Springer Series in
CHEMICAL PHYSICS 66

Springer
Berlin
Heidelberg
New York
Barcelona
Hong Kong
London
Milan
Paris
Singapore
Tokyo

Springer Series in

CHEMICAL PHYSICS

Series Editors: F. P. Schäfer J. P. Toennies W. Zinth

The purpose of this series is to provide comprehensive up-to-date monographs in both well established disciplines and emerging research areas within the broad fields of chemical physics and physical chemistry. The books deal with both fundamental science and applications, and may have either a theoretical or an experimental emphasis. They are aimed primarily at researchers and graduate students in chemical physics and related fields.

63 **Ultrafast Phenomena XI**
Editors: T. Elsaesser, J.G. Fujimoto, D.A. Wiersma, and W. Zinth

64 **Asymptotic Methods in Quantum Mechanics**
Application to Atoms, Molecules and Nuclei.
By S.H. Patil and K.T. Tang

65 **Fluorescence Correlation Spectroscopy**
Theory and Applications
Editors: R. Rigler and E.S. Elson

66 **Ultrafast Phenomena XII**
Editors: T. Elsaesser, S. Mukamel, M.M. Murnane, and N.F. Scherer

Series homepage – http://www.springer.de/phys/books/chemical-physics/

Volumes 1–62 are listed at the end of the book

T. Elsaesser S. Mukamel M.M. Murnane
N.F. Scherer

Ultrafast Phenomena XII

Proceedings
of the 12th International Conference,
Charleston, SC, USA,
July 9–13, 2000

With 468 Figures

Springer

Professor T. Elsaesser
Max-Born Institut für Nichtlineare Optik
und Kurzzeitspektroskopie
Max-Born-Str. 2A
12489 Berlin, Germany

Professor M.M. Murnane
University of Colorado,
JILA
Boulder, CO, USA

Professor S. Mukamel
University of Rochester,
Chemistry Department
Rochester, USA

Professor N.F. Scherer
University of Chicago,
Chemistry Department
Chicago, IL, USA

Series Editors:

Professor F.P. Schäfer
Max-Planck-Institut für Biophysikalische Chemie
D-37077 Göttingen-Nikolausberg, Germany

Professor J.P. Toennies
Max-Planck-Institut für Strömungsforschung
Bunsenstrasse 10
D-37073 Göttingen, Germany

Professor W. Zinth
Universität München,
Institut für Medizinische Optik
Öttingerstr. 67
D-80538 München, Germany

ISSN 0172-6218

ISBN 3-540-41211-5 Springer-Verlag Berlin Heidelberg New York

Library of Congress Cataloging-in-Publication Data applied for

Die Deutsche Bibliothek – CIP-Einheitsaufnahme
Ultrafast phenomena ...: proceedings of the ... international conference ... - 4 (1984)-. - Berlin ; Heidelberg ; New York ; Barcelona ; Hong Kong ; London ; Milan ; Paris ; Singapore ; Tokyo : Springer, 1984
Früher u.d.T.: Picosecond phenomena
12. Charleston, SC, USA, July 9–13, 2000. - 2000
(Springer series in chemical physics ; 66)
ISBN 3-540-41211-5

Springer-Verlag Berlin Heidelberg New York
a member of BertelsmannSpringer Science+Business Media GmbH

Typesetting: Camera-ready copies by the editors
Cover concept: eStudio Calamar Steinen
Cover production: *design & production* GmbH, Heidelberg

Printed on acid-free paper SPIN: 10777895 57/3141 XO - 5 4 3 2 1 0

Preface

This volume is a collection of papers presented at the Twelfth International Conference on Ultrafast Phenomena held at Charleston, SC, from July 9 to 13, 2000. The Ultrafast Phenomena Conferences are held every two years and provide a forum for discussion of the latest results in ultrafast optics and their applications in science and engineering. A total of more than 200 papers was presented, reporting progress in the technology of generating and characterizing ultrashort pulses, including new techniques for shortening the duration of laser pulses, for stabilizing their absolute phase, and for improving tunability over broad wavelength ranges, output powers and peak intensities. Ultrafast spectroscopy, including techniques like time-resolved X–ray diffraction and two-dimensional spectroscopy, continues to give new insights into fundamental processes in physics, chemistry and biology. Control and optimization of the outcome of ultrafast processes represent another important field of research. There is an increasing number of applications of ultrafast techniques, for instance in material diagnostics and processing and in imaging and microscopy. The enthusiasm of the participants, among them many students, the high quality of the papers and the nice conference site made the conference successful and pleasant.

Many people contributed to the success of the conference. The members of the international program committee reviewed the submissions and put the final program together. The staff of the Optical Society of America deserves special thanks for making the meeting arrangements and running the meeting smoothly. We would like to thank the Optical Society of America for sponsorship, and also acknowledge support from Coherent, Lucent Technologies, Newport, Positive Light, Spectra Physics Lasers and Thorlabs.

Berlin, Germany — *Thomas Elsaesser*
Rochester, USA — *Shaul Mukamel*
Boulder, USA — *Margaret Murnane*
Chicago, USA — *Norbert Scherer*
September 2000

Contents

Part I: Coherent Control

Part II: Lasers for Ultrashort Pulses

Part III: Pulse Characterization, Shaping and Measurement Techniques

Part IV: THz Generation and Measurement

Part V: Imaging and Microscopy

Part VI: X-Ray Generation and Diffraction, Higher Harmonics, High Intensity Physics

Part IX: Isolators, Metals and Superconductors

Part X: Electron Dynamics and Charge Transfer at Interfaces and in Liquids

Part XII: Dynamics in Liquids

Part XIV: Ultrafast Chemistry

Part XV: Biology

Part I

Coherent Control

Fig. 1. Kent Wilson (1937–2000), in Torrey Pines State Park, Del Mar, California, August 1999

The Odyssey of Kent Wilson: Holding Molecules in the Light

Dudley Herschbach

Department of Chemistry and Chemical Biology, Harvard University,
12 Oxford Street, Cambridge, MA 02138, USA
E-mail: herschbach@chemistry.harvard.edu

1 Introduction

Anyone pursuing laser chemistry is indebted to Kent Wilson as an inspiring architect of the field. Many are now using innovative laser techniques conceived by him to probe the molecular dynamics of chemical and biochemical reactions. Many will know too that, as a remarkably wide-ranging physical chemist, he also contributed much seminal work, both theoretical and experimental, to several other fields, among them economics, engineering, computer science, environmental science, archaeology, and educational filmmaking.

His legacy to science over the past four decades is immense, but his death on 27 March at the age of 63 deprived him of fully enjoying three major projects brought to fruition in the last few years by his group at the University of California, San Diego. Each required development of powerful new apparatus, exemplifying Wilson's architectural vision. One machine achieves "quantum control" of photochemical processes using iterative feedback to automatically tailor optimal laser excitation pulses. Another, the culmination of long effort, enables direct observation of atomic motions by means of ultrafast x-ray diffraction and spectroscopy. Most recently, a laser microscope has been built that can obtain very high resolution images of living tissue without killing the cells. These remarkable achievements have opened up exhilarating frontiers.

Those privileged to know Kent personally cherished him as wonderfully excitable and creative in his work but also ardent and devoted as a friend, eager to share his enthusiasms and adventures. His intense affection and admiration for his students, colleagues, and family was warmly reciprocated. A reunion of the Wilson group, held in 1996 to celebrate Kent's 60th birthday, produced a book of reminiscences by group members [1]. The book is amply illustrated with photos, many of parties, outings, or trips Kent had initiated, and conveys vividly the exuberant, joyful spirit he fostered in work and play. In a dedication, Kent wrote:

> Freud said that the two most important ingredients of a good life are work and love. I am privileged to have had a life's work which I love, which is to me deeply meaningful and sustaining, an endless source of fascination, excitement and challenge. The best and most important part of this life's work is you, the people I have worked with. I take great pride in you and your lives; to me, you are a second family. You have enriched my life by working with me and now by coming back to gether again in a reunion to celebrate ourselves and our years together.

There is a Quaker saying, "I'm holding you in the light". With people as well as molecules, Kent Wilson fulfilled this admirably in his life.

We are fortunate to have from Kent himself an autobiographical essay, "Summing Up", published last December in a special issue of the Journal of Physical Chemistry honoring him [2]. It is a poignant "love letter to posterity", with candid and wise observations on his life in science, his joys and regrets. I urge everyone concerned with science, whether a young person contemplating a career, a mature scientist concerned about better managing a research group, or a cultural scholar, to read Kent's essay. In this paper, I will augment portions of his essay and a brief obituary article [3]. Chiefly, I'll describe episodes from his graduate work with my group and his early years at San Diego. Now distant history, these stories may yet be instructive as harbingers of the enterprising, zestful style that emerged in his career and grew to a crescendo in his recent work.

2 Beginnings

Kent Wilson was born on 14 January, 1937 in Philadelphia and grew up, as he liked to say, "inside an experiment". This was located in a nearby village, Bryn Gweled Homesteads, a mostly Quaker community which his parents had helped to found, designed to foster communal living. Encouraged by prominent scientists in the village, Kent by the age of eight had made a rudimentary radio receiver. Later, his family spent winters in Washington, D.C., where his father headed a Quaker lobbying organization working for disarmament, civil rights, and combating poverty. Kent became an avid visitor to the Smithsonian Museums and the Library of Congress; he credited that with "saving my academic career from ruin" because he "did very little of the assigned homework in school".

At Harvard College, Kent began as a government major, focused on history and political science. However, he was captivated by a chemistry course given by the fabled Leonard Nash. That led Kent to switch his major to chemistry and physics. After completing a bachelor of arts degree in 1958, he went to the University of Strasbourg for a year and wrote a Diplome thesis in economics. Deciding then to pursue graduate study in chemistry, he came to the University of California at Berkeley. Susceptible to evangelical fervor, Kent joined my fledgling research group there soon after we began crossed molecular beam experiments of reactions of alkali atoms with halogen compounds [4].

Young scientists pursuing work with molecular beams now are blissfully unaware how rudimentary were the tools available back then, before the advent of lasers and modern computers and before vacuum chambers and all sorts of elegant gear could be ordered from a catalog. The prime detection method, limited to alkali atoms and alkali halide molecules, relied on surface ionization by hot filaments. This took advantage of a somewhat mysterious differential sensitivity (discovered by Sheldon Datz and Ellison Taylor). A tungsten filament ionized efficiently both nonreactively scattered atoms and reactively scattered halide molecules, whereas a platinum alloy filament responded almost only to atoms. Thus, the difference of readings taken on the tungsten and platinum alloy filaments (often small!), permitted measurement of the angular distributions of the alkali halide product. After initial success with reactions of alkali atoms with alkyl iodides (work of George Kwei, Jim Norris, and Jim Kinsey), however, my group encountered great difficulty in extending the experiments to analogous reactions with diatomic halogens and other molecules. Such molecules seemed to "poison" the platinum alloy filament, causing its signals to drift and fluctuate intolerably.

Kent undertook to improve the detection method in two ways. He built a quadrupole mass spectrometer (designed by Gilbert Brink) from scratch. It was a major job for a graduate student long before such instruments could be bought. He was justly proud of his instrument, as seen in the photo on the following page. The motivation was to extend the chemical scope of surface ionization by mass analyzing the ions emitted by the filaments. Kent also built an auxiliary apparatus with which he carried out an extensive study of surface ionization efficiency for a wide range of filament materials and operating conditions. Both his mass spectrometer and ionization study proved quite valuable in later work. Here, I note just one aspect indicative of Kent's alert, questioning approach. In the literature on surface ionization, much ado was made of the fact that bursts of potassium ions emerge as background even from well-aged tungsten filaments in high vacuum, without exposure to incident beams. This was attributed to a similarity in the ionic radii of potassium and tungsten. Kent looked for and found other alkali ions were also emitted. He had the satisfaction of confirming that the relative yields did not correlate with ionic radii but simply matched abundances in the earth's crust, as he had suspected.

Another episode exemplifies Kent's characteristically optimistic outlook about what he might accomplish in the future. The summer following his first year in graduate school, he went off with a student group on a trip of 7 weeks or so to the Soviet Union, wanting to see first-hand how a communistic economy worked. After his return, box after box of books arrived at our lab. These contained more than a hundred classic texts of modern physics and chemistry, Russian translations that Kent had found he could buy very cheaply. He proudly assembled them in a large bookcase above his desk. Thereafter, I happened to come across a Russian article in need of translation. When I took it to Kent, I discovered he could not read Russian; he had blithely assumed that he'd learn how someday.

At the time my group moved to Harvard, in the fall of 1963, we were still struggling with detector poisoning. Fortunately, Kent's studies of surface ionization led him to a paper (by Eugene Toux and John Triscka) which showed us how to "immunize" the filament. Armed with such black art, Kent and others of my group were able to examine many more alkali reactions.

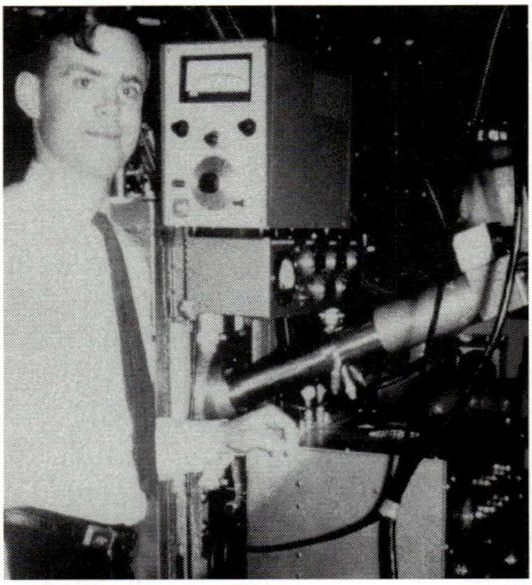

Fig. 2. Kent Wilson as a gaduate student, with the mass spectrometer he had built; at Harvard, Spring 1964

By the next spring, we had established that reactions with diatomic halogens proceeded by a "stripping" mechanism, essentially opposite to the "rebound" process found three years before for methyl iodide. The paper reporting this, with Kent as first author, was submitted just a few weeks after a seemingly definitive theoretical paper, based on extensive classical trajectory calculations, had appeared with the conclusion that all reactions would go by a rebound mechanism.

Kent went on to survey reactions with carbon tetrachloride and several other polyhalide compounds. Thereby he mapped out generic aspects of a transition between the rebound and stripping limits [4]. This revealed how the reaction dynamics could be correlated with electronic structure of the reactant molecules. His doctoral thesis, completed in the fall of 1964, presented his abundant data and described in exemplary detail the design, construction, and operation of what was then state-of-the-art apparatus.

3 Career at San Diego

In 1965 Kent gladly returned to California and began teaching and building his own laboratory at San Diego. In his research, he tackled the challenging task of developing experiments to elucidate the dynamics of molecular photodissociation by observing the distributions in angle and velocity of the dissociation products. The techniques of ordinary spectroscopy put almost exclusive emphasis on stable excited states, but in molecules there exist many more dissociative excited states. Kent's technique of photofragment spectroscopy opened up a vast lode of information about such states and repulsive intermolecular forces, thereby also contributing to understanding energy release in a large family of chemical reactions [4].

His success was hard won. The impetus for photofragment spectroscopy came from Dick Zare's doctoral thesis, also done in my group in 1964, which treated the basic theory. However, feasibility estimates made by presuming use of the continuous light sources then available had discouraged us from undertaking photodissociation experiments. Kent realized that a pulsed light source offered substantial advantages and bravely set to work. In the reunion book [1], a superb chapter from the 1970 thesis of Kent's first graduate student, George Busch, is reprinted. It describes the long series of failures confronted by Kent and his team, their inventive responses and their ultimate triumph, achieved only in late 1968 when a pulsed solid-state laser with decent conversion efficiency for doubling became available.

Although his later research explored quite diverse areas, Kent's scientific leitmotif remained molecular dynamics. In the 1980s, by combining ultrafast laser experiments with computer simulations he was able to markedly advance understanding of reaction dynamics in solution, and bring out connections to gas phase dynamics. In the 1990s, his ultrafast x-ray laser apparatus revealed unanticipated dynamical aspects of the transition between solid and liquid phases, in a phenomenon termed "nonthermal melting". In striving for his ultimate goal of "quantum control", Kent arranged to conduct "conversations" with molecules via laser light pulses, thereby enabling the molecules to teach experimenters how to negotiate a desired chemical choreography of atomic motions.

For Kent, his teaching and mentoring were adventures in both intellectual and social dynamics. A unique part of his research group was dubbed by him the "Senses Bureau". For this he recruited first-year undergraduates, "young enough so that they do not yet know the meaning of the word impossible". Over 35 years, they created original software and hardware for computer animation and produced many educational films and interactive "virtual reality" devices used widely in high schools, colleges and medical schools. A classic film explicated protein synthesis with the aid of 200 dancers, poetry, and a jazz-rock band. The Bureau also had a large role in Kent's studies of air pollution in California and Mexico, and in the use of chemical archaeology to elucidate trade routes in Africa.

In mid-career Kent became increasingly frustrated with the cost in time and tribulation of applying for grants to support his research. Too often, that prohibits work that requires building expensive, unprecedented apparatus. On a sabbatical leave in 1985, Kent came up with an extraordinary solution, discovering "a mapping between parts of physics and economics". Boldly betting on his theory, over the next few years he invested in the stock markets of 25 countries, especially developing nations in financial chaos. This succeeded so well that he was able to resign all his federal grants, thereafter supporting from his own funds a sizable research group as well as equipping his laboratory with the finest laser and computer instrumentation he could buy or build.

4 Benediction

Late in 1998, Kent learned he had incurable, late stage prostate cancer. For the remaining 14 months of his life he dealt with this as his "most challenging experiment", organizing a team of biochemists to work in collaboration with physicians pursuing new therapeutic approaches. With the support of his wife Lana and their daughters Tasha and Maya, and many friends, he maintained a "balance of hope and realism", making arrangements for others to continue key lines of his research.

In the closing paragraph of his farewell essay [2], Kent describes his joy in visiting wildflowers along a favorite trail, akin to his pleasure in the blooming of his scientific projects and progeny. Late last August, he escorted me over that trail, happily introducing his favorite flowers. That was when I took the photo prefacing this article. With it, I wanted to hold Kent "in the light". I also had in mind a poem by Max Planck [5]. It offers an apt benediction for Kent, who so earnestly pursued "that great central mystery of the connection of mathematics and physics to what happens in the world of matter". In a letter to another scientist, Planck wrote:

What you have picked, what I have picked
These we will bind together,
Entwining thus a fair bouquet
From gifts we give each other.

References

1. *Wilson Group Reunion*, 1965–1996, University of California, San Diego
2. K.R. Wilson, J. Phys. Chem. A **103**, 10022, 1999
3. D. Herschbach, Nature **405**, 902, 2000
4. D. Herschbach, Angew. Chem. International Ed. Engl. **26**, 1221, 1987
5. Quoted (in English translation) by James Murphy in his introduction to *Where is Science Going?* by Max Planck, 1933, reprinted by Oxbow Press, Connecticut, 1981

The Kent Wilson Group in the 1990s

Bern Kohler*

The Ohio State University, Department of Chemistry, 100 W. 18th Ave.,
Columbus OH 43210, USA.
E-mail: kohler@chemistry.ohio-state.edu

*Postdoctoral scientist in the Wilson Group from 1992-95.

1. Introduction

Scholarly publications form an enduring, but incomplete record of the life of a scientist. Sadly, this record preserves almost nothing in the way of human qualities. I am grateful for this opportunity to share some of my memories of Kent Wilson. All of us can recall one or two extraordinary individuals, whose confidence in us and whose steady encouragement helped us to succeed. Kent was just this kind of person to his many students, postdocs, colleagues, and former collaborators. He was an extraordinary mentor, and a generous friend.

2. Paths less traveled

Kent loved travelling, whether it was a quick trip across the border to nearby Tijuana, or a journey to Timbuktu. He preferred the remote and exotic, and loved to tell the tales that are the reward for visiting those places. Many former group members can recall Kent's stories about eating durian, a foul-smelling (but apparently tasty) fruit from the jungles of Southeast Asia. One day at lunchtime, Kent surprised everyone with a ripe durian, which he claimed to have bought at a local Chinese market. Kent gathered the group around him, and cut up the fruit with great ceremony. While most of us held our noses, Kent enjoyed every bite.

Kent was an avid walker and many in the group have fond memories of treks with him high above treeline in the German Alps or in the mountains of Southwestern Colorado. Kent was equally fond of the sandy and fragrant paths through the rare Torrey Pine trees near his home. In science, Kent was an explorer who constantly made his own paths to fabulous new territories. He was ahead of his time in so many areas. Pioneering efforts in photofragment spectroscopy in the 1970s were followed by equally pioneering work in computer graphics, chemical archaeology, and air pollution monitoring. Kent showed 3D-movies of his work back when the rest of us were making do with transparency pens for our presentations.

Like many great explorers, Kent showed marvelous tenacity and steadfastness of purpose when it came to launching new voyages which others had deemed too risky. Kent's ideas rarely fell within the current paradigms, and always seemed to push current thinking (and current technology) to the limit. Kent's work in quantum control, the subject of this tribute session dedicated to his memory, is a good example. Since the invention of the laser, the possibility of controlling molecules with coherent light has been a siren's song for many. Twenty years ago, Berens and Wilson published an eloquent and visionary statement in the conference proceedings of the second of this series of meetings that anticipated much of the group's future course [1],

> "The accessible time range for picosecond experimental measurements is now short enough, and that for molecular dynamics theoretical computation is now long enough, to significantly overlap. This overlap of theory and experiment can, at least in principle, be exploited to discover the trajectories of atoms during chemical processes in solution, for example rotational and vibrational relaxation, rotational and vibrational dephasing, and chemical reactions."

The above passage heralded a dramatic switch in the group's research focus from experiment to theory. Figure 1 shows a timeline of Kent Wilson's publications categorized as experimental or theoretical. For most of the 1980s the Wilson Group was a small group of theorists who made significant contributions to condensed-phase chemical reaction dynamics by computer simulation, and laid the foundations for what was to come.

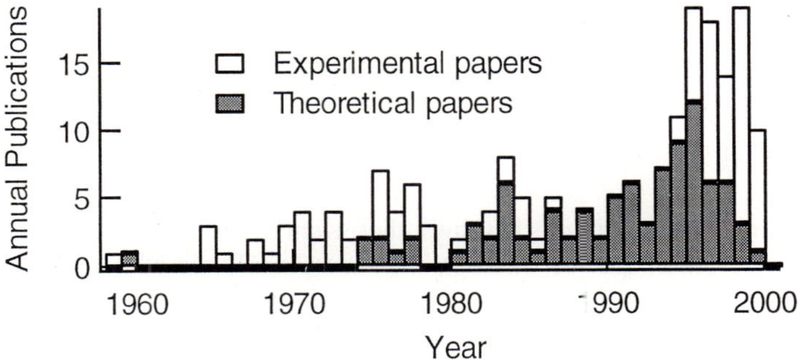

Fig. 1. Timeline of Kent R. Wilson's scientific publications (through June 2000).

3. Unifying experiment and theory

At the start of the 1990s, powerful theoretical tools were in place for calculating the optimal light fields for controlling small quantum mechanical systems. In addition, solid-state femtosecond lasers became available that could reliably synthesize these fields. Kent had grand plans for bridging the gulf that often separates experiment from theory [2]. The transition from theory back to experiment is an expensive one, and Kent realized that the traditional channels of support would respond too slowly and too cautiously to keep pace with his dreams. Kent's solution to this problem is legendary [3]. He made millions of dollars via global investment, and by the early 1990s was ready to begin.

Thus, began the third, and, perhaps, most remarkable phase of the Wilson Group. Beginning in 1992, the group grew rapidly as Kent assembled a team of researchers with expertise that spanned experiment and theory. Having started in experiment, and then switched to theory, Kent was finally able to unite the two (Figure 1). Kent's legendary enthusiasm for challenging science and his unique solution to the customary financial constraints attracted an enormous number of applications to join his group from some of the best young chemists and physicists from around the world. The group members benefited immensely from each other, and from the steady stream of visiting senior scientists. UCSD's location in La Jolla, California was of course an additional, powerful draw, particularly in the wintertime!

Like all successful expedition leaders, Kent was relentlessly optimistic, and at times irritatingly confident in the ultimate success of his group's quests. The surge in experimental papers around 1995 (Figure 1) documents the success of these efforts. First, the work in quantum control came to fruition [4], and, more recently, extraordinary work has been accomplished in time-resolved x-ray diffraction [5,6]. At this conference, sessions were held on both topics. Today, these new territories no longer appear quite as inaccessible as they did a decade ago. Kent Wilson would be the first to acknowledge that this was not a solo voyage, but resulted instead from the parallel efforts of many other explorers from around the world. Still, Kent Wilson's spirit stands out for its boldness. As many of us in ultrafast science set our compasses toward these new lands, I have to wonder where Kent's dreams would be leading him today.

4. Farewell

Sadly, the timeline in Figure 1 shows that Kent's life ended just as he was achieving some of the greatest success of his career. Those who spoke with him in recent years know firsthand the great joy his group brought him. For many of us, and I think for Kent himself, as great as the science was, the group was something much greater. It is fitting then to end with a photograph. Figure 2 shows Kent with some of his group members on the beach at Torrey Pines, a short distance from his laboratories at the University of California, San Diego. Although this photograph was taken in 1995, I believe that anyone who ever worked with Kent can recall the warmth of a similar gathering. In his floppy straw hat and loose-fitting shirt, Kent is celebrating the accomplishments of his group members and talking of new voyages. And he is smiling.

Fig. 2. Prof. Kent Wilson (center, in white shirt and straw hat) and some of the members of his research group on Torrey Pines Beach, California, July 1995 (photo by B. Kohler).

References

1 P. H. Berens and K. R. Wilson, in *Picosecond Phenomena II*, Edited by R. Hochstrasser, W. Kaiser, and C. V. Shank, Springer-Verlag, Berlin, 1980; p. 247.

2 Y. J. Yan, B. Kohler, R. E. Gillilan, R. M. Whitnell, K. R. Wilson and S. Mukamel, In *Ultrafast Phenomena VIII*, Edited by J.-L. Martin, A. Migus, G. A. Mourou, and A. H. Zewail, Springer-Verlag, Berlin, 1993; p. 8.

3 Some details can be found in Kent Wilson's remarkable essay "Summing Up", K. R. Wilson, *J. Phys. Chem. A* **103**, 10022, 1999.

4 B. Kohler, J. L. Krause, J. L., F. Raksi, K. R. Wilson, V. V. Yakovlev, R. M. Whitnell, Y.-J. Yan, *Acc. Chem. Res.* **28**, 133, 1995.

5 C. Rose-Petruck, R. Jimenez, T. Guo, A. Cavalleri, C. W. Siders, F. Ráksi, J. A. Squier, B. C. Walker, K. R. Wilson, and C. P. J. Barty, *Nature* **398**, 310, 1999.

6 C. W. Siders, A. Cavalleri, K. Sokolowski-Tinten, C. Tóth, T. Guo, M. Kammler, M. Horn von Hoegen, K. R. Wilson, D. von der Linde, and C. P. J. Barty, *Science* **286**, 1340, 1999.

Algorithms for Closed Loop Ultrafast Control of Quantum Dynamics

Herschel Rabitz

Department of Chemistry, Princeton University, Princeton NJ 08544, USA
E-mail: hrabitz@princeton.edu

Abstract. Learning control techniques have a special role to play in the manipulation of quantum dynamics phenomena. The unique capabilities of quantum systems making them amenable to learning control are (a) the ability to have very large numbers of identical systems for submission to control, (b) the high duty cycle of laboratory laser controls, and (c) the ability to observe the impact of trial controls at ultrafast time scales. Various learning algorithms have been proposed to guide this control process. The present paper will discuss these proposals, as well as some new perspectives.

1. Introduction

There is rapidly growing interest in the control of quantum systems for a variety of applications with atoms, molecules, and condensed phases[1]. Most cases under consideration employ tailored ultrafast laser electric fields $\varepsilon(t)$ as the control, with hopefully, the capability of driving the system from a chosen initial state $\psi(0)$ to a desired final state $\psi(T)$ at time T. As quantum systems naturally exhibit complex motions, it is exceedingly difficult to intuit the necessary structure of the control field $\varepsilon(t)$ capable of producing successful manipulations of the quantum system. Despite the difficulties with theoretically designing quantum controls, much insight has been learned from such efforts, with the most important point being that successful control demands full cooperation between the temporal field structure $\varepsilon(t)$ and the dynamical capabilities of the quantum system.

The difficulties outlined above for controlling quantum systems naturally suggest considering closed loop learning control[2]. Quantum control has some especially attractive features when approached as a learning problem. First, it is often easy to produce many samples, with each starting in the same state, and the duty cycle of going from one control to the next can be many thousands or more per second[3]. The action of each control, in turn, can be observed by a second rapid probe, and the overall process repeated with a new sample at an exceedingly high rate under computer control[4]. Thus, one can envision performing thousands or more distinct learning control experiments on comfortable laboratory time scales. An enabling technical feature of quantum learning control is the ability to create rather arbitrarily shaped control fields $\varepsilon(t)$ through advanced laser pulse shaping technology[3], including the ability to switch from one control field to another at the rate of $\sim 10^3 - 10^6$/sec. The last key element of the control process is the nature of the learning algorithm in the closed loop, which is the focus of this paper. In principle, any of a variety of algorithms could be considered, but one requirement is that they be operationally fast, as they will be directly employed in the laboratory with real-time ongoing experiments.

Under rather general conditions, the non-linear nature of controlled quantum dynamics suggests that there will also be multiple control fields $\varepsilon(t)$ giving comparable results. The overall learning control process aims to discover at least one of these good fields. Section 2 of this paper will discuss the nature of the currently considered learning algorithms, and open up the prospect of a new family with special features. Section 3 will introduce a new non-linear learning algorithm. Finally, section 4 presents brief concluding remarks.

2. Quantum Learning Control Algorithms

The control of a quantum system is normally expressed[5] as an optimization problem by specification of a cost functional $J = J_Q + J_\varepsilon$. Here, J_Q refers to the costs associated with steering the quantum system as close as possible to a desired outcome for the target operator O expectation value $\langle \psi(T)|O|\psi(T)\rangle$, and J_ε refers to any relevant costs associated with the magnitude or form of the control field. The goal is minimization of J with respect to the control $\varepsilon(t)$. The field is often expressed in terms of its Fourier transform $\tilde{\varepsilon}(\omega)$, which is more easily amenable to manipulation in the laboratory[5]. At present, there is no information suggesting that the control of quantum systems lends itself to a special class of algorithms, and as such, a variety have been considered as summarized below.

Gradient Algorithms. Nominally, the employment of gradient algorithms in laboratory optimization calls for caution, as the gradient $\delta J / \delta(t)$ must be measured in the laboratory where inherent errors will be introduced. A very large number of experiments may be performed with nominally the same variation $\delta\varepsilon(t)$ of the functional response J; thus, the ensemble average $\langle \delta J/\delta\varepsilon(t) \rangle$ may be observed to drive down the noise contributions. This procedure has been explored in simulations[6], where it was shown to be quite effective, and this technique deserves further analysis and attention.

Genetic Algorithms. The use of genetic algorithms (GA) for quantum learning control[4,7] introduces a stochastic-type element in the search process, to more thoroughly explore for viable control field $\varepsilon(t)$ solutions. However, a potential difficulty with these algorithms is the need to work with large numbers of control variables, ranging in the hundreds to thousands, typically coming from a discretization of the field Fourier transform $\tilde{\varepsilon}(\omega)$ in keeping with how the fields are generated in the laboratory[3]. Thus, the optimization algorithm (genetic or otherwise) must deal with a very high dimensional space of control variables. Under normal conditions, GA's have been reported to critically slow down with such large numbers of variables. All of the current laboratory learning experiments in quantum systems have employed GA's involving approximately 100 generations and families of 50 members. Despite the very large number of control variables, good quality results have been obtained[8,9] with GA's.

Regardless of the algorithm employed for quantum learning control, an ancillary desire is to deduce as much as possible about the dynamical mechanism of control from the structure of the optimal field. In order to ensure a reliable physical interpretation of the control in terms of mechanism, it is essential that the field $\varepsilon(t)$ or $\varepsilon(\omega)$ contain no extraneous features that have essentially no impact on the quantum observable[10]. In the context of GA's, this implies that "genetic pressure" needs to be applied against the turning of those quantum control knobs that have little impact on the control output. Simulations have shown dramatic field structure differences, arising from placing genetic pressure against extraneous field components[4].

Linear Control Maps. Although quantum observables typically draw on the control field $\varepsilon(t)$ in a non-linear fashion, one may consider the generation of approximate input-output (IO) relationships based on a *sequence* of laboratory-generated linear mappings. As an illustration, the simplest case occurs with temporal experimental data $O(t) = \langle \psi(t)|O|\psi(t)\rangle$, 0 = t = T. Thus, let $\varepsilon^s(t)$ be the current best estimate of the control field at the s-th iterative map level producing the data track $O^s(t)$ as an approximation to the optimal or desired track $O^d(t)$. As a further step towards the desired track, an improved field $\varepsilon^{s+1}(t) = \varepsilon^s(t) + \delta\varepsilon^s(t)$ can be obtained from inverting the forward mapping relationship

$$\delta O^s(t) = \int_0^t M^s(t,t')\delta\varepsilon^s(t')dt' \tag{1}$$

where $\delta O^s(t) = O^d(t) - O^s(t)$. This process would then be repeated to form $\varepsilon^{s+2}(t)$, etc., until acceptable convergence is reached approaching the desired track $O^d(t)$. The forward map $M^s(t,t')$ may be determined by inputting a set of trial fields $\varepsilon_i^s(t)$, $i = 1,2,\ldots$, and observing their impact $O_i^s(t)$, i = 1,2,.... At this juncture, quantum learning control with linear maps has been successfully tested under simulation for quantum systems[11].

3. Non-Linear Maps for Learning Control and the Identification of Cooperative Control Features

Section 2 summarized the current algorithmic technology being considered for the control of quantum systems in the laboratory. Thus far, only genetic-type algorithms have been implemented[8,9] in the laboratory, and some very encouraging results have been obtained. GA's, when successful, only provide the control field, and not much insight into the physical basis behind the control field structure. A step towards addressing the latter important issue is evident in the learning algorithm involving the generation of linear maps discussed above. The presence of very large numbers of control variables implies that better optimization procedures may be necessary to effectively find the controls. The next step of forming non-linear mappings may prove to be valuable for the field of quantum control[13].

To appreciate the perspective involved with non-linear control maps, it is useful to consider the control field as broken into a set of N variables $\varepsilon \equiv (\varepsilon_1,...,\varepsilon_N)$. The observable $O(\varepsilon) = O(\varepsilon_1,...,\varepsilon_N) = \langle \psi(T)|O|\psi(T)\rangle$ can be viewed as a function over the N-dimensional space of input controls. The full structure of such spaces in quantum systems is not known; however, it expected that $O(\varepsilon)$ will be a reasonably behaved function over at least some local, or even rather global, domain in the space of controls $\varepsilon \equiv (\varepsilon_1,...,\varepsilon_N)$. The number of variables N can be as large as $\sim 10^2 - 10^3$. The objective is to determine a series of maps $O^{\ell}(\varepsilon)$, $\ell = 1,2,\ldots$, where each is quantitatively valid over some domain $\mathcal{D}^{\ell}$ in the space of controls.

The analysis above produces a high dimensional function representation problem[13]. Arguments from the statistical analysis of many diverse problems suggests that even rather large domains may exist, where only a modest number of experiments (i.e., not growing exponentially with N) are sufficient to determine the function representation as a map. In developing such maps, it is useful to consider the degree of cooperativity amongst the control variables $\varepsilon \equiv (\varepsilon_1,...,\varepsilon_N)$, which naturally leads to a hierarchical expansion of the output observable $O^{\ell}(\varepsilon)$ as

$$O^{\ell}(\varepsilon) = O_0^{\ell} + \sum_i O_i^{\ell}(\varepsilon_i) + \sum_{i<j} O_{ij}^{\ell}(\varepsilon_i,\varepsilon_j) + \ldots + O_{1,2,\ldots,N}^{\ell}(\varepsilon_1,\ldots,\varepsilon_N) \qquad (2)$$

Here, O_0^{ℓ} is the zero-th order constant mean map output, while $O_i^{\ell}(\varepsilon_i)$ is a first order independent, and likely non-linear, response of the map output to the action of the i-th control variable ε_i acting alone. Similarly, the term $O_{ij}^{\ell}(\varepsilon_i,\varepsilon_j)$ gives the second order response due to the non-separable cooperative role of the variables ε_i and ε_j. Finally, the last term $O_{1,2,\ldots,N}^{\ell}(\varepsilon_1,\ldots,\varepsilon_N)$ closes the expression with any residual full N-dimensional cooperativity. Using the laboratory data on the left-hand of Eq. (2), it is necessary to determine the relevant individual terms on the right-hand side. This may be conveniently done by defining the component functions $O_0, O_i(\varepsilon_i), O_{ij}(\varepsilon_i,\varepsilon_j)$, to be fully orthogonal, both within and between each order through a suitable metric[12].

Given the ability to determine the component functions on the right-hand side of Eq. (2), in terms of the laboratory data, an overall optimal learning control algorithm may be set up. Although there is extra overhead in determining the maps, they may be utilized a large number of times by the optimizer in the learning algorithm, as the maps can be exceedingly fast to evaluate[12]. Furthermore, the structure of the maps reveals how the control variables act alone or cooperatively (i.e., by examining $O_i(\varepsilon_i)$ and $O_{ij}(\varepsilon_i,\varepsilon_j)$, respectively), to reach the objective. Several simulations have been carried out with this algorithm on finite-dimensional quantum systems of up to 10 discrete states and 18 control variables in the frequency domain[12]. Excellent results were obtained with maps using up to either first or second order terms in Eq. (2). The story told by the maps, at times varied from that found by simple kinematic analysis of the field structure alone. That is, the non-linear relationship between the input control field and the output quantum observable lead to subtle roles for the input control variables.

4. Conclusions

This paper presented the current state of learning algorithm development for controlling quantum systems, with special emphasis given to a new algorithm based on non-linear maps in section 3. Besides utility in quantum mechanics, the algorithm presented in section 3 also should have application to broader control problems in engineering.

Acknowledgements. The author acknowledges support from the Department of Defense (DoD) and the National Science Foundation (NSF).

References

1 H. Rabitz, R. de Vivie-Riedle, M. Motzkus, and K. Kompa, *Science*, **288**, 824-828 (2000).
2 K. More, *Iterative Learning Control for Deterministic Systems* (Springer, London, 1993).
3 H. Kawashima, M.M. Wefers, and K.A. Nelson, *Ann. Rev. Phys. Chem.*, **46**, 627-656 (1995).
4 R.S. Judson and H. Rabitz, *Phys. Rev. Lett.*, **68**, 1500 (1992).
5 A.P. Peirce, M. Dahleh, and H. Rabitz, *Phys. Rev. A*, **37**, 4950 (1988); S. Shi and H. Rabitz, *Chem. Phys.*, **139**, 185 (1989); R. Kosloff, S.A. Rice, P. Gaspard, S. Tersigni, D.J. Tannor, *Chem. Phys.*, **139**, 201 (1989).
6 P. Gross, D. Neuhauser, and H. Rabitz, *J. Chem. Phys.*, **98**, 4557 (1993).
7 G.J. Tóth, A. Lörincz and H. Rabitz, *J. Chem. Phys.*, **101**, 3715 (1994).
8 A. Assion, T. Baumert, M. Bergt, T. Brixner, B. Kiefer, V. Seyfried, M. Strehle, G. Gerber, *Science,* **282**, 919 (1998).
9 C.J. Bardeen, V.V. Yakovlev, K.R. Wilson, S.D. Carpenter, P.M. Weber, W.S. Warren, *Chem. Phys. Lett.,* **280**, 151 (1997).
10 H. Rabitz and W. Zhu, Optimal Control of Molecular Motion: Design, Implementation, and Inversion, *Accts. of Chem. Res.*, in press.
11 M.Q. Phan and H. Rabitz, *J. Chem. Phys.*, **110**, 34 (1999).
12 E. Weiss, J. Geremia, and H. Rabitz, Nonlinear Input-Output maps for Quantum Control, *J. Chem. Phys.*, submitted.
13 H. Rabitz and Ö. Alis, *J. Math. Chem.*, **25**, 197 (1999).

Control of Quantum Dynamics by Adaptive Femtosecond Pulse Shaping

M. Bergt, T. Brixner, B. Kiefer, M. Strehle, and G. Gerber

Physikalisches Institut, Universität Würzburg, 97074 Würzburg, Germany
E-mail: gerber@physik.uni-wuerzburg.de

Abstract. Femtosecond laser pulses are modified in a pulse shaper and optimized iteratively by a learning evolutionary algorithm in order to reach specific optimization goals in coherent control experiments. The significance of the electric field structure is investigated by reducing the search parameter space. Correlation experiments with second-harmonic generation (SHG) reveal a high degree of problem complexity.

1. Introduction

Control of chemical reactions, i.e., selective cleavage or formation of chemical bonds, has been a fascinating prospect since the invention of the laser 40 years ago. Various theoretical approaches have been proposed [1,2] for achieving such control. For small molecules, experimental realizations have been successful (see Refs. in [3,4]). For large molecules, however, these approaches are difficult or impossible to apply. This is because such control schemes require knowledge of molecular potential energy surfaces, which for complex molecules are usually not accurately known. We present the experimental realization of a different approach first suggested theoretically by Judson and Rabitz [5]. The goal of this scheme is to maximize an experimental result such as cleavage of a specific chemical bond by modifying adequately a femtosecond laser pulse. The outcome of the experiment is used as feedback in a learning algorithm which mimics the biological evolution. Depending on the optimization goal, the algorithm decides which laser pulses out of a pool of individuals (laser pulses with different pulse shapes) are better adapted to the given problem. These "survivors", modified by mutation and crossover, build a new generation of individuals which are again tested in the experiment. Through iteration, the fitness of the survivors increases until the optimization goal is reached. In our setup, the phase of amplified femtosecond laser pulses is modified using a liquid-crystal display (LCD, 128 pixels) pulse shaper.

2. Results and Discussion

Our initial efforts to demonstrate the application of this optimization scheme focused on the automated compression of femtosecond laser pulses [6,7]. In general, ultrafast laser pulses, produced by a chirped-pulse amplification (CPA) system are not completely bandwidth-limited. In order to achieve such pulses,

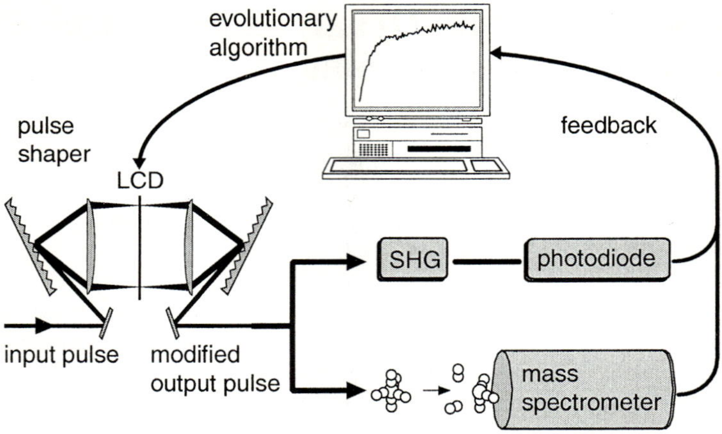

Fig. 1. Experimental setup. A computer-controlled pulse shaper modifies the spectral phase of femtosecond laser pulses. The experimental result (second harmonic generation (SHG) or photodissociation product yield) obtained with the tailored laser pulses is used as feedback in an evolutionary algorithm. The laser field is improved iteratively until an optimal result is found

considerable efforts must be made in day-to-day alignment of the laser components. We therefore apply the optimization scheme, using second harmonic generation (SHG) of the laser pulse in a BBO crystal as feedback signal (see Fig. 1). The efficiency of this process is an experimental signal which measures the degree of compensation of the higher order chirp components of the laser pulse. Within about 100 generations of the learning algorithm, the complex phase-modulated laser pulse is compressed to its bandwidth limit. No knowledge of the input pulse phase is needed, and the compression is achieved in a totally automated fashion.

Just as this technique is capable of removing phase contributions from laser pulses, it is also able to introduce them. We present the first example where such an optimization procedure is used to control the photodissociation product yield of a complex molecule. In contrast to the femtosecond laser pulse compression experiment (where the optimization goal is a bandwidth-limited laser pulse), the shape of the laser pulse necessary for optimizing particular reaction channels is not known. The $CpFe(CO)_2Cl$ molecule, which incorporates different types of chemical bonds, is used as a prototype [3]. The optimization goal chosen was to control the relative quantum yield of two different photodissociation channels. In the first case, the product is $CpFe(CO)Cl^+$, where one Fe-CO bond is broken. In the second case the product channel leading to $FeCl^+$ is chosen. As such, the algorithm was directed to maximize and, in a second step, to minimize the branching ratio $CpFe(CO)Cl^+/FeCl^+$. The resulting product ratios of this control experiment are shown in Fig. 2. Significant optimizations can be achieved in both cases; i.e., the product ratio can be controlled between 5:1 in the maximization experiment and 1:1 in the minimization experiment. The electric fields E(t) calculated from the optimal phase functions and the laser pulse spectrum are also shown in Fig. 2. Both fields exhibit a very complex structure with clearly visible

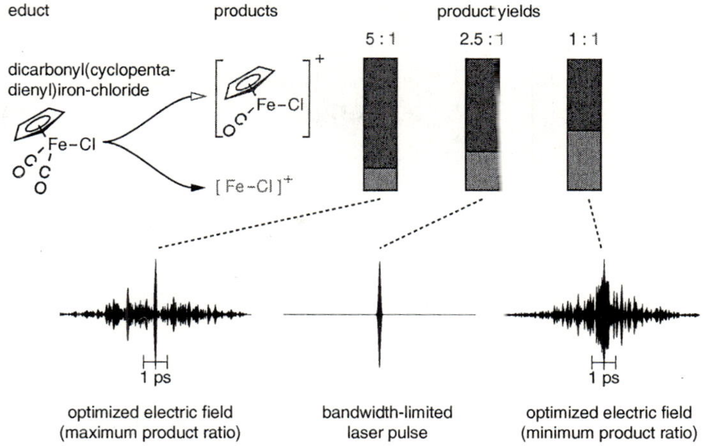

Fig. 2. Relative $CpFe(CO)_2Cl$ photodissociation product yields. Starting from the parent molecule, two different bond breaking reactions lead to chemically different final products. The branching ratio $CpFeCOCl^+/FeCl^+$ is maximized as well as minimized and the resulting optimized electric fields are also shown

differences for the two optimization channels. Applying a bandwidth-limited laser pulse leads to a branching ratio of 1:2.5, which lies between the optimized results.

These experiments clearly demonstrate that the applied method of feedback-controlled optimization is capable of controlling the outcome of a photodissociation reaction. No information about the involved molecules or the reaction mechanism is needed in order to achieve this goal. Since the electric fields are quite complex, the question of the significance of these features arises. In the following, we want to explore the degree of complexity of these types of control experiments.

A first issue to be discussed is whether it is necessary to search the optimum in the enormous search space spanned by the 128 individual optimization parameters, i.e., the 128 pixels of the LCD. Therefore we reduced the search space by coupling the phase of adjacent pixels. In a first step, every second pixel was driven with the same voltage value as its neighbor. In a second step, the constraint was such that the LCD array was divided in sequences of four identically phased neighboring pixels. The resulting 64, or 32, respectively, individual parameters were then optimized to maximize again the $CpFe(CO)Cl^+/FeCl^+$ ratio. The evolution of the fitness curves is shown in Fig. 3. It can be inferred that all the 128 individual optimization parameters are needed to reach the best optimization result. When the search space is reduced, the optimization procedure is slower and the optimum ratio is smaller. This is an indication that the topology of this multi-dimensional search space is very complex and only the full set of optimization parameters is sufficient for reaching the best optimization results.

The question now arises if it can be proved experimentally that these types of optimization experiments are very complex and that it is in general not sufficient to use simple one-parameter control schemes such as a pure variation of the temporal laser pulse intensity. A first hint is given already by the electric fields in Fig. 2. The bandwidth-limited laser pulse, which has the highest peak intensity of

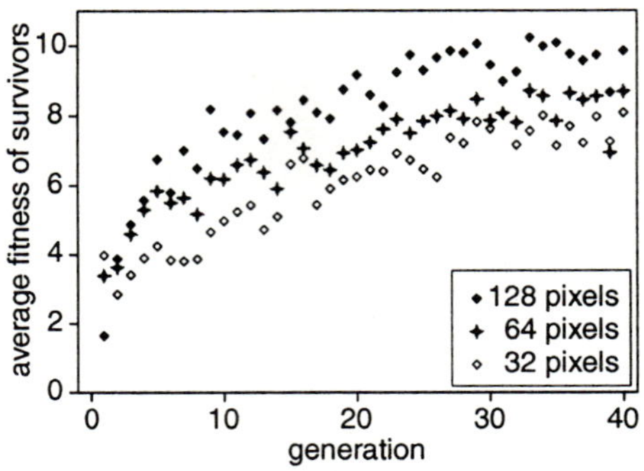

Fig. 3. Evolution of the fitness function with reduced search space

the three pulses used, leads to an intermediate branching ratio. The two optimal pulses both have lower peak intensities arising from the introduction of complex phase modulation. In one case the lower peak intensity leads to an increase of the product ratio whereas in the other case it leads to a decrease. It appears that there is no simple correlation between pulse intensity and optimization result because the detailed phase structure is important.

For a more detailed examination, a correlation experiment was performed. The idea is to try to map the coherent control experiment of unknown complexity onto a simpler control experiment whose search space topology is known. Since we wanted to investigate intensity dependencies, we chose to measure the amount of SHG light and simultaneously the product yield for each laser pulse shape during the optimization procedure. If the outcome of the product yield/ratio depends upon the laser pulse intensity in the same way as does SHG (i.e., the higher the intensity, the higher the efficiency), the two signals (SHG, product yield/ratio) have to be correlated. On the other hand, if the two signals are not correlated the molecular control experiment requires a specific phase function and not simply the correct laser pulse intensity. In a first correlation experiment, the direct ionization of the parent molecule $CpFe(CO)_2Cl$ was maximized. Figure 4a shows that the ionization of the parent molecule is directly correlated to the SHG process. It is known from previous fs-experiments on organometallic molecules that the ionization process of these molecules strongly depends on the pulse duration [8]. The longer the pulse duration, the higher is the degree of fragmentation. Therefore the ion yield of the parent molecule $CpFe(CO)_2Cl$ increases for shorter laser pulses, as does the SHG intensity. In this test experiment, it was therefore possible to map the search space topology of the unknown control experiment onto the topology of the well-understood SHG process.

A completely different behavior is observed for the maximization of the $CpFe(CO)Cl^+/FeCl^+$ ratio (see Fig. 4b). There is no clear correlation between product ratio and SHG efficiency. The complexity of this control experiment is therefore much higher than the complexity of the SHG process. A variety of laser pulse shapes with different SHG signals and therefore different pulse intensities are capable of producing the same product ratio. Or, from a different perspective, fixing the pulse intensity alone is not sufficient to reach the optimal branching ratios. Specifically phase-shaped laser pulses are required instead. This result

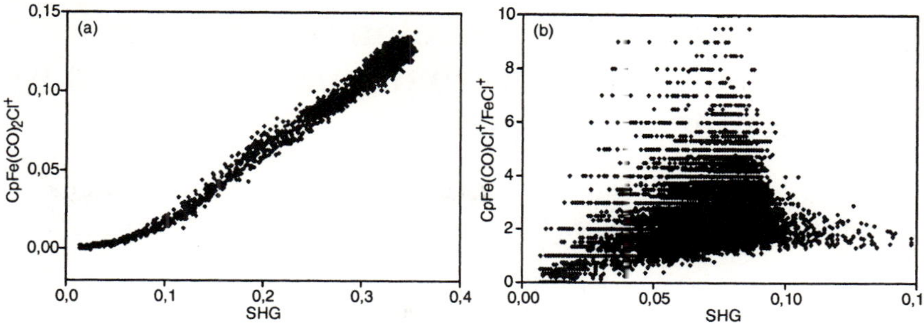

Fig. 4. Correlation diagram. The amount $CpFe(CO)_2Cl^+$ ion yield (a) and the $CpFe(CO)Cl^+/FeCl^+$ product ratio (b) versus the corresponding SHG signal is plotted for each individual laser pulse shape during the two optimization experiments

confirms the prior observation that the best product ratios are reached by the optimization of all 128 parameters individually.

3. Conclusion

Feedback-controlled optimization of the photodissociation reaction of a complex organometallic molecule has been demonstrated. No prior knowledge of the potential energy surfaces of the parent molecule or its fragments is needed in order to control the branching ratio between two different photodissociation channels. If the evolutionary optimization algorithm uses the full set of the available optimization parameters, the best branching ratios are achieved. The high degree of problem complexity was determined by examining correlations with second-harmonic production.

References

1 P. Brumer and M. Shapiro, Chem. Phys. Lett. **126**, 541, 1986.
2 D. J. Tannor, R. Kosloff, and S. A. Rice, J. Chem. Phys. **85**, 5805, 1986.
3 A. Assion, T. Baumert, M. Bergt, T. Brixner, B. Kiefer, V. Seyfried, M. Strehle, and G. Gerber, Science **282**, 919, 1998.
4 M. Bergt, T. Brixner, B. Kiefer, M. Strehle, and G. Gerber, J. Phys. Chem. **103**, 10381, 1999.
5 R. S. Judson and H. Rabitz, Phys. Rev. Lett. **68**, 1500, 1992.
6 T. Baumert, T. Brixner, V. Seyfried, M. Strehle, and G. Gerber, Appl. Phys. B **65**, 779, 1997.
7 T. Brixner, M. Strehle, and G. Gerber, Appl. Phys. B **68**, 281, 1999.
8 L. Bañares, T. Baumert, M. Bergt, B. Kiefer, and G. Gerber, J. Chem. Phys. **108**, 5799, 1998.

Optimal control of two-photon transitions: bright and dark femtosecond pulses designed by a self-learning algorithm

T. Hornung, R. Meier, D. Zeidler, K.-L. Kompa, D. Proch, and M. Motzkus*

Max-Planck-Institut für Quantenoptik, P.O. Box 1513, D-85740 Garching/Germany
*E-mail: mcm@mpq.mpg.de

Abstract. Coherent quantum control of the nonresonant 3s-5s excitation in sodium is demonstrated. "Bright" or "dark" pulses are designed by an evolutionary algorithm which evaluates a signal from the experiment to control a phase mask.

1. Introduction

The synthetization of arbitrarily shaped fs pulses by spectral filtering to drive a quantum mechanical system towards a target state requires detailed knowledge of the Hamiltonian which, except for the simplest molecules, is generally unknown. As an alternate strategy, Rabitz *et al.* [1] have thus suggested to establish an interactive dialogue between the experiment and a self-learning mathematical algorithm which iteratively optimizes a suitable merit function and eventually produces the parameters of the „ideal" spectral filter. The scientific literature since then reports several successful implementations of this principle [2-4].

As a demonstration of effective feedback-control this paper describes how an appropriately tailored spectral phase of the pump pulse can either maximize or minimize the population transferred by the nonresonant two-photon process (3s→→5s). This is, at least to the best of our knowledge, the first demonstration of *dark pulses* that have been designed by a self-learning genetic algorithm which „listens" to the experiment.

2. Experimental Methods

Sodium in a heat-pipe oven was excited by phase-modulated 100 fs pulses at λ = 598nm (close to the 3s→→5s resonance) with a power density at the focus of $\approx 1\times 10^{11}$ W/cm^2. The modulation was realized with a liquid crystal (LC) mask in a 4-f arrangement. The evolutionary strategy sets out from a random seed to adjust 48 vectors of 128 parameters (LC voltages). The heat pipe cell buffered with 10mbar of Ar was operated at T=520K to eliminate pulse propagation effects.

The population of the 5s target level optically decays to 3p, or undergoes collisional relaxation to the 4p state. Both levels are monitored separately via their fluorescence to the 3s ground state at 589nm and 330nm, respectively.

3. Results

Population of the 5s level is not necessarily due to a two-photon absorption since the spectral width of the pump pulses permits the competitive (1+1)-photon excitation via 3p (at 589nm). Meshulach *et al.* [5] have given a theoretical treatment of the quantum control of multiphoton transitions by shaped ultrashort pulses which excludes strong field effects. Their model served to interpret an experiment which recorded the observed 4p-fluorescence following the 3s→→5s transition as a function of the π step position induced by the SLM and showed one maximum around ≈602nm in the midst of two symmetrically arranged minima. The conclusion clearly suggests a nonresonant two-photon process at λ ≈602nm.

The optimization routine processes the signal intensity of 4p-fluorescence. In contrast to 4p-, the concomitantly induced 3p-fluorescence is drastically diminished in the absence of the wavelength which matches the one-photon 3s-3p resonance. This represents an additional hint that 5s, as the precursor to 4p, is accessed nonresonantly, rather than by a (1+1)-sequence. The power dependence of the 4p- and the 3p-fluorescence signals reveals 4p as being exclusively fed from 2-photon excited 5s. The 3p state draws to some extent from 5s, but is predominantly pumped in a resonant single step.

The nonresonant two-photon interaction of an ultrashort pulse with a two-level system induces a transition with a probability S_2 [5]:

$$S_2 = \left| \int A\left(\frac{\omega_0}{2}+\Omega\right)\cdot A\left(\frac{\omega_0}{2}-\Omega\right)\cdot \exp\left\{ i \left[\underbrace{\left[\Phi\left(\frac{\omega_0}{2}+\Omega\right) + \Phi\left(\frac{\omega_0}{2}-\Omega\right) \right]}_{\text{interference term}} \right] \right\} \cdot d\Omega \right|^2$$

where ω_0 is the energy of the 3s→→5s transition (301nm). Two-photon events are induced by all pairs of photons which satisfy the condition $\omega_1 + \omega_2 = \omega_0$. The detuning of frequencies ω_1, ω_2 from $\omega_0/2$ is denoted by Ω. Control of the excitation process is exercised via the interference term. Maximization of S_2 is obviously achieved if the interference term vanishes, which describes the minimum-duration transform limited pulse. This solution is not singular, however, since any tailored pulse with the same power spectrum $A(\omega)$ but with an antisymmetric phase function, $\Phi(\omega_0/2-\Omega) = -\Phi(\omega_0/2+\Omega)$, will produce the same result. This is counterintuitive since longer, i.e. less intense, pulses should be less effective in transferring population. So-called *dark pulses* which nullify the two-photon pumping probability are induced as long as $\Phi(\Omega) = \cos(\beta\cdot\Omega)$. The total of solutions, discriminated by virtue of the parameter β is symmetric with respect to the center frequency $\omega_0/2$.

The task to maximize or cancel S_2 was left to an evolutionary strategy embedded in a feedback loop. The *Ansatz* for the phase filter, $\Phi(n) = a\cdot\cos(b\cdot n + c)$ with n as the variable numbering the LC pixels, and a, b, and c as free parameters to be optimized proved still manageable but sufficiently general to comprise the solutions given in [6]. The phase filters which were retrieved are either symmetric (extinction), or antisymmetric (enhancement of fluorescence) with reference to the center frequency $\omega_0/2$.

Animated by the question of the existence of additional solutions of altogether different character we thus expanded the parameterization to 128 sample points, each falling between 0 and 2π as before. Fig. 1 documents the convergence towards the dark (bottom) and the bright pulse (top). In accordance with theory an antisymmetric phase function promotes population enhancement. No likewise obvious symmetry properties, however, characterize the suppression of two-photon pumping. Repeated dark-pulse optimizations produced identical experimental results but varying phase functions which bore no resemblance with the prediction of theory. The shaped pulses show a complex phase- and amplitude-time structure of comparable duration (≈2ps). Rather than their peak power it is their phase distribution which produces qualities such as „bright“ or „dark“.

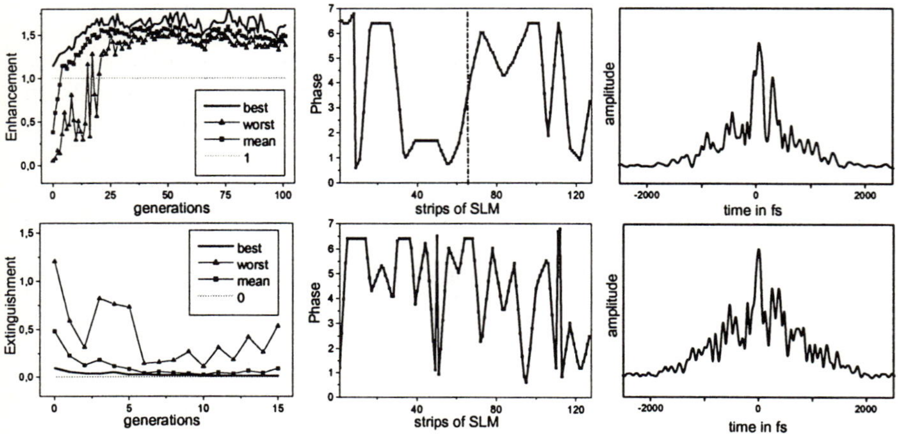

Fig. 1. Convergence of the 128-parameter search for dark (bottom) and bright (top) pulses. Left: Convergence of normalized fluorescence intensity. Dashed lines mark „no signal“ (0) and „unshaped reference pulse“ (1). Middle: Phase structure of bright pulse showing antisymmetry. Right: Pulse shape in the time domain.

4. Conclusions

Designed pulses which achieve distinct control of a two-photon transition were iteratively found by an unbiased evolutionary algorithm which communicated with the experiment. The search of the feedback algorithm in an extended parameter space produced in addition to known analytic solutions novel phase structures which are not intuitively understandable and call for further theoretical studies.

References

1 R. S. Judson, H. Rabitz, Phys. Rev. Letters 68 (1992) 1500
2 C. J. Bardeen, et al., Chem. Phys. Letters 280 (1997) 151
3 A. Assion, et al., Science 282 (1998) 919
4 T. C. Weinacht, J. Ahn, P. H. Bucksbaum, Nature 397 (1999) 233
5 D. Meshulach, Y. Silberberg, Phys. Rev. A 60 (1999) 1287

Feedback optimization of molecular states using a parameterization in frequency and time domain

T. Hornung, R. Meier, and M. Motzkus*

Max-Planck-Institut für Quantenoptik, Hans Kopfermann Strasse 1, D - 85741 Garching, Germany
*E-mail: mcm@mpq.mpg.de

Abstract: Combining arbitrary pulse-shaping with degenerate four-wave-mixing (DFWM) in a feedback controlled experiment, parameterization in the time domain is demonstrated giving direct evidence of the applied control process in a learning loop.

1. Introduction

A feedback loop, consisting of a computer controlled fs pulse modulator and an evolutionary algorithm, allows to optimize a certain physical objective in a molecule without prior knowledge of the potentials [1-3]. The obtained optimal solution of the electric field which forces the molecule into this certain target state has typically a complex structure which is difficult to interpret in terms of the induced control processes.

The optimization usually takes place by changing the amplitude and phase of certain parts of the spectrum of the fs laser pulse. In this contribution we present a parameterization of the shaped fs pulse in the time domain which directly represents the pulse structure and allows to reduce the number of active parameters. This leads to faster convergence of the algorithm and gives direct insight into the applied control mechanism.

In order to compare the optimization in the frequency domain with one in the the time domain we have chosen K_2 in the gas phase which is excited and probed in a femtosecond three pulse degenerate four-wave-mixing (3-DFWM) setup. 3-DFWM can be used to monitor [4] and control the excited as well as the ground state dynamics of molecules[5]. We combined DFWM with arbitrary pulse-shaping extending the possibilities to influence this non-linear process by multiple pulses. In addition, our results on K_2 show that control of ground and excited state vibrational dynamics can be further influenced by phase-locked pulse trains.

2. Experimental setup

100 fs laser pulses at 825 nm are split into three beams having each an energy of 50 nJ / pulse arranged in a forward BOXCARS geometry. The beams are focused

into a heat pipe filled with potassium and argon (buffer gas) heated to a temperature of 360°C. The FWM signal is detected with a PMT and using a scanning monochromator as bandpass. One of the three beams is sent through an all-reflective pulse shaper with a phase and amplitude modulating spatial light modulator (SLM) at its Fourier plane. An evolutionary algorithm automated this setup controlling the shaper according to a feedback signal received from the experiment. Altogether this setup represents a learning-loop.

3. Results and Discussion

For the feedback signal a 5 ps long DFWM transient (FFT window) was acquired and Fourier transformed resolving the two peaks corresponding to ground (X) and excited (A) electronic state vibrations. The feedback signal was attained by computing the difference between these two vibrational peak intensities and adding a mean of the subsoil of the Fourier data. It was necessary to acquire the FFT window for time separations τ far out the shaping window (Fig. 1).

It was shown that pulse trains can excite one vibrational mode of molecules embedded in a matrix, a process known as Impulse Stimulated Raman Scattering (ISRS) [6], or even transfer population into one vibrational level [7]. Furthermore some influence of phase-locked pulses on DFWM was reported [8].

Choosing adequate parameterizations we could restrict the learning-loop to these promising mechanisms and at the same time make the genes, the evolutionary algorithm operates on, of short length.

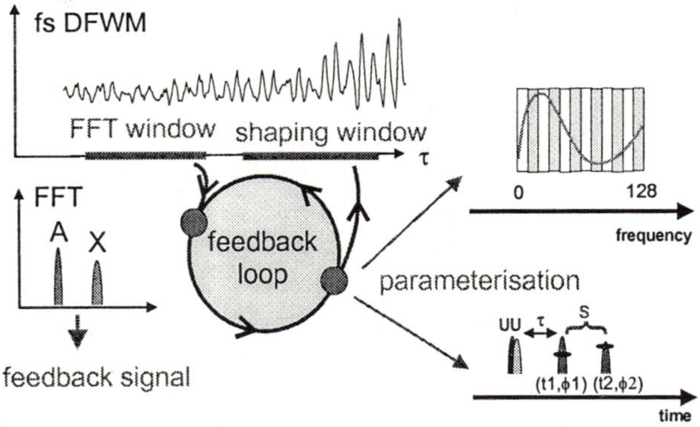

Fig. 1. The feedback signal is derived from the transient by FFT and evaluating the difference in peak heights between A and X. This number serves as feedback to an evolutionary algorithm which uses genes either representing frequency or time domain.

Pulse trains with a constant interpulse separation can be shaped most easily by binding the modulator pixels together into a spectral phase function of periodic type $a \cdot \sin(b \cdot x+c)$. In this case the genes are of length 3, since they are built from frequency domain floating point values a,b and c. In fig 2a the optimized pulse train, maximizing X state vibrations, is shown. The shaped pulse has an interpulse separation of exactly 2 times the vibrational periods of the X state (720 fs). This measurement presents for the first time a proof that control of molecules in the gas phase can be achieved through ISRS (to be published).

Phase-locked double pulses are modulated by quite complex phase and amplitude mask patterns, allowing no easy functional representation. In contrast double pulses can be easily parameterized in the time domain. Here the genes consist of only four parameters, two for each pulse representing its phase and time of occurrence. A double pulse of time separation 560 fs with a phase difference of 2.3 rad optimizes A as shown in Fig.2b. A pulse pair with a separation of 785 fs and 4.4 rad optimizes X (not shown). The smallest time step the shaper could realize in our setup was 80 fs, so an exact matching of the interpulse separation to the vibrational periods of the A and X state dynamics was not possible. Despite this fact the molecular vibration can still be restricted either to the excited or ground state, if the phase in ISRS is used as a control knob. An additional measurement shows, that varying only the phase relationship in a pulse pair can yet control the ratio between A and X vibrations.

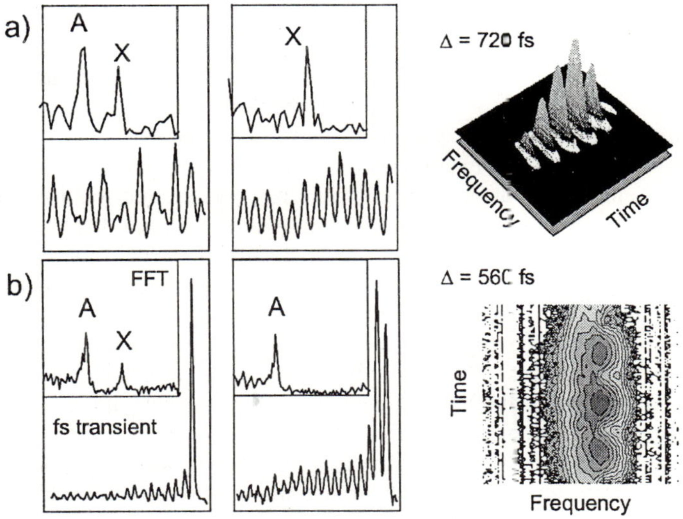

Fig. 3. Optimization using genes in (a) frequency and (b) time domain.

References

1 R. S. Judson, H. Rabitz, Phys. Rev. Lett. 68 (1992) 1500
2 C. J. Bardeen, et al., Chem. Phys. Lett. 280 (1997) 151
3 A. Assion et al., Science 282 (1998) 919
4 M. Schmitt, G. Knopp, A. Materny, W. Kiefer, Chem. Phys. Lett. 280 (1997) 339
5 I. Pastirk, et al., Faraday Discussions 113 (1999) 1
6 H. Kawashima, M. Wefers, K. Nelson, Annu. Rev. Phys. Chem. 46 (1995) 627
7 K. Sundermann, R. de Vivie-Riedle, J. Chem. Phys. 110 (1999) 1896
8 W. de Boeij, M. Pshenichnikov, D. Wiersma, Chem. Phys. Letters 238 (1995) 1

Controlling and Probing Impulsively Induced Ground State Vibrational Dynamics

Erez Gershgoren[1], Sanford Ruhman[1], Jiri Vala[2], Ronnie Kosloff[2].

[1] Dept. of Physical Chemistry, and the Farkas Center for Light Induced Processes, the Hebrew University, Jerusalem 91904, Israel. E-mail: sandy@fh.huji.ac.il.
[2] Dept of Physical Chemistry, and the Fritz Haber Center for Molecular Dynamics, the Hebrew University, Jerusalem 91904, Israel. E-mail ronnie@fh.huji.ac.il.

Abstract. Wavelength, intensity and chirp are used to control impulsive generation and detection of I_3^- ground state vibrational coherences. Intensity harmonic switching, sub- $1/v^2$ dependence of second harmonic T^*_2(vib), and phase space mapping of vibrational coherences by chirped probing are demonstrated.

Resonant impulsive optical excitation of molecular species not only generates compact wavepackets on the excited state, but also drills coherent dynamical holes on the ground electronic surface. This ubiquitous phenomenon was investigated in our lab for UV photolysis of I_3^- ions, and demonstrated to be a valuable spectroscopic tool for following ground state dynamics in real time [1]. In those experiments dephasing of the symmetric stretching fundamental was measured. No modulations related to other vibrations, or higher harmonics of the symmetric stretch were observed. To analyze our results, a novel coordinate dependent two level system scheme was devised. It provided detailed insight into molecular response to impulsive excitation at intensities extending beyond the perturbative limit, in terms of a textbook model. Here we report a renewed investigation of this process using tunable 20-30 fsec laser pulses, to fully explore how pulse intensity, frequency and chirp can control the coherence imposed on ground state vibrations, and probe its evolution.

Model calculations detect intensity effects on the structure of the ground state hole, even at mild pump fluences, indicating an increase in the hole's momentum upon an increase of pump power. To demonstrate this effect we present simulated coherent holes generated for a 30 fsec excitation pulse centered at 360 nm, for a one dimensional model of I_3^-. Integration of the quantum mechanical equations of motion is conducted for a model consisting of ground and excited state sections along the symmetric stretch alone. A Wigner representation of residual ground state population following photoexcitation of a room temperature distribution is presented. Only the dynamic part of $\rho(x,p)$ is plotted, by subtracting ρ_{eq} from the density. This results in a zero norm density, which captures all the dynamical aspects of the subsequent evolution.

The upper part of the Figure below exhibits dynamical holes for 4 pumping fluences. As time evolves, this hole rotates in the designated direction around the

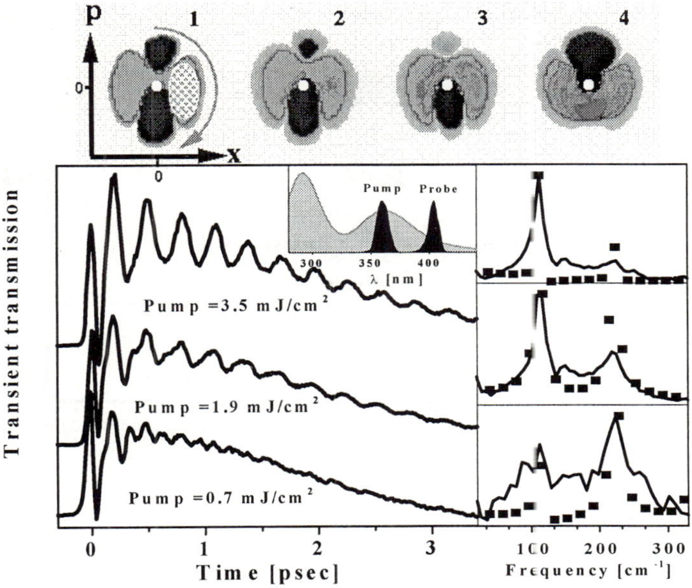

origin (white dot). Black and gray portions of the hole designate negative and positive densities respectively. At the lowest fluence, the hole consists of alternating positive and negative quadrants. Alternatively stated, at low intensities, impulsive excitation in the center of a band drills a nearly zero momentum hole, consisting of two counter propagating holes. A 400 nm pulse probes the evolving hole at its outer turning point, as demonstrated by the designated observation window. For each period (2π rotation in phase space), it thus detects 2 minima and 2 maxima, and its absorption is modulated at $2\nu_{vib}$. A moderate increase in pump intensity causes a rapid gain in hole momentum, reshaping it as a single cycling minimum, leading to detection of the fundamental frequancy predominantly.

The lower panel of figure 1 demonstrates the fine agreement between theory and experiment. Experiments were conducted with a home made multipass amplified Ti;sapphire laser system coupled to a commercial OPA (TOPAS) producing ~20 fsec pulses throughout the visible. Frequency doubling of these pulses in thin BBO crystals provided both pump and probe. Laser spectra are shown along with the absorption of I_3^- in ethanol solution in the insert. The pump/probe data are depicted on the left, with the appropriate Fourier analysis of the oscillatory components shown to the right. A Fourier analysis of the numerical simulations using the same fluence levels is superimposed on the frequancy analysis of the experimental data, demonstrating the outstanding agreement between the simulated and experimental results. In both the excitation probabilities ran from 5 – 30 % in the tested fluence range.

The effect of chirped pump on the RISRS process has been both theoretically studied [2], and experimentally tested [3]. Here we deal with the effect of chirp in probing pulses on the detection of ground state dynamics, as demonstrated in the following figure [4]. The residual chirp was controlled by changing the insertion of a pair of suprasil dispersion compensating prisms, and changed in jumps of 250 fsec^2. The pump at 400 nm generates a dynamic hole near the outer turning point

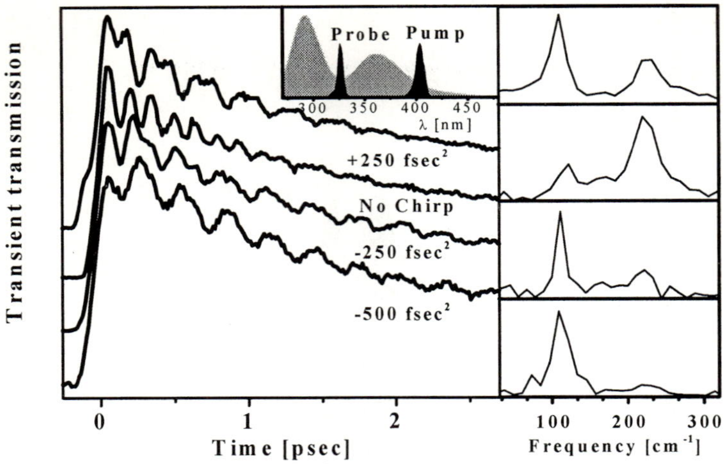

of the ground state density, leading to a single cycling hole in phase space. Positioned near the midpoint between the two I_3^- absorption peaks, 325 nm probes simultaneously detect the ground state density at the inner and outer turning points. Thus, the signature of even a single evolving hole in the ground state should predominantly appear at twice the symmetric stretching frequency. While this is the case for transform limited probe pulses, even slight amounts of residual chirp ($\pm$ 270 fs^2), which change the pulse duration less than 30%, shift the modulations to the first harmonic. This is apparent also in the Fourier transforms of the modulations. Closer scrutiny of the data shows however that the first harmonic modulations obtained through opposite signed chirp are precisely antiphased.

Conclusions of this study are A) Even at mild excitation fluences, dramatic intensity effects are observed on ground state coherences and their signature in transmission. Agreement of experiment and theory stresses utility of nonperturbative simulation approaches when light induced effects on the material system are required. B) Slight residual chirp of probing pulses can equally cause dramatic changes in the temporal absorption modulations caused by such coherences, and serve to map the dynamic hole position in phase space. C) These and other methods for suppressing modulations at the fundamental frequency boost the accuracy of characterizing vibrational dynamics of higher harmonics and other weak modulation features. The T_2^* ratio between the first and second harmonics of the symmetric stretch fall well below the expected 4:1. More study will be required to resolve this issue.

References

1 G. Ashkenazi et.al., Adv. Chem. Phys., **100**, 229-317(1997).
2 S. Ruhman and R. Kosloff, J. Opt. Soc. Am. **B**, **7**, 1748-1752 (1990).
3 C. J. Bardeen et. al., J. Phys. Chem. A, **102**, 2759-2766 (1998).
4 R. Zadoyan et. al., Chem. Phys., **233**, 353-365 (1999).

Dynamics and Coherent Control of Condensed Phase Vibrational Coherences

Valeria D. Kleiman, Dale McMorrow, and Joseph S. Melinger

Naval Research Laboratory, Code 6820, Washington DC 20375, USA. Tel: 202-767-2447.
E-mail: valeriak@ccs.nrl.navy.mil

Abstract: The dynamics and control of condensed phase vibrational coherences are studied using electronically resonant transient birefringence and dichroism. The chirp direction and mode of detection yield qualitatively different responses in the birefringence and dichroism signals.

In recent years the study and manipulation of molecular dynamics using phase-modulated excitation pulses has generated great attention[1]. The creation of non-stationary wavepackets using ultrafast excitation pulses has lead to the experimental manipulation of wavepacket dynamics in the gas phase[2]. Recently, there has been a growing interest in using phase properties of ultrashort pump pulses to influence the excitation of molecules in solution[3].

In this paper we explore the properties of vibrational coherences in condensed phase molecules created with phase-modulated pump pulses. Measuring either the real or the imaginary components of the third-order susceptibility created by both transform limited and chirped pulse excitation, we investigate specific response functions, which are related to the created vibrational coherences.

The transient anisotropy induced by femtosecond pulse excitation on the dye IR125 was measured using the OHD-RIKES experimental setup[4]. Phase modulation on the pump pulse was introduced through a pair of prisms positioned on the pump beam pathway. Transform limited (TL) pulse profiles yield a 35 fsec (FWHM) pulsewidth, in good agreement with the experimentally determined bandwidth of 515 cm^{-1}. The phase-modulated pulses have cross correlations of 75 and 82 fs for positive (PC) and negative chirp (NC), respectively.

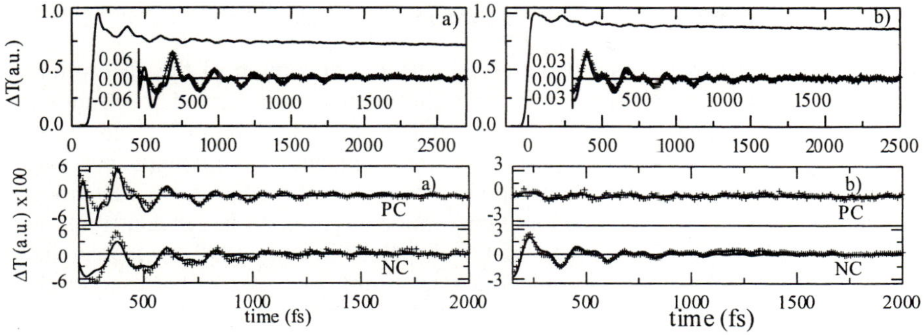

Fig. 1. Normalized transient signals for birefringence (a) and dichroism (b). Top panels: insets shows data (crosses) after subtraction of the long-lived contributions and fits (solid lines). Bottom panels: excitation with with PC and NC pulses.

Figure 1 (top) shows the birefringence and dichroism responses excited by a near TL pulse. The data were analyzed by first subtracting the exponential contributions, and then fitting the oscillatory portion of the response to a sum of damped sinusoids. These sine waves arise from intramolecular vibrational degrees of freedom, which yield the vibrational coherences.

Analysis of birefringence transients reveals 4 vibrational modes, with the most prominent contributions at 141 cm^{-1} and 298 cm^{-1}. Also evident is a low frequency component at ~20 cm^{-1}, and a higher frequency mode at 417 cm^{-1}. The dichroism data generated by near TL pulses appears similar to the birefringence, although the relative amplitude of the oscillatory components is smaller. Fitting to a sum of damped sinusoidal waves yields the 141cm^{-1} and 298 cm^{-1} oscillatory components.

To better understand the source of the oscillatory components, the birefringence measurement was repeated with the probe beam dispersed by a grating before detection. Figure 2 shows the birefringence transients obtained for each spectral

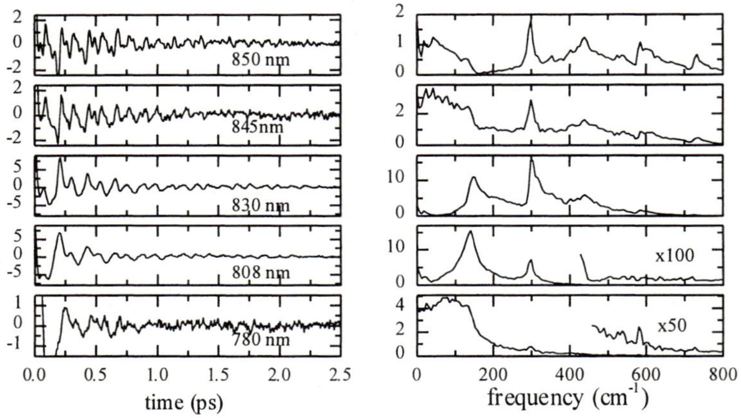

Fig. 2. Birefringence signal for selected probe frequencies. Left panel shows transients corresponding to each probe frequency; right panel shows the corresponding FFT spectrum.

region together with the corresponding spectra. These transients and spectra show a pronounced dependence on the probe frequency. The Fourier transforms were obtained on the reduced datasets after subtracting a long time (ps) exponential contribution and a fast decay ($\tau \sim 140$ fs).

The frequency-dispersed spectrum (808) near the maximum of the absorption spectrum (810nm) is qualitatively similar to the frequency-integrated spectrum of a TL pulse, with strong vibrational activity at 141 cm^{-1} and 298 cm^{-1}. In contrast, at the wings of the probe spectrum, new high-frequency components become evident. These new components correspond to vibrational modes at *c.a.* 420 cm^{-1}, 585 cm^{-1} and 730 cm^{-1}. The broad 420 cm^{-1} mode has a much larger contribution to the FT spectrum at the red end of the dispersed spectra (850 nm), than its contribution to the TL frequency-integrated spectrum. The spectra corresponding to the blue side (780 nm) of the absorption band show the presence of the 585 cm^{-1} and 730 cm^{-1} peaks, although with a much weaker contribution.

Figure 1 (bottom) shows the oscillatory components corresponding to phase-modulated excitation pulses (for frequency integrated detection). The changes due to PC or NC excitation are quite pronounced, and go beyond the trivial degradation of modulation depth due to the longer excitation pulses. For the birefringence measurement, the 417 cm^{-1} mode, which is present in both the TL and PC data, makes no contribution to the signal generated with a NC pulse. While the amplitude of the 298 cm^{-1} mode remains unchanged, the amplitude of the 141 cm^{-1} mode is diminished by 30%. When frequency components corresponding to excited state vibrations are excited with NC pulses they should appear as weaker contributions to the total signal, since the blue-to-red sweep is ordered to maximize the coupling back to the ground state. This argument, together with its very fast decay (τ=130 fs), suggests that the 417 cm^{-1} oscillation arises from an excited state mode. In contrast, the dichroism signal shows oscillatory contributions in the NC data that are nearly identical to those obtained with TL excitations. For PC excitation, however, the oscillations are nearly completed attenuated, with only a slight contribution from the 298 cm^{-1} mode (one order of magnitude smaller).

In summary, for a TL excitation pulse, the percentage of the total signal that is due to vibrational coherences is at least a factor of two larger in the birefringence than in the dichroism. For the transform-limited pulse, four modes are observed in the birefringence response, and only two in the dichroism (when measured with a frequency integrated probe pulse). Finally, for excitation with PC pulses, the vibrational coherences are nearly completely attenuated in the dichroism, while remaining prominent in the birefringence.

One possible explanation for these results is that the birefringence and dichroism contributions to $\chi^{(3)}$ preferentially sample different regions of the potential energy surfaces coupled by the laser field. For example, the dichroic component should have a frequency response that is similar to that of a transient absorption/stimulated emission process. For shifted potential energy surfaces, the peak of the frequency response should occur in the region where wavepacket broadening is at a maximum, whereas the high and low frequency wings will sample the wavepacket near the turning points. Since we measure the dichroic response integrated over all probe frequencies one may expect this measurement to have preferentially sampled the central region of the coupled surfaces, thus, leading to an attenuation of the coherent oscillations. The birefringence component should have a different frequency response, possibly with larger weighting near the turning points of the coupled surfaces. If this were the case one would expect a larger contribution from the coherent oscillations to the total birefringence signal.

References

1 R. J. Gordon and S. A. Rice, *Ann. Rev. Phys. Chem.* **48**, 601 (1997).

2 B. Kohler, V. Yakovlev, J. Che, J. L. Krause, M. Messina, and K. R. Wilson, *Phys. Rev. Lett.* **74** (17), 3360 (1995) ;.. G. Gerber and T. Baumert, *Is. J. Chem.* **34**, 103 (1994).

3 V.D. Kleiman, D. McMorrow, and J.S. Melinger *World Scientific* (2000); V.D. Kleiman, S.M. Arrivo, J.S. Melinger, and E. J. Heilweil, *Chem. Phys.* **233**, 207 (1998); C. J. Bardeen, Q. Wang, and C.V. Shank, *J. Phys. Chem.* **102**, 2759 (1998).

4 W.T. Lotshaw, D. McMorrow, N. Thantu, J.S. Melinger and R. Kitchenham, *J. Ram. Spec.* **26**, 571 (1995).

Spatiotemporal Coherent Control

Richard M. Koehl and Keith A. Nelson

Department of Chemistry, M.I.T., Cambridge, MA 02139. Email: kanelson@mit.edu

Abstract. Spatially shifted and temporally delayed pulses and pulse pairs are used for phase-coherent manipulation of phonon-polariton wavepackets as they propagate at light-like speeds. Spatiotemporal coherent control is conducted manually or through automated spatiotemporal pulse shaping, and is guided by spatiotemporal imaging of the propagating responses.

Ultrafast optical control over coherent molecular and material responses has generally been conducted through the use of excitation light fields with specified time-dependent amplitude and phase profiles, but not with time-varying spatial profiles. This is adequate for control over localized responses, but it does not allow the excitation light to follow and continue to manipulate a propagating response as it moves through its host medium. Here we report "spatiotemporal" coherent control over phonon-polaritons (mixed lattice vibrational/electromagnetic waves) that move at light-like speeds through polar crystals. Ultrashort pulses that drive coherent phonon-polaritons through impulsive stimulated Raman scattering arrive at specified times and sample locations to manipulate selected responses.

Results of the simplest demonstration are shown in Fig. 1. At $t = 0$ ps, an amplified ultrashort 800-nm pulse focused to a line in a $LiTaO_3$ crystal generates counterpropagating, "ripple"-like phonon-polariton wavepackets which are monitored through spatiotemporal imaging. [1] Selected images are shown in column (a), wavepacket amplitudes in (b), and intensities in (c). At $t = 2.13$ ps, a second optical pulse of equal intensity generates a second pair of wavepackets. The timing and position of the second pulse are optimized for amplification of the rightward moving wavepacket $|v'\rangle$ which has approximately twice the amplitude and four times the intensity of the initial wavepacket $|v\rangle$. The two leftward moving wavepackets are not spatially or temporally coincident and do not interact.

In the next examples, the two spatially separated excitation beams are passed through different regions of a binary "grating" phase mask pattern. The grating diffracts each beam into ±1 diffraction orders which are focused by a two-lens telescope onto the sample. This yields two excitation interference patterns at the sample, temporally delayed and spatially separated from each other. Results are shown in Fig. 2. In (a), selected images of traveling phonon-polariton wavepackets generated by the first pulse pair at $t = 3.35$ ps are shown. The leftward-propagating wavepacket is canceled at approximately $t = 20.83$ ps by the second pulse pair, while the rightward-propagating wavepackets evolve without interference. Figure 2b shows a condensation of the images in (a) and many more, averaged over the vertical dimension and placed one above the other to produce a vertical time axis.

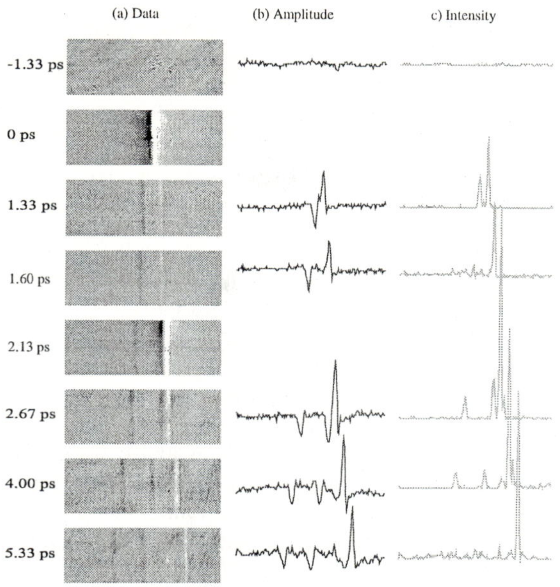

Fig. 1. Coherent wavepacket addition. Two spatially and temporally separated ultrafast optical pulses generate and amplify a propagating terahertz-band coherent state. See text. (a) Representative phonon-polariton images. The spatial scale is 1.5 mm. (b) Amplitudes of each wavepacket, calculated by averaging along the vertical dimension of the data represented in column (a). Values are not shown for 0 and 2.13 ps because the incident optical pulses generate a strong electronic response. The opposite signs are an artifact of the imaging method. (c) Approximate intensities of wavepackets calculated from column (b).

Figure 2c shows results of a similar experiment with the relative timing of the pulse pairs adjusted for constructive rather than destructive wavepacket interference, yielding amplification rather than cancellation.

Finally, a recently improved form of automated spatiotemporal pulse shaping [2] has been used for a preliminary demonstration of automated spatiotemporal coherent control with many pulses. Successive pulses were made to arrive at the sample at nearby locations, yielding a round excitation spot that moved along the sample at the phonon-polariton speed of about c/6. Each spot generated a ripple-like phonon-polarion response radiating outward [1], the front part of which was amplified by the sequence of pulses to produce a visible image. In contrast, excitation waveforms that were shaped similarly but with slightly different timing did not match the phonon-polariton speed and yielded no amplification.

Spatiotemporal coherent control and its automation will have important applications in nonlinear lattice dynamics, exciton as well as phonon-polariton manipulation, terahertz spectroscopy, and ultrafast optical signal processing.

Acknowledgements. Supported in part by National Science Foundation grant no. CHE97-13388 and Army Research Office grant no. DAAG55-98-1-0431.

References

[1] R. M. Koehl, S. Adachi, and K. A. Nelson, "Direct visualization of collective wavepacket dynamics," J. Phys. Chem. A. **103**, 10260-10267 (1999).
[2] R. M. Koehl, T. Hattori, K. A. Nelson, "Automated spatial and temporal shaping of femtosecond pulses," Opt. Comm. **157**, 57-61 (1998).

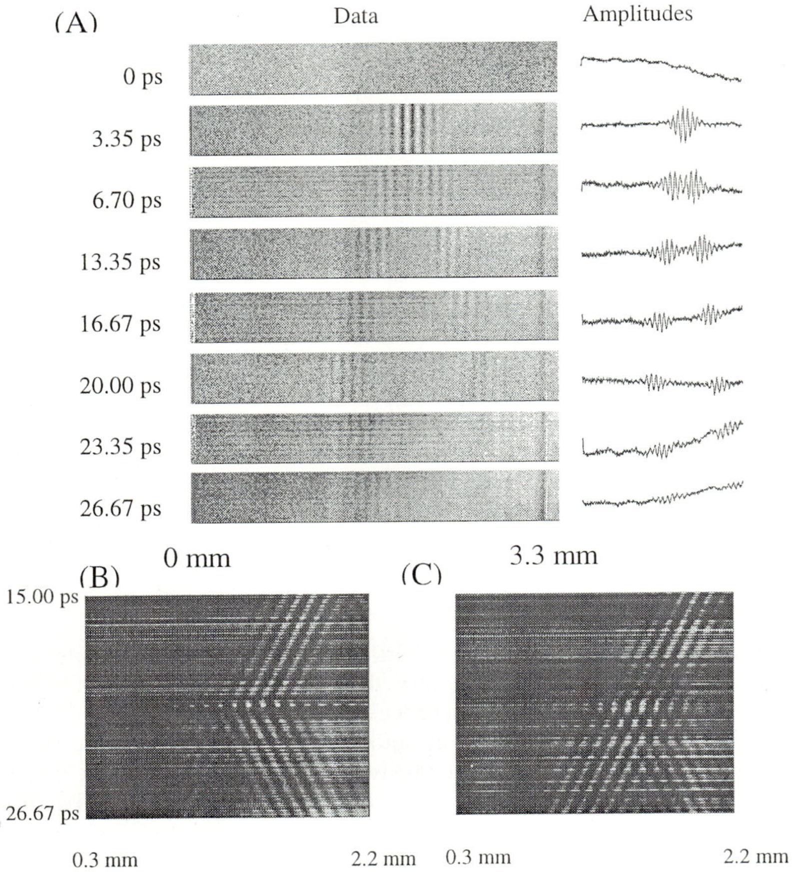

Fig. 2. Wavepacket cancellation and amplification. (a) Images showing cancellation by a second excitation pulse pair of the leftward-propagating wavepacket generated by a first pulse pair. (b) Condensation of images like those in (a), showing times and locations near the second excitation event. (c) Same as (b) but with timing of the second pair changed to yield constructive rather than destructive interference, and wavepacket amplification rather than cancellation. See text.

Coherent Control and Nonlinear Interactions of Semiconductor Cavity Polaritons

Y. –S. Lee[1], A. V. Maslov[2], D. S. Citrin[2], and T. B. Norris[1]

[1] Center for Ultrafast Optical Science, The University of Michigan, Ann Arbor, MI 48109-2099, USA
E-mail: yunslee@eecs.umich.edu
[2] Department of Physics and Materials Research Center, Washington State University, Pullman, WA 99164-2814, USA

Abstract. We demonstrate coherent control of the cavity-polaritons of a quantum-well semiconductor microcavity in a two-color scheme. Evidence for a coherent two-photon biexciton transition and excitation induced dephasing as the dominant nonlinear optical process is given.

Cavity polaritons are the coupled modes of a quantum well exciton and the cavity photon mode of a microcavity. When the exciton and field modes are degenerate, the lower and upper normal modes are the symmetric and antisymmetric superpositions of the exciton and cavity modes. We use a pair of phase-locked optical pulses (with relative delay τ_{12}) to excite the lower polariton branch, thereby controlling the reflection of a probe pulse tuned to the upper branch via the nonlinearity of the exciton transition which underlies both modes. Coherent control of the exciton population in the system is demonstrated by varying the relative phase of the two pump pulses, and is used to measure the dephasing time of the lower normal mode. Studies of the spectral lineshape at different probe delays provide insight into the mechanism of the excitonic nonlinearity.

The sample in these experiments has two $In_{0.04}Ga_{0.96}As$ QW's positioned at the antinodes of a $3\lambda/2$ cavity formed by two 99.6 %-reflectivity distributed Bragg reflectors (14 (16.5) periods of GaAs/AlAs on top (bottom) mirrors). The pump and probe pulses were generated by spectral filtering the output of an 82-MHz mode-locked Ti:sapphire laser producing 75-fs pulses centered at 831 nm. The pump (probe) pulse was resonant with the lower (upper) normal mode, with pulse durations of 190 fs. The pump pulse was split into two excitation pulses with variable delay using a Michelson interferometer with a piezoelectric translator in one arm to control the relative phase between the two pulses. The pump and probe beams were overlapped on the sample at near normal incidence. Probe reflectance spectra and the time-resolved probe reflectivity at the reflectivity minimum of the upper mode were measured at 10 K.

Figure 1 shows the probe reflectivity spectra when the probe delay was set between, and after the two pump pulses for (a) co-circular and (b) cross-circular pump and probe polarizations. The pump pulse phase difference was π. Due to the cavity enhancement, the amplitude of the differential reflectivity (DR) signal $\Delta R/R$ was quite large - up to 20 % - which is at least three orders of magnitude larger than the signal obtained from a bare QW without a cavity. The qualitative form of the reflectivity spectra varies considerably with the pump and probe polarizations.

When both pump and probe are co-circularly polarized, the first pump pulse induces a broadening in the upper mode spectrum, and an increase in the reflectivity at line center. Following the second pulse, the reflectivity spectrum nearly recovers due to the coherent destruction of the exciton population. The recovery is not complete because of the incoherent population dephased between the two pump pulses. When the pump and probe are cross-circularly polarized, there is initially a substantial shift of the reflectivity minimum to higher energy. Following the second pump pulse, again there is a substantial recovery of the spectrum.

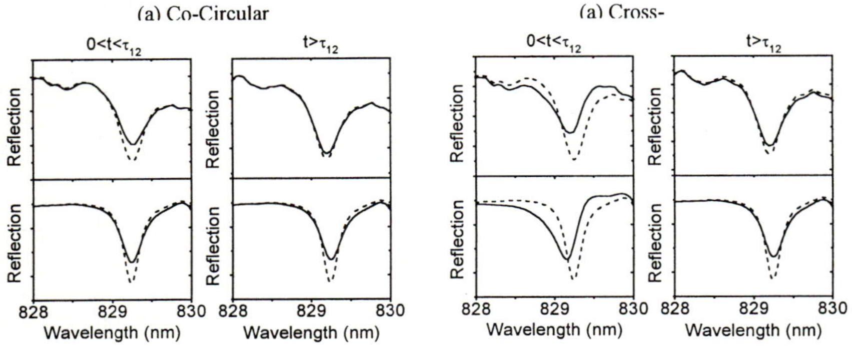

Fig. 1. Measured (top) and calculated (bottom) temporal evolution of DR spectra for (a) co-circular and (b) cross-circular pump-probe polarization at $0<t<\tau_{12}$ and $t>\tau_{12}$ when single pump pulse fluence per pulse is 6.8 $\mu J/cm^2$ and τ_{12} is 0.9 ps. Dotted lines are probe spectra without pump pulses.

Qualitatively, the broadening observed in the reflectivity spectra may be understood as the consequence of EID [1]. As the pump pulses modulate the exciton density and hence the exciton linewidth, the width of the upper normal mode is modulated. The spectral blue shift seen in the cross-circularly polarized case, however, can only be explained if there is a transient increase in the oscillator strength of the excitons underlying the normal modes seen by the probe pulse. When the sample is excited by a single pump pulse, the blue shift is observed to last only as long as the cavity lifetime, indicating that the nonlinear response depends only on the instantaneous field at the position of the QW's. An increase in oscillator strength due to a biexciton transition has been proposed previously [2]; the difference in the model used here is that the oscillator strength increase occurs only during the presence of the optical field. A coherent two-photon biexciton transition (CTBT) opens a new channel for optical absorption, thus it increases the oscillator strength seen by the probe field. A simple phenomenological model can reproduce the main features of the experimental results. We calculate the propagation of optical pulses through the microcavity-accounting for the details of the layered dielectric structure-using a 1D finite-difference time-domain method. The exciton response is modeled as a two-level system, where the optical Bloch equations have been modified to incorporate the

EID and CTBT effects. EID is included by assuming that the dephasing rate is proportional to the total number of excitons ($T_2=T_2^0/(1+\alpha N_{ex})$). If the pump and probe beams are cross-circularly polarized, the CTBT effect is included by increasing the oscillator strength for the excitons created by the probe as $d=d_o(1+\beta|E|^2)$ where $|E|^2$ is the optical intensity at the QW position. The results of the simulation are shown in Fig. 1, and the main experimental features are reproduced. Since the CTBT term is proportional to the optical intensity, the spectral shift lasts only for the cavity lifetime. At long time delays, or for co-circular polarization, the nonlinear signal is entirely explained by EID.

The coherent control of the exciton population was further demonstrated by fine-tuning the temporal delay $\Delta\tau_{12}$ (and hence the phase) between the two pump pulses. The probe delay was fixed at 15 ps following the first pump pulse, so that the DR signal corresponds to the final incoherent population remaining in the system. The dephasing time (T_2) of the lower mode polariton can be obtained from the decay time of the interference oscillation amplitude. Figure 2 shows the measured and fitted results of the oscillation amplitude vs. τ_{12} for cross-circular polarization. The exponential decay times are 2.7 and 6.4 ps for pump fluence per pulse of 6.8 and 0.7 μJ/cm^2 respectively, which indicates the strong EID at high excitation. At low excitation limit, the QW exciton linewidth (0.7 meV) corresponds to the exciton dephasing time of 2 ps. Thus, our measurement indicates that T_2 of the lower mode polariton is much longer than of the exciton.

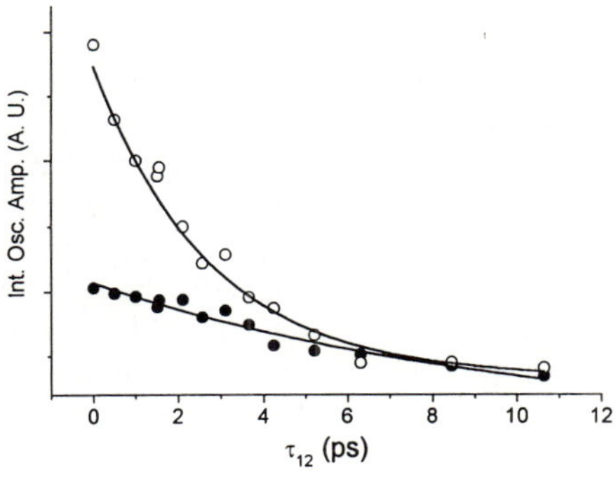

Fig. 2. The temporal decay of the interference oscillation for pump fluence per pulse of 6.8 (open circle) and 0.7 (closed circle) μJ/cm^2

Acknowledgements. The Michigan portion of the work was supported by AFOSR and NSF. The Washington portion was supported by AFOSR, ONR, and NSF via DMR 9705403.

References

1 H. Wang et al., Phys. Rev. Lett. 71, 1261 (1993).
2 X. Fan et al., Phys. Rev. B 57, R9451 (1998).

Coherent Control of XUV Radiation

Randy Bartels[1], Sterling Backus[1], Ivan Christov[2], Lino Misoguti[1], Gleb Vdovin[3], Erik Zeek[1], Margaret M. Murnane[1], and Henry C. Kapteyn[1]

[1] JILA, University of Colorado, Campus Box 440, Boulder, CO 80309-0440, USA
E-mail: bartels@jila.colorado.edu
[2] Department of Physics, Sofia University, Bulgaria
[3] Delft University of Technology, 2600 GA Delft, The Netherlands

Abstract. We demonstrate that using phase-only control of ultrafast pulses, we can selectively enhance the brightness of high-harmonic generation by over an order of magnitude, and generate near transform-limited soft x-ray pulses.

1. Introduction

In this work, we demonstrate for the first time coherent control of a very high-order nonlinear process[1]. By altering the shape of an ultrafast laser pulse using phase-only control with a deformable mirror, and by adaptive control of the pulse shape using an evolutionary algorithm, we can both enhance the brightness and control the spectral characteristics of high-order harmonic generation (HHG). We demonstrate an enhancement in peak brightness by over an order of magnitude, as well as enhancing a single harmonic order compared to its adjacent orders. We can also spectrally narrow a given harmonic order, resulting in the probable generation of ultrashort x-ray pulses with near transform-limited temporal pulse shape. This work represents the first time that adaptive or "learning" control of a high-order nonlinear process has been experimentally demonstrated. Surprisingly, the optimal laser pulse shapes differ only slightly from the transform-limited case (21fs versus 18fs), but have significant phase variations on the leading edge of the optimized pulse.

2. Experimental Methods and Results

For our experiment, pulses are generated by an amplified Ti:Sapphire laser system capable of generating transform-limited 15fs pulses with mJ energy levels.[2] A zero-dispersion stretcher folded by a micro-machined deformable mirror manipulates the phase of the amplified pulses. When the mirror is deformed, the spectral phase (and thereby the temporal phase and profile) are modified. The shaped pulses are injected into a hollow-core fiber and generate HHG radiation in the XUV region. [3] The generated light is coupled into a flat-field grating spectrometer and the spectral data is collected by an x-ray CCD camera.

Optimization of the HHG is accomplished using an evolutionary strategy which begins with a set of random pulse shapes whose associated HHG spectra are measured. Those pulses that produce the brightest, most spectrally pure, and most energetic x-rays are retained, duplicated, and perturbed; then the HHG spectra of the new trial shapes and best previous are measured and the process repeats. Prior to each optimization of the HHG radiation, the output laser pulses are compressed to the transform-limit using SHG intensity as a fitness function [2]; the corresponding HHG spectrum is the baseline of the optimization (Fig. 1, It. 0). Fig. 1 shows an optimization increasing the brightness of the 27th harmonic by a factor of 8 without increasing the brightness of the neighboring harmonic orders.

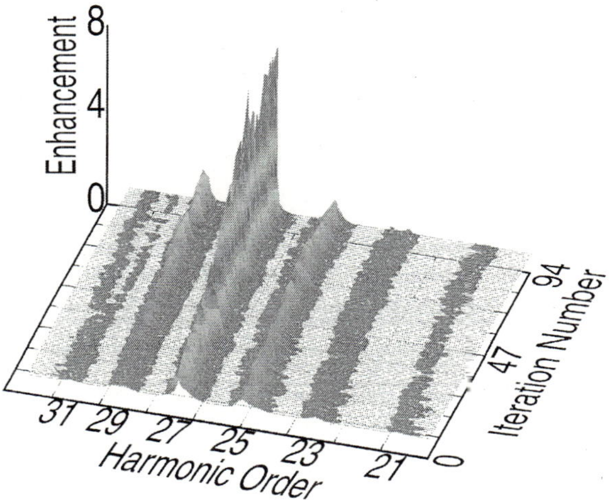

Fig. 1. Optimization of a single (27th) harmonic in argon while suppressing adjacent harmonics. The HHG spectrum for the best pulse of each iteration

Figure 2 shows the laser pulse shapes corresponding to the initial (it. 0) and final (it. 94) HHG spectra shown in Fig. 1. The conversion efficiency of the x-ray generation process optimizes for a non-transform-limited pulse in stark contrast to recent experimental evidence from a number of groups that has shown low-order, perturbative, non-linear processes (i.e. second-harmonic generation) have a maximum conversion efficiency for a transform-limited pulse.[1,2,4] Subtle pulse shape changes resulting from the nonlinear phase distortion on the leading edge produces a substantial and beneficial change in the output energy, brightness (3x-11x), and spectrum of the HHG radiation in all noble gasses tested (2-95 torr), under conditions with and without phase matching of the HHG process [3].

To explain the unexpected result that very slight changes in laser pulse shape can so dramatically enhance and select individual harmonics, we consider the semi-classical rescattering model of HHG. Electrons returning to the core with a specific return energy generate particular harmonics on each half-cycle. The continuum x-ray emission generated by a particular cycle of the laser pulse reinforces constructively or destructively with different parts of the continuum generated by adjacent cycles. Controlling the nonlinear phase distortion on the leading edge of the pulse manipulates the phases of the optimized harmonic x-ray bursts so that they add together more constructively, increasing the brightness of the harmonic emission. Increasing the coherence between particular frequencies from the x-ray bursts of each half-cycle constitutes a new intra-atomic phasematching process.

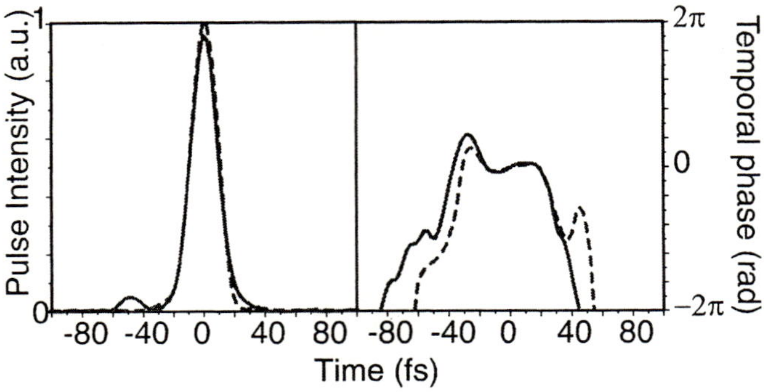

Fig. 2. Amplitude and phase of the laser pulses corresponding to Fig. 1: [left] initial transform-limited pulse (dashed) and final optimized pulse (solid); [right] transform-limited phase (dashed) and optimized phase (solid)

3. Conclusions

In summary, this work represents the first demonstration of control of a very high-order nonlinear process. We demonstrate significantly increased enhancement and selectivity of high-order harmonic generation, as well as the generation of near-transform-limited harmonics. We achieve this using a new type of phasematching, which allows us to channel energy into a selected harmonic. This result has immediate utility in the probing of dynamics of chemical and material systems, because it provides a way to select a harmonic without temporally broadening it. The result is a bright, quasi-monochromatic, and highly spatially-coherent soft x-ray light source for use in techniques such as photoelectron spectroscopy and pectromicroscopy, time-resolved x-ray studies of material and chemical systems, and time-resolved holographic imaging.

References

1 Bartels, R., *et al*, Shaped-pulse optimization of coherent emission of high-harmonic soft x-rays, *Nature* (13 July 2000).

2 Zeek, E., *et al.* Adaptive pulse compression for transform-limited 15fs high-energy laser pulse generation. *Optics Letters* , (April 15, 2000).

3 Rundquist, A., *et al.* Phase-matched Generation of Coherent Soft X-rays. *Science* **280**, 1412-1415 (1998).

4 Yelin, D., Meshulach, D. & Silberberg, Y. Adaptive femtosecond pulse compression. *Optics Letters* **22**, 1793-5 (1997).

Enhancement of Kα yield from femtosecond laser produced plasmas by automated control of plasma parameters

Christian Ziener, Ingo Uschmann, Gregor Stobrawa, Thomas Feurer, Heinrich Schwoerer, Roland Sauerbrey

[1] Institut für Optik und Quantenelektronik, Friedrich-Schiller-Universität Jena, Max-Wien-Platz 1, Germany
E-mail: ziener@qe.physik.uni-jena.de

Abstract. We demonstrate the automated optimization of Kα emission from silicon, titanium and cobalt, generated in a multi-terawatt femtosecond laser produced plasma. The plasma parameters were adjusted by a newly developed laser prepulse device.

1. Introduction

Femtosecond laser produced plasmas are a unique source of fast electrons and ions, of discrete x-ray line emission and bremsstrahlung as well as even neutrons [1, 2, 3]. For Kα-emission the duration of the generated x-rays is as short as the laser pulse, leading to x-ray intensities hardly accessible by other techniques. For various applications of Kα x-rays in time resolved x-ray diffraction, x-ray spectroscopy or x-ray lithography, the efficiency of the emission needs to be optimized. Since the Kα lines are created by accelerating electrons in the plasma and stopping them in the cold target bulk, the plasma parameters have to be adjusted for the optimal output of the desired line.

It is shown that the implementation of automated control techniques in fact leads to an optimization of the fs Kα-emission by controlling the properties of the fs laser produced plasma in a feedback loop [4].

The parameters which are accessible in the experiment are the atomic number Z of the target, the energy and duration of the main laser pulse, the focal spot size and time delay, energy and intensity of a laser prepulse. This prepulse determines the target status at the time the main laser pulse interacts with it. The preplasma conditions can strongly influence the intensity of the observed lines [5]. In our experiment we observed and optimized Kα-emission from silicon (Z=14), titanium (Z=22) and cobalt (Z=27) with Kα photon energies of 1.7 keV, 4.5 keV and 6.9 keV respectively.

2. Experimental Methods

The optimization was performed by a newly developed prepulse unit which was incorporated in the Jena multi TW Ti:sapphire laser system.

The prepulse unit is set up in a Mach-Zehnder interferometer arrangement with two 9:1 beam splitters, a micrometer translation stage driven by a stepping motor in one arm of the interferometer, two half wave plates and a polarizer. The orientation of the wave plates also can be controlled by stepping motors. The half wave plates and the polarizer allow for adjusting the energy of each arm separately by more than one order of magnitude.

To guarantee the spatial and directional overlap of the recombined beams one of the mirror mounts is motorized. This mirror was automatically adjusted for optimal overlap of pre- and main pulse in the focal plane at the position of the target for every position of the translation stage. This procedure is crucial, since the optical path length from the unit to the target is 15m and the angular accuracy required by the focusing optics is 15μrad. Finally the automatic mirror control assures the overlap of the 10μm x 12μm foci within 1.5μm for (negative) delay times between 0 and 300ps.

The Kα radiation was monochromatized with toroidally bent crystals of a horizontal bending radius of 150mm and a vertical bending radius of 106.4mm [6]. For silicon and cobalt Kα the quartz (10-10) was used in the 1. respectively in the 4. reflection order as well for titanium the Si (311). The x-ray detector consists of a GaAsP-Schottky photodiode protected with a 7μm thick Be window. The x-ray source and the photodiode were placed on the Rowland circle of the crystals.

3. Results and Discussion

Table 1 summarizes the parameters of the Kα lines, the used Bragg crystals and the achieved gain.

Table 1. Summary of the observed Kα-lines, the bent crystals and the gain achieved by the prepulse.

element	Z	Photon energy / keV	Crystal-reflection	Number of photons	Maximum gain	Optimal delay /ps
Silicon	14	1.740	Quartz (10-10)	5.8×10^9 x gain	2.2	150
Titanium	22	4.505	Si (311)		1.5	110
Cobalt	27	6.930	Quartz (40-40)		4.6	25

In Figure 1 the results for the variation of the prepulse delay for all three Kα-lines are plotted in detail. We plot the efficiency of the Kα-emission against the delay between the two pulses. The efficiency was normalized to the value without prepulse. Obviously the effect is largest for the Co Kα-emission at 6.9 keV (triangles). Each point is an average of six laser shots and the experiment was repeated several times on different days. At 50 ps delay of the main pulse with respect to the prepulse the efficiency of the generation of this line is enhanced by more than a factor of four. The intensities on the target were 5×10^{17} W/cm^2 and 5×10^{15} W/cm^2 for main and prepulse, respectively, corresponding to the maximum throughput of both arms of the prepulse unit. The maximum enhancement of the

1.7 keV silicon Kα-line (squares) was achieved with a 150 ps delay of the prepulse and reached a value slightly larger than two.

The 4.5 keV titanium Kα-line (circles) could only be enhanced by a factor of 1.5 at about 100 ps delay. A more detailed analysis of the results shows two separated maxima at different delay times for Si and Ti and a reduction of efficiency at short times between zero and 30 ps for all elements.

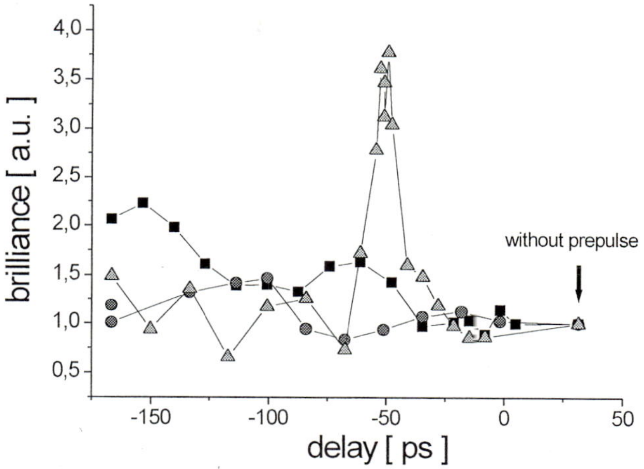

Fig. 1. Efficiency of Kα emission of Si (squares), Ti (circles) and Co (triangles) depending on the delay of the laser prepulse. Intensities were 5 x 10^{17} W/cm^2 and 5 x 10^{15} W/cm^2 for main and prepulse respectively.

The Kα yield depending on the prepulse delay was also measured for various prepulse and main pulse intensities. Because of the number of parameters we performed a feedback controlled optimization of the silicon Kα-emission. The diode signal of 5 laser shots was averaged and used as a feedback signal for an algorithm which consists of a combination between a simplex downhill and a simulated annealing algorithm. The algorithm was capable of finding the optimal set of parameters within a few minutes.

References

1 T. E. Cowan et al., in *Physical Review Letters*, **84**, 903, 2000.
2 K. W. D. Ledingham et al., in *Physical Review Letters*, **84**, 899, 2000.
3 G. Pretzler et al., in *Physical Review E*, **58**, 1165, 1998.
4 T. Feurer, in *Applied Physics B*, **68,** 55, 1999.
5 S. Bastiani et al., in *Physical Review E,* **56**, 7179, 1997.
6 T. Missalla et al., in *Review of Scientific Instruments,* **70**, 1288, 1999.

Part II

Lasers for Ultrashort Pulses

Challenges and Limitations on Generating Few Cycle Laser Pulses directly from Oscillators

F. X. Kärtner[1], U. Morgner[1], R. Ell[1], Ch. Jirauschek[1], G. Metzler[1], T. R. Schibli[1], Y. Chen[2], H. A. Haus[2], E. P. Ippen[2], J. Fujimoto[2], V. Scheuer[3], G. Angelow[3] and T. Tschudi[3]

[1] High-Frequency and Quantum Electronics Laboratory,
Universität Karlsruhe, Kaiserstraße 12, 76128 Karlsruhe, Germany
E-mail: F.Kaertner@etec.uni-karlsruhe.de

[2] Massachusetts Institute of Technology, Cambridge, MA 02139, USA
E-mail: ippen@mit.edu

[3] Technische Hochschule Darmstadt
E-mail: Volker.Scheuer@Physik.TH-Darmstadt.de

Abstract. Sub-two-cycle optical pulses with bandwidths in excess of 500 nm have been generated directly from a Ti:sapphire laser using double-chirped mirror pairs and exploiting dispersion-managed soliton formation in the cavity. For the first time sub-two-cycle pulses directly from the laser are generated with broad enough spectra to detect and control the evolution of the phase between carrier and envelope using interference between second and third harmonic of the fundamental spectrum.

1. Introduction

Because of the limited gain bandwidth of Ti:sapphire of about 200 nm, and the limited bandwidth and dispersion control available by current laser mirrors, pulses with spectra covering one octave and width less than 5 fs have been generated by external compression at low repetition rates [1,2]. Spectral broadening by self phase modulation in microstructure fiber [3], generating spectra from about 500 nm to 1500 nm just from the pulse train of a standard modelocked Ti:sapphire oscillator has been demonstrated. However, because of the complicated phase structure on these pulses pulse compression has yet to be demonstrated. Intracavity spectral broadening is advantageous in a sense that the spectral phase is cleaned up each roundtrip by the pulse shaping provided by the saturable absorber action, making the extracavity pulse recompression more successful. Directly from a Ti:sapphire oscillator pulses in the range of 5-6 fs have been generated recently [4,5]. The drawback of using intracavity spectral broadening is the requirement in high mirror reflectivity and well controlled dispersion over the desired bandwidth.

In this contribution, we demonstrate a Ti:sapphire laser oscillator employing double-chirped mirror pairs (DCMPs) and CaF_2-prisms for dispersion compensation and a second focus in a glass plate for additional spectral broadening. By exploiting dispersion managed soliton formation [6] in the laser ultrabroad spectra extending over 500 nm and even over one octave on a logarithmic scale are generated directly from a laser oscillator. Well shaped spectra extending over one octave are important for further progress in ultrahigh resolution optical coherence tomography, for control of the phase between carrier and envelope and for frequency metrology [7,8].

2. Ultrabroad Double-Chirped Mirror Pairs

For sub-10 fs pulse generation directly from the oscillator dispersion compensating mirrors with an extended high-reflectivity range of more than the 200 nm of standard quarter-wave dielectric Bragg-mirrors are required. This has been achieved by the invention of chirped mirrors [9] and double-chirped mirrors (DCM) [10]. Recently, we demonstrated double-chirped mirror pairs (DCMPs) which allow for GDD compensation from 600 nm to 1200 nm while being highly reflective over the full octave [11]. The mirrors type I and type II should be used in subsequent pairs to minimize oscillations in the GDD curve. Fig. 1a show the designed reflectivity and GDD of type I and type II mirrors. The GDD of both mirrors show strong oscillations. However, the average GDD of both mirrors in the design is rather smooth over the full octave. Fig. 1b shows the measured

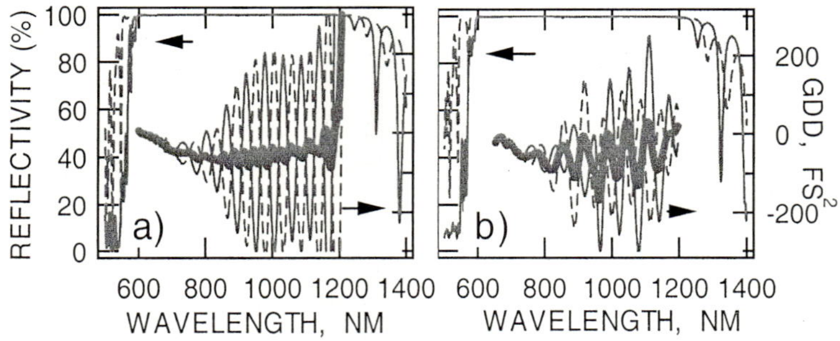

Fig. 1. (a) Designed reflectivity and GDD of type I (solid curves) and type II (dashed curves) mirrors. The thick line shows the average GDD of type I and type II mirrors; (b) shows the measured results of a fabricated mirror pair.

reflectivities and GDD of the fabricated mirrors of type I and II, using ion-beam sputtering. Because of fabrication tolerances only a cancelation of GDD oscillations by a factor of 3 has been achieved, instead of a factor of 10 as is shown in the design. As is shown below these fabricated mirror pairs are already good enough to allow for considerable progress in short pulse generation towards the single-cycle regime.

3. Sub-Two-cycle Laser Set-up and Results

The laser set-up is shown in Fig. 2. The 2.7 mm thin Ti:sapphire crystal is pumped by a frequency doubled Nd:YVO_4 laser. Two prisms made from CaF_2 allow for dispersion fine tuning. The separation between the prisms is adjustable from 95 to 115 cm, leading altogether to a repetition rate of about 65 MHz. The

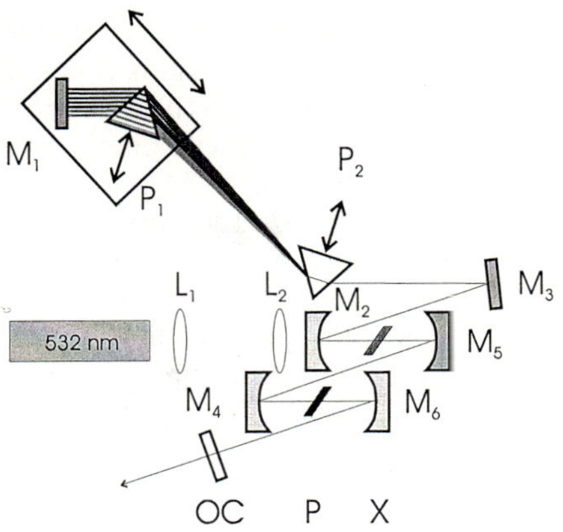

Fig. 2. Laser set-up: M_1, M_3 and M_5 are type I DCMs and M_2, M_4 and M_6 are type II DCMs; P_1 and P_2 are CaF_2-prisms, P is a 2 mm thick BK7-plate, X the 2.7 mm thick Ti:sapphire crystal and OC is a quarter wave 1% output coupling mirror made from $ZnSe/MgF_2$.

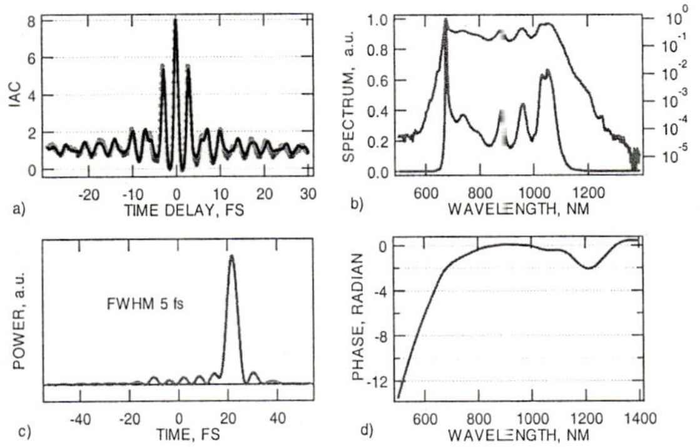

Fig. 3. a) Measured (solid) and retrived (dotted) line. b) Measured output spectrum on a linear and logarithmic scale. c) Retrived intensity envelope versus time. d) Retrived spectral phase.

output coupling mirror is a broadband quarter wave mirror made from $ZnSe/MgF_2$ with a center transmission of 1 %. The pulses from the laser are recompressed externally by means of four bounces on DCMPs and a sequence of CaF_2-prisms. The pulses are detected by a symmetrical second-harmonic interferometric autocorrelator using metal-coated thin beamsplitters and thin wedged KDP doubling crystal. Figs. 3a and b show the measured interferometric autocorrelation and spectrum. Figs. 3c and d show the retrieved intensity of the pulse in the time domain and the spectral phase using a phase retrival technique demonstrated recently [12] and resulting in a FWHM of 5 fs. The recomputed interferometric autocorrelation from the retrived pulse is also shown in Fig. 3 a and agrees very well with the measured one. The retrieved phase shows that the extracavity dispersion for the shorter wavelengths is not yet optimized. Therefore, further reduction in pulse width seem to be possible by a refined external dispersion compensation sheme. In a laser with one focus only, the pulse energy is limited by the requirement for optimum KLM action, which is also limiting the amount of SPM per round-trip and, therefore, the spectral broadening within one roundtrip. To overcome this limitation, we placed a second focus in the cavity. The dispersion in the Ti:sapphire crystal is balanced by the dispersion upon reflection on the DCMP. The BK7 plate in the second focus shows as much dispersion as the Ti:sapphire crystal. In the pedestal the spectrum, see Fig. 3b, extends from 600 to 1300 nm, resulting in a pulsewidth of only 4.3 fs, if a flat phase is assumed. The structure in the spectrum is due to the remaining GDD oscillations and increased output coupling in the wings. Despite the large oscillations in the GDD caused by fabrication tolerances and the imbalanced GDD oscillations of Mirror M_6, the spectrum is relatively smooth, which can be explained by the strong KLM action, which continuously cleans up the pulse shape and the enhanced SPM due to the second focus. Due to the strong SPM per round-trip a large nonlinear phase shift per round-trip is generated, so that subsequent pre- and post pulses generated upon reflection of the pulse on the DCMs due to GDD-oscillations dephase rapidly. The center wavelength was slightly shifted from the Ti:sapphire gain maximum to 825 nm, corresponding to an optical period of 2.75 fs.

4. Phase-Control of 5 fs Pulses

Subsequently, the pulse train is launched into a Mach-Zehnder type interferometer. The short wavelength part of the spectrum at 660 nm is doubled by a critically phase matched Type I BBO to 330 nm. The long wavelength part at

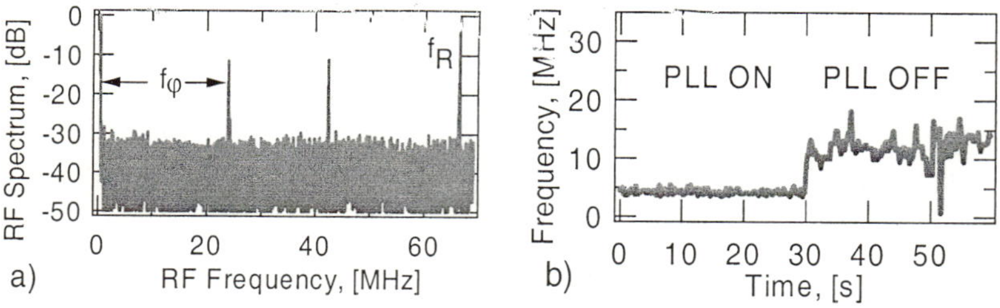

Fig. 4. a) RF signal from the interference. The fundamental repetition rate is about 68 MHz. The phase-slip beat signal is 20dB above shot noise. b) Locking of the phase-slip signal to an rf-oscillator.

990 nm is first doubled with a 0.5 mm thick noncritically phasematched $KNbO_4$-crystal and subsequently sum-frequency mixed with the remaining fundamental in a 1 mm thick critically phase matched BBO-crystal generating also light at 330 nm. The signals from the two arms interfere at a beam splitter and the beat is

detected via a photomultiplier tube behind a 10 nm wide spectral filter at 330 nm and a spatial filter. The beat signal is displayed in Figure 4 a. The beat signal at $f\varphi=(\varphi/2\pi)f_{R,,}$ where $f_{R,,}$ is the repetition rate of the laser, indicates the repetition rate with which the electrical field of the pulse is reproduced due to the difference between phase and group delay in the cavity. The beat frequency can easily be shifted by changing the insertion of a prism or by tilting the prismatic end mirror. In the free running system the phase-slip frequency changes slowly (on a millisecond time scale) in a 10 MHz window, see Fig 4 b. A phase-locked-loop can lock the phase-slip frequency to an external rf-oscillator, which opens up possibilities for new experiments in nonlinear optics that depend on the electric field rather than on intensity.

5. Conclusions

Double-chirped mirror pairs spanning one octave with high reflectivity and well controlled dispersion lead to the generation of phase controlled 5fs pulses directly from an oscillator. Further improvement in the external dispersion compensation should allow for 4.5 fs pulses close to one and a half cycle in width. This opens up new experiments in nonlinear optics, and frequency metrology.

Acknowledgements. The authors greatly acknowledge the fabrication of the broadband output coupler by M. Lederer, B. Luther-Davis and A. Boiko.

References

1 A. Baltuska, Z. Wei, M. Pshenichnikov, and D. Wiersma, Opt.Lett. **22**, 102, 1997.
2 M. Nisoli, S. De~Silvestri, O. Svelto, R. Szipöcs, K. Ferenz, Ch. Spielmann, S. Sartania, F. Krausz, Opt. Lett. **22**, 522, 1997.
3 J. Ranka, R. Windeler, A. Stentz, Opt. Lett. **25**, 25 (2000).
4 U. Morgner, F. X. Kärtner, S. H. Cho, Y. Chen, H. A. Haus, J. G. Fujimoto, E. P. Ippen, Opt. Lett. **24**, 411, 1999.
5 D. H. Sutter, G. Steinmeyer, L. Gallmann, N. Matuschek, F. Morier-Genoud, U. Keller, V. Scheuer, G. Angelow, T. Tschudi, Opt. Lett. **24**, 631, 1999.
6 Y. Chen, F. X. Kärtner, U. Morgner, S. H. Cho, H. A. Haus, J. G. Fujimoto, E. P. Ippen, JOSA B **16**, 1999, 1999.
7 D. J. Jones, S. A. Diddams, J. K. Ranka, A. Stentz, R. S. Windeler, J. L. Hall, S. T. Cundiff, Science 288, 636, 2000.
8 A. Poppe, A. Apoloski, G. Tempea, Ch. Spielmann, F. Krausz, T. Udem, R. Holzwarth, T. W. Hänsch, to appear in Phys. Rev. Lett.
9 Szipöcs and K. Ferencz, C. Spielmann, and F. Krausz, Opt. Lett. **19**, 201, 1994.
10 F. Kärtner, N. Matuschek, T. Schibli, U. Keller, H A. Haus, C. Heine, R Morf, V. Scheuer, M. Tilsch, and T. Tschudi, Opt. Lett. **22**, 831, (1997).
11 F. Kärtner, U. Morgner, T. Schibli, J. G. Fujimoto and E. P. Ippen, Proceedings of the Conference on Lasers and Electro-Optics, 2000, paper CThE5.
12 A. Baltuska, A. Pugzlys, A.S. Pshenichnikov, D.A. Wiersma, B. Hoenders, H. Fewerda, in Ultrafast Optics Conference, paper Th10, Ascona, Switzerland (1999).

Extremely Flexible and Accurate Chirp-Compensation for 75-MHz Repetitive Glass-Fiber Output of a More-Than 100-THz Bandwidth -- Generation of a-Few-Optical-Cycle Transform-Limited Pulses --

Shinki Nakamura[1*], Liming Li[2], Naoki Karasawa[1], Ryuji Morita[1], Hidemi Shigekawa[3], and Mikio Yamashita[1]

[1] Department of Applied Physics, Hokkaido University and CREST, Japan Science and Technology Corporation (JST), Kita-13, Nishi-8, Kita-ku, Sapporo 060-8628, JAPAN
[2] Department of Photonics Material Science, Chitose Institute of Science and Technology and CREST, JST, 758-65 Bibi, Chitose 066-8655, JAPAN
[3] Institute of Applied Physics, University of Tsukuba and CREST, JST, 1-1-1 Tennodai, Tsukuba 305-8573, JAPAN
[*] Present address: Department of Media and Telecommunications Engineering, Ibaraki University, 4-12-1, Nakanarusawa, Hitachi 316-8511, JAPAN
E-mail: nakamura@dmt.ibaraki.ac.jp, mikio@eng.hokudai.ac.jp

Abstract. 92-fs, 75-MHz-repetitive pulses from a simple laser-oscillator glass-fiber system were compressed to 7.1-fs transform-limitedly by a chirp compensator consisting of a prism-pair and a spatial-phase-modulating pulse shaper that *in-situ*-adjusts independently and accurately any order dispersions.

Recently, optical pulses in the 4-5 fs regime have been generated by external pulse compression [1] with an amplifier at the repetition rate of 1 kHz, and directly from Ti:sapphire lasers at the repetition rate of 100 MHz [2]. For their chirp compensations, the combinations of prism-pairs, grating-pairs and chirped mirrors have been employed [1,2]. However, the inter-dependence among the group-delay dispersion (GDD), the third-order dispersion (TOD) and the higher-order dispersions as well as the bandwidth limitation of these techniques make it difficult to generate a-few-optical-cycle *transform-limited* pulses. To overcome these problems, we have developed a 4.9-fs, 1-kHz-repetition pulse compressor based on a novel hybrid chirp compensation of a prism pair and a spatial-light-modulating pulse shaper [3]. In this paper, we demonstrate for the first time that this compensation technique is applied to compression to 7.1-fs transform-limited pulses at a 75-MHz repetition rate from a fused-silica fiber in a simple, non-amplified optical source. Furthermore, *in-situ* accurate measurements of the TOD dependence of interferometric autocorrelation traces with keeping GDD constant are demonstrated without realignment of any optics.

In our experimetal setup, the 12-fs, 10-nJ pulses at the center wavelength of 800 nm were generated from a mode-locked Ti:sapphire laser. The 12-fs pulses were coupled into a 2.5-mm silica fiber by a 36x reflective objective which introduced no additional GDD and TOD. A single-mode fused silica fiber with a core diameter of 2.7 μm was used. The output from the 2.5-mm fiber was collimated by another reflective objective and measured by a spectrometer. At the input pulse power of 127 kW, the spectrum was broadened owing to the dispersive self-phase modulation effect and the broad-spectral pulses had the duration of 92 fs. This chirped pulse was first pre-compressed by a pair of fused silica prisms with a separation length of 59 cm. The output pulse from the prism-pair was coupled into a pulse-shaping apparatus [4], which consisted of a pair of 300 lines/mm gratings and a pair of concave 200-mm focal-length spherical silver mirrors. A programmable 256-pixel spatial-light-modulator (SLM) with the total effective width of 27 mm was placed on the Fourier plane.

The 92-fs chirped pulses from the 2.5-mm fiber were compressed to the Fourier transform limited pulses by compensating for GDD and TOD. The GDD and TOD of a silica fiber output were measured to be 342 fs^2 and 187 fs^3 at 800 nm, respectively, by the second-harmonic generation frequency-resolved optical gating. The GDD by the fiber was canceled by using the prism pair, whose GDD and TOD are -342 fs^2 and -479 fs^3 respectively. Considering the SLM's material dispersion (GDD = 165 fs^2 and TOD=125 fs^3), we applied the GDD of -150 fs^2 and TOD of +220 fs^3 to the SLM, which gave the total GDD to be 15 fs^2 and TOD to be 53 fs^3.

The shortest pulse width after complete phase control corresponded to the 7.1-fs pulse duration calculated by the inverse Fourier transform (IFT) of the broad-spectrum from 670 to 930 nm as shown in Fig.1. We confirmed that the dispersion evaluation is correct by measuring the interferometric autocorrelation traces as a function of TOD under the constant GDD = -150 fs^2 as shown in Fig. 2. It shows that GDD = -150 fs^2 and TOD = +220 fs^3 are optimum, and even the small change in TOD greatly affects the temporal pulse profile.

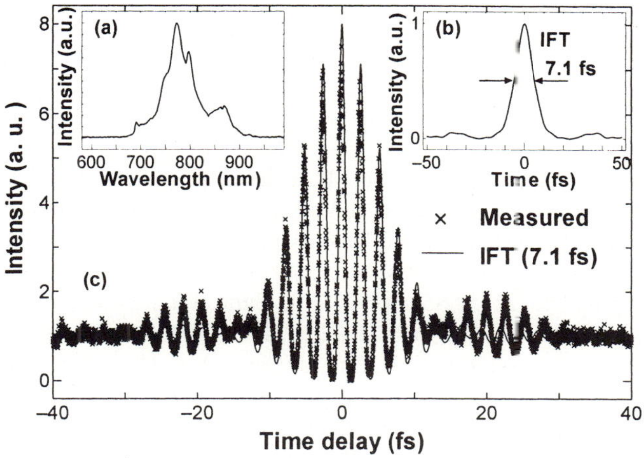

Fig. 1. Measured interferometric autocorrelation trace of the Fourier-transform-limited pulse (×). The solid curve is a fit of calculation from IFT of the pulse spectrum (inset (a)) with FWHM of 7.1 fs (inset (b)).

In conclusion, we performed compression to 7.1-fs transform-limited pulses at the 75-MHz repetitive rate using a prism pair and a 256-pixel SLM in a pulse-shaping configuration. The accurate TOD characteristic of interferometric autocorrelation traces of more-than 100-THz bandwidth pulses was measured while keeping GDD constant.

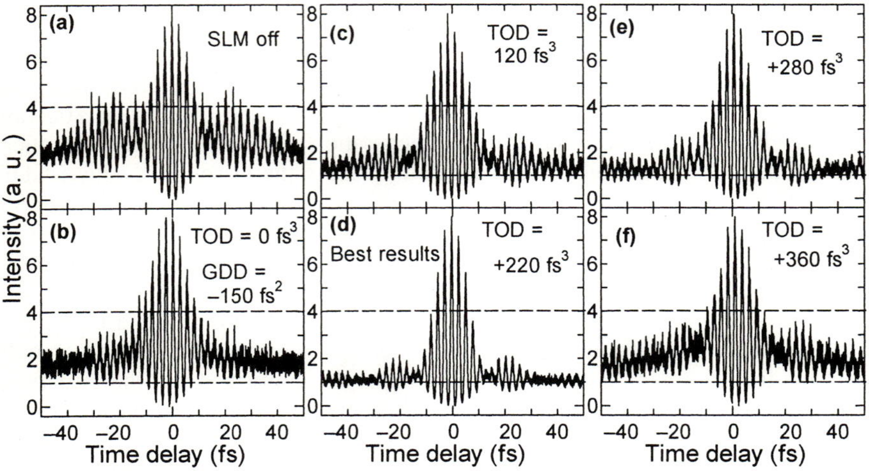

Fig. 2. TOD dependence of measured interferometric autocorrelation traces while keeping GDD = -150 fs^2 constant.

References

1 Z. Cheng, G. Tempea, T. Brabec, K. Ferencz, C. Spielman and F. Krausz, in *Ultrafast Phenomena XI*, (Springer-Verlag, Berlin, 1998), pp. 8-10.; A. Baltuška, Z. Wei, M. Pshenichnikov, and D. Wiersma, Opt. Lett. **22**, 102-104 (1997).; M. Nisoli, S. D. Silverstri, O. Svelto, R. Szipocs, Ch. Spielmann, S. Sartania, F. Krausz, Opt. Lett. **22**, 522-524 (1997).; A. Shirakawa, I. Sakane, M. Takasaka, and T. Kobayashi, Appl. Phys. Lett. **74**, 2268-2270 (1999).

2 U. Morgner, F. X. Kärtner, S. H. Cho, Y. Chen, H. A. Haus, J. G. Fujimoto and E. P. Ippen, Opt. Lett. **24**, 511-513 (1999).; H. Sutter, G. Steinmeyer, L. Gallmann, N. Matuschek, F. Morier-Genoud, and U. Keller, Opt. Lett. **24**, 631-633 (1999).

3 Lin Xu, Naoya Nakagawa, Shinki Nakamura, Ryuji Morita, and Mikio Yamashita, in *Conference on Lasers and Electro-Optics, San Francisco, California, May 11, 2000*, CLEO 2000 Technical Digest Series (Optical Society of America, Washington, DC, 2000), paper CThE6, pp. 412-413.

4 A. M. Weiner, D. E. Leaird, J. S. Patel, and J. R. Wullert, IEEE J. Quantum Electron. **28**, 908-920 (1992).

14-fs pulses at 1.3 μm generated from an all-solid-state Cr:forsterite laser

C. Chudoba[1], J.G. Fujimoto[1], E.P. Ippen[1], H.A. Haus[1], U. Morgner[2], F.X. Kärtner[2], V. Scheurer[3], G. Angelow[3], and T. Tschudi[3]

[1] Department of Electrical Engineering and Computer Science and Research Laboratory of Electronics, Massachusetts Institute of Technology, Cambridge, MA 02139 USA
E-mail: chudoba@mit.edu
[2] High-Frequency and Quantum Electronics Laboratory, University of Karlsruhe (TH), D-76131, Germany
[3] Institute for Applied Physics, TH Darmstadt, D-64289, Germany

Abstract. 14-fs pulses at 1.3 μm were generated by an all-solid-state, Kerr-lens mode-locked Cr:forsterite laser using double-chirped mirrors. The spectral bandwidth (FWHM) was 250 nm with an 80 mW average power and 100 MHz repetition rate.

Ultrashort, broadband light sources at 1.3 μm are important for studies of fibers and photonic devices as well as for biomedical imaging techniques, such as two-photon microscopy and optical coherence tomography. We report the generation of 14 fs pulses by an all-solid-state Cr:forsterite laser, using dispersion compensating double-chirped mirrors (DCM). The laser spectrum covers wavelengths of 1230-1580 nm with a full-width-half maximum of 250 nm. Previous investigators have reported pulse durations of 20 fs with bandwidths of 90 nm [1, 2]. To the best of our knowledge, our results represent the shortest pulses and broadest bandwidths ever produced by a Cr:forsterite laser.

A schematic of the laser is shown in Figure 1. The laser uses a standard z-fold cavity-design. The Cr:forsterite crystal is placed between two curved mirrors with 100 mm radius of curvature. The crystal is 5x5x5 mm^3 Brewster-cut and cooled to 263° K. A frequency diode pumped Nd-YAG is used as the pump source and an optical isolator is used in the pump beam to prevent feedback. For an incident power of 6 W, the crystal absorbs 4.3 W, corresponding to an absorption coefficient of 2.4 cm^{-1}. The output coupler has a 1% transmission with a 210 nm bandwidth. All mirrors in the cavity are DCMs except for the output-coupler OC. This setup results in 5 bounces on DCMs per round-trip. Two Brewster-cut, high dispersion PBH71-prisms are used for adjusting the dispersion compensation. The apex-to-apex prism separation is 0.05 m. The prisms are mounted on translation stages for control of both the prism insertion and the prism separation. To start mode-locking we scan the prism separation slightly Soft aperture Kerr-lens mode-locking gives stable pulse formation and no additional hard aperture is needed. The beam path is enclosed by plexiglas tubes and the cavity is purged by dried nitrogen to avoid dispersion from the water absorption band at 1.45 μm.

Intracavity dispersion compensation achieves pulse shaping by dispersion-managed mode-locking [3]. Balancing the dispersion from the gain and prism materials is particularly challenging in the 1.3 μm wavelength range because the

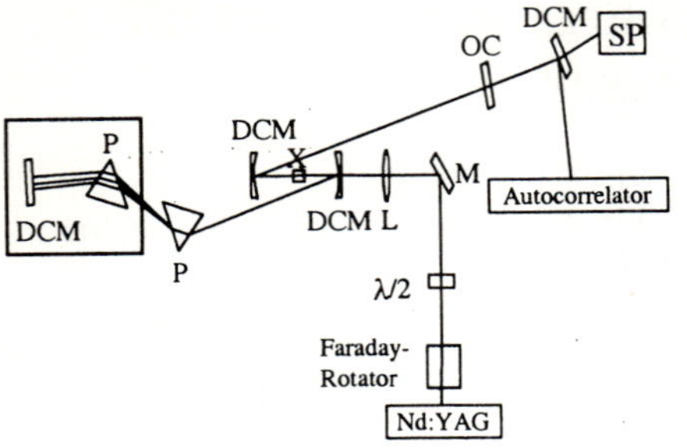

Fig. 1. Cr:forsterite laser with DCMs, PBH71 prisms P, crystal X, output-coupling mirror OC and focusing lens L. M is a standard dielectric mirror.

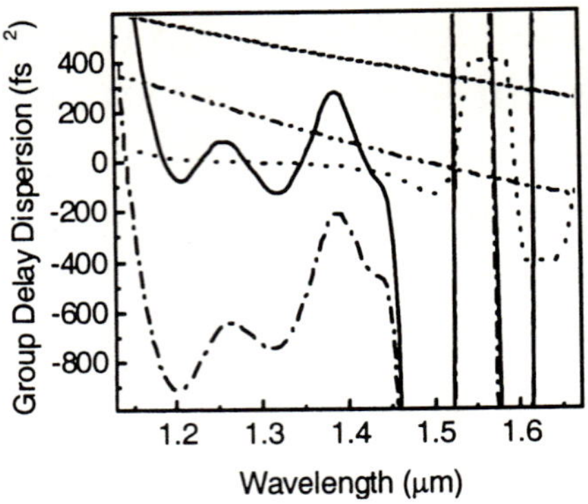

Fig. 2. The DCMs (dash dot) in combination with 4 mm PBH71 (dashed) balance the dispersion of the Cr:forsterite crystal (dash dot dot) and the output-coupling mirror (dotted). The solid line is the resulting total intracavity dispersion.

zero dispersion wavelength of most materials occurs in this wavelength range. Near the zero dispersion wavelength, higher-order dispersion becomes the dominant process limiting pulse durations. Higher order dispersion is difficult to compensate over an extended wavelength range with DCMs. Therefore, it is important to minimize higher order dispersion in the cavity by reducing the crystal length and choosing an optimum prism material. We used PBH71 (SF58) as prism material with a zero dispersion wavelength of 2 μm. This zero dispersion wavelength is at much longer wavelengths than the lasing wavelength and results in a favorable ratio of second to third order dispersion in the lasing range, enabling the design of broader bandwidth DCMs (Fig.2).

Since Cr:forsterite is a low gain material, high reflectivity, low loss mirrors are important. The reflectivity of the DCMs is greater than 0.998 from 1.1-1.5 μm, resulting in a bandwidth of 400 nm. At shorter wavelengths of <1.10 μm, the reflectivity decreases rapidly in order to have high transmission of the pump beam at 1.064 μm. The DCMs consist of 48 layers and were fabricated by ion-beam sputtering, the state-of-the-art technique for dielectric multi-layer deposition [4].

The dispersion contribution of the laser rod, the prism material, the output-coupling mirror and the DCMs inside the laser cavity are shown in Fig. 2. The group delay dispersion of the DCMs and the output-coupling mirror were measured by whitelight interferometry. Since the prism separation is small, only the material insertion contributes significantly to the intracavity dispersion. Higher order dispersion from the prism material is minimized by the insertion of only 4 mm PBH71 per round-trip. The DCMs in combination with the PBH71 prisms balance the dispersion in the range from 1.2-1.48 μm. This dispersion compensation results in the broad spectrum shown in Fig. 3 and enables the generation of 14-fs pulses as shown in Fig. 4.

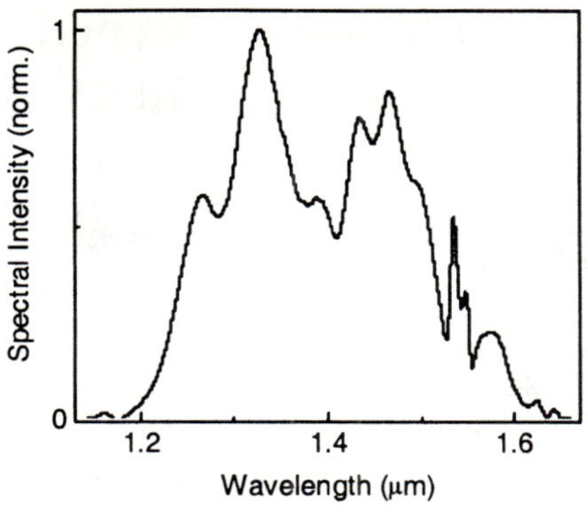

Fig. 3. Output spectrum of the Cr:forsterite laser showing 250 nm FWHM bandwidth.

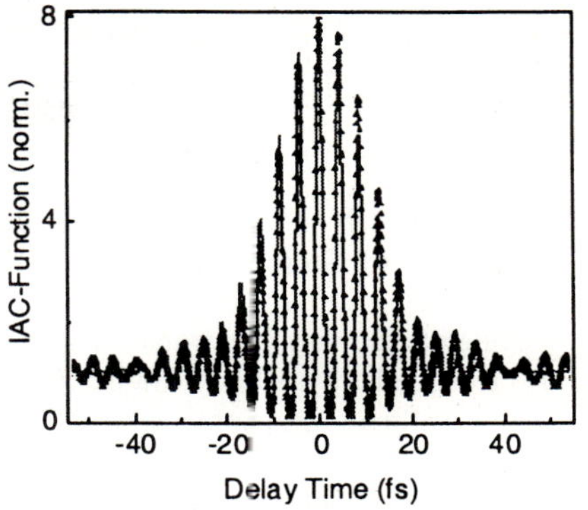

Fig. 4. Interferometric autocorrelation measurement (triangles). The solid line is a $sech^2$-fit with a FWHM of 14 fs.

The pulse durations were measured using an interferometric autocorrelator [5] with counter-oriented beamsplitters. An extracavity DCM compensated for the dispersion of the 6.35 mm thick CaF_2-substrate of the output-coupling mirror and the 1 mm thick beamsplitters in the autocorrelator. The peaks in the output spectrum correspond well with regions of net negative intracavity dispersion caused by the GDD oscillations of the DCMs. Stronger Kerr-lens mode-locking or reduced GDD oscillations of the DCMs are required to minimize the modulation of the spectrum.

In summary, we generated 14-fs pulses covering the spectral range from 1220-1580 nm produced by a Cr:forsterite laser. This record pulse duration is achieved by using specially designed DCMs for improved intracavity dispersion compensation in the 1.3 μm wavelength regime.

Acknowledgements. C. Chudoba acknowledges support from the Deutsche Forschungsgemeinschaft (DFG). This research was supported by the Air Force Office of Scientific Research contract F4920-98-1-0139 and the Medical Free Electron Laser Program, Office of Naval Research contract N00014-94-1-0717.

References

1. Z. Zhang, K. Torizuka, T. Itatani, K. Kobayashi, T. Sugaya, and T. Nakagawa, *IEEE Journal of Quantum Electronics*, 1997. **33**(11): p. 1975-81.
2. Z. Zhang, K. Torizuka, T. Itatani, K. Kobayashi, T. Sugaya, T. Nakagawa, and H. Takahashi, *Opt. Lett.*, 1998. **23**(18): p. 1465-7.
3. Y. Chen, F.X. Kärtner, U. Morgner, S.H. Cho, H.A. Haus, E.P. Ippen, and J.G. Fujimoto, *J. Opt. Soc. Am. B*, 1999. **16**(11): p. 1999-2004.
4. M. Tilsch, V. Scheurer, J. Staub, T. Tschudi, *Proc. SPIE*, 1994. **2253**: p. 414.
5. C. Spielmann, X. Lin, and F. Krausz, *Applied Optics*, 1997. **36**(12): p. 2523-5.

Smooth dispersion compensation: novel chirped mirrors with suppressed dispersion oscillations

L. Gallmann[1], N. Matuschek[1], D.H. Sutter[1], V. Scheuer[2], G. Angelow[2], T. Tschudi[2], G. Steinmeyer[1], and U. Keller[1]

[1] Institute of Quantum Electronics, Swiss Federal Institute of Technology
ETH Hönggerberg – HPT, CH-8093 Zürich, Switzerland
E-mail: gallmann@iqe.phys.ethz.ch
[2] Institute for Applied Physics, TH Darmstadt
D-64289 Darmstadt, Germany
E-mail: Volker.Scheuer@physik.th-darmstadt.de

Abstract. We present a novel design approach for chirped mirror dispersion compensation, suitable for octave-filling spectra with strongly reduced dispersion oscillations. This method is experimentally demonstrated on a mode-locked Ti:sapphire laser. From the full amplitude and phase characterization of the resulting ultrabroadband pulses, we obtained a full width at half maximum duration of 5.8 fs.

1. Introduction

A major limitation in chirped mirror design arises due to interference between light reflected at different penetration depths inside the mirror structure. This results in residual oscillations in the group delay dispersion (GDD) which ultimately limits pulse shortening. Unfortunately, there is always a trade-off between GDD-oscillations and reflection bandwidth. To date, the shortest pulses directly generated from a Ti:sapphire laser have a duration below 6 fs [1, 2]. Double-chirped mirrors (DCMs) [3] were used for higher order dispersion compensation over a bandwidth of 180 THz with GDD-oscillations of about 50 fs^2 (peak-to-peak) or 15 fs^2 (rms). Our new DCM design increases the dispersion compensating bandwidth to 220 THz with GDD oscillations of only 5 fs^2 (peak-to-peak) or 2 fs^2 (rms).

2. Chirped mirrors and impedance matching

So far, the DCM technique has been the most promising way to keep dispersion oscillations at a minimum. Double chirping means that, in addition to the local Bragg wavelength, the local coupling of the incident wave to the reflected wave is independently chirped. This corresponds to an adiabatic matching of the impedance. A theoretical DCM is used as an excellent starting design, requiring only local numerical optimization [4]. However, the strong index discontinuity at the air-coating interface still gives rise to unwanted interference, which results in oscillations of the group delay with wavelength similar to a Gires-Tournois interferometer. These detrimental dispersion oscillations are reduced by the

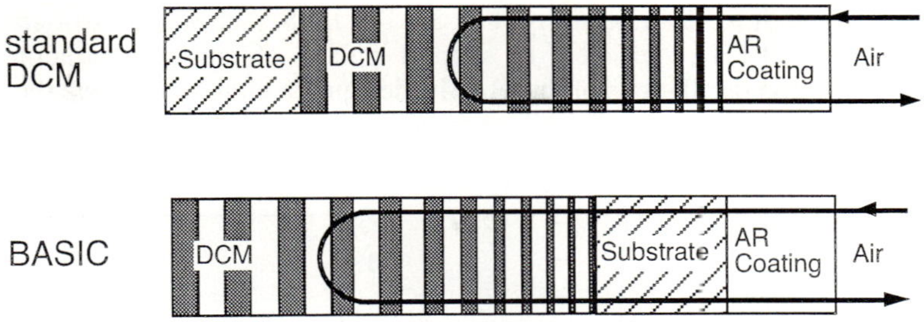

Fig. 1. Schematic comparison of a standard DCM design (top) and a BASIC chirped-mirror design (bottom)

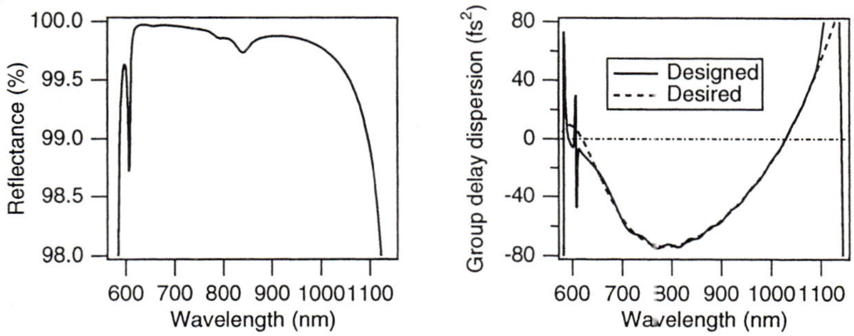

Fig. 2. Reflectance (left) and group delay dispersion (right) of the BASIC DCM design

antireflection (AR) coating on top of the high reflector stack, matching the impedance of air to the high or low index material in the front part of the mirror. To suppress dispersion oscillations sufficiently for the use inside a mode-locked Ti:sapphire laser, a reflectivity of 10^{-4} or less is required for the AR coating. For broadband AR coatings in general, one has to trade bandwidth for residual reflectivity, which effectively limits the usefulness of the direct DCM approach to about 200 THz bandwidth.

Our novel approach to mirror dispersion compensation, in principle, totally eliminates dispersion oscillations by geometrically separating the AR coating and the chirped mirror stack. The chirped mirror is coated on the back of a very thin substrate (Fig. 1). Perfect impedance matching between the back-side coated (BASIC, [5]) chirped mirror and the substrate is accomplished if the index of the substrate is chosen identical to the index of the high or low index material of the coating. Detrimental interferences between the top side of the substrate and the BASIC chirped mirror are avoided by making the two surfaces noninterfering. This can be achieved, e.g., by wedging the substrate or using different radii of curvature for the two surfaces. An AR coating may be used on the front surface to reduce losses. Other than in the conventional design, however, imperfections of the AR coating do not affect the dispersion properties of the chirped mirror. Not being limited by dispersion oscillations, we designed a BASIC chirped mirror that supports the wavelength range from 610 nm to 1100 nm for dispersion compensation with >99% reflectivity (Fig. 2). The BASIC chirped mirror covers a bandwidth of 220 THz, sufficient to support pulses of sub-5-fs duration. The residual dispersion oscillations of this BASIC design are significantly smaller than those of conventional DCMs of similar bandwidth.

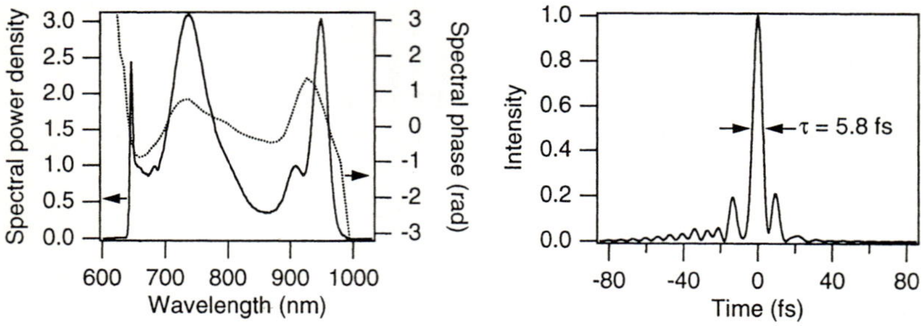

Fig. 3. Spectrum and spectral phase (left) and temporal intensity profile (right) of a 5.8 fs pulse obtained with BASIC DCMs and measured with spectral phase interferometry for direct electric-field reconstruction (SPIDER)

3. Experimental results and conclusion

With two mirrors fabricated according to the design shown in Fig. 2 and one additional conventional DCM inside the laser cavity, we were able to generate pulses as short as 5.8 fs (Fig. 3). This combination of mirrors was necessary to cancel out some residual dispersion oscillations due to the limited layer deposition accuracy (about 0.2 nm rms per layer). It is to be expected that deposition accuracy will be improved to about 0.1 nm rms in the near future, allowing to fully exploit the potential of the BASIC design approach. This technique should allow for smooth octave-spanning dispersion compensation.

Acknowledgements. This work has been supported by the Swiss National Science Foundation.

References

1 D. H. Sutter, G. Steinmeyer, L. Gallmann, N. Matuschek, F. Morier-Genoud, U. Keller, V. Scheuer, G. Angelow, and T. Tschudi, Opt. Lett. 24, 631-633, 1999.

2 U. Morgner, F. X. Kärtner, S. H. Cho, Y. Chen, H. A. Haus, J. G. Fujimoto, E. P. Ippen, V. Scheuer, G. Angelow, and T. Tschudi, Opt. Lett. 24, 411-413, 1999.

3 F. X. Kärtner, N. Matuschek, T. Schibli, U. Keller, H. A. Haus, C. Heine, R. Morf, V. Scheuer, M. Tilsch, and T. Tschudi, Opt. Lett. 22, 831-833, 1997.

4 N. Matuschek, F. X. Kärtner, and U. Keller, IEEE J. Quantum Electron. 35, 129-137, 1999.

5 N. Matuschek, L. Gallmann, D. H. Sutter, G. Steinmeyer, and U. Keller, U.S. patent application priority date Feb. 11, 2000.

Dispersion Management over one Octave with Tilted-Front-Interface Chirped Mirrors

Gabriel Tempea and Ferenc Krausz

Institut für Photonik, Technische Universität Wien
Gußhausstr. 27/387, A-1040 Vienna, Austria
E-mail: tempea@tuwien.ac.at

Abstract. The spectral range over which a chirped mirror is capable of introducing tailored dispersion can be extended to one octave by tilting the front (air) interface of the coating with respect to other interfaces of the multilayer. In addition, the elimination of adverse interference effects also results in a significant smoothening of the group delay dispersion versus wavelength.

Dispersive ("chirped") mirrors (CMs) [1] allow controlling the frequency-dependence of the group delay by means of the penetration depth of different spectral components. In order to achieve e.g. negative group delay dispersion (GDD), wave packets carried at longer wavelengths have to suffer increased delay by having to penetrate deeper into the coating before getting reflected as compared to short-wavelength wavepackets. However, at the dielectric-air interface fairly wavelength-independent reflection of about 3% occurs. Interference occurs between rays reflected at the air-interface and rays reflected deeper in the coating, leading to a strong distortion of the reflectance and particularly of the phase characteristics of the mirror. This adverse effect can be alleviated by using an antireflection (AR) coating [2] or a narrow bandstop filter [3] on top of the reflecting structure. Efficient suppression of spurious resonances calls for a reflectance significantly less than 10^{-3} at the air interface. AR coatings or bandpass filters can meet this requirement only over a limited spectral range, restricting the bandwidth of CMs usable for dispersion control to ≈ 150 THz at 800 nm [2,3]. In addition, the GDD-versus-wavelength curve is often plagued by large fluctuations approaching the nominal value of GDD caused by residual interference effects.

In this paper we demonstrate that the above-mentioned drawbacks can be overcome by tilting the front interface (the interface to the incident medium, which is most commonly air) with respect to other interfaces of the multilayer structure, as sketched in Fig. 1. In this manner adverse interference between the beam reflected at the air-dielectric interface (dotted line) and those reflected by internal interfaces (dashed line) can be completely eliminated and requirements to the AR coating on the tilted-front surface substantially relaxed. To keep the overall reflectance high, the front-surface reflectance can be reduced below 0.2 % over 500-1000 nm without introducing notable phase distortions by using an AR coating consisting of merely 13 layers.

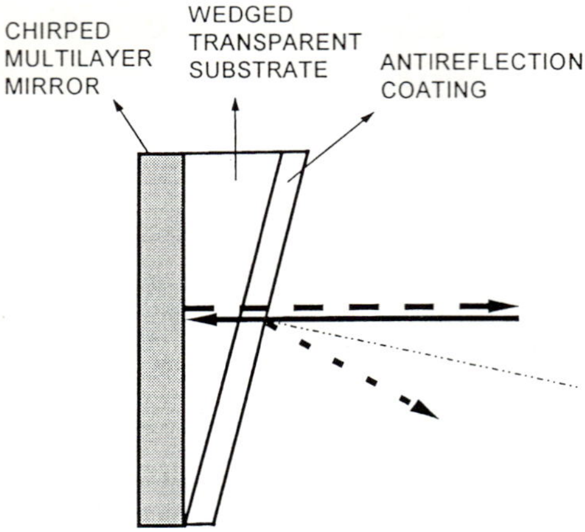

Fig. 1. Simplified schematic of a TFICM, coated on a thin wedged glass substrate and illuminated from the substrate-side. The substrate introduces additional positive dispersion, reducing the overall negative GDD of the mirror. However, the wedged plate can be as thin as 100-200μm. Alternatively, the multilayer can be coated on a thick optical substrate and than the thin AR-coated wedged plate applied on the mirror, using an index-matching fluid.

To demonstrate the power of this concept, we have designed a tilted-front-interface chirped mirror (henceforth briefly TFICM), which *provides dispersion control up to (and including) the fourth order over one octave*, between 500 and 1000 nm (Fig. 2). Employing several reflections off such a mirror, the spectral phase imposed on a broadband wave packet upon travelling several millimeters in e.g. fused silica and tens of centimeters in air can be compensated with an accuracy sufficient for the generation of pulses with durations of around 3 fs in the single-cycle regime. A first attempt has already been made to manufacture this mirror. The squares in Fig. 2 depict GDD-data measured by white light interferometry [4]. Deviations of the measured GDD-curve from the designed one are due to manufacturing errors, which need to be reduced in the future. The mirror consists of 68 SiO_2 and TiO_2 layers deposited on a thin wedged fused silica substrate. The need for the comparatively high number of layers is primarily related to negative third-order dispersion (TOD) required for compensating the material TOD of opposite sign. If only second-order dispersion were to be introduced, less layers were required and the absolute value of GDD could be increased. However, in order to push the performance of pulse compression techniques towards the single-cycle limit, control of higher-order dispersion is essential.

With the problem of impedance matching to air being inherently resolved, the choice of the starting multilayer structure becomes less critical than in the case of conventional CMs. As a result, even a linearly chirped initial structure leads to fast convergence of the computer optimization algorithm to the target specifications using the TFI concept. The GDD-versus-wavelength curve used to be highly sensitive to manufacturing errors, because small deviations from the nominal thickness of the top layers can easily destroy the impedance matching implemented by conventional methods [1-3]. TFI designs are significantly more stable, because impedance mismatch between the internal multilayer structure and the (tilted) air-mirror interface is eliminated independently of the layer structure. Statistical error sensitivity analysis shows that the fluctuations of the GDD-versus-wavelength curve of a conventional CM introducing a nominal GDD of $-45fs^2$

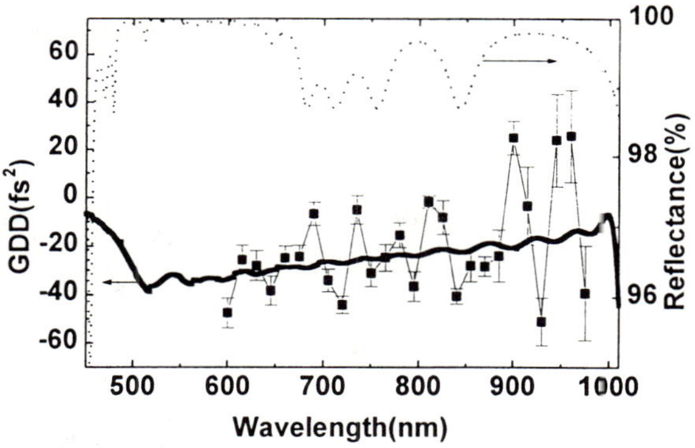

Fig. 2. Reflectance (dots) and GDD (full line) of a TFI-CM consisting of 68 SiO_2 and TiO_2 layers. The mirror precisely compensates the dispersion introduced by a physical path of .45mm in fused silica plus 15cm in air. The GDD could not be measured bellow 600nm because of bandwidth limitation sin our white light interferometer

over 300 nm at 800 nm have a relative amplitude of 100% for a maximum absolute deviation of 0.5 nm from the nominal layer thickness. By contrast, the maximum amplitude of the GDD oscillations is only 30% for a TFI mirror meeting the same specifications.

In conclusion, perfect impedance matching of CMs to the environment can be achieved over an arbitrary broad spectral range by tilting the front interface with respect to internal interfaces of the multilayer. As a result, *the bandwidth of CMs drawing on the TFI concept is only limited by technological issues* (number of layers that can be coated and accuracy of thickness control). Additional benefits emerging from the TFI concept include i) reduced sensitivity to manufacturing errors, ii) suppression of the higher-order reflection bands (i.e. higher transmittance outside the HR band), and iii) potentially fluctuation-free dispersion characteristics. Combined with well-established spectral-broadening techniques [5,6], these mirrors pave the way towards single-cycle optical pulse generation.

References

1 R. Szipöcs et al., Opt. Lett. 19, 201, 1994.
2 F. X. Kärtner et al., Opt. Lett. 22, 831, 1997.
3 G. Tempea et al., IEEE JSTQE 4, 193, 1998.
4 W. H. Knox et al., Opt. Lett. 13, 574, 1988.
5 A. Baltuska et al., Opt. Lett. 22, 102, 1997.
6 Z. Cheng et al., in Ultrafast Phenomena XI, J.G. Fujimoto, D.A. Wiesma, W. Zinth, eds. (Springer), 8, 1998.

A prism-pair-formed pulse shaper compresses optical pulses to the 6-fs regime

Lin Xu, Naoya Nakagawa, Ryuji Morita, and Mikio Yamashita

Department of Applied Physics, Hokkaido University and CREST, Japan Science and Technology Corporation (JST), Sapporo 060-8628, Japan
Email: lin.xu@physics.gatech.edu

Abstract. We experimentally demonstrate the first programmable optical pulse compression to 6 fs using a prism-pair-formed pulse shaper with a spatial light modulator. This method manifests the potential to compress pulses to the one-cycle regime.

1. Introduction

Typically, prism-pairs, grating-pairs, chirped mirrors and the combinations of these elements are utilized for dispersion compensation to generate ultrashort optical pulses. However, these methods have the drawbacks concerning the independent control of different orders of dispersion and the limited bandwidth due to using gratings and chirped mirrors. They will definitely prevent optical pulse compression down to the one-cycle regime. Pulse shaping technique manifested the capability of independent arbitrary phase control [1,2]. Recently, an 11-fs pulse was generated by using a spatial light modulator (SLM) in a pulse shaper [3]. More recently, we have demonstrated 5-fs pulse compression with a programmable SLM in a grating-pair-formed pulse shaper [4]. In this pulse shaper, the bandwidth of the diffractive gratings is broader, but its limitation still exists. In addition, if the pulse spectrum is broader enough to cover one octave range, which can be produced by induced-phase modulation (IPM) technique and by self-phase modulation (SPM) in a microstructure fiber [5,6], respectively, the diffraction gratings in the pulse shaper will introduce complicate phase controlling. One of methods to solve this problem is to utilize prisms instead of gratings. In this paper, we present a prism-pair-formed pulse shaper with a programmable SLM for chirp-compensation. We experimentally demonstrate a 6-fs pulse compression using this method for the first time.

2. Experiment and results

The prism-pair-formed pulse shaper is shown in the dashed frame of Fig.1. It is consisted of a pair of highly dispersive Brewster TF5 prisms (P3 and P4), and a pair of concave 200-mm focal-length spherical mirrors (M10 and M11). The prisms are placed at the focal planes of the concave mirrors to form a 4-*f* system. And a commercial 128-pixel liquid crystal SLM (Meadowlark Optics) is located on the Fourier plane of the pulse shaper, where frequencies are spatially dispersed.

In the experiment, a 400-μJ, 30-fs pulse is launched into an argon-filled capillary fiber with an inner diameter of 140 μm and a length of 60 cm. At gas pressure of 2.0 bars, a white light continuum (WLC), which extends from 500 nm to 1000 nm, is generated due to the dispersive SPM effect. Fig.2 (a) shows the measured spectrum. The pulse duration is also measured to be of 177 fs (full width at half maximum, FWHM). The WLC beam is directed into a pair of prisms (BK7) formed pre-compressor, which is utilized to compensate for the main linear chirp of WLC and material dispersion of TF5 prisms in the pulse shaper. The separation length between two prisms is set to 65 cm.

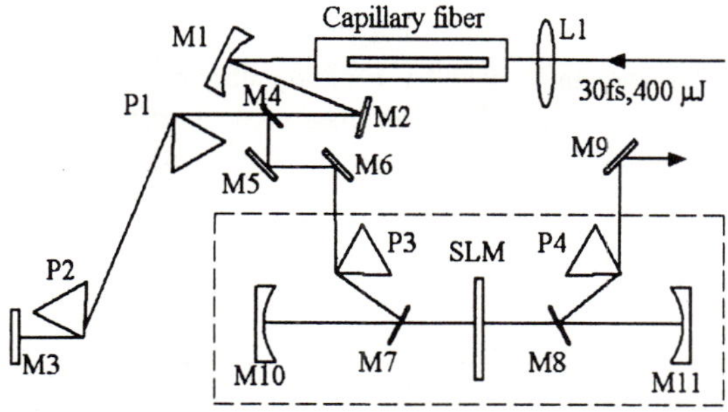

Fig. 1. Layout of experimental setup. M2-M9, silver-coated plane mirrors; M1, M10, and M11, Al-coated spherical mirrors, R = -400 mm; L1, spherical lens, f =300 mm; P1, P2, Brewster-angle cut BK7 prisms; P3, P4, Brewster-angle cut TF5 prisms. The pulse shaper is shown within the dashed frame.

We couple the pre-compensated WLC into the pulse shaper. In order to realize the programmable phase control of SLM, the frequency distribution on the SLM plane is numerically evaluated. The nonlinear phase dispersion applied on the SLM can be written as:

$$\phi_{SLM}(\omega_j) = \frac{1}{2!}\frac{d^2\phi}{d\omega^2}\bigg|_{\omega_0}(\omega_j - \omega_0)^2 + \frac{1}{3!}\frac{d^3\phi}{d\omega^3}\bigg|_{\omega_0}(\omega_j - \omega_0)^3 + \ldots ,$$

where j=1,2…128 (pixel number). $d^2\phi/d\omega^2|\omega_0$ and $d^3\phi/d\omega^3|\omega_0$ represent group velocity dispersion (GVD) and third-order dispersion (TOD) at the center frequency ω_0, respectively. A computer controls the phase on the SLM. When the quadratic phase (GVD) and cubic phase (TOD) are set to be of –330 fs^2 and 2000 fs^3, respectively, the shortest optical pulse is observed. The interferometric autocorrelation trace is shown in Fig. 2(b). The compressed pulse spectrum is depicted in the inset. The direct Fourier transform (FT) of the spectrum gives the pulse duration of 5.6fs (FWHM), and the fitting on the assumption of a $sech^2(t)$ intensity profile shows the pulse duration of 6 fs. The well agreement between the measured autocorrelation and FT-constructed autocorrelation indicates the smallest residual phase errors existing. In our experiment, the generated shortest optical pulse is limited by the compressed pulse spectrum. Comparing the compressed pulse spectrum (inset of Fig.2 (b)) with the generated WLC spectrum (Fig. 2 (a)), we observed the short wavelength components of the compressed pulse were cut by the second BK7 prism in the pre-compressor due to its smaller size. This limitation can be removed by using large size prisms. As a result, a shorter optical pulse can be generated by the compression technique in the near future if the forth-order dispersion (FOD) and the higher-order dispersion are compensated for.

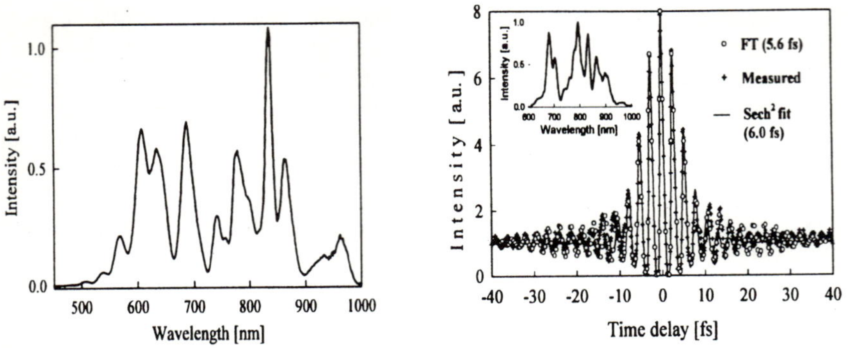

Fig. 2. (a) The WLC spectrum generated from an argon-filled capillary fiber pumped by a 30-fs, 400-μJ 1-kHz pulse centered at 780 nm. (b) The measured interferometric autocorrelation trace when applied GVD = -330 fs^2 and TOD = 2000 fs^3 (see text).

3. Summary

In summary, a prism-pair-formed pulse shaper with a commercial-SLM has been demonstrated for programmable chirp compensation for the first time. The shortest optical pulse of 6-fs has been compressed when the quadratic phase and the cubic phase have been placed to be of –330 fs^2 and 2000 fs^3, respectively, on the SLM. The prism-pair-formed pulse-shaping compressor has unique features compared with its grating-pair partner. It has the significantly large bandwidth and lower losses while employing the Brewster angles. Due to the unique features, this method provides the possibility to compress an optical pulse down to the one-cycle regime.

References

1. A. M. Weiner, D. E. Leaird, J. S. Patel, and J. R. Wullert, IEEE J. of Quantum Electro., **28**, 908, 1992.
2. A. Efimov, C. Schaffer, and D. H. Reitze, J. Opt. Soc. Am. B **12**, 1968, 1995.
3. D. Yelin, D. Meshulach, and Y. Silberberg, Opt. Lett. **22**, 1793, 1997.
4. L. Xu, N. Nakagawa, S. Nakamura, R. Morita and M. Yamashita, in Technical Digest of Conference on Lasers and Electro-Optics (CLEO) 2000, CThE6, 412.
5. N. Karasawa, R. Morita, H. Shigekawa and M. Yamashita, Opt. Lett. **25**, 183, 2000.
6. J. K. Ranka, R. S. Windeler and A. J. Stentz, Opt. Lett. **25**, 25, 2000.

Correcting the Failure of the Slowly Varying Amplitude Approximation for Short Pulses

Deborah Alterman[1] and Jeffrey Rauch[2]

[1] Dept. of Applied Mathematics, Box 526, University of Colorado, Boulder, CO 80309-0526, USA
[2] Dept. of Mathematics, University of Michigan, Ann Arbor, MI 48019, USA

Abstract. The slowly varying amplitude assumption, a useful simplification, is invalid for few cycle pulses. We offer a rigorously justified alternative, valid for short pulses, which leads to qualitatively different behaviors.

When studying the propagation of electromagnetic waves, both computations and analysis are simplified by assuming a plane wave solution $\mathbf{E}(t,\mathbf{x}) = \mathbf{E}_0 e^{-i\omega(\mathbf{x}/c-t)}$. To describe wavetrains with decaying tails, $\mathbf{E}_0$ is replaced by an amplitude function $\mathbf{A}(t,\mathbf{x})$. For high frequencies the phase $e^{-i\omega(\mathbf{x}/c-t)}$ oscillates rapidly compared to changes in $\mathbf{A}$. Diffraction appears over longer times, a third variable scale. Introduce the small dimensionless parameter $\varepsilon = \mathcal{O}(\lambda)$, relative to the natural length scale. The envelope function $\mathbf{A}(\varepsilon t, t, \mathbf{x})$ captures changes on the slow time scale $T = \varepsilon t$. Equations for $\mathbf{A}(T,t,\mathbf{x})$ are simpler than the original equations and computing $\mathbf{A}$ does not require resolution on lengths scales of $\mathcal{O}(\varepsilon)$, $\mathcal{O}(1)$ and $\mathcal{O}(1/\varepsilon)$.

The development of ultrafast lasers in the last decade raises questions about the slowly varying amplitude assumption. These lasers typically pass a point in times measured in femtoseconds and contain only a few cycles or even only half a cycle. In such cases this assumption is inappropriate.

This problem has been addressed in two ways. The full Maxwell's equations can be studied as in [1]-[4]. This approach yields insight into the differences between short pulses and wavetrains, but requires costly computations which resolve the solution on the small ε length scale. Another line of inquiry, as in [5]-[10] uses asymptotics to derive new approximate solutions valid for short pulse initial data. Sometimes weaker assumptions about pulse length or amplitude evolution replace the slowly varying assumption.

Our work assumes an initial short pulse profile, and derives and rigorously justifies a new approximate solution using only a few physically reasonable assumptions [11]. This short pulse approximation leads to simpler equations and computations which need not resolve all three length scales.

The differences between the slowly varying amplitude approximation and the short pulse approximation emerge in the wave equation in 2D plus time

$$\partial_t^2 u - \partial_{x_1}^2 u - \partial_{x_2}^2 u = 0 \,. \tag{1}$$

Consider wavetrain initial data having slowly varying amplitude multiplying a rapidly oscillating phase, given by $u(\varepsilon, 0, x_1, x_2) = f(x_1, x_2)e^{ix_1/\varepsilon}$ and $\partial_t u(\varepsilon, 0, x_1, x_2) = 0$.

Explicit solution on the Fourier side shows that $u(\varepsilon, t, \mathbf{x}) \approx A(\varepsilon t, x_1 - t, x_2)e^{i(x_1-t)/\varepsilon}$ where $A(T, x_1, x_2)$ satisfies the Schrödinger equation

$$\partial_T A = -\frac{i}{2}\partial_{x_2}^2 A \qquad A(0, x_1, x_2) = f(x_1, x_2) \,. \tag{2}$$

Assuming a short pulse initial profile leads to a different equation. This initial data decays rapidly in the propagation direction and hence can be represented in the form $u(\varepsilon, 0, x_1, x_2) = f(x_2, x_1/\varepsilon)$ and $\partial_t u(\varepsilon, 0, x_1, x_2) = 0$ where $\lim_{|\varphi|\to\infty} f(x_2, \varphi) = 0$. This represents a pulse of width $\mathcal{O}(\varepsilon)$ in x_1 and width $\mathcal{O}(1)$ in x_2. Such initial data better describe the output of an ultrafast laser with beam width of $\mathcal{O}(1)$. For short pulses $u(\varepsilon, t, \mathbf{x}) \approx B(\varepsilon t, x_2, (x_1 - t)/\varepsilon)$ where $B(T, x_2, \varphi)$ satisfies a *different* linear equation.

$$\partial_T \partial_\varphi B = -\frac{1}{2}\partial_{x_2}^2 B \qquad B(0, x_2, \varphi) = f(x_2, \varphi) \tag{3}$$

The wavetrain approximation and the short pulse approximation to solutions of (1) lead to two different profile equations, (2) and (3). Numerical studies show that these different equations lead to qualitatively different descriptions. We used a single cycle initial pulse shape $u(\varepsilon, 0, x_1, x_2) = e^{-(x_1/\varepsilon)^2}e^{-x_2^2}\sin(x_1/\varepsilon)$, which yields an explicit formula for the wavetrain approximation. The approximations are functions of the variable $(x_1 - t)/\varepsilon$ so computations need not be scaled to ε. After propagation for time $\varepsilon t = 1$, two very different pictures emerge.

Compared to the wavetrain approximation, the short pulse approximation has increased intensity at the leading edge of the pulse and a split parabolic tail behind the pulse, while the wavetrain approximation shows more spread in the transverse, or x_2 direction. These features appear in Figure 1 and have been observed experimentally in THz pulses in free space. [12].

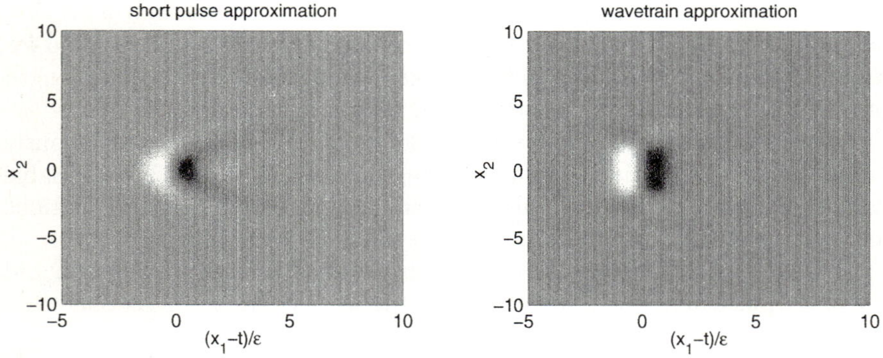

Fig. 1. A comparison of the two approximations to solutions of (1) at time $\varepsilon t = 1$

Like the slowly varying amplitude approximation, the short pulse approximation can also be applied in many settings. N wave propagation equations in d spatial variables with order J nonlinearity can be written in the form

$$\partial_t \mathbf{u} + \sum_{j=1}^{d} A_j \partial_{x_j} \mathbf{u} + \mathbf{\Phi}(\mathbf{u}) = 0 \tag{4}$$

with $N \times N$ matrices A_j. Under certain physically reasonable assumptions [11], the following method produces an approximate solution $\mathbf{U}(\varepsilon, t, \mathbf{x})$ in the short pulse approximation, valid for ε small compared to the scale of $\mathbf{x}$.

Choose a wavenumber $\mathbf{k}_0$ and frequency $\omega = \omega(\mathbf{k}_0)$ so that the matrix $L_0 = \omega I + \sum_{j=1}^{d} k_j A_j$ is singular. Define $\pi = \pi(\mathbf{k}_0)$ to be the orthogonal projection operator which projects into the kernel of L_0. Let $p = 1/(J-1)$, the power in ε of solutions where diffractive effects and nonlinear effects appear on the same time scale. The short pulse initial profile can be described

$$\mathbf{u}(\varepsilon, 0, \mathbf{x}) = \varepsilon^p \, \mathbf{f}(\mathbf{x}, (\mathbf{k}_0 \cdot \mathbf{x})/\varepsilon) \tag{5}$$

The function $\mathbf{f}(\mathbf{x}, \varphi)$ satisfies $\pi \mathbf{f}(\mathbf{x}, \varphi) = \mathbf{f}(\mathbf{x}, \varphi)$ and $\pi \mathbf{U}(\varepsilon, t, \mathbf{x}) = \mathbf{U}(\varepsilon, t, \mathbf{x})$. Define the group velocity $\mathbf{v}$ by $\mathbf{v} := (v_1, \ldots, v_d)$ where $v_j = -\frac{\partial \omega}{\partial k_j}\big|_{\mathbf{k}=\mathbf{k}_0}$. Define the second order differential operator $\mathbf{R}(\partial_{\mathbf{x}})$ by

$$\mathbf{R}(\partial_{\mathbf{x}}) := -\frac{1}{2} \sum_{l,m=1}^{d} \frac{\partial^2 \omega}{\partial k_l \partial k_m}\bigg|_{\mathbf{k}=\mathbf{k}_0} \frac{\partial^2}{\partial x_l \partial x_m} \, .$$

Let $\mathbf{\Phi}_J(\mathbf{u})$ denote the first nonzero term in the Taylor expansion of $\mathbf{\Phi}(\mathbf{u})$ about $\mathbf{0}$. An approximate solution to equations (4) and (5) is then given by $\mathbf{U}(\varepsilon, t, \mathbf{x}) = \varepsilon^p \mathbf{U}_0(T, t, \mathbf{x}, \varphi)\big|_{T=\varepsilon T,\, \varphi=(\mathbf{k}_0 \cdot \mathbf{x} - \omega t)/\varepsilon}$ where $\mathbf{U}_0(T, t, \mathbf{x}, \varphi) = \mathbf{F}(T, \mathbf{x} - \mathbf{v}t, \varphi)$ and the function $\mathbf{F}(T, \mathbf{x}, \varphi)$ is determined by the initial value problem (6), which generalizes (3).

$$\partial_T \partial_\varphi \mathbf{F} + \mathbf{R}(\partial_{\mathbf{x}})\mathbf{F} + \partial_\varphi \pi \mathbf{\Phi}_J(\mathbf{F}) = 0 \qquad \mathbf{F}(0, \mathbf{x}, \varphi) = \mathbf{f}(\mathbf{x}, \varphi) \tag{6}$$

References

1. R. W. Ziolkowski: Phys. Rev. A **39** (1989)
2. R. W. Hellwarth, P. Nouchi: Phys. Rev. E **54** (1996)
3. C. V. Hile, W. L. Kath: J. Opt. Soc. Am. B **13** (1996)
4. S. Feng, H. G. Winful, R. W. Hellwarth: Phys. Rev. E **59** (1999)
5. J. E. Rothenberg: Opt. Lett. **17** (1992)
6. E. Esarey, P. Sprangle, M. Pilloff, J. Krall: J. Opt. Soc. Am. B **12** (1995)
7. T. Brabec, F. Krausz: Phys. Rev. Lett. **78** (1997)
8. M. A. Porras: Phys. Rev. E **58** (1998)
9. A. E. Kaplan: J. Opt. Soc. Am. B **15** (1998)
10. J. K. Ranka, A. L. Gaeta: Opt. Lett. **23** (1998)
11. D. Alterman, J. Rauch: Phys. Lett. A **264** (2000)
12. Z. Jiang, X.-C. Zhang: Opt. Express **5** (1999)

Precise Control of the Pulse-to-Pulse Carrier-Envelope Phase in a Mode-Locked Laser

David J. Jones[1], Scott A. Diddams[1], Jinendra K. Ranka[2], Robert S. Windeler[2], Andrew J. Stentz[2], John L. Hall[1], and Steven T. Cundiff[1]

[1] JILA, University of Colorado and National Institute of Standards and Technology, Boulder, CO 80309-0440 USA
[2] Bell Laboratories, Lucent Technologies, 700 Mountain Avenue, Murray Hill, NJ 070974

Abstract. Using a coherent frequency domain technique, we lock the relative pulse-to-pulse carrier-envelope phase of a 15-fs pulse produced by a mode-locked Ti:Sapphire laser to various rational integer fractions of 2π.

With generation of optical pulses approaching the single cycle regime [1], control over the phase of the carrier with respect to the pulse envelope is becoming increasingly relevant, particularly in extreme nonlinear optics. A number of strong field processes in nonlinear optics have been either predicted and/or observed to be dependent on the relative carrier-envelope phase [2] including high harmonic/soft x-ray generation [3–5], attosecond pulse generation [6], and above-threshold ionization [7]

The physical origin of this relative carrier-envelope phase is the difference between the cavity-averaged group velocity (v_g) and phase velocity (v_p). After each round trip, the phase slip between the carrier and envelope of subsequent pulses is,

$$\Delta\phi = \left[\frac{l}{v_g} - \frac{l}{v_p}\right]\omega \bmod (2\pi), \tag{1}$$

where l is the cavity length and ω is the carrier frequency of the pulse. In general, $\Delta\phi$ is not zero and moreover fluctuations in the laser's operational parameters (e.g. cavity length, intracavity power, dispersion) vary v_g and v_p by different amounts, effectively randomizing the relative carrier-envelope phase from pulse to pulse emitted by an unstabilized mode-locked laser.

In this work we demonstrate, using both temporal cross-correlations and frequency domain measurements, precise control over the the pulse to pulse relative carrier-envelope phase ($\Delta\phi$) of a Kerr-Lens mode-locked Ti:Sapphire laser. [8] To obtain this phase control, we employ stabilization techniques that were first developed when mode-locked lasers were utilized in optical frequency metrology applications. [9,10] Further discussion of this emerging application of ultrafast lasers is included in the end of this paper.

The definition of $\Delta\phi$ is shown schematically in Fig. 1. It is important to note $\Delta\phi$ is not the absolute carrier-envelope phase of the pulse, rather it is the pulse to pulse carrier-envelope phase. The absolute phase, ϕ, of the mth pulse is given by $\phi(m) = \phi_o + m\Delta\phi$ where ϕ_o is unknown constant phase offset and m is an integer labeling individual pulses. Using the frequency domain technique described below, $\Delta\phi$ can be stabilized. However, the absolute carrier-envelope phase cannot determined. A similar, but refined measurement or an additional technique, such as above threshold ionization [2], is necessary to measure and control ϕ_o.

In the frequency domain $\Delta\phi$ shows up as a frequency offset (δ) of the fs laser's n^{th} frequency comb mode (f_n) from a harmonic of the pulse repetition rate f_{rep} (i.e., $f_n = nf_{rep} + \delta$, see Fig. 1). The offset frequency and $\Delta\phi$ are related via $2\pi\delta = \Delta\phi f_{rep}$. By stabilizing both δ and f_{rep}, we control $\Delta\phi$. The comb spacing is locked to a precision microwave oscillator, while δ is stabilized using the self-referencing technique described below.

Our experimental setup is described in detail elsewhere. [8] To phase lock f_{rep}, a high harmonic of the 90 MHz repetition rate is mixed with a microwave oscillator, thereby producing an error signal that is fed back to the mirror-mounted tilting PZT inside the laser cavity. As shown in Fig 1, a direct measure of the offset frequency δ is obtained by forming a heterodyne beat between a mode on the visible end of the spectrum (f_{2n}) and the second harmonic of a mode from the near infrared ($2f_n$). Subsequently, a digitally programmable error signal is formed between δ and $\frac{k}{16}f_{rep}$ and then is supplied to the PZT to translate the cavity mirror The variable k is a programmable integer $0 \leq k < 16$ thus allowing us to lock the relative carrier envelope phase from 0 to 2π in discrete steps of $\pi/8$. By directly referencing the carrier-envelope phase evolution of the output pulse train to the pulse repetition rate itself (which is a locked quantity), a pulse train with constant

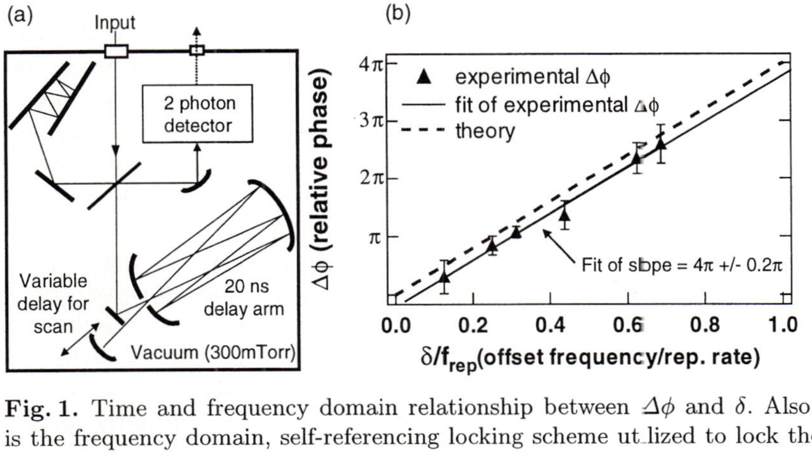

Fig. 1. Time and frequency domain relationship between $\Delta\phi$ and δ. Also shown is the frequency domain, self-referencing locking scheme utilized to lock the offset frequency δ

carrier-envelope phase can be easily obtained with a pulse picker driven at the appropriate subharmonic of f_{rep}. Subsequent integration of such a phase stable oscillator into a low repetition rate, high pulse energy amplifier system [12] can then be easily implemented. In order to obtain the full optical octave spectral bandwidth required for this locking scheme, we employ a 10 cm piece of air-silica microstructure fiber [11] that, in our case, generates continuum spanning 260 to 590 THz (510 to 1125 nm in wavelength).

To measure $\Delta\phi$ in the time domain, we perform a time-averaged interferometric cross-correlation [13] between the pulse i and pulse $i+2$ shown in Fig 2(a). A multi-pass cell in one arm of the correlator is used to generate the required 20 ns differential delay. To obtain a well-formed interferogram, the mirror curvatures and their separations were chosen to mode-match the output from both arms. To minimize the possible phase shift imbalance between the two correlator arms, all mirrors in the correlator have identical coatings except the PZT-mounted scanning mirror. Furthermore, each arm has an equal number of bounces. The entire correlator is in a vacuum chamber held below 300 mTorr to reduce the dispersive effects of air. The second order cross-correlation is measured using a two photon technique by reflective focusing of the recombined beam with a 10-cm radius of curvature spherical mirror onto a window-less GaAsP photodiode.

We determine $\Delta\phi$ by fitting the fringe peaks of the interferogram to a correlation function assuming a Gaussian pulse envelope. From the fit parameters, we determine the center of the envelope and compare it with the phase of the underlying carrier. A fit of the fringe peaks assuming a hyperbolic secant envelope produced nearly identical results. A plot of the experimentally determined relative phases for various offset frequencies, along with a linear fit of the averaged data, is given in Fig. 2(b). Also shown is the expected relation $\Delta\phi = 4\pi\delta/f_{rep}$ (the extra factor of two results from correlating the pulse i and pulse $i+2$). We attribute the phase offset between experiment and theory is an artifact from the cross-correlation. Specifically, it is most likely due to a mirror-reflection induced phase imbalance in the correlator. The experimental slope is within 5% of the theoretically predicted value and demonstrates precise control over the pulse to pulse carrier-envelope phase.

Although our time-domain cross-correlations clearly show successful phase control, a measurement of the phase stability can also be made in the frequency domain by counting the offset frequency δ. At each of the different lock points, we counted δ with a gate time of 0.1 s. A maximum uncertainty of 3.6 Hz was measured, giving an uncertainty in $\Delta\phi$ of 0.19 mrad. During the counting runs, a maximum excursion of 5.6 Hz in δ was observed, corresponding to a carrier-envelope phase slippage of 0.56 cycles/sec. Results of a more thorough examination of the phase stability will be presented elsewhere. There are a number of possibilities why our time domain phase measurements have a higher degree of uncertainty. The correlator has a significantly smaller amount of averaging due to both its shorter effective gate time as well as its

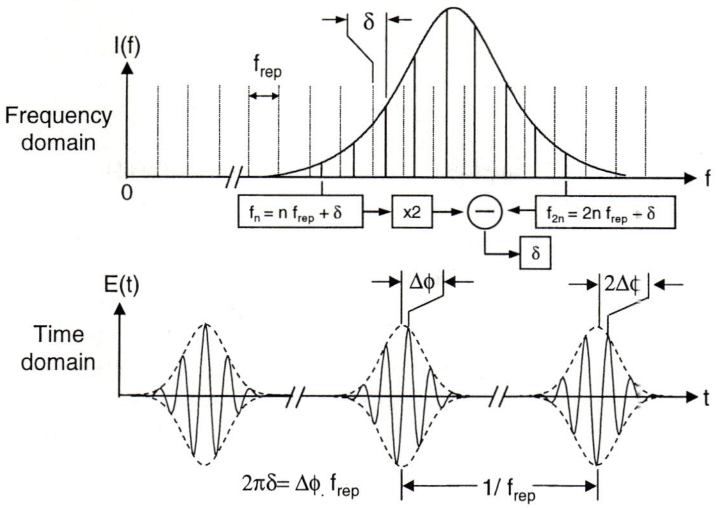

Fig. 2. (a) Cross correlator for measuring $\Delta\phi$ between pulse i and pulse $i+2$. Except for the PZT-mounted scanning mirror, all other mirrors have similar coatings to minimize phase imbalance. The entire correlator was placed in a vacuum chamber and pumped down to 300 mTorr to minimize the dispersive effects of air. (b) Plot of the relative carrier-envelope phase vs. the offset frequency (normalized to the pulse repetition rate). A slope of 4π is expected. As indicated, a linear fit of our averaged data produces this slope with a 5% uncertainty. The offset in phase from expected relation is attributed to a phase imbalance in the cross-correlator

measurement of only pulse i and $i+2$. In contrast the frequency domain measurement is inherently an average over all possible pulse combinations within the measurement time. Fluctuations in the correlator itself also contribute to the uncertainty.

In the experiments described above, it could be argued that the observed shift between carrier and envelope is simply an artifact of reshaping the pulse envelope due to the tilting of one of the laser's mirrors. Our locking method is coherently based on the repetition rate so the pulse shape is required to be stable. In other words, any reshaping of the pulse envelope would accumulate from pulse to pulse, eventually leading to an instability of the oscillator. Furthermore, we have monitored the pulse spectrum and the width of the cross-correlation. Neither is observed to vary systematically within the measurement bandwidth (~1 s), which indicates that spectral filtering and subsequent pulse reshaping is not occurring on this time scale. With small tilt angles it is evident that the phase shift at the cavity end mirror is very well approximated by a linear function such that only the group delay is modified, while higher order dispersion remains fixed at its steady state operating value. The phase delay is also modified, though by a different amount than the group delay. This differential change is the key point. Depending on the frame of reference we are either moving the carrier underneath the pulse envelope (if one is moving in a frame with the pulse envelope) to obtain phase synchronization or vise versa. In either case, the result is the same: successful control of the carrier-envelope phase. Although the intracavity pulse shape is dynamic, the output of the laser consists of a constant average carrier frequency that is amplitude modulated by a constant, repetitive envelope. Our experiments demonstrate that the relative phase delay between the carrier and envelope can be accurately controlled to provide for the first time a pulse train in in which every k^{th} pulse is truly identical.

This work also has tremendous impact on the area of optical frequency metrology. Because of their superior stabilities, optical frequency standards have long been recognized as attractive tools for precision measurements in fundamental physics. However, a longstanding obstacle in precision optical frequency metrology is the difficulty of generating widely tunable optical oscillators and determining their absolute frequencies with respect to the 9.193 GHz cesium microwave standard. When the frequency offset δ of the octave-spanning mode comb can be controlled in addition to the mode spacing, one obtains an array of millions of stable oscillators spanning the entire visible and near infra-red, all of which are directly referenced to the cesium standard. [8,10] In this regard, a phase-stabilized mode-locked laser can also be described as an optical frequency synthesizer, which elegantly solves the problem of measuring virtually any optical frequency. It is anticipated that when used in this manner phase-stabilized femtosecond lasers will play a key role in defining new optical standards in addition to providing the "clockwork" for future all-optical frequency standards.

References

1. U. Morgner *et al*: Opt. Lett. **24**, 411, 1999.
   D. H. Sutter *et al*: Opt. Lett. **24**, 631, 1999.
2. T. Brabec and F. Krausz: Rev. Mod. Phys. **72**, 545, 2000.
3. C. G. Durfee III *et al*: Phys. Rev. Lett. **83**, 2187, 1999.
4. A. de Bohan *et al*: Phys. Rev. Lett. **81**, 1837, 1998.
5. G. Tempea, M. Geissler, and T. Brabec: J. Opt. Soc. of Am. B **16**, 669, 1999.
6. I. P. Christov *et al*: Phys. Rev. Lett. 78 1251, 1997.
7. P. Dietrich, F. Krausz and P. B. Corkum: Opt. Lett. **25**, 16, 2000.
8. D. J. Jones *et al*: Science, **288**, 635, 2000.
9. J. Reichert *et al*: Opt. Comm. **172**, 59, 1999.
10. S. A. Diddams *et al*: PRL **84**, 5102, 2000.
11. J. Ranka *et al*: Opt. Lett. **25**, 25, 2000.
12. C. G. Durfee III *et al*: IEEE J. Sel. Topics Quant. Elec., **4**, 395, 1998.
13. L. Xu *et al*: Opt. Lett. **21**, 2008, 1996.

Carrier Envelope Offset Phase Stabilization for Few-Cycle Nonlinear Optics

G. Steinmeyer[1], D.H. Sutter[1], U. Keller[1], J. Stenger[2], and H.R. Telle[2]

[1] Inst. of Quantum Electronics, Swiss Federal Institute of Technology
ETH Hönggerberg HPT, CH-8093 Zürich, Switzerland
E-mail: sguenter@iqe.phys.ethz.ch
[2] Physikalisch-Technische Bundesanstalt (PTB)
PO-Box 3345, D-38023 Braunschweig, Germany
E-mail: harald.telle@ptb.de

Abstract. We report methods for the measurement and stabilization of the relative phase between carrier and envelope of an ultrashort optical pulse, based on sum frequency generation and heterodyning of a modelocked laser spectrum with optical transfer oscillators.

1. Introduction

Nonlinear optics has been traditionally described in the framework of the slowly-varying envelope approximation (SVEA). With the advent of few-cycle pulse sources, however, a breakdown of the SVEA is expected for processes like high-harmonic generation and multiphoton ionization. Consequently a dependence of the conversion efficiency of nonlinear optical processes on the carrier envelope offset (CEO) phase difference φ_{CEO} (see Fig. 1) is expected. Moreover, several proposed schemes for the generation of single attosecond pulses inherently rely on the CEO phase control.

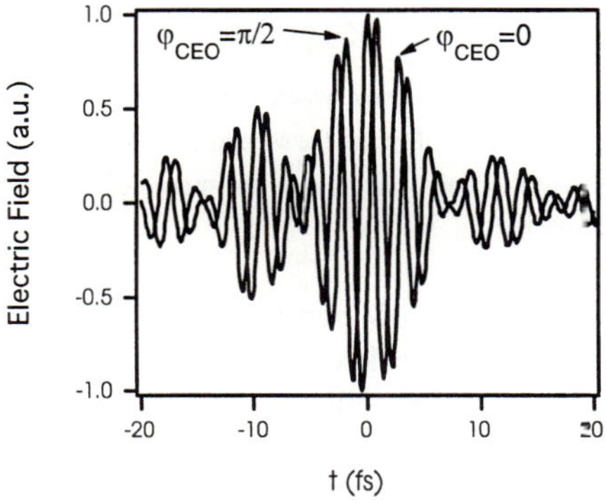

Fig. 1. Reconstructed electric field of a 5.9-fs pulse from a Ti:sapphire laser [1] for two different values of the CEO phase (dashed and solid line). A difference between the phase and group velocities inside the laser cavity causes the carrier-envelope offset to slip from pulse to pulse, resulting in a temporally varying phase difference φ_{CEO}.

In principle, the dependence of conversion efficiency on φ_{CEO} could be used directly for a stabilization scheme. However, the conversion efficiency is also highly susceptible to amplitude fluctuations of the laser source, in particular if a high-order nonlinear optical process is employed. We propose alternative routes to phase control [2], relying on sum frequency generation and heterodyning of an ultrabroadband sub-10-fs laser spectrum and cw transfer oscillators. Using an octave-spanning supercontinuum spectrum generated in fibers, one of the schemes proposed in [2] has been recently demonstrated for CEO phase control [3, 4]. Extending on that we also show how the CEO phase of more narrowband sources can be measured and stabilized.

2. Methods for CEO phase control

The spectrum of an ultrafast laser pulse consists of a comb of frequencies, which are separated by the repetition rate of the laser f_{rep} (Fig. 2). In a recent experiment the separation of these modes was confirmed constant to a degree of relative uncertainty of less than 10^{-15} [5]. Theoretically, this uniform comb can be extended to zero frequency and the positive mode closest to zero is defined as the CEO frequency ν_{CEO}. A nonvanishing value of ν_{CEO} is a manifestation of the difference between the group and phase velocities or alternatively, it represents the difference in the round-trip times of the carrier and the envelope inside the oscillator cavity. The goal of a CEO stabilization scheme is to lock ν_{CEO} to zero, thereby locking the carrier phase to the envelope phase and rendering all modes integral multiples of f_{rep}.

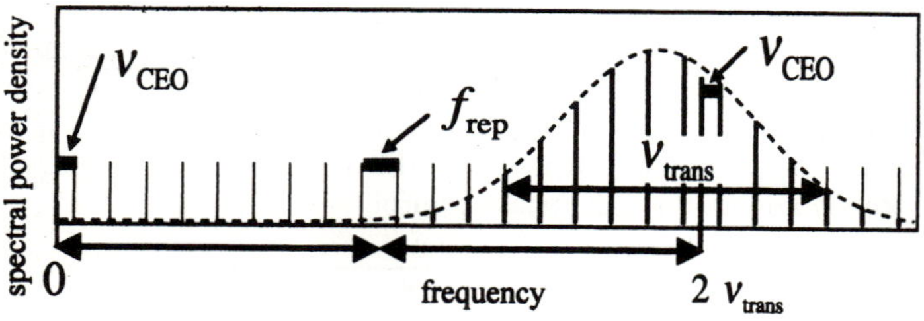

Fig. 2. Measuring ν_{CEO}. The residual frequency when the modes from the mode-locked laser are theoretically extended to zero is the ν_{CEO}. Mixing the second harmonic of the transfer oscillator with the next highest mode of the mode-locked laser allows for measurement of the ν_{CEO} if $\nu_{transfer}$ is an integral number of the mode spacing f_{rep}.

We employ the heterodyning of cw transfer laser oscillators with a Ti:sapphire laser oscillator within a phase-locked loop to control the CEO with sub-femtosecond precision. The transfer oscillator can be used, e.g., for a collective up-conversion process from the low-frequency to the high-frequency part of the spectrum (Fig. 3). With active control of the cavity length, the frequency shift provided by the transfer oscillator can be set to an integral number of modes such that $\nu_{trans} = nf_{rep}$

With this first phase-lock, the frequency difference between the second harmonic of the transfer oscillator and the nearest Ti:sapphire mode becomes equal

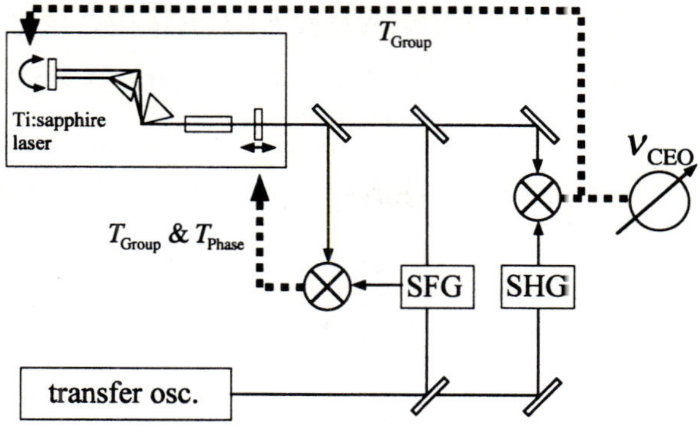

Fig. 3. Setup for the measurement and control of the CEO for a scheme like the one depicted in Fig.2. SFG: sum frequency generation; SHG: second-harmonic generation; crossed circles: heterodyning; dashed lines: phase-locked loops; solid lines: optical paths.

to ν_{CEO}. With intracavity control of the group delay independent of phase delay, ν_{CEO} can be kept constant or set to zero with a second phase-lock [6]. While this scheme requires a mode-locked spectrum with a relative width $\Delta\nu/\nu \geq 0.4$, this constraint can be further relaxed by additional frequency interval bisection steps [7], trading experimental simplicity for the required width of the frequency comb.

3. Conclusion

The ability to control the relative phase of the carrier of a laser pulse within its envelope invites new applications of sub-10-fs pulses. Providing a direct phase-coherent link between a primary microwave standard and an optical frequency substantially simplifies precision optical frequency measurements [3]. In turn, more accurate measurements of fundamental physical constants and high-precision spectroscopy could lead to more precise atomic clocks [5]. Moreover, phase-control opens up an avenue to a completely new regime of nonlinear optics, beyond the slowly-varying envelope approximation.

References

1. L. Gallmann et al., *Opt. Lett.* **24**, 1314 (1999).
2. H. R. Telle et al., *Appl. Phys. B* **69**, 327 (1999).
3. D. J. Jones et al., *Science* **288**, 635 (2000).
4. A. Apolonski et al., *Phys. Rev. Lett.* **85**, (2000).
5. T. Udem et al., *Opt. Lett.* **23**, 1387 (1998).
6. T. Udem et al., *Phys. Rev. Lett.* **82**, (1999).
7. H. R. Telle et al., *Opt. Lett.* **15**, 532 (1990).

Sub-10-fs light pulses with stabilized carrier-envelope phase: optical waveform synthesis

Andreas Poppe[1], Ronald Holzwarth[2], Alexander Apolonski[1,3], Gabriel Tempea[1], Christian Spielmann[1], Theodor W. Hänsch[2], and Ferenc Krausz[1]

[1] Institut für Photonik, Technische Universität Wien, Gusshausstr. 27 / 387, A-1040 Wien, Austria; E-mail: ferenc.krausz@tuwien.ac.at.
[2] Max-Planck-Institut für Quantenoptik, Hans-Kopfermann-Str. 1, D-85748 Garching, Germany.
[3] Institute of Automation and Electrometry, SB RAS, Novosibirsk 630090, Russia.

Abstract: Electronic control of the carrier-envelope phase in a 24-MHz train of 1-MW, sub-6-fs laser pulses has been achieved with simultaneous reduction of the pulse energy noise to unprecedented levels. Few-cycle femtosecond optical waveforms can now be synthesized with electric and magnetic fields reproducible to within known phase error (< 0.3 rad) for the first time.

1. Introduction

State-of-the-art ultrafast optical techniques allow the generation of intense light pulses comprising merely a few field oscillation cycles. These pulses open the way to exploring and exploiting nonlinear light-matter interactions in the strong-field regime, where the electric field $E(t) = A(t)\cos(\omega_0 t + \varphi)$ and at relativistic intensities also the magnetic field rather than the intensity envelope $I(t) \propto |A(t)|^2$ govern the evolution of the irradiated atomic system. As a consequence, processes, such as few-cycle-driven high harmonic generation and the predicted emergence of isolated attosecond XUV pulses sensitively depend on the *absolute phase (or carrier-envelope phase)* φ of few-cycle laser pulses [1].

Laser mode locking does not result in absolute-phase-stabilized pulses, because of a difference between the round-trip phase delay and group delay in a laser cavity [2]. Access to and control of φ was restricted to a measurement of its pulse-to-pulse variation $\Delta\varphi = \varphi_{n+1} - \varphi_n - 2k\pi$ (with the positive integer k chosen to yield $0 < \Delta\varphi < 2\pi$) until recently [2]. This proved insufficient for generating femtosecond pulses with stabilized φ because nonlinear effects translate small pulse energy fluctuations into minor fluctuations of $\Delta\varphi$, which can accumulate to a substantial carrier-envelope phase jitter over many round trips in the laser [2].

The consequence of $\Delta\varphi$ in the frequency domain is an offset of the equidistant comb of phase-locked modes by an offset frequency $f_{ceo} = (\Delta\varphi/2\pi)f_r$ [3], yielding a frequency comb $f_n = f_{ceo} + nf_r$, where f_r is the pulse repetition rate and f_{ceo} is the frequency at which the carrier slips through the envelope in the output pulse train, i.e. the same waveform reproduces itself in the output, and has been hence referred to as the carrier-envelope-offset frequency [4,5]. Heterodyning different parts of the mode-locked spectrum can yield a beat signal oscillating at f_{ceo} [4,5].

This carrier-slip beat signal has recently been generated by broadening the mode-locked spectrum of a Ti:sapphire laser in a single-mode fiber to cover a full octave and beating the frequency-doubled output of the infrared spectral wing with the high-frequency tail of the fundamental spectrum [6]-[9]. This beat signal can be easily shown to oscillate at f_{ceo} and, when carefully filtered, may allow selecting pulses with reproducible waveforms from the output pulse train [8]. Locking f_{ceo} to an external electronic oscillator opens not only the way to realizing a "frequency ruler" for precision optical frequency metrology [5,6,9], but also improves the accuracy of fixing φ in pulses selected from the train for further amplification. In Ref. [6], a servo loop translating and tilting a cavity mirror allowed reducing the jitter of φ to some 140 cycle slips within one second.

In this paper we report phase-locking of f_{ceo} to a local oscillator with a servo loop controlling the intracavity pulse energy W in a Ti:sapphire laser via the pump power [9] utilizing the sensitive dependence of $\Delta\varphi$ on W [2]. The system presented here allows full electronic control of the evolution of φ *without cycle slipping* over extended periods of time in few-cycle light pulses and results, in addition, in a *reduction of pulse energy noise to unprecedented levels*. These advances open the way to *synthesizing few-cycle optical waveforms with precisely reproducible electric and magnetic fields*.

2. Mode locking over an octave

Fig. 1 shows the schematic of our optical pulse generator consisting of a Kerr-lens-mode-locked mirror-dispersion-controlled Ti:sapphire oscillator including an intracavity telescope (not shown in Fig. 1) and a standard-single-mode-fiber/chirped-mirror-compressor, which is described in more detail in Ref. [8].

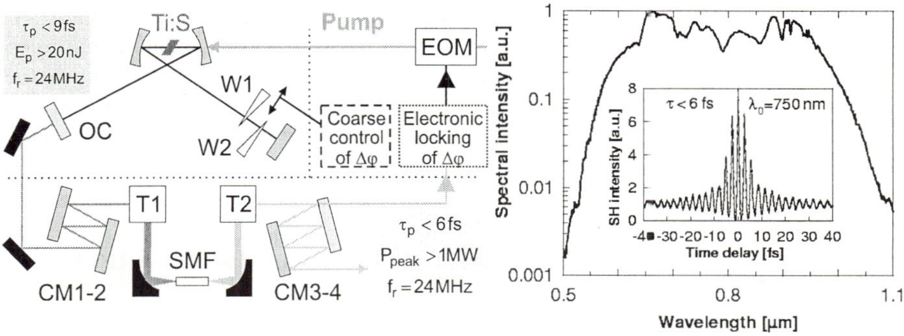

Fig. 1 Schematic of the few-cycle optical pulse generator (a), spectrum (b) and interferometric autocorrelation (inset) of the output pulses.

The system delivers sub-6-fs pulses (see inset in Fig. 1b) with a peak power exceeding 1 MW at a repetition rate of 24 MHz. The wings of the spectrum extend from 500 to 1100 nm, spanning a full octave (Fig. 1b). Dichroic chirped mirror CM3 transmits the spectral components near 1080 nm and 540 nm towards the optical system creating the beat signal oscillating at f_{ceo}.

3. Generating the carrier-slip beat signal

The carrier-slip beat signal oscillating at f_{ceo} has been generated in a nonlinear interferometer schematically illustrated in Fig. 2. Dichroic mirror DM transmits the infrared components at 1080 nm, which are frequency-doubled in a 1-mm-thick BBO crystal and recombined with the 540-nm beam propagating through the other arm of the Mach-Zender interferometer at beam splitter BS (SM stands for silver mirrors) [10].

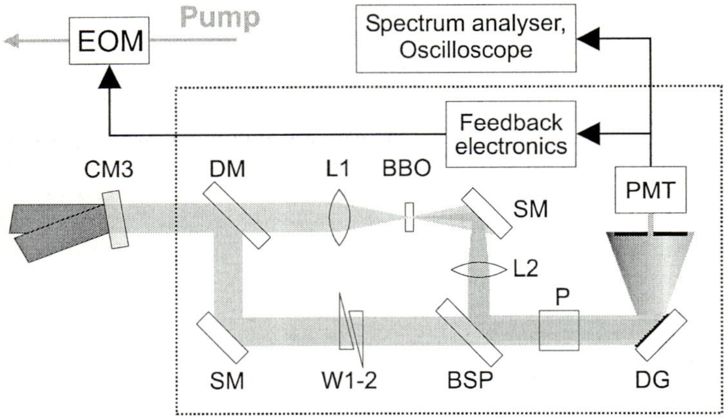

Fig. 2 Schematic of the optical setup generating the carrier-slip beat signal at ≈ 540 nm.

Spatial and temporal overlap of the two wave packets are ensured by the telescope formed by lenses L1-2 and by the wedged glass plates W1-2, respectively. Interference between the orthogonally polarized 540-nm wave packets is enforced by a polarizer (P) aligned to yield maximum beat signal at f_{ceo}, 30 dB above the noise level in 100 kHz bandwidth, sufficient for phase locking [9]. After spectral filtering by a diffraction grating (DG) and a slit, the combined 540-nm radiation is detected by a photomultiplier (PMT).

4. Electronic control of the carrier-envelope phase evolution

We have phase locked the carrier-slip beat note to 1 MHz (synthesized by a stabilized local oscillator) by adjusting the pump power with an electro-optic modulator (EOM) [9]. This servo loop makes use of the intracavity nonlinearities translating a change of the intracavity pulse energy W into a change of $\Delta\varphi$ [2]. Coarse adjustment of f_{ceo} is performed manually by translating a thin fused-silica wedge (W2 in Fig. 1a). The rf spectrum of the stabilized carrier-slip beat note is shown in Fig. 3. The depicted trace represents the average of 20 scans, providing conclusive evidence of the long-term stability of the phase locking [10]. (With the phase locking switched off, f_{ceo} fluctuates over the full frequency range shown in Fig. 3 during the time it takes to record 20 frequency scans.)

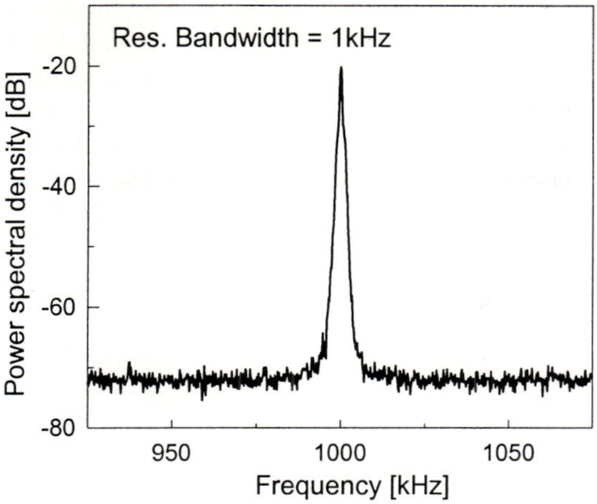

Fig. 3 Power spectrum of the in-lock carrier-slip beat note.

We have also monitored the ratio of the frequency of the in-lock beat signal to that of the local oscillator with an rf frequency counter to assess the quality of phase locking. Up to the maximum counter gate time of 10 s deviation of the two frequencies is to within the 0.1 Hz resolution of the counter. This indicates that *no random cycle slip of the carrier through the envelope occurs* over periods as long as 10 s (and probably significantly longer) [10].

5. Measurement of the jitter of the carrier-envelope phase: optical waveform synthesis

Precision frequency counting yielded that $\varphi(T)$ is firmly locked to the electronically defined $\varphi_{electr}(T) = 2\pi\, f_{local} T$ to within a cycle over more than 10 s in the generated few-femtosecond pulse train. However, we are unable to quantify the *sub-cycle jitter* of $\varphi(T)$, which is indispensable for reliable optical waveform synthesis. To gain access to this important quantity, we utilize that the parameter primarily controlled by the servo loop, namely the pulse energy, i) can be directly and permanently monitored and ii) is connected to the pulse-to-pulse shift in φ by a known (measured) relationship [10]. As a consequence, the carrier-envelope phase jitter defined as the root-mean-square deviation of $\varphi_r(T)$ from $\varphi_{electr}(T)$ that accumulates over a measurement time T_m due to (controlled or random) variation of the pulse energy can be determined from the power spectral density $S_W(f)$

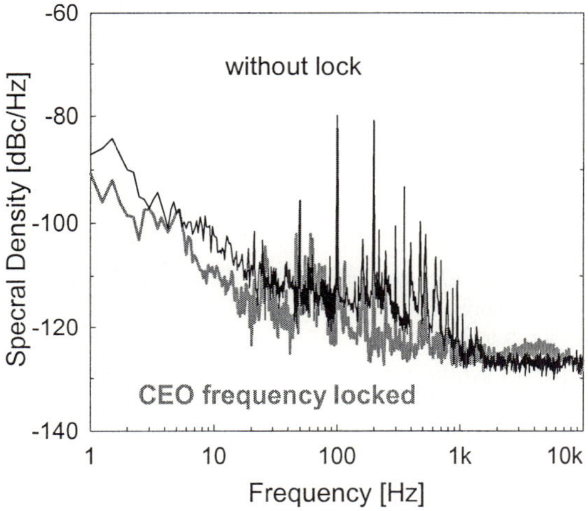

Fig. 4 Pulse energy noise spectrum of the Kerr-lens-mode-locked Ti:sapphire laser with the carrier-envelope phase locked (thick line) and unlocked (thin line).

of pulse energy variations [2,10]. Fig. 4 contrasts $S_W(f)$ of the carrier-envelope-phase-locked system with that observed when the servo loop was interrupted. With the f_{ceo} phase-locked the noise is significantly reduced, yielding an unprecedented rms energy fluctuation below 0.01% and hence fields variations below 3% within the frequency band of 1 Hz – 100 kHz. Even more importantly, the amplitude noise is notably suppressed at frequencies beyond 1 kHz.
Calculation of the energy-noise-induced jitter of φ for the corresponding measurement time of approximately 1 ms [10] yields a value of less than 0.3 radian, indicating that the servo loop responds to such small random phase shifts and hence keeps the carrier-envelope phase jitter below this level.

6. Summary

We have demonstrated cycle-slip-free electronic control of the evolution of the carrier-envelope phase in few-cycle light wave packets and measured its sub-cycle jitter, opening the way to synthesizing optical waveforms with precisely reproducible electric and magnetic fields for the first time. After amplification, these few-cycle optical waveforms will allow controlling high-field processes on the time scale of the light oscillation period. Anticipated impacts include the reproducible generation of isolated attosecond soft-X-ray pulses by few-cycle-driven high harmonic emission.

Acknowledgements. This work has been sponsored by the Austrian Science Fund, Grant Y44-PHY and within the framework of the Special Research Program F016. C.S. acknowledges financial support from the Austrian Academy of Science (APART fellowship).

References

1. T. Brabec and F. Krausz, Rev. Mod. Phys. **72**, 545 (2000).
2. L. Xu *et al.*, Opt. Lett. **21**, 1259 (1996).
3. T. W. Hänsch, private communication.
4. H. R. Telle *et al.*, Appl. Phys. B **69**, 327 (1999).
5. J. Reichert *et al.*, Opt. Commun. **172**, 59 (1999).
6. D. J. Jones *et al.*, Science **288**, 635 (2000).
7. A. Poppe *et al.*, in *Conference on Lasers and Electro-Optics*, OSA Technical Digest (OSA, Washington DC, 2000) Postdeadline paper CPD1
8. A. Apolonski *et al.*, Phys. Rev. Lett. **85**, 740 (2000).
9. R. Holzwarth *et al.*, "An optical frequency synthesizer for precision spectroscopy," submitted for publication.
10. A. Poppe *et al.*, "Few-cycle optical waveform synthesis," submitted for publication.

Frequency Domain Control of Femtosecond Pulse Trains with Fabry-Perot Reference Cavities for Optical Frequency Metrology

R. Jason Jones and Jean-Claude Diels

Department of Physics and Astronomy, University of New Mexico, 800 Yale Blvd. NE, Albuquerque, NM 87131
E-mail: jonron@unm.edu and jcdiels@unm.edu

Abstract. We demonstrate control and stabilization of the carrier frequency, phase, and repetition rate of femtosecond light pulses from a Kerr-lens mode-locked Ti:sapphire laser utilizing a Fabry-Perot reference cavity for use in precision optical frequency measurements.

1. Introduction

Ultra-short pulse laser systems are making a dramatic impact in the fields of optical frequency metrology and nonlinear optics, which require precise control over the pulse frequency and phase [1-3]. Using a Fabry-Perot reference cavity and a novel combination of actuators, we have actively stabilized the comb of frequencies emitted from a mode-locked femtosecond laser for use in high-precision direct optical frequency measurements. Without the need for a separate electronic radio frequency reference, this technique also offers the possibility of synthesizing a comb of highly stable radio frequencies directly from optical transitions. The stability of the femtosecond frequency comb relative to the reference cavity was characterized by measuring the beatnote between the components of the comb and a CW laser locked to the same reference cavity. The dispersion of the imperfect reference cavity was also measured using this technique.

In order for the quasi-monochromatic wavepackets that make up the pulse to stay together in the laser resonator, the nonlinear phase shift associated with the mode-locking mechanism must compensate for the cavity linear dispersion to second order and higher. This results in a spectrum for the mode-locked laser which consists of a series of equally spaced modes separated by the inverse of the pulse repetition time. The Fabry-Perot cavity provides an infinite number of reference modes to which one can simultaneously stabilize all the modes of the femtosecond comb [4]. In the time domain picture, the carrier frequency and phase of the incident pulses from the laser are being locked to the cavity in such a way that they constructively interfere with the pulse stored in the reference cavity. It is in this way that we can think of the Fabry-Perot cavity as providing an absolute frequency, phase, and repetition rate reference for the pulse train.

2. Experimental Methods

To stabilize the more than 500,000 longitudinal modes of the Kerr-lens mode-locked (KLM) Ti:sapphire laser to the reference cavity, two separate feedback loops were used to independently stabilize the average frequency of the fs comb and the mode spacing. Using a grating to disperse the pulses reflected from the reference cavity, an error signal was derived from two different spectral regions separated by ~28 THz using the standard Pound-Drever-Hall technique commonly used for single frequency lasers. (See Fig. 1) In this case, however, each error signal was the composite signal from the ~20,000 modes detected in each spectral region.

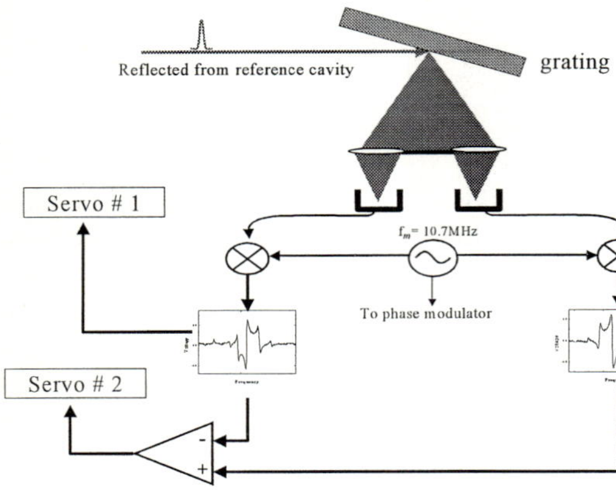

Fig. 1. Schematic of the detection system used to obtain error signals for the average frequency (Servo #1) and the repetition frequency, or mode spacing (Servo #2).

The stabilization to the first error signal was performed using an acousto-optic modulator (AOM) placed just outside the laser cavity in a double pass configuration. (The average frequency of the AOM can be kept centered using a piezo to slowly translate an intra-cavity prism.) Since the AOM changes only the position of the frequency comb and not its spacing, the remaining noise seen at the second error signal is due to the inherent noise in the mode spacing of the laser.

The second error signal is obtained by amplifying the difference between the error signals obtained from the two extreme spectral regions as shown in Figure 1. The difference of these two error signals provides a discriminator for locking the repetition rate of the mode-locked laser to that of the reference cavity, *independent* of the average frequency of the laser (to first order). This signal can be applied to tilt the end mirror of the laser resonator at the prism dispersed end of the cavity to adjust the group delay of the pulse for slow corrections of the repetition rate (< 1kHz) as first demonstrated by Reichert et. al. [5]. A larger bandwidth can be achieved in this second servo loop by sending the higher frequency components of the error signal simultaneously to a piezo-mounted mirror and the AOM to quickly adjust the cavity length while keeping the average frequency of the fs comb constant.

To measure the stability of the fs comb locked to the reference cavity, a CW Ti:sapphire laser was also locked to the same reference cavity. The beatnote between the CW laser and different components of the mode-locked laser were recorded. The beatnote between the CW laser and two modes of the fs comb is shown in Figure 2. The bandwidth of the beatnote was consistently less than 10 kHz. when measured in different regions of the laser spectrum.

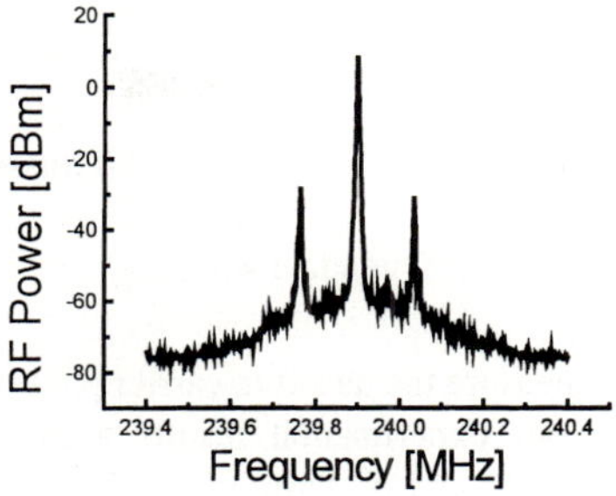

Fig. 2. Beating between CW laser and m+1 and m-1 modes of the femtosecond comb. The large peak in the center corresponds to the repetition rate of the mode-locked laser.

The low frequency beatnote between the CW laser and the mth mode of the mode-locked laser was counted on a frequency counter with a gate time of 0.1 seconds. The Allan deviation computed from this data is shown in Figure 3.

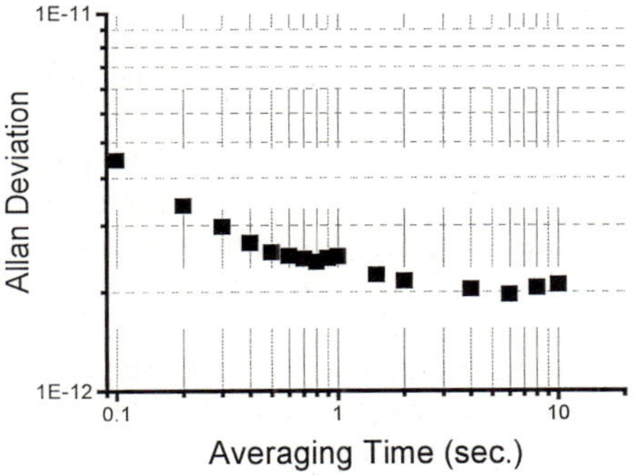

Fig. 3. Allan deviation of optical frequency of the mth mode of the femtosecond comb. Gate time: 0.1 seconds.

The use of the reference cavity with the mode-locked laser can enable the synthesis and coherent linking of stable radio frequencies derived from optical transitions when the reference cavity length is stabilized to an atomic transition for long term stabilization.

References

1 Th. Udem, J. Reichert, R. Holzwarth, and T.W. Haensch, in *Optics Letters,* Vol. 24, 881, 1999.

2 D.J. Jones, S.A. Diddams, J.K. Ranka, A. Stentz, R S. Windeler, J.L. Hall, and S.T. Cundiff, in *Science,* Vol. 288, 635, 2000.

3 M. Niering, R. Holzwarth, J. Reichert, P. Pokasov, M. Weitz, Th. Udem T.W. Haensch, P. Lemonde, G. Santarelli, M. Abgrall, P. Laurent, C. Salomon, and A. Clairon, in *Physical Review Letters,* Vol. 84, 5496, 2000.

4 R.J. Jones, J.-C. Diels, J. Jasapara, and W. Rudolph, in *Optics Communications,* Vol. 175, 409, 2000.

5 J. Reichert, R. Holzwarth, Th. Udem, and T.W. Haensch, in *Optics Communications,* Vol. 172, 59, 1999.

Nonlinear Optical Method for Determining the Absolute Carrier Phase of a Laser Pulse

M. Mehendale, S. A. Mitchell, D. M. Villeneuve and P. B.Corkum

Steacie Institute for Molecular Sciences, National Research Council of Canada, 100 Sussex Drive, Ottawa, *K1A 0R6 Canada*
Tel.:(613) 990-0947, Fax: (613) 991-3437, E-mail: Manjusha.Mehendale@nrc.ca

Abstract. We present a nonlinear optical technique to measure the absolute carrier phase of a sub-ten-femtosecond visible pulse. We compare our experimental results with our numerical model, which reveals that the interference of 2nd and 3rd harmonics of a 4.5 fs pulse is sensitive to the absolute phase.

In materials, the envelope of a laser pulse moves with the group velocity and the oscillation with the phase velocity so the phase ϕ of the oscillation with respect to the envelope changes. Diffraction also changes ϕ. Therefore measuring the envelope cannot give information about the carrier envelope phase.

All sufficiently nonlinear optical phenomena are phase dependent. The phase dependence arises from the interference of multiple pathways to the same final state [1]. We present a method to measure the carrier envelope phase of a sub-5 femtosecond visible pulse using the sensitivity of the interference between different harmonics to ϕ. Our work relates to the method already developed to fix the carrier phase of a pulse train from femtosecond oscillators [4]. In contrast, our method can determine, in a single laser shot, the actual value of ϕ. It also relates to proposals for measuring it during multiphoton ionization [2,3]

We develop a numerical model in which we generate 2nd and 3rd harmonics of a short visible pulse by all combinations of fundamental frequencies. Fig. (1)

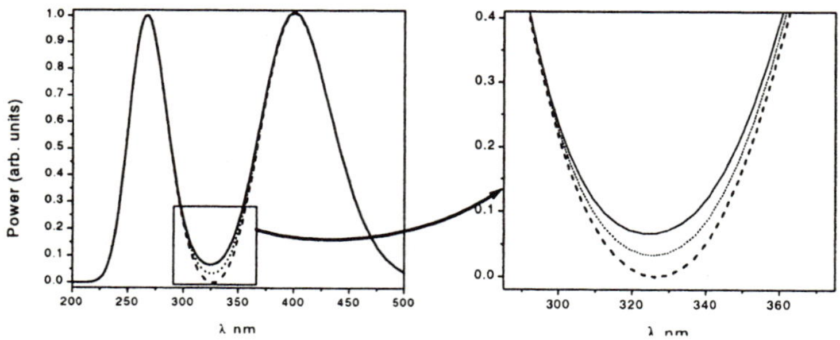

Fig. 1. Calculated 2nd and 3rd harmonic power spectrum of a 4.5 fs transform limited Gaussian pulse for carrier phases of 0 (solid), $\pi/2$ (dotted), π (dashed).

shows 2nd and 3rd harmonics spectra calculated for different values of ϕ for a 4.5 fs full width half maximum transform limited pulse with central wavelength of 800 nm. In the overlap region the interference is constructive or destructive depending on the carrier envelope phase of the short pulse. This is illustrated in Fig. (1), which shows the interference for carrier phases of 0, $\pi/2$ and π. Measuring the interference enables us to measure the carrier envelope phase.

Our laser pulse is not yet short enough to produce interference. Instead we demonstrate that single-shot measurements can be made with 10 fs pulses available in our lab. These pulses are generated by spectral broadening of 50 fs, 500 μJ pulses from a regenerative amplifier [4] using hollow core fiber self phase modulation followed by compression after passing through two pairs of fused-silica prisms of 20° apex angle in a double pass geometry. This geometry enables the compensation of only the overall 2nd order dispersion limiting the pulse to a duration of 10 fs.

We make predictions by taking an ideal 4.5 fs pulse and applying 3rd and 4th order dispersion as calculated for the dispersive optical elements in our setup. Fig.(2b) shows the predicted spectrum under our experimental conditions. It shows that there is no significant spectral overlap, due to the residual higher order chirp.

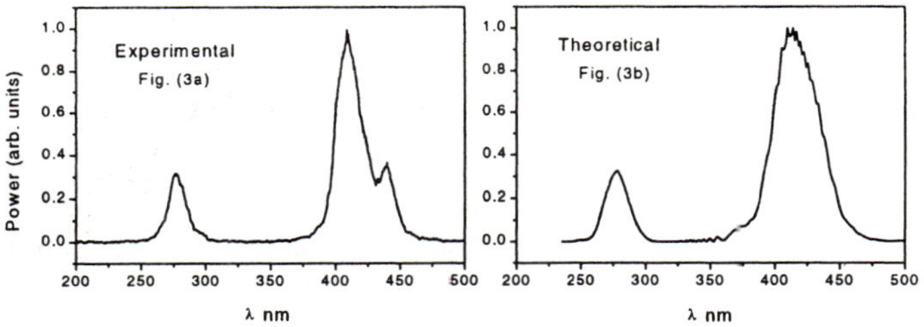

Fig. 2. Power spectra of 2nd and 3rd harmonics of a pulse with bandwidth corresponding to a 4.5 fs Gaussian pulse and with 3rd and 4th order of dispersion.

The interference between the 2nd and 3rd harmonics of a short pulse was achieved using the process of near surface harmonic generation at the surface of Si (001) [5]. The *p*-polarized short pulses were incident on a sample of Si(001) in air at 45° angle. The *p*-polarized 3rd harmonic emerged along the direction of reflection of the incident pulse. The fundamental pulse and it's 3rd harmonic passed through a 10 μm-BBO crystal where the 3rd harmonic interfered with the 2nd harmonic generated from the BBO crystal. The dispersion in the BBO crystal of the third harmonic is very small and doesn't harm the interference of the two harmonics. The spectrum was recorded using a spectrometer and a diode array detector monitored by a computer. The single shot measurements were performed with the same setup by changing the repetition rate of the laser to 50 Hz and making the exposure time to be 20 ms. Fig.(2a) shows the single shot spectra of 2nd and 3rd harmonic of our 10 fs fundamental pulse at 800 nm. We compare this with Fig.(2b), which shows a plot of 2nd and 3rd harmonics of a 10 fs pulse generated by our numerical model. The model assumes realistic 3rd order dispersion = -830 fs^3 and 4th order dispersion = -8635 fs^4. The comparison shows that our numerical model successfully represents our experimental results.

We now show that the interference on which our method is based is sensitive to 2nd order chirp as well. Fig. (3) shows the spectral interference with 2nd order dispersion of 35 fs^2 which is equivalent to the chirp introduced by a 1 mm thick fused silica plate. The two curves in Fig. (3) show the effect of a change in the carrier phase of π. Phase sensitivity is still present but is now written in the spectral position of constructive and destructive interference.

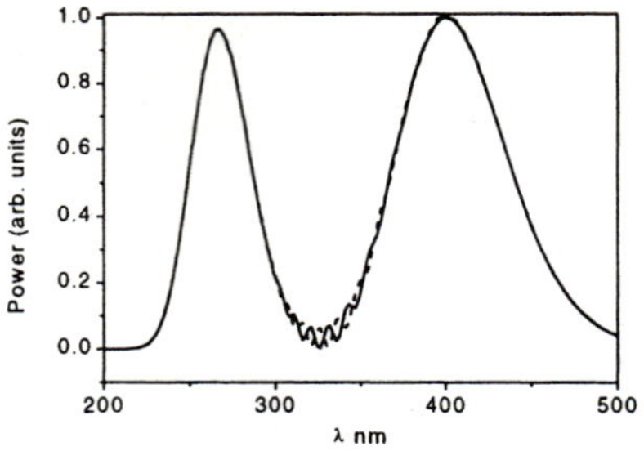

Fig. 3. Calculated 2nd and 3rd harmonic power spectrum of the 4.5 fs pulse used in Fig. (1) but with linear dispersion of 35 fs^2 for carrier phases of 0 (dashed), π (solid).

Fig. (2) and Fig. (3) illustrate the importance of controlling the pulse in the spectral extremes. We plan to optimize the pulse by introducing feed back control for pulse compression. Although we have concentrated on using 2nd and 3rd harmonics, the interference of higher harmonics is phase sensitive. For example, overlap between 3rd and 4th harmonics would enable one to measure the carrier envelope phase of transform limited pulses even longer than 4.5 fs.

The carrier envelope phase of an oscillator can now be fixed so that it does not change from pulse to pulse in a mode-locked train [6,7]. It is not a fixed quantity but changes during pulse propagation, amplification, focussing, etc. Hence measuring it at the end of this process will be essential. Our approach to measuring the carrier envelope phase is influenced by the concepts of coherent control [1]. Just as measurement of the carrier envelope phase is related to coherent control, carrier envelope phase is a new parameter for coherent control.

References:

1 P. Brummer and M. Shapiro, Ann. Rev. Phys. Chem 43, 257 (1992)
2 E. Cormier and P. Lambropoulos, Eur. Phys. J. D 2, 15 (1998)
3 P. Dietrich, F. Krausz and P. B. Corkum, Opt. Lett. 25, 16 (2000)
4 M. Nisoli, S. De Silvestri, O. Svelto, R. Szipocs, K. Ferencz, Ch. Spielmann, S. sartania, and F. Krausz, Opt. Lett. 22, 522 (1997)
5 J. E. Sipe, D. J. Moss and H. M. van Driel, Phys. Rev. B, 35, 1129 (1987)
6 David J. Jones, Scott A. Diddams, Jitendra K. Ranka, Andrew Stentz, Robert S. Windeler, John L. Hall and Steven T. Cundiff, Science, 288,635 (2000)
7 F. Krausz et al, private communication

Generation of Relativistic Intensity Pulses at 300-Hz Repetition Rate

H. Wang[1], O. Albert[2], J. Nees[1], D. Liu[1], G. Mourou[1], and Z. Chang[1]

[1] Center for Ultrafast Optical Science, University of Michigan, 2200 Bonisteel Blvd., Ann Arbor, MI 4810-2099, USA
E-mail: hwang@umich.edu
[2] Laboratoire d'Optique Appliquée, CNRS-UMR 7639, ENSTA, Ecole Polytechnique, 91761 Palaiseau cedex, France

Abstract. By focusing 21-fs laser pulses down to a 1.6-μm spot with aberration-compensated optics, we demonstrated for the first time that a peak intensity of 4×10^{18} W/cm^2 can be produced at 300-Hz repetition rate.

1. Introduction

Relativistic intensity refers to a new regime in optics where the nonlinear character of laser-matter interaction is dominated by the relativistic motion of the electrons characterized by a ponderomotive energy of greater than 0.5 MeV, the energy of the electron at rest [1]. It is a fundamentally new optical regime, where the intensity for 1-μm light must be greater than 10^{18} W/cm^2. It extends the traditional realm of ultrafast optics investigation—limited so far to the valence and core electrons, with characteristic energy of a few electron volts, and which has dominated the ultrafast field for 25 years—to the nuclear regime. This intensity level has been possible because of the technique of CPA combined with the refinement in short-pulse generation [2]. However, this novel regime could previously be reached only with relatively large and expensive lasers working at relatively low repetition rates, i.e., single-shot to 10 Hz [3,4]. Recently, we showed that by combining ultrashort pulses with a large numerical aperture, f# = 1, and deformable mirror (DM), a spot size of ~1μm and relativistic intensities could be reached at 1 kHz with much smaller and less expensive systems with good stability [5]. In this contribution, we are showing that intensities close to 10^{19} W/cm^2 are now becoming accessible at 300-Hz repetition rate. Just as, 30 years ago, high-repetition lasers made possible the study of ultrafast reactions in the visible regime, the possibility of studying, at very high repetition, physical phenomena with mega-electron-volt characteristic energy will give a second wind to the field of ultrafast phenomena.

2. Experimental Methods and Results

The schematic diagram of the Ti:sapphire laser system is shown in Fig. 1. The front end is a Kerr-lens, mode-locked, Ti:sapphire oscillator (FemtoSource

Scientific Pro; FemtoLasers GmbH) that can generate sub-10-fs pulses. The pulses are then stretched to 40 ps by an all-reflective grating stretcher. A Faraday isolator is used after the stretcher to isolate the oscillator from any possible feedback. After the Faraday isolator is a Pockels cell, which selects 300-Hz seed pulses. The seed pulses are then injected into a multipass Ti:sapphire pre-amplifier.

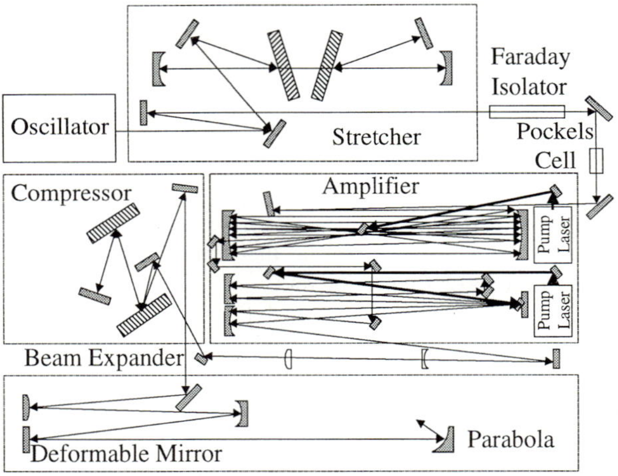

The pre-amplifier has seven passes and consists of only two curved mirrors. Pulse energy of 1 mJ can be obtained for a pump energy of 9.5 mJ. The beam is then injected into the second amplifier, which is a four-pass Ti:sapphire amplifier as shown in Fig. 1. A folding mirror is used to make the design simple and compact. Over 3.5-mJ pulse energy can be obtained for a pump energy of 17 mJ.

After the beam is amplified, it is compressed by a grating compressor. The output pulse energy is 2.5 mJ at 300 Hz. An autocorrelation measurement shows that the pulse has a FWHM of 21 fs, assuming a sech^2 pulse shape.

In order to generate high intensity greater than 10^{18} W/cm^2, a f/1 parabola is used to focus the beam. A DM is used to correct the wave-front distortion from the laser system and to precompensate the aberration caused by the focusing optics. The DM is a 2-inch-diameter silver mirror with 37 electrostrictive actuators (from Xinetics). In most traditional adaptive optical systems, one determines the optimal shape of the DM that will compensate for wave-front aberrations by first using a wave-front sensor to measure the aberrated wave front, thus yielding a wave-front error correction, which can then be fed to the DM to produce a flat wave front [6]. Because it is extremely difficult to measure the wave front at the focus of a high-numerical-aperture parabola, we have designed a scheme that will determine the optimal correction by the DM without the necessity to perform a wave-front measurement [7]. The scheme that we implement here uses nonlinear optics in conjunction with machine learning through an evolutionary algorithm described below. Strong nonlinear effects occur at the focus spot of an ultrashort pulse. A stronger nonlinear signal generated at the focus corresponds to a smaller focal spot. Thus we can obtain diffraction-limited focusing by maximizing the nonlinear signal using the DM without the necessity to know anything about the wave front of the laser on the parabola.

We implement the nonlinear correction scheme by using second-harmonic generation (SHG) in a thin β-barium borate crystal at the focal spot. The SHG signal is used as the feedback for the computer-controlled DM. The shape optimization of the DM is done by genetic algorithm (GA). Figure 3 shows the close-to-diffraction-limited focal spot with the deformable mirror. The focal spot had a diameter of 1.6 μm at FWHM and 78% of the total energy is inside the $1/e^2$ radius, which leads to a peak intensity of 4×10^{18} W/cm^2. The actual peak intensity should be much higher due to the aberration of the microscope objective in our imaging system, which is significant but not taken into account in our measurement.

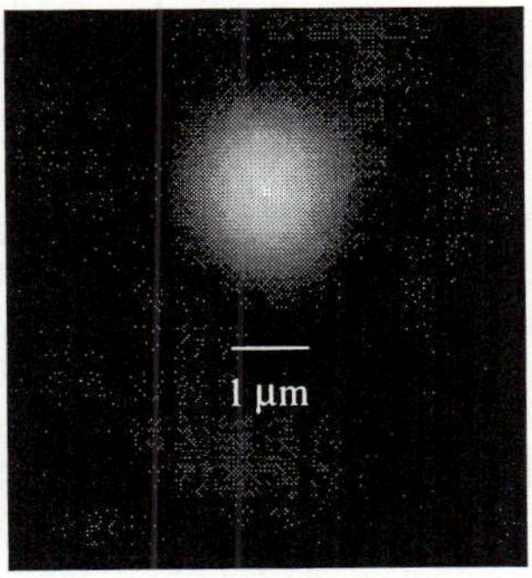

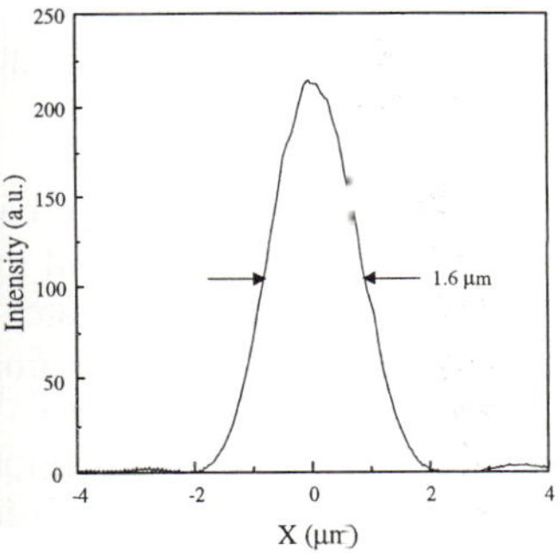

3. Conclusion

We have shown that by correcting the wavefront distortion with adaptive optics, a spot size of ~1 μm and an intensity close to 10^{19} W/cm^2 could be reached with femtosecond lasers operating at high repetition rate.

References

1 M. D. Perry and G. Mourou, Science **264,** 917, 1994.
2 D. Strickland and G. Mourou, Opt. Commun. **56**, 219, 1985.
3 C. N. Danson, L. Barzanti , Z. Chang et. al., Opt. Commun. **103**, 392, 1993.
4 Perry M.D. et. al., Opt. Lett. **24**, 160, 1999.
5 O. Albert et. al., Opt. Lett., to be published.
6 F. Druon et. al., Opt. Lett. **23**, 1043, 1998.
7 O. Albert et. al., Opt. Lett. **25**, 52, 2000.

Reflection-double-pass, Ti:sapphire continuous-wave amplifier delivering 5.77-W average power, 82-MHz repetition rate, 100-fs pulse

Zhenlin LIU, Hidetoshi MURAKAMI, Toshimasa KOZEKI, Hideyuki OHTAKE, and Nobuhiko SARUKURA

Institute for Molecular Science (IMS), Myodaiji, Okazaki 444-8585, Japan
Tel: +81 564 55 7477 Fax: +81 564 55 7218 E-mail: murakami@ims.ac.jp

Abstract: A confocal, reflection-double-pass, Ti:sapphire continuous-wave amplifier with a small-signal gain of 4.2 has been invented. Femtosecond pulses with an 82-MHz repetition rate from a mode-locked Ti:sapphire laser are amplified to 5.77-W average power through an amplifier pumped by three cw green lasers, and the extraction efficiency reaches 10.6 %.

There have been two approaches in the continuous effort to obtain high-average-power, high-repetition-rate femtosecond pulses, power scaling of mode-locked oscillators [1-6] and the master oscillator and power amplifier (MOPA) scheme [7]. The scaling of the average power from the oscillator is restricted from the thermal management in the gain medium. Various schemes have been proposed to solve this problem, including the cylindrical pumping scheme [3, 4] and the multiple gain-module scheme in a single laser cavity [5, 6]. Previously, we achieved 3.4-W average power from a mode-locked Ti:sapphire laser oscillator with an additional intra-cavity amplifier [6]. Furthermore, successful continuous-wave (cw) amplification of the mode-locked laser output has only been reported in $Nd:YVO_4$ [7]. However, such successful cw amplification of femtosecond pulses had never been accomplished in a broad-band-width, low-gain medium like Ti:sapphire. In this paper, we will report a confocal, reflection-double-pass, Ti:sapphire cw amplifier with a small-signal gain of 4.2 and 10.6% extraction efficiency. This system produces successfully 5.77-W average-power, 82-MHz repetition-rate, femtosecond pulses.

The experimental set-up of the confocal, reflection-double-pass Ti:sapphire cw amplifier is illustrated in Fig. 1. The femtosecond seed pulses at 782 nm with an 82-MHz repetition rate were produced by a regeneratively mode-locked Ti:sapphire laser (Tsunami). The input beam with an average power of 1.7 W was passed through the isolator to avoid unnecessary feed back, then guided to the cw amplifier unit. The 1.3-cm optical path Ti:sapphire amplifier crystal with a 1-cm^{-1} absorption coefficient at 532 nm was designed to have a Brewster surface on one end, and a dichroic coating on the other flat end. The gain crystal was attached to the water-cooled copper heat sink with the water temperature of 7 degrees centigrade. The input beam was focused with a 10-cm-focal-length lens with antireflection coating, and passed through the Brewster surface to the flat surface of the gain medium. The beam was then reflected back with a small horizontally deviated angle of about 5 degrees and collimated with the same lens. A dichroic flat mirror with high reflectivity at 780 nm and high transmission at 532 nm was

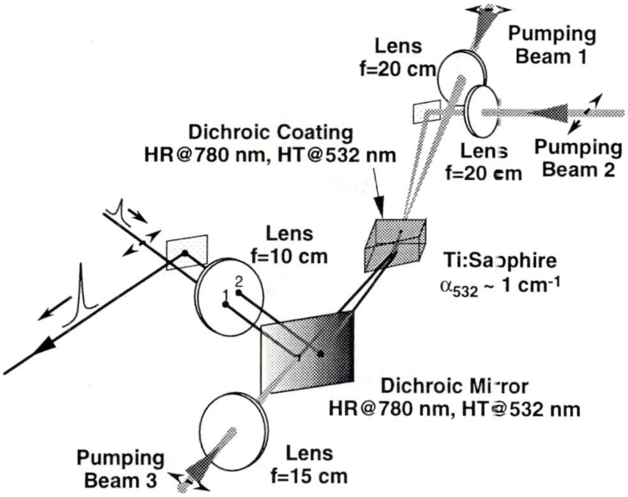

Fig. 1. Experimental set-up of the confocal, reflection-double-pass Ti:sapphire cw amplifier. Two large-frame Ar-ion lasers and one all-solid-state green laser are installed as pumping sources.

inserted between the focusing lens and the Brewster surface of the crystal to introduce the third pumping beam. Reflection-double-pass amplifier advantages over the confocal, conventional double-pass or multi-pass amplifiers with Brewster-cut crystals are not only the compactness and implementation ease, but also the efficient coupling of the pumping beams. In the reflection-double-pass amplifier, the input beam has a focal point near the flat end of the crystal, and the pumping power is the highest at the focal point of the signal beam.

Two large-frame, all-line Ar-ion lasers and one all-solid-state green laser are installed as pumping sources of the cw amplifier. Two Ar-ion laser pumping beams were focused separately with 20-cm focal-length lenses to overlap with each seed beam pass through the flat back surface of the Ti:sapphire crystal. The third pumping beam from the 10-W, all-solid-state green laser was focused with a 15-cm focal-length lens from the flat bending mirror side through the dichroic mirror to overlap with the second amplification pass. This design produces an almost perfect beam overlapping of the pumping beams and seed beams. The pumping-power dependence of the output power of the amplifier is shown in Fig. 2(A). The maximum average output power reaches 5.77 W. To our knowledge, this is the highest average power of high-repetition-rate femtosecond pulses at around 800 nm. The power extraction efficiency for the absorbed pumping power reached 10.6 %. This efficiency is almost comparable to that of the cw Ti:sapphire laser oscillator. The input-power dependence of the output power for full pumping power is shown in Fig. 2(B). The small-signal gain was 4.2 times that of the transmitted pulse without pumping. This gain factor almost equals that of a single-stage Ti:sapphire pulsed amplifiers pumped by the second harmonics of Q-switched Nd:YAG lasers. After amplification, the Ti:sapphire laser beam was guided to a double-pass, dispersion-compensating compressor. The compressor consisted of an SF6 Brewster prism pair separated by 46.5 cm and a roof reflector with a dielectric coating for a small vertical displacement for the second pass in

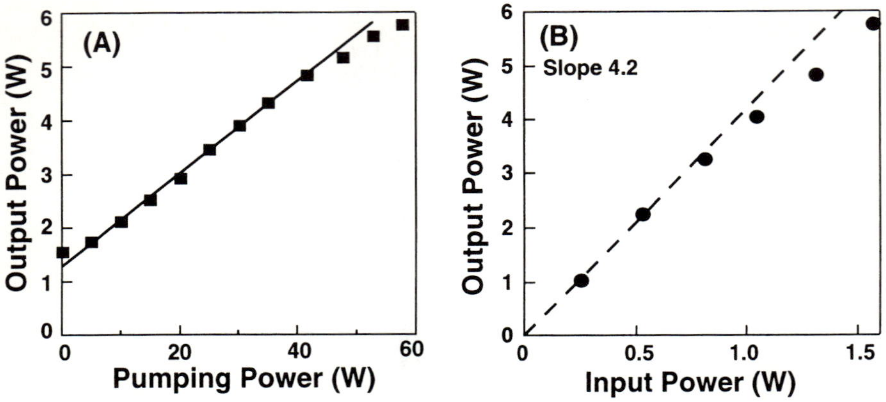

Fig. 2. (A) Input-power dependence of the output power, and the small-signal gain was 4.2. (B) Pumping-power dependence of the output power of the amplifier. The maximum average output power reaches 5.77 W.

the compressor. The amplified 190-fs pulse was compressed to a 100-fs pulse assuming a $sech^2$ pulse shape with a nearly transform-limited time and band-width product of 0.38. The output power was limited only by the available pumping power or the number of pumping lasers. Basically, this cw-amplifier unit is cascadable for further average power scaling.

In conclusion, we have successfully constructed a confocal, reflection-double-pass Ti:sapphire cw amplifier with a small-signal gain of 4.2, 5.77-W average power, and 10.6 % extraction efficiency [9]. Such high-average-power and high-repetition-rate femtosecond optical pulses will open up new application possibilities including parametric frequency conversion, THz-emitter driver [8], and material processing. This cw-amplification scheme will also be widely applicable to other low-gain and laser-pumped gain media.

References

1 W. S. Pelouch, P. E. Powers, and C. L. Tang, Opt. Lett. **17**, 1070 (1992).
2 D. Kopf and U. Keller, Opt. Let. **22**, 99 (1997).
3 D. Kopf and U. Keller, in Advanced Solid-State Lasers, Technical Digest (Optical Society of America, Washington, DC), pp. 126 (1997).
4 J. Aus der Au, F. H. Loesel, F. Morier-Genoud, M. Moser, and U. Keller, Opt. Let. **23**, 271 (1998).
5 G. J. Spuhler, T. Sudmeyer, R. Paschotta, U. Keller, in Advanced Solid-State Lasers, Technical Digest (Optical Society of America, Washington, DC), paper TuC2 (2000).
6 Z. Liu, S. Izumida, S. Ono, H. Ohtake, and N. Sarukura, Appl. Phys. Lett. **74**, 3622 (1999).
7 J. D. Kafka, M. L. Watts, and J. W. Pieterse, J. Opt. Soc. Am. B **12**, 2147 (1995).
8 N. Sarukura, H. Ohtake, S. Izumida, and Z. Liu, J. Appl. Phys. **84**, 654 (1998).
9 Z. Liu, H.Murakami, T. Kozeki, H. Ohtake, and N. Sarukura, Appl. Phys. Lett. **76**, 3182 (2000).

Chirped pulse amplification for ultraviolet femtosecond pulses using Ce:LiCAF crystal

Zhenlin Liu[1], Toshimasa Kozeki[1,2], Yuji Suzuki[1,2], Nobuhiko Sarukura[1,2], Kiyoshi Shimamura[3], Tsuguo Fukuda[3], Masahiro Hirano[1], Hideo Hosono[1,4]

[1] Japan Science and Technology Corporation (JST), C-1232, KSP, Takatsu-ku, Kawasaki, Kanagawa 213-0012, Japan. E-mail: z-liu@net.ksp.or.jp
[2] Institute for Molecular Science (IMS), Myodaiji, Okazaki 444-8585, Japan
[3] Tohoku University, 2-1-1 Katahira, Aoba-ku, Sendai 980-8577, Japan
[4] Tokyo Institute of Technology, Nagatsuta, Midori-ku, Yokohama 226-8503, Japan

Abstract. Chirped pulse amplification in ultraviolet region is demonstrated using broadband Ce:LiCAF laser medium. The 4-pass amplifier has 370-times gain and delivered 6-mJ, 290-nm pulses. After dispersion compensation, output pulses can be compressed down to 115 fs.

There have been continuous efforts for obtaining high peak power femtosecond ultraviolet pulses. A series of excellent works had been reported using KrF excimer amplifiers up to multiterawatt level [1]. There are two approaches for this purpose. One is the nonlinear frequency conversion of Ti:sapphire amplifier system [2]. The other is the amplification of 248-nm pulses with KrF excimer laser amplifier [1]. In case of the nonlinear frequency conversion from near-infrared, femtosecond pulses, the output energy is limited to a few millijoule level obviously by the available size of tripling or quadrupling crystals. With KrF excimer laser amplifier, this limitation was moved away by sacrificing the tunability. The shortest possible pulse duration for KrF amplifier is estimated to be 29-fs from its bandwidth. On the other hand, Ce:LiCAF laser medium invented by Dubinskii et al was known to have broad tunability at least from 281 nm to 315 nm corresponding to 3-fs bandwidth [3]. Additionally it has the attractively high saturation fluence of 115 mJ/cm^2. Previously, the available crystal size was at most 1-cm diameter that was not sufficiently large for the high-peak-power laser system. For the future TW Ce:LiCAF chirped pulse amplification (CPA), we have grown Ce:LiCAF crystal with 5-cm diameter successfully [4]. This medium can be pumped efficiently by the fourth harmonics of Nd:YAG lasers. In the case of Ce:LiCAF oscillator, 60-mJ, 289-nm pulses were generated by 230 mJ, 266-nm pumping [5]. Also in the Ce:LiCAF power amplifier, 18% extraction efficiency was demonstrated previously for 1-ns pulses [6]. Moreover, owing to the recent development of new nonlinear crystal like CLBO [7] and $Li_2B_4O_7$ [8], the conversion efficiency from the second harmonics of Nd:YAG lasers to the fourth harmonics is as high as 50%. From these energy efficiency respects, the Ce:LiCAF CPA laser system aiming for the high peak power system is feasible enough. In this presentation, we will report a very first chirped pulse amplification in ultraviolet region using this Ce:LiCAF. The 4-pass amplifier has 370-times gain and delivered 6-mJ, 290-nm pulses. After dispersion compensation, output pulses can be compressed down to 115 fs.

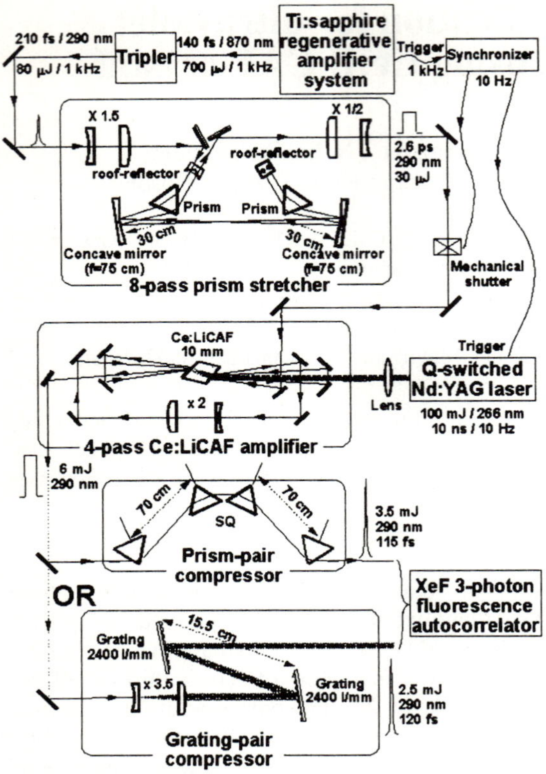

Fig. 1. Experimental setup of the Ce:LiCAF CPA laser system.

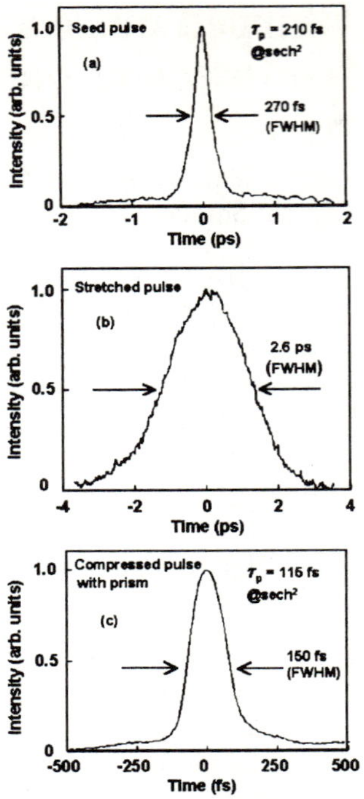

Fig. 2. Pulsewidths measured by three-photon fluorescence autocorrelator of (a) compressed seed pulse; (b) stretched pulse; and (c) compressed amplified pulse.

The experimental set-up of the Ce:LiCAF CPA system is illustrated in Fig. 1. The femtosecond seed pulses at 290 nm with 1-kHz repetition-rate were provided by the frequency tripling of the Ti:sapphire regenerative amplifier output. The duration of 290-nm, femtosecond pulses were evaluated by the XeF three-photon fluorescence method that is capable to measure down to around 50-fs [9]. The duration of the seed was 210 fs, that could be compressed to 140 fs assuming a $sech^2$ pulse shape with 0.5 time and bandwidth product after dispersion compensation by the quartz prism pair (Fig. 2a). The seed pulses without any dispersion compensation were guided to the 8-pass quartz-prism and reflector stretcher to expand the duration to 2.6 ps avoiding spatial chirping and aberration (Fig. 2b). 33-µJ stretched seed pulses were guided to the modified bow-tie style 4-pass Ce:LiCAF amplifier. 100-mJ, 266-nm, 10-ns pumping pulses were provided from a synchronously operated 10-Hz, Q-switched Nd:YAG laser. The 1 x 1 x 1-cm^3 Brewster-cut Ce:LiCAF crystal with 82% absorption for the pump beam was used for this amplifier stage. The pumping beam was softly focused to 4-mm diameter. The diameter of the seed beams were 1.5 mm for the first and the second passes, and expanded to 3-mm for the third and fourth passes to avoid deep gain quenching. The optical delays for the each pass were approximately 1.1 ns, 3.6 ns, and 1.9 ns. The maximum output energy was 6-mJ for this amplifier. The power extraction efficiency for the absorbed pumping power reached 7.3%.

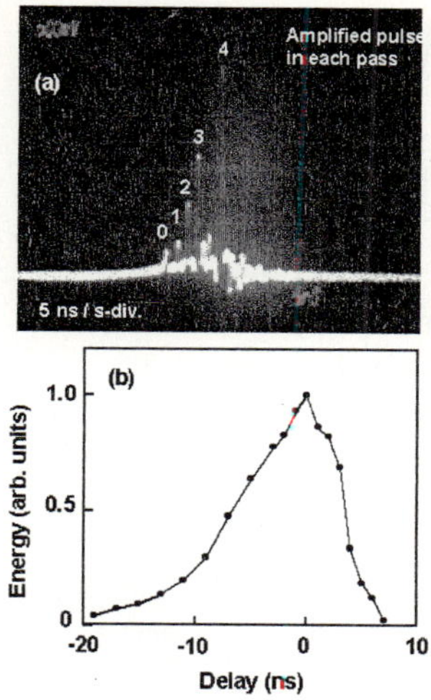

Fig. 3. (a) Gain factors for each pass in the 4-pass amplifier. (b) Output energy dependence on the delay of seed and pumping pulses.

The total gain factor is 370 for the transmission pulse. Estimated from Fig. 3a, the gain factors for each pass are 7, 3, 2, and 7. The gain factor difference should be due to the spatial and temporal overlap quality of the pump and signal beams. The allowance of the relative delay of the input 290-nm pulse and pump pulse was approximately a few nanoseconds (Fig. 3b). This is large enough for the accuracy of the present timing control jitter with 0.2 ns. After the amplification, the beam was guided to the dispersion-compensating, double-pass, quartz Brewster prism-pairs separated by 70-cm, or 15.5-cm apart 2400L/mm grating pair. The amplified pulse was compressed down to 115fs (Fig. 2c) and 120 fs duration for each case assuming the $sech^2$ pulse shape with time and bandwidth products of 0.66 and 0.69. The estimated B-integral for this CPA system is less than 5 even including compressor prism part. Therefore, the pulse-stretching ratio should be adequate for this amplification level. Presently, the output power was only limited by the available pumping power.

In conclusion, we have successfully demonstrated chirped pulse amplification in ultraviolet region for the first time using broadband Ce:LiCAF laser medium. The 4-pass amplifier has 370-times gain and delivered 6-mJ, 290-nm pulses. After dispersion compensation, output pulses can be compressed down to 115 fs. Future scaling of this Ce:LiCAF laser system will open up new category of high peak power lasers in ultraviolet region.

References

1 Y. Nabekawa, K. Kondo, N. Sarukura, K. Sajiki and S. Watanabe, Opt. Lett. **18**, 1922, 1993.

2 C. P. J. Barty, T. Guo, C. Le Blanc, F. Raksi, C. Rose-Petruck, J. Squire, K. R. Wilson, V. V. Yakovlev, and K. Yamakawa, Opt. Lett. **21**, 668, 1996.

3 M. A. Dubinskii, V. V. Semashko, A. K. Naumov, R. Y. Abdulsabirov, and S. L. Korableva, J. Mod. Opt. **40**, 111, 1993.

4 K. Shimamura, N. Mujilatu, K. Nakano, S. L. Baldochi, Z. Liu, H. Ohtake, N. Sarukura, and T. Fukuda, J. Crystal Growth **197**, 896, 1999.

5 Z. Liu, K. Shimamura, K. Nakano, N. Mujilatu, T. Fukuda, T. Kozeki, H. Ohtake, and N. Sarukura, Jpn. J. Appl. Phys. **39**, L466, 2000.

6 N. Sarukura, Z. Liu, H. Ohtake, Y. Segawa, M. A. Dubinskii, R. Y. Abdulsabirov, S. L. Korableva, A. K. Naumov, and V. V. Semashko, Opt. Lett. **22**, 994, 1997.

7 Y. K. Yap, K. Deki, Y. Kagebayashi, M. Horiguchi, Y. Mori, and T. Sasaki, OSA TOPS, Advanced Solid-State Lasers, vol. 19, pp. 380, 1998.

8 R. Komatsu, T. Sugawara, K. Sassa, N. Sarukura, Z. Liu, S. Izumida, Y. Segawa, S. Uda, T. Fukuda, and K. Yamanouchi, Appl. Phys. Lett. **70**, 3492, 1997.

9 N. Sarukura, M. Watanabe, A. Endoh, and S. Watanabe, Opt. Lett. **13**, 996, 1988.

Parabolic Pulses from Yb:fiber Amplifiers: A New Paradigm for High Power Ultrashort Pulse Generation

J. D. Harvey[1], J. M. Dudley[1], V. I. Kruglov[1], B. C. Thomsen[1], M. E. Fermann[2]
[1] Dept of Physics, University of Auckland,, Auckland, New Zealand
E-mail: j.harvey@auckland.ac.nz
[2] IMRA America, 1044 Wooddridge Avenue, Ann Arbour, Michigan 48105, USA
E-mail: fermann@imra.com

Abstract. We report parabolic pulse generation from a high gain Yb:doped fiber amplifier. After self-similar propagation in standard optical fiber and subsequent pulse compression, 70 fs pulses with 80 kW peak power are generated.

1. Introduction

Optical fibre amplification of mode locked seed pulses has the potential for generating a convenient, compact source of high power ultrashort pulses, by harnessing the potential for integration, and the high optical efficiencies of fibre amplifiers. When coupled with well-known techniques for pulse compression, fiber amplifiers have already demonstrated the generation of sub-100 fs pulses with 10's of kW peak power [1]. However, at these high intensities, severe pulse distortions such as optical wave breaking are usually observed [2], imposing stringent limitations on amplifier and compressor design.

In this paper we report a new technique for generating high power sub-100 fs pulses based on the compression of parabolic pulses generated from a high gain normal dispersion fibre amplifier [3]. Based on the theoretical analysis of the nonlinear Schrödinger equation (NLSE) with gain, we show that parabolic pulses are exact asymptotic solutions of the NLSE with gain, and propagate in the amplifier self-similarly subject to exponential scaling of amplitude and temporal width. In addition they posses a strictly linear chirp. These "optical similaritons" represent a new class of solution to the NLSE, with their linear chirp facilitating efficient pulse compression.

2. Theoretical Analysis

Consider the evolution of pulses in an optical amplifier in the absence of gain saturation and for pulses with spectral bandwidths less than the amplifier bandwidth. In this case, propagation is described by the NLSE with gain [2]:

$$i\frac{\partial A}{\partial z}=\frac{1}{2}\beta_2\frac{\partial^2 A}{\partial T^2}-\gamma|A|^2A+i\frac{g}{2}A \tag{1}$$

Applying self-similarity analysis [4] to (1) for g • 0 and β_2>0, yields an exact asymptotic (z→•) solution given by:

$$A(z,T) = A_0 \exp[gz/3]\,(1-[T/T_0(z)]^2)^{1/2} \exp[i\varphi(z,T)]\ ,\ |T| \le T_0(z),$$

where $A_0 = 0.5(gE_{IN})^{1/3}(\gamma\beta_2/2)^{-1/6}$, $T_0(z) = 3g^{-2/3}(\gamma\beta_2/2)^{1/3}E_{IN}^{1/3}\exp(gz/3)$

and $\varphi(z,T) = 3\gamma A_0^2(2g)^{-1}\exp(2gz/3) - g(6\beta_2)^{-1}T^2$. (2)

This solution represents a pulse with parabolic intensity profile, and a quadratic temporal phase (i.e. a linear chirp). This pulse is an asymptotic solution with exponentially increasing width and amplitude towards which any input pulse evolves with sufficient propagation distance. Gaussian input pulses having a range of pulse durations (FWHM) from 100 fs – 5 ps, but fixed energy E_{IN} = 12 pJ were propagated numerically in a 6 m long fiber amplifier with realistic parameters corresponding to Yb:doped fiber: γ= 5.8 × 10^{-3} $W^{-1}m^{-1}$, β_2 = 25 × $10^{-3}ps^2m^{-1}$, g = 1.9 m^{-1} [1]. Figure 1(a) compares the evolution of the pulse obtained from simulations with the analytic prediction given by (2). The evolution approaches the asymptotic limit in all cases. Fig. 1(b) shows the output characteristics for the input 200 fs pulse, illustrating the excellent agreement (over 10 orders of magnitude) between the simulation results (circles) and the expected asymptotic profile from (2) (dashed line). We note also that the exponentially decaying "skirts" on the pulse shown in Figure 2b and the associated chirp of these features are also in agreement with a more complete self similarity analysis, although it is unlikely that these features could be detected experimentally

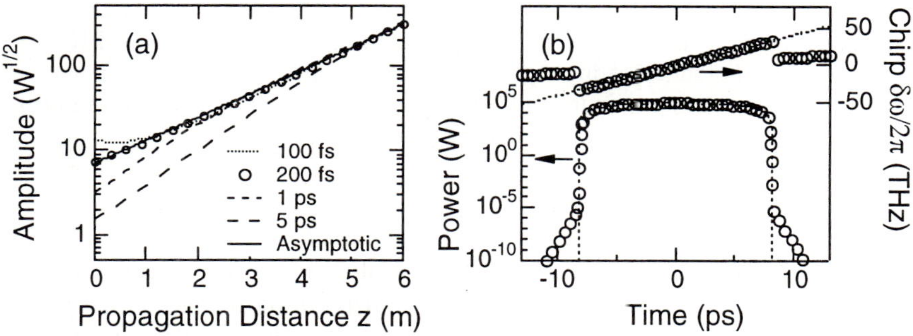

Fig. 1 (a) Simulated evolution for different input pulses (indicated in caption) compared to theoretical prediction (solid line). (b) compares NLSE output (circles) with asymptotic parabolic pulse solution I(dashed line)

3. Experiment

Our experiments used 12 pJ, 200 fs pulses at 1.06 μm and at a repetition rate of 63 MHz injected into a 3.6 m length of Yb:doped fiber co-directionally pumped at 976 nm. Characterization of the output pulses was carried out using FROG based on second-harmonic generation (SHG) in a KDP crystal [5]. FROG measurements were carried out on the pulses directly after the Yb:doped fiber amplifier, as well as after subsequent propagation in 2 m of standard undoped single mode fiber (SMF). The solid lines in Fig. 2(a) show the measured intensity and chirp for the amplifier output with g = 1.9 m^{-1}. These results are compared with NLSE simulations (circles) and the predicted asymptotic characteristics (short dashes). Both the measured intensity and chirp are in good agreement with the results of NLSE simulations, but more significantly, there is agreement over two orders of magnitude with the asymptotic parabolic pulse predicted by (2). To emphasize the parabolic nature of these pulses, the figure also includes a $sech^2$ fit to the measured intensity profile (long dashes).

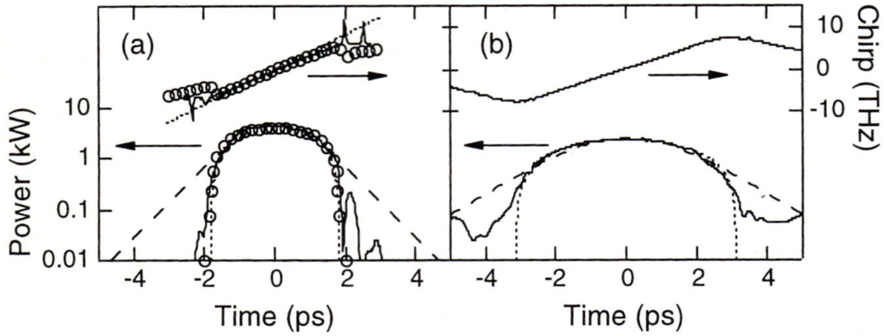

Fig. 2: (a) Parabolic pulses from amplifier: measured (solid); simulations (circles); asymptotic solution (short dashes). (b) Parabolic pulses after SMF propagation: measured (solid); parabolic fit (short dashes). The long dashes show a $sech^2$ fit.

Fig. 2(b) shows how these pulses retain their parabolic profile after self-similar propagation in 2 m normally-dispersive undoped single mode fiber (SMF). To demonstrate the potential of high power parabolic pulses in ultrafast optics, we used a dispersive grating pair for pulse compression, obtaining a minimum pulse duration of $\Delta\tau = 68$ fs with a corresponding peak power of 80 kW.

4. Conclusions

In conclusion, we have developed a theoretical treatment of the amplification of pulses in high gain fiber amplifiers which predicts the formation of high power parabolic similariton pulses from any input pulse. This is the first experimental demonstration of the existence of similariton pulses and their application to pulse compression. We expect that they will find wide application in ultrafast optics.

References

1. M. E. Fermann *et al.* Opt. Lett. **24**, 1428 (1999).
2. G. P. Agrawal, *"Nonlinear Fiber Optics,"* Academic Press, San Francisco (1995).
3. D. Anderson et al., J. Opt. Soc. Am. B **10**, 1185 (1993); K. Tamura and M. Nakazawa, Opt. Lett. **21**, 68 (1996).
4. G. I. Barenblatt, *"Scaling, Self-similarity, and Intermediate Asymptotics,"* Cambridge University Press, Cambridge (1996).
5. R. Trebino *et al.*, Rev. Sci. Instrum. **68**, 3277 (1997).

Generation, Amplification and Characterization of Tunable Visible Ultrashort Shaped Pulses

Howe-Siang Tan, Warren S. Warren, and Elmar Schreiber

Center for Ultrafast Laser Applications and Department of Chemistry, Princeton University, Princeton NJ 08544-1009, USA
E-mail: eschreib@princeton.edu

Abstract. By producing non-collinear optical amplification (NOPA) with an internal acousto-optic modulator (AOM), we generate amplified shaped ultrashort pulses (~5μJ, <50fs) tunable from 500 to 700nm with up to 50THz bandwidths. The shaped pulses are fully characterized by a variation of the spectrally and temporally resolved up-conversion technique (STRUT).

1. Introduction

Our particular goal is to obtain shaped pulses in the highly interesting visible spectral region. Shaping and control of ultrashort laser pulses is of growing interest and its use has been demonstrated in a wide variety of optical applications, including quantum control, high speed communications, and material characterization. The common approaches include: a programmable spatial liquid crystal arrays [1] or acousto-optic modulator (AOM) [2,3] to alter the dispersed pulse spectrum based on a Fourier transform technique. An AOM with collinear acousto-optic interaction [4] and electrostatically deformable mirrors [5] have also been used. In [1,2,5] the pulse shaping takes place after amplification, whereas in [3,4] and in our approach it is performed prior to amplification.

2. Pulse Generation and Amplification

To generate ultrashort laser pulses with broad wavelength tunability in the visible, we utilized the NOPA scheme as introduced by Riedle [6]. The NOPA is pumped by an amplified Ti:sapphire laser system (1kHz repetition rate) capable of producing nearly transform-limited 90 to 120fs pulses with up to 2.5mJ output energy per pulse, however, just 10% is sufficient to run the NOPA with high stability. As seen in fig. 1, the NOPA consists of two separate beampaths, one for the white light generation (pumped with ~10μJ) and another for the parametric amplification.

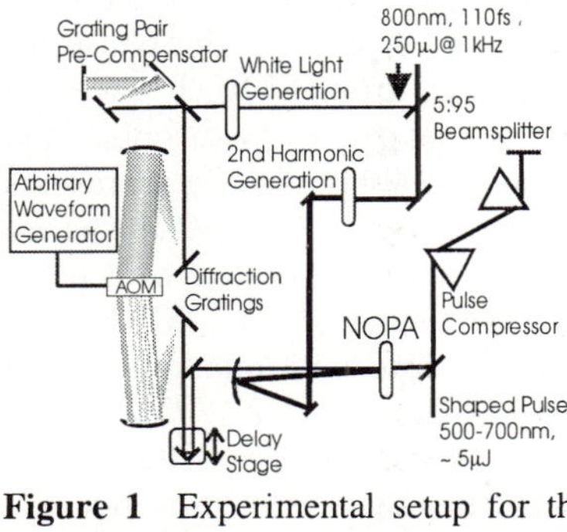

Figure 1 Experimental setup for the generation and amplification of shaped ultrashort pulses in the visible regime.

A white light continuum pulse is generated in a 3mm sapphire crystal and sent through a Bragg regime AOM pulse shaper (TeO_2 crystal, 10mm thick) [2]. Most of its quadratic dispersion has been pre-compensated by a diffraction grating pair.

At the pulse shaper, the spectrum of the white light pulse is spatially dispersed by a diffraction grating and imaged by a concave mirror onto the AOM crystal (40mm) placed in the Fourier plane. An arbitrary waveform generator supplies a shaped radio frequency modulation $f(t)=exp[i\alpha(2\pi T)^2+i\beta(2\pi T)^3]$ to the AOM crystal to generate the acoustic waves. Hence, the AOM imposes a position-dependent phase and amplitude modulation on the dispersed spectrum, which is refocused onto a further grating. This recombines the modulated spectrum and thus generates a shaped white light pulse. In the present study, we used an acoustic wave with $\alpha=29MHz^2$ and $\beta=26MHz^3$ to compensate for residual quadratic and cubic (TeO_2:~5500fs^3/cm (@600nm)) material dispersions, respectively.

The pulse shaper can be tuned to shape the spectral range of choice (bandwidth • 50THz) of the white light pulse. This shaped white light pulse provides the seed pulse which is amplified to ~5μJ via a non-collinear optical parametric (type I) process in a BBO crystal by a 100μJ pump pulse (@400nm). To ensure efficient amplification, a small amount of quadratic chirp is introduced to the pulse. This, and any additional chirp acquired from the pulse's propagation through the BBO crystal is finally compensated by a pair of fused silica prisms.

3. Temporal and Spectral Pulse Characterization

A variation of (STRUT) [7] that provides information about the pulse's amplitude and phase is used to characterize the generated shaped pulses. An amplified shaped pulse is cross-correlated with an unshaped reference pulse with a relatively narrow bandwidth (~4THz) centered at 800nm. The sum frequency spectrum generated in a type I non-linear optical process in a 0.3mm thick BBO crystal is collected as a function of delay time between shaped and reference pulse resulting in a time-wavelength correlation (STRUT) trace. A 25μJ pulse sampled from the regenerative amplifier provides the reference pulse.

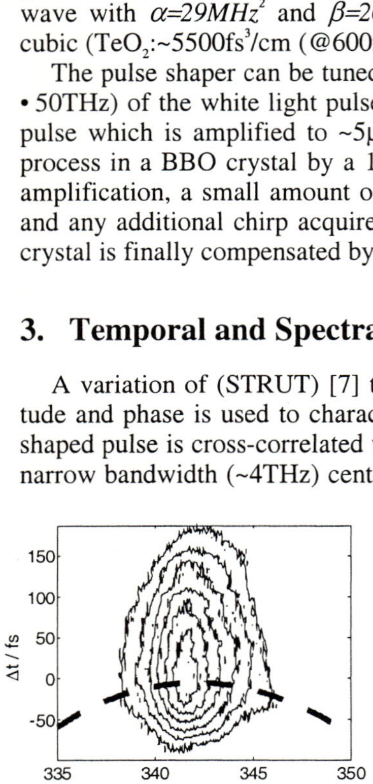

Figure 2 Contour plot of a STRUT trace generated by a sum frequency process between a shaped pulse centered at 600nm and a reference pulse at 800nm. The phase profile of this chirped pulse is retrieved by obtaining the negative delay-time Δt for the maxima at each wavelength λ, as is idealized by the dashed line. The actual retrieved curve is depicted in Fig 3(left panel).

Figure 2 shows a typical STRUT trace for a shaped pulse and fig. 3 presents the phase information as retrieved from STRUT traces of two differently shaped pulses with cubic and quadratic phase shifts. While fig. 3a represents the phase characterization of a pulse shaped with a cubic phase shift, generated with parameters $\alpha=28.70$MHz^2 and $\beta=26.24$MHz^3, fig. 3b depicts a pulse shaped with quadratic phase shift with $\alpha=36.08$MHz^2 and $\beta=22.14$MHz^3.

Due to the thickness of the BBO crystal used, the phase-matching condition is not fulfilled for the whole spectral range of the shaped pulse. Hence, the phase characterization could not be performed over the full spectral band-

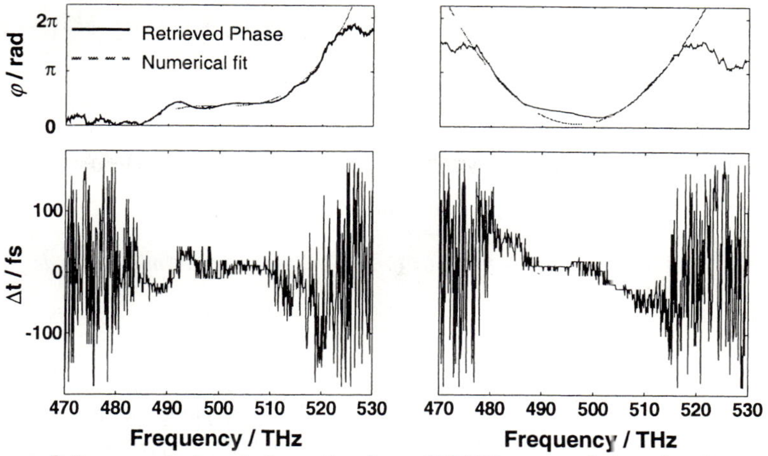

Figure 3 Recovered phase information from STRUT traces of shaped pulses centered at 600nm. Bottom panels depict the first derivative of phase φ. In the top panels, solid lines represent the retrieved phases. The dashed lines are fits to determine numerically the amount of phase shift imparted. Right panels: a pulse with cubic phase shift of ~$8000fs^3$ (recovered from the STRUT trace of fig. 2). Left panels: a pulse with quadratic phase shift of ~$550fs^2$.

width of the pulse. A major limitation of our amplification scheme is the length of the shaped pulse that is restricted by the length of the pump pulse.

4. Conclusions

We have demonstrated experimentally the successful amplification and characterization of shaped pulses in the visible regime using AOM technique. First simple examples of shaped amplified pulses with quadratic and cubic phase shifts are shown. The presented system shall be used for optimal (coherent) control spectroscopy of complex molecules in the highly interesting spectral range from 480 to 750nm. Therefore, as a next step we plan to implement feedback control by an evolutionary algorithm.

Acknowledgements. This work is supported by The New Jersey Commission on Science and Technology and the Air Force Office of Scientific Research. The authors would like to thank E. Riedle (LMU Munich), P. Dietrich (FU Berlin) and M. Motzkus (MPIQ Garching) for many helpful advices.

References

1 A.M. Weiner, et. al., *IEEE J. Quantum Elec.*, **28** (1992) 918.
2 M.A. Dugan, J.X. Tull, and W.S. Warren, *J. Opt. Soc. Am. B*, **14** (1997) 2348.
3 M.R. Fetterman et. al., *Opt. Exp.*, **3** (1998) 366.
4 P. Tournois, *Opt. Com.* **140** (1997) 245; F. Verluise et al. *Opt. Lett.* 25 (2000) 575.
5 E. Zeek et al ., Opt. Lett. **24** (1999) 493.
6 T. Wilhelm, J. Piel, and E. Riedle, *Opt. Lett.* **22** (1997) 1494.
7 J.K. Rhee et al., *J. Opt. Soc. Am.* **B 13** (1996) 1780.

Synthesis of Supershort UV-Pulses Using Phase-Locked Raman Side-Band Generation

G.Korn, A.Nazarkin, M.Wittmann

Max-Born-Institut für Nichtlineare Optik und Kurzzeitspektroskopie,
Max-Born-Str. 2A, D-12489 Berlin, Germany
Phone : +49-30-6392-1277, Fax : +49-30-6392-1289, E-Mail : korn@mbi-berlin.de

Abstract. We demonstrate a new technique of fs-pulse generation employing ultrafast modulation of a laser field phase by impulsively excited molecular rotational or vibrational motion with subsequent temporal compression. An ultrashort pump pulse at 800 nm performs impulsive excitation of a molecular gas in a hollow waveguide and a weak delayed probe pulse at 400 nm is scattered on the temporal oscillations of its dielectric index. The resulting sinusoidal phase modulation of the probe pulse allows pulse temporal compression using both positively and negatively dispersive elements. The potential of this new method is demonstrated by the generation of a periodic train of 5.5 fs pulses at 400 nm which are the shortest ever generated in the UV.

1. Introduction

In the last few years, the traditional technique of ultrashort pulse generation based on Kerr self-phase modulation and subsequent pulse compression with dispersive elements has been considerably improved [1] and enables presently the generation of the shortest pulses in the visible lasting only two optical periods [2]. The difficulty in controlling the phase of the generated broad spectrum and the lack of suitable, easy-to-handle dispersive elements for pulse compression, however, set certain limitations on the potential of this technique for sub-fs pulse generation. Recently, considerable interest has been focused on the alternative mechanisms, particularly, on the idea of Fourier synthesis of sub-fs waveforms using phase-synchronized lines of stimulated Raman scattering (SRS) [3-5]. In our recent paper [6], we have shown that synchronization of the phases of the SRS-lines can be achieved for the SRS-process occurring in a preliminary impulsively excited medium. An ultrashort pump pulse prepares a Raman medium impulsively (with the pump pulse duration shorter than the molecular vibrational period [7,8]), and a weak probe pulse is scattered on the coherently oscillating molecular vibrations or rotations, which results in the generation of multiple side-band Raman components. Due to the temporal separation of the pumping and the scattering process, the deteriorating effects of self-phase modulation, medium ionization etc. on the phases of the components are completely eliminated in this linear scattering regime. The process can be treated as an ultrafast modulation of the injection field phase by molecular vibrations or rotations. Since an isolated rotational (or vibrational) mode of the molecule causes a periodical modulation of the probe pulse phase, the transformation of the produced phase modulation into amplitude one can be done using either negative or positive group-delay dispersive elements (GDD) [9].

In this paper the results of temporal compression of the output side-band Raman spectra generated in an impulsively excited gas-filled hollow waveguide are reported. Even with the positive GDD-compensation, we were able to generate periodic sequences of sub-6fs pulses at 400nm, which are to our knowledge the shortest pulses ever produced in the UV. The potential of this new fs-pulse generation technique is examined using different rotationally and vibrationally-active Raman gases for the ultrafast phase modulation.

2. Experimental Results and Discussion

In our experiments we used a Ti:sapphire chirped pulse amplification laser system at 1 kHz repetition rate producing 25 fs pulses at 800 nm on a 1 mJ level. About 0.2 mJ of the fundamental were used to excite the gaseous medium. The other part was frequency doubled in a type-I BBO (thickness 1 mm) to generate the probe-pulses at 400 nm with durations of ca. 100 fs and an energy of some 10 μJs. The experiments were carried out in a 1 m long hollow waveguide to elongate the interaction region and to ensure the spatial overlap of pump- and probe beams. The emerging radiation at 400 nm also possesses excellent beam quality due to the guiding in the waveguide.

In the first series of compression experiments, excitation of pure vibrational degrees of freedom was used for the ultrafast phase modulation. An ideal object for such investigations is the SF_6 molecule. This highly symmetric molecule has no rotational Raman spectrum and its vibrational mode A_{1g} (period T=43 fs, Ω =775 cm^{-1}) is much stronger than other two modes. As a result, only A_{1g}-mode is effectively excited by the nonselective pumping mechanism. The coherent molecular vibration in gaseous SF_6 was prepared impulsively by a single intense 25 fs pulse at 800 nm (ca. 300 μJ). (For this purpose the pulse duration of the exciting pulse needs to be shorter than the molecular vibrational period.) The probe pulse at 400 nm was delayed (200 fs) and sent through the excited medium. In Fig.1a we present results on compression of the output laser radiation transmitted through the impulsively excited SF_6. The dashed line in Fig.1a represents the output pulse shape at 400 nm. without the Raman excitation. The corresponding output Raman side-band spectrum (three Stokes and anti-Stokes lines) is shown in Fig.1b, the dashed line represents the input spectrum at 400nm. We observed a periodic sequence of compressed laser pulses following with the period of the excited vibrational mode A_{1g}. Interestingly, the use of both positive and negative group-velocity dispersion (GVD) for the compression was found to be as appropriately efficient. By adjusting carefully the net GVD with chirped mirrors (providing negative GVD at 400 nm) and varying the light pass in air together with the exit window of the waveguide housing (positive GVD), we measured UV pulses as short as 5.8 fs (solid line of Fig.1a) in the positive net GVD region. The corresponding compression ratio is seen to be higher than 10.

Rotational Raman scattering is also suitable for the technique of fs-pulse generation. This was demonstrated by using gaseous H_2 and N_2 for the ultrafast phase modulation. For H_2 (actually, we had a mixture of 75% ortho-H_2 and 25% para-H_2), we observed a periodic train of 15 fs pulses with the distance between the pulses (57 fs) corresponding to the rotational period of ortho-H_2. In the case of N_2, impulsive excitation with sub-ps pulses led to a formation of a coherent superposition of rotational lines which caused a rather complicated modulation law of the probe pulse phase. We found that for the ultrafast phase modulation process to be effective it is necessary that the number of simultaneously excited (rotational or vibrational) modes would be possibly small (practically 1-:-2).

3. Conclusion

In conclusion, we have described a new technique of fs-pulse generation which employs impulsively excited molecular rotational and vibrational motion to modulate the dielectric index of a probe laser pulse. Propagation of the probe pulse in a Raman medium results in a sinusoidal phase-modulation which can be transformed into periodic pulse sequences using both negative and positive GDD. The method was demonstrated by the generation of compressed pulse trains in SF_6 (vibrational Raman scattering) and H_2 (rotational Raman scattering) in the region of positive GDD-compensation. The 5.5 fs-pulses at 400 generated with this technique are the shortest ever reported in the UV.

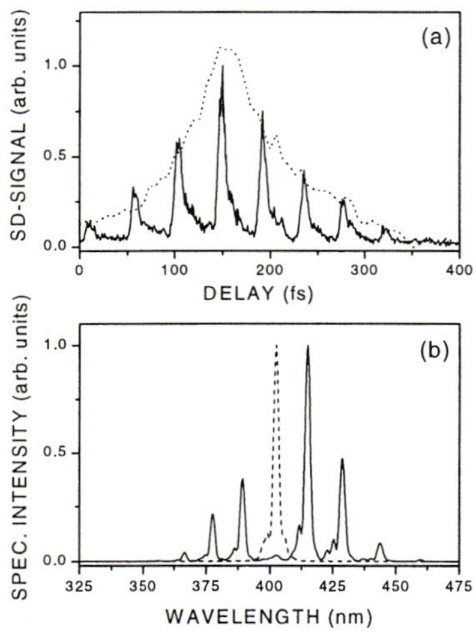

Fig.1. Temporal structure (a) and spectrum (b) of the compressed pulse at 400 nm preliminary transmitted through impulsively excited SF_6. The dotted line in (a) and dashed line in (b) show correspondingly the pulse shape and the pulse spectrum without impulsive medium excitation.

The authors thank Prof. T. Elsaesser for the support and stimulating discussions. We are also grateful for the support from the Deutsche Forschungsgemeinschaft.

References

[1] M. Nisoli, S. De Silvestri, and O. Svelto, Appl. Phys. Lett. 68, 2793 (1996) O. Dühr, E.T.J. Nibbering, G. Korn, G. Tempea and F. Krausz, Opt. Lett. 24, 34 (1999)
[2] M. Nisoli, S. De Silvestri, O. Svelto, R. Szipöcs, K. Ferencz, C. Spielmann, S. Sartania, and F.Krausz, Opt. Lett. 22, 522 (1997)
[3] S. Yoshikawa and T. Imasaka, Opt. Commun. 96, 94 (1993)
[4] T. Mukai, R. Wynands, and T. W. Hänsch, Opt. Commun. 95, 71(1993)
[5] H. Kawano, Y. Hirakawa, and T. Imasaka, IEEE J. Quant. Electr. 34, 260(1998)
[6] A. Nazarkin, G. Korn, M. Wittmann, and T. Elsaesser, Phys. Rev. Lett. 83, 2560 (1999)
[7] S. De Silvestri, J. G. Fujimoto, E. P. Ippen et al Chem. Phys. Lett., 116, 146 (1985)
[8] S. Ruhman, A. Joly and K.Nelson, IEEE J. Quantum Electron. 24, 460 (1988)
[9] M. Wittmann, A. Nazarkin, and G. Korn, Phys. Rev. Lett.,84, 5508 (2000)

Generation and measurement of ultrafast tunable VUV light

L. Misoguti[1], C.G. Durfee[2], S. Backus[1], M. M. Murnane[1], and H. C. Kapteyn[1]

[1] JILA-Physics Department, University of Colorado and NIST, Boulder CO 80309, USA
E-mail: misoguti@jila.colorado.edu
[2] Department of Physics, Colorado School of Mines, Golden CO 80401, USA.

Abstract. We report the efficient generation of phase-matched tunable VUV light from 80-300nm by wave mixing and harmonic generation in a gas-filled hollow-core fiber. Distinct phase-matching and cascaded processes are observed.

1. Introduction

In this work we investigate the generation of ultrafast tunable deep-UV and VUV light using nonlinear processes in a gas-filled hollow-core fiber. Using ultra-short pulses, we observe two distinct regimes of phase-matching - the first relying on mode propagation in the capillary, while the second is dominated by dispersion near the resonance lines of the gas. Self-phase modulation of the fundamental pulses makes it possible to generating very broad-bandwidth pulses in the deep-UV and VUV that should be compressible to very short pulse duration. Cascaded processes also allow us to generate broadband VUV light efficiently.

2. Experimental Methods

In recent work, we demonstrated the use of hollow-core waveguides [1, 2] as an effective method for phase-matched frequency conversion of ultrashort pulses into the ultraviolet [3] and the soft x-ray regions of the spectrum [4]. In a gas-filled capillary, phase-matching over broad bandwidths can be obtained by controlling the phase velocity of the light. This is achieved by balancing the mode dispersion of the capillary with the dispersion of the gas. The wave vector for light at a wavelength λ propagating in a fiber of diameter a is -

$$k(\lambda) = \frac{2\pi}{\lambda} + \frac{2\pi P\delta(\lambda)}{\lambda} - \frac{u_{nm}^2\lambda}{4\pi a^2}, \qquad (1)$$

where P is the pressure of the gas, $\delta(\lambda)$ is related the index of refraction of the gas by $n=1+P\delta(\lambda)$, and u_{nm} is a discrete constant that varies with the propagation mode. Alternatively, phase matching can also be achieved on the short-wavelength side of the resonance line of an atom [5]; however, the bandwidth is limited by the rapidly varying dispersion in the vicinity of the resonance lines [6].

In these experiments, we implemented a number of two-color frequency-mixing schemes in a 125-175 μm capillary waveguide, using 400 nm pump pulses, and either 800 nm fundamental, or tunable light from an OPA at 1100-

2900 nm, as the idler. The output UV light is given by $\omega_{UV}=2\omega+2\omega-\omega_{idler}$ for the simple four-wave mixing processes, or this process is repeated for cascaded UV generation

3. Results and Discussion

Fig. 1 shows intense signals generated at 240nm using parametric four-wave mixing ($\omega_{UV}=2\omega+2\omega-\omega_{idler}$). As the pressure within the waveguide is tuned, we observe two effects. First, the output UV signal (Fig. 1(b)) is optimized at distinct pressures that correspond to phase-matched conversion into distinct spatial modes of the waveguide. Second, at high pressures, the effect of nonlinear frequency broadening of the pump or signal wavelengths is evident. The phase of the 240nm light is likely to be well behaved, as we recently demonstrated by compressing 267nm pulses to 8fs [7]. The bandwidth is sufficient to support a 6fs pulse.

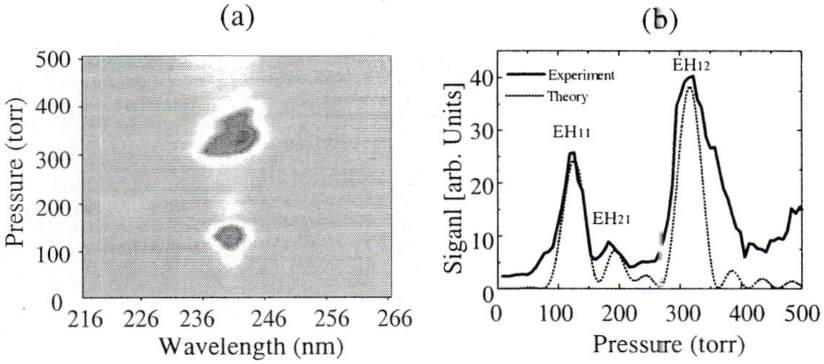

Fig. 1: Generation of broadband 240nm light by four wave mixing ($2\omega+2\omega-\omega_{idler}$) in Ar; (a) spectrum; (b) intensity as function of pressure.

We also generated light in the VUV using higher-order nonlinear processes such as six-wave mixing, third- and fifth harmonic generation, as well as cascaded four-wave mixing. In the VUV, broadening of the spectrum of the pump has a dramatic effect on the generated signal, as shown in Fig. 2. We observe clear indications that in some regions, resonant phase matching dominates frequency conversion processes, limiting the pulse duration to tens of fs. However, far from such resonances, very short-duration pulses can be generated. Finally, we observe very efficient generation of deep UV light at 200nm (4ω) and 160nm (5ω) through 1- and 2-stage cascaded processes [8]. Here we first generated 3ω light very efficiently (30%) using four wave mixing $3\omega=2\omega+2\omega-\omega$. The resultant 267nm light was then used as the seed for a second four-wave mixing step, giving rise to light at either 160nm or 200nm. Although many process can contribute to such cascaded schemes, careful comparison of theory and experiment allows us to identify the likely high-order mixing processes. For example, light at 5ω is generated mainly by cascaded difference-frequency four-wave mixing processes: $5\omega = 2*(3\omega)-\omega$, $5\omega = 4\omega+2\omega-\omega$ and $5\omega = 4\omega+3\omega-2\omega$. We believe that this is the first time cascaded processes have been used to upshift ultrashort-pulse light efficiently into the UV.

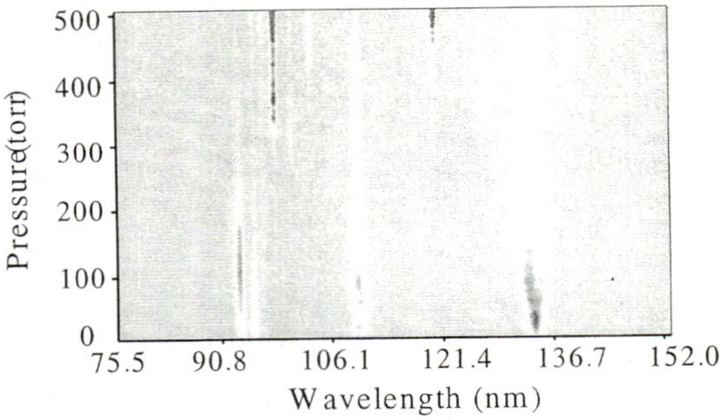

Fig. 2: Several discrete wavelengths generated simultaneously due to extreme phase modulation and broadening of the pump light at 400nm. The narrow lines are as a result of resonant phase matching, while light at 133nm (6ω) is non-resonantly phase-matched, and has a very broad 10nm bandwidth, capable of supporting a 2.5fs pulse.

4. Conclusions

In summary, through guided-wave frequency conversion we generate light throughout the VUV, from 80 to 300nm. We observe intense four-wave mixing signals at 214-267nm, with 5-30% conversion efficiency. Intense spectral broadening of the pump light leads to the generation of broadband UV light, with the potential to produce 2fs pulses in the VUV. The combined effects of resonant dispersion and self-phase modulation allow us to simultaneously generate light at a number of wavelengths, from 85 to 120nm (10-15eV). Finally, cascaded processes promise to significantly enhance conversion efficiencies.

Acknowledgements. The authors gratefully acknowledge support from the Department of Energy and the National Science Foundation.

References

1 *E. A. J. Marcatili and R. A. Schmeltzer,* Bell System Technical Journal, *vol. 43, p. 1783, 1964.*

2 *M. Nisoli, S. De Silvestri, and O. Svelto,* Applied Physics Letters, *vol. 68, p. 2793, 1996.*

3 *C. G. Durfee, S. Backus, M. M. Murnane, and H. C. Kapteyn,* Optics Letters, *vol. 22, p. 1565, 1997.*

4 *A. Rundquist, C. G. Durfee III, S. Backus, C. Herne, Z. Chang, M. M. Murnane, and H. C. Kapteyn,* Science, *vol. 280, p. 1412, 1998.*

5 *J. F. Reintjes: Nonlinear optical parametric processes in liquids and gases, Orlando, FL: Academic Press, 1984.*

6 *R. Mahon, T. J. McIlrath, V. P. Myerscough, and D. W. Koopman,* IEEE Journal of Quantum Electronics, *vol. QE-15, p. 444, 1979.*

7 *C. G. Durfee, S. Backus, H. C. Kapteyn, and M. M. Murnane,* Optics Letters, *vol. 24, p. 697, 1999.*

8 *L. Misoguti, C. G. Durfee, H. C. Kapteyn, and M. M. Murnane, submitted to Physical Review Letters.*

Part III

Pulse Characterization, Shaping and Measurement Techniques

Autocorrelation Measurement of Femtosecond Optical Pulses Using Two-Photon-Induced Photocurrent in a Photomultiplier Tube

Toshiaki Hattori, Yoshitsugu Kawashima, Masahiro Daikoku, Hideyuki Inouye, and Hiroki Nakatsuka

Institute of Applied Physics, University of Tsukuba, Tsukuba, Ibaraki, 305-8573, Japan
E-mail: hattori@bk.tsukuba.ac.jp

Abstract. Very sensitive intensity autocorrelation measurement of 15-fs, 800-nm pulses from a Ti:sapphire laser has been achieved by using two-photon-induced photocurrent in a photomultiplier tube.

1. Introduction

Recent progress in intensity autocorrelation measurements of femtosecond optical pulses based on two-photon-induced processes in photodetectors, such as photodiodes (PDs) and light-emitting diodes, has made the pulse-width measurement of ultrashort pulses very simple and easy [1]. Sensitivity of these methods are generally better than the conventional technique based on second-harmonic generation in a nonlinear crystal. However, more sensitive measurements are required in many areas of application of femtosecond optical pulses, such as pulse shaping and ultrafast spectroscopy. Here, we present our recent results of very sensitive autocorrelation measurement of 15-fs optical pulses based on two-photon-induced photoemission in a photomultiplier tube (PMT).

PMTs are very sensitive photodetectors because of the secondary-emission multiplication of photoelectrons. Use of a PMT, therefore, as a nonlinear detector in autocorrelation measurements should lower the required input light power to a considerable extent. Autocorrelation measurements of picosecond optical pulses have already been reported [2,3]. However, no studies have been reported on the femtosecond response of the two-photon-induced processes in PMTs, or on autocorrelation measurements of 800-nm light from Ti:sapphire lasers, which are widely used as femtosecond pulsed light sources.

2. Experiment

Attenuated output of a Ti:sapphire laser generating 15-fs, 800-nm pulses at a repetition rate of 89MHz was sent to a Michelson-type interferometer. The output of the interferometer was loosely focused by a concave mirror with a focal length of 200 mm onto the photocathode of a PMT with a spot size of 80 μm. Two-photon-induced photocurrent of the photodetector was measured by scanning the delay time of one arm in the interferometer. Two PMTs were used in the experiment: 1P28 (Hamamatsu Photonics), with a reflection-mode Cs_3Sb photo-cathode and R2557 (Hamamatsu Photonics), with a transmission-mode Na_2KSb photocathode. Both of these PMTs have one-photon response spectra similar to that of a GaAsP PD, which was used by Ranka *et al.* in the autocorrelation measurement based on the two-photon-induced photocurrent in the PD [1]. A GaAsP PD was also used in the present experiment for comparison.

Interferometric autocorrelation traces obtained with the GaAsP PD at two values of input power are shown in Fig. 1. Figure 1(a) shows an autocorrelation trace with a very good signal-to-noise ratio, which was obtained with 70 mW incident light. On the other hand, with incident light of 500 μW, traces obtained with the PD were not only very noisy but also deformed by a considerable contribution of one-photon-induced photocurrent, as shown in Fig. 1(b).

With similar input power of 410 μW, very good autocorrelation trace was obtained with the PMT as shown in Fig. 2. No broadening of the pulse is observed in the figure.

Signal traces obtained with a PMT R2557 show tails which decay at time constant of 270 fs, as shown in Fig. 3. This suggests existence of an intermediate state in the two-photon transition, and the decay time observed here is attributed to the relaxation time of the inter-mediate state. It has been re-ported that the bandgap energies of the photocathode materials Cs_3Sb and Na_2KSb are 1.6 eV and 1.0 eV, respectively, although the sum of the

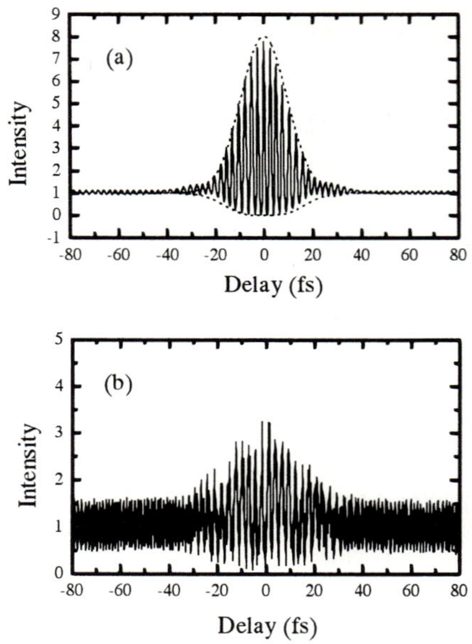

Fig. 1. Interferometric autocorrelation trace of 15-fs, 800-nm pulses from a Ti:sapphire laser obtained by measurement of two-photon-induced photocurrent in a GaAsP photodiode at input power of (a) 70 mW and (b) 500 μW.

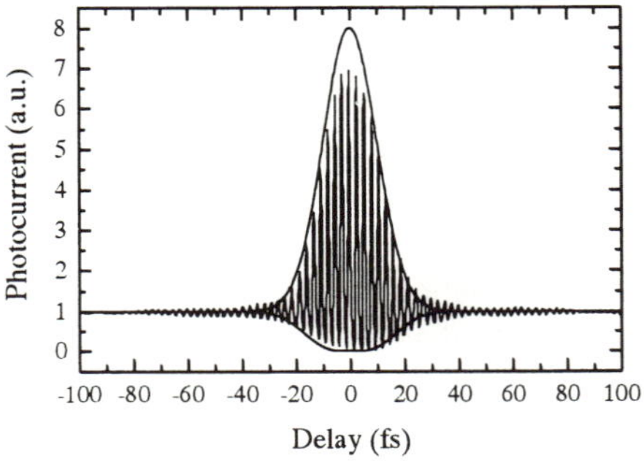

Fig. 2. Interferometric autocorrelation trace of 15-fs, 800-nm pulses from a Ti:sapphire laser obtained with a 1P28 photomultiplier tube at input power as low as 410 μW. Simulated envelopes of the autocorrelation trace of 15-fs Gaussian pulses are also shown.

bandgap energies and the electron affinities are 2.0 eV for both materials [4]. These facts explain the observed results. Since the 800-nm incident light has photon energy of 1.55 eV, it can induce interband transition with one photon in Na_2KSb, and electrons reach the vacuum level by two-photon absorption via a two-step process. Simulation by perturbation calculation based on the optical Bloch equation showed the intermediate state lifetime of 270 fs and the dephasing time less than 5 fs [5].

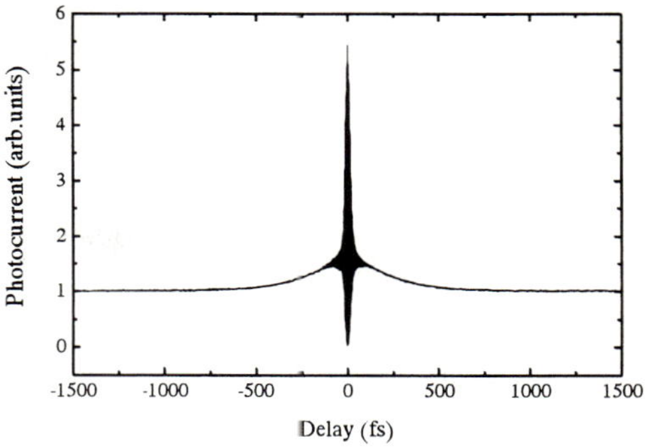

Fig. 3. Signal trace obtained with a PMT R2557. It has tails with a decay time of 270 fs, which can be regarded as the lifetime of the intermediate state of the two-photon-induced process.

3. Conclusions

In summary, very sensitive autocorrelation measurement of 15-fs, 800-nm optical pulses has been achieved by using two-photon-induced photoemission processes in a PMT [6]. Ultrafast relaxation of electrons excited in an intermediate state of the two-step two-photon transition was observed in a different type of PMT with the same measurement setup [5].

References

1 J. K. Ranka, A. L. Gaeta, A. Baltuska, M. S. Pshenichnikov, and D. A. Wiersma, Opt. Lett. **22**, 1344 (1997).
2 W. R. Bennett, Jr., D. B. Carlin, and G. J. Collins, IEEE J. Quantum Electron. **QE-10**, 97 (1974).
3 Y. Takagi, Appl. Opt. **33**, 6328 (1994).
4 W. E. Spicer, J. Appl. Phys. **31**, 2077 (1960).
5 T. Hattori, Y. Kawashima, M. Daikoku, H. Inouye, and H. Nakatsuka, Jpn. J. Appl. Phys. **39**(8), (2000) in press.
6 T. Hattori, Y. Kawashima, M. Daikoku, H. Inouye, and H. Nakatsuka, to be published in Jpn. J. Appl. Phys.

Precision and accuracy of ultrashort optical pulse measurement using SPIDER

Ellen M. Kosik, Matthew E. Anderson, Luís E. E. de Araujo, Christophe Dorrer, and Ian A. Walmsley

The Institute of Optics, University of Rochester, Rochester, NY 14627
e-mail: kosik@optics.rochester.edu

Abstract: We have examined the accuracy and precision of pulse reconstruction using Spectral Phase Interferometry for Direct Electric-field Reconstruction, and identified methods for improving the sensitivity of the technique.

Spectral Phase Interferometry for Direct Electric-field Reconstruction (SPIDER) is an embodiment of spectral-shearing interferometry in which two replica pulses are combined with a third stretched (chirped) pulse in a nonlinear optical process. The resulting two frequency-shifted replica pulses have a spectral shear dependent on the time delay between the incident pulse pair and the amount of chirp on the stretched pulse. The spectrum of the replicas is measured using a spectrometer and a spectral interferogram is recorded on a photodiode array. The spectral and temporal intensities and phases can be easily extracted from this interferogram through a simple, non-iterative inversion routine [1]. The SPIDER method is very robust and can be used to measure a wide range of pulses. The method has been demonstrated at many wavelengths from the infrared to the blue [2]; it can measure pulses from oscillators or amplifiers[1,3]; it can run in real time [3] and in single shot configurations [4]; and has been implemented for pulses with durations from 6 fs [5] to 500 fs[6].

Fundamental to the usefulness of any measurement tool is how well it performs in real laboratory settings, in particular how accurate and precise it is in the presence of noise. Precision, the ability of the meaurement to return the correct electric field, was analyzed by simulating noisy interferograms and using the SPIDER algorithm to reconstruct the temporal pulse shape. Three types of error were used in this study, additive noise, multiplicative noise, and quantization. The accuracy of the device, the ability of the device to return the same results from several data sets, was also analyzed via simulation. Due to the fact that the spectral sampling of the detector is much finer than the spectral shear, the measured SPIDER interferogram, corresponding to a single pulse, contains redundant data. This redundancy can be used to calculate several determinations of the electric field of the pulse and determine an accuracy function, thereby helping to track sources of noise or systematic errors in the measurement process. Our analysis has shown the SPIDER method capable of very accurate and very precise reconstructions in the presense of noise (Fig. 1,2). Roughly, for a signal to noise ratio on the interferogram of 10, the precision of the reconstructed phase is better than 0.03 radians, and the accuracy is better than 2%. (The precision is measured by the intensity-weighted rms temporal phase and the accuracy by the rms pulse shape deviation respectively.).

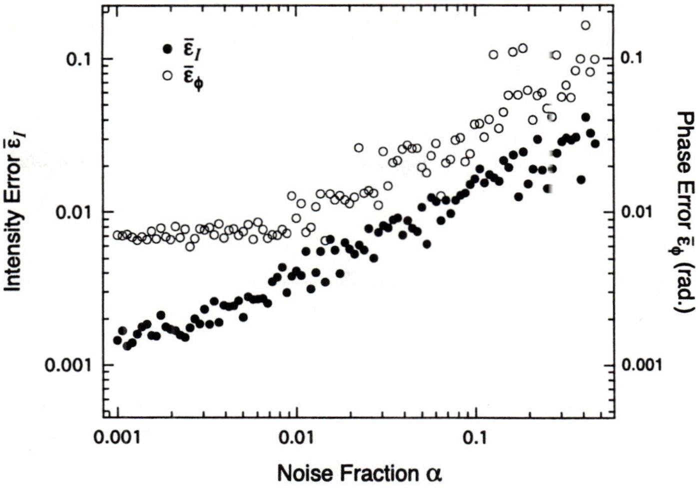

Fig.1. Error in reconstructed temporal intensity (solid circles) and phase (open circles) versus noise fraction for the case of additive noise. The error measure was the time integrated difference in input and reconstructed intensity and phase respectively.

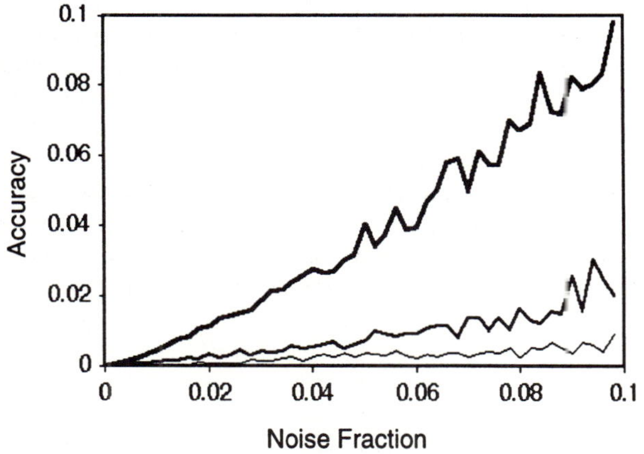

Fig.2. Accuracy of reconstructed electric field versus noise fraction for three sizes of filter width in the time domain. The thinest line correspond to the smallest window, the thickest line to the largest window. The error measure was the time integrated squared difference between the input and reconstructed temporal electric fields, using a simple interferogram.

We have also identified techniques which improve the sensitivity of the SPIDER method. As there is a one-to-one relation between the spectral phase of the input pulse and the spectral phase of the upconverted pulse, it is possible to use a narrow-band non-linear crystal and to scan the upconverted portion of the input field to retrieve the spectral phase on small portions, as illustrated in Fig.3. The global spectral phase is then obtained directly by stitching the different

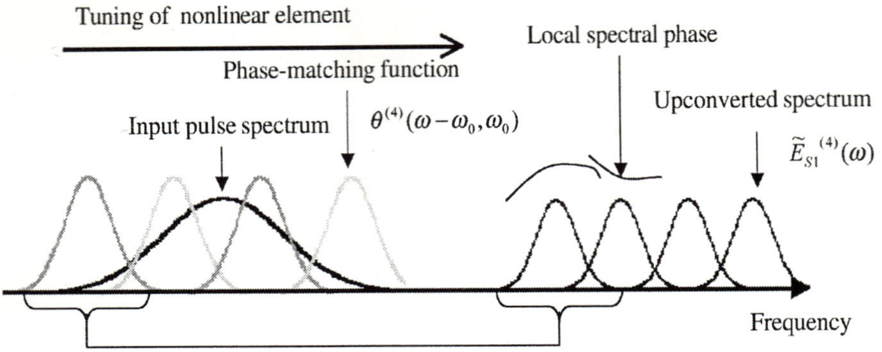

Fig. 3. Even when the nonlinear process has a narrow phase-matching bandwidth, the unique mapping of the local spectral phase from the input to the shifted spectrum allows accurate reconstruction of the local phase. The global phase (to within a constant) is obtained by tuning the phase-matching function and stitching the locally-recovered spectral phase functions.

determinations of the spectral phase. A second method of increasing the sensitivity of the SPIDER method is by homodyning the electric field of one of the upconverted replicas with a strong uncharacterized pulse. By measuring the phase difference between each of the replicas and the same homodyne pulse identical information to the direct SPIDER technique can be extracted. Because a strong homodyne pulse can be obtained using a small part of the energy of the pulse to be characterized, the technique is still self referencing but gives an enhancement in sensitivity (equivalent to a factor of 5 decrease in the energy of the input pulse in our preliminary experiment).

SPIDER is an easy to implement, self-referencing pulse characterization technique with a robust, non-iterative inversion algorithm. It provides accurate and precise measurements of a wide array of optical pulses in realistic laboratory conditions, and allows simple modification for increased sensitivity.

References

1 C. Iaconis, I. A. Walmsley: Opt. Lett. **23**, 792 (1998); C. Iaconis, I. A. Walmsley: IEEE J. Quantum Electron. **35**, 501 (1999).

2 C. Iaconis, I. A. Walmsley: *OSA Annual Conference*, (Baltimore, MD, 1998).

3 T. M. Shuman, M. E. Anderson, J. Bromage, C. Iaconis, L. Waxer, I. A. Walmsley: Optics Express **5**, 134 (1999).

4 C. Dorrer , B. de Beauvoir , C. Le Blanc , S. Ranc , J. P. Rousseau , P. Rousseau , J. P. Chambaret , F. Salin: Opt. Lett., **24**, 1644 (1999); C. Dorrer, Opt. Lett., **24**, 1532 (1999).

5 L. Gallmann, D. H. Sutter, N. Matuschek, G. Steinmeyer, U. Keller, C. Iaconis, I. A. Walmsley: Opt. Lett. **24**, 1314 (1999).

6 Christophe Dorrer, C. Le Blanc, B. Wattellier, T. Zou, C. Sauteret, A. Migus, personal communication.

Highly SimplifiedUltrashort Pulse Measurement

Patrick O'Shea[1], Mark Kimmel , Xun Gu, and Rick Trebino[2]

[1] School of Physics, Georgia Institute of Technology, Atlanta, GA 30332-0430
E-mail: pat@socrates.physics.gatech.edu
[2] E-mail: rick.trebino@physics.gatech.edu

Abstract. A *thick* nonlinear crystal can replace the usual thin crystal and spectrometer, and a Fresnel biprism can replace the beam splitter and delay line in a FROG device, yielding a remarkably simple single-shot ultrashort-pulse intensity-and-phase measurement device with no sensitive alignment parameters and significantly greater sensitivity.

1. Introduction

The measurement of ultrashort laser pulses remains an arduous task. The autocorrelator, the device of choice for decades, is complex, comprising many components. The pulse must be split into two; the two pulses must then be spatially and temporally overlapped in a carefully aligned second-harmonic-generation (SHG) crystal; and the pulse relative delay must be scanned while maintaining alignment. Alignment involves many sensitive degrees of freedom and has great potential for error. Worse, bandwidth considerations require a very thin SHG crystal, which is expensive and difficult to align, obtain, and handle, and which severely limits device sensitivity.

Frequency-Resolved Optical Gating (FROG)[1], which yields the full pulse intensity and phase, requires adding a spectrometer to an autocorrelator apparatus, further complicating an already complicated apparatus. And FROG is the simplest proven intensity-and-phase measurement device.

As a result, we have developed a remarkably simple, full-intensity-and-phase pulse-measurement device that 1) *uses no spectrometer*, 2) *uses no beam-splitter or delay line*, 3) avoids the problematic thin crystal and instead uses a more convenient, *thick* SHG crystal, 4) is considerably more sensitive, 5) involves no sensitive alignment parameters, and 6) comprises only a few simple optical components. We call this new device GRating-Eliminated No-nonsense Observation of Ultrafast Incident Laser Light E-fields (GRENOUILLE). It yields traces essentially identical to those of SHG FROG, and hence yields the full pulse intensity and phase.

2. Grating-Eliminated No-nonsense Observation of Ultrafast Incident Laser Light E-fields

FROG involves spectrally-resolving a pulse that has been time-gated by itself. GRENOUILLE (Fig. 1) combines both of these operations in a single thick SHG crystal. It uses the finite phase-matching band-width of the SHG crystal and the angular variation of the crystal's phase-matched wavelength as a spectral filter. A cylindrical lens focuses the beam tightly enough to yield a large range of crystal incidence angles, and the SHG signal is detected vs. angle (i.e., wavelength) in the vertical direction.

GRENOUILLE also involves replacing the entire beam-splitter/delay-line with a Fresnel biprism, which splits the beam into two, mapping the relative beam delay onto horizontal position at the crystal, which is imaged onto the camera, where the signal is detected vs. position (i.e., delay) in the horizontal direction. So the resulting signal at the camera will be an SHG FROG trace with delay running horizontally and wavelength running vertically. Thus, GRENOUILLE yields a trace identical to that of SHG FROG, but with considerably less effort and considerably greater sensitivity.

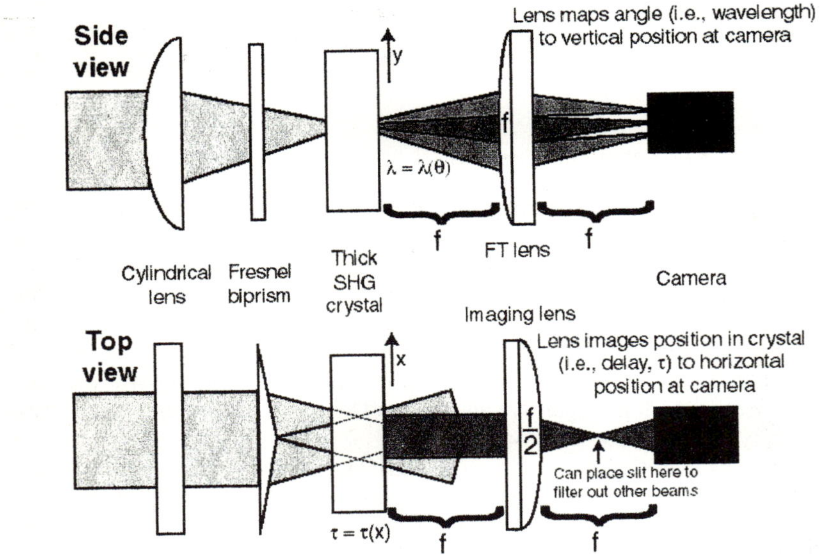

Fig. 1. GRENOUILLE: A highly simplified ultrashort-pulse measurement device. A Fresnel biprism is used to replace the beam splitter and delay line. It also utilizes a thick SHG crystal acting as *both* the nonlinear-optical element and the spectrometer. The wavelength that achieves phase-matching varies linearly with crystal incidence (and exit) angle in the vertical plane. Focusing the beam into the crystal yields a large beam divergence in the vertical direction, so each wavelength of the pulse phase-matches for some angle. A Fourier-transforming cylindrical lens with vertical power maps crystal vertical output angle (and hence wavelength) onto vertical position at the camera. Meanwhile, an imaging cylindrical lens with horizontal power maps delay onto horizontal position at the camera to achieve a full FROG trace of SHG signal pulse energy vs. delay and wavelength.

The key issue in GRENOUILLE is the thickness of the crystal. It must be thick enough that group-velocity *mismatch* (GVM) between the pulse wavelength and its second harmonic cause it to act as a wavelength filter with enough spectral resolution to resolve the details of the pulse spectrum: GVM L >> τ_p, where τ_p is the pulse length. But it must not be too thick, or group velocity *dispersion*

(GVD) between wavelengths present in the pulse will cause the pulse to spread in time, distorting it: GVD L $\ll \tau_c$, where τ_c is the pulse coherence time. A crystal length (L) that satisfies both conditions will exist if:

$$GVM / GVD \gg TBP$$

where TBP is the pulse time-bandwidth product ($\sim\tau_p / \tau_c$). For a near-transform-limited pulse (TBP ~ 1), this is easy because GVM $\gg$ GVD for all but near-single-cycle pulses.

3. GRENOUILLE Experiment

We first tested GRENOUILLE using a 6-mm thick BBO crystal (about 30 times too thick for standard autocorrelators or FROG) using a standard beam splitter and delay line. We used a test pulse comprising the sum of two linearly chirped (~40-fs) pulses separated by about a pulse length—an ideal test pulse for GRENOUILLE because it has fine structure in both delay and frequency. The theoretical and experimental (single-shot) SHG FROG traces are shown in Figs. 2a and b, respectively. Our experimental GRENOUILLE trace is shown in Fig. 2c. Note the very good agreement among the traces.

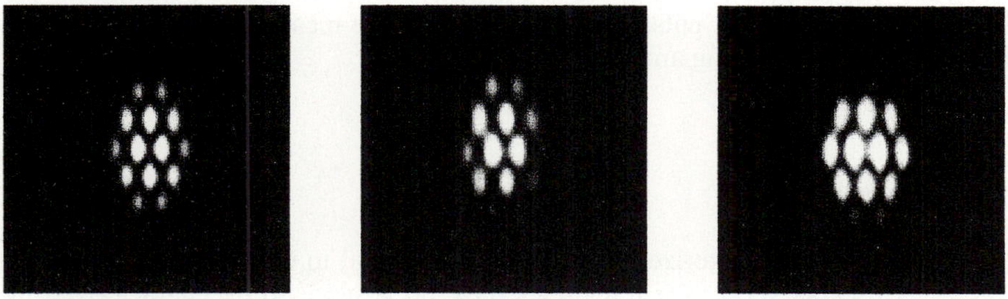

Fig. 2. a. Theoretical SHG FROG trace (intensity vs. frequency and delay) for a pulse comprising the sum of two linearly chirped Gaussian-intensity pulses separated by about one pulse length. b. Experimental single-shot SHG FROG trace of this pulse. c. Experimental GRENOUILLE trace for this pulse. In all traces, delay runs horizontally, and frequency runs vertically.

4. Conclusions

We feel that GRENOUILLE is ideal for most ultrafast laser diagnostics. Only a few simple optical elements are required and no sensitive alignment is required. It is also extremely compact and more sensitive than other pulse diagnostics, including even those that do not yield the full intensity and phase.

5. References

1 R. Trebino, K. W. DeLong, D. N. Fittinghoff, J. N. Sweetser, M. A. Krumbügel, and D. J. Kane, "Measuring Ultrashort Laser Pulses in the Time-Frequency Domain Using Frequency-Resolved Optical Gating," Review of Scientific Instruments, vol. 68, pp. 3277-3295, 1997.

Time-gated FROG: A new technique for studying the buildup of optical pulse field in mode-locked ultrafast lasers

Y.S. Hsing[1], J. M. Hsieh[2], and Ci-Ling Pan.[1]

[1] Institute of Electro-Optical Engineering National Chiao Tung University 1001 Ta Hsueh Rd. Hsinchu, Taiwan, 30010

E-mail: clpan@cc.nctu.edu.tw

[2] National Nano Devices Laboratory 1001 Ta Hsueh Rd. Hsinchu, Taiwan, 30010

E-mail: jmshieh@mail.ndl.nctu.du.tw

Abstract. The amplitude and phase of the evolving optical pulse field during the buildup of mode-locked pulses in ultrafast lasers are measured for the first time by combining time-gating and FROG techniques..

1. Introduction

The ability to characterize ultrashort pulses is vital to the advance of ultrafast lasers. Pulses as short as a few cycles have been successfully characterized by either Frequency Resolved Optical Gating (FROG) or Self-referencing spectral interferometry (SPIDER). [1,2] To the best of our knowledge, the transient state of optical pulse field (i.e., amplitude and phase) as the laser evolves from noise to steady state mode-locked pulses have not been reported. Previous studies on the buildup dynamics in ultrafast lasers mostly rely on the buildup time as determined by the buildup of the second harmonic signal of the pulse train. Further insight can be obtained by analysis of the transient autocorrelation traces. [3] Neither technique provides phase information on the optical pulse field, which should also play an important role in the buildup process. In this work, we report for the first time a new technique that allows dynamic optical pulse field (both amplitude and phase) measurement.

2. Experimental Methods

There are two main part of the experiment, one is FROG and the other is time gating system.

This is shown in Fig. 1. We use two phase-locked optical choppers to gate the laser output that was fed to a FROG apparatus. The timing diagrams of the choppers are shown in Fig. 2. The intra-cavity chopper (chopper 1) periodically blocks the laser at 500 Hz. The extra-cavity chopper (chopper 2) opens a time window of about 20 μs for the chopped beam. The relative jitter of the opening times of the two time-windows is about 1 μs. By adjusting the phase difference between the choppers, we obtain the FROG trace of the transient laser output at a certain delay time relative to the laser onset. The buildup dynamics of a self-starting passively mode-locked Ti: sapphire laser with a saturable Bragg Reflector (SBR) [4] was investigated. The average output power of the laser is 300mW. It generates ≈150 fs pulses at 82 MHz.

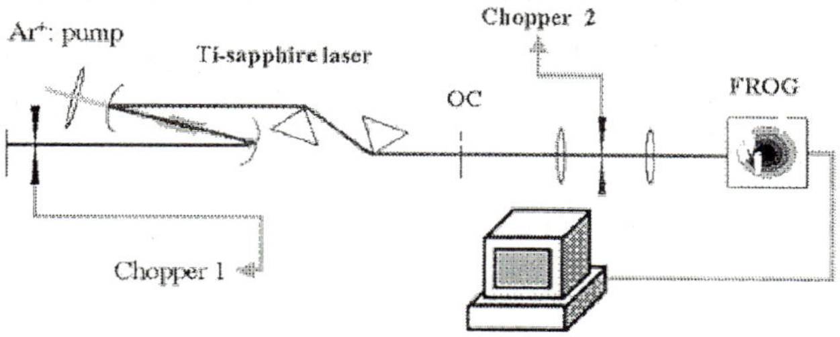

Fig.1 The experimental setup of buildup dynamics.

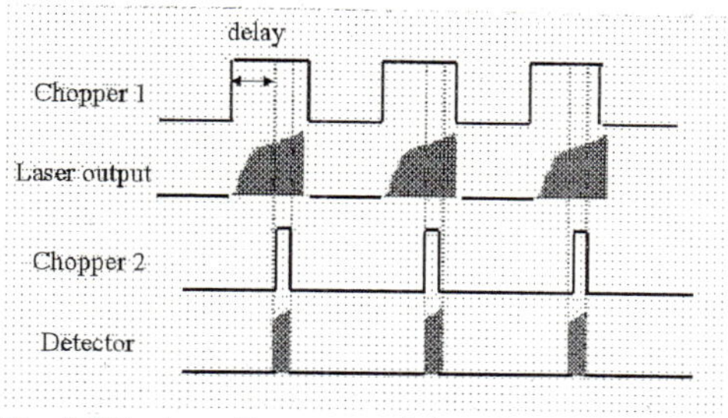

Fig.2 Timing diagram of the choppers.

3. Conclusions

The buildup dynamics of a self-starting passively mode-locked Ti: sapphire laser with a saturable Bragg Reflector (SBR) [4] was investigated. The average output power of the laser is 300mW. It generates ≈150 fs pulses at 82 MHz. In Fig. 3-5we show the typical FROG trace and retrieved intensity and phase of the laser output at the delay time 126, 187, and 841 μs with respect to the laser onset. These correspond to the initial, mid-point, and final stages of the buildup process respectively. The FWHM durations of the steady pulses (Fig. 5) are 150 fs, in agreement with direct intensity autocorrelation results. We observed highly asymmetric pulse field during the early stage of buildup (with high-order chirp). In the steady state, the pulse is almost symmetric. This is attributed to the use of the SBR.

In summary, time-gated FROG is proposed and demonstrated for the first time. Initially results on the transit temporal phase evolution during the buildup of a passively mode-locked Ti: sapphire/SBR laser are presented.

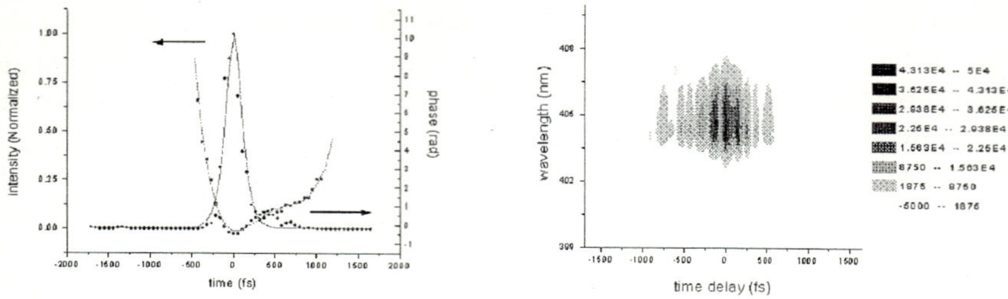

Fig.2 The FROG trace and retrieval field at 126 μs respect to the laser onset.

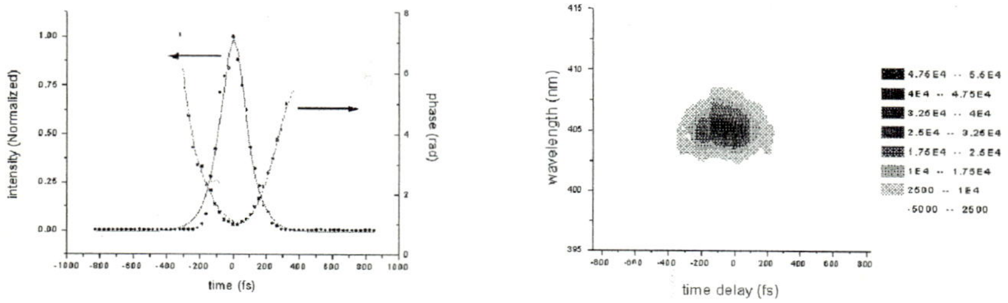

Fig.3 The FROG trace and retrieval field at 187 μs respect to the laser onset.

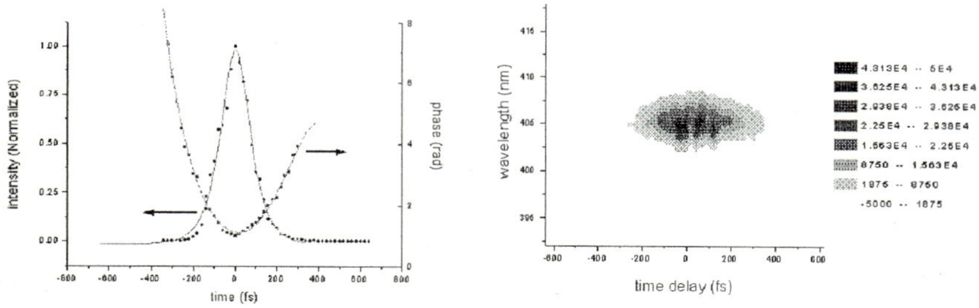

Fig.4 The FROG trace and retrieval field at 841 μs respect to the laser onset.

References

1. **K. W. Delong, D. N. Fittinghoff, and R. Trebino**, "Practical issues in ultrashort-laser-pulse measurement using Frequency-Resolved Optical Gating, "IEEE. J. Quantum. Elect **32,** pp 1253-1264 (1994)
2. **C. Iaconis, I. A. Walmsley**, "Spectral phase interferometer for direct electric-field reconstruction for ultrashort optical pulse," Opt. Lett **23**, pp 792-794(1998)
3. **J. M. Shieh, T .M. Twu, and C. L. Pan**, "The effects of intracavity dispersion on the starting dynamics of CW passively mode-locked Ti: sapphire/DDI lasers," Opt. Lett. **21**, pp 1058-1060 (1996)
4. **T. A. Liu, C. L. Pan,** "Effect of wavelength on optical nonlinearities and ultrafast carrier dynamics of strained quantum well saturable Bragg reflector," Opt. Comm'98 Taiwan.

Measuring the intensity and phase of ultrabroaband continuum

Lin Xu[1], Mark W. Kimmel[1], Patrick O'Shea[1], Rick Trebino[1], Jinendra K. Ranka[2], Robert S. Windeler[2], and Andrew J. Stentz[2]

[1] School of Physics, Georgia Institute of Technology, Atlanta, GA 30332, USA
E-mail: rick.trebino@physics.gatech.edu
[2] BellLaboratories,Lucent Technologies,700 MountainAve.,Murray Hill,NJ 07974 USA

Abstract. Dithering the nonlinear crystal in Cross-correlation FROG (XFROG) allows the measurement of the intensity and phase of ultrabroadband (up to 1200-nm wide) continuum, generated using recently developed microstructure optical fiber.

1. Introduction

It has recently become possible to generate ultrabroadband continuum from 400 to 1600 nm with a TEM_{00} spatial mode and nearly flat spectrum. This involves propagating <1-nJ, 100-fs 800-nm pulses through newly developed microstructure optical fiber, which can be nearly *dispersion-free at 800 nm*.[1] The continuum has a stable spectrum, but it has unknown residual spectral phase.[1]

To measure its spectral phase, i.e., its intensity and phase vs. time, FROG [2] and other such techniques require a nonlinear-optical medium with an immense phase-matching bandwidth. Baltuska, et al.[3] have measured 2.5-cycle pulses with spectra from 600 to 1000 nm using second-harmonic-generation (SHG) FROG with a 10-µm-thick crystal. But no known SHG medium achieves the huge bandwidth required to measure this continuum (See Fig. 1).

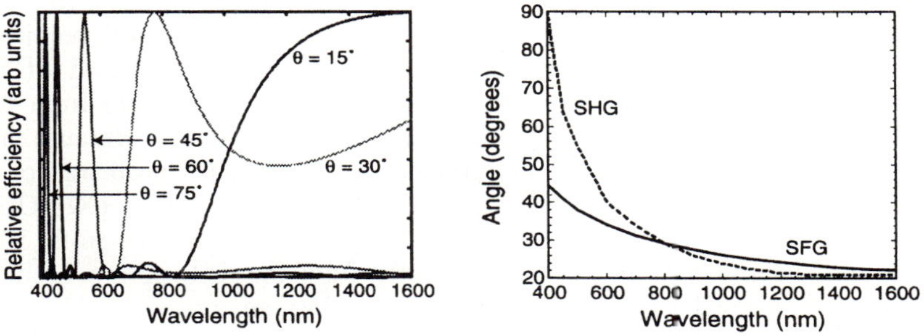

Figure 1. Left: Phase-matching efficiency vs. wavelength for 10-µm-thick BBO for angles ranging from 15° to 75° (not including effects such as nonlinearity dispersion). Note that angles ranging from ~20° to ~90° are required for the continuum's bandwidth. (KDP does not even phase-match wavelengths below ~500 nm.) Right: BBO phase-matching angle vs. wavelength for SHG and SFG with an 800-nm reference pulse.

2. Cross-correlation FROG (XFROG) using a dithered crystal

We have found that it is possible to increase the bandwidth of an autocorrelator or FROG by angle-dithering the crystal. Because the crystal phase-matching angle varies with wavelength, angle-dithering the crystal during a pulse measurement can increase the range of wavelengths that achieves phase-matching during the measurement period. This increases the effective bandwidth of the crystal.

We verified that this approach works using pulses with less bandwidth before using it for the difficult case of the ultrabroadband continuum. Here we use near-transform-limited ~30-fs, 800-nm, ~1-nJ oscillator pulses. Figure 2 shows FROG traces taken with a thin crystal with sufficient bandwidth, a thick crystal with insufficient bandwidth, and the same thick crystal but dithered over a range of about 15°. Note the agreement between the thin-crystal and the dithered-thick-crystal traces and pulse field.

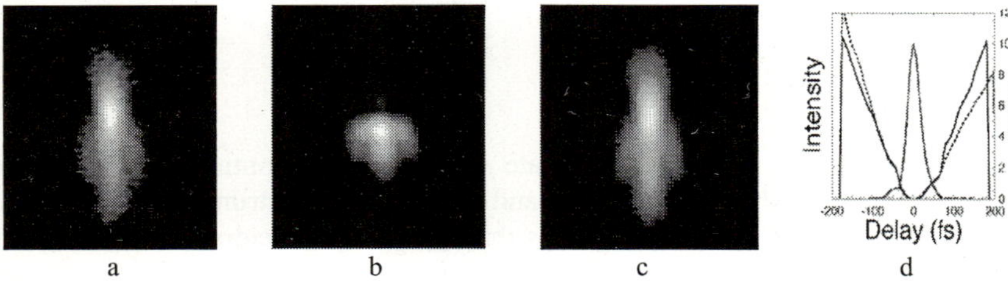

Figure 2. a. Experimental SHG FROG trace of a ~30-fs pulse using a 100-μm thick KDP crystal. b. Same, but using a too thick (1-mm) BBO crystal (note the severe trace cropping in frequency due to insufficient phase-matching bandwidth). c. Same, but with a *dithered* too thick (1-mm) BBO crystal (note that the trace is now complete and no longer cropped; and it agrees well with the trace in (a)). d. Retrieved intensities and phases vs. time for the pulse yielding the traces in (a) and (c), and which agree well.

And we have also found that it is possible to *reduce* the bandwidth required for the measurement by using sum-frequency generation (SFG) instead of SHG. Performing an SHG-based measurement of a pulse with wavelengths from 400 to 1600 nm requires generating wavelengths from 200 to 800 nm. Because crystal dispersion increases drastically near 200 nm, blue is especially problematic. This difficulty can be reduced by performing *Cross-correlation FROG* (XFROG),[4] which uses SFG with a known narrower-band gate pulse (here a ~30-fs, 50-nm-bandwidth, 800-nm-wavelength pulse directly from the Ti:Sapphire oscillator). SFG generates signal wavelengths from ~260 to ~550 nm, a considerably reduced bandwidth when measured in terms of the *signal* bandwidth, which is, in fact, the relevant quantity. Indeed, the required range of angles is much less (See Fig. 1).

We have combined the dithered-crystal technique above with the XFROG technique to measure the continuum generated by sending ~1 nJ through 16 cm of microstructure fiber. Using a 1-mm-thick BBO crystal (whose GVD-induced chirp is small compared to that of the continuum here) dithered by about 20°, we obtain the very broadband XFROG trace shown in Fig. 3. Interestingly, this trace shows

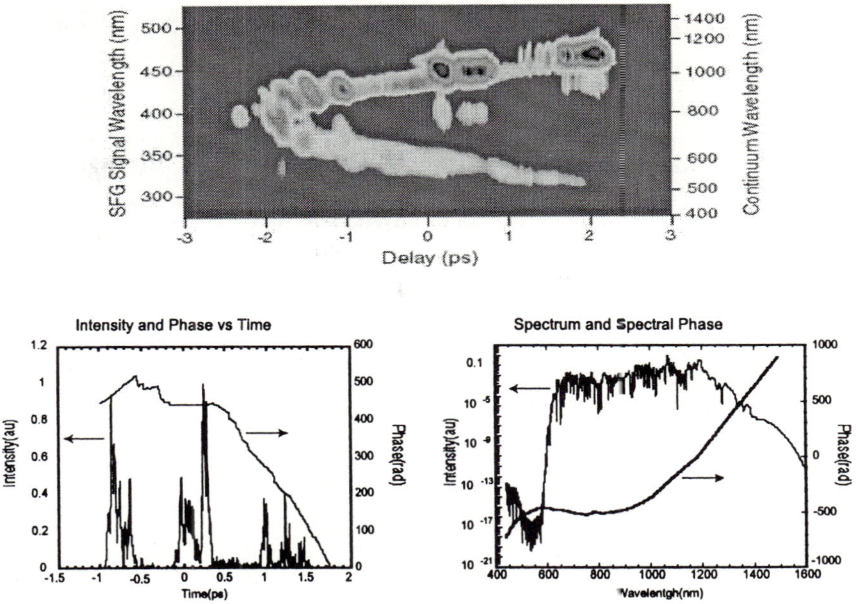

Figure 3. Top: Experimental XFROG trace of the continuum pulse, showing its parabolic delay vs. wavelength (or cubic phase vs. wavelength). Bottom: The continuum retrieved intensity and phase vs. time and frequency.

that the continuum exhibits very strong cubic spectral phase (which would not be observable using standard techniques).

3. Conclusions

We have used XFROG with an angle-dithered BBO crystal to measure the continuum generated using microstructure optical fiber and have shown that the continuum has approximately cubic spectral phase.

References

1 J. K. Ranka, R. S. Windeler, and A. J. Stentz, presented at CLEO, CPD8, 1999; J. K. Ranka, R. S. Windeler, and A. J. Stentz, Opt. Lett., **25**, 25, 2000.

2 R. Trebino, K. W. DeLong, D. N. Fittinghoff, J. N. Sweetser, M. A. Krumbügel, and D. J. Kane, Review of Scientific Instruments, **68**, 3277, 1997.

3 A. Baltuska, M. S. Pshenichnikov, and D. A. Wiersma, IEEE J. of Quantum Electron., **35**, 459, 1999.

4 S. Linden, J. Kuhl, H. Giessen, Opt. Lett., **24**, 569, 1999.

Amplitude and Phase Measurement of Mid-IR Femtosecond Pulses Using XFROG

D. T. Reid, P. Loza-Alvarez, C. T. A. Brown, T. Beddard and W. Sibbett

J. F. Allen Research Laboratories, School of Physics and Astronomy, University of St Andrews, North Haugh, St Andrews, Fife KY16 9SS, UK
E-mail: derryck@st-andrews.ac.uk

Abstract. We describe a cross-correlation-based frequency-resolved optical gating (XFROG) technique for simultaneously measuring the amplitude and phase of two ultrashort pulses with different wavelengths but derived from a common modelocked oscillator.

1. Introduction

Frequency-resolved optical gating (FROG) is a versatile technique for measuring ultrashort pulses across a wide range of wavelengths, and common approaches for CW-modelocked pulses utilise a detector array to record the spectrum of the second-harmonic generation (SHG) autocorrelation signal. In certain spectral regions (eg. near-UV, mid-IR) implementing SHG-FROG is impractical, either because no sensitive array detectors exist at the SHG wavelength or because crystal absorption and phasematching considerations are restrictive. A solution to this problem is often available by using a sum- [1, 2] or difference-frequency [3] FROG geometry to generate an optical mixing signal in a more convenient wavelength region. Previous work using a cross-correlation FROG (XFROG) geometry has relied on mixing the unknown pulse with a reference pulse which had been previously characterised by a conventional FROG geometry [4]. Here we describe a simultaneous measurement of two unknown pulses, one in the near-IR and the other in the mid-IR, with the only *a priori* knowledge of the pulses being their conventionally-measured wavelength spectra. The technique is an extension of the earlier "TREEFROG" method [1] originally used to characterise two pulses of the same wavelength but with different spectral phase. Finally, we note that other techniques have been applied for IR pulse measurement, including SHG-FROG which was used to measure 5 μm pulses from a free-electron laser [5].

2. Experimental description

Our technique simultaneously measures two pulses, one at a wavelength in the mid-infrared and generated by a PPLN-based synchronously-pumped femtosecond optical parametric oscillator (OPO) and the other the OPO pump pulse originating from a self-modelocked Ti:sapphire laser. The experimental arrangement is shown in Fig. 1.

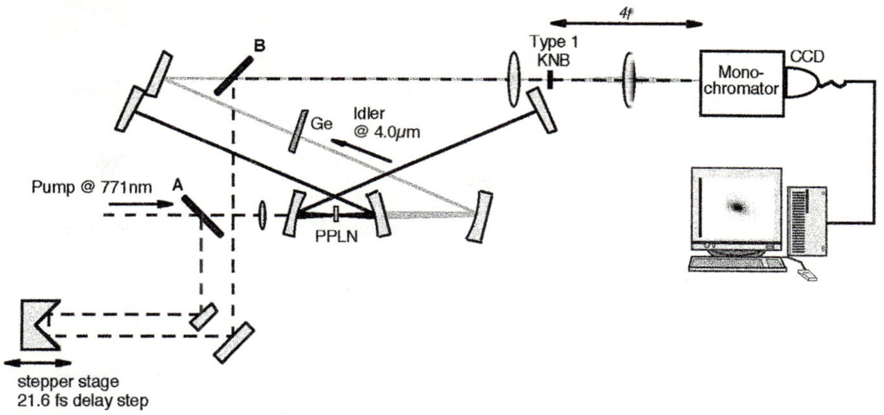

Figure 1. Experimental configuration. The pump pulse is sampled at beamsplitter A (uncoated 45° microscope slide) and, after a variable delay, is recombined with the OPO idler pulse at dichroic beamsplitter B (high reflectivity @ 800 nm / high transmission @ 5000 nm).

Retrieval of the pulse data from the measured XFROG trace is achieved using an adapted version of the principal components generalised projections algorithm [7] and by applying alternate pump / idler spectral constraints as described in Ref. [1].

3. Results and Discussion

Experimental data are shown in Fig. 2 with the measured and retrieved XFROG traces shown in Fig. 2(a) and 2(b) respectively. The contour lines are directly proportional to the XFROG signal however the grayscale shading corresponds to the square-rooted signal. Significantly, the measurement required only 1.2 mW (800 μW) of average pump (idler) power. The high detectivity is explained by the large effective nonlinear coefficient of the $KNbO_3$ crystal (d_{eff} ~ 9 pm/V) and suggests that it should be possible to measure idler pulses with only a few tens of μW average power. The results presented in Fig. 2(c, d) show that both pulses are chirped, with the signs of the chirps consistent with the normal (anomalous) dispersion experienced by the pump (idler) pulses. The chirp of the pump pulses is explained experimentally by their transmission after the laser though a Faraday optical isolator and that of the idler pulses by negative group-velocity dispersion contributed in the mid-infrared by the OPO mirror substrates and measurement focussing optics. The insets to Figs. 2(c) and 2(d) show the retrieved idler and pump spectral intensities together with the measured idler and pump spectra.

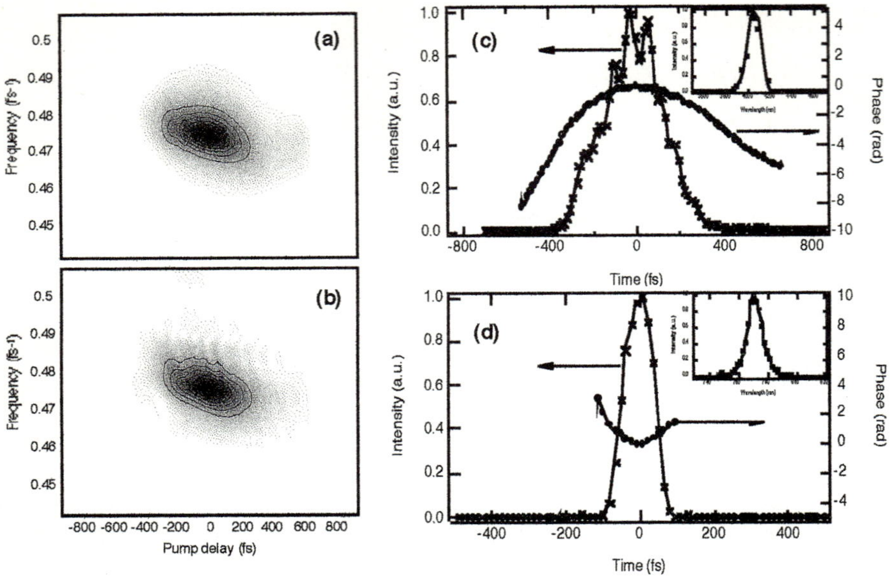

Figure 2 (a) Measured and (b) retrieved XFROG traces for a pump (idler) at 771 nm (4.0 μm); (c) retrieved idler and (d) retrieved pump pulse data.

4. Conclusions

We have demonstrated a practical, sensitive measurement technique for ultrafast mid-IR pulses which is readily extendable to other wavelengths and $\chi^{(2)}$ processes.

Acknowledgements. This work was funded by the UK EPSRC.

References

1 K.W. DeLong, R. Trebino and W.E. White, J. Opt. Soc.Am.B 12, 2463 (1995)
2 H. Giessen (Private Communication)
3 S. Linden, J. Kuhl and H. Giessen, Opt. Lett. 24, 569 (1999)
4 S. Linden, H. Giessen and J. Kuhl, Phys. Status. Solidi B - Basic Res., 206, 119 (1998)
5 B.A. Richman, M.A. Krumbugel and R. Trebino, Opt. Lett. 22, 721 (1997)
6 Spectragen Inc., 1700 South Mount Prospect Road, Des Plaines, IL 60018
7 D. J. Kane, IEEE J. Sel.Top. Quantum. Electron.4, 278 (1998)
8 D. J. Kane, G. Rodriguez, A. J. Taylor and T. S. Clement, J. Opt. Soc. Am. B. 14, 935 (1997)
9 D. T. Reid, W. Sibbett, J. M. Dudley, L. P. Barry, B. Thomsen and J. D. Harvey, Appl. Opt. 37, 8142 (1998)

Rapid retrieval of ultrashort pulse amplitude and phase from a sonogram trace

I. G. Cormack, W. Sibbett and D. T. Reid

School of Physics and Astronomy, University of St Andrews, Fife, KY16 9SS, UK
E-mail: igc2@st-andrews.ac.uk

Abstract. By measuring the sonogram of an ultrashort pulse we have demonstrated a robust and sensitive characterization method with acquisition rates as high as 0.5 Hz. Retrieved data are compared to independent experimental measurements.

1. Introduction

Methods by which ultra-fast pulses are measured and characterised have matured greatly over the past years. Techniques such as SHG-FROG [1] and SPIDER [2] have made it possible to accurately measure the amplitude and phase of an ultrashort pulse. We have developed an alternative technique that can rapidly and accurately characterize ultra-fast pulses and which has many advantages over the methodologies mentioned above.

Our technique involves recording the sonogram of the pulse [3], rather than the spectrogram as in the case of the SHG-FROG system. A sonogram is a map of group delay against frequency made using a spectral bandpass filter G(ω), whose width is comparable to that of the pulse, E(ω).

$$I_{sonogram}(t,\Omega) = \left| \int_{-\infty}^{\infty} E(\omega) G(\omega - \Omega) e^{-i\omega t} d\omega \right|^2 \tag{1}$$

Whereas the SHG-FROG technique is gated within the time domain and requires the use of a doubling crystal and spectrometer to measure the spectrum of the pulse autocorrelation, the sonogram technique is gated within the frequency domain. This major difference enables us to take full advantage of two-photon detectors such as a photodiode or LED [4] to directly measure the sonogram trace. Two photon detectors are very inexpensive and ideal for ultrashort pulse measurement due to their sensitivity, broad detection bandwidth and polarisation insensitivity. An additional advantage which the sonogram has over SHG-FROG is its ability to determine the sign of the retrieved temporal phase. This is significant because it provides exact knowledge of the direction of the pulse in time.

As in the SHG-FROG technique, once the sonogram trace is experimentally obtained it must be processed by a retrieval algorithm to deduce the pulse characteristics. Over recent years however, the speed and robustness of retrieval algorithms have improved greatly [5] and combining this with the steadily increasing speed of computers, a point is reached where the time taken to retrieve a trace is insignificant compared to the time it takes to acquire the trace experimentally. Thus we have developed a rapid method for experimentally acquiring sonograms. Using our fast retrieval algorithm (>100 iterations per second on a Pentium II 450Mhz personal computer) we can currently demonstrate acquisition rates as high as 0.5Hz.

2. Experimental Methods

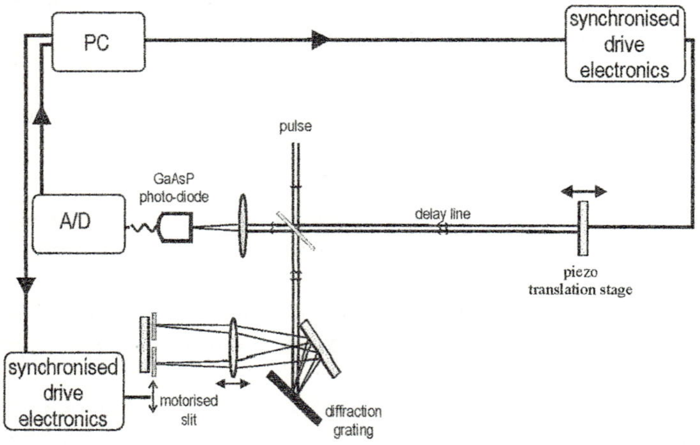

Fig. 1. The experimental setup used to acquire the sonograms

The sonogram is acquired (Figure 1) by equally splitting the pulse in a Michelson interferometer. One interferometer arm varies the pulse delay by means of a piezo translation stage, while the other consists of a frequency filter. Using white light interferometry [6], the dispersion within the filter has been accurately characterised enabling precise positioning of the lens to give a near dispersionless filter. The pulses from the delay arm and the frequency filtered arm are then recombined on a GaAsP photodiode which acts as a two-photon detector and a cross-correlation signal is obtained. The sonogram is constructed from many cross-correlation profiles, each recorded with a different frequency component (ie. different slit position).

Precise control of the slit and piezo translation stage, as well as the acquisition of the two-photon signal, is achieved using a combination of an analogue-digital / digital-analogue expansion card within a personal computer and a specially designed digital signal generator. This device uses the clock signal from the expansion card to produce two different waveforms. The first waveform controls the piezo stage and consists of a digitally constructed sine-wave with a period equal to 1024 clock signals. The second waveform controls the movement of the slit by a trigger output pulse once every sine-wave period. Having precise control over the movement of the slit and the piezo stage allows one sonogram to be acquired every two seconds. The data acquisition by the computer continues in the background allowing the calibration and retrieval of the previously acquired sonogram to be carried out, the result of which is displayed on the monitor at a refresh frequency of 0.5Hz.

3. Results and Discussion

To emphasize how robust this pulse characterization setup is, retrieved data was compared to that obtained experimentally. Typical retrieved data show excellent agreement with the independently measured pulse spectrum (Figure 2a) and intensity autocorrelation (Figure 2b).

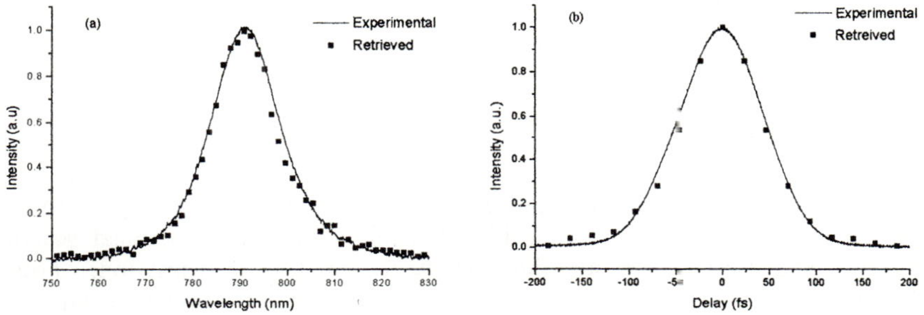

Fig. 2. Comparison of experimental and retrieved (a) spectrum, (b) intensity autocorrelation

4. Conclusions

The potential for rapid and sensitive acquisition, combined with robust, inexpensive and flexible implementation makes the sonogram technique an ideal candidate for a real time, user friendly pulse characterization technique.

References

1 K W. DeLong, R. Trebino, J. Hunter, W. E. White, J. Opt. Soc. Am, B **11**, 2206-2215 (1994)
2 C. Iaconis, I. A. Walmsley, Opt. Lett, **23**, 792-794 (1998)
3 V. Wong, I. A. Walmsley, J. Opt. Soc. Am. B, 14, 944-949 (1997)
4 D. T. Reid, M. Padgett, C. McGowan, W. E. Sleat, W. Sibbett, Opt. Lett, **22**, 233-235 (1997)
5 D. T. Reid, IEEE J. Quantum Elecron, **35**, 1584-1589 (1999)
6 K. Naganuma, K. Mogi and H. Yamada, Opt. Lett. **9**, 393-395 (1990)

Simultaneous two-dimensional Space and Time Measurement of Ultrashort Optical Pulses based on Spatial-Spectral Interferometry.

Takasumi Tanabe, Nobutastu Takei, Hiroshi Tanabe, and Fumihiko Kannari

Department of Electronics and Electrical Engineering, Keio University, 3-1-1 Hiyoshi Kohoku-ku Yokohama, 223-8522, JAPAN
E-mail: takasumi@kinet.or.jp

Abstract. We proposed a diagnostics to measure temporal and spatial profiles of ultrashort optical pulses based on spatial-spectral interferometry. With this measurement technique we directly observed space-time coupling at a Fourier pulse shaper.

1. Introduction

Arbitrary shaping of ultrashort pulses is a powerful tool for applications of ultrashort laser pulses such as ultrafast sensing, controlling chemical reaction, coherent control of matters or high-density information transmission. Fourier pulse shapers consisting of a pair of grating and lens in a 4-f optical arrangement have been widely used to meet such requirement. The complex amplitude of an input spectrum is modulated with Fourier masks, which are generally formed with a programmable spatial light modulator (SLM). This pulse shaping technique can supply almost arbitrary pulse shapes in both amplitude and phase within the resolution defined by the input pulsewidth and up to the pulsewidth defined mainly by the Fourier mask resolution. However, the temporally shaped pulses always exhibit unwanted spatial shift leading to a poor spatial profile. Furthermore, the shaped waveform shows a different spatial frequency at different time. This is the time-space coupling caused at 4-f Fourier shaping arrangement.

Since the wave front of input pulse is always tilted after the diffractive grating at the input side of the 4-f shaper, any newly added spatial frequency components at the Fourier plane by the shaping mask also causes shaping in the spatial beam profile at the output [1].

So far, several measurement techniques of ultrashort pulses were developed. Among those methods, Frequency Resolved Optical Gating (FROG)[2] or Spectral Phase Interferometry for Direct Electric-field Reconstruction (SPIDER)[3] have been frequently used for reconstructing complex amplitude of pulses. FROG uses iterative calculations from a two-dimensional spectrogram, whereas SPIDER can obtain both intensity and phase distribution without any iterative calculations from only a one-dimensional fringe image. That leads to a quick reconstruction of waveforms. Both methods use either second or third order nonlinearlity without known reference pulses. On the other hand, Spatial-Spectral interferometry (SSI)[4] reconstructs both intensity and phase distribution without using any nonlinear devices when a known reference pulse is given. To characterize the temporal phase from a spectral fringe pattern, the Fourier-transform method of fringe-pattern analysis [5] is used. Though these pulse measurement techniques show a remarkable progress in their resolution, accuracy, and reconstruction speed, they always assume a uniform spatial frequency for pulses measured.

In this paper we propose a two-dimensional measurement; time and one space dimension, based on SSI to directly measure the space-time distribution of the shaped pulses whose spatial frequency varies in time.

2. Space-time coupling of a pulse shaper

The temporal and spatial distribution of a shaped pulse is given by the equation:

$$e_{\text{out}}(x,t)=\frac{1}{\lambda f\sqrt{2\pi}}\exp(i\varpi t)\cdot\left[\iint\tilde{e}_{in}(-x',\Omega)M\left(\frac{2\pi(x-x')}{\alpha\lambda f}\right)\exp\left\{i\Omega\frac{t+\beta(x-x')}{\alpha}\right\}d\Omega dx'\right] \tag{1}$$

where the time and space coupling is determined by the constant α/β, which is given by

$$\frac{\beta}{\alpha}=\frac{\lambda}{cd\cos\theta_i}. \tag{2}$$

Here, c is the velocity of light, d is the groove interval of the grating and λ is the wavelength and θ_i is the incident angle of the beam to the grating.

In order to discuss the performance of the space and time measurement technique based on the SSI, we compared the experimental results with numerical calculations of shaped pulses. Fig. 1 (a) shows the calculated pulse when alternate phase mask with 0 and π phase shifts is given. The pixel number of the mask is 128, and the spectral width of the input pulse is 45 nm (FWHM). The spatial pattern linearly shifts from the center at the wings in the time.

3. Two dimensional Spatial-Spectral Interferometry and Experimental results

Experimental setup consisting of a pulse shaper and an SSI is shown in Fig.2. A couple of mirror is placed to rotate the beam profile by 90 degrees, since the space-time coupling effect appears only in the identical direction of the dispersion of the gratings.

In SSI, both signal and reference beams are horizontally dispersed by the grating and interfere on the CCD plane. Also, the reference beam is slightly tilted, instead of setting a time delay between the signal and reference pulses, so that the fringe pattern exhibits also spatial information on the vertical axis in addition to the frequency information on the horizontal axis. By applying the Fourier-transform method for fringe-pattern analysis along the vertical axis, both amplitude and phase distributions of spatial domain are obtained. By repeating this process to each frequency (horizontal axis) component, the pulse is characterized in both time and space.

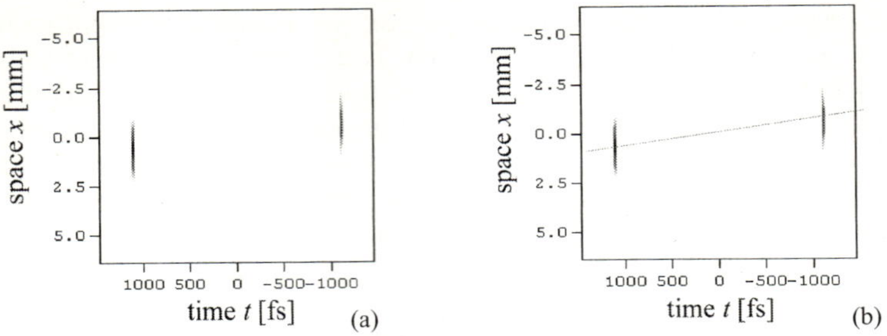

Fig. 1. (a) Calculated space-time distribution of shaped pulse. The calculated profile tilt is β/α=5.678×10^{-4} fs/mm. (b) Experimental result. The profile tilt agrees well with the calculation (shown in the figure with a solid line).

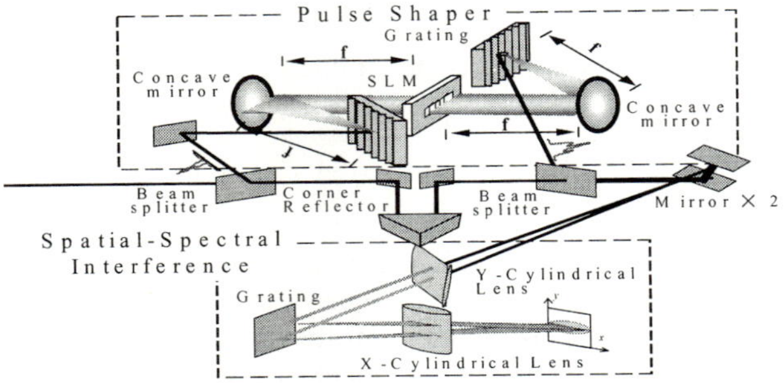

Fig. 2. Experimental setup. The spatial profile of the shaped waveform is rotated by 90 degrees with a pair of mirror.

The experimental result is shown in Fig.1 (b). The space-time distribution shows an excellent agreement with the calculation. Time resolution is determined by the observed spectral bandwidth as Δt=27 fs.

4. Conclusions

Using a two-dimensional SSI, we obtained simultaneously two-dimensional ultrashort pulse profiles in space and time. The shaped ultrafast pulse from a 4-f pulse shaper was measured, and the space-time coupling was experimentally demonstrated. The profile tilt shows a good agreement with the calculation.

This two-dimensional profile measurement is very useful to observe both amplitude and phase variations in the space together with the temporal profile for optical beams propagated some optical nonlinear media.

References

1 M. M. Wefers and K. A. Nelson, *IEEE J. Quantum Electron.*, Vol. 32, 161, 1996.
2 D. J. Kane and R. Trebino, *IEEE J. Quantum Electron.*, Vol. 29, 571, 1993.
3 C. Iaconis and I. A. Walmsley, *IEEE J. Quantum Electron.*, Vol. 35, 501, 1999.
4 D. Meshulach, D. Yelin, and Y. Silberg, *J. Opt. Soc. Am. B*, Vol. 14, 2095, 1997.
5 M. Takeda, H. Ina, and S. Kobayashi, *J. Opt. Soc. Am.*, Vol. 72, 156, 1982.

Reliability of Fourier-transform spectral interferometry

Nadia Belabas[1], Christophe Dorrer[2], Jean-Pierre Likforman[1] and Manuel Joffre[1]

[1] Laboratoire d'Optique Appliquée, ENSTA – Ecole Polytechnique, 91761 Palaiseau Cedex, France
E-mail : manuel.joffre@polytechnique.fr
[2] The Institute of Optics, University of Rochester, Rochester, New-York 14627, USA
E-mail: christophe.dorrer@polytechnique.org

Abstract. We demonstrate the incidence of several issues detrimental to the reliability of spectral interferometry: frequency calibration, frequency interpolation, spectral resolution and aliasing. Taking these issues into account results in a more accurate retrieved spectral phase.

1. Introduction

Fourier-transform spectral interferometry (FTSI) is a widely used technique allowing to retrieve the difference in spectral phase between two ultrashort pulses [1,2]. It relies on the measurement of the spectrum resulting from the interference between two ultrashort pulses separated by a fixed time delay τ, $E_1(t)$ and $E_2(t-\tau)$. The numerical treatment of the data involves an inverse Fourier transform of the interference spectrum into time domain, followed by a truncation in order to extract the correlation product and finally a Fourier transform back into frequency domain [1]. The complex quantity $E_1^*(\omega)E_2(\omega)\exp(i\omega\tau)$ is thus obtained, whose argument yields $\Delta\varphi(\omega)+\omega\tau$. The difference in spectral phase between the two pulses, $\Delta\varphi(\omega)$, can then be obtained simply by subtracting off the linear phase $\omega\tau$ resulting from the time delay between the two pulses.

2. Frequency calibration of the spectrometer

One of the most simple steps of FTSI, namely subtracting $\omega\tau$ from the calculated phase, turns out to be particularly demanding on the accuracy of the spectrometer calibration [3]. Indeed, for a value of the time delay of a few picoseconds, $\omega\tau$ takes values much greater than $\Delta\varphi(\omega)$ and even a small error in the determination of ω will result in large errors in the retrieved phase. This is illustrated in Fig.1 which shows the spectral phase retrieved for different values of the time delay between the two pulses E_1 and E_2. Although these curves should all have been identical to the phase difference $\Delta\varphi(\omega)$ between the two pulses, they are observed to differ dramatically due to the calibration error. Note that the incidence of the calibration error increases with increasing values of the time delay, so that the spectral phase obtained for $\tau = 1$ ps is rather close to the correct value (see Fig. 2(b)). The data shown in Fig. 1 actually provide a simple technique to measure the calibration error [3]. Once the frequency axis was properly calibrated using this technique, we found a spectral phase independent of the time delay used for the measurement.

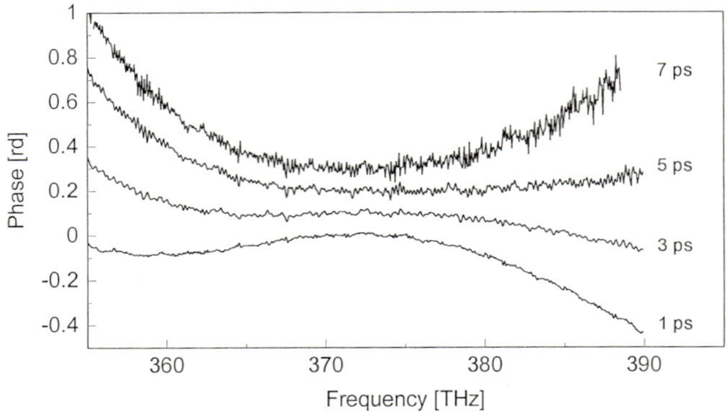

Fig. 1. Spectral phase difference between two pulses retrieved using spectral interferometry for different values of the time delay between the two pulses. The main term consisting of a phase varying linearly with frequency has been subtracted off. The curves have been shifted vertically for sake of clarity.

3. Effect of the uneven spacing in frequency domain

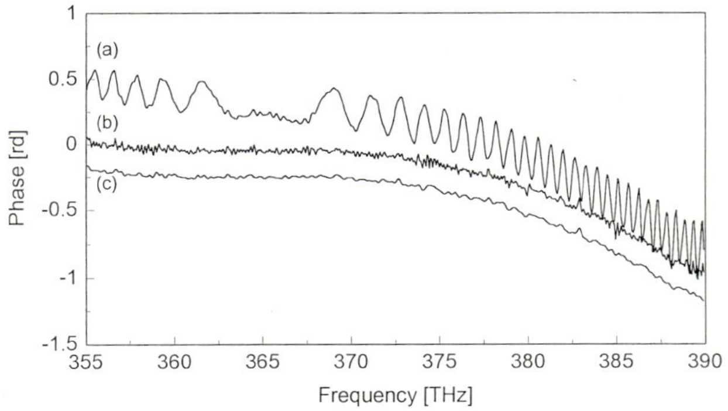

Fig. 2. Spectral phase retrieved using spectral interferometry after a proper calibration of the spectrometer. The different curves correspond to the various techniques described in the text. The time delay was fixed to $\tau = 5$ ps.

The first step of FTSI involves Fourier transforming the spectral data into time domain. However, experimental data are usually available as an array of points regularly spaced in wavelength domain. In contrast, Fast Fourier Transform (FFT) algorithms require data points regularly spaced in frequency domain. One simple approach is to interpolate the experimental data so as to numerically generate the required array of points regularly spaced in frequency domain. Fig. 2(a) shows the spectral phase obtained when using such an interpolation scheme. A significant oscillatory component is observed, which can be attributed directly to the interpolation error (more important for large values of the time delay). Therefore, other techniques should be preferred, such as using a FFT directly into the Fourier conjugate of the experimental data (b), or the zero-padding technique (c).

4. Spectrometer resolution

The finite spectral resolution of the spectrometer results in a simple multiplication in the reciprocal domain. This will result in a faster apparent decay of the measured field in time domain, particularly detrimental when FTSI is used for analyzing photon-echo emissions. However, after proper characterization of the resolution function in reciprocal domain, the spectral resolution can be corrected, thus increasing the useful time range of spectral interferometry and improving the reliability of the data obtained in time domain.

5. Conclusion

FTSI is subject to several experimental limitations which should be carefully taken into account in order to determine the reliability of the retrieved spectral phase [4]. We propose here an improved procedure taking into account the issues raised above : The frequency calibration must first be performed with great accuracy and corrected. The data are then directly Fourier transformed using a FFT algorithm and divided by the spectrometer response. The relevant Fourier components are kept, so that the complex data can be obtained after a Fourier transform back into wavelength domain. Only then can the phase be interpolated in frequency domain, so that a last Fourier transform can provide the proper data in time domain when needed.

References

1 L. Lepetit, G. Chériaux and M. Joffre, J. Opt. Soc. Am. B **12**, 2467 (1995).
2 D. N. Fittinghoff, J.-L. Bowie, J. N. Sweetser, R. T. Jennings, M. A. Krumbugel, K. W. DeLong, R. Trebino and I. A. Walmsley, Opt. Lett. **21**, 884 (1996); Opt. Lett. **21**, 1313 (1996).
3 C. Dorrer, J. Opt. Soc. Am. B **16**, 1160 (1999).
4 C. Dorrer, N. Belabas, J.-P. Likforman, M. Joffre, *Spectral resolution and sampling issues in Fourier-transform spectral interferometry*, J. Opt. Soc. Am. B (2000).

Direct Measurement of Spectral Phase of Femtosecond Pulses using Optical Parametric Effect

Kazunori Naganuma and Shigeo Ishibashi

NTT Photonics Laboratories, 3-1 Morinosato-wakamiya, Atsugi 243-0198, Japan
E-mail: ngnm@aecl.ntt.co.jp

Abstract. We present a new class of spectral interferometry that uses the optical parametric effect for the noniterative complete characterization of optical pulses. Due to the inherent broadband character of the effect, a frequency-folding interferometer can measure femtosecond pulses without relying on conventionally used ultra-thin crystals, thus enabling more accurate evaluation of weaker pulses.

Introduction. Direct measurement of the spectral phase is needed to avoid ambiguity about the numerical convergence in iterative pulse-reconstruction calculations [1-3]. With this in mind, the frequency-shearing interferometer was proposed, where a spectrum is superimposed on a frequency-shifted replica of itself. For femtosecond pulses, the scheme called SPIDER [4] using dual sum-frequency generations (SFGs) has been devised and applied to sub-6-fs pulses [5].

Here we propose a different class of spectral interferometry called a frequency-folding interferometer (FFI), in which a pulse spectrum is superimposed on its frequency-reversed replica. While the replica can be generated by four-wave mixing or optical parametric effect, the FFI using the latter has a broadband character that makes it especially suited to measurements in femtosecond regime.

Theory of FFI using Optical Parametric Effect. We consider an optical parametric crystal through which an optical signal wave propagates along with a narrow-band strong wave having a frequency ν_3 nearly twice that of the signal (second-harmonic pump wave). Coupling occurs between any pair of frequency components (ν_1, ν_2) in the signal spectrum satisfying $\nu_1 + \nu_2 = \nu_3$. Solving the standard three-wave coupled equation neglecting pump depletion results in the following output spectrum expressed in photon flux, $s_i = I_i/(h\nu_i)$ $(i = 1,2)$:

$$s_1(L) = (1+m^2)s_1 + m^2 s_2 + 2m\sqrt{1+m^2}\sqrt{s_1 s_2}\,\sin(\phi_1 + \phi_2 - \phi_3 + \delta)\,. \tag{1}$$

When parametric gain g_0L is small, mixing coefficient m and phase bias δ are approximated by $(g_0L/2)\mathrm{sinc}(\Delta kL/2)$ and $\Delta kL/2$, respectively, which show the effect of the phase-matching bandwidth of the crystal. The measurement bandwidth, defined as an interval between the zeros of m around the degenerate frequency $\nu_3/2$, is calculated for type I phase matching as $\Delta\nu = (\pi DL/2)^{-1/2}$. Such a dispersion ($D$)-governed bandwidth is a remarkable contrast to those of conventional SHG and SFG which are governed by the group delay mismatch [6].

Consequently the bandwidth is much larger, which alleviates the necessity of using ultra-thin (< 100 μm) crystals even in the femtosecond regime.

The spectrum expressed by Eq. (1) carries interference phase of a form $\phi_1 + \phi_2 - \phi_3$. We denote this as a symmetrized phase because it is always symmetrical in terms of the degenerate frequency $\nu_3/2$. Such a symmetrized form originates from the nature that the spectrally-reversed replica participating in the interference is phase-conjugated at the same time.

The symmetrized phase is only sufficient in deducing spectral phases that are symmetrical around $\nu_3/2$. To be general, we need another symmetrized phase taken for a slightly different pump frequency $\nu_3'/2$. On a frequency mesh with the step size of $|\nu_3 - \nu_3'|/2$, the spectral phase $\{p_k\}$ can be noniteratively calculated from a couple of symmetrized phases $\{\sigma_k\}$ and $\{\sigma_k'\}$:

$$\begin{aligned} p_n &= \sigma_n/2, & p_{n'} &= \sigma'_{n'}/2 \\ p_{n-k} &= \sigma_{n-k} - p_{n'+(k-1)}, & p_{n'+k} &= \sigma'_{n'+k} - p_{n-(k-1)} \end{aligned} \qquad (2)$$

where σ_n and $\sigma'_{n'}$ are values at the respective degenerate frequency, and $|n - n'| = 1$ holds by definition.

Experimental Setup and Measurement Results. We built the preliminary FFI apparatus shown in Fig. 1 to test the proposed method. The coherent ($\phi_3 = \text{const.}$) SH pump wave is obtained by the SHG of the signal itself. For this, we used a long (8-mm) periodically poled $LiNbO_3$ (PPLN) crystal, with which we intentionally exploited the limited SHG bandwidth to attain a narrow-band (0.7-nm FWHM from 1.5-μm fundamental) pump wave. To maintain a relative phase between the pump (ν_3) and signal (ν) waves, the apparatus was equipped with the standard interferometer stabilization scheme using a He-Ne laser. The waves were collinearly focused onto an optical parametric crystal, which was identical to the other PPLN except that it was 2 mm long to retain enough bandwidth (57.2 THz).

We can adjust the pump phase ϕ_3 with a phase controller (PC: actually a wedged glass plate). By varying the glass thickness, ϕ_3 changes because of the refractive index difference between the ν_3 and $\nu_3/2$ waves. For each pump frequency, we recorded three spectra, s_a, s_b and s_c in which ϕ_3 was changed by 90° by using the PC. From the set of quadrature data, the interference phase is calculated as $\tan^{-1}((s_a - s_c)/(2s_b - s_a - s_c))$ in a way completely independent of mixing coefficient m.

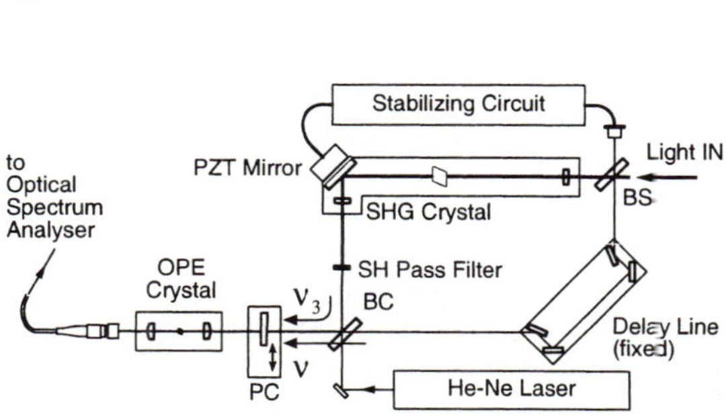

Fig. 1. Setup of frequency-folding interferometer based on quadrature phase detection

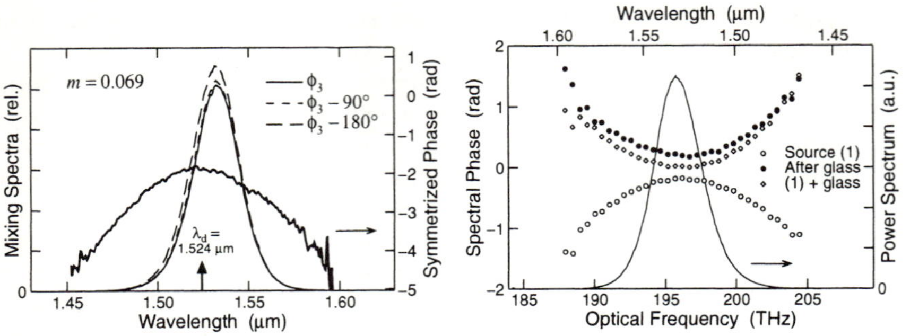

Fig. 2. Measured symmetrized phase (left) and spectral phase (right) for a femtosecond Cr:YAG laser

Figure 2 shows the measured results for pulses from a femtosecond Cr:YAG laser (50-mW average power and 180-MHz repetition). As shown in the quadrature spectra data (left figure), an interference modulation as small as several percent was sufficient for the phase detection down to far spectral wings. Besides the data shown for 1.524-µm λ_d, another symmetrized phase for 1.528-µm λ_d was measured. The degenerate wavelength λ_d was altered by changing the temperature of SHG crystal to change the pump wavelength.

The spectral phase (open circles in the right figure) calculated from these symmetrized phases clearly shows that the laser pulses were slightly down-chirped. We subsequently conducted a similar measurement on pulses after passing them through an SF glass block (filled circles). In this case the chirp was reversed. This spectral phase agrees well with that (diamonds) predicted from the original phase and from the measured dispersion data of the glass, demonstrating the validity and accuracy of these measurements.

Conclusion. We proposed and demonstrated the frequency-folding interferometer (FFI) as a new type of spectral interferometry aimed at the direct and precise detection of the spectral phase. Due to its broadband nature, it potentially eases the tradeoff between time resolution and sensitivity and facilitates the detailed characterization of the low-power signals used in telecommunication.

References

1 K. Naganuma, K. Mogi and H. Yamada, IEEE J. Quantum Electron. **25**, 1225, 1989.
2 J. Paye, IEEE J. Quantum Electron. **30**, 2693, 1994.
3 R. Trebino, K. W. DeLong, D. N. Fittinghoff, J. N. Sweetser, M. A. Krumbügel and B. A. Richman, Rev. Sci. Instrum. **68**, 3277, 1997.
4 C. Iaconis and I. A. Walmsley, Opt. Lett. **23**, 792, 1998.
5 L. Gallmann, D. H. Sutter, N. Matuschek, G. Steinmeyer, U. Keller, C. Iaconis and I. A. Walmsley, Opt. Lett. **24**, 1314, 1999.
6 A. M. Weiner, IEEE J. Quantum Electron. **19**, 1276, 1983.

Ultrashort Pulse Characterization by Frequency-Resolved Pump-Probe

Maxim S. Pshenichnikov, Andrius Baltuška, Foppe de Haan, and Douwe A. Wiersma

Ultrafast Laser and Spectroscopy Laboratory, University of Groningen, Nijenborgh 4, 9747 AG Groningen, The Netherlands
E-mail: M.S.Pshenichnikov@chem.rug.nl

Abstract. We present a novel method for measuring amplitude and phase of ultrashort optical pulses. The technique makes use of frequency-resolved pump-probe geometry that is widely used in femtochemistry and femtobiology. The experimental realization of the method is demonstrated.

In recent years, frequency-resolved optical gating (FROG) technique has been widely applied for ultrashort pulse characterization [1]. A number of outstanding features such as experimental simplicity, uniqueness of the retrieved amplitude and phase, and independent data checks make FROG an invaluable tool in ultrafast spectroscopy. In general, the FROG technique is based on homodyne detection that calls for appreciable pulse intensity. This requirement may not be met in many spectroscopic applications, where the high transition dipole moment and thermal load on the sample impose substantial limitations on the pulse energy. It has been shown [2,3] that optically heterodyne-detected (OHD) FROG has a number of advantages over homodyne-based FROG schemes.

In this Contribution we introduce a new member to the FROG family: frequency-resolved pump-probe (FRPP). Unlike OHD FROG [2,3], which requires additional polarizers and measurements of both in-phase and in-quadrature signals FRPP directly employs pump–probe geometry widely used in ultrafast spectroscopic experiments. For the vast majority of experiments in the liquid phase, which involve the use of transparent solvents with fast electronic response, FRPP pulse characterization can be conducted on a bare solvent prior to introducing the sample molecules into it. Noteworthy, no other pulse characterization technique provides such an implicit adequacy of "what-you-measure-is-what-you-use". We also demonstrate the application of the novel FRPP technique to the output of a Ti:sapphire laser.

The schematic of the set-up is that of a conventional FRPP experiment. Two replicas of the input pulse, delayed by the time τ, are crossed in a medium with instantaneous 3rd-order nonlinearity. Behind the medium, one of the beams is detected by an OMA that registers the FRPP signal related to both pulses (e.g., employing a lock-in or synchronous detection) [4]:

$$S_{FRPP}(\Omega,\tau) \propto -\mathrm{Re}\left[E^{*}(\Omega) \int \left| E(t+\tau) \right|^{2} E(t) e^{i\Omega t} dt \right] \quad (1)$$

where $E(t)$ and $E(\Omega)$ denote the pulse electric field and spectrum, respectively.

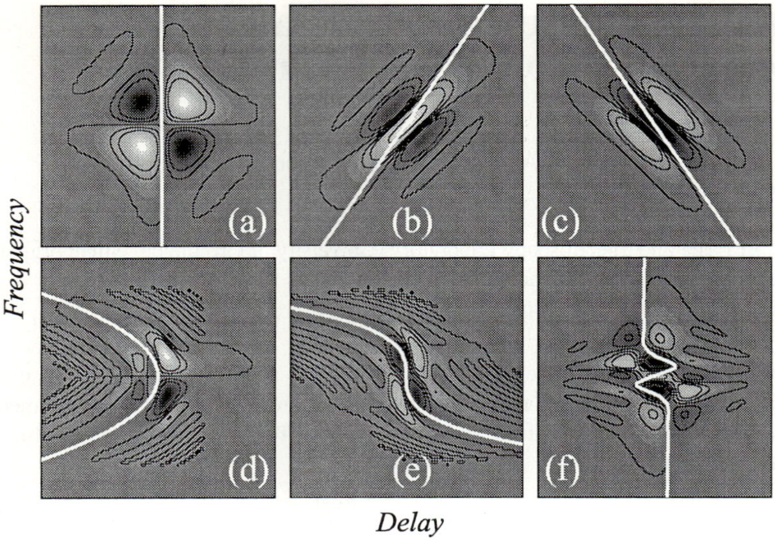

Fig.1. Examples of FRPP traces: spectrum-limited pulse (a), quadratic positive (b) and negative (c); cubic negative (d); quartic negative (e) spectral phases and self-phase modulation case (f). White curves, drawn along the respective group delays, demonstrate the intuitiveness of HD FROG technique. The amplitude of the traces changes from negative (white color) to positive (black color) values.

Figure 1 presents several examples of FRPP traces of different pulses. As it is usually the case for the $\chi^{(3)}$-based techniques, the traces, in general, are not symmetric around $\tau = 0$ and therefore the direction of time flow is unambiguously determined. Furthermore, the FRPP traces are highly intuitive: a curve drawn along the crests and the valleys follows the group delay, i.e. the first derivative of the spectral phase (white curves in Fig.1). In particular, the trace corresponding to a spectrum-limited pulse, consists of four distinct quadrants with well-defined boundaries (Fig.1a).

An essential and powerful feature of the FROG technique is the ability to check experimental data via the so-called marginals. FRPP is no exception in this respect. The peculiarity of FRPP lies in the fact that both temporal and spectral marginals are equal to zero [3]. The time marginal indicates whether spectral filtering has occurred in the process of recording the FRPP trace, while the frequency marginal is an extremely sensitive tool to verify the instantaneous nature of the nonlinearity.

The signal enhancement in the FRPP technique due to heterodyning allows characterization of low-intensity pulses using the 3rd-order nonlinearity and, as a consequence, the perfect phase matching conditions [5]. Therefore, the demand for extremely thin nonlinear medium, as is the case of the second-harmonic generation FROG, is substantially relaxed in FRPP. The proposed technique essentially requires the short nonlinear length $L_{NL} = c/\omega n_2 I(t)$, which has several important advantages (e.g., negligible material dispersion, no requirements on the intensity distribution, retained intuitiveness) over other methods [5,6].

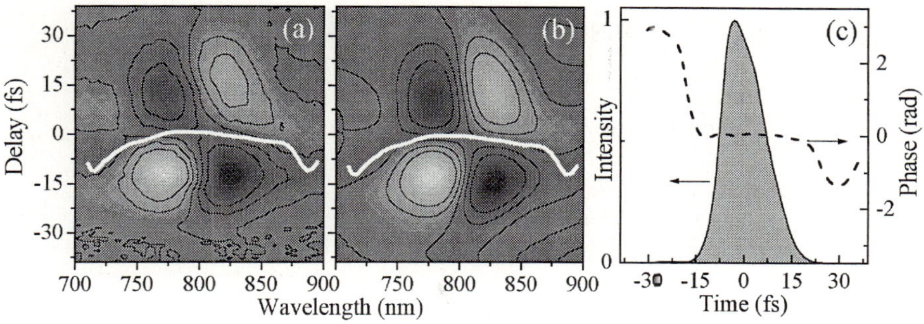

Fig.2. Experimental (a) and reconstructed (b) FRPP traces for a pulse the intensity and phase of which are shown in (c). The gray scale in (a), (b) is identical to that in Fig.2. White curves represent the recovered group delay.

The retrieval algorithm is based on obtaining the best match for the measured and computed FRPP matrices. The *rms* difference between the two traces is minimized by a standard nonlinear least-squares fit algorithm. To accelerate the algorithm convergence, we represent the pulse amplitude and phase by a cubic spline drawn through few equally-spaced points (nodes). The algorithm is initiated with a modest amount of nodes (typically 7-11) that have random values. Automatic incrementing of the amount of the nodes used to avoid algorithm stagnation due to coarse sampling also provides a self-adjustment to any pulse complexity.

Figure 2 shows the experimental and reconstructed FRPP traces obtained from a 100-μm water sample. The temporal intensity and phase of the pulses used in the experiments are given in Fig.2c. Since the overall water response is extremely fast and dominated by electronic hyper-polarizability [7], it is well justified to treat the nonlinearity as instantaneous. As can be judged from Fig.3, the reconstructed FRPP trace reproduces fairly well the essential features of the experimental pattern.

References

1. K.W. DeLong, R. Trebino, J. Hunter, and W.E. White, J. Opt. Soc. Am. B **11,** 2206 (1994); D.N. Fittinghoff, K.W. DeLong, R. Trebino, and C.L. Ladera, J. Opt. Soc. Am. B **12**, 1955 (1995).
2. R. Trebino, K.W. DeLong, US Patent #5,530,544.
3. J.A. Gardecki, S. Constantine, Y. Zhou, and L.D. Ziegler, JOSA B **17**, 6652 (2000).
4. M.D. Levenson and S.S. Kano, *Introduction to Nonlinear Laser Spectroscopy* (Academic Press, New York, 1988).
5. M.D. Thomson, J.M. Dudley, L.P. Barry, and J.D. Harvey, Opt. Lett. **23**, 1582 (1998).
6. M.A. Franco, H.R. Lange, J.-F. Ripoche, B.S. Prade, and A. Mysyrowicz, Opt. Comm. **140**, 331 (1997).
7. E.W. Castner, Y.J. Chang, Y.C. Chu, G.E. Walrafen, J. Chem. Phys. **102**, 653 (1994).

Attosecond Cross Correlation Technique

Armin Scrinzi, Michael Geissler, and Thomas Brabec

Institut für Photonik, Technische Universität Wien,
Gußhausstr. 27/387, A-1040 Vienna, Austria
E-mail: scrinzi@tuwien.ac.at; brabec@tuwien.ac.at

Abstract. A technique for attosecond pulse detection by cross-correlation with a strong laser pulse is presented and theoretically analyzed. The method is linear in the xuv intensity, which ensures high efficiency and applicability for wavelengths to below 10nm.

1. Introduction

The experimentally observed spectral width [1] of xuv harmonics generated by intense femtosecond laser pulses suggests that attosecond (as) pulses have already been produced. However, direct observation of pulses with as-duration has not been possible so far. Autocorrelation techniques [2] that are applied for the time measurement of femtosecond laser pulses cannot be extended to the xuv-regime because of the absence of efficient nonlinear processes. For cross-correlation [3] no suitable sub-femtosecond time scales have been identified to date. Here we propose to use as the basic time scale for cross-correlation the rapid oscillations of the electric field of an intense laser pulse. The nonlinear process which correlates the laser electric field with the xuv-as pulse is one-(xuv)photon ionization in the presence of the laser electric field. The photoionization signal depends linearly on the xuv-as pulse intensity, which provides for high efficiency and therefore, allows realization of this technique at short wavelengths with currently available experimental parameters.

2. Results and Discussion

For the description of the attosecond cross correlation (ACC) technique we assume that the as-pulse is created by high harmonic generation (HHG) with a few-cycle Ti:S laser pulse [4] in a noble gas. The few-cycle laser pulse is split into two parts, where one part is focused into a gas target for harmonic generation. The high energy end of the harmonics corresponds to a single as-pulse [4] and is selected by a suitable combination of filters or by a multi-layer mirror. The as-pulse and the second part of the laser pulse are then simultaneously focused into a singly ionized He plasma, which is generated by an intense circularly polarized laser prepulse. With circular polarization non-sequential double ionization [5] of He is suppressed. The photon energy of the xuv-pulse is chosen smaller than the transition energy between the He^+ 1s and 2p states, see Fig. 1. The peak electric field strength of the laser pulse must be large enough to suppress the Coulomb barrier below the 2p state of the He^+ ion. Then, the as-pulse generates a maximum number of He^{++} ions, when it coincides with a peak of the laser electric field, while He^{++} production is lowest at zero field strength. Without xuv radiation the number of He^{++} ions remains negligible. The depth of modulation of the He^{++} yield with respect to the relative delay between the two pulses gives a measure of the as-pulse duration.

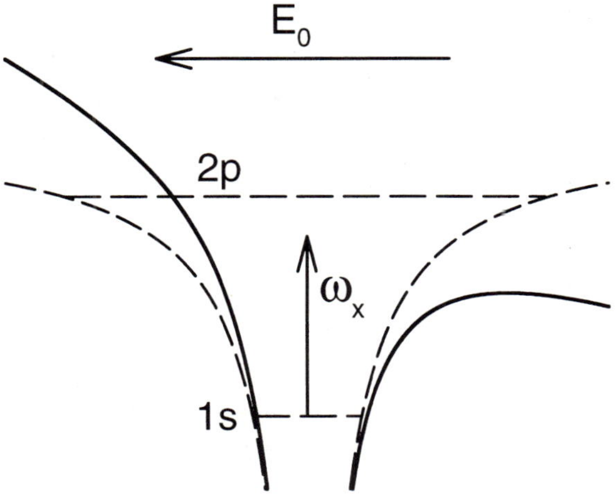

Fig. 1. Principle of attosecond cross correlation measurement: The strong electric laser field E_0 suppresses the Coulomb potential of a He^+ ion. The solid and dashed lines refer to the ion potential at zero and peak laser field, respectively. At the peak field strength an as-pulse can single-photon ionize He^+. The xuv photon energy is chosen smaller than the energy difference between the 1s and 2p state (40.8eV).

Ionization yield and time resolution of the ACC technique are shown in Fig. 2, which was obtained by solving the three dimensional Schrödinger equation for He^+. In the inset of Fig. 2 the He^{++} yield for the combined field of a 5fs Ti:S laser pulse and of a 600as xuv-pulse is depicted versus delay time between the two pulses. The peak intensity and the center wavelength of the laser pulse and of the xuv pulse are $I_0=10^{15}W/cm^2$, $\lambda_0=800nm$, $I_x=10^{12}W/cm^2$, $\lambda_x=36.5nm$, respectively. The modulation depth is approximately constant over a delay time of 2 optical cycles around the laser pulse center. Outside the range of constant modulation, the laser electric field is too weak to suppress the Coulomb barrier and ionization drops rapidly. A single attosecond pulse can be distinguished from a train of attosecond pulses, because the delay time over which the modulation depth remains constant depends on the number of peaks in the pulse train.

The modulation depth of the He^{++} ion yield as a function of as-pulse duration is depicted in the main part of Fig. 2. Measurement of the He^{++} ion yield in combination with the data of Fig. 2 determines the duration of attosecond pulses. The xuv-pulse parameters used for the calculation are easily realizable with state of the art table top laser systems. The He^{++} ion yield of $He^{++}/He^+ > 10^{-4}$ is well above the detection limit. Since the yield decreases only linearly with xuv-intensity, even lower xuv peak intensities can be used. Finally, for the detection of as-xuv pulses at shorter wavelengths the ion charge must be increased, such that the ground state energy remains below the single photon ionization threshold.

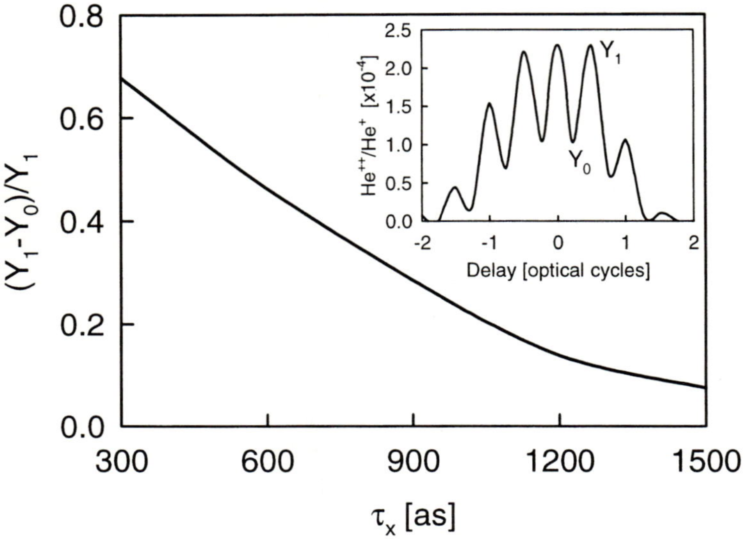

Fig. 2. Inset: He^{++} yield versus relative delay between attosecond pulse and laser pulse; for the parameters see the text. Main part of Fig. 2: modulation depth of the ionization probability (Y1-Y0)/Y1 versus as-pulse duration τ_x.

Acknowledgements. This work was supported by the Austrian Science Fund, Y142-TPH and by the special research program F016.

References

1 T. Brabec and F. Krausz, Rev. Mod. Phys. **72**, 545 (2000).

2 Y. Kobayashi, O. Yoshihara, Y. Nabekawa, K. Kondo, and S. Watanabe, Opt. Lett. **21**, 417 (1996).

3 T. E. Glover, R. W. Schoenlein, A. H. Chin, and C. V. Shank, Phys. Rev. Lett. **76**, 2468 (1996).

4 I. P. Christov, M. M. Murnane, and H. C. Kapteyn, Phys. Rev. Lett. **78**, 1251 (1997).

5 B. Walker et al., Phys. Rev. Lett. **73**, 1227 (1994).

Unbalanced multiphoton autocorrelation techniques for fs pulse measurements in the near ir

S. Marzenell, P. Mahnke, and R. Beigang

Fachbereich Physik, Universität Kaiserslautern, D-67663 Kaiserlautern, Germany
E-mail: beigang@physik.uni-kl.de

Abstract. We report on balanced and unbalanced two- and three-photon interferometric autocorrelation measurements of femtosecond pulses in the near and mid infrared. The influence of pulse asymmetries and phase modulations is shown experimentally and theoretically.

1. Introduction

Interferometric autocorrelation techniques are routinely used for ultrashort pulse measurements [1,2]. However, autocorrelation measurements do not provide information about pulse shapes and asymmetries as they are always symmetric. Recently it was shown [3] that unbalanced three-photon intensity autocorrelation measurements are well suited for the determination of pulse shapes of ultrashort pulses. In this contribution we demonstrate how unbalanced two- and three-photon interferometric autocorrelation techniques can provide additional information about ultrashort pulses.

2. Theory

An autocorrelation function $g^n(\tau)$ of n-th order can be written as:

$$g^n(\tau) = \int_{-\infty}^{\infty} I(t) \cdot I^n(t+\tau)dt$$

For high orders of n $(n \rightarrow)$ $g^n(\tau)$ will approach the original function $I(t)$ as lim $I^n(t) = (t)$ is the delta-function. This can be achieved using an unbalanced autocorrelation set-up with $I_1 = kI_2$ $(0 < k \leq 1)$. Neglecting terms of higher orders of k one obtains for a two- and three-photon intensity autocorrelation, respectively:

$$g_I^2(\tau) \approx 1 + 4k^2 \cdot \int_{-\infty}^{\infty} I(t) \cdot I(t+\tau)dt \qquad g_I^3(\tau) \approx 1 + 9k^2 \cdot \int_{-\infty}^{\infty} I^2(t) \cdot I(t+\tau)dt$$

Only for a three-photon autocorrelation information about a possible pulse asymmetry can be obtained. However, if an *interferometric* autocorrelation is performed the upper envelope of the autocorrelation functions have the form:

$$g_E^2(\tau) \approx 1 + 4k \cdot \int_{-\infty}^{\infty} E^3(t) \cdot E(t+\tau)dt \qquad g_E^3(\tau) \approx 1 + 6k \cdot \int_{-\infty}^{\infty} E^5(t) \cdot E(t+\tau)dt$$

In this case already a two-photon autocorrelation contains information about the pulse shape. With increasing order of the nonlinear process the measured pulse shape becomes more comparable to the original pulse shape. If one compares g_I^3 and g_E^2 it turns out that a two-photon interferometric autocorrelation contains more information about a possible asymmetry than a three-photon intensity autocorrelation assuming Fourier limited pulses. Calculated unbalanced two-photon interferometric autocorrelation functions for a double pulse are shown in Figure 1a) for different values of *k*. For $k = 1$ the autocorrelation is symmetric whereas with decreasing *k* the autocorrelation function approaches the shape of the original pulse.

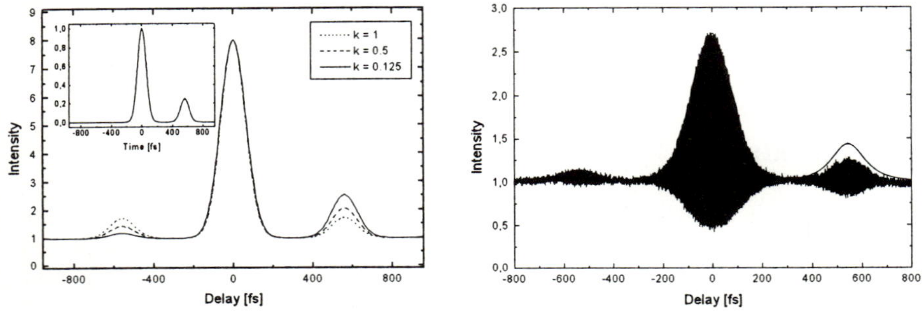

Fig. 1.
a) Calculated upper envelope of unbalanced interferometric two-photon autocorrelation functions for a double pulse and different values of k

b) Interferometric two-photon autocorrelation (k = 0.26) at 800 nm using a GaAsP photodiode. The solid line represents the field envelope of the original pulse

3. Experimental Results

Autocorrelation measurements were performed in a typical Michelson set-up in the wavelength range from 800 nm to 8 µm using different photo-diodes as the nonlinear element. Depending on the wavelength and photodiode two- or three-photon processes were applied. Table 1 shows the wavelength ranges for various photodiodes which have been tested in our experiments. The ultrashort pulses were provided from a synchronously pumped mode-locked $AgGaSe_2$ optical parametric oscillator (OPO) pumped by a CTA OPO around 1.5 µm [4]. The signal and idler pulses from the $AgGaSe_2$ OPO were tunable between 2 µm and 8 µm. It was possible to measure on-line interferometric autocorrelations with all diodes listed in Table 1. In the case of InAs the photodiode had to be cooled with a thermoelectric cooler to –30 °C. Typical experimentally observed interferometric three-photon autocorrelations at 1.5 µm are shown in Figure 2.

The influence of the ratio between the intensities in the unbalanced arms was measured at 800 nm for a two-photon autocorrelation with a GaAsP photodiode

Table 1. Photodiodes for two- and three-photon autocorrelations

Photodiode	Two-photon absorption	Three-photon absorpption
GaAsP	680 – 1360 nm	1380 – 2040 nm
Si	1100 – 2100 nm	2200 – 3300 nm
InGaAs	1800 – 3500 nm	(3600 – 5400 nm)
InAs	3800 – 7600 nm	-

for an asymmetric double pulse (see Figure 1b)). It can clearly be seen that the envelope of the autocorrelation approaches the original pulse shape with decreasing k. With an intensity autocorrelation (balanced or unbalanced) there would be no information about the asymmetry. The experimental results were in good quantitative agreement with the calculations.

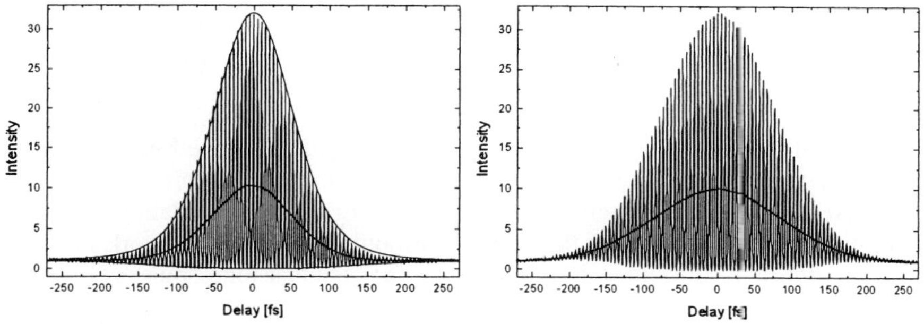

Fig. 2. Interferometric three-photon autocorrelation at 1.5 μm using a GaAsP photodiode: a) unchirped and b) chirped pulse

4. Conclusions

In summary we have investigated balanced and unbalanced interferometric autocorrelation techniques using direct two- and three-photon absorption in photodiodes to measure pulses in the wavelength range from 800 nm to 8 μm. From the upper envelope of the unbalanced interferometric autocorrelation function the pulse shape was deduced. The evaluation of high order interferometric autocorrelation measurements which can easily be performed with commercially available photodiodes can thus provide additional information about ultrashort pulses in a wide wavelength range.

References

1 see. e. g. D. T. Reid and W. Sibbett. E&L notes: Suppl. To Optics & Photonics News **9**, 1998.

2 J. K. Ranke, A. L. Gaeta, A. Baltuska, M. S. Pshenichnikov, and D. A. Wiersma, Opt. Lett. **22**, 1344, 1997.

3 P. Langlois and E. P. Ippen, Opt. Lett. **24**, 1868, 1999.

4 S. Marzenell, R. Beigang, and R. Wallenstein, Appl. Phys. **B69** 423, 1999.

Unbalanced Third-Order Correlations for Characterizing the Intensity and Phase of Femtosecond Pulses

Jeffrey W. Nicholson, Mark Mero, Jayesh Jasapara, and Wolfgang Rudolph

Department of Physics and Astronomy, University of New Mexico, Albuquerque, NM 87131, USA
E-mail: jeffn@phys.unm.edu, wrudolph@unm.edu

Abstract. The intensity and phase of femtosecond pulses can be retrieved from measurements of third-order intensity correlations and the pulse spectrum using a numerical algorithm. Unbalancing the interferometer makes the retrieval algorithm sensitive to the direction of time, and improves the retrieval of certain pulse shapes. The sensitivity of the retrieval to direction of time as a function of the degree of imbalance in the correlator is investigated.

A variety of techniques such as FROG [1] and SPIDER [2] have been introduced to characterize the intensity and the phase of ultrashort pulses. One of the earliest techniques relied on measurements of the second-order autocorrelation, the pulse spectrum and the spectrum of the second-harmonic [3]. Recently, work has been done to retrieve the intensity and phase of femtosecond pulses from an unbalanced second order intensity correlation combined with the pulse spectrum [4]. This measurement, called PICASO (Phase and Intensity from Correlation And Spectrum Only), used a Michelson interferometer unbalanced by a dispersive element to help resolve the ambiguity in the direction of time that is typical for second order measurements. The measurement proved to be a robust method for characterizing ultrashort pulses. Such techniques can be combined with multi-photon current detectors allowing for compact and simple devices with minimal intrinsic dispersion.

Unbalanced third order correlations show a temporal asymmetry [5]. Here we describe PICASO that is based on the measurement of a third order correlation (TOC) , where unbalancing with an amplitude and phase filter makes the retrieval intrinsically sensitive to the direction of time. In addition, unbalancing the TOC improves the retrieval of known difficult pulse shapes such as pulses with a symmetric spectrum and cubic spectral phase.

The retrieval algorithm works by assuming an initial spectral phase of the pulse. Together with the measured pulse spectrum this yields an initial guess for the complex electric field of the pulse which is used to calculate the TOC. The calculated correlation, I_r, is compared to the measured TOC, I_m, and an error (root-mean square deviation), Δ, is obtained:

$$\Delta = \sqrt{\frac{1}{N}\sum_{i=1}^{N}\left[I_m(\tau_i) - I_r(\tau_i)\right]^2} \qquad (1)$$

Here N is the number of sampling points taken at successive delays τ_i. The retrieval error Δ is minimized using a multi-dimensional search algorithm that iteratively improves the guess for the spectral phase.

To record TOC's, fs pulses were focused onto a glass slide for third-harmonic generation (THG) [6]. A thin (180 nm) Ta_2O_5 film increased the signal by more than a factor of 10 as compared to the SiO_2 substrate while confining the effective interaction length and thus eliminating dispersive effects. The TH was filtered with an interference filter and recorded with a PMT. Focusing a few mW average power with a parabolic mirror resulted in sufficiently high signal to noise ratios.

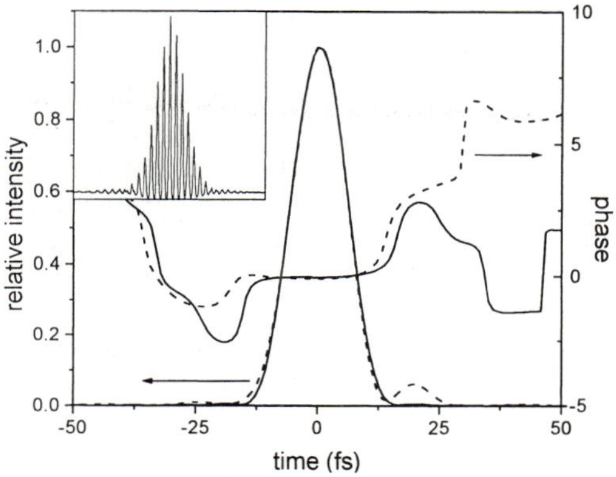

Fig. 1. Pulse retrieval from an experimentally measured spectrum and an unbalanced third-order correlation. The solid (dashed) lines correspond to focusing with a parabolic mirror (microscope objective). The inset shows a balanced TOC (parabolic mirror).

A typical balanced TOC of pulses from a Ti:sapphire oscillator is shown in the inset of Fig.1. To unbalance the correlator a ND filter (50 % transmission, 1.1 mm BK7 substrate) was introduced in one arm of the correlator. Correlations with the parabolic focusing mirror and a microscope objective were recorded and the pulses retrieved, cf. Fig.1. For the microscope objective we pre-compensated the second- and most of the third-order dispersion with a combination of prisms and a third-order dispersion (TOD) mirror. The small satellite is a result of uncompensated TOD and illustrates the sensitivity of PICASO.

An important aspect of any pulse reconstruction is its ability to resolve ambiguities (i.e. direction of time and sign of chirp). We investigated in more detail the ability of the unbalanced TOC to distinguish between the correct and time-reversed solutions. Synthetic pulses were designed that had an intensity and phase profile comparable to pulses from the microscope objective, for comparison with experimental data. The retrieval was allowed to stagnate either at the local minimum of the correct or the time-reversed pulse depending on the initial guess. The Δ values for both pulses were then plotted as a function of filter transmission, T, see Fig.2. As expected the same retrieval error occurs for balanced TOC's (T = 1). Increasing the imbalance results in an increased error difference, which allows the retrieval of the correct pulse without ambiguity. As with the unbalanced second-order PICASO [4] the phase response of the filter is crucial to resolve the chirp for pulses that are nearly symmetric in time.

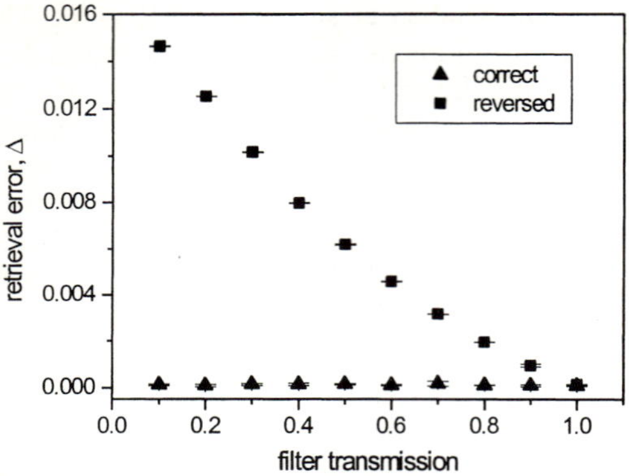

Fig. 2. Retrieval error, Δ, as a function of the imbalance in the correlator.

While for few-cycle pulses the full interferometric correlation (IC) can be used as a retrieval target, for longer pulses this is prohibitive because of the large N that is necessary to resolve the fringes. On the other hand, the intensity correlation does not have phase information that is contained in higher frequency components of the IC. There are also indications that Δ obtained with the IC is less sensitive to pulse shape variations than the intensity correlation and higher frequency components. This suggests that the inclusion of additional components in the retrieval should further increase the accuracy, for both second and third order correlations.

In summary, PICASO with an unbalanced third-order correlation is an accurate pulse characterization technique that is inherently sensitive to the direction of time. It is well suited for use with multi-photon current detectors in compact and inexpensive correlators for real-time pulse retrieval. Finally, use of a single shot geometry for the correlation coupled with a 2-dimensional detector array could allow for simultaneous measurements of the correlation as a function of beam radius. This would be of importance in experiments where coupling between space-time coordinates is expected, e.g. plasma wakefield acceleration.

Acknowledgements. We thank the Laser Zentrum Hannover (Dr. Ristau) for providing the Ta_2O_5 coatings. This project was supported in part by NFBI and NSF (PHY-9601890).

References

1 D. J. Kane and R. Trebino, in *Opt. Lett.* Vol. 18, 823, 1993.
2 C. Iaconis and I.A. Walmsley, in *Opt. Lett.*, Vol. 23, 792, 1998.
3 J.C. Diels, J.J. Fontaine, I.C. McMichael, and F. Simoni, in *Appl. Opt.*, Vol. 24, 1270, 1985. K. Naganuma, K. Mogi, and H. Yamada, in *IEEE J. Quant. Electron.*, Vol. 25, 1225, 1989.
4 J.W. Nicholson, J. Jasapara, W. Rudolph, F. Omenetto, and A.J. Taylor, in *Opt. Lett.*, Vol. 24, 1774, 1999.
5 P. Langlois and E.P. Ippen, in *Opt. Lett.*, Vol. 24, 1868, 1999.
6 D. Meshulach, Y. Barad, and Y. Silberberg, in *J. Opt. Soc. Am., Vol. B* 14, 2122, 1997.

Spectral phase correlator for coded waveform recognition using second harmonic generation

Z. Zheng[1], A. M. Weiner[1], K. R. Parameswaran[2], M. H. Chou[2,3] and M. M. Fejer[2]

[1] 1285 EE Building, Department of Electrical and Computer Engineering, Purdue University, West Lafayette IN 47907-1285, USA
E-mail: zzheng@ecn.purdue.edu

[2] E. L. Ginzton Laboratory, Stanford University, Stanford CA 94305-4085, USA
E-mail: krishp@stanford.edu

[3] Current address: Tellium Inc., 2 Crescent Place, Oceanport NJ 07757-0901, USA

Abstract. We report a novel spectral phase correlator scheme for coded ultrashort pulses using SHG in long PPLN crystals. Results using bulk crystals and waveguides show high contrast ratios and the potential for very low power operation.

In second harmonic generation (SHG) of ultrashort optical pulses, the spectral structure of the pump pulse, along with effects like group velocity mismatch (GVM), can significantly affect the outcome of the second harmonic process. Here we study SHG of spectrally phase coded optical pulses in long nonlinear crystals under the large GVM condition, a condition that is traditionally avoided in ultrafast optics. Our results show that, under the low conversion condition, the SHG output can be directly related to the spectral correlation function of the pump field spectrum due to the spectral filtering effect caused by the GVM, and, thus, SHG in long nonlinear crystals can serve as a nonlinear optical 'spectral phase correlator'. This phenomenon can find applications in the proposed ultrashort pulse optical code-division multiple access (CDMA) systems [1, 2], where a nonlinear device is needed to distinguish between correctly decoded signals and incorrectly decoded multi-access interference from different multiplexed channels which carry different spectral coding information.

In ultrashort pulse SHG using a very long nonlinear crystal of length L, large GVM between pump and second harmonic (SH) pulses leads to strong narrowing of the phase matching bandwidth. Therefore, P_{SHG} is dominated by the spectral component at the center of SH spectrum. This relationship can be further simplified, if we assume that the original pump spectrum is flat-topped and unchirped and is then divided into 2N `bins' with an equal frequency width, each of which is coded with a phase shift Φi (i=-N,...,-1,1,...,N) (as shown in Fig. 1). When there is no depletion and the intrapulse group velocity dispersion at the fundamental wavelength is negligible [3], P_{SHG} is given by [4]:

$$P_{SHG} = 2\pi\Gamma^2 L\left|\int_0^\infty E_1(\Omega')E_1(-\Omega')d\Omega'\right|^2 \Big/ |\alpha| \propto \Gamma^2 L\left|\sum_{i=1}^{N}\exp(-j\Phi_i)\exp(-j\Phi_{-i})\right|^2 \Big/ |\alpha| \qquad (1)$$

i.e., the SHG signal is proportional to the square of the magnitude of the correlation of two code words $C_1 = \{\Phi_{-1}\ldots\Phi_{-N}\}$ and $C_2 = \{\Phi_1 \ldots \Phi_N\}$, as in Fig. 1, respectively. Here Γ is the nonlinear coupling coefficient, α is the GVM parameter, Ω' is the angular frequency detuning from the center frequency, and E_1

is the field envelope spectrum function of the fundamental. In the proposed CDMA system, C_1 would be applied at the transmitter and C_2 at the receiver; C_1 and C_2 would be taken from a code set with low cross correlations. Then, the SHG spectral correlator can distinguish different categories of waveforms based on the correlation of C_1 and C_2. It is noted that P_{SHG} scales linearly with L, so that highly efficient SHG is possible using long crystals [5]. It should also be noted that the same functional form as Eq. (1) also governs the transition probability for two-photon absorption (TPA) with no intermediate resonance in atoms [6]. Therefore, our nonlinear optical but classical experiments have an interesting relationship to the TPA quantum coherent control experiments in [6].

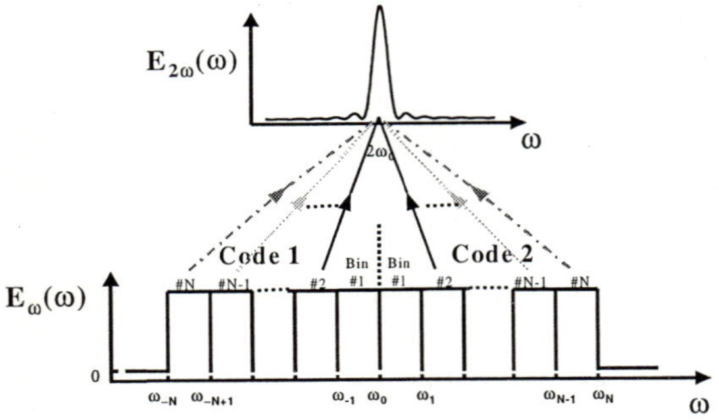

Fig. 1. Scheme of optical phase coding of ultrashort pulses.

Our experimental setup consisted of a passively mode-locked femtosecond fiber laser, a fiber-pigtailed femtosecond pulse shaper [2, 7] with a 128-pixel amplitude/phase liquid crystal modulator (LCM) array, periodically poled lithium niobate (PPLN) crystals, and a spectrometer. The output spectrum from the pulse shaper was ~18 nm wide centered at 1560 nm, and the PPLN phase matching wavelength was tuned to this wavelength. The transmittance/phase function of the LCM was controlled to flatten the spectral amplitude and compensate any residual spectral phase chirp simultaneously with the coding process. In our experiments, we measured the SHG output for C_1 set to a fixed code word from a binary code set with known correlation properties, with C_2 varied among all the code words in the set.

In our first experiments [4], we used a 20-mm long commercial bulk PPLN crystal. The first code set was the length-7 M-sequence code, which has an auto-/cross-correlation ratio of 7. This should lead to a contrast ratio of 49 between matched and unmatched codes. Shown in Fig. 2(a), for uncoded pulses, the SHG has a bandwidth of ~ 0.3 nm due to the large (6 ps) GVM. For coded pulses with $C_1=C_2$, the SHG peak is comparable to that of the uncoded case. For each of the six other cases where $C_1 \neq C_2$, the SHG at the center wavelength is suppressed by at least a factor of 50, as expected, although the spectra have different shapes at other wavelengths. As a second case, we studied a set of length-8 Hadamard codes, which are orthogonal and should suppress the SHG signal more strongly, ideally to zero (Fig. 2(b)). The seven traces for $C_1 \neq C_2$ show very weak SHG. Compared with the uncoded case, the SHG at the center frequency was suppressed by at least 600 for all unmatched codes and by > 1500 in some traces.

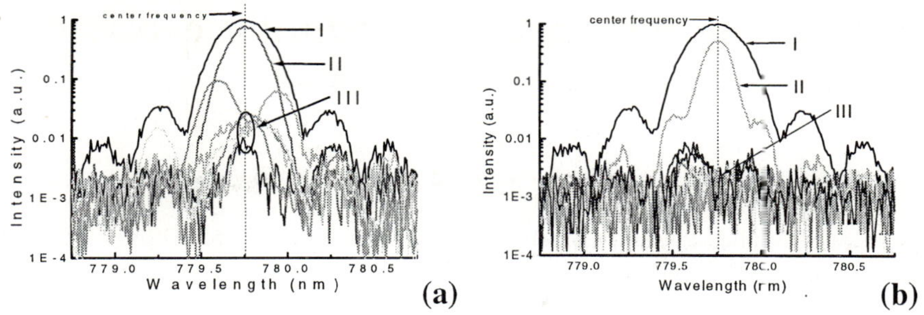

Fig. 2. SHG spectra measured under different spectral phase coding situations. I: no phase coding. II: matched phase codes ($C_1=C_2$). III: all unmatched phase codes ($C_1 \neq C_2$) (a)—Length-7 M-sequence codes; (b)—Length-8 Hadamard codes.

In our second experiment, we used a PPLN waveguide [8], which had a 55.5-mm-long SHG section with a uniform grating period of 14.75 μm. The SHG bandwidth was ~ 0.3 nm FWHM. Our preliminary waveguide data show an internal SHG conversion efficiency of ~ 50%/pJ for uncoded pump pulses, which is ~ 500 times higher compared to our bulk PPLN experiments. This boost in efficiency and the narrow SHG bandwidth enabled direct real-time detection at low pump powers using a high-speed avalanche photodiode (APD) responsive to light at 780 nm. Here the length-8 Hadamard codes were used. For coded pulses with $C_1=C_2$, real-time detection was possible with an average pump pulse energy of ~ 0.25pJ, which shows the high sensitivity of the nonlinear detection process. For the cases where $C_1 \neq C_2$, the SHG signal was significantly suppressed by at least 50 times. Further improvement in the suppression ratio should be possible.

In summary, we demonstrate a novel spectral correlation scheme with high contrast and high sensitivity between coded ultrafast optical waveforms. Our scheme applies ideas from coherent control to classical nonlinear optics, which provides the possibility of pulse-by-pulse processing at GHz rates.

References

1 J. A. Salehi, A. M. Weiner and J. P. Heritage, in *J. Lightwave Technol.*, Vol. 8, 478, 1990.

2 H. P. Sardesai , C. C. Chang and A. M. Weiner, in *J. Lightwave Technol.*, Vol. 16, 1953, 1998.

3 M. A. Arbore, O. Marco and M. M. Fejer, in *Opt. Lett.*, Vol. 22, 865, 1997.

4 Z. Zheng and A. M. Weiner, in *Opt. Lett.*, Vol. 25, 984, 2000.

5 A. M. Weiner, A. M. Kan'an and D. E. Leaird, in *Opt. Lett.*, Vol. 23, 1441, 1998.

6 D. Meshulach and Y. Silberberg, in *Nature*, Vol. 396, 239, 1998.

7 A. M. Weiner, D. E. Leaird, J. S. Patel and J. R. Wullert, II, in *IEEE J. Quantum Electron.*, Vol. 28, 908, 1992.

8 M. H. Chou, I. Brener, M. M. Fejer, E. E. Chaban and S. B. Christman, in *IEEE Photon. Technol. Lett.*, Vol. 11, 653, 1999.

Femtosecond nonlinear polarization evolution from cascade quadratic processes

X. Liu, F. O. Ilday, K. Beckwitt, F. W. Wise

Department of Applied Physics, Cornell University, Ithaca, NY 14853, USA
E-mail: fwise@ccmr.cornell.edu

Abstract. We experimentally demonstrate strong amplitude modulation of 100 fs pulses with nonlinear polarization evolution based on type-I second-harmonic generation. Applications of this process to mode-locking of bulk and fiber lasers is promising and will be discussed.

Nonlinear polarization evolution (NPE) provides intensity-dependent transmittance or reflection of light beams, which is essential for many applications. NPE based on cubic nonlinearities was first demonstrated by [1], and is now commonly employed in mode-locked fiber lasers [2, 3]. NPE can also be implemented using nonlinear phase shifts ($\Delta\Phi_{NL}$) produced by quadratic nonlinear interactions [4, 5, 6] which can potentially be larger than with cubic nonlinearities. Furthermore, the sign of the phase shift is controlled. Phase retardation generated by employing phase-matched type-II second-harmonic generation (SHG) was demonstrated by Lefort and Barthelemy [7] and has been successfully applied to the mode-locking of a Nd:YVO_4 laser [8] to produce 2.8 ps pulses. However, group-velocity mismatch (GVM) between the ordinary (o) and extraordinary (e) input pulses practically restricts the technique to applications involving pulses ~1 ps in duration or longer.

The prospective application to mode-locking femtosecond bulk and fiber lasers motivates the consideration of NPE in quadratic media using a type-I interaction, which eliminates the effects of GVM between the input pulses. The use of a type-I quadratic process for NPE was proposed theoretically by Saltiel *et al.* [9], but without any consideration of the effects of GVM between the fundamental and harmonic pulses. This GVM has several undesirable effects on phase shifts produced in the cascade process. In previous work [10, 11], we have established that these effects can be reduced by appropriate choice of the phase mismatch.

In our experiment, the input pulse is incident on a second-harmonic generation (SHG) crystal that is oriented with the incident polarization at an angle α with respect to the normal to the principal plane. Under phase-mismatched conditions, o component at the fundamental frequency accumulates a nonlinear phase shift, but the e component does not. Passage through a 45° Faraday rotator and reflection from a mirror rotates the polarizations of the original o and e components, and the new o component accumulates another nonlinear phase shift on the return through the SHG crystal. The net phase retardation $\Delta\Phi_{NL}$ between the components produces polarization rotation that is converted to amplitude modulation by a polarizer. This arrangement fully compensates for the linear phase delay and group-velocity delay between the o and e waves, which is essential for NPE with ultrashort pulses [12]. Neglecting linear losses, the transmittance of light through the polarizing beam splitter can be written as

$$T = 1 - \sin^2(2\alpha)\,\sin^2(\Delta\Phi_{NL}/2) \qquad (1)$$

To a first approximation [6],

$$\Delta\Phi_{NL} \approx -\,(\Gamma^2L^2/\Delta kL)\,(\Delta I/I) \qquad (2)$$

where $\Delta k = k_{2\omega} - 2k_{\omega}$ is the phase mismatch, L is the crystal length, I is the total intensity, and ΔI is the intensity difference between the o and e components.

Nearly transform-limited pulses with energy up to 20 μJ and duration 80 fs at 1.53 μm were produced by an optical parametric amplifier. The quadratic medium was a 17-mm-long barium metaborate (BBO) crystal cut for type-I SHG at ~1.55 μm and we set $\alpha = 28°$ to maximize the nonlinear transmission. Since GVM can severely degrade the phase shifts on the 100 fs time scale, the phase mismatch must be chosen correctly. Following the criterion given in [11] we conclude that $|\Delta kL| \geq 10\pi$ is necessary in the present application. A separate advantage is the reduction of the energy loss due to SHG at large phase mismatch.

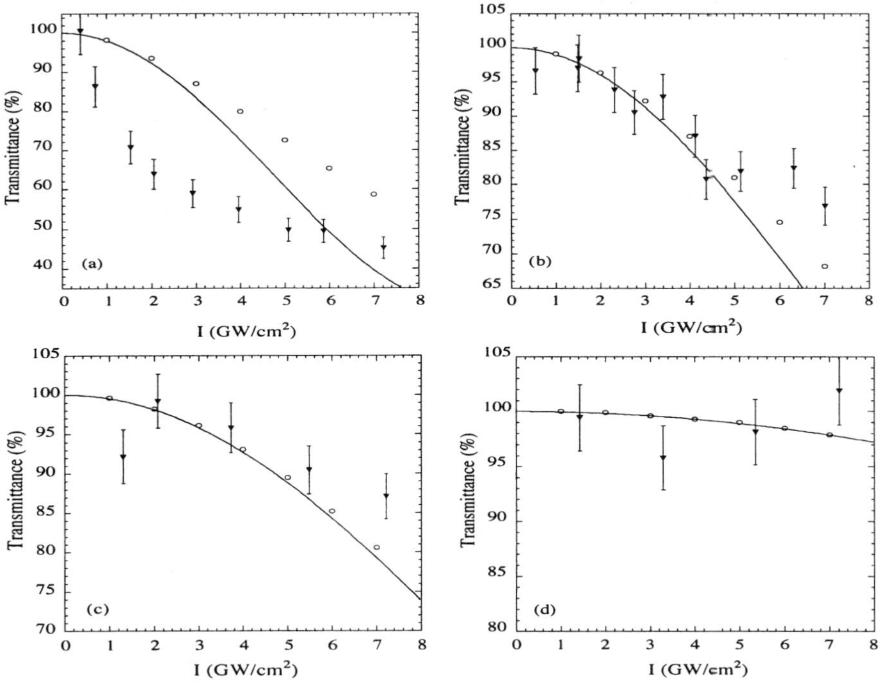

Fig. 1. Experimental (triangles) and calculated transmittance using analytic expression (line) and numerical simulations (circles) at phase mismatches of (a) -14π, (b) -20π, (c) -30π and (d) -120π.

The measured nonlinear transmittance is shown in Fig. 1, along with the results of approximate analytical calculations based on Eq. (2) and numerical solutions to the one-dimensional coupled wave equations. The experimental and calculated results for phase mismatches of ΔkL = -14π, -20π, -30π, and -120π agree overall. The measurement at -120π can be viewed as a control experiment since the effect is negligible. For I = 7GW/cm^2, a total nonlinear phase shift $\Delta\Phi_{NL}$~0.7π is expected. From the measured broadening of the pulse spectrum we infer an experimental value in excellent agreement with the calculated one. Also, from Eq. (1), we expect T ≈ 0.44, in agreement with the measured transmittance of 0.45.

For mode-locking of fiber lasers, we envision implementing the type-I NPE described here using periodically-poled lithium niobate in a waveguide structure. Assuming a 20-mm-long waveguide with 50 μm^2 guiding area, the pulse energy to achieve $\Delta\Phi_{NL}$ = π with 200 fs pulses at 1.55 μm is estimated to be ~ 100 pJ. This approach is thus quite promising. In addition to the amplitude modulation, it will be possible to use the positive (negative) nonlinear phase shifts for soliton-like pulse shaping with anomalous (normal) GVD.

In conclusion, we have demonstrated strong intensity-dependent transmittance of femtosecond pulses by the use of NPE based on type-I phase mismatched SHG. This approach has the advantage of 100 fs response time with respect to type-II NPE. Compared to NPE in cubic media, type-I NPE benefits from lower power requirements to achieve a given rotation of polarization, and also produces less background nonlinear phase shifts. We anticipate that using linearly polarized light in type-I NPE will be advantageous for mode-locking polarization maintaining fibers. Possible applications include all-optical switching and modulation, optical limiting, pulse shaping in addition to mode-locking of ultrafast lasers.

References

1. R. H. Stolen, J. Botineau, and A. Ashkin, Opt. Lett. **7**, 512 (1982).
2. M. Hofer, M. E. Fermann, F. Haberl, M. H. Ober, and A. J. Schmidt, Opt. Lett. **16**, 502 (1991).
3. M. E. Fermann, M. J. Andrejco, Y. Siilberberg, and M. L. Stock, Opt. Lett. **18**, 894 (1993).
4. H. J. Bakker, P. C. M. Planken, L. Kuipers, and A. Lagendijk, Phys. Rev. A. **42**, 4085 (1990).
5. N. R. Belashenkov, S. V. Gagarskii, M. V. Inochkin, Opt. Spectrosc. **66** 1383, (1989).
6. R. DeSalvo, D. J. Hagan, M. Sheik-Bahae, G. Stegeman, E. W. Van Stryland, and H. Vanherzeele, Opt. Lett. **17**, 28 (1992).
7. L. Lefort and A. Barthelemy, Opt. Lett. **20**, 1749 (1995).
8. V. Couderc, F. Louradour, and A. Barthelemy, Opt. Commun. **166**, 103 (1999).
9. S. Saltiel, K. Koynov, and I. Buchvarov, Appl. Phys. B **63**, 371 (1996).
10. L. Qian, X. Liu, and F. Wise, Opt. Lett. **24**, 166 (1999).
11. X. Liu, L. Qian, and F. W. Wise, Opt. Lett. **24**, 1777 (1999).

Femtosecond polarization detection using high-speed pulse shaping

Dorine Keusters, Peifang Tian, and Warren S. Warren

Center for Ultrafast Laser Applications, Princeton University, Princeton, NJ, 08544
E-mail: wwarren@princeton.edu

Abstract. We demonstrate a method for femtosecond, phase sensitive detection of optical polarization, using an acousto-optic pulse shaper to create a sequence of up to several hundred phase coherent pulses. Essential to this method is the ability of the acousto-optic pulse shaper to update the phase relation of the pulses in the sequence on a nanosecond timescale. The method is demonstrated by measuring the optical free induction decay of rubidium vapor, and can be particularly useful for experiments involving very low or very high optical densities. It can easily be extended to multidimensional spectroscopy.

1. Introduction

Direct, phase sensitive measurement of magnetization began the transformation of NMR from a physicist's toy into a broadly applicable technique. Similarly, a fast and convenient method for phase-sensitive detection of the induced optical polarization on a fs timescale could have significant impact on phase coherent optical spectroscopy. For example, multidimensional techniques [1] might be able to unravel more complex dynamics than simple pump-probe experiments, [2]. In addition, measurement of the free induction decay (FID), while equivalent to measuring the spectrum in NMR, provides additional information in optical experiments, particularly in systems with nonstationary dynamics [3]. Phase sensitive detection of polarization on a fs timescale has been demonstrated [4,5] but the approaches to date are technologically difficult - particularly if the sample has very few absorbers (so phase matching does not collimate the polarization) or very many absorbers (so propagation effects and scattering dominate).

In this paper, we demonstrate a different method for femtosecond phase-sensitive polarization detection. We start with a previously demonstrated concept [5,6]: an extra phase-coherent detection pulse applied at time τ transfers part of this polarization present at that time into excited state population. This population can easily be detected by measuring the total fluorescence emitted by the sample, or by measuring the total absorption. Modulating the phase of this added pulse modulates the transferred polarization, which permits its detection. What is new here is that we use an acousto-optic pulse shaper (AOPS) to accomplish this modulation automatically and at very high speed (>>10^6 Hz), exploiting what is often considered a *disadvantage* of the technique.

2. Experimental Method

The AOPS shapes each pulse from a 80 MHz modelocked laser (Coherent Vitesse, 20 fs pulsewidth when transform-limited, 3 nJ/pulse) into a sequence of pulses with well-defined delays. Changing the RF frequency of the AOPS changes the pulse delay [7], by about 231 fs/MHz; applying N simultaneous sine waves creates N pulses [8]. However, due to acoustic wave propagation, the phase relation between the pulses changes systematically from one sequence to the next. For example, if the AOPS is driven by sine waves at 200 and 210 MHz, each modelocked pulse will create a two-pulse sequence with 2.3 ps delay, but the phase difference between the two pulses will go through one full cycle in 100 ns(=1/(10 MHz)). This means, in turn, that the component of the fluorescence or absorption modulated at 10 MHz will reflect both the amplitude and phase of the polarization 2.3 ps after the first pulse. This can be extracted by lock-in detection using the beat frequency between the two sine waves as a reference.

To demonstrate the technique, the optical free induction decay (OFID) of rubidium atoms (D-lines at 794.76 nm and 780.03 nm) was measured in two different ways. First we produced one excitation-pulse and one detection pulse, and varied the delay τ between these pulses to obtain the complete OFID signal. Transmission through the Rubidium-cell (10 cm, 300°C) is detected by a photodiode. The parts of the signal oscillating in phase and in quadrature with the RF reference are plotted as a function of delay in Fig. 1a. The beating can be clearly seen, but direct Fourier transformation gives a spectrum that is shifted by about 50 nm. This shift arises because the reference travels directly to the lock-in amplifier, but the RF-wave has to travel at an acoustic speed through the crystal. In our geometry, the propagation delay is 5.58 μs. This translates into a frequency shift in the optical domain. If this extra frequency is subtracted, the spectrum shown in Fig. 1b is obtained. Within the accuracy of the experiment (0.5 nm for a total delay of 4 ps), the two absorption peaks at 780.4 and 794.4 nm agree with the values published for Rb in literature.

In the second method, a pulse sequence was generated, consisting of one large excitation pulse and a few hundred detection-pulses with different delays τ. (Fig. 2) Since the phase-difference between the excitation pulse and each of the detection pulses oscillates at its own RF beat frequency, the signal arising from each excitation-detection-pair can be separated. The complete OFID signal (Fig. 1c) is obtained by Fourier transforming the transmission through the Rb cell and

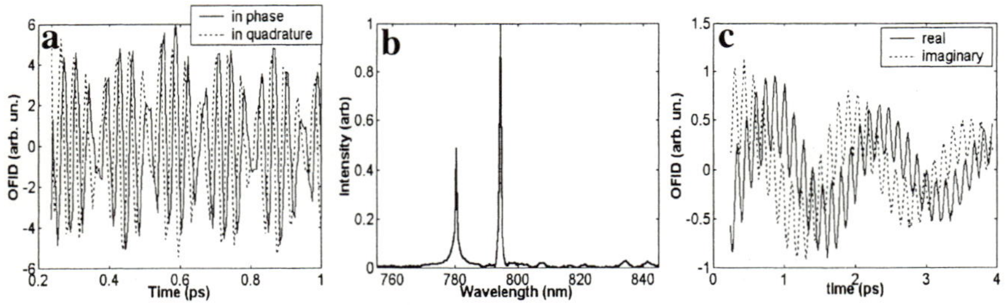

Fig 1. (**a**). First ps of OFID signal obtained in step by step method. Note the high oscillation-frequency caused by the delay between the arrival of the reference and the signal at the lock-in amplifier. (**b**). Spectrum obtained from (a) after correcting for delay between reference and signal. (**c**) OFID signal obtained in 'single shot' method. Note the signal oscillates at the proper frequency, which is smaller than in (a), because in this method no reference is used. Fourier transform of (c) gives a spectrum similar to (b) without any corrections.

selecting only the components oscillating at the proper beat frequencies of the RF sine waves. Here the OFID signal appears at the proper frequency, which is much smaller than in Fig. 1a, because in this method no reference is used. Fourier transformation gives a spectrum similar to Fig. 1b with peaks at 780.1 and 795.6mnm.

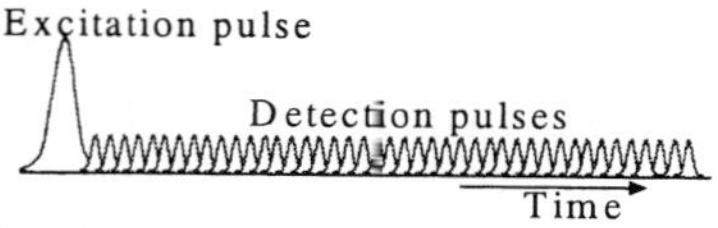

Fig. 2 Pulse sequence used in 'single shot' OFID measurement.

Note that spurious signals arising from interaction with two detection pulses will not interfere with the OFID measurement if the RF-frequency and amplitude of the detection pulses is chosen properly. The second method is faster and more convenient than the first, since the complete OFID is measured in a 'single shot' without any adjustments to the optical setup or RF waveform. The first method, however, offers a much higher sensitivity.

3. Conclusion

A technique has been developed for femtosecond phase sensitive detection of optical polarization, which exploits the fast pulse shaping capabilities of an acousto optic pulse shaper. The technique was demonstrated by measuring the optical free induction decay of Rb, but extension to multidimensional techniques is straightforward. The lack of phase matching requirements and immunity to scattering makes it a useful tool for spectroscopy involving very low and very high optical densities respectively.

Acknowledgements. Help from Wolfgang Wagner and Hua Cao is greatfully acknowledged. Financial support came from AFOSR and the State of New Jersey

References

1. P. Hamm, M. Lim, F. Degrado, R.M. Hochstrasser, J. Chem. Phys. **112** (4), 1907, 2000; J.D. Hybl, A.W. Albrecht, S.M. Gallagher Faeder D.M. Jonas, Chem. Phys. Lett. **297,** 307, 1998, W.M. Zhang, V Chernyak, S. Mukamel, J. Chem. Phys. **110** (11), 5011, 1999
2. S.M. Gallagher Faeder, D.M. Jonas, J. Phys. Chem. A **103** (49), 10489, 1999; D. Keusters, H.S. Tan, W.S. Warren, J. Phys. Chem. A **103** (49), 10369, 1999.
3. S. Mukamel, *Principles of Nonlinear Optical Spectroscopy*, Oxford Univ. Press, 1995.
4. S.M. Gallagher, A.W. Albrecht, J.D. Hybl, B.L. Landin, B. Rajaram, D.M. Jonas, J. Opt. Soc. Am. B, **15** (8) 2338, 1998; L. Lepetit, G. Chériaux, M. Joffre, J. Opt. Soc. Am. B, **12** (12), 2467, 1995; W.P. de Boeij, M.S. Pshenichnikov, D.A. Wiersma, Annu. Rev. Phys. Chem. **49,** 99, 1998.
5. N.F. Scherer, R.J. Carlson, A. du Matro, A.J. Ruggiero, V. Romero-Rochin, J.A. Cina, G.R. Fleming, S.A. Rice, J. Chem. Phys. 95 (3), 1437, 1991
6. W.S. Warren, A.H. Zewail, J. Chem. Physs. 75 (12) 5956, 1981; F. Spano, M. Haner, W.S. Warren, Chem. Phys. Lett. 135 (1-2)97, 1987
7. W. Yang, D. Keusters, D. Goswami, W.. Warren, Opt. Lett. 23 (23), 1843, 1998
8. J.X. Tull, M.A. Dugan, W.S. Warren, Adv. Magn. Opt. Res., **20,** 1, 1997.

Tunable third order dispersion of a prismatic pulse compressor

K. Osvay[1], P. Dombi[1], A.P. Kovács[1], J. Klebniczki[1], G. Kurdi[2] and Z. Bor[2]

[1] Department of Optics and Quantum Electronics, University of Szeged, P.O.Box 406, Szeged 6701, Hungary
E-mail: osvay@physx.u-szeged.hu
[2] HAS Research Group on Laser Physics, Dóm tér 9, Szeged 6720, Hungary

Abstract. The wavelength dependent thermal refractive index provides an extra degree of freedom for the adjustment of higher order dispersion of a prismatic pulse compressor. The effect allows for fine tuning of the intracavity dispersion of ultrafast oscillators.

1. Introduction

For generating laser pulses shorter than 15 fs in ultrashort pulse oscillators, it is vital to compensate for the intracavity dispersion to the highest possible order. The most frequently used intracavity prismatic pulse compressor can cancel the group delay dispersion (GDD) but leaves uncompensated third order dispersion (TOD), which can be usually reduced by chirped mirrors. However, due to the inherent nature of these ultrafast mirrors and also the inaccuracy in the production of these complicated dielectric layers, there is always some residual TOD in the cavity which sets a limit for the generation of the shortest possible pulses.

In this paper we propose a new method for fine tuning of intracavity dispersion. It is shown that the change of the material temperature of the intracavity prism pair results in a change of the ratio of GDD to TOD.

2. A thermal prism pair compressor

It is widely known that the phase shift as well as its higher order derivatives (GDD, TOD) due to a prism pair can be tuned simply by changing the position of either of the prisms. Upon this procedure eventually the ratio of the total glass path in the prisms l_g and the prism apex distance l_p is varied. It can be directly deduced from Ref. [1], that in general the specific phase shift ϕ_S (i.e. the phase shift per unit prism apex distance) is

$$\phi_s = \frac{\phi}{l_p} = \frac{2\pi}{\lambda}\cos\theta(\lambda) = \frac{2\pi}{\lambda}\sqrt{1-\left(R\frac{\cos\alpha_1\cos\beta_0}{\sin\varphi}\right)^2}\,, \tag{1}$$

where $R=l_g/l_p$ and the angles $\alpha_1(\lambda)$ and $\beta_0(\lambda)$ are defined by the angle of incidence at the first prism α_0, the prism apex angle φ and the refractive index n [2].

Increasing R from zero both GDD and TOD are negative but change sign typically around $0.03<R<0.06$. In the case of intracavity dispersion control, when mainly the dispersion of the laser crystal is to be compensated, it has a consequence that the prism pair operates at an optimal R which provides full compensation for GDD and the least residual negative TOD. Once R has been set, there is no more degree of freedom to change the ratio of GDD to TOD.

The refractive index of the prism material practically depends not purely on wavelength but also on temperature. Thus, varying the temperature of the prism pair only, the changes of phase shift at different wavelengths will unequally depend on the temperature. Hence, at a given R the ratio of GDD and TOD is expected to be sensitive of temperature. These effects can be calculated from the thermal refractive index $n=n(\lambda,T)$. Since for a given material the dispersion at a certain temperature T_0 and the thermal change of refractive index $\eta(=\Delta n/\Delta T)$ are separately known, the approximation $n(\lambda,T)=n(\lambda,T_0)+\eta(\lambda)(T-T_0)$ was used [2].

3. Results

The values of specific thermal GDD and specific thermal TOD have been calculated at 850 nm for Brewster prism pairs as a function of the ratio of total glass path and prism apex distance (Fig.1.). The material of the prisms has been chosen as fused silica [3,4] and SF10 [5], both of which have positive η and represent materials with low and high dispersion, respectively, while FK54 [5] is one of the rare example of optical glasses with negative η. From the figures it can be established that i) the specific GDD changes much less with temperature than the specific TOD compared to their absolute values and ii) if $\eta>0$ then both the specific GDD and TOD change sign at different R's but in the case of $\eta<0$ their signs remain positive.

Typical spectral dependence of the specific phase derivatives (group delay (GD), GDD and TOD) are shown for fused silica and SF10 in Fig.2. It may be worth noting that the specific thermal GDD and even the specific thermal TOD vary much more rapidly for SF10 than for fused silica. The strong dependence of higher order dispersion on wavelength may be an attractive feature for compensation also for sophisticated phase shifts such as the one generated upon noncollinear optical parametric amplification of ultrashort pulses [6,7].

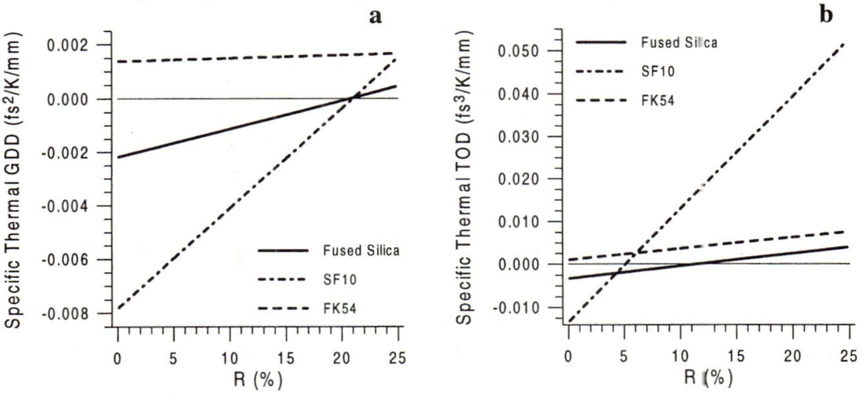

Fig. 1. Temperature change of specific GDD (a) and specific TOD (b) vs. ratio of total glass path and prism apex distance at 850 nm.

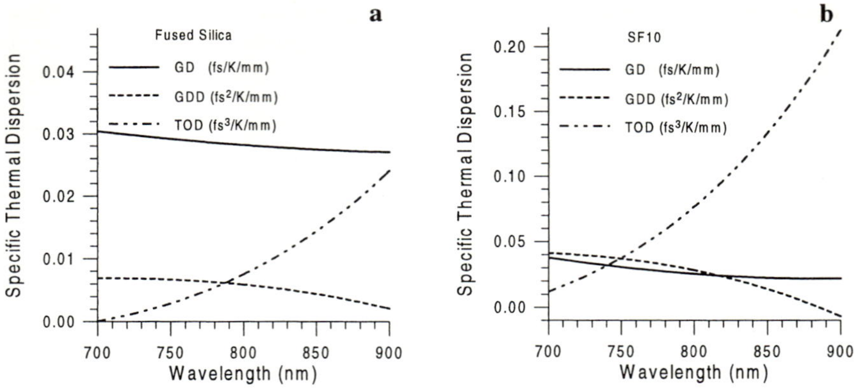

Fig. 2. Spectral change of specific thermal dispersions of fused silica (a) and SF10 (b).

4. Discussion

The higher order dispersion control proposed in this paper can be realised basically in two ways: either the temperature of the existing intracavity prism pair is changed resulting in a 20-30% improvement of the ratio of GDD and TOD or a second thermal prismatic compressor is constructed. The latter case is more complicated but allows for a more accurate adjustment. For example, setting up the thermal SF10 prism pair to R=0.051, where the specific thermal TOD is zero and heating the prism pair, it is only the GDD that is varied at a constant total specific TOD of -12.78 fs^3/mm. In the special case of warming by 12.2 °C, this system exhibits no GDD at all. The other way around is also working: the adjustment of the thermal prism pair to R=0.21 results in zero specific thermal GDD at the trade in a constant GDD of 24.1 fs^2/mm. Once this thermally constant GDD of the thermal prism pair is cancelled by the original intracavity pulse compressor, any change of the temperature of the "thermal" compressor gives rise to the change of the total specific TOD by a rate of 0.05 fs^3/mm/K.

Acknowledgements. This project has been supported by OTKA #T33018 and by FKFP #2029/97. During this research Z.Bor and K.Osvay have held the Szëchenyi Professorship of the Ministry of Education of Hungary.

References

1 O. E. Martinez, J. P. Gordon, R. L. Fork, J. Opt. Soc. Am. A **1**, 1003, 1984
2 K. Osvay et al., submitted to Appl. Phys. B
3 I. H. Malitson, Appl. Opt. **2**, 1103, 1963
4 S. De Nicola, G. Carbonara, A. Finizio, G. Pierattini, Appl. Phys. B **58**, 133, 1994
5 *Optisches Glas*, The Catalogue of Schott Glaswerke, Mainz, Germany, 1980
6 A. Shirikawa, I. Sakane, T. Kobayashi, Opt. Lett. **23**, 1292, 1998
7 I. N. Ross et al., Appl. Opt. **39**, 2422, 2000

Nonlinear Dispersion Properties Of Sub-wavelength Photonic Crystals

S. Vijayalakshmi(1), Y. Zhang(1), Gul Yaglioglu(2), R. Dorsinville(2), C. W. White(3), and H. Grebel(1)

(1) Nonlinear Nanostructures laboratory, Electrical and Computer Engineering, New Jersey Institute of Technology, Newark, NJ-07102
Phone: (973) 596-3538, Fax: (973) 596-6513
E-mail: grebel@njit.edu

(2) Electrical Engineering Department,
City College and Graduate Center of The City University of New York, New York, NY-10031
E-mail: dorsin@ee.ccny.cuny.edu

(3) Oak Ridge National Laboratory, Oak Ridge, TN-37831

Abstract:

Nonlinear pulse compression and pulse broadening have been demonstrated with a coherent array of sub-micron silica spheres embedded with silicon nanoparticles ('nano with nano').
OCIS codes: (320.7110) Ultrafast Nonlinear Optics

Summary

The most common pulse compression scheme is a combination of a frequency chirp and broadening with a dispersion compensator. The frequency chirp and broadening is achieved by propagation in a nonlinear medium, most notably a Kerr medium. Periodic structures, such as, phase gratings, have been used as dispersion media needed for the compensation process [1]. Photonic crystals embedded with nonlinear material such as, silicon nanoclusters, bear the potential of incorporating the two requirements in one structure. The fcc structure of silica based opaline sample was ion implanted with Si ions. Following annealing, 5-6-nm clusters of silicon were formed within a 0.3-μm thick region, as determined by SEM and Raman spectroscopy [2]. The structures have been shown to be highly nonlinear [3], reaching index changes on the order of $|\delta n_{NL}|>0.1$ at λ=0.8-μm, and strongly dispersive [4]. Unlike Bragg fibers [5], we employ here a transverse scheme, namely, the dispersion is achieved via *transverse Bragg reflections* [6] while maintaining non-resonance conditions along the propagation direction.

The optical wave propagation in photonic crystal is rather complicated. An insight into the wave propagation within a transverse structure may be obtained as follows. The wave propagation consists of many spatial modes. The transverse propagation constant of the *q*th mode may be in resonance with the periodic structure, namely, $\beta_x^{(q)} \sim k_0 n_0 \sin(\theta) = q(\pi/\Lambda)$, where Λ is the transverse pitch. From this equation we may calculate that the incident angles for transverse Bragg condition are 19^0 and 43^0 for a non-implanted opaline structure and an opal containing 5-6-nm Si clusters, respectively. The coupling strength, with coupling, $|\kappa| = \pi \Delta n / \lambda_0 \cos\theta$, Δn=0.5, λ=0.8-μm and interaction length L=10-μm, is $\kappa L \sim 30$ for a 10-μm thick opal.

The propagation constant in the direction of propagation is $\beta_z^{(q)}=\{k_0^2n^2(0)-[\beta_x^{(q)}]^2\}^{1/2}$. The time of arrival per unit length of a spectral line, ω, of mode number, q, is,

$$\tau(\omega,q)=\partial\beta_z^{(q)}/\partial\omega\sim(1/c)[n+(1/8)(q\lambda_0/\Lambda)^2/n]. \tag{1}$$

Here we used, n, the effective linear refractive index and, $k_n=\omega n_0/c=2\pi/\lambda_n$. Consider a 100-fs pulse. The delay-time introduced by the modal dispersion is $\delta\tau=\tau(\omega,q+1)-\tau(\omega,q)=(2q+1)$0.2-fs/μm. Shown in Fig. 1 is the broadening of the pulse by a non-implanted and implanted opaline samples as a function of angle. As predicted, a large broadening occurs at the transverse Bragg angle. Based on the previous arguments, the pulse propagation is made via multi-spatial modes.

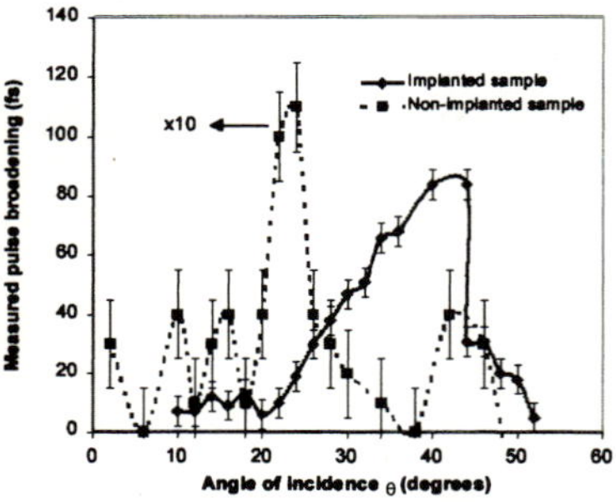

Fig. 1. Pulse broadening as a function of incident angle, θ. The theoretical estimation of the Bragg peaks are 19^0 and 43^0 for non-implanted and Si implanted samples, respectively

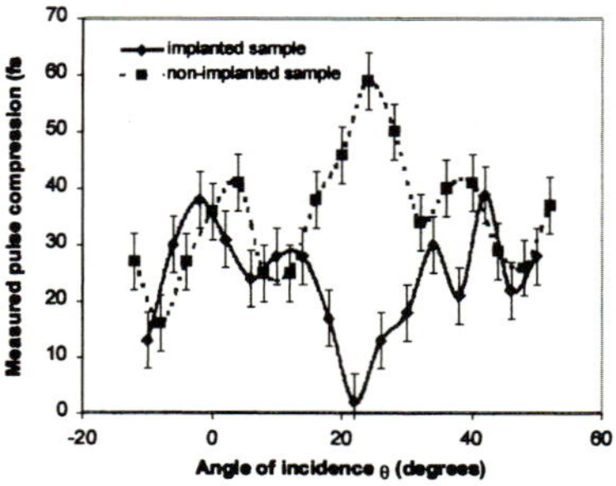

Fig. 2. Pulse compression for a pre-chirped pulse at 265-fs as a function of the incident angle, θ, for implanted and non-implanted 10-μm thick opaline samples. The implanted region was 0.3-μm

The opal structure may compensate for frequency chirp introduced by external means. Such pulse 'compression' for a 10-μm thick non-implanted opaline samples is shown in Fig. 2. The incoming pulse of 100-fs from a Ti:sapphire laser was positively chirped, without a change in the spectral width, using a 15-cm long, water-filled cuvette. The pulse duration after the water cuvette was 265-fs and the corresponding re-compression was 60-fs. The compression for the implanted sample was not as effective as the non-implanted sample. The reason is that the implanted region distorts the incident phase for the non-implanted segment thereby, resulting in a non-optimized effect. At low power levels (see also Fig. 1) the broadening from the combined effects is actually maximized. Also, owing to the negative signs of both linear and nonlinear dispersions one may expect a substantial broadening.

Nonlinear properties may be incorporated upon replacing n(I) by, $n=n_0+\gamma I$; here I is the intensity distribution of the propagating pulse, commonly taken to be of gaussian shape, $I=I_0exp(-2t^2/\tau_p^2)$. Self-phase modulations will broaden the pulse spectrum and include a spectral chirp owing to the large index changes. The linear dispersion will determine the relative time-of-arrival of the tail and peak of the pulse, which corresponds to the relative time of arrival of the high and low frequencies, respectively. The nonlinear effect was indirectly demonstrated by measuring the pulse broadening of a 100-fs input pulse as a function of intensity. In Fig. 3 we show changes in pulse broadening as a function of pulse energy at two angles: (a) at the angle of maximum pulse compression and (b) at an angle of minimum compression. Note that $\kappa L \sim 0.2$ for the nonlinear region. One may observe a saturation of index change for large energy densities. We also added the flat response of a non-implanted, 10-μm thick opaline structure. The latter demonstrates that, in general, the silica-based opaline structure does not contribute to the self-phase modulations. The small dip near 250 μJ/cm^2 may indicate a true compression owing to the positive nonlinearity of the silica spheres; this is still under investigation. One should expect that as the nonlinear chirp increases, its effect on the pulse broadening decreases. This is because the nonlinear chirp contributes a term like $A/(1+A^2)$, where A is proportional to the instantaneous frequency [7].

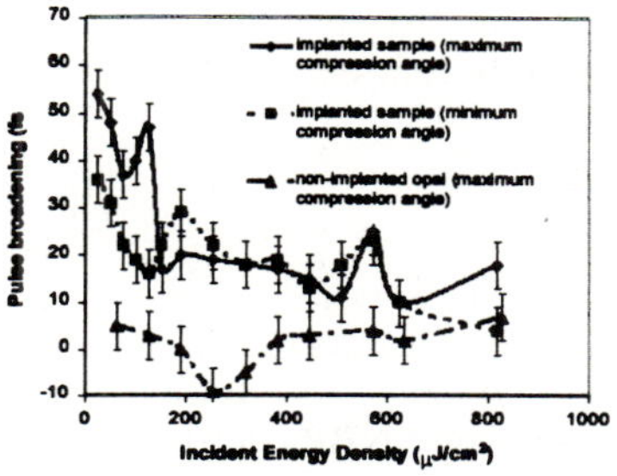

Fig. 3. Change in pulse broadening as a function of pulse energy density

In summary, we have shown that by incorporating nano silicon clusters (a nonlinear medium) within an opaline structure and employing a transverse diffraction scheme we were able to manipulate ultra-short pulses.

Acknowledgement: Research sponsored at NJIT in part by the Oak Ridge National Laboratory, managed by UT-Battelle, LLC, for the U.S. Department of Energy under contract DE-AC05-00OR22725 and by the NSF-CIRE (#9872777) and NSF-IGERT (#9874275) programs at CCNY.

References

1. E. B. Treacy, "Optical pulse compression with diffraction gratings", *IEEE J. QE.*, **QE 5**, 454 (1969)
2. Y. Zhang et al, "Linear properties of Si nanoclusters embedded in an ordered array of silica spheres", submitted to *Appl. Phys. Letts*, November, 1999.
3. M. Ajgaonkar, Y. Zhang H, Grebel and C. W. White, "Nonlinear optical properties of a coherent array of sub-micron SiO_2 spheres (opal) embedded with Si nanoparticle", *Appl. Phys. Letts*, **75**, 1532 (1999).
4. J. M. Tobias and H. Grebel, "Self-imaging in photonic crystals at sub-wavelength range", *Opt. Letts.*, 24, 1660 (1999).
5. B. Eggleton and R. Slusher, "Bragg grating solitons", *Phys. Rev. Letts*, **76**, 1627 (1996).
6. H. Grebel and W. Zhong, "Holographic interconnects: Transverse Bragg waveguides", *Opt. Letts.*, **18**, 1123 (1993).
7. A. Yariv, "*Optical Electronics in Modern Communication*", Oxford University Press, New York, 5th ed., p 225.

Carrier-induced optical pulse shaping in a semiconductor nonlinear Bragg reflector

K. Ogawa* and Y. Matsui

The Femtosecond Technology Research Association, 5-5 Tokodai, Tsukuba, Ibaraki 300-2635, Japan, *E-mail: kenogawa@festa.or.jp

Abstract. Carrier-induced pulse shaping in a semiconductor nonlinear Bragg reflector is presented. Absorption saturation and two-photon absorption produce positive pulse chirping, and the pulse chirping is compensated by negative group-delay dispersion in the Bragg reflector.

1. Introduction

Semiconductor nonlinear Bragg reflectors (SNBRs) consist of periodic layers in which optical nonlinearities are distributed. Coupling of incident optical waves with distributed optical nonlinearity in SNBRs leads to the formation of gap solitons.[1,2] This type of solitons have been interested mainly for applications to all-optical switching. All-optical switching was investigated experimentally in a semiconductor periodic waveguide, based on gap solitons generated by carrier-induced nonlinearity under nanosecond laser pulse excitation.[3] In ultrafast optics, solitons have also played a distinct role in optical pulse shaping. In this paper, carrier-induced optical nonlinearity and pulse propagation in an SNBR are characterised by measurements on pump-probe reflectance and time-wavelength pulse specrogram.

2. Semiconductor nonlinear Bragg reflector

The SNBR investigated consisted of 7-nm InGaAs quantum wells (QWs) periodically distributed in 100-nm InP/109-nm InGaAs optical confinement Bragg layers.[4] Cross-section of the SNBR is depicted in the lower-left inset in Fig. 1. The Bragg layers were epitaxially stacked on an InP substrate. Unlike the case with high-reflectance NBR for all-optical switching by the gap solitons, optical resonance occurred at the centre of the SNBR forbidden gap even in linear-response (low-intensity) limit. This optical resonance was produced by a phase shift of π in the refractive-index periodicity, compared with that of the ordinary high-reflectance Bragg reflectors.[4] The thickness of the QWs was adjusted to tune the lowest exciton resonance to the peak wavelength (1500 nm) of the optical resonance at room temperature. Such dual resonance was effective in the enhancement of carrier-induced optical nonlinearity and produced significant optical phase modulation as described below.

3. Results and Discussion

Pulse shaping in the SNBR was achieved by the two processes: (i) generation of pulse chirping by carrier-induced self-phase modulation and (ii) chirp compensation by group delay dispersion. Carrier-induced optical nonlinearity was characterised by pump-probe reflectance measurements. Pulse autocorrelation traces and time-wavelength pulse spectrograms were then measured to investigate changes in pulse shape after the reflection. All the measurements were performed at room temperature.

Carrier-induced optical nonlinearity in the SNBR was characterised by pump-probe measurements in reflection geometry. The pump and probe pulses were generated from an 80-MHz optical parametric oscillator (OPO). The pulses from the OPO were in the transform limit with 190-fs pulse duration and 15-nm spectral width. The pulse peak wavelength was 1500 nm. The energy of the pump and probe pulses was 1.2 nJ and 100 fJ, respectively. The pulses were focused in 30-μm diameter, which lead to pump photon density of 1×10^{15} cm^{-2} per pulse.

Pump-induced reflectance change obtained as a function of probe delay time is plotted in Fig. 1. In the initial stage of photoexcitation, probe reflectance was increased by absorption saturation (AS) in the QWs. At the peak of pump pulses, instantaneous two-photon absorption (TPA) in InP and InGaAsP layers overwhelmed the AS. Carriers were accumulated in the InP and InGaAsP layers after the TPA and caused electronic-bandgap reduction (BR) which formed a slowly decaying absorption background after the photoexcitation. More than 20 % absolute reflectance change was produced by the AS and the TPA during the photoexcitation in the SNBR.

Such a large reflectance change leads to significant modulation of the optical phase of the pump pulses reflected from the SNBR. The reflected pump spectrum is presented in the upper-right inset in Fig. 1.The reflected pump spectrum consisted of multiple peaks and was broadened more than twice of the width of the incident pump pulses. Spectrally resolved pump-probe transient reflectance was also measured to investigate spectral modulation of the pump pulses. The pump-probe reflectance spectrogram showed that the pump pulses were positively chirped in the process of carrier-induced phase modulation: the peak of the pump-probe reflectance was delayed by about 500 fs with a wavelength shift from 1510 nm to 1480 nm.

Autocorrelation traces of pump pulses were obtained in the geometry illustrated in the inset of Fig. 2. The linearly polarised incident pulses were passed through a polarisation beam splitter (PBS), circularly polarised after a quarter-wavelength plate (QWP) and focused normal onto the sample (the SNBR or a reference metal mirror). The reflected pulses were

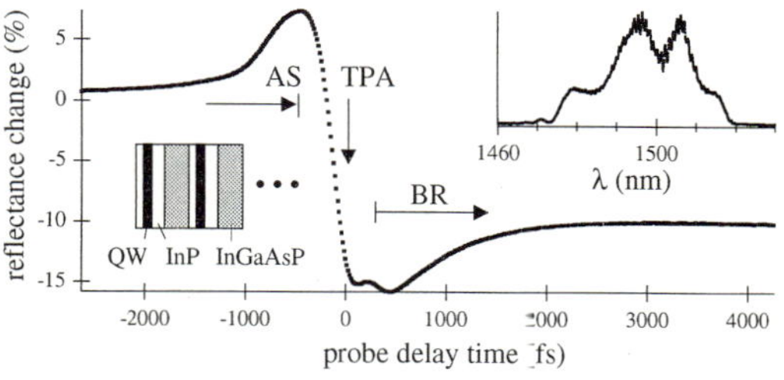

Fig. 1. Pump-induced reflectance change. Insets: spectrum of reflected pump pulses (upper right) and cross-section of the SNBR (lower left).

picked up by the PBS and evaluated by the TPA autocorrelation.

Autocorrelation traces from the SNBR were compared with the traces of reference pulses from a metal mirror as shown in Fig. 2. The width of the autocorrelation trace from the SNBR was 254 fs, reduced to 74 % in comparison with that of the reference pulses. The effect of the narrowing was weaker when the incident pulse energy was decreased. The reflected pulses from the SNBR showed autocorrelation traces almost same as those of the incident pulses at pulse energy below 0.2 nJ.

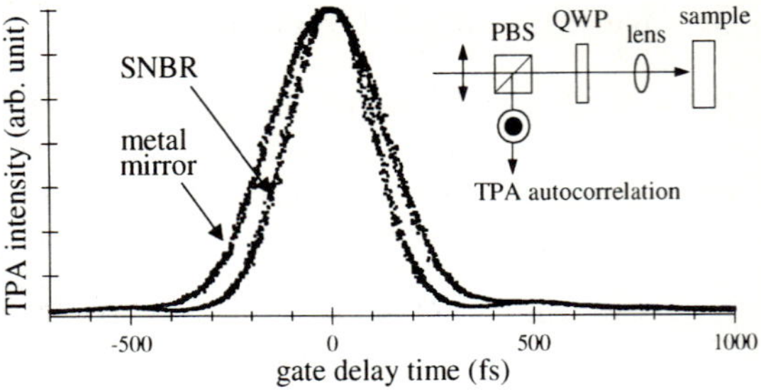

Fig. 2. TPA autocorrelation traces of reflected pulses from the SNBR and a metal mirror. Inset: measurement geometry.

The pulse narrowing in the SNBR was analysed by TPA pulse spectrograms.[5] The spectrogram of the reflected pulses from the SNBR with 1.2-nJ pulse energy is presented and compared with the reference spectrogram in Fig. 3. The pulses from the SNBR were broadened in spectral domain while they were compressed in time domain. Transfer-matrix analysis of the SNBR provided group-delay dispersion of 5 fs/nm at 1500 nm. This negative group-delay dispersion has the opposite sign to the positive pulse chirp induced by the carrier nonlinearity and leads to the compensation of the pulse chirping.

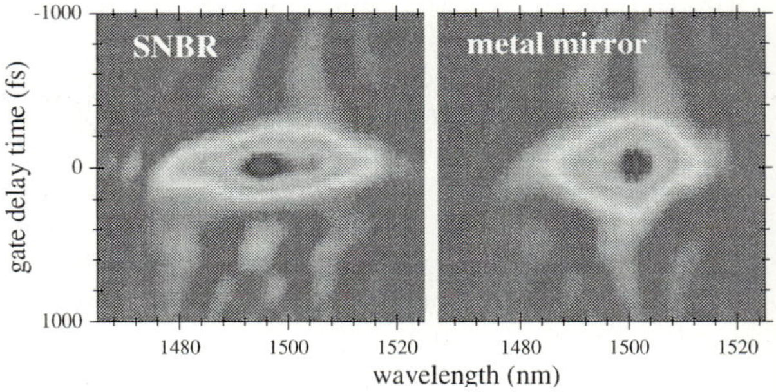

Fig. 3. TPA spectrograms of reflected pulses from the SNBR (left) and a reference metal mirror (right).

Acknowledgements. This work was supported by the New Energy and Industrial Technology Development Organization (NEDO).

References

1 W. Chen and D. L. Mills, Phys. Rev. Lett. **58**, 160 (1987).
2 C. Martijin de Sterke and J. E. Sipe, Phys. Rev. A**42**, 2858 (1990).
3 N. D. Sankey, D. F. Prelewitz and T. G. Brown, Appl. Phys. Lett. **60**, 1427 (1992).
4 K. Ogawa and Y. Matsui, Appl. Phys. Lett. **74**, 2569 (1999).
5 K. Ogawa and M. D. Pelusi, submitted to Optics Express.

20-fs pulse shaping with a 512-element phase only liquid crystal modulator

H. Wang[1], Z. Zheng[1], D. E. Leaird[1] , A. M. Weiner[1], T. A. Dorschner[2], J. J. Fijol[2], L. J. Friedman[2], H. Q. Nguyen[2] and L. A. Palmaccio[2]

[1]School of Electrical and Computer Engineering, Purdue University, West Lafayette, IN 47907-1285
[2]Raytheon Company, 131 Spring Street, Lexington, MA 02421

Abstract: We report pulse shaping experiments in 20-fs range using a 512-element phase-only liquid crystal modulator array. The results demonstrate the potential for significant improvement compared to 128-element liquid crystal modulators previously used for pulse shaping.

OCIS codes: (320.5540) Pulse shaping, (230.3720) Liquid-crystal devices.

1. Introduction

Pulse shaping technique is a powerful tool for ultrafast scientific studies as well as in optical communications [1]. For now, liquid crystal modulators (LCMs) [2,3], acousto-optic (AO) modulators [4] and deformable mirrors [5,6] are generally used in programmable pulse shaping. LCMs are currently the most widely used modulators in femtosecond pulse shapers. Their use for shaping of sub-15 fs pulses was demonstrated several years ago [7]. However, previous devices were limited to 128 pixels. Here we report experiments in the 20-fs regime, which for the first time demonstrate pulse shaping using a 512-element phase-only LCM. Our results show the potential for high-accuracy pulse shaping with significantly increased complexity.

2. Experimental setup

As in Fig. 1, in our pulse shaper, we used two 600 grooves/mm diffraction gratings and two 300mm radius spherical mirrors. The incident angle onto the grating was about 28 degrees, and the diffraction angle was about 0 degrees.

The LCM is based on an advanced LCM array technology developed for beam steering in laser radar applications [8]. It is a phase-only modulator array with 512 elements. The center-to-center spacing of adjacent pixels was 25μm for a total

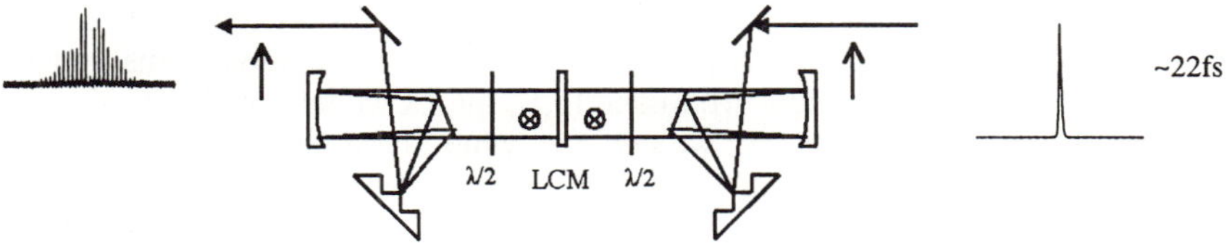

Fig. 1, layout of the pulse shaper

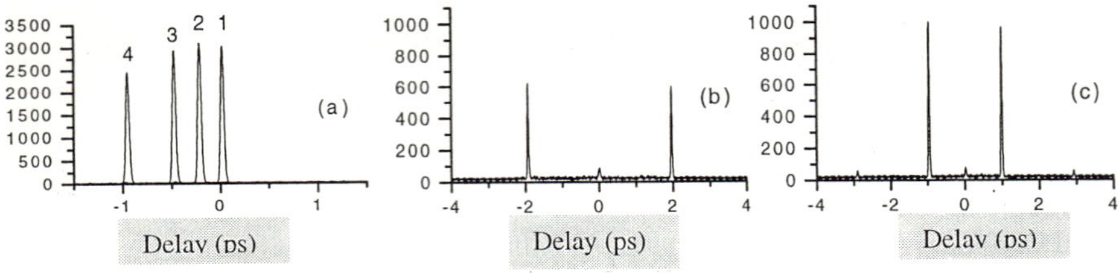

Fig.2, Cross-correlation measurements of shaped pulses

(a) temporally shifted pulse by linear phase modulation: pulse #1-no modulation, pulse #2-$\pi/16$ per pixel phase change, pulse #3-$\pi/8$ per pixel, pulse #4-$\pi/4$ per pixel.
(b)(c) 0-π-0-π periodical modulation,(b) two pixels per digit, (c)four pixels per digit.

spatial aperture of 13mm. The frequency difference between adjacent pixels was 130GHz, and the LCM covered the spectrum from 728nm to 870nm.

We used ~22fs pulses from a mode-locked Ti:Sapphire laser with 60nm FWHM spectral width centered at 798nm. Because of the large bandwidth, it is essential to take into account the nonlinearity of the pixel number vs. optical frequency relationship [9] in order to obtain good results while making use of the increased pixel count.

3. Results

Fig. 2(a) shows the cross-correlation data of pulses with different linear phase ramps applied (total phase shifts of 0, 32π, 64π, and 128π across the LCM, respectively). The linear phase ramps shift the pulse in time domain proportional to the slope of the modulation. If a 128π spectral phase shift was imposed on a 128-element LCM, it would be π phase change per pixel. The undersampling of the phase ramp would lead to a severe distortion. No significant pulse distortions were observed in any of the data in Fig. 2(a). The advantage of having 512 pixels for this application is a 4X increase in the spectral resolution and temporal window as compared to that with a 128-element modulator array.

Fig. 2(b) and 2(c) show the results of periodic 0-π-0-π modulations. In Fig. 2(b), each 'bit'(0 or π) covered two pixels, while in Fig. 2(c) each 'bit' covered four pixels. Two pulses were generated and distributed symmetrically about zero delay, and the distance between them was inversely proportional to the period of modulation function. So in Fig. 2(c) the LCM works like a 128-pixel LCM.

Fig. 3 shows the results with periodic M-sequence phase modulations. M-sequences are pseudo-noise digital sequences [10]. The length-7 M-sequence we used is (0 0 1 0 1 1 1); the length-15 M-sequence is (0 0 0 1 0 0 1 1 0 1 0 1 1 1 1). We applied 0 phase for "0" and π phase for "1". We can see pulse trains under an envelope [1] in Fig. 3. In Figs. 3(b,e) and (c,f), the LCM emulated 256-pixel and 128-pixel LCMs respectively. As we can see, when the number of pixels per bit increases, the intervals between the pulses become smaller, and also the time window becomes narrower. Therefore the results using all 512 pixels (Fig. 3(a,d)) show an improved time window compared to the 128-pixel case(Fig. 3(c,f)).

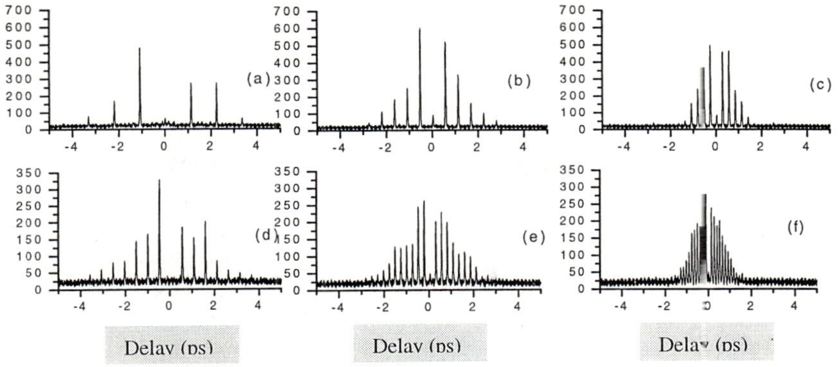

Fig. 3, Cross-correlation measurements of shaped pulses

(a)-(c): Periodical length-7 M-sequence modulation, (d)-(f): Periodical length-15 M-sequence modulation. In (a) and (d) one pixel per digit, in (b) and (e) two pixels per digit, in (c) and (f) four pixels per digit.

4. Conclusion

In summary, we have reported the first femtosecond pulse shaping experiments using a 512-element liquid crystal phase modulator. The increased pixel account allows substantially enhanced pulse shaping resolution compared to the 128-element LCMs commonly available. We have also performed preliminary adaptive pulse shaping [11] experiments, in which we have demonstrated that new patterns may be applied to the LCM every 90 ms, which is significantly faster than most current LCMs. This should translate into faster convergence in adaptive pulse shaping routines.

Reference

1. A. M. Weiner, Prog. Quantum Electron. 19, 161-238 (1995).
2. A. M. Weiner, D. E. Leaird, J. S. Patel, and J. R. Wullet, IEEE J. Quantum Electron. 28, 908-920(1992).
3. M. M. Wefers, and K. A. Nelson, Opt. Lett. 20, 1047(1995).
4. M. A. Dugan, J. X. Tull, and W. S. Warren, J. Opt. Soc. Am. B 14, 2348-2358(1997)
5. P. J. Delfyett, L. T. Florez, N. Stoffel, T. Gmitter, N. C. Andreadakis, Y. Silberberg, J. Heritage, and G. Alphonse, IEEE J. Quantum Electron Electron. 28, 2203(1992)
6. E. Zeek, R. Bartels, M. M. Murnane, H. C. Kapteyn and S. Backus, Opt. Lett. 25, 587(2000).
7. A. Efimov, C. Schaffer, and D. H. Reitz, J. Opt. Soc. Am. B. 12, 1968-1980(1995).
8. P. F. McManamon, T. A. Dorschner, D. L. Corkum, L. J. Friedman, D. S. Hobbs, M. Holz, S. Liberman, H. Q. Nguyen, D. P. Resler, R. C. Sharp, E. A. Watson, IEEE Proc. 84, (2) 268-298, 1996.
9. L. Wang, A. M. Weiner, Opt. Comm., 167, 211-224(1999).
10. J. G. Proakis, "Digital Communications" (third edition), McGraw-Hill Inc., ISBN 0-07-051726-6, 724-726(1995).
11. D. Meshulach, D. Yelin, Y. Silberberg, Opt. Comm. 138, 345-348 (1997).

Coherently controlled shaping of ultrafast electric field-transients in the mid-infrared

F. Eickemeyer[1], R.A. Kaindl[1], M. Woerner[1], A.M. Weiner[1,2]

[1] Max-Born-Institut für Nichtlineare Optik und Kurzzeitspektroskopie, D-12489 Berlin, Germany, E-mail: woerner@mbi-berlin.de
[2] Permanent address: School of Electrical and Computer Engineering, Purdue University, West Lafayette, Indiana 47907

Abstract: Amplitude and phase-controlled mid-infrared field transients are generated by phase-matched difference frequency mixing of near-infrared pulses shaped using a liquid crystal modulator. The generated electric field transients are directly measured using ultrafast free-space electro-optic sampling.

Coherent control experiments have found increased interest in recent years. In particular, such experiments can improve our understanding of chemical reactions and dynamical processes in semiconductor quantum structures [1,2]. Many fundamental processes take place at energies corresponding to the mid-infrared spectral range. So far, however, it was not possible to carry out coherent control experiments in this range because a suitable light source for shaped mid-infrared pulses is still lacking. Here we experimentally demonstrate the first coherently controlled shaping of ultrafast electric field-transients in the mid-infrared.

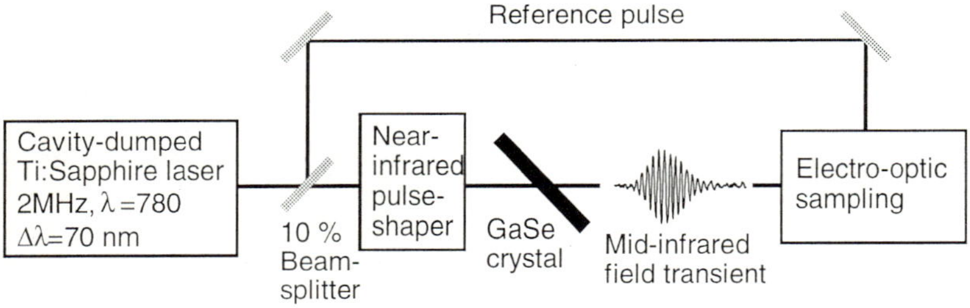

Fig. 1. Experimental Setup

The experimental setup is schematically shown in Fig. 1. Near-infrared pulses centered at a wavelength of 780 nm with a large bandwidth of 70 nm are generated in a cavity-dumped, modelocked Ti:sapphire oscillator [3]. Shaping of the mid-infrared pulses is obtained here in a two-stage scheme: first, the beam is passed through a programmable pulse shaper which consists of a liquid crystal modulator array placed in a zero dispersion stretcher unit [4]. In this way, amplitude- and phase-shaped near-infrared pulses are obtained. In the second step, these pulses are used to generate ultrafast mid-infrared electric field transients by phasematched difference-frequency mixing in a 0.5-mm-thick GaSe crystal [5].

A complete characterisation of the shaped mid-infrared electric-field transient is obtained by free-space electro-optic sampling, recently developed by Zhang et al. [6]. For this, the shaped mid-infrared pulse is focused onto a 10-μm thick ZnTe crystal. A 17-fs near infrared reference pulse taken as a reflection from the oscillator is focused onto the same position. The electro-optic modulation of the reference pulse induced by the ultrafast Pockels effect is detected using a pair of balanced photodiodes. By varying the time delay between the optical pump and probe pulses, the temporal waveform of the electric field transient is directly sampled.

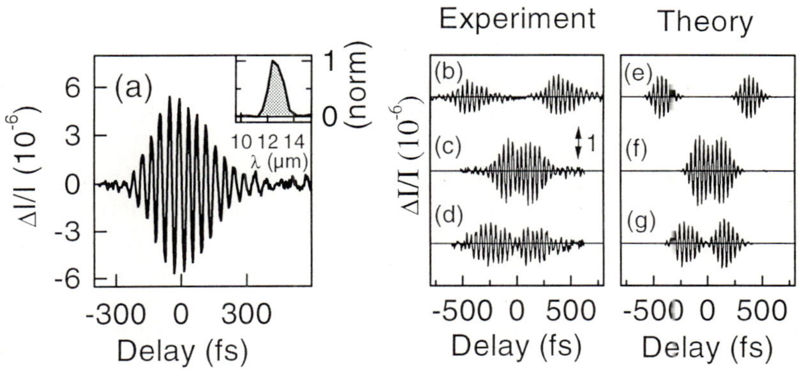

Fig. 2. (a) Mid-infrared electric field at 12.5 μm wavelength generated from an unshaped pump pulse. Inset: Fourier transform of the waveform giving the pulse spectrum. Shaped mid-infrared waveforms (left column shows experiment, right column theory) for different temporal separations of the pump pulses, (b,e) 800 fs, (c,f) 200 fs and (d,g) 60 fs.

Figure 2 (a) shows the electric-field of an unshaped mid-infrared pulse at a center wavelength of 12.5 μm. The corresponding spectrum as obtained by Fourier transformation of the waveform is shown in the inset. The high sensitivity of electro-optic sampling allows a complete electric field characterization of mid-infrared pulses even on a sub-pJ level.

In Fig. 2 (b) the mid-infrared waveform generated by two near-infrared pulses with a mutual separation of 800-fs is shown. As expected one obtains two mid-infrared pulses with the same distance as the pump pulses. What happens, when the distance between the two near-infrared pulses is lowered and ultimately comes into the range of the mid-infrared pulse length? The result of such measurements is given in Figs. 2 (c) and 2 (d). For a near-infrared pulse separation of 200 fs we already observe a joint electric field transient due to the overlap of the two generated mid-infrared pulses. However, a further reduction of the separation of the two input pulses down to 60 fs leads to a surprising result: We observe again two independent mid-infrared pulses with 500 fs separation.

This unexpected behaviour can be explained as follows: When the two pump pulses are widely separated then each of the two input pulses generates its own infrared field by difference frequency mixing of the respective ordinary with the extraordinary polarization component of each pulse. As the separation of the near-infrared pulses approaches the mid-infrared pulselength the generated mid-infrared pulses merge into a single pulse [Fig. 2 (c) and (f)]. If the separation of the two pump pulses is further decreased then a second mid-infrared pulse occurs [Fig. 2 (d) and (g)] which is generated by the extraordinary component of one input pulse interacting with the ordinary component of the second input pulse meeting a second time due to the different group velocities.

The coherently controlled shaping of mid-infrared electric field transients is fully confirmed by model calculations shown in Figs. 2(e,f,g). Our theoretical model is described in detail in Ref. [3]. We observe a good agreement of the experiment and theory which promises the generation of more complex electric field patterns using appropriately shaped input pulses.

In conclusion, we report the first generation of **shaped** high repetition-rate femtosecond mid-infrared pulses. We used a pulse shaper to influence near-infrared pulses from a cavity-dumped Ti:Sapphire oscillator and a GaSe crystal for difference frequency mixing these pulses to generate the mid-infrared. As an example we generated a series of double pulses with different temporal distance. We detected them with an electro-optic technique and found that the results can very well be predicted with our theoretical model. This now enables us to shape mid-infrared pulses in a controlled manner which can be used for coherent control experiments in the mid-infrared wavelength range.

References

1 Assion, A. et al., " Control of Chemical Reactions by Feedback-Optimized Phase-Shaped Femtosecond Laser Pulses," Science, Vol. **282**, 919-922 (1998).

2 Baumberg, J.J. et al., " Ultrafast coherent carrier control in quantum wells," J. Opt. Soc. Am. B, Vol. **13**, 1246-1250 (1996).

3 M. S. Pshenichnikov et al., "Generation of 13-fs, 5-MW pulses from a cavity-dumped Ti:sapphire laser," Opt. Lett. **19**, 572-574 (1994).

4 A. M. Weiner et al., "Programmable Shaping of Femtosecond Optical Pulses by Use of 128-Element Liquid Crystal Phase Modulator," IEEE J. Quantum Electron., Vol. **28**, 908-919 (1992).

5 R.A. Kaindl et al., "Broadband phase-matched difference frequeny mixing of femtosecond pulses in GaSe: experiment and theory," Appl. Phys. Lett. **75**, 1060-62 (1999).

6 Q. Wu and X.-C. Zhang, "Free-space electro-optic sampling of mid-infrared pulses," Appl. Phys. Lett. **71**, 1285-86 (1997).

Spatial Replenishment and Self–Focusing Dynamics of Ultrashort Laser Pulses

K.D. Moll and Alexander L. Gaeta

School of Applied and Engineering Physics, Cornell University,
Ithaca, NY 14853, USA
E-mail: a.gaeta@cornell.edu

Abstract. We investigate the collapse dynamics of an ultrashort pulse. Our results indicate that multiple collapse events can occur and that supercontinuum generation (SCG) is due to the formation of an "optical shock" wave at the point of collapse.

1 Introduction

The nonlinear propagation of an ultrashort pulse through a transparent medium can result in dramatic changes to its temporal, spatial, and spectral properties. The primary process responsible for these changes is self-focusing, which causes the pulse to be compressed in space resulting in a corresponding increase in the peak intensity. For input powers that are above the critical power P_{cr} for self-focusing, steady-state theoretical analysis predicts that the pulse will undergo catastrophic collapse. For experiments in the long-pulse (>1 ns) regime, this behavior results in plasma formation which can lead to filamentation [1] and damage [2]. For femtosecond pulses, theoretical and experimental work has shown that the effects of dispersion in the medium can completely alter the dynamics of the self-focusing process. For example, even for input powers significantly above the critical power for self-focusing, the effects of dispersion can halt the explosive increase in the peak intensity through the process of pulse splitting [3,4].

2 Theoretical Investigation of Self-Focusing Dynamics

We theoretically studied the nonlinear dynamical behavior of the self-focusing of femtosecond pulses both near and above the point at which the pulse effectively undergoes catastrophic collapse. Even for pulses substantially longer than a single optical cycle, our results indicate that the dynamics can only be described by a model that goes beyond the slowly-varying envelope approximation [5]. The model we propose includes a self-steepening and space-time focusing term as well as plasma formation. We show that as the pulse approaches the collapse point, a steep edge is formed at the back of the pulse (i.e., an optical "shock wave") and is accompanied by a large phase jump

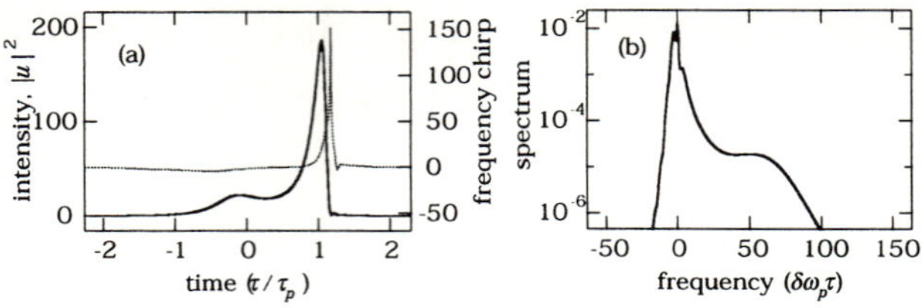

Fig. 1. (a) Temporal profile of the on-axis intensity and frequency chirp for the case in which the input power is near the threshold for catastrophic collapse, and (b) the corresponding total pulse spectrum.

[see Fig. 1(a)]. The resulting pulse spectrum [see Fig. 1(b)] exhibits a broad blue-shifted pedestal with a sharp cut-off, which is in good qualitative agreement with the experimentally observed SCG spectrum [6]. At higher powers, plasma defocusing limits the collapse. However, the pulse may self-focus to multiple collapse points provided sufficient energy remains.

It has been proposed (see [2]) that SCG is a result of plasma generation. Our theoretical treatment without space-time focusing and shock terms results in the formation of a steep edge at the front of the pulse which produces a broad red-shifted pedestal, clearly showing that mechanisms beyond plasma formation are necessary for SCG.

3 Experimental Observations of Spatial Dynamics

In our studies of the spatial dynamics of self-focusing, we use the beam from a Ti:sapphire multi-pass amplifier generating 50-fs pulses which is focused into sapphire and fused-silica samples. A microscope and digital camera are used to observe the light from the beam that is scattered in the transverse direction. As the pulse energy is increased, the location of the collapse point is seen to move towards the front face of the sample as predicted by numerical simulations [7]. At higher powers, the pulse refocuses and forms a second

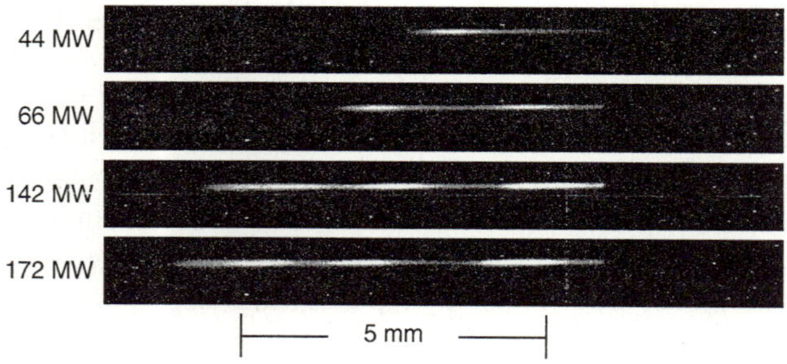

Fig. 2. Formation of multiple foci as the pulse energy is increased.

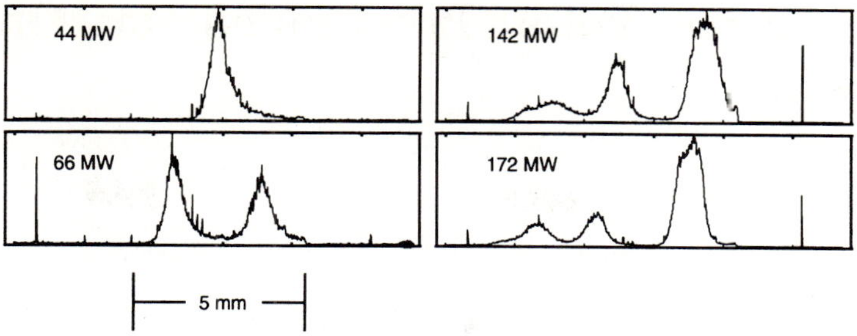

Fig. 3. Line–outs of the photos of multiple replenishment for various peak powers.

collapse point [see Fig. 2]. Similar spatial replenishment is predicted in the propagation of ultrashort pulses through air [8]. Further increase in the pulse energy results in the distance between the two collapse points to decrease. Additional collapse events also occur at still higher powers. Figure 3 displays line-outs of the photos acquired with a digital camera as the input power is increased.

4 Conclusion

We have investigated a model which demonstrates that the SCG process is a result of self-steepening and space-time focusing which cause an "optical shock" to develop in the medium. We have also seen that multiple collapse events may occur when there is sufficient power for self-focusing to overcome the effects of plasma defocusing at previous collapse points.

References

1. R. Y. Chiao, E. Garmire, and C. H. Townes: Phys. Rev. Lett. **13**, 479 (1964); Y. R. Shen and M. T. Loy: Phys. Rev. A **3**, 2099 (1971); E. Yablonovitch and N. Bloembergen: Phys. Rev. Lett. **29**, 907 (1972).
2. N. Bloembergen: J. Quantum Electron. **QE-10**, 375 (1974).
3. J. E. Rothenberg: Opt. Lett. **17**, 583 (1992); P. Chernev and V. Petrov: Opt. Lett. **17**, 172 (1992); G. G. Luther, A. C. Newell, J. V. Moloney, and E. M. Wright: Opt. Lett. **19**, 789 (1994).
4. J. K. Ranka, R. W. Schirmer, and A. L. Gaeta: Phys. Rev. Lett **77**, 3783 (1996); S. A. Diddams, H. K. Eaton, A. A. Zozulya, and T. S. Clement: Opt. Lett. **23**, 379 (1998).
5. J. K. Ranka and A. L. Gaeta: Opt. Lett. **23**, 534 (1998).
6. A. Brodeur and S.L. Chin: Phys. Rev. Lett. **80**, 4406 (1998).
7. J.H. Marburger: Prog. Quant. Electr. **4**, 35 (1975).
8. M.Mlejnek, E.M. Wright, and J.V. Moloney: Opt. Lett. **23**, 382, (1998); A.Talebpour, S. Petit, and S.L. Chin: Opt. Comm. **171**, 285 (1999).

Femtosecond Pulse Propagation in Optical Fibers

Fiorenzo G. Omenetto, Dzmitry Yarotski and Antoinette J. Taylor

Materials Science and Technology Division, MST-10, MSK764
Los Alamos National Laboratory, Los Alamos NM 87545,
USA
E-mail: omenetto@lanl.gov, dzmitry@lanl.gov, ttaylor@lanl.gov

Abstract: Femtosecond pulse propagation at 1.55 μm through a 90-meter link of single-mode optical fiber is analyzed. Observation and characterization of pulse evolution as a function of energy, distance, soliton formation and soliton self-frequency shift are shown.

Ultrafast pulse propagation in optical fibers offers challenges both in the basic and applied sciences. The short duration of the optical pulses makes higher order nonlinear effects dominant affecting deeply the study of short pulse propagation and its related phenomena such as soliton formation, phase dynamics and pulse deformation.

A variety of techniques that recover the intensity and phase of the pulse have been developed in recent years [1-3], driven by the progressively decreasing duration of pulses down to just a few optical cycles. The availability of a complete ensemble of information (knowledge of the electric field's temporal envelope, phase and spectral properties) in the context of fiber optics, provides a unique probe into ultrashort pulse propagation dynamics and nonlinear effects in fibers.

By adapting Frequency Resolved Optical Gating (FROG) [2] to the infrared wavelengths (λ=1.55 μm) of the third window for optical communications we were able to demonstrate the detection and accurate characterization of shaped femtosecond pulses [4] and describe in detail the phase dynamics leading to the formation of femtosecond optical solitons in single-mode fibers [5,6].

This paper describes new experiments that have been performed to obtain more information on pulse behavior for different fiber lengths and pulse energies. The approach is briefly outlined as follows: pulses having duration τ~170 fs generated by an optical parametric oscillator tuned at λ= 1.55 μm (E=3.5 nJ at 80 MHz) are attenuated and coupled into a link of conventional single-mode optical fiber (Corning SMF-28). After propagation, the output pulses from the fiber are sent to a single-shot SHG-FROG. The pulse energies are carefully varied before being launched into a fixed length segment of fiber and FROG traces are correspondingly detected for the indiviual energy values. The fiber is then cut (10 m per cut in the case described) and the process repeated. This allows *(i)* to perform energy-dependent analysis while keeping the fiber length fixed and *(ii)* to trace the propagation of the pulse for a fixed energy through the length of the fiber. While the former approach has been used to assess soliton formation dynamics [5] indirectly by varying the effect of the nonlinearity through the variation of the input pulse energy, the latter provides a means to follow the propagation of the pulse (and its phase and spectral evolution) through the fiber, thereby looking "directly" at the pulse dynamics during propagation and enabling the characterization of various fiber parameters. By examining the ensemble of data we can trace the collective dependence of the pulse duration as a function of pulse energy launched into the fiber. An observable shortening of the pulse duration can be detected for input pulse powers around 18 mW, independently of the fiber length at hand and oscillations in the pulse duration can be detected. As chirped solitons propagate, the dispersive energy associated with the initial chirp gradually leaves the soliton in a series of periodic bursts accompanied by characteristic oscillations in the peak amplitude of the pulse. These dynamics were described theoretically in 1984 by Desem and Chu [6] and originate from the pulse redistributing its energy within the center and the wings of the pulse, eventually settling to the predicted soliton waveform. In addition, the pulse is experiencing a self-frequency shift as its duration gets shorter. This effect was first observed by Mitschke and Mollenauer and is a transfer of energy from the higher to the lower frequency components of the soliton pulse due to the Raman effect [7]. This shift is quite evident in the FROG data where the trace can be seen moving towards the longer wavelength region of the spectrogram as the power of the input pulse is increased. The results that illustrate a sampling of the observed collective frequency shift are illustrated in figure 1. The largest shift is observed, as expected, for the longest segment of fiber and is in excess of 12% of the fundamental wavelength. The dependence of the self-shifting on the peak power (i.e. temporal duration) of the pulse is also evident, with more shift occurring per unit length as the pulse duration gets shorter.

As the pulse is shifted, the soliton nature of the pulses is maintained. The solitonic nature of the frequency shifted pulses can be investigated by observing the phase profile of the pulse. As an example, we show in figure 2, the

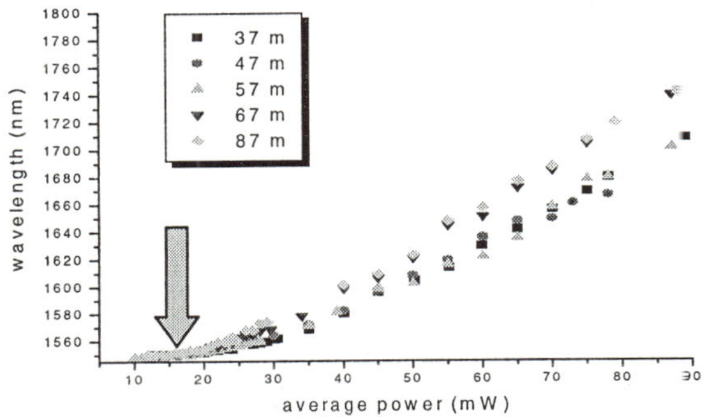

Fig.1. Wavelength shift of the output pulses for a set of fiber lengths. The appearance of a continuous shift from the fundamental value of 1550 nm starts becoming visible for input average powers around 16 mW.

deconvolution of a FROG trace obtained from a strongly shifted (100 nm) pulse for a length of Z=67 m. The reconstructed pulse duration is τ=220 fs and its phase profile is flat across the pulse, a signature of soliton formation. By individual analysis of the reconstructed FROG traces, a more accurate analysis of the oscillation leading to asymptotic soliton formation can be performed. The global analysis is underway at the time of writing and has verified the propagation of a stable soliton at low energy with flat phase throughout the entire fiber link.

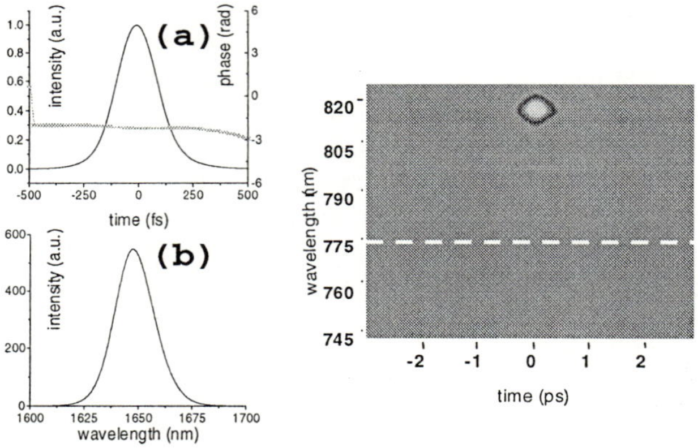

Fig.2. Reconstructed (a) electric field and phase profile and (b) spectrum for a strongly shifted pulse. The phase is noticeably flat across the pulse. The recovered time value is τ=220 fs. Also shown on the right is the corresponding shifted SHG-FROG trace

In conclusion, application of phase sensitive measurement techniques to this class of problems can be of great help in gaining key insight on femtosecond pulse propagation dynamics. Nonlinear effects such as Raman–induced soliton self-frequency shift can be observed and the global temporal and spectral evolution of femtosecond pulses can be traced. The phase characteristics which are accessible will permit accurate characterization of these and other nonlinear effects which affect short pulse propagation.

References

1. C. Iaconis and I. A. Walmsley, Opt. Lett, **23**, 792 (1998)
2. R. Trebino, K. W. DeLong, D. Fittinghoff, J. Sweetster, M. A. Krumbugel and D. Kane, Rev. Sci. Instr., **68**, 3277 (1997)
3. J. W. Nicholson, J. Jasapara, W. Rudolph, F. G.Omenetto and A. J. Taylor, Opt. Lett., **24** (23), 1774, 1999
4. F. G. Omenetto, J. W. Nicholson and A. J. Taylor, Opt. Lett. , **24** (23), 1780, 1999
5. F. G. Omenetto, B. Luce, D. Yarotsky, and A. J. Taylor, Opt. Lett., **24**, 1392, 1999
6. C. Desem and P. L. Chu, Opt. Lett, **11** (4), p. 248, 1986
7. F. M. Mitschke and L. F. Mollenauer, Opt. Lett., **11** (10), 659, 1986

Microcavity-enhanced surface-emitted second-harmonic generation for signal processing in the ps and fs regimes

Todd G. Ulmer[1], Brian R. Washburn[1], Anthony J. SpringThorpe[2], Carl M. Verber[1], and Stephen E. Ralph[1]

[1] School of Electrical and Computer Engineering, Georgia Institute of Technology, Atlanta, Georgia 30332-0250 USA
E-mail: stephen.ralph@ece.gatech.edu
[2] Nortel Networks Microelectronics, Dept. 5C14, Ottawa ON K2H 8E9, Canada

Abstract. Resonant cavity enhancement of second-harmonic generation using ps and sub-ps pulses is experimentally evaluated and numerically analyzed. We assess the spectral filtering/cavity lifetime of the microcavity, and demonstrate that significant enhancements are possible for pulses as short as 150 fs.

1. Introduction

Signal processing applications such as high-speed optical time-division demultiplexing [1] require pulse widths on the order of 1 ps or less. In this regime, the large spectral content of the pulses presents serious challenges for the exploitation of nonlinear effects in cavities. Enhancement of collinear second-harmonic generation (SHG) by resonating either the fundamental or the SH has long been appreciated [2]. Recently, these enhancements have also been applied to the surface-emitted geometry (SESHG) [3-4], where fundamental pulses counter-propagating in a waveguide produce SH that propagates vertically. Previous work has been limited to the quasi-cw case, where the pulse bandwidth is small. Here we report on sub-ps pulses, where the bandwidth can be much larger than the spectral width of the cavity resonance. Thus, the spectral filtering effect of the cavity must be considered. Furthermore, an optimum cavity finesse exists that is dependent on the spectral content of the pulses, the size of the cavity, and the absorption. Preliminary experimental results for an AlGaAs structure with a dielectric top mirror agree qualitatively with the numerical analysis, and suggest that the enhancement can be significant even for 150-fs pulses.

2. Theory

We consider a multilayer waveguide that is quasi-phase matched in the vertical direction [1, 3]. The resonant cavity is formed by distributed Bragg reflectors (DBRs) on the top and bottom surfaces of the waveguide. Combining the approaches in [2] and [5], we find the frequency-dependent SH resonant enhancement to be

$$\frac{P_t^{(2\omega)}}{P_0^{(2\omega)}} = \frac{\left(1-|r_t|^2\right)\left(1+|r_b|\exp(-\alpha d/2)\right)^2}{\left(1-|r_t||r_b|\exp(-\alpha d)\right)^2 + 4|r_t||r_b|\exp(-\alpha d)\sin^2\frac{\theta}{2}} \tag{1}$$

where $P_t^{(2\omega)}$ is the SH power transmitted through the top mirror, $P_0^{(2\omega)}$ is the SH power generated in one direction in the absence of the resonator, r_t and r_b are the top and bottom SH field reflectivities, α is the average power absorption coefficient for the cavity, d is the distance between the mirrors, and θ is the round-trip phase shift in the cavity. The round-trip phase shift includes the penetration depth into the Bragg reflectors [6], which can be large for low-contrast mirrors.

3. Results and Discussion

The SH efficiency is determined experimentally in a rear-facet reflection geometry, where a single pulse is coupled into the waveguide and allowed to reflect through itself at the opposite waveguide facet to generate SH. The multilayer structure incorporates an AlGaAs lower DBR and waveguide core, as described in [1]. The top dielectric DBR has a measured reflectance of 80%. Using a waveguide sample with a top antireflection coating as a reference, we measure an enhancement of 20 for a 1-ps FWHM fundamental and 2 for a 150 fs fundamental. This is shown in Fig. 1a and 1b.

We have implemented a numerical optimization by varying the top mirror reflectance with all other cavity parameters fixed. The lower mirror is an AlGaAs DBR with a power reflectance of >99% so that most of the SH is emitted through the top mirror rather than into the substrate. The number of top DBR layers is then varied for given values of the attenuation. For simplicity, we have assumed that dispersion of the refractive index and $\chi^{(2)}$ are zero. Results are shown in Fig. 1. The results are normalized to the case where $r_t = 0$ for comparison with the experimental data, and thus the total cavity enhancement is up to a factor of four larger than shown.

4. Conclusions

We have shown that resonant cavity enhancement of sub-ps surface-emitted second-harmonic pulses can be large. With optimization of the cavity parameters, enhancements of up to 28 (×4) for 1-ps fundamental pulses and up to 5.5 (×4) for 150-fs pulses are possible.

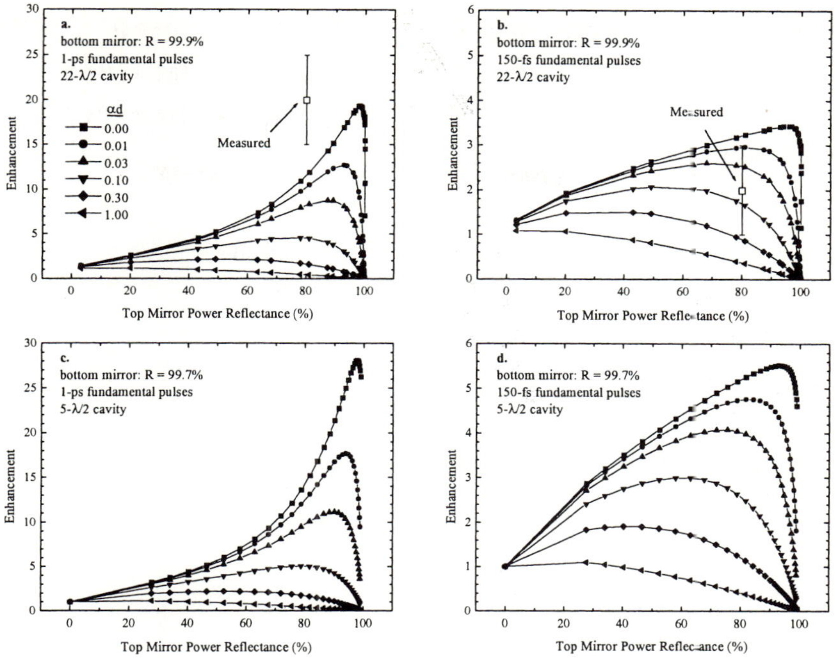

Fig. 1. Calculated resonant enhancement of second-harmonic power as function of top mirror reflectance for (a.) 1-ps FWHM 1.55-μm fundamental pulses in a 22-λ/2n cavity, (b.) 150-fs pulses in a 22-λ/2n cavity, (c.) 1-ps pulses in a 5-λ/2n cavity, and (d.) 150-fs pulses in a 5-λ/2n cavity. The cavity absorption is given by αd. All results are normalized to the power for $r_t = 0$, and thus the enhancement shown is relative to the case where only the lower mirror is present.

References

1 T. G. Ulmer, M. C. Gross, W. S. Astar, P. W. Juodawlkis, B. R. Washburn, A. J. SpringThorpe, R. P. Kenan, C. M. Verber, and S. E. Ralph, in Materials and Devices for Photonic Circuits, SPIE 44th annual meeting, Denver, CO, July 21-22, 1999.

2 A. Ashkin, G. D. Boyd, and J. M. Dziedzic, IEEE J. Quantum Electron. **QE-2**, 109-124 (1966).

3 R. Lodenkamper, M. L. Bortz, M. M. Fejer, K. Bacher, and J. S. Harris, Jr., Opt. Lett. **18** 1798-1800 (1993).

4 S. Janz, Y. Beaulieu, A. Fiore, P. Bravetti, V. Berger, E. Rosencher, and J. Nagle, Opt. Express **2** 462-470 (1998).

5 D. Vakhskoori, J. Appl. Phys. **70** 5205-5210 (1991).

6 D. I. Babic and S. W. Corzine, IEEE J. Quantum Electron. **28** 514-524 (1992).

Femtosecond Time-Resolved Fluorescence Spectroscopy Utilizing Optical Kerr Shutter: Direct Observation of a Temporal Fluorescence Band Shape

Tetsuhiko Nagahara[1], Yasuo Kanematsu[2], and Tadashi Okada[1]

[1] Department of Chemistry, Osaka University, Toyonaka, 560-8531, Japan
E-mail: nagahara@chem.es.osaka-u.ac.jp, okada@chem.es.osaka-u.ac.jp
[2] Venture Buisiness Laboratory, Osaka University, Suita, 565-0871, Japan
E-mail: kanematu@ mls.eng.osaka-u.ac.jp

Abstract. A novel femtosecond fluorescence spectrum measurement system using optical Kerr shutter has been developed. Wide range of 400nm wavelengths are obtained at once. Temporal changes in fluorescence spectra of a styryl dye DASPI are demonstrated.

1. Introduction

Time-resolved fluorescence (TRF) measurements in femtosecond regime have been intensively investigated since fluorescent probe molecules are often used to study solvation dynamics, which may control chemical reactions in solution phase [1,2]. Solvation dynamics in this regime should be further investigated by quantitative evaluation of its time-resolved spectral shape, which reflects not only diffusion process but also vibrations of the system including solvent molecules.

Optical Kerr shutter (OKS) has been known as the most effective method to obtain temporal sections of ultrafast phenomena such as TRF. Duguay and Hansen used CS_2 liquid as a Kerr material [3], which has instantaneous electronic optical Kerr effect (OKE) response as well as picosecond slow response due to molecular reorientation. Nonresonant-type glasses, however, has exclusively electronic response. Utilizing the glass plate as a Kerr material, a novel OKS was constructed. Our TRF measurement system using the combination of OKS and liq. N_2 cooled CCD spectrograph enables us to obtain TRF spectra "directly". Much more reliable TRF spectra to evaluate the spectral shape would be obtained in comparison with conventional methods such as fluorescence upconversion [4] or transient gain spectrum measurement by pump-probe technique [5]. In both cases, sophisticated processing is needed to generate reliable TRF spectra from acquired data.

2. Experimental Methods

An amplified Ti:Sapphire laser (150fs FWHM, 900mW, 200kHz, 830nm) was used for TRF measurements; its output of second harmonics (ave. power 10mW) is used for sample excitation, the rest fundamental attenuated (ave. power 20mW) is optically delayed and used for gating, respectively.

Details of apparatus are illustrated in Fig. 1. Nonresonant-type inorganic glass, PBH21 (OHARA glass) was used as Kerr material. Efficiency of OKS is known to be comparable to that of CS_2 [6]. The OKE response of this glass itself was checked by optical heterodyne detected OKE experiment using 20fs Ti:sapphire oscillator. As shown in Fig. 2., OKE response is exclusively electronic. Therefore, time resolution of this apparatus strongly depends on the geometry of sample excitation, fluorescence collimation, and fluorescence light focusing onto Kerr material. In order to avoid spatial dispersion that makes response worse, the excited volume should be as small as possible. For this reason, excitation pulse is focused by a lens which has short focal length of 50mm onto a 300μm thick flowing quartz cell containing sample solutions; a Cassegranian is used for collecting and collimating fluorescence light; both fluorescence light and optically delayed gating pulse are focused by a off-axis parabolic mirror (focal length: 25mm) onto thin glass plate (thickness: 300μm). Additionally, a thin 50μm diameter pinhole is attached in front of the glass plate to avoid photons from different optical paths. Although this thin glass plate and pinhole reduce transmitted fluorescence intensity, still a few percent of total fluorescence can be collected. In order to reduce group delay dispersion (GDD) of this system, a thin film polarizer (thickness: 2mm) is used. In order to avoid reabsorption, dye concentration was adjusted to obtain an absorbance around 0.2 at 415nm.

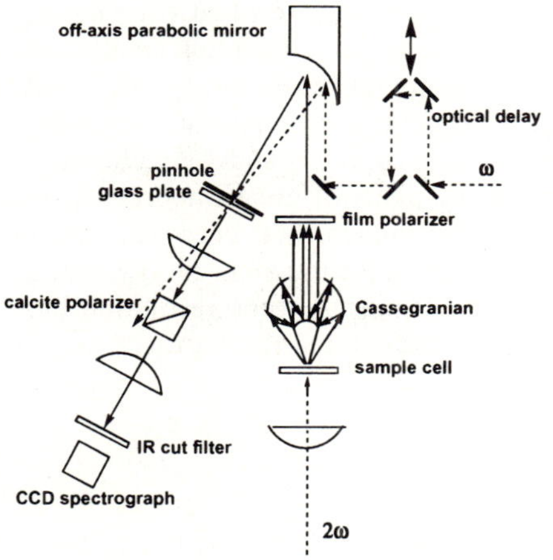

Fig. 1. Schematic illustration of apparatus

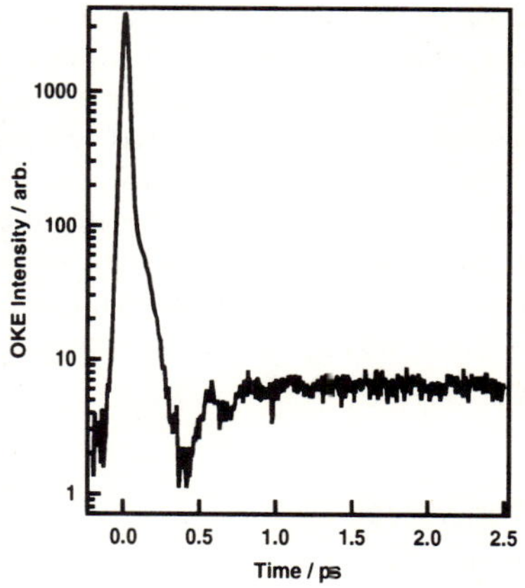

Fig. 2. OHD-OKE signal of PBH21

3. Results and Discussion

Fig. 3 illustrates the TRF spectra of DASPI in methanol after excitation. In spite of raw data, TRF spectra have been clearly seen without any processing. From these data, availability and reliability of this method might be easily recognized. As we mentioned, this method is reliable to evaluate spectral shape without consuming much time for data acquisition and processing. As a result,

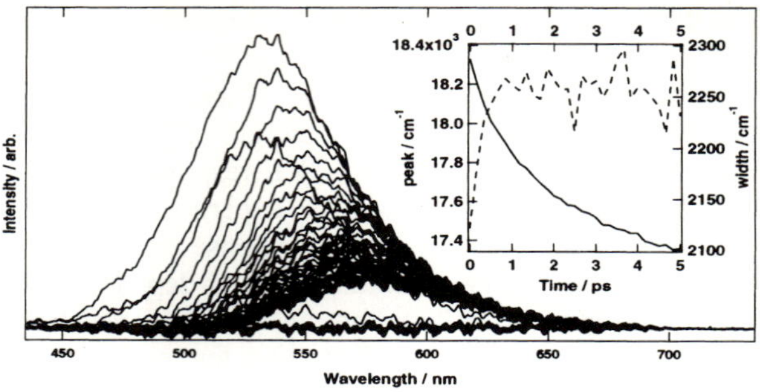

Fig. 3. Time-resolved fluorescence spectra in methanol (167fs / step), peak position (inset, solid line), and bandwidth (inset, dashed line) against time in cm^{-1}.

peak position and bandwidth against time (Fig.3, inset), which indicate relaxation of average energy and relaxation of dispersion around average energy respectively, were easily detected.

Due to the pulse width of laser used, time resolution was not sufficient to detect the fast initial changes of the spectra, however, temporal peak shift was clearly observed (Fig.3 inset, solid line). From spectral shift correlation function, corresponding relaxation times for our experiment were τ_1=0.73ps (30%) and τ_2=6.6ps (60%). The instrumental time resolution was determined to be c.a. 200fs by measuring either the cross-correlation signal between residual excitation and gate pulses or signal of Raman line in pure solvent. The apparatus enabled us to obtain wide range of spectra at once with this temporal resolution of 200fs, which totally restricted by laser pulse width used. To get higher temporal resolution, an apparatus using cavity-dumped 13fs laser as a light source is under developing.

Detection of TRF spectra in femtosecond regime is of great importance, since the fluorescence dynamics relate directly to the fundamental chemical reactions such as electron/charge transfer, proton/hydrogen-atom transfer, energy transport as well as isomerization reaction in solution. Making the time resolution higher will enable us to perform the detailed analysis of the initial rapid processes. This would be done soon.

References

1 S. J. Rosenthal, X. Xie, M. Du, and G. R. Fleming, J. Chem. Phys. **95**, 4715 (1991)

2 K. Yoshihara, K. Tominaga and Y. Nagasawa, Bull. Chem. Soc. Jpn. **68**, 696 (1995)

3 M.A. Duguay, J. W. Hansen, Appl. Phys. Lett. **15**, 192 (1969)

4 J. Shah, IEEE J. Quantum Electron. **24**, 2, 276 (1988)

5 D. Bingemann and N. P. Ernsting, J. Chem. Phys. **102**, 7, 2691 (1995)

6 (a) Y. Kanematsu, H. Ozawa, I. Tanaka, and S. Kinoshita, J. Lumin. **87-89**, 917 (2000); (b) S. Kinoshita, H. Ozawa, Y. Kanematsu, I. Tanaka, N. Sugimoto, and S. Fujiwara, to be published in Rev. Sci. Instrum. (2000)

Part IV

THz Generation and Measurement

An intense broadband terahertz source based on a novel four wave rectification process

D. J. Cook, J.-X. Chen and R. M. Hochstrasser

Department of Chemistry, University of Pennsylvania, 231 S. 34th St., Philadelphia, PA 19104-6323. USA
E-mail: hochstra@sas.upenn.edu

Abstract. The fundamental and second harmonic pulses from an ultrafast laser are mixed in a gaseous medium to generate an intense, (~ 1 kV cm^{-1}) broadband (≥ 4 THz) pulse through a process which resembles third order optical rectification.

1. Introduction

We have developed a new method for the generation of intense broadband THz pulses, which resembles four wave rectification (FWR)[1]. In the process of four wave rectification, ultrafast pulses with frequencies ω_0 and 2 ω_0 interact; the carrier frequencies of three input fields add to zero, and the non-linear response is driven by the product of the field envelopes of the input pulses. In one possible mechanism for FWR, the nonlinear polarization induced by the input waves acts as the source for the radiated THz. This is essentially the inverse of the previously reported method of Terahertz Field Induced Second Harmonic generation (TFISH) [2], in which a THz field and an optical field at ω_0 interact to generate 2 ω_0 ± ω_{THz}. The THz pulses generated by the FWR process were initially observed in ambient air; subsequent measurements in a variety of gases have obtained signals of similar intensity. The measured bandwidth and peak field of these pulses are sufficiently great (i.e. comparable or greater than other sources) that this method can be employed as a practical laboratory THz source.

2. Experimental Methods

In the experiments presented here, pulses were used from a regeneratively amplified Ti:sapphire laser system at a repetition rate of 1 kHz. In order to generate the THz radiation, 150 μJ, 65 fs, 800 nm pulses are focused through a beta barium borate (BBO) crystal and then to a beam waist in which the 800 nm light is mixed with the second harmonic generated by the BBO crystal. Following the focus, an off axis paraboloid was used to collimate the radiated THz. The majority of the 400 nm and 800 nm light passes through a hole in the paraboloid, and a Si window rejects the residual optical energy and transmits the THz. The THz is then focused into a ZnTe crystal and detected using the EO detection method.

In a further effort to elucidate the THz generation mechanism, a gas cell was used to study the dependence on the gas composition and pressure. N_2, O_2, and the noble gases were studied, and the absolute pressure was varied from vacuum to 5 atmospheres. In order to ensure proper phase locking (see below) the BBO crystal was placed inside the gas cell. Polycrystalline diamond was used for the exit window in order to transmit both the generated THz and residual optical pulses.

3. Results and Discussion

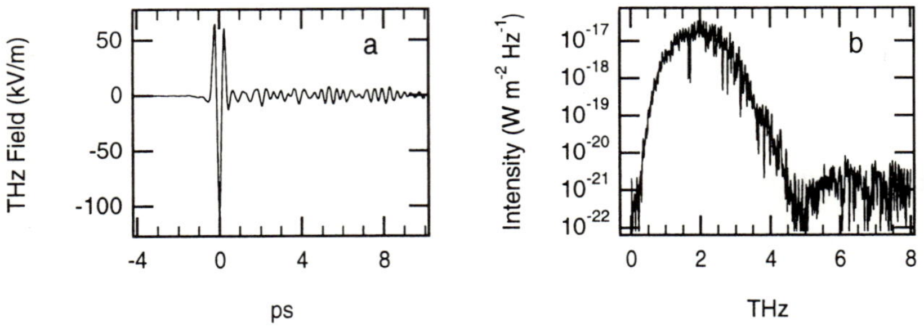

Fig. 1. (a.) The time dependent THz field from FWR in ambient air, and (b.) the associated power spectrum. The spectrum at frequencies above 3 THz is perturbed by the 5.3 THz transverse optical phonon resonance in the ZnTe crystal used for the EO detection.

Fig. 1a shows the time dependent THz pulse generated by the FWR process in ambient air. The Fourier transformed power spectrum is depicted in Fig. 1b. The field is estimated by assuming the effects of the linear response of the ZnTe EO crystal are negligible over the length of the crystal. If the area of the THz beam waist in the EO crystal is known, the energy of the THz pulse can then be estimated. For the pulse in Fig. 1, the fluence in the ZnTe crystal is estimated to be 2.7 nJ cm^{-2}. The assumption of a diffraction limited THz beam waist (≈250 μm) yields a detected energy ~ 1 pJ.

Phase Dependence. A THz signal generated by a FWR process will depend on the relative phase of the fundamental and second harmonic pulses. The phase dependence can be readily illustrated by considering the third order polarization:

$$P_i^{(3)}(t) = \gamma_{ijkl}\, E_j^{2\omega_0}(t)\, E_k^{\omega_0}(t)\, E_l^{\omega_0}(t). \tag{1}$$

Associating E_k and E_l with the 800 nm pulse (ω_0), $E_k(t) = E_l(t) = 1/2\ \mathcal{E}_{800}(t)\, e^{i\omega_0 t} + c.c.$ Likewise, $E_j = 1/2\ \mathcal{E}_{400}(t)\, e^{i(2\omega_0 t+\varphi)} + c.c.$, where φ is the phase difference between the two pulses. The product of the fields results in a $4\omega_0$ term, a $2\omega_0$ term, and a rectified term. Noting that the 400 nm radiation is generated by SHG from the 800 nm pulse, $\mathcal{E}_{400}(t) \propto \mathcal{E}_{800}^2(t)$, and $P_i^{(3)}(t) \propto \mathcal{E}_{800}^4(t)\cos\varphi$.

Effects of Gas Composition and Pressure. The pressure dependent data exhibit a clear threshold at low pressure and saturation at high pressure. Oscillatory behavior is also observed and is associated with the aforementioned phase dependence; between the BBO crystal and the focus, the 400 nm and 800 nm pulses acquire a differential phase shift which is associated with the dispersion of the intervening gas. The maximum signal is not greatly affected by the nature of the gas. However, the appearance and saturation pressures depend on the gas composition. There appears to be a correlation between the ionization potential and the appearance and saturation pressures. These data therefore indicate a plasma driven process. In this regard, the mechanism for THz generation may be related to previous experiments in which THz radiation has been generated from plasmas using single color high field pulses[3], and also to experiments in which two colors at ω_0 and $2\omega_0$ generate THz through quantum interference of photocurrents in semiconductors[4]. However, other aspects such as the spectrum and polarization dependence are predicted by a third order FWR description.

4. Conclusions.

The intensity of the THz radiation generated by the FWR process (in the gaseous plasma) is surprisingly strong. Rectification of shorter pulses in these nonpolar gases with no far IR absorption could lead to enhanced bandwidth, with no spectral limitations associated with self absorption in the medium used to generate the THz radiation. Furthermore, if three different frequencies, one (ω') near $2\omega_0$ and two (ω'' and ω''') near ω_0, are used for the input pulses, frequency generation at $\omega' - \omega'' - \omega'''$ in the hitherto problematic region between 100 cm^{-1} and 1000 cm^{-1} should be, in principle, possible.

Acknowledgements. This work was supported by the NIH, and the NSF, and by the Research Resource grant RR13456.

References

1 D. J. Cook and R. M. Hochstrasser, Opt. Lett. *in press*, (August, 2000).
2 D. J. Cook, J. X. Chen, E. A. Morlino, and R. M. Hochstrasser, Chem. Phys. Lett. **309,** 221 (1999).
3 H. Hamster, A. Sullivan, S. Gordon, W. White, and R. W. Falcone, Phys. Rev. Lett. **71,** 2725 (1993).
4 D. Côté, J. M. Fraser, M. DeCamp, P. H. Bucksbaum, and H. M. van Driel, Appl. Phys. Lett. **75,** 3959 (1999).

THz generation by photo-ionization of electrically biased air

Torsten Löffler, Frank Jacob, and Hartmut G. Roskos

Physikalisches Institut, Johann Wolfgang Goethe – Universität Frankfurt am Main, Robert-Mayer-Str. 2-4, D-60054 Frankfurt am Main, Germany
E-mail: t.loeffler@physik.uni-frankfurt.de

Abstract. We present an experimental investigation of the generation of far-infrared (terahertz) pulses by photo-ionization of electrically biased air with amplified laser pulses. The generation process is examined in terms of the radiation pattern and signal dependence on the electric bias field, on the laser pulse energy and on the gas pressure.

1. Introduction

THz emission from photo-ionized unbiased noble gases was firstly reported by Hamster et. al. [1,2]. In his experiments, the emission of the electromagnetic transients results from the ponderomotive forces within the laser focus inducing an ultrafast charge separation in the plasma which leads to emission of THz pulses along a cone at an angle of approximately 45° with respect to the propagation direction of the laser beam. We have recently reported that a significantly increased THz emission intensity in the forward direction can be achieved, when a static electric bias field is applied to the air plasma [3].

2. Experimental and Results

A CLARK CPA 2001 laser system provides laser pulses of 150 fs duration at a wavelength of 775 nm with a repetition rate of 1 kHz. Pump pulses with an energy of 500 μJ (chopped at 500 Hz) are focused into the gap between two electrodes separated by 1 mm from each other. The THz radiation emitted from the laser-generated plasma is collected by an off-axis paraboloidal mirror with either 30° or 70° collection cone angle β as defined in Fig. 1. (a) and focused by a second one onto a 1-mm-thick <110> ZnTe crystal for electro-optic [4] detection. 5 scans with a lock-in time constant of 30 ms are averaged.

In Fig. 1. (b), the peak-to-peak amplitude of the THz radiation collected with a paraboloid with $\beta = 30°$ is depicted as function of the view angle α. In agreement with the work of Hamster et. al. [1,2], the radiation without applied field occurs predominantly along a cone at an angle α of approximately 45°. In agreement with symmetry considerations no radiation is emitted in the forward direction. In contrast, the maximum emission with applied bias field occurs in forward direction which is consistent with idea of a current surge model. The emission cone is significantly wider than 90°, which is our experimentally accessible range.

This large cone is a result of the small (sub-wavelength) size of the air plasma emitter. An increase of in the collection cone angle β to 70° by using a paraboloidal with a shorter focal length results in an increase of the THz amplitude by a factor of 2 and in a slight shift of the spectrum towards lower frequencies.

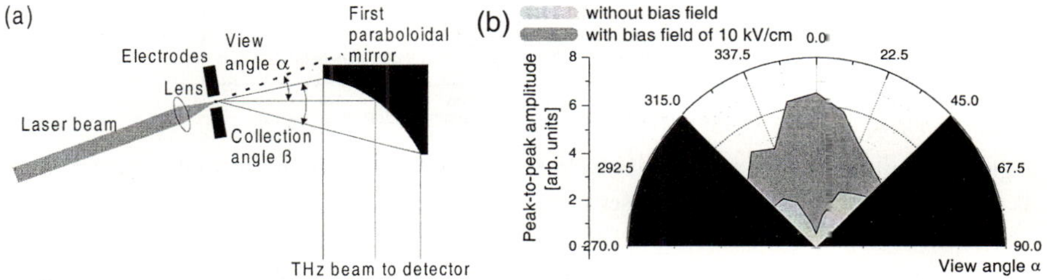

Figure 1. (a) Definition of view angle α and collection angle β. (b) Angular pattern of the THz field from an air plasma without bias (light gray) and of the additional signal induced by a 10 kV/cm bias field (dark gray). Black: region not accessible in the experiment.

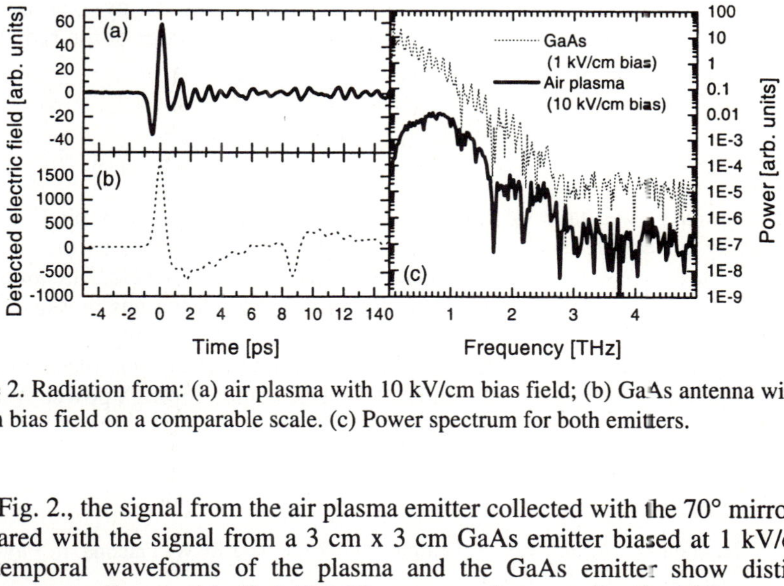

Figure 2. Radiation from: (a) air plasma with 10 kV/cm bias field; (b) GaAs antenna with 1 kV/cm bias field on a comparable scale. (c) Power spectrum for both emitters.

In Fig. 2., the signal from the air plasma emitter collected with the 70° mirror is compared with the signal from a 3 cm x 3 cm GaAs emitter biased at 1 kV/cm. The temporal waveforms of the plasma and the GaAs emitter show distinct differences as discussed in Ref. [3]. The spectra show a roll-off which cuts the noise level at about 3 THz. This is a result of the detector bandwidth being limited by the phase mismatch of the THz pulse and the optical signal within the 1-mm-thick ZnTe crystal [5]. Also water-vapor absorption lines can be identified. Although the absolute amplitude of the air plasma signal in the time domain is a factor of 20 weaker, the spectral data show that the signal-to-noise ratio above 1 THz is comparable with that of the GaAs emitter. The increased noise floor of the GaAs source is a result of both a decreased lock-in sensitivity and electromagnetic distortion by the charging and discharging of the GaAs.

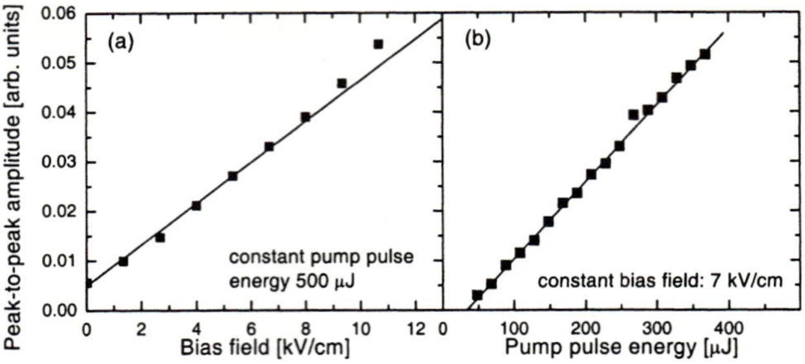

Figure 3. Amplitude of the detected electric field of the THz radiation from the air plasma as a function of: (a) bias field; (b) laser pump pulse energy.

In Fig. 3., the variation of the peak-to-peak amplitude of the THz signal with the bias field (a) and the laser pulse energy (b) is depicted. A linear relationship is found in both cases [3].

By increasing the absolute gas pressure from 1 bar to 5 bar we found an increase of the THz amplitude by 15 %. Additionally, the higher breakthrough voltage at 5 bar allows us to raise the maximum bias field from 10 kV/cm to 80 kV/cm. This results in an increase of the THz amplitude by one order of magnitude [6]. No significant changes of the spectral properties of the radiation is observed.

3. Conclusion

In summary, we have demonstrated that THz radiation from photo-ionized electrically biased air has comparable signal-to-noise properties as the radiation from a biased GaAs emitter in the spectral range from 1 to 3 THz. The amplitude of the signal from the gas plasma rises superlinearly with the gas pressure, when the maximum applicable bias field for each pressure setting is applied. We expect this to continue up to high pressures. Thus the high-pressure gas plasma emitter has the potential to became a very interesting THz source for high power systems.

References

1 H. Hamster, A. Sullivan, S. Gordon, W. White, and R. W. Falcone, in *Phys. Rev. Lett.*, Vol. 71, 2725 (1993).
2 H. Hamster, A. Sullivan, S. Gordon, and R. W. Falcone, in *Phys. Rev. E,* Vol. 49, 671 (1994).
3 T. Löffler, F. Jacob, and H. G. Roskos, in *Appl. Phys. Lett.,* Vol. 77, (17 July 2000) in press.
4 Z. G. Lu P. Campbell, and X.-C. Zhang, in *Appl. Phys. Lett.,* Vol. 71, 593 (1997).
5 P. Y. Han, and X.-C. Zhang, in *Appl. Phys. Lett.,* Vol. 73, 3049 (1998).
6 data not shown, to be published elsewhere.

Coherent THz emission from optically pumped intersubband plasmons in parabolic quantum wells

R. Bratschitsch, T. Müller, R. Kersting, G. Strasser, and K. Unterrainer

Institute for Solid State Electronics, Technical University Vienna, Floragasse 7, A-1040 Vienna, Austria
EMail: rudolf.bratschitsch@tuwien.ac.at

Abstract. We report on few-cycle THz emission from modulation-doped parabolic quantum wells which are optically excited by near-infrared femtosecond laser pulses. The observed THz emission corresponds to the intersubband plasmon of the parabolic quantum well. We identify the excitation mechanism of the intersubband plasmon and hence THz emission to be ultrafast field screening.

1. Introduction

Recently, Sekine et al. [1] reported on the optically pumped emission of THz radiation from grating-coupled *intra*subband plasmons in a doped single quantum well. We present experiments which show that *modulation doped* parabolic quantum wells (PQWs) emit coherent THz radiation corresponding to the *inter*subband plasmon when excited by near infrared femtosecond laser pulses.

2. Experimental Methods

The samples used in the experiments are modulation doped GaAs/AlGaAs PQWs, with widths in the range of 1200 - 2000 Å and carrier sheet densities of 1.7×10^{11} - 5×10^{11} cm^{-2}. We perform THz autocorrelation (AC) measurements where two temporarily delayed visible femtosecond laser pulses hit the sample. The emitted THz radiation is collected by parabolic mirrors and detected by a bolometer. The Fourier transform of the recorded AC signal then gives the spectrum of the coherent radiation emitted by the source. The samples are mounted in a continuous flow cryostat to cool them to approximately 5 K and the whole setup is purged with nitrogen gas to avoid absorption of the THz radiation by water vapor.

3. Results and Discussion

Figure 1 shows an AC trace of a modulation doped PQW (W=1400 Å, $n_{2D}=5\times10^{11}$ cm^{-2}) excited by 780 nm (τ_{FWHM} = 80 fs) laser pulses. The density of the optically

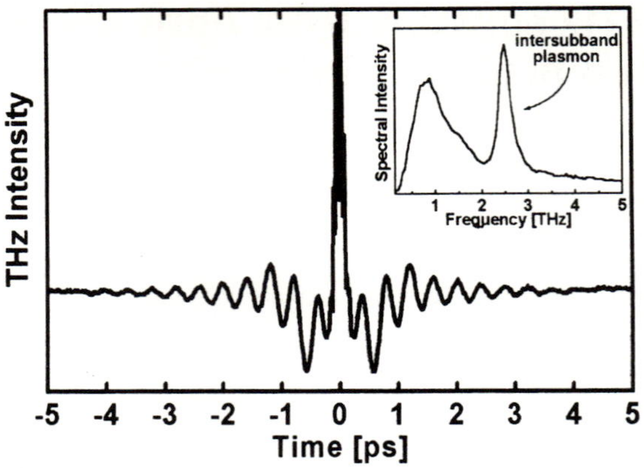

Fig. 1. THz autocorrelation signal of the 1400 Å PQW excited by 780 nm laser pulses (T = 5 K). Inset: Fourier transform of the recorded AC.

generated carriers is kept well below the carrier density inside the PQW due to the modulation doping.

The spectrum of the emitted THz radiation (inset of fig. 1) consists of two components, a broad one around 0.8 THz and a narrow one (FWHM: 0.3 THz) with a center frequency of 2.55 THz. These two emission peaks can be observed within a wide range of excitation wavelenghts (815-760 nm).

While the broadband component varies in frequency, the emission at 2.55 THz doesn't change. The low frequency broadband component is found with all the different PQW samples and shows roughly the same frequency dependence, i.e. it is independent of the PQW sample structure. The origin of the broadband component is due to THz generation at the surface of the sample [2].

The narrowband emission results from the oscillation of the carriers inside the PQW [3]. FTIR absorption and THz-Time Domain Spectroscopy measurements [4] show nearly the same resonance frequency of 2.2 THz which is the characteristic frequency of the intersubband plasmon of this PQW. The width of the THz emission line is as narrow as 0.2 THz for the 2000 Å PQW sample.

The excitation mechanism for the intersubband plasmon is due to screening of the surface depletion field by the electron-hole pairs injected by the ultrafast laser pulse [5]. In this way the electrons inside the quantum well experience a kick and begin to oscillate with their eigenfrequency (Fig. 2). This is supported by the fact that the oscillation can be excited over a large wavelength range of the femtosecond pulses (815-760 nm). This large range implies that the excitation mechanism is clearly a non-resonant phenomenon, in contrast to THz quantum beat experiments. The difference between the two excitation mechanisms - quantum beats and ultrafast field screening - is also discernible when we perform

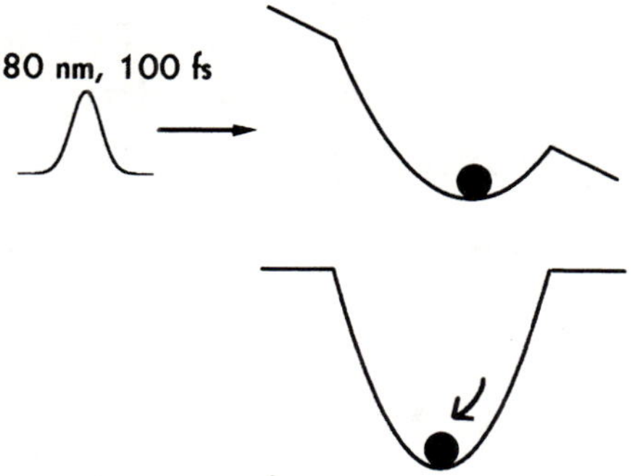

Fig. 2. Schematic drawing of the excitation mechanism: an oscillation of the carriers in the parabolic potential is initiated by ultrafast field screening.

THz emission experiments on an *undoped* PQW of the same width (L = 1400 Å). In this case no THz emission associated with the PQW can be observed, i.e. no quantum beats can be excited.

4. Conclusions

We have demonstrated optically driven THz emission from intersubband plasmons in modulation-doped parabolic quantum wells. The excitation mechanism is due to screening of the surface depletion field by electron-hole pairs injected by an ultrafast laser pulse. Due to this non-resonant excitation mechanism, a completely symmetric nanostructure can emit optically driven THz radiation. THz emission due to quantum beats can be excluded since we observe no THz radiation from an identical but undoped PQW. The combination of the designability of the transition frequency, the narrowband emission, and the absence of any processing of the sample make modulation-doped PQWs attractive and easy-to-use THz emitters.

References

1 N. Sekine, K. Yamanaka, K. Hirakawa, M. Voßebürger, P. Haring-Bolivar, and H. Kurz, Appl. Phys. Lett **74**, 1006, 1999.

2 X.-C. Zhang, B. B. Hu, J. T. Darrow, and D. H. Auston, Appl. Phys. Lett. **56**, 1011, 1990.

3 R. Bratschitsch, T. Müller, R. Kersting, and K. Unterrainer, Appl. Phys. Lett. **76**, 3501, 2000.

4 R. Kersting, R. Bratschitsch, G. Strasser, K. Unterrainer, and J. N. Heyman, Opt. Lett. **25**, 272, 2000.

5 R. Kersting, K. Unterrainer, G. Strasser, H. F. Kauffmann, and E. Gornik, Phys. Rev. Lett. **79**, 3038, 1997.

Narrow-Band Terahertz Wavetrains Generated by Optical Rectification in Periodically-Poled Lithium Niobate

T. Meade[1], Y.-S. Lee[1], V. Perlin[1], H. Winful[1], T.B. Norris[1], and A. Galvanauskas[2]

[1] Center for Ultrafast Optical Science, 2200 Bonisteel Blvd., University of Michigan, Ann Arbor MI 48109, USA
E-mail: tmeade@umich.edu
[2] IMRA America, 1044 Woodridge Ave., Ann Arbor MI 48105, USA

Abstract. We demonstrate a method for generating multi-cycle terahertz electromagnetic waveforms by optically rectifying femtosecond pulses in periodically-poled lithium niobate. For a 1.2 mm crystal with 30 μm domains, a 1.80 THz wave with 0.07 THz bandwidth is generated. The crystal domain structure largely determines the THz waveform.

Recently, experiments to generate narrow-bandwidth terahertz (THz) radiation have attracted considerable attention. Some of this work relies on controlling interference between two continuous-wave lasers (photomixing), which is limited by antenna response [1]. Other research has utilized pulse-shaping techniques to narrow the bandwidth of the THz signal, which is limited by the efficiency of one's pulse-shaper [2,3]. In contrast, we demonstrate a technique by which the nonlinear crystal itself is engineered to generate the desired THz waveforms by optical rectification.

The process of optical rectification has been widely applied to single-domain crystals such as ZnTe, GaAs, and GaP to produce THz-bandwidth single-cycle electromagnetic pulses. In optical rectification, the nonlinear polarization induced in the medium is proportional to the intensity envelope of the pulse $|E(t)|^2$. In the far field, the radiated field associated with this polarization is proportional to $\partial^2(|E(t)|^2)/\partial t^2$. For optical pulses on the order of 100 femtoseconds, rectification gives rise to waveforms with THz frequencies [4]. Due to the velocity mismatch in lithium niobate between the optical and THz beams, an optical pulse will lead the THz wave by one optical pulse-width after propagating a walk-off length $l_w = c\tau_p/(n_{THz} - n_{opt})$ through the crystal. The fundamental idea behind our approach is to reverse the polarity of the domains of a periodically poled crystal on a scale $l_d \approx l_w$, so that as the optical pulse travels through the crystal, it induces a nonlinear polarization that is essentially a convolution of the crystal's domain structure with the pulse intensity envelope [5]. The radiated THz waveform then corresponds to the domain structure of the crystal. Under such circumstances, one may envision a model of the polarization induced by the optical beam as shown in Figure 1.

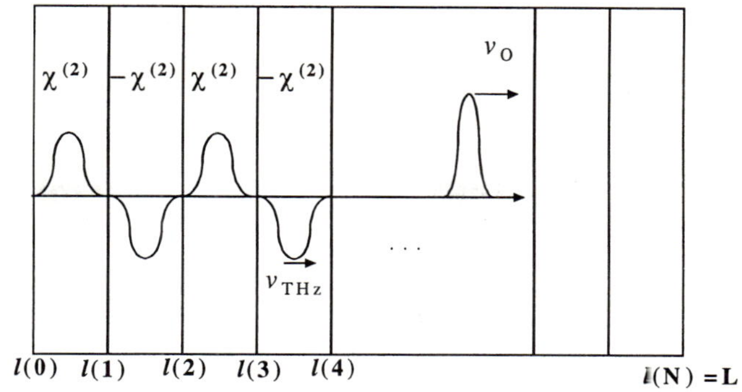

Fig. 1. Schematic diagram of the nonlinear polarization induced in the PPLN crystal.

For our experiments, we used a z-cut periodically-poled lithium niobate (PPLN) crystal, 1.2 mm long with a domain length of 30 microns. We performed the experiments with regeneratively amplified 150-fs pulses at 800 nm at a repetition rate of 250 kHz [6]. The pulse energy incident on the sample was 400 nJ, focused to a spot size of roughly 100 microns. The THz signal was measured electro-optically using a 1 mm thick ZnTe crystal [7] and was modulated at 50 kHz for lock-in detection.

At room temperature, we have measured the optical index of lithium niobate to be $n_{opt}=2.3$; the index at THz frequencies has been reported to be $n_{THz}=5.2$ [8]. If the sample is perfectly periodic, we expect a narrow bandwidth waveform to be generated, with a peak frequency of $f_f=c/[2l_d(n_{THz}-n_{opt})]$. In figure 2(a) we show the THz waveform taken with the PPLN sample at room temperature. The data show decaying oscillations, peaked at 1.67 THz with a bandwidth of 0.13 THz, as shown by the power spectrum in figure 2(b).

It has been demonstrated that absorption loss due to phonons reduces at lower temperatures [9]. We therefore investigated the properties of the PPLN at 13 K. A representative time trace of the THz field under these conditions is shown in figure 2(c), and its power spectrum in 2(d). A low-frequency peak is visible in figure 2(d). Part of the polarization radiates in the backward direction and then reflects off the front crystal surface into the forward direction. This wave has a peak frequency given by $f_b=c/[2l_d(n_{THz}+n_{opt})]$. The extra propagation distance accounts for the time lag between the high-frequency forward-propagating wave and the low-frequency backward-propagating wave. Detailed measurements of the low-frequency signal remain a subject of further study.

The difference between the room temperature and low temperature scans is dramatic. The absorption loss for the forward propagating THz wave throughout the PPLN crystal is negligibly small at low temperatures. We measured the frequency of the forward propagating wave to be 1.80 THz with a bandwidth of 0.07 THz. In this low-loss case, the bandwidth essentially follows $\delta f/f \sim 2/N$, where N is the number of domains in the crystal. Because the absorption loss is so small here, the use of even longer crystals is feasible; using a 7.2 mm crystal, a

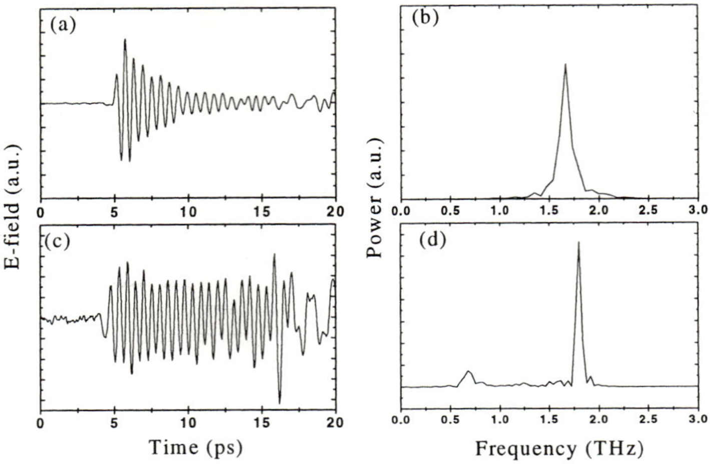

Fig. 2. (a) Time trace of the THz field at room temperature; (b) power spectrum of 2(a); (c) time trace of THz field at 13 K; (d) power spectrum of 2(c).

bandwidth of 18 GHz has been obtained at low temperature. The discrepancy between the peak frequencies at high and low temperatures can be attributed to several factors. At low temperatures, the crystal shrinks and so our domains shorten, thereby increasing the peak THz frequency. Additionally, the real part of the indices of refraction may be temperature dependent.

We have demonstrated engineerable THz waveforms in poled nonlinear media. Crystals poled in a manner where the domain size increases as one moves laterally across the crystal could also provide a source for tunable multi-cycle THz fields, which may prove useful for spectroscopic applications.

References

1 E. R. Brown, K. A. McIntosh, K. B. Nichols, and C. L. Dennis, Appl. Phys. Lett. **66**, 285, 1995.
2 Y.Q. Liu, S.G. Park, and A.M. Weiner, IEEE Jour. Sel. Top. Quant. Electr **2**, 709, 1996.
3 J. Bromage, S. Radic, G.P. Agrawal, C.R. Stroud, P.M. Fauchet, and R. Sobolewski., Opt. Lett. **22**, 627, 1997.
4 L. Xu, X.-C. Zhang, and D.H. Auston, Appl. Phys. Lett. **61**, 1784, 1992.
5 Y.-S. Lee, T. Meade, V. Perlin, H. Winful, T.B. Norris, and A. Galvanauskas, Appl. Phys. Lett., **76**, 2505, 2000.
6 T.B. Norris, Opt. Lett. **17**, 1009, 1992.
7 Q. Wu and X.-C. Zhang, Appl. Phys. Lett. **68**, 1604, 1996.
8 H.J. Bakker, , S. Hunsche, and H. Kurz, Phys. Rev. B **50**, 914, 1994.
9 H. Ito, K. Kawase, and J. Shikata, IEICE Trans. Electron. **E81-C**, 264, 1998.

Optimum geometrical conditions for femtosecond-laser-irradiated, magnetic-field enchanced, THz-radiation InAs emitter and its saturation effect in high magnetic field

Hideyuki Ohtake, Shingo Ono, Masahiro Sakai, Zhenlin Liu, Hidetoshi Murakami, and Nobuhiko Sarukura

Institute for Molecular Science (IMS), Myodaiji, Okazaki 444-8585, Japan
Tel: +81 564 55 7477   Fax: +81 564 55 7218   E-mail: ohtake@ims ac.jp

Abstract. Thermal-receiver detectable THz radiation is generated from the femtosecond-laser-irradiated, magnetic-field enchanced InAs emitter, by optimization of the geometrical layout and magnetic field direction. Moreover, geometrically dependent saturation of the radiation and spectral shift are observed.

There have been numerous necessities for intense, compact, and simple THz-radiation sources. Zhang et al. reported enhancement of THz-radiation power due to quadratic magnetic-field dependence of THz-radiation power from GaAs irradiated with a femtosecond laser [1]. From the practical point of view, the advantage of using InAs as a THz emitter is approximately one order higher efficiency of THz-radiation power compared with the GaAs case due to its smaller effective mass [2]. In this letter, we report saturation of THz-radiation power from femtosecond-laser-irradiated InAs in a high magnetic. Additionally, the clear incidence-angle dependence of the THz-radiation power and spectrum was also observed. As a result, THz radiation can be observed by a room-temperature operated pyroelectric thermal receiver without even using a liquid-helium-cooled bolometer.

The experimental setup is almost similar to that of Ref. 4. An 82 MHz repetition-rate mode-locked Ti:sapphire laser delivers nearly transform-limited 70-fs pulses at 800 nm. The sample is undoped bulk InAs with a (100) surface. With a split-coil superconducting magnet with cross-room-temperature bores can provide a magnetic field up to 5 T, five different optical geometries, as illustrated in Fig. 1, are compared by changing the magnetic field, magnetic-field direction, and excitation laser incident angle. A liquid-helium-cooled bolometer is provided for detecting the power of the total radiation. The THz radiation polarizes almost horizontally, as shown in Fig. 2(a). From lower magnetic field to near 2 T, the THz-radiation power shows quadratic magnetic-field dependence [4]. Moreover, in the case of the magnetic field parallel to the surface and the laser incident angle 45 degree to the surface normal (G-1), the saturation at around 3 T and the reduction of the radiation power above 3 T are observed. On the contrary, clear saturation is not observed in the case of the G-5, as shown in Fig. 2(b). In the case of the G-5 geometrical layout, the polarization of the THz radiation changed dramatically with increasing magnetic field. Photoexcited electrons are accelerated on the plane that is perpendicular to the propagation direction of the THz radiation in the G-5 geometrical layout. Since the polarization of THz radiation reflects the projection

Geometry	Magnetic field direction	Saturation field (T)	Relative power
G-1 (Laser, THz, B ⊗, InAs)	⊗	+3.2	1 (max)
	●	-3.0	0.77
G-2 (B, InAs, Laser, THz)	N→S	+3.2	0.11
	S←N	-3.1	0.10
G-3 (InAs, B, Laser, THz)	No radiation was observed.		
G-4 (Laser, B, THz, InAs)	S↑N	+4.8	0.67
	N↓S	-4.7	0.67
G-5 (Laser, THz, B, InAs)	N→S	+5.0 >	0.70
	S←N	-5.0 >	0.68

Fig. 1. Geometry dependence of THz-radiation power. The saturation of THz-radiation power is observed in the upper two cases.

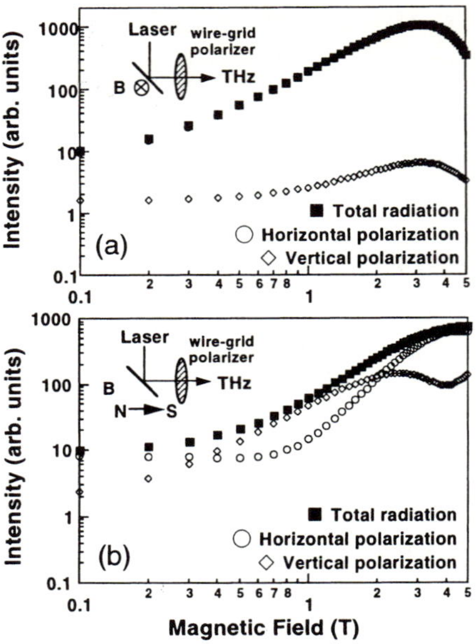

Fig. 2. Magnetic-field dependence of THz-radiation power. (a) for the G-1 and (b) for the G-5 geometrical layouts, respectively.

of the carrier accelerating direction to each direction, both components of polarization should be observed. The saturation of THz-radiation power is observed only in the G-1 and G-2 geometrical layouts. Additionally, in the G-1 geometrical layout, we have observed the remarkable magnetic-field direction dependence of the THz-radiation power. This result also originates from the difference of the electron acceleration direction. We are currently working to clarify the saturation mechanism with a combination of time-resolved spectroscopy and complicated magnetic direction dependence of THz-radiation power and spectra.

Above mentioned experiments demonstrated that the THz-radiation power from an InAs with the magnetic field direction parallel to the semiconductor surface was higher than that from other geometries. Therefore, the only remaining parameter for geometrical optimization was the incidence angle of the excitation laser. THz-radiation power at each incidence angle was measured using a 1.7-T electromagnet by rotating the InAs and the bolometer simultaneously. The Brewster angle was measured to be 78 degrees for InAs around the 800-nm wavelength region. The angular dependence of the THz-radiation power is shown in Fig. 3. The THz-radiation power is the highest at nearly Brewster angle, and this information is quite important for designing a practical THz-radiation source. The highest THz-radiation power was almost five times of the 0-degree incidence-angle power. This is due to the higher photoexcited carrier density that generates THz radiation from the higher absorption efficiency of excitation power. Due to this optimization, the THz radiation from such an emitter is intense enough to be detected by a pyroelectric thermal receiver. The pyroelectric thermal receiver has several advantages. Besides the simplicity of this receiver, there is no need to gate the signal for obtaining the spectra. Moreover, it can be operated at room temperature.

For practical applications, it is also necessary to know the geometrical dependence of the THz-radiation spectra. To extract the features of the THz-radiation spectra in various cases, the center frequency of the THz radiation was defined as the integral average of the frequency. We can see the asymmetrical peak shift with respect to the magnetic field direction due to the difference of the carrier acceleration direction induced by changing the magnetic field direction [5]. The peak shift of THz-radiation spectrum becomes larger for higher incidence angles as shown in Fig. 4, because of the asymmetry of the carrier motion. In contrast to the earlier Brewster geometry excitation work by Weiss et al. [6], it should be noted that both spectrum shape and intensity were changed by flipping the magnetic field direction [5].

In conclusion, we have found saturation of THz-radiation power at around 3 T. This magnetic field will be easily achieved by our permanent magnet for the compact source [6]. Moreover, we found that the optimum geometrical conditions are the magnetic field direction parallel to the semiconductor surface and the Brewster incidence angle of the excitation laser. In this condition, the THz radiation can even be observed by a pyroelectric thermal receiver. Unlike time-gated detectors, there is no requirement of timing adjustment. Such a qualitative advance in emitters will certainly widen the application field of THz radiation.

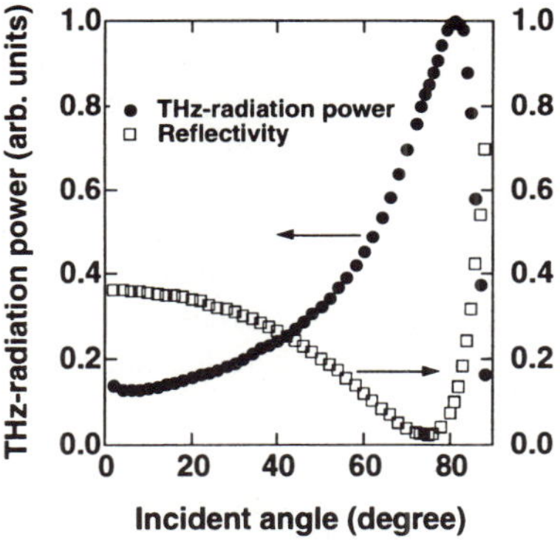

Fig. 3 Angular dependence of 800-nm excitation laser reflectivity and THz-radiation intensity from InAs.

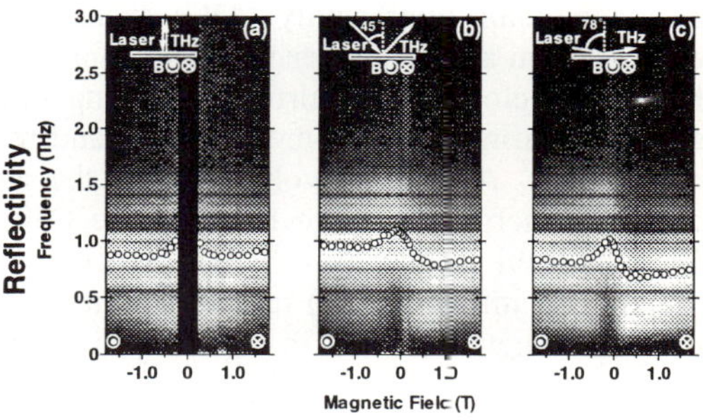

Fig. 4 Two-dimensional plot for THz-radiation spectra on different magnetic fields up to 1.7 T. Polarization of excitation laser is horizontal. Incidence angles of the excitation laser are (a) 0 degrees, (b) 45 degrees and (c) 78 degrees (Brewster angle).

References

1 X.-C. Zhang, Y. Lin, T. D. Hewitt, T. Sangsiri, L. E. Kingsley, and M. Weiner, Appl. Phys. Lett. **62**, 2003 (1993).

2 N. Sarukura, H. Ohtake, S. Izumida, and Z. Liu, J. Appl. Phys. **84**, 654 (1998).

3 For example, D. M. Caughey and R. E. Thomas, Proc. IEEE **55**, 2192 (1967).

4 H. Ohtake, S. Ono, M. Sakai, Z. Liu, T. Tsukamoto and N. Sarukura, Appl. Phys. Lett. **76**, 1398 (2000).

5 C. Weiss, R. Beigang, and R. Wallenstein, Conference on Lasers and Electro-Optics 1999, paper CTuK14.

6 S. Ono, T. Tsukamoto, M. Sakai, Z. Liu, H. Ohtake, N. Sarukura, S. Nishizawa, and A. Nakanishi, Rev. Sci. Instrum. **71**, 554 (2000).

Tunable Coherent Far Infrared Radiation Emission from Biased Semiconductor Microcavities

R. Harel [(†)], I. Brener [(††)], L. N. Pfeiffer, K. W. West, J. M. Vandenberg, S. G. Chu and J. Wynn

Bell-Laboratories, Lucent Technologies, 600 Mountain Ave, Murray Hill, NJ, 07974
E-mail: roey@lucent.com

Abstract. A tunable coherent far infrared radiation source from a biased semiconductor microcavity is demonstrated. The radiation is emitted after resonant excitation of the cavity polaritons by a short laser pulse and its energy is tuned by applying different bias fields.

1. Introduction

Semiconductor microcavity (MC) structures consist of quantum wells (QWs) embedded in a cavity layer which is confined by two high-reflectivity distributed Bragg reflector (DBR) mirrors. A strong interaction between the QWs excitons and the confined photon mode occurs and results in the formation of two cavity polaritons [1,2]. Applying a voltage across the sample polarizes the QW excitons and coherent energy exchange between the QW exciton and the MC mode, at the vacuum Rabi frequency (VRS) [3] is accompanied by an oscillating dipole that radiates far infrared (FIR) radiation [4]. In addition, an applied external bias across the MC structure results in a reduction of the VRS due to the reduction of the exciton oscillator strength [5].

2. Experimental Setup and Microcavity Structure

The experimental setup used in order to measure the FIR radiation is based on the free space electro-optic sampling detection technique [6] where the FIR radiation induced polarization change of a weak probe beam as it propagates through an electro-optic crystal. The polarization change is transformed into an intensity change by measuring the probe beam intensity after passing it through a polarizer.

The MC structure under study was grown on an n^+-doped GaAs wafer. It consists of 10 102Å/108Å AlAs/GaAs QWs embedded at the antinode of the confined MC mode in a 5λ cavity layer. The top and bottom DBR mirrors had 13 and 25 periods of λ/4 $AlAs/Al_{0.1}Ga_{0.9}As$, respectively. In the bottom DBR mirror, the last period close to the cavity layer, was left nominally undoped, while the following 4 periods were n^+-doped to a level of $3\cdot10^{16}$ cm^{-3} and the last 20 periods were n^+-doped to a level of 10^{18} cm^{-3}. The cavity layer and the top DBR mirror were nominally undoped. Reflection spectrum measured from the MC structure at T=4K, θ=45deg and under no external bias shows E_{VRS}=9.05±0.3meV.

3. Experimental results

Fig. 1(a) shows the THz time-domain spectrum observed at T=4K, θ=45deg and applied bias of V=-16Volt. The excitation spot position on the wafer corresponds to a resonance between the MC mode and the QW exciton. From the measured THz signal, a frequency of ν=1.94±0.02THz corresponding to a FIR radiation energy of E_{THz}=hν=8.01±0.09meV is extracted. The VRS obtained from the simultaneously measured reflection spectrum is E_{VRS}=8.1±0.05meV. Fig. 1(b) presents the FIR radiation spectrum intensity.

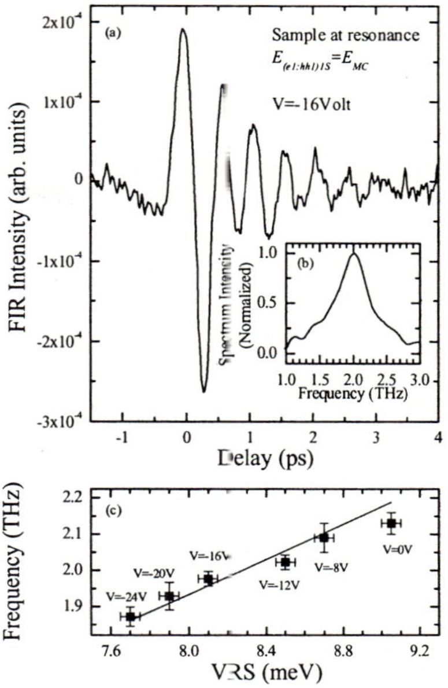

Figure 1. (a) THz time-domain signal measured at T=4K and θ=45deg under applied voltage of V=-16volt. (b) Spectrum intensity calculated from (a). (c) FIR radiation frequency as a function of E_{VRS} (dot) for different applied bias voltage. The line presents the relation ν=E_{VRS}/h.

The dots in Fig. 1(c) show the extracted FIR radiation frequency as a function of the VRS for different applied bias voltages. At each voltage, the exciton energy is different due to the quantum confined Stark effect (QCSE) and compensation was done by exciting different spot positions on the wafer at which the MC mode energy shifts by the same amount. This results in a resonance condition between the MC mode and the QW exciton at all applied bias voltages. As can be seen, a monotonic decrease of the FIR radiation frequency and of the VRS energy with the applied electric field is clearly observed. This is due to the reduced excitonic oscillator strength under applied bias voltage [5].

Fig. 2 presents the THz time-domain signal measured for different applied fields for an illumination spot position where a resonance between the MC mode and the QW exciton occurs at V=0. Although these spectra were taken for different small bias values, the exciton shift, due to the QCSE, is very small, resulting in a resonant condition for all cases. The FIR radiation resulting from the instantaneous creation of free electron-hole polarization (which appears at t=0ps) disappears in the case of zero bias. This result indicates that there is no built-in electric field in the sample, and that the conduction and valence bands (in the cavity region) are flat. At low bias fields, very weak FIR emission was observed, and this is attributed to the small exciton dipole moment. Even without external bias (Fig. 2(c)) oscillations in the FIR emission spectrum are observed, but they are time shifted relatively to those observed with an applied bias.

4. Model

In order to explain the FIR radiation frequency dependence on the applied bias voltage, we present a model based on the solution of the time-dependent

Schrödinger equation [7]. From the model, the FIR radiation electric field is given by:

$$E_{THz}(t) \propto \alpha\beta \cdot \cos(\Omega t) \cdot \langle ex|\mathbf{r}|ex\rangle \qquad (1)$$

Here, α and β are the fraction of the upper and lower polariton population, Ω is the VRS and $\langle ex|\mathbf{r}|ex\rangle$ is the exciton dipole moment. As can be seen, the FIR radiation energy is predicted to be equal to the VRS. The solid line in Fig. 1(c) presents the linear relationship ν=VRS/h. An excellent agreement between the measured FIR radiation frequency and the VRS is achieved.

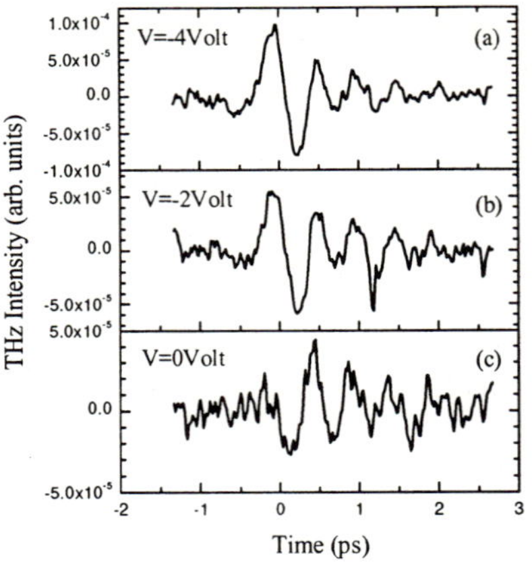

Figure 2. THz time-domain signal measured at T=4K and θ=45deg for different applied bias. The excitation position was chosen such that a resonance condition exists at V=0Volt.

A question arises regarding the FIR radiation source under no applied bias voltage. The experimental results presented in Fig. 2 show that there is no residual electric field inside the cavity (confirmed by the disappearance of the THz time-domain peak at t=0ps). This leads to the conclusion that the induced dipole is from nonlinear polarization inherent to GaAs ($\chi^{(2)} \neq 0$) [8]. We calculate the nonlinear dipole using the nonlinear parameters of bulk GaAs. After considering the MC structure reflection, the cavity enhancement and the cavity polariton broadening, a nonlinear dipole of μ=1.5Å was extracted [7]. This value is similar to the excitonic dipole moment induced in a 100Å GaAs QW by an applied dc electric field of about 1kV/cm. This electric field corresponds to an applied bias of about -3V and, as can be seen in Fig. 2, the FIR intensity is at the same order of magnitude as in the case of an applied bias of V=2V.

In conclusion, we studied the electric field effect on the properties of the coherent FIR radiation emission from a semiconductor MC. We found that the FIR radiation energy decreases monotonically with increasing applied bias and is equal to the VRS. Without external bias, a weak FIR radiation was observed and we attribute this radiation to difference frequency mixing in the cavity layer between the two confined photon modes at the cavity polaritons energy.

Acknowledgements: R. H. acknowledges the support of the US-Israel Binational Science Foundation (BSF), Jerusalem, Israel.

(†) Also at the Solid State Institute, Technion - Israel Institute of Technology, Haifa, Israel.
(††) Present address: Tellium Inc., 2 Crescent Place, P.O. Box 91, Oceanport, NJ, 07757.

1. C. Weisbuch *et al.* Phys. Rev. Lett. **69**, 3314 (1992).
2. R. Houdré *et al.* Phys. Rev. Lett. **73**, 2043 (1994).
3. T. B. Norris *et al.* Phys. Rev. B **50**, 14663 (1995).
4. R. Harel *et al.* CLEO/QELS, Baltimore (1999), paper JThC2
5. T. A. Fisher *et al.* Phys. Rev. B **51**, 2600 (1995).
6. Q. Wu and X. -C. Zhang, Appl. Phys. Lett. **67**, 3523 (1995).
7. R. Harel *et al.* Phys. Stat. Sol. (a) **178**, 365 (2000).
8. J. B. Khurgin, J. Opt. Soc. Am. B, **13**, 2129 (1996)

Amplitude and Phase Resolved Detection of Tunable Femtosecond Pulses with Frequency Components beyond 100 THz

A. Brodschelm, F. Tauser, R. Huber, J. Y. Sohn, and A. Leitenstorfer[a]

Physik-Department E11, TU München, James-Franck-Straße, 85748 Garching, Germany
[a] E-mail: aleitens@ph.tum.de

Abstract. The electric field amplitude and phase of ultrashort transients generated by phase-matched optical rectification in GaSe is detected via ultrabroadband electro-optic sampling. Full tunability is achieved between 6 μm and 40 μm with pulse durations below 50 fs. Frequency components approaching the near infrared regime are observed.

We report on both widely tunable generation and field-resolved detection of femtosecond electromagnetic transients. The center frequency of the THz pulses is readily varied over an extraordinarily wide interval from 10 THz to 45 THz [1]. Employing phase-matched optical rectification of ultrashort laser pulses in GaSe crystals with a thickness between 20 μm and 90 μm, we successfully generate bandwidth-limited pulses shorter than 50 fs. The electric field transients are detected with a time resolution of 10 fs and a high dynamic range. Calculated amplitude and phase spectra based on a plane-wave model agree well with the measured data [1,2]. The energy of the generated THz pulse saturates with the thickness of the GaSe crystal exceeding approximately 100 μm.

We start with 10 fs laser pulses at a repetition rate of 64 MHz an energy of 15 nJ and a spectral width of 103 nm (51 THz) at a central wavelength of 780 nm. The major part of the laser intensity is focused on a z-cut GaSe crystal. The crystal can be tilted by the phase matching angle θ about a horizontal axis perpendicular to the direction of the incident beam. $\theta = 0$ denotes normal incidence. A half-wave plate is used to rotate the polarization by 45° to provide both ordinarily and extraordinarily polarized components in order to satisfy phase matching. The multi-THz radiation, emitted in forward direction, is collimated by a gold coated off-axis parabolic mirror. The ultrashort transients are focused on a 10.3 μm thick <110> oriented electro-optic crystal of ZnTe. Changes of the polarization state of a time delayed probe pulse due to induced birefringence in ZnTe are analyzed in a detection scheme operating at the shot-noise level. The difference ΔI between the photocurrents of two balanced photodiodes is proportional to the applied THz electric field present in the electro-optic crystal during propagation of the probe pulse [3,4]. An extraordinarily low noise level of 3×10^{-9} $Hz^{-1/2}$ and a high dynamic range of 5×10^{3} are achieved in our ultrabroadband electro-optic THz system.

At small phase matching angles θ, the electric field of the emitted transients oscillates relatively slowly and displays significant chirp. The corresponding spectrum of the transient at $\theta = 2°$ extends from approximately 100 GHz to 60 THz, i.e. from 3 mm to 5 μm, with only a small gap between 5 THz and 7 THz, the Reststrahlen band of GaSe [1]. At higher values of θ, a very short burst of rapid oscillations at early times is found in addition to a low-frequency component extending over several ps. The spectral components below 10 THz vanish while the amplitude above 10 THz increases considerably. At sufficiently large angle the spectrum splits in two separated peaks: A narrow line at 8 THz to 10 THz shifts slightly to the red with increasing θ while a broad maximum at higher frequencies experiences a strong blue-shift. This may be understood as a phase-matched nonlinear-optical $\chi^{(2)}$ process, mixing different Fourier components of our broadband laser-spectrum [1].

In a plane wave model incorporating the dispersion of the first and second order polarizabilities of GaSe, it is important to go beyond the slowly varying amplitude approximation for the temporal envelope of the pulse. A single integral describes the generated THz electric field E_{THz} at the exit face of the crystal in terms of the laser field E_L [1,2]. After solving this integral one has to take into account frequency dependent reflection losses at the crystal surfaces.

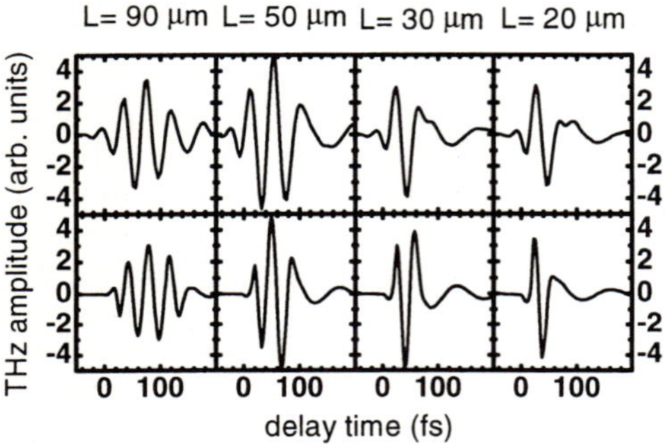

Figure 1. THz transients generated in GaSe crystals of various thickness L at θ = 50°. Upper row: Measurement. Lower row: Simulation

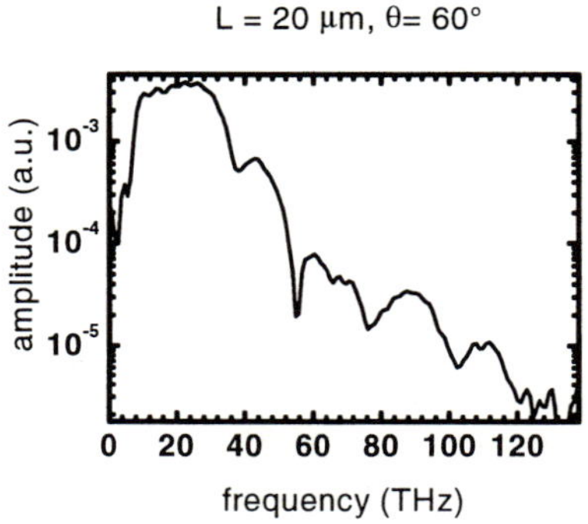

Figure 2. Measured amplitude spectrum of a transient emitted by a L = 20 μm GaSe crystal at an angle of θ = 60°. Detector: 10.3 μm ZnTe.

Measured (top) and calculated (bottom) transients emitted from GaSe crystals with varying thickness L between 90 μm and 20 μm at a phase matching angle θ of 50° are depicted in Fig. 1. The duration of the pulse at a central frequency of 25 THz becomes shorter with decreasing crystal thickness, ranging from a few oscillation periods down to single cycle pulses. Interestingly, the peak electric field amplitudes (approximately 1 kV/cm) are comparable in all four cases. The small deviations between the calculated transients and the measured data are explained by the phase distortions introduced by the THz optical system and the frequency response of the 10.3 μm ZnTe electro-optic detector. In Fig. 2, the amplitude spectrum of the transient emitted from the L = 20 μm GaSe specimen at θ = 60° is depicted. Frequency components above 100 THz (λ = 3 μm) are observed, approaching the near-infrared spectral regime.

Fig. 3 shows a simulation of the energy of the generated THz pulse as a function of crystal thickness. For perfect phase matching over the entire bandwidth, a quadratic increase of pulse energy with L would be expected. This behavior is observed only for very thin crystals (L < 10 μm), as seen in the logarithmic inset of Fig. 3. The pulse energy saturates at a thickness above approximately 100 μm. The effective conversion length inside the crystal is limited due to the group velocity mismatch of the ordinary and extraordinary polarized components of the pump-pulse (e.g. GVM ≈ 100 fs/mm at for ω_{THz} = 18 THz). For L > 200 μm, we predict strongly chirped pulses with a constant bandwidth due to the dispersion of GaSe.

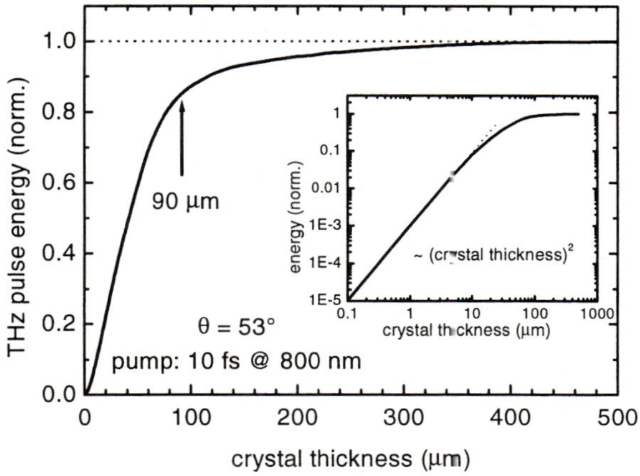

Figure 3. Calculated energy of the emitted THz pulse as a function of the thickness L of the GaSe crystal. Note the logarithmic scaling of the inset

We have recently extended our setup in order to perform coherent control and pump-probe experiments on vibrational and electronic resonances in the mid infrared. Femtosecond transmission measurements in highly excited semiconductors are under way. The high dynamic range and wide tunability of our multi-THz system provides access to a new class of amplitude and phase resolved studies on ultrafast phenomena.

References

1 R. Huber, A. Brodschelm, F. Tauser and A. Leitenstorfer, Appl. Phys. Lett. **76**, 3191 (2000).

2 R. A. Kaindl, F. Eickemeyer, M. Woerner and T. Elsaesser, Appl. Phys. Lett. **75**, 8 (1999).

3 Q. Wu and X.-C. Zhang, Appl. Phys. Lett. **68**, 1604 (1996), ibid. **70**, 1784 (1997), ibid. **71**, 1285 (1997).

4 Leitenstorfer, S. Hunsche, J. Shah, M. C. Nuss and W. H. Knox, Appl. Phys. Lett. **74**, 11 (1999), Phys. Rev. Lett. **82**, 5140 (1999), Phys. Rev. B 61, 16642 (2000)

Ultrabroadband detection of THz radiation by LT-GaAs photoconductive antenna gated with 15-fs laser pulses

Shunsuke Kono, Masahiko Tani, Ping Gu, and Kiyomi Sakai

Kansai Advanced Research Center, Communications Research Laboratory, MPT
588-2, Iwaoka, Nishi-ku, Kobe, 651-2492, Japan
E-mail: kono@crl.go.jp

Abstract. Ultrabroadband detection of terahertz radiation is demonstrated with a low-temperature-grown GaAs photoconductive antenna gated with 15-fs laser pulses. The observed spectra extends up to 20 THz and shows an absorption band between 7~10 THz due to the phonon resonance in GaAs substrate of the detector.

1. Introduction

The generation and detection of terahertz (THz) radiation using a photoconductive (PC) antenna gated with ultrashort optical pulses have been studied extensively during the last decade. However, it has been believed that the bandwidth of PC antenna is restricted within several THz due to the carrier lifetime and the RC time constant. The highest frequency detected with PC antennas reported so far was 5 THz [1,2]. On the other hand, it was reported that PC antennas with a long carrier lifetime was able to detect THz radiation, as well as that with a sub-picosecond carrier lifetime, with only slight degradation in detection bandwidth [3,4]. The detection of THz radiation with a PC antenna with a long-carrier-lifetime substrate is considered to be due to the rising edge of carriers injected by the short laser pulses. This suggests that the detection bandwidth is possibly extended by using shorter laser pulses. In this article, we report an ultrabrodband detection of THz radiation by using a low-temperature-grown (LT) GaAs PC antenna gated with 15-fs laser pulses. The detected radiation frequency exceeded 20 THz. This is the highest frequency detected with PC antennas reported so far.

2. Experimental Methods

A mode-locked Ti:sapphire laser generated 12-fs laser pulses with 75 MHz repetition rate. The laser pulses were negatively chirped by a multiple reflection between a pair of group-velocity compensated dielectric mirrors outside of the laser cavity. The laser beam was divided into pump (110 mW) and probe beam (10 mW) by a 1-mm-thick glass plate beam splitter. The pump beam was focused by an off-axis parabolic mirror (f=51 mm) onto a (100)-cut SI-InP wafer with an incident angle of 45°. The THz radiation emitted in the reflection angle of the

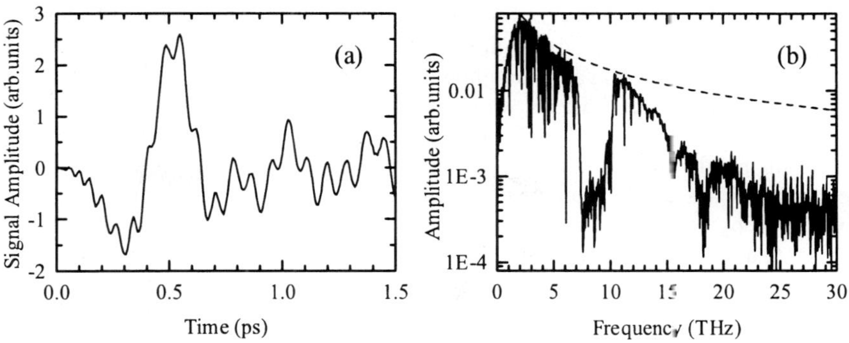

Fig. 1. (a) The detected THz waveform from (100) InP, and (b) its Fourier transformed spectrum for a longer time window. The dashed line in the spectrum indicates a circuit response function with an RC time constant of 0.2 ps.

pump beam was collected and focused onto a PC antenna detector with a pair of parabolic mirrors. The PC antenna was a 30-μm long dipole antenna with 5-μm gap at the center, fabricated on an LT-GaAs. The carrier lifetime of the photoconductive substrate was estimated to be 1.4 ps from a transient photoreflectance measurement. The probe beam was focused onto the photoconductive gap of the antenna by a reflection-type objective lens to avoid broadening of the optical pulses. The timing of the pump and probe pulses was changed by a delay-line in the probe beam path consisting of a corner reflector and a high-precision motorized translation stage. The pulse width at the emitter surface and the PC detector was estimated to be about 15 fs (FWHM). The current signal from the PC antenna was amplified by a low noise current amplifier and then detected by a lock-in amplifier referenced to the frequency (2 kHz) of a mechanical chopper in the pump beam path.

3. Results and Discussion

Figure 1(a) shows the detected waveform of the THz radiation from the (100) InP emitter for a single scan. The time constant of the lock-in amplifier was 0.1 sec, and the scanning time for the 3-ps time widow was about 3 minutes with a 1-μm step resolution (=6.7 fs) in the delay-line translation stage. The waveform in Fig. 1(a) shows a single cycle THz pulse near 0.5-ps delay time, whose width at half of the peak amplitude is about 0.2 ps. There are very fast oscillations superposed to this large cycle. The shortest oscillation period is about 45 fs. These fast oscillations are attributed to the optical rectification effect on the InP emitter. The ringings after the main peak pulse are due to dispersion and absorption in GaAs substrate of the detector, and absorption by water vapor in the ambient air. Figure 1(b) shows the Fourier transformed spectrum of the waveform in Fig. 1(a) for a longer time widow. The detected frequency exceeds 20 THz. This is the broadest bandwidth detected with an LT-GaAs PC antenna reported so far The absorption band observed between 7 and 10 THz is due to the phonon resonance in the 0.4-

mm thick GaAs substrate (ν_{TO}=8.0 THz, ν_{LO}=8.6 THz). Many sharp lines observed are due to the water vapor absorption.

It is extraordinary that a PC antenna with ~1 ps carrier lifetime has a broadband response more than 10 THz. However, it is known that a PC antenna with a very long carrier lifetime (>100 ps) can detect THz radiation with almost the same detection bandwidth of that with a subpicosecond carrier lifetime [4]. The fast response of a slow PC antenna originates from a fast rise of photocurrent on excitation by the short laser pulse. The same discussion will apply to our case. Other than carrier lifetime, the RC time constant of the antenna should be considered to explain the frequency response of the signal current from the PC antenna. In a previous report [5] we estimated a RC time constant for the same type of antenna to be on the order of subpicosecond (~0.2 ps). This finite RC time constant will restrict the bandwidth of the PC antenna and reduce the responsivity at high frequencies. For example, with a 0.2 ps RC time constant the cut-off frequency is about 0.8 THz. The responsivity of the antenna at 5 THz and 10 THz decreases to 25% and 13% of that at the cut-off frequency, respectively. The circuit response function in frequency domain calculated for a RC time constant of 0.2 ps is indicated by a dashed line in Fig. 1(b), which qualitatively agrees with the spectral profile in the frequency range lower than 5 THz. Although the reduction of the responsivity due to the carrier lifetime should be taken into account in the frequency region higher than 5 THz, this slow decay of the RC frequency response function can be one of the reasons of our ultrabroadband detection with the PC antenna.

Our result shows that the PC sampling has broad detection bandwidth comparable to the EO sampling when the PC antenna is gated with short enough laser pulses. In the EO sampling one of the draw-back is the strong absorption and dispersion at the phonon resonance. In the PC sampling, the phonon absorption of the antenna substrate can be avoided by growing LT-GaAs on other materials such as high resistivity silicon (ν_{TO}=15.5 THz).

4. Conclusions

We have demonstrated ultrabroadband detection of THz radiation with LT-GaAs PC antenna with 15-fs gating laser pulses. The frequency spectrum extended more than 20 THz. This is the widest bandwidth achieved with a PC antenna reported so far.

References

1 N. Katzennellenbogen and D. Grischkowsky, *Appl. Phys. Lett.*, **61**, 840 (1992).

2 M. Tani, R. Fukasawa, H. Abe, S. Matsuura, K. Sakai, and S. Nakashima, *J. Appl. Phys.*, **83**, 2473 (1998).

3 F. G. Sun, G. A. Wagoner, and X.-C. Zhang, *Appl. Phys. Lett.*, **67**, 1656 (1995).

4 M. Tani, K. Sakai and H. Mimura, *Jpn. J. Appl. Phys.*, **36**, L1175 (1997).

5 Shuji Matsuura, Masahiko Tani and Kiyomi Sakai, *Appl. Phys. Lett.*, **70**, 550 (1997).

A Simple Terahertz Spectrometer

Timothy F. Crimmins, Michael J. Gleason, David W. Ward, and Keith A. Nelson

Department of Chemistry, MIT, Cambridge, MA 02139. Email: kanelson@mit.edu

Abstract: A simple terahertz spectrometer with only visible light in/out and with no terahertz optics or free-space terahertz wave propagation is demonstrated. Single-cycle or tunable narrowband THz radiation can be generated and used.

The far-IR spectral region is rapidly emerging as an important component in the spectroscopic study and optical imaging of a very wide range of samples extending from simple and complex liquids to biological materials to integrated circuits. [1] Spectroscopy in this terahertz frequency regime is especially potent in liquids and partially ordered solids, where it probes properties on length and time scales intermediate between those characteristic of bulk low-frequency dielectric responses and of local intermolecular or interionic motions. Conventional dielectric spectroscopy rarely reaches frequencies over 100 GHz, and crucial components of the dynamics of normal and supercooled liquids, mixed ferroelectric crystals, and other complex materials have remained elusive.

Ultrafast spectroscopy in the terahertz frequency range typically involves generation of the THz radiation through femtosecond laser illumination of photoconducting dipole antennas or nonlinear optical crystals to generate broad bandwidth, coherent bursts of far-IR output with substantial spectral density from 100 GHz to 5 THz or higher. This THz radiation is typically focused, manipulated, and directed through free space and to a sample using far-IR optics, and finally detected using a second photoconducting dipole antenna or other nonlinear element which is gated by a variably delayed femtosecond pulse. The main drawbacks of this approach are its requirements for manipulation of far-IR radiation through space, for far-IR optics, and for fabrication of photoconducting dipole antennas. In addition, complex non-Lorentzian spectra may be more easily analyzed through examination with tunable, narrow-band THz radiation than with broadband radiation consisting of one or very few cycles.

Figure 1 depicts an alternate apparatus for conducting terahertz spectroscopy. The setup uses only visible light pulses as inputs and outputs, and it requires no free-space propagation of far-IR radiation, no far-IR optics, and no photoconducting antennae. Conveniently tunable, narrowband THz radiation or broadband single-cycle radiation can be generated and used for spectroscopy. In this apparatus, femtosecond pulses in the visible spectrum (in our case amplified Ti:sapphire output) are used to irradiate a first crystal of $LiTaO_3$ or similar material, driving a mixed lattice vibrational-electromagnetic wave called a phonon-polariton, via impulsive stimulated Raman scattering (ISRS). [2] The phonon-polariton response propagates through the crystal at the speed of far-IR radiation in the crystal (in $LiTaO_3$, about 1/6 the speed of light in air) to the crystal edge, at which its electromagnetic component is partially coupled out of the crystal and into the sample. The THz radiation propagates through the sample and into a second crystal, in which it regenerates phonon-polaritons of the original

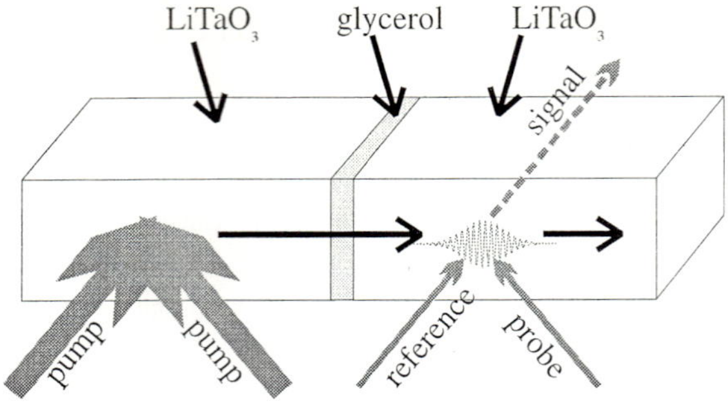

Fig. 1. Diagram of a simple terahertz spectrometer. Terahertz-frequency waves are generated in the first crystal, propagated through a liquid sample layer, and detected via heterodyned diffraction in a second crystal.

frequency and wavelength. The phonon-polaritons are detected through coherent scattering of visible probe pulses that are appropriately shifted spatially and delayed temporally.

In our initial experiments, we used crossed pulses to generate an interference pattern whose wavelength determines the phonon-polariton wavelength and the corresponding frequency (typically in the 0.5-5 THz range) according to the phonon-polariton dispersion relation. The use of diffractive optics (a phase mask pattern consisting of parallel trenches etched into a glass substrate) to generate the two excitation pulses yields a large overlap region with many fringes, thereby generating a narrowband phonon-polariton response. The phonon-polariton frequency is conveniently and rapidly tuned by simply inserting a different phase mask pattern (among the 20 patterns etched into a single glass substrate) into the beam path. Phonon-polaritons in the second $LiTaO_3$ crystal were monitored through heterodyned coherent scattering of a variably-delayed, visible probe pulse. Heterodyne detection and spatially separated detection make further use of the phase mask for generation of the probe and reference beams.

Preliminary data from the viscous liquid glycerol at room temperature are plotted in Fig. 2. The responses from phonon-polaritons which have and have not propagated through the liquid sample layer (for the latter, the sample apparatus was translated such that the excitation and probing beams are all within the same crystal) are plotted together for comparison. The attenuation and temporal shifting of the signal caused by propagation through the liquid can be determined from data of this sort, revealing the complex dielectric response of the liquid at the terahertz frequencies selected.

Since in this apparatus the terahertz radiation never leaves the crystal-sample-crystal assembly and never propagates through free space, only visible light is used, and no far-IR optics or antenna detectors are necessary. The crystal-sample-crystal assembly can be considered the core of a simple THz spectrometer that requires only visible light inputs and outputs. The entire assembly measures roughly 1 cm in each dimension and can be used for temperature-dependent

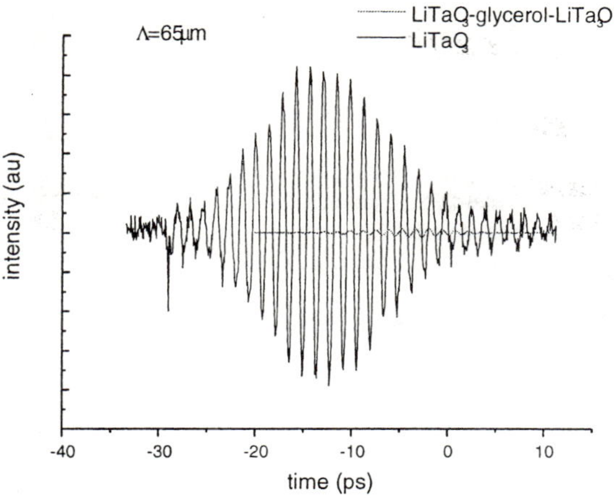

Fig. 2. Terahertz response of glycerol. The lower amplitude data show signal from THz waves generated in a $LiTaO_3$ crystal, passed through a layer of glycerol, and detected in a second $LiTaO_3$ crystal. Plotted for comparison is signal from phonon-polaritons generated and detected within the same crystal, with the same spatial separation between excitation and detection regions.

measurements. Solid samples could be used, and thin liquid layers could be inserted between the $LiTaO_3$ crystals and the sample for index matching.

It may prove possible to extend the method for nonlinear THz spectroscopy. The crystal edges could be tailored for focusing of the THz radiation to achieve optimal intensity levels. An intense THz excitation pulse and a subsequent THz pulse or narrowband probe could be generated in the first crystal and detected in the second. Alternatively, an intense narrowband THz excitation field could be used for hole-burning or related nonlinear measurements. In general, the form of the THz excitation field can be tailored for selected nonlinear spectroscopic measurements or other uses based on the temporal and spatial characteristics of the visible radiation used to produce it. Finally, time-dependent responses could be probed directly within the sample.

Acknowledgements. Supported in part by National Science Foundation grant nos. CHE97-13388 and DMR-9710140.

References

[1] X.-C. Zhang, "Free space THz optoelectronics," http://www.rpi.edu/~zhangxc/ (1999).
[2] R. M. Koehl, S. Adachi, and K. A. Nelson, "Direct visualization of collective wavepacket dynamics," J. Phys. Chem. A. **103**, 10260-10267 (1999).

Ultrafast Applications of THz Waveguides

D. Grischkowsky, G. Gallot[1], S.P. Jamison[2] and R.W. McGowan[3]

School of Electrical and Computer Engineering and Center for Laser and Photonics Research, Oklahoma State University, Stillwater, Oklahoma 74078, USA
E-mail: danielg@thzsun.ecen.okstate.edu
[1]Present address, Laboratoire d'Optique Quantique, CNRS, Ecole Polytechnique, 91128 Palaiseau cedex, France.
[2]Present address, Department of Physics and Applied Physics, University of Strathclyde, Glasgow, UK.
[3]Present address, Imation, Oakdale, Minnesota 55128, USA

Abstract. Quasi-optical techniques have been used to efficiently couple freely propagating subpsec pulses of THz electromagnetic radiation into rectangular metal waveguides. We have observed dispersive, low-loss propagation over the frequency band from 0.65 to 3.5 THz with typical waveguide cross-section dimensions of the order of 250 μm and lengths of 25 mm. We demonstrate both theoretically and experimentally that it is possible to achieve efficient TE_{10} single-mode coupling and propagation in a suitably sized rectangular waveguide for an incoming focused, linearly polarized THz pulse with a bandwidth covering many octaves in frequency and overlapping more than 35 waveguide modes. The possible applications enabled by these results will be discussed.

1. Introduction

The THz frequency range located midway between microwaves and visible light presents a new frontier containing an abundance of technical applications and fundamental research problems. Consequently, THz research solutions and techniques can come from either optics or microwaves or in many cases combinations of both. Because of the developing nature of THz capability, at the present time, one-of-a-kind laboratory, optoelectronic THz transmitters and receivers provide interim solutions enabling both fundamental research and applications [1,2].

One outstanding problem has been the guided wave propagation of THz radiation, and the associated efficient coupling between guided and freely propagating THz waves. Recently, efficient, broad-band coupling of freely propagating pulses of THz radiation into circular metal waveguides has been demonstrated [3]. This work also reported the highest guided-wave performance to date, obtained with a 240 μm diameter stainless steel waveguide over the frequency range from 0.8 to 3.5 THz and with a power absorption coefficient of less than 1 cm^{-1}. This absorption is much less than that of coplanar and microstrip transmission lines [3]. However, for circular waveguides the undistorted propagation of ps THz pulses was shown not to be possible, because of the extreme dispersion near the cut-off frequencies overlapped by the pulse bandwidth covering many octaves in frequency.

Here, we present an experimental study of the application of quasi-optical techniques to couple freely propagating pulses of THz radiation into rectangular brass metal waveguides. Even though the input spectrum overlaps the cut-off frequencies of more than 35 rectangular waveguide modes, the calculated coupling coefficients and experiments show that the linearly polarized incoming THz pulses only couple significantly into the the TE_{10} and TM_{12} modes [4]. We demonstrate theoretically and experimentally that it is indeed possible to achieve TE_{10} single-mode coupling and propagation in a suitably-sized rectangular waveguide for an incoming focused, linearly polarized THz pulse with a bandwidth covering many octaves in frequency. In addition, we have achieved broadband pulse propagation with a power absorption coefficient of less than 0.5 cm^{-1}, the lowest loss figure to date.

2. Experimental Set-Up

The experimental set-up for broadband THz waveguides shown in Fig. 1 consists of an optoelectronic transmitter and receiver along with beam shaping and steering optics. A detailed description of the entire system has been previously published [1,2]. In the standard THz-TDS setup the sample is placed at the THz beam waist located at the confocal position between the two parabolic reflectors. The THz beam system has symmetric confocal optics with respect to the center line of the system between the transmitter and receiver. This situation gives a unity, frequency-independent, power coupling efficiency between the transmitter and receiver (see Eq. 2.16, Ref. 5). The transmitting antenna is at the focus of the silicon lens, which collimates the frequency independent far-field pattern of the antenna into an approximately Gaussian beam with a 1/e beam waist diameter of 5 mm and with a plane wave phase front. This is beam waist W1. This beam waist is in turn in the focal plane of the transmitter paraboloidal mirror, which focuses the THz beam to the second beam waist W2 with beam diameters proportional to wavelength and with plane wave phase fronts. In the standard system the THz optical train is identical (but reversed) from this point on to the receiver. For the waveguide experiments an additional silicon lens is placed at the beam waist W2, thereby producing a third frequency independent beam waist W3 with a 1/e amplitude waist diameter of 200 μm and a plane wave phasefront. This third beam

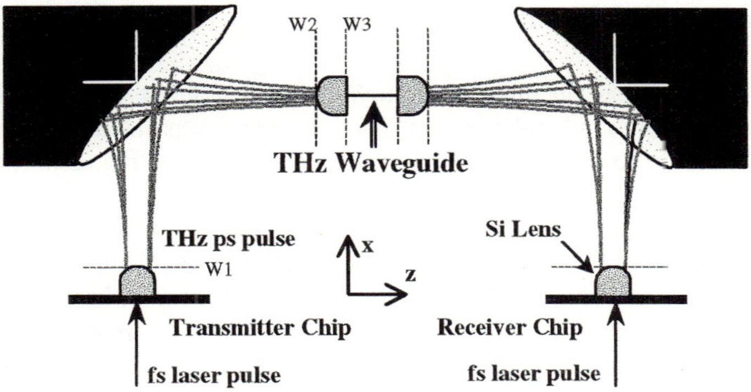

Fig. 1. The optoelectronic THz-TDS system incorporating quasi-optical coupling to the THz waveguide. The generated THz pulse is linearly polarized in the figure plane and along the X-axis at the waveguide entrance

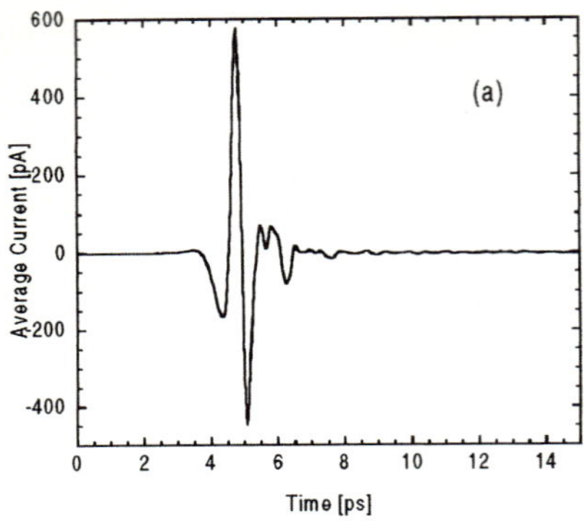

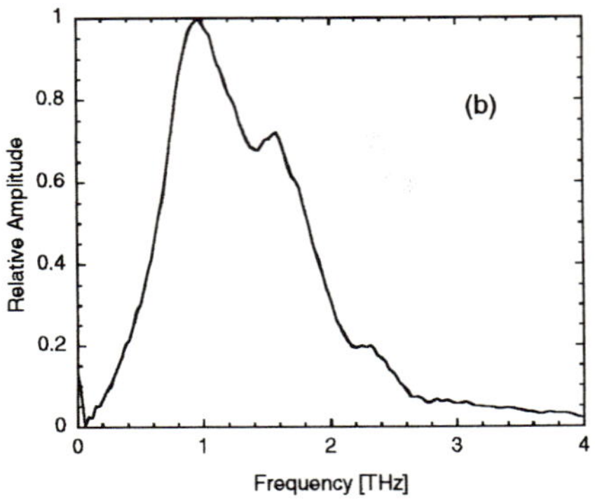

Figure 2: (a) Measured reference THz pulse. (b) Relative amplitude spectrum of the reference pulse. The small oscillations apparent in (a) at approximately 6 ps are due to reflections between the confocal silicon lenses.

waist overlaps the waveguide entrance face. The output face of the waveguide is again in the focal plane of an identical silicon lens with identical but reversed optics to the receiver. Considering the earlier work [3,4] that used stainless steel circular waveguides stainless steel (SS), much longer waveguides of high conductivity metals could be used, for example copper and brass. Copper has a conductivity 36 times larger than SS, giving an absorption coefficient 6 times smaller for Cu than for SS. Brass is also a good choice with a conductivity slightly less than copper. The rectangular waveguides discussed here have been fabricated in our laboratory by cutting grooves in brass plates. Using a 250 μm jewelers saw blade, a precise groove is cut on a 25 mm-square, 6 mm-thick brass plate. Another brass plate is tightly connected to the grooved plate to complete the waveguide. The entrance and exit side of the guide are carefully milled to obtain high-precision flat and burr-free entrance and exit faces. In order to study the influence of the waveguide dimension on the relative coupling efficiencies, several waveguides of different size were cut on the same 25 mm long plate.

3. Experimental Results

The reference pulse is obtained by removing the waveguide and moving the two silicon lenses to the confocal position, where they are separated by their common focus W3. Figure 2a shows such a reference THz pulse, and Fig. 2b shows that the useful amplitude spectrum extends from 0.1 to 4 THz. Via our understanding of the coupling of the incoming linearly polarized, focused THz pulse with a plane wave phase front to the waveguide modes, we have shown that it is possible to design a rectangular waveguide for which the THz pulse couples to only a single mode of the waveguide [4]. As will be shown below, we have experimentally demonstrated this conclusion by observing single mode propagation over the frequency range from 0.7 to 4 THz for an optimal 250x125 μm rectangular brass waveguide 25 mm long. The calculated frequency dependent loss, phase velocity and group velocity for the three important modes of this rectangular waveguide

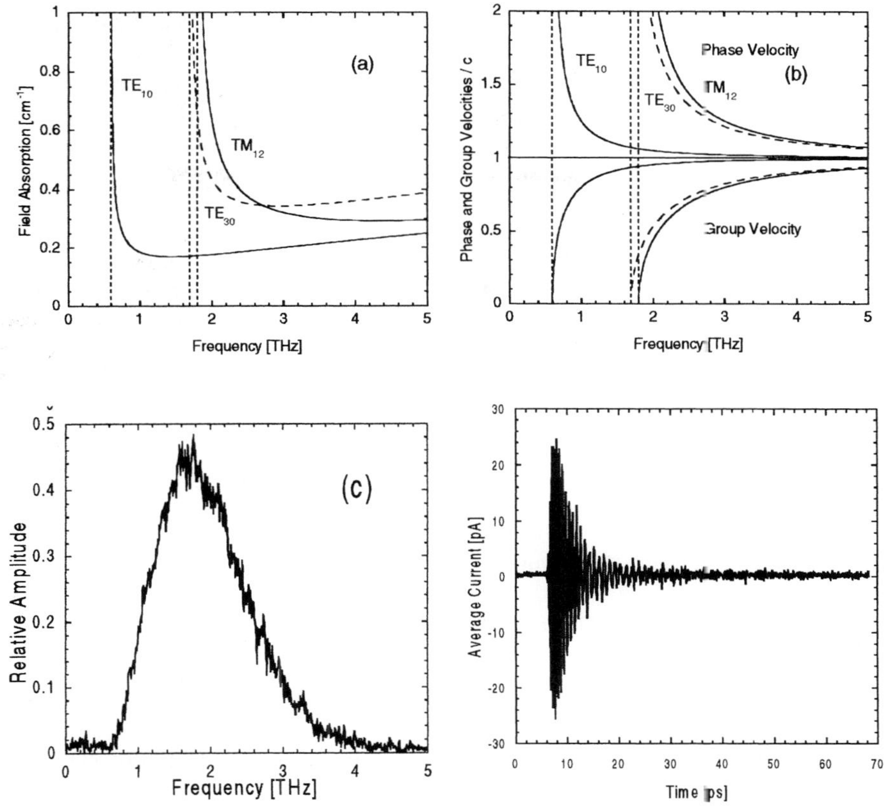

Fig. 3. (a) Calculated field absorption coefficient and (b) the phase and group velocities for the dominant 3 modes in the air-filled 250x125 μm rectangular brass waveguide. (c) Relative amplitude spectrum of the transmitted THz pulse. through the 25 mm long, brass waveguide. (d) Measured transmitted THz pulse. Single-mode propagation involves the TE_{10} mode.

are shown in Figs. 3a and 3b, where TE_{10} is the excited mode for the single mode propagation.

Figure 3c presents the amplitude spectrum of the measured THz pulse transmitted through the 25 mm long, 250x125 μm waveguide, shown in Fig. 3d. The transmitted pulse shows regular oscillations with some irregular structure on the trailing edge of the pulse due to water vapor in the path of the THz beam. Although the pulse has been significantly broadened by the group velocity dispersion of the TE_{10} mode shown in Fig. 3b, the resulting 13 ps, 1/e pulsewidth is significantly less than the corresponding 40 ps, 1/e pulsewidth for the circular guide, obtained earlier [3], due to the difference between single and multi-mode propagation. These results are significant in that they enable THz pulse propagation in the waveguide with a single-valued, analytic propagation vector.

Single mode propagation makes possible the application of waveguide THz time-domain spectroscopy (THz-TDS), e.g., surface-specific absorption measurements. For an air-filled waveguide, the observed frequency-dependent absorption would determine the difficult-to-measure, frequency-dependent conductivity. Our waveguide theory for thin film measurements has shown an enhancement of measurement sensitivity of up to 50 times that of a comparative single pass reflective measurement [4]. The waveguide THz-TDS technique also appears to be ideal for the THz (far-infrared) study of low-volume, high-pressure, long-path waveguide sample cells containing high-pressure, toxic or reactive gases and liquids. From experimental measurements made on the absorption of water, we believe that for a similar polar molecule, THz-TDS has the sensitivity to characterize the absorption of nanogram quantities of samples of toxic or precious gases contained in a THz waveguide. As a final comment, similar to optical fibers these THz waveguides can preserve the high intensity focus of a lens over the long path of the waveguide and thereby enable the observation of nonlinear effects.

Acknowledgements. This work was partially supported by the National Science Foundation and the U.S. Army Research Office.

References

1. M. van Exter and D. Grischkowsky, IEEE Trans. Microwave Theory Tech. **38**, 1684, 1990.
2. D. Grischkowsky, S. Keiding, M. van Exter and Ch. Fattinger, J. Opt. Soc. B, **7**, 2006, 1990.
3. R. W. McGowan, G. Gallot, and D. Grischkowsky, Optics Lett. **24**. 1431,1999.
4. G.Gallot, S.P. Jamison, R.W. McGowan and D. Grischkowsky, J. Opt. Soc. Am. B, **17**, 851, 2000.
5. J. Lesurf, *Millimeter-wave Optics, Devices and Systems*, Adam Hilger, Bristol and New York, 1990.

Variable Angle Impulse Ranging and Image Reconstruction of Dielectric Cylinders

M. T. Reiten, D. Grischkowsky, R. A. Cheville

School of Electrical Engineering and Center for Laser & Photonics Research, 202 Engineering South, Oklahoma State University, Stillwater, OK 74078, USA
E-mail: kridnix@okstate.edu

Abstract A new bistatic terahertz impulse ranging system has permitted isolation of the surface wave loss and dispersion on a dielectric cylinder. Direct, frequency dependent comparisons of surface wave loss and propagation velocity are compared to Mie theory. The ability to measure returned pulse trains phase coherently over a range of angles permits imaging through wavefront reconstruction. Preliminary results on image reconstruction are presented.

1. Introduction

Since the early work of Clebsch and Mie [1] analytic solutions for electromagnetic scattering have been approached in the frequency domain by solving for fields at boundaries. An alternative method of extracting the physical mechanisms of scattering is to spatially localize the excitation energy through use of a short duration pulse. The temporal resolution permits the experimental isolation of processes in time which are difficult to extract from boundary value methods such as propagation of surface waves [2] and Gouy phase shifts [3]. The contributions of various physical mechanisms to the total scattered field depend strongly on geometry-dominant mechanisms.

Through direct measurement of time resolved surface waves over a variable range of angles has enabled a direct measurement of complex index of refraction independent of other mechanisms such as radiation or coupling [2]. These results demonstrate enhancement of surface wave contributions for angles near backscattering. The ability to measure scattered pulses in the time domain with picosecond resolution at a range of well-defined angles permits wavefront reconstruction algorithms to be used for THz bandwidth imaging.

2. Experimental Methods

The impulse ranging system utilizes picosecond duration (near delta function) electromagnetic pulses to illuminate the target with time gated, coherent measurement of the scattered field over a 1 THz bandwidth [2,4]. The angle θ between the source and the detector can be continuously adjusted from 21° to 180°. The target is a high dimensional accuracy alumina rod of $1.5 \pm .001$ mm.

The time dependent scattered electric field is shown in Figure 1(a). Temporally isolated scattering mechanisms are clearly visible; a specular reflection striking the surface of the cylinder at glancing incidence (**S**), the surface or creeping wave radiation (Θ), and a wave propagating through the target (**T**). The surface wave arises for the portion of the incidence pulse which strikes the cylinder at near grazing incidence, couples to the surface, and travels around the outside of the target (inset Figure 1(a)). Due to the curvature of the cylinder this surface wave continually loses power to radiation.

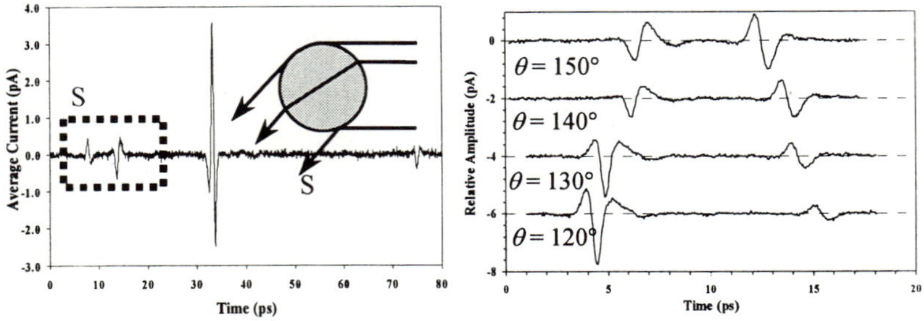

Fig. 1. a) Impulse response of alumina cylinder at θ = 150°. b) Specular reflection, T, and surface waves, Θ, at various θ.

3. Results and Discussion

Fourier transforming the time resolved data, the detected electric field is given by scattered from a cylinder of radius, r, at an angle, θ is given by:

$$E_{rcv}(\omega,\theta) = R(\omega)E_{surf}(\omega,\theta) = R(\omega)E_{surf}(\omega,0)e^{-in_{eff}(\omega)k_o r\theta} \quad (1)$$

where $E_{rcv}(\omega,\theta)$ is the signal detected at the receiver, $E_{surf}(\omega,\theta)$ is the electric field of the surface wave, and $R(\omega)$ describes the out-coupling of radiation. The complex index of refraction, $n_{eff}(\omega)$ determines the propagation of the surface wave through $k_s = n_{eff}k_o$.

The loss of radiation from the surface wave tangentially from the surface into space is responsible for the pulse labeled Θ in Figure 1(a). Measuring the surface wave radiation at oblique angles using the ranging system described above, the detected surface wave is effectively isolated from paths propagating through the target dielectric [4]. To measure surface wave propagation characteristics we experimentally vary θ over the range the range 120° ≤ θ ≤ 150° with the surface wave path lengths differing by $d_{surf} = r\Delta\theta$. Specular reflection (**S**) and surface wave measurements (Θ) are shown in Figure 1(b)

The temporally isolated surface waves were Fourier transformed, and surface wave field measured independent of coupling mechanisms:

$$\frac{E_{sig}(\omega,\theta+\Delta\theta)}{E_{ref}(\omega,\theta)} = \frac{R(\omega)E_{surf}(\omega,0)e^{-in_{eff}k_o r(\theta+\Delta\theta)}}{R(\omega)E_{surf}(\omega,0)e^{-in_{eff}k_o r\theta}} = e^{-in_{eff}(\omega)k_o r\Delta\theta} = e^{-ik_{surf}(\omega)d_{surf}} \quad (2)$$

E_{ref} is the isolated surface wave measured at θ = 150°and E_{sig} are the signal spectra taken at the signal angles of 140°, 130°, and 120°. This notation is equivalent to the notation previously used [2] to determine k_s. Figure 3 shows the real part of the effective index, $n(\omega)=\mathrm{Re}[n_{eff}]$, and the field loss coefficient, $\alpha(\omega) = \mathrm{Im}[n_{eff}]k_o$, measured as a function of frequency. The real index is within 4% of the previously published results [5] while the loss component is consistently higher (≈

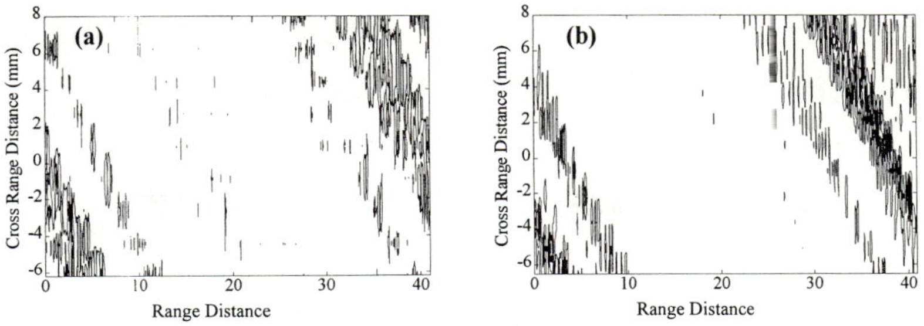

Fig. 2. Image from measured ranging data over 90 degree aperture compared to simulated data over aperture.

9 cm^{-1}) indicating that additional mechanisms are involved with paths crossing interior chords of the target.

The ability to phase coherently measure the scattered field over a range of angles leads to the ability to reconstruct images. Using a wavefront reconstruction algorithm it is possible to measure both range and cross range position of scattering features from a dielectric cylinder. These are shown in Figure 2 for 9 angular measurements taken over the range 49° to 139°. Figure 2(a) is a contour plot of the measured scattering center positions, while figure 2(b) shows the calculated image using numerical data obtained from a Mie scattering algorithm.

4. Conclusions

A continuously variable angle THz impulse ranging system has made possible the first direct measurement of the surface wave propagation characteristics in the THz frequency range. The deviation from the frequency independent loss obtained from previous measurements highlights the importance of as yet unidentified coupling mechanisms involved with surface waves. This system can be used for remote THz imaging. The authors acknowledge the support of the Army Research Office and National Science Foundation.

References

1 N. A. Logan, Proc. IEEE Vol. 53, 773, 1965.

2 R. A. Cheville, R. W. McGowan, and D. Grischkowsky, Phys. Rev. Lett. Vol. 80, 269, 1998.

3 R. W. McGowan, R. A. Cheville, and D. Grischkowsky, Appl. Phys. Lett. Vol. 76, 670, 2000.

4 R. A. Cheville, R. W. McGowan, and D. Grischkowsky, IEEE Trans. Ant. Prop. Vol. 45, 1518 1997.

5 R. W. McGowan, R. A. Cheville, and D. Grischkowsky, IEEE Trans. Microwave Theory Tech. Vol. 48, 417, 2000.

Dielectric Property Measurement of Sub-micron Thin Film by Differential Time-domain Spectroscopy

Kwang-Su Lee, Jin Young Kim[1], Jeffrey Fortin[1], Zhiping Jiang, Ming Li, Toh-Ming Lu[1], X.-C. Zhang

Department of Physics, Rensselaer Polytechnic Institute, Troy, NY 12180, USA
Email: zhangxc@rpi.edu
[1]CAIST, RPI

Abstract. Differential time-domain spectroscopy allows the dielectric property measurement of sub-micron films in the frequency rage from GHz to THz. We present measurement of dielectric property of sub-micron thin film.

1. Introduction

To increase the process speed of semiconductor devices, attempts to replace SiO_2 layers to low dielectric constant thin films are widely undertaken. The electronic device industry is pushing semiconductor devices operating at hundreds GHz and even at THz frequency range. For optimized operation of the devices, it will be essential to characterize the dielectric properties of thin films of dielectric materials or semiconductors in this frequency range. The dielectric properties are quite different from those of bulk due to the presence of fine grains, mechanical stresses, formation of interfacial layers, or rough interfaces [1]. Unfortunately, there are limits to characterizing dielectric properties of sub micron thin films in this frequency range because of very small interaction length with a normal incident of electromagnetic wave. At most, a vector analyzer can measure dielectric constant up to the range of a few ten GHz [2]. Several methods are proposed to measure dielectric constant of thin films. However, those methods are very time consuming or thickness limited for the measurement [3]. In this paper, a promising method to measure the dielectric constant of sub-micron thin film is proposed and differential time-domain measurement of dielectric property of a sub-micron thin film is presented.

A THz coherent wave generated in GaAs is used for the experiment when a 100 fs laser is illuminate on its surface. If the THz beam is incident on a thin film on a substrate with n_2 of refractive index, d of thickness of the thin film, and n_3 of refractive index of the substrate, a transmitted electric filed due to multi-reflection is given by [4]

$$E_t(\omega) = \frac{t_{12}\, t_{23}\, \exp\left[i\delta(\omega) - \alpha(\omega)d/2\right]}{1 - r_{21}\, r_{23}\, \exp\left[2i\delta(\omega) - \alpha(\omega)d\right]} E_0(\omega) \tag{1}$$

where E_0, t_{12}, t_{23}, r_{21}, r_{23} are incident electric field, transmission, reflection coefficients between interfaces and d, $\delta(\omega)$, $\alpha(\omega)$ are thickness, phase shift, the absorption coefficient of thin film medium, respectively. On the other hand, a reference electric field incident on the substrate is given by $E_{ref}(\omega) = t_{13} E_0(\omega)\exp(i\omega d / c)$. If defined as $E_{diff}(t) \equiv E_t(t) - E_{ref}(t)$ and assumed $d << c/\omega$, $\delta(\omega) << 1$, the ratio of $E_{diff}(\omega)/E_{ref}(\omega)$ is given by

$$\frac{E_{diff}(\omega)}{E_{ref}(\omega)} \approx i\frac{\omega}{c}d\left[n_2 - 1 + \frac{(n_2 - n_1)(n_2 - n_3)}{(n_1 + n_3)}\right] - \alpha(\omega)d\left[1 + \frac{(n_2 - n_1)(n_2 - n_3)}{(n_1 + n_3)}\right] \quad (2)$$

By solving this equation on n_2 in terms of $E_{diff}(\omega)/E_{ref}(\omega)$, d, n_1, n_3 and denoting $A(\omega) \equiv \frac{c}{\omega d}\left|E_{diff}(\omega)/E_{ref}(\omega)\right|$, the equation on n_2 is given by

$$n_2(\omega) = \sqrt{[1 + A(\omega)](n_1 + n_3) - n_1 n_3} \quad (3)$$

If $A(\omega)$ is measured, n_2 can be deduced. For most cases, the dielectric constant of a thin film is given by $\varepsilon(\omega) = n(\omega)^2$

2. Experimental Methods

For the experiment, a Ti: sapphire laser (Tsunami) with 100 fs in pulse width, 800 nm in central wavelength, and 82 MHz in repetition rate is used. The laser light is divided into two beams, a pump and a probe. The pump laser beam is illuminated on a GaAs inclined surface to generate THz. The generated THz is collimated and focused by two pairs of parabolic mirrors. The probe beam is coincident with the THz wave to detect a polarization change due to Pockel's effect of THz. The sample is designed to be a half coated with thin film on the substrate and the other half uncoated. It was mounted on a shaker (galvanometer) with a frequency of 66 Hz at the focal point of THz beam to measure E_{diff}. This shaking frequency is fed by a function generator and synchronized with a lock-in amplifier. E_{ref} can be measured by blocking the THz beam with a metal plate at the thin film coated part.

3. Results and Discussion

Differential time-domain spectroscopic signals were measured in several thin film samples, including 1.8 μm free standing, 1.8 μm and 300nm Parylene-N films on silicon substrates. Fig.1 shows a measured differential THz waveform of 300nm Parylene-N film. The measured amplitude of $E_{diff}(t)$ is 0.2 % of $E_{ref}(t)$. Time resolved waveforms of both E_{diff} and E_{ref} of thin films on the silicon substrate show reflection waveforms due to internal reflections in the substrates. E_{diff} waveforms from reflected second pulses in both thin films on the substrates are observed to be unexpectedly high in amplitudes. The frequency dependence of dielectric constants are obtained directly from Eq. (6) with Fourier transformed E_{diff} and E_{ref}. The reference waveform ranges from 200 GHz to 1.6 THz in

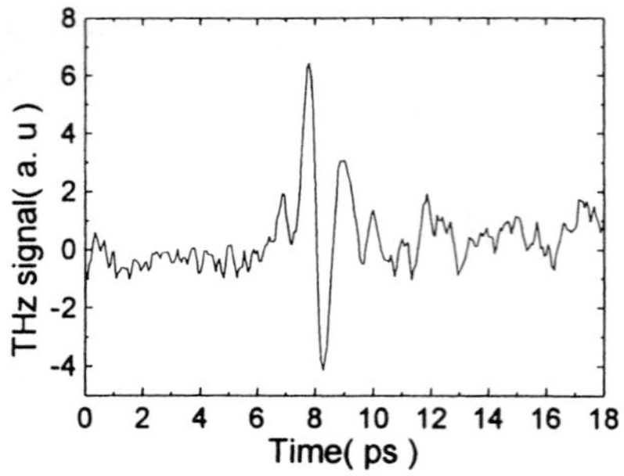

Fig. 1 Differential THz waveform of 300nm Parylene-N thin film.

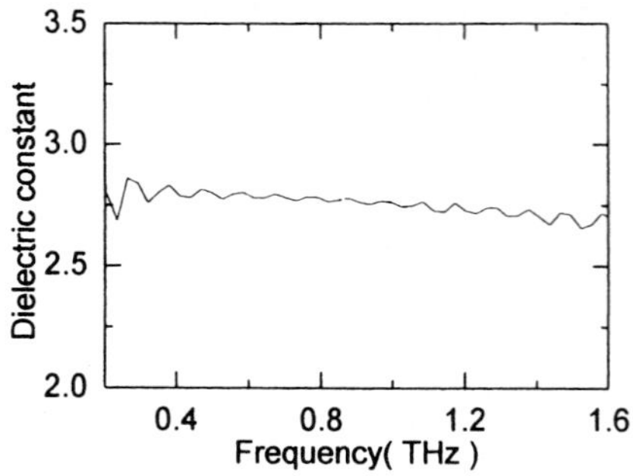

Fig. 2 Dielectric constant of 300nm Parylene-N thin film.

frequency and has maximum amplitude at 0.75THz in the frequency spectrum. The maximums of differential waveforms are shifted to higher frequencies and have a maximum at 0.9 THz.

For the deduction of dielectric constant, the second reflected waveforms are not counted. Fig.2 shows dielectric constant of 300nm Parylene-N thin film. The 300nm film shows a lower dielectric constant than that of bulk material. The detailed origin of the lower dielectric constant of these Parylene-N thin films is not systematically investigated. However, as mentioned in the earlier part, the presence of fine grains, mechanical stresses, formation of interfacial layers or rough interfaces are considered to influence its dielectric constant by part.

4. Conclusions

The differential time-domain spectroscopic method for the dielectric property measurement of sub-micro thin film in GHz to THz frequency range is presented. The dielectric constant of sub micron thin film is measured. The 300nm Parylene-N thin film shows different dielectric properties than that of bulk material.

Acknowledgements. This work was supported by the National Science Foundation and Army Research Office.

References

1 V. Craciun and R. K. Singh, Appl. Phys. Lett. 76, 1932 (2000).

2 D. Grischkowsky, Soren Keiding, Martin van Exter, and Ch. Fattinger, J. Opt. Soc. Am. B. 7, 2006 (1990).

3 M. Li, G. C. Cho, T.-M. Lu, X.-C. Zhang, S.-Q. Wang and J.T. Kennedy, Appl. Phys. Lett. **74**, 2113 (1999).

4 M. Born and E. Wolf, *Principles of Optics*, 6th edition, (Cambridge University Press, 1980), chapter 7.6.

Excitation Dynamics beyond the Slowly-Varying Envelope Approximation

R. Kersting[1], R. Bratschitsch[2], G. Strasser[2], and K. Unterrainer[2]

[1] Department of Physics, Rensselaer Polytechnic Institute, Troy, NY 12180, USA
E-mail: kerstr@rpi.edu
[2] Institute for Solid State Electronics, TU Wien, Floragasse 7, A-1040 Wien, Austria

Abstract. We present THz measurements of nonadiabatic excitation dynamics in semiconductor heterostructures which are recorded with a time-resolution shorter than the oscillation time of the exciting light.

1. Introduction

During the past decades the excitation dynamics of two-level systems have been the subject of a variety of time-resolved optical experiments. When interacting with a light field, the coherent response of a dipole transition can be treated using the density formalism which considers the Hamiltonians of the system, of the interaction, and of the relaxation. In general, the optical Bloch equations are solved using the slowly-varying envelope approximation (SVEA) [1]. This leads to such coherent phenomena as optical nutation, Rabi-oscillations and photon echoes. Here, we report experiments performed with a time-resolution shorter than the period of the excitation field. For the first time, we fully resolve the excitation dynamics on a time-scale beyond the SVEA and observe the nonadiabatic following of electrons.

2. Experimental Method

The recent developments of coherent sources emitting far-infrared light pulses have made sub-cycle time-resolutions possible. In our experiments we use THz light pulses with tailored envelopes generated by femtosecond laser excitation of n-doped GaAs [2]. Few-cycle broad-band THz pulses (0.5 THz – 4.0 THz) are used to excite electronic intersubband transitions in semiconductor heterostructures. After transmission the THz pulses are time-resolved by mixing them with nearly half-cycle reference pulses [3].

The experiments were performed on electronic intersubband transitions in modulation-doped parabolic quantum wells (PQWs). The 1,200 Å and 1,400 Å wide GaAs/AlGaAs PQWs have equally spaced transitions of 3.0 THz and 2.2 THz, respectively. Semi-transparent Schottky gratings are fabricated on top of the structures to couple the exciting THz pulse to the intersubband transition. Application of a negative bias to the Schottky grating depletes the structure of electrons and therefore enables the control of the intersubband absorption and dispersion. Using this electronic control for time-resolved modulation measurements gives a signal which is proportional to the macroscopic polarization of the excited quantum well states. The signal is proportional to the expectation value of the electron motion [4].

3. Results and Discussion

Figure 1a) shows the response of electrons in a PQW structure when driven nearly resonantly by a few-cycle THz pulse. During excitation the macroscopic polarization follows the driving light field. The amplitude of the polarization increases gradually during the presence of the exciting THz pulse. When the driving pulse has decayed, the electrons continue to oscillate and undergo free induction decay due to scattering. Off-resonance experiments give further insight into the nonadiabatic electron dynamics during excitation. Figure 1b) shows the response of a system with an intersubband transition frequency of 3.0 THz. It is driven by a THz pulse having a center frequency of 1.6 THz. Even in this strongly off-resonant case the electrons follow the excitation field despite the discrete nature of their quantum states. After the decay of the driving THz field, the electrons change their oscillation frequency from 1.6 THz to their eigenfrequency of 3.0 THz.

The experimental data are analyzed using model calculations based on an ansatz, which considers the nonadiabatic following and takes into account the shape of the exciting pulse. Due to the low excitation density, the model enables both, the calculation of the polarization as well as of the population of the excited state [4]. Figure 2 shows the agreement between calculated and experimental data for the off-resonant case. During excitation a build-up and decay of the population of the upper state is observed. The increased population can be understood in terms of the Heisenberg picture. During interaction with the exciting field the electrons are driven into a virtual state which is a superposition of both the upper

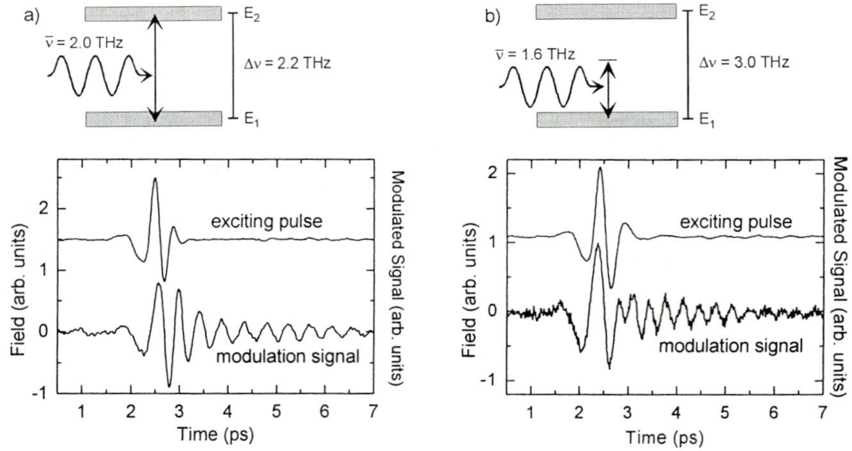

Fig. 1. Resonant (a) and nonresonant (b) excitation of intersubband transitions. The upper curves show the exciting THz pulses and the lower curves the modulation signal. The Jablonski diagrams indicate the excitation process.

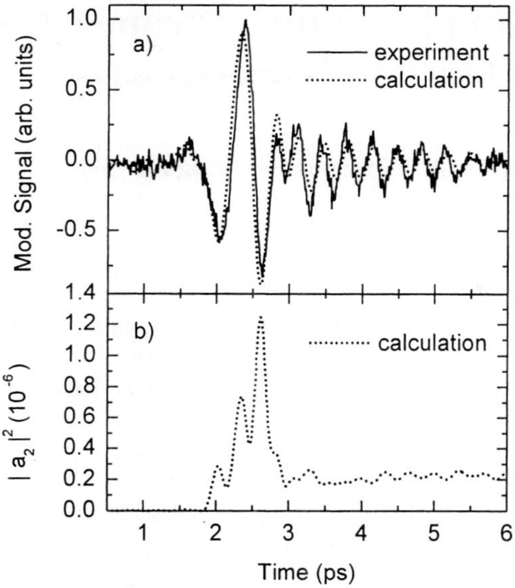

Fig. 2. : Model calculation of polarization (a) and population of the upper level (b). The solid line shows the experimental data, the dotted lines show the calculated data.

and the lower state wave functions. This population decays with the decay of the driving field. However, it does not vanish completely due to the small spectral overlap between the exciting THz pulse and the intersubband transition.

It is worth mentioning that the observed nonadiabatic phenomena are invariant to scaling of the wavelengths. The reported dynamics show the fundamental nature of optical excitations in matter. Besides these fundamental aspects, the observed dynamics enable novel electro-optic devices such as THz modulators [5].

References:

1 Meystre and M. Sargent III, in *Elements of Quantum Optics*, Springer, New York, 1998.
2 R. Kersting, K. Unterrainer, G. Strasser, H.F. Kauffmann, and E. Gornik, Phys. Rev. Lett. **79**, 3038, 1997.
3 J.N. Heyman, R. Kersting, and K. Unterrainer, Appl. Phys. Lett. **72**, 644, 1998.
4 R. Kersting, R. Bratschitsch, G. Strasser, and K. Unterrainer. Opt. Lett. **25**, 272, 2000.
5 R. Kersting, G. Strasser, and K. Unterrainer. Electr. Lett. **36**, 1156, 2000.

THz-Pulse Studies of Superluminal Propagation in Frustrated Total Internal Reflection

John J. Carey, Justyna Zawadzka, Dino A. Jaroszynski and Klaas Wynne

University of Strathclyde, Dept. of Physics and Applied Physics, Glasgow G4 ONG, UK
URL: http://dutch.phys.strath.ac.uk/

Abstract. Tunneling of photons in frustrated total internal reflection has been studied in the time domain with single-cycle THz pulses. Superluminal phase, group and energy velocities have been observed. Diffractive spreading rules out superluminal communication.

Waves propagating, for example, along interfaces, through waveguides below the cutoff frequency or through structures with a photonic band gap, are evanescent. The wavevector of an evanescent field has one or more imaginary elements resulting in an exponential decay of the field amplitude with distance in one or more spatial directions. The quantum-mechanical equivalent of evanescent-wave propagation is tunneling in which the wave function decays inside a classically forbidden barrier. [1] The most curious phenomenon of tunneling is superluminal propagation in which it may appear that signals travel faster than the speed of light in vacuum.

We will describe a study of Frustrated Total Internal Reflection (FTIR) in which evanescent electromagnetic waves travel in the air gap between two prisms. It has been shown using an indirect technique [2] that light travels superluminally in this gap. New data will be presented [3] demonstrating superluminal propagation entirely measured and analyzed in the time domain using single-cycle terahertz pulses. It is found that the pulses are advanced more than ever seen before: More than a pulse width. Theoretical analysis shows that in FTIR part of the incoming pulse travels backwards in time apparently violating the principle of causality.

The laser system used is a Ti:sapphire oscillator and regenerative amplifier producing 150-fs pulses at a 250-kHz repetition frequency resulting in 600-mW average power at 800 nm. The beam from this laser is split in two as it enters the THz setup. One is the gate beam, which is used for detection by electro-optic sampling with a <110> cut 1-mm thick ZnTe crystal. The other 78% is incident on a photoconductive dipole antenna consisting of a 3x5-mm^2 piece of LT-GaAs. With 2-kV bias applied, a 0.8-ps THz pulse with a 1-mm center wavelength is produced. Scanning the delay of the gate beam at 6 Hz gives a real-time electric-field trace, which is averaged for 5-10 minutes to give a signal-to-noise ratio >100.

Two right-angle Teflon prisms (θ_c=44.4^0±0.2^0) of 40-mm side length are placed in the beam with variable air spacing. THz-field traces with the gap between the prisms set to zero ($L = 0$) are directly compared with traces at nonzero gap. Translating the second prism away from $L = 0$ results in the creation of an air gap between the prisms and the removal of a column of air outside the prism pair. If the speed of light in air were independent of the presence of the prisms, the two experiments (at $L = 0$ and $\neq 0$) would give the same result apart from optical pathlength changes due to refraction. Any changes in the optical pathlength are indicated in these experiments as a relative delay.

Several series of experiments were performed in which L was varied between 0 and 20 mm, the incidence angle was varied between 35^0 and 55^0, and with p- and s-polarization of the THz beam. The figure shows data for angles of incidence where the waves in the gap are evanescent. In this case, it is found that the transmission varies from unity at low frequencies to zero at high frequencies (~60 cm^{-1}) and that the transmission spectrum narrows with increasing gap. In both the evanescent and non-evanescent cases, the advance as measured by the group or centroid delay [4] is on the order of a few picoseconds and larger than the pulse width.

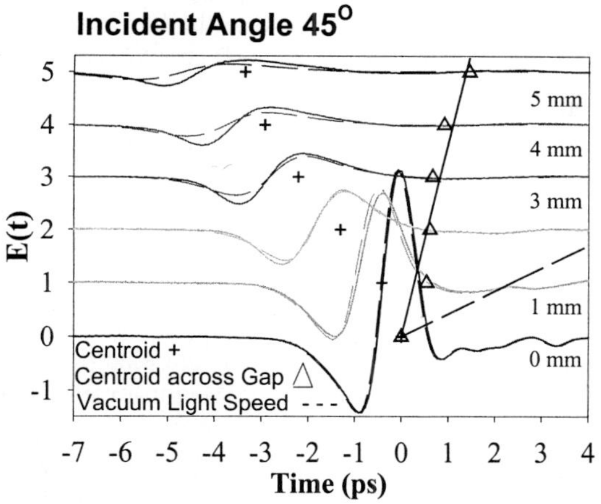

The raw data show temporal advances larger than a pulse width for both evanescent and non-evanescent waves. In the evanescent case, the temporal advance across the gap is in fact much larger than apparent because the waves have to travel through more prism material as the gap is increased. In the non-evanescent case, the measured temporal advance is because the diffracted beam travels through less material in the second prism as the gap is increased. In the non-evanescent case, the centroid of the pulses is found to cross the gap at speed (0.99±0.01) c. In the evanescent cases, the pulses cross the gap at superluminal speed, consistent with the theoretical time-response function.

The theory used to describe these data [3] appears to show that the principle of causality does not apply in FTIR. However, in the case of non-evanescent wave propagation, any apparent violation of the principle of causality is due to a trivial shadow effect. However, in the case of evanescent-wave propagation, the time-response function has a value at negative time (extending to negative infinity). To obtain this superluminal effect, the pulses have to travel through prism material in which the pulses are delayed with respect to free space. However, in principle, the gap could be made arbitrarily large and the superluminal advance would outstrip the delay due to the prism material. The transmission of evanescent waves is poor but this could be compensated for by using a powerful input pulse.

Do the results presented here imply that superluminal communication is possible and causality violated in a non-trivial way? If the angle of incidence is chosen close to but larger than the critical angle, the signal will travel superluminally with transmission close to unity. However, to have a single angle of incidence, the signal beam has to be perfectly collimated necessitating an infinitely extended beam. This requires infinitely large prisms and hence an infinitely large delay compared to free space. In practice therefore, the prisms must have a finite extent resulting in a distribution of incidence angles. In our

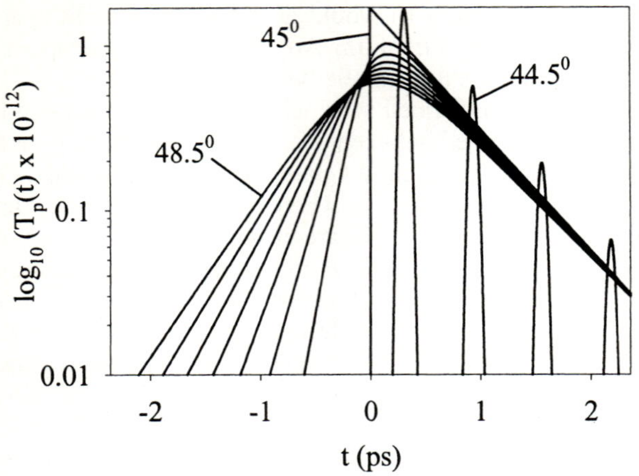

Plot of the theoretical time-response function for FTIR. The graphs show the response in the second prism when a delta-function pulse reaches the first interface at $t = 0$. $n = \sqrt{2}$ (critical angle 45^0), $L = 1$ mm ($d/c = 2.4$ ps) and $\theta = 44.5, 45, \ldots 48.5^0$.

experiment, we measured a beam width of 1.5cm implying that the electric-field traces in the figure may be "contaminated" with as much as 25% non-evanescent waves. Non-evanescent waves have transmission close to unity for any gap whereas the evanescent waves decay exponentially. Thus, for any *average* angle of incidence $\theta > \theta_c$, the beam still has components with $\theta < \theta_c$, which will end up dominating the signal for sufficiently large gap. Therefore when the gap becomes large enough to make this superluminal communication scheme useful, the signal will be dominated by non-evanescent waves.

Acknowledgements. We acknowledge the financial support of EPSRC and the technical support of Andy Hughes. JZ gratefully acknowledges financial support from an Overseas Research Students Award Scheme.

References

1. Chiao, R.Y. and A.M. Steinberg, *Tunneling Times and Superluminality.* Progress in Optics, 1997. **37**: p. 345-405.
2. Balcou, P. and L. Dutriaux, *Dual Optical Tunneling Times in Frustrated Total Internal Reflection.* Physical Review Letters, 1997. **78**: p. 851-854.
3. Carey, J.J., *et al.*, *Non-Causal Time Response in Frustrated Total Internal Reflection.* Physical Review Letters, 2000. **84**: p. 1413-1434.
4. Wynne, K., *et al.*, *Tunneling of Single-Cycle Terahertz Pulses through Waveguides.* Optics Communications, 2000. **176**: p. 429-435.

Part V

Imaging and Microscopy

Optical Coherence Tomography for Biomedical Imaging

James Fujimoto[1], Wolfgang Drexler[1], Xingde Li[1], Ingmar Hartl[1], Christian Chudoba[1], Uwe Morgner[2], Franz Kärtner[2], Costas Pitris[1], Erich Ippen[1], and Mark Brezinski[3]

[1] Department of Electrical Engineering and Computer Science, Research Laboratory of Electronics, Massachusetts Institute of Technology, Building 36-345, 77 Massachusetts Avenue, Cambridge, MA 02139, USA

[2] High-Frequency and Quantum Electronics Laboratory, University of Karlsruhe (TH), D-76131, Germany

[3] Brigham and Women's Hospital, Harvard Medical School, Boston, MA, USA

Abstract. Optical coherence tomography (OCT) enables imaging of tissue pathology in situ and in real time. Using ultrashort pulse lasers, ultrahigh resolutions of ~1 μm as well as spectroscopic imaging are possible. OCT can function as a type of optical biopsy because it enables imaging tissue microstructure in situ without the need to excise and process a specimen as in conventional histopathology. OCT can be used to image pathology in tissues where biopsy is impossible.

1. Introduction

Optical coherence tomography (OCT), is an emerging technology for biomedical imaging. OCT enables the high resolution *in vivo* and *in situ* imaging of internal tissue microstructure and pathology in a variety of medical fields [1-4]. OCT imaging is analogous to ultrasound except that it uses low coherence light and interferometry to perform precise measurements of the echo delay of reflected or backscattered light from tissue. Cross sectional images are measured by scanning the transverse position of the incident optical beam, resulting in a two dimensional array which represents the backscattering of the tissue in a cross sectional plane. OCT is a powerful imaging modality because it permits the imaging of tissue *in situ* without the need for excision and processing as in conventional biopsy and pathology.

The axial resolution of OCT is governed by the bandwidth of the light source used for imaging. In typical OCT systems superluminescent diodes are used, yielding resolutions of 10-15 μm. Femtosecond KLM solid state-lasers are attractive light sources for OCT because they generate extremely short coherence length light with single spatial mode and high power [5-8]. Ti:sapphire and Cr:forsterite lasers have been used for *in vitro* and *in vivo* imaging with 4 μm and 6 μm longitudinal resolution, respectively [3-5]. Recently, resolutions of 1 μm have been achieved using ultrabroad bandwidth Ti:sapphire generating pulse durations of ~2 cycles [6,7].

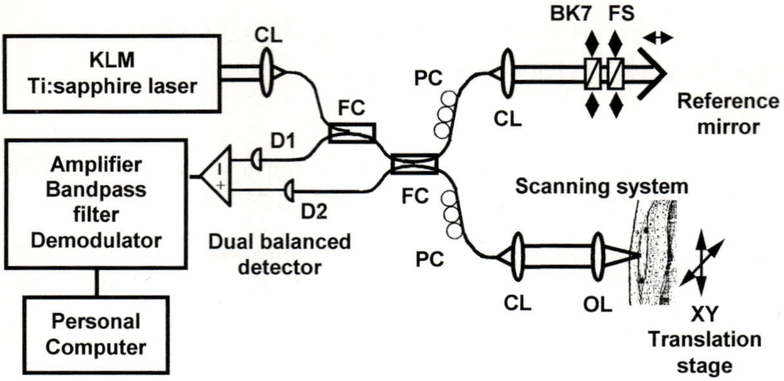

Figure 1. Ultrahigh resolution fiber OCT system using a KLM modelocked Ti:sapphire laser. The interferometer is dispersion balanced, optically and electronically optimized to support broad optical bandwidths.

2. Methods

Using a state of the art femtosecond Kerr-lens modelocked Ti:sapphire laser as a low coherence light source, an ultrahigh resolution, spectroscopic OCT system has been developed [7,8]. This laser emits sub two optical cycles pulses, corresponding to bandwidths of over 350 nm centered at 800 nm. High performance is achieved using specially designed double chirped mirrors with high reflectivity bandwidth and controlled dispersion response, in combination with low-dispersion calcium-fluoride prisms for intracavity dispersion compensation. By carefully matching dispersion, minimizing chromatic aberrations, and optimizing optics, fibers, electronics and data acquisition, a fiber optic OCT system has been developed, being able to support 260 nm and enabling *in vivo* imaging with axial resolutions of 1 µm in biological tissues. In addition, the Ti:sapphire laser output spectrum can be filtered or post processing algorithms can be applied, so that adjustable optical bandwidths at wavelengths in the near infrared can be generated, and spectroscopically resolved OCT imaging can be performed.

3. Results

OCT can be applied to a wide range of biomedical applications [1-4]. Figure 2 shows an ultrahigh resolution cross-sectional tomograms of the *Xenopus laevis* tadople *in vivo*. The image has an axial resolution of 1 µm.

Figure 2. Ultrahigh resolution OCT image of African frog tadpole (Xenopus laevis) in vivo showing subcellular level resolution. (1x5μm, longitudinal x transverse; 0.75x0.5mm; 1700x1000 pixels) Multiple cells of various sizes are clearly shown. Bar: 100μm.

Figure 3 shows an ultrahigh resolution cross-sectional tomogram of the retina of a normal human volunteer compared to a standard resolution OCT image. An axial resolution of ~3 μm has been achieved. This is an improvement by a factor of ~ 4-5 over standard, 10-15 μm resolution, OCT retinal imaging, and to our knowledge, the highest OCT resolutions achieved for ophthalmic imaging. Both tomograms were recorded in the same eye at approximately the same site, in the foveal region over a transverse length of approximately 4 mm, consisting of 100 x 500 pixel and 600 x 1300 pixel sampling, i.e. 30 μm x 7 μm (top) and 7 μm x 1 μm (bottom) pixel spacing. Ultrahigh resolution OCT offers an unprecedented view and quantification of retinal morphology with structures such as the retinal nerve fiber layer, retinal pigment epithelium, inner and outer plexiform layers, and the photoreceptor layer. In other medical specialties, the combination of OCT with catheters, endoscopes, or needles can be used to image both intraluminally and in solid organs.

Spectral and functional information can be obtained by measuring the full interference signal instead of the envelope data only using digital signal post processing [8]. This method is analogous to Fourier transform spectrometery. A full spectra may be obtained for each pixel in the OCT image. In order to display the spectral features in an image, the "center of mass" of the spectra was calculated and mapped to a false color scale. Hue, saturation, luminance (HSL) color space were used to map the backscattering intensity into the saturation and the spectral center of mass into the hue, keeping luminance constant. This permits the intensity and spectral shift of the backscattered light to be visualized. Figure 4 shows the conventional ultrahigh resolution OCT image (left) and the spectroscopic OCT image (right) of the same site of the mouse retina *in vivo*. In the spectroscopic OCT image (right), red hue indicates a long wavelength enhancement of the backscattered light, while green hue is a short wavelength enhancement. The spectroscopic image shows enhanced differentiation between different layers of the retina which arise from their different optical scattering properties. Functional information, such as blood oxygenation or tissue hydration, may also be extracted.

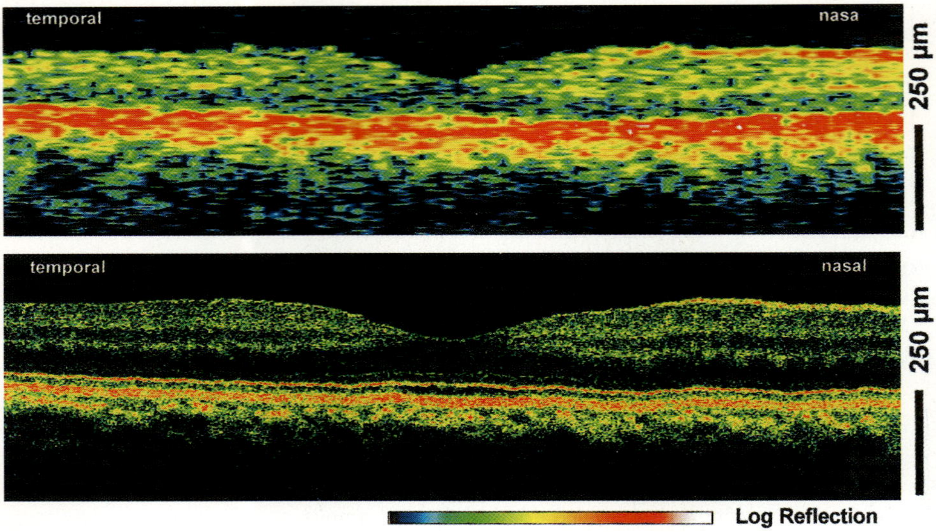

Figure 3. OCT image foveal region of the retina of a normal human subject with standard (top, 10 μm axial resolution) and ultrahigh resolution (bottom, 3 μm axial resolution). This unprecedented image resolution will significantly improve visualization of intraretinal structures and should have a powerful impact on clinical ophthalmology. (with J. Schuman, New England Eye Center)

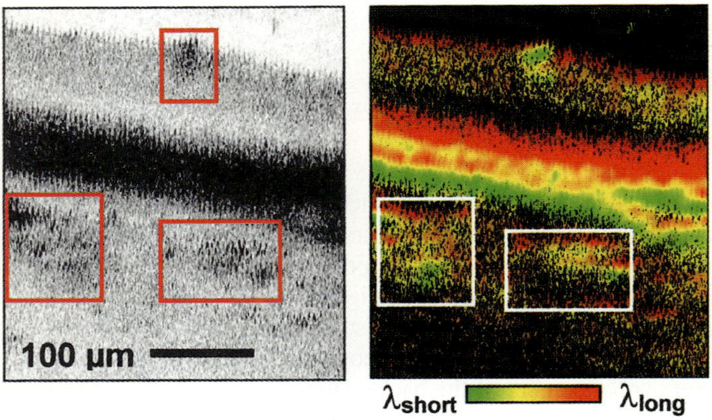

Figure 4. Spectroscopic OCT imaging of the mouse retina *in vivo*. Spectroscopically resolved imaging can be a powerful method for enhancing the differentiation of structures and assessing their functional state. This data is preliminary and the motion artifact in the image can be greatly reduced by faster image acquisition or image processing.

4. Conclusions

OCT is an emerging imaging modality which functions as a type of "optical biopsy," to image tissue pathology *in vivo* and *in situ.* Ultrahigh resolution OCT imaging using femtosecond laser sources achieves a quantum leap in performance over current OCT systems. Femtosecond laser sources also enable broad bandwidth spectroscopic imaging. Spectroscopically resolved OCT can enhance image contrast by differentiating tissues according to their scattering properties and also promises to permit the functional imaging of tissue. OCT imaging has the potential for early diagnosis and monitoring of many different diseases and can be applied to a wide range of clinical applications ranging from cancer detection to surgical guidance. The availability high performance of OCT will strongly depend on the development of low cost, compact femtosecond laser sources or other low coherence light sources.

Acknowledgements. W. Drexler was visiting from the University of Vienna. This research was supported by the Air Force Office of Scientific Research contract F4920-98-1-0139 and the Medical Free Electron Laser Program, Office of Naval Research contract N00014-94-1-0717.

References

1 D. Huang, E. Swanson, C.P. Lin C, J.S. Schuman, W.G. Stinson, W. Chang, M.R. Hee, T. Flotte, K. Gregory, C.A. Puliafito, and J.G. Fujimoto, Science **254**, 1178, 1991.

2 G. J. Tearney, M. E. Brezinski, B. E. Bouma, S. A. Boppart, C. Pitris, and J. G. Fujimoto, Science, **276**, 2037, 1997.

3 M. E. Brezinski and J. G. Fujimoto, IEEE Journal of Selected Topics in Quantum Electronics, **5**, 1185-92, July/August 1999.

4 J. G. Fujimoto, C. Pitris, S. Boppart, and M. Brezinski, Neoplasia, **2**, 9-25, January 2000.

3 B. E. Bouma, G. J. Tearney, S. A. Boppart, M. R. Hee, M. E. Brezinski, J. G. Fujimoto, Optics Letters **20**, 1486, 1995.

4 B. E. Bouma, G. J. Tearney, I. P. Bilinsky, B. Glubovic, and J. G. Fujimoto, Optics Letters **21**, 1839, 1996.

5 S. A. Boppart, B. E. Bouma, C. Pitris, J. F. Southern, M. E. Brezinski, and J. G. Fujimoto, Nature Medicine **4**, 861, 1998.

6 U. Morgner, F. X. Kartner, S. H. Cho, Y. Chen, H. A. Haus, J. G. Fujimoto, E. P. Ippen, V. Scheuer, G. Angelow, and T. Tschudi, Optics Letters **24**, 411, 1999.

7 W. Drexler, U. Morgner, F. X. Kartner, C. Pitris, S. A. Boppart, X. D. Li, E. P. Ippen, and J. G Fujimoto, Optics Letters **24**, 1221, 1999.

8 U. Morgner, W. Drexler, F. X. Kartner, X. D. Li, C. Pitris, E. P. Ippen, and J. G. Fujimoto, Optics Letters **25**, 111, 2000.

Ultrahigh resolution OCT using continuum generation in an air-silica microstructure optical fiber

I. Hartl[1], X.D. Li[1], C. Chudoba[1], R. Ghanta[1], T. Ko[1], and J.G. Fujimoto[1]
J.K. Ranka[2], R.S. Windeler[2] and A.J. Stentz[2]

[1] Department of Electrical Engineering and Computer Science and Research Laboratory of Electronics, Massachusetts Institute of Technology, 77 Mass. Avenue, Cambridge, MA 02139, hartl@mit.edu
[2] Bell Laboratories, Lucent Technologies, 700 Mountain Avenue, Murray Hill, NJ 070974

Abstract: We demonstrate ultrahigh resolution optical coherence tomography using continuum generation in an air-silica microstructure fiber. A broadband OCT system was developed, supporting a bandwidth of 370 nm at 1.3 μm center wavelength. We achieved longitudinal resolutions of 2.6 μm in air, or ~2 μm in tissue. To our knowledge this is the highest longitudinal OCT resolution demonstrated in this wavelength range and the first application of this new light source for OCT. We also describe the application of this technique for imaging biological tissue *in vivo*.

Optical coherence tomography (OCT) is an emerging technology for micron-scale cross-sectional imaging of biological tissue [1]. The longitudinal resolution of OCT images is inversely proportional to the optical bandwidth of the light source. Superluminescent diodes are used in conventional OCT systems and typically yield 10-15 μm longitudinal resolutions. OCT imaging with resolutions of 1 μm at 800 nm [2] and 5.1 μm at 1300 nm [3] in biological tissue have been demonstrated using a Kerr-lens mode-locked Ti:sapphire laser with double-chirped mirrors and the self phase modulation broadened spectrum of a Kerr-lens mode-locked Cr:forsterite laser, respectively.

High nonlinearity, air-silica microstructure fibers [4] can generate an extremely broadband continuum using low energy femtosecond pulses. Ultrahigh resolution OCT in the spectral region between 1.2 μm and 1.5 μm requires extremely broad bandwidths because of the $\lambda^2/\Delta\lambda$ dependence of the longitudinal resolution. However, this spectral region is of particular interest for OCT because of the high penetration depth in biological tissue.

We demonstrated ultrahigh resolution OCT using an air-silica microstructure fiber as a light source. We launched unamplified 100 fs pulses from a commercially available femtosecond laser (Coherent Mira) into the microstructure fiber. Dramatic nonlinear effects are observed in this fiber which result in the generation of an extremely broadband continuum extending from 390 nm to 1600 nm [4]. Since the full bandwidth of the continuum is too broad to allow focussing without chromatic aberration and too broad to be detected by a photodiode made from a single semiconductor, we used only the near-infrared part of the light source for OCT imaging.

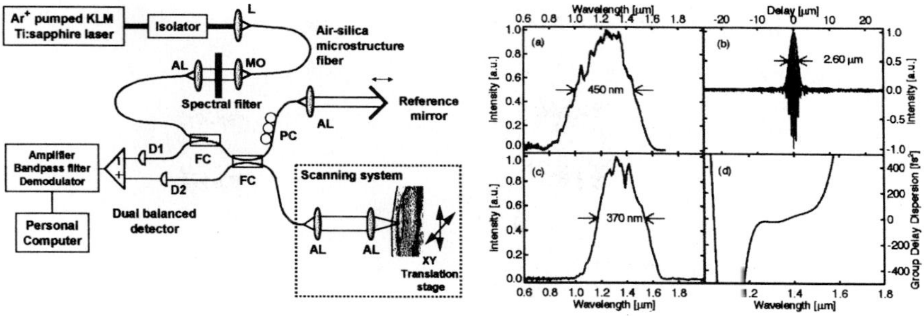

Fig. 1. Left: Schematic diagram of experimental setup. L: f=3.5mm 0.65 NA best-form lens, MO: 0.85 NA 60x microscope objective, FC: 3dB 2x2 fiber coupler, AL: f=10mm chromatically corrected custom-designed 0.3 NA lens, D1,D2: InGAs photodiodes. A: air gap. PC: polarization controller. **Right:** (a) Typical spectrum of the continuum after spectral filtering and coupling to a single-mode fiber measured at the entrance port of the second fiber coupler. (b) Interference fringes recorded by using a silver mirror as sample. The full width at half maximum was determined to 2.6 µm. Detected optical spectrum (c) and group delay dispersion mismatch of the interferometer arms (d) obtained from the intensity spectrum of the Fourier transform of the interferometric signal (b).

A schematic of the experimental setup is shown in Fig1 (left). The output of a KLM Ti:sapphire laser, centered at 803 nm, is focussed into the air-silica microstructure fiber. The fiber continuum is collimated, spectrally filtered and focussed into the OCT setup, which uses custom-designed lenses and broadband 3dB fiber couplers (FC), designed for 1300 nm center wavelength.

Fig 1a shows the spectrum of the light source after spectral shaping, coupling into the single-mode fiber and passing the first fiber coupler. The performance of the OCT system was measured by placing a silver mirror at the sample position. The interference signal obtained and its envelope is shown in Fig 1b. We obtained a free space longitudinal resolution of 2.6 µm, corresponding to ~2 µm in tissue. The detected optical spectrum (Fig 1c) was calculated by Fourier transforming the interferometric signal. The bandwidth was reduced to approximately 370 nm due to the loss of the short wavelength part of the spectrum. This reduction of the bandwidth is mainly caused by the poor spectral sensitivity of the InGAs photodiodes at shorter wavelength. The sensitivity of the system was experimentally measured to be -100dB by attenuating the signal in the sample arm with neutral density filters.

We demonstrated the feasibility of *in vivo* ultra high resolution imaging by using the cheek pouch of a Syrian hamster, a well-established animal model for studies of cancer progression. Fig. 2 shows an *in vivo* image of the cheek pouch which shows the epithelium, connective tissue and muscle layers at ultrahigh resolution.

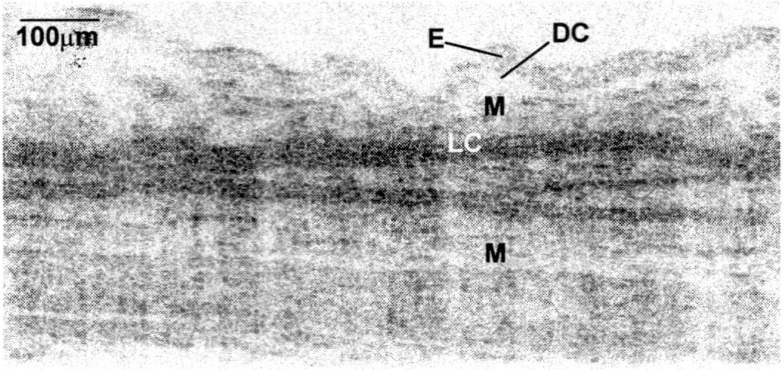

Fig. 2. *In vivo* ultrahigh resolution (~2 μm x 6 μm, longitudinal x transverse resolution) OCT image of a Syrian hamster cheek pouch. A layered structure showing epithelial layer (E), dense connective tissue (DC), muscle layers (M) and loose connective tissue (LC) is clearly visible. This image consists of 1000 transverse scans with 2000 pixels per scan, covering an area of 1 mm by 0.5 mm. The incident power was 0.5 mW.

In conclusion, we demonstrated the application of a white light continuum generated in an air-silica microstructure fiber as a light source for ultrahigh resolution OCT imaging. We achieved a longitudinal resolution of ~2 μm in biological tissue at a center wavelength of 1.3 μm. This light source may be engineered into a compact design; suitable for clinical application, by using commercially available, all-solid state, integrated Ti:sapphire lasers or fiber lasers.

Acknowledgements. We wish to thank Dr. Wolfgang Drexler and Costas Pitris for technical assistance and helpful discussions. This research was supported by the Air Force Office of Scientific Research contract F4920-98-1-0139 and the Medical Free Electron Laser Program contract N00014-94-1-0717.

References

1 D. Huang, E.A. Swanson, C.P. Lin, J.S. Schuman, W.G. Stinson, W. Chang, M.R. Hee, T. Flotte, K. Gregory, C.A. Puliafito, and J.G. Fujimoto, Science 254, 1178-1181, (1991).

2 W. Drexler, U. Morgner, F.X. Kärtner, C. Pitris, S.A. Boppart, X.D. Li, E.P. Ippen and J.G. Fujimoto, Optics Letters **24**, 1221 (1999).

3 B.E. Bouma, G.J. Tearney, I.P. Bilinsky, B. Golubovic, and J.G. Fujimoto, Optics Letters **21**, 1839-1841, (1996).

4 J.K. Ranka, R.S. Windeler and A.J. Stentz, Optics Letters **25**, 25 (2000). J.K. Ranka, R.S. Windeler and A.J. Stentz, CLEO 1999 Paper CPD8 (1999).

Intracellular water diffusion probed by femtosecond nonlinear CARS microscopy

Eric O. Potma, Wim P. de Boeij, and Douwe A. Wiersma

Ultrafast Laser and Spectroscopy Laboratory, Materials Science Centre, Department of Chemistry, University of Groningen, Nijenborgh 4, 9747 AG Groningen. The Netherlands
e-mail: potma@chem.rug.nl

Abstract: We report on a nonlinear coherent anti-Stokes Raman microscope system based on a high repetition rate femtosecond cavity-dumped visible optical parametric oscillator. This microscope enables real-time mapping of water concentration gradients in single living cells at high spatial resolution.

1. Introduction

The use of short pulse excitation in microscopy has been recognized as an extremely powerful extension in advanced optical imaging [1,2]. The major benefit of femtosecond pulses is found in the availability of high peak powers at benign average powers. Recently, nonlinear optical techniques, like second- and third harmonic generation [3-5] and coherent anti-Stokes Raman scattering [6-8], have been adopted in microscope configurations. Among the these techniques, the method of coherent anti-Stokes Raman scattering (CARS) is particularly promising for biological imaging since it allows spectrally selective probing of specific biochemical species without the need for extrinsic markers. Above all, CARS microscopy provides an additional contrast method for otherwise optically transparant media, such as water. Water is the major molecular constituent of biological cells and comprises about 70 percent of the average cell volume weight. Since water molecules constitute the dynamical matrix in which all cellular biochemical and biophysical processes occur, knowledge of intracellular hydrodynamics is of vital importance in understanding the functioning of living cells. Cellular hydrodynamics are usually modelled by the water membrane permeability P_d and the intracellular water diffusion constant D_w [9], whose determination in living cells is far from trivial. Although nuclear magnetic resonance (NMR) techniques are generally used to measure transient water concentrations in biological samples [10-12], these methods do not provide the connection between hydrodynamics and morphology of single cells at high spatial resolution.

In this contribution we report on the optical registration of dynamical water distributions in living cells based on vibrational CARS contrast in a laser scanning microscope. Signals are derived directly from the vibrational active OH-stretch mode of the water molecules. Whereas earlier studies were hampered by low spatial resolution and long integration times, implementation of the femtosecond laser technology permits imaging of water distributions at microscopic dimensions with rapid acquisition times. We show that by using CARS microscopy, spatio-temporal water distributions can be directly visualized, for the first time, in single living cells.

Alternatively, the motional properties of intracellular water on ultrashort timescales are studied. Using a novel technique based on optical Kerr effect (OKE) spectroscopy, the rotational diffusion of water molecules is succesfully monitored in single cells. Application of this nonlinear technique contributes to elucidate the physical characteristics of the aqueous intracellular domain.

2. Experimental Setup

The beam-scanning CARS microscope is constructed around a high repetition rate femtosecond multicolour lightsource based on a cavity-dumped visible optical parametric oscillator (OPO)[13]. The layout of the setup is depicted in figure 1.

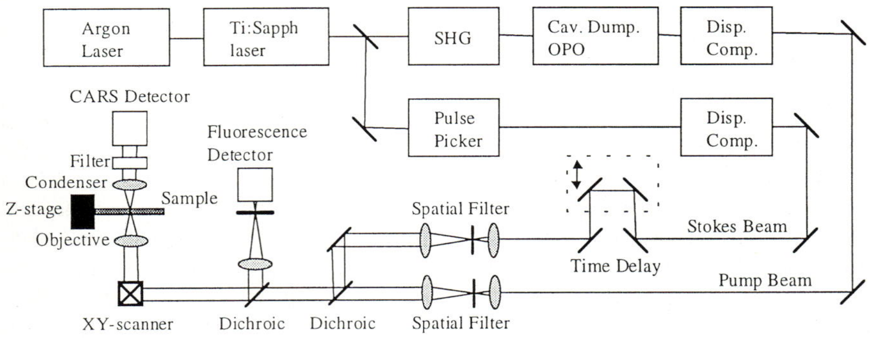

Fig. 1. Experimental layout of the femtosecond nonlinear CARS microscope. SHG: second harmonic generation unit; Cav. Dump. OPO: Cavity-dumped Optical Parametric Oscillator; Disp. Comp. Prism dispersion compensator

The OPO is pumped by the second harmonic (400 nm) of a mode-locked femtosecond Ti:sapphire laser and delivers 30 fs pulsed excitation within the tuning range of 570-660 nm at variable repetition rates. The radiation of the OPO (pump-probe) as well as the residual fundamental beam of the Ti:sapphire oscillator (Stokes) is used in the CARS process. The current system is capable of addressing molecular vibrations in the range of 2400-4500 cm^{-1}, which includes the C-H and O-H stretch vibrations. The femtosecond pulsed radiation gives rise to high peak intensities at the focal spot that promote the nonlinear CARS process. The high repetition rate light source enables real-time CARS mapping of intracellular hydrodynamics of single living cells. We have configured our microscope such that two-photon excited (TPE) fluorescence (epi-mode) can be simultaneously detected along with the (transmission-mode) CARS signal. Average powers were kept low in order to minimize thermal and photo-induced damage to the cells.

3. Intracellular water distribution and diffusion

In figure 2 the CARS image of a living *Dictyostelium discoideum* cell is depicted. Signals derived from the aqueous compartments in the cell compare with the signal intensities measured in the extracellular environment. Hydrophobic regions like the plasma membrane as well as intracellular structures show up as dark regions in the CARS image. The morphology of the cell as deduced from the CARS recording is confirmed in the two-photon excited (TPE) autofluorescence image that was recorded simultaneously. With the lasers tuned to the vibrational transition around 3300 cm^{-1} no resonant contribution from heavy water (D_2O) is observed. Consequently, in order to visualize the mobility of intracellular water distributions, heavy water solutions may be used as a contrast agent to resolve dynamical distributions of water.

The cells were kept in a perfusion chamber that was initially filled with aqueous buffer. The interior and exterior of a single cell, that was allowed to attach to the glass slide, was imaged by scanning the focal spot along a line parallel to the equatorial axis of the cell as a function of time. At a given moment, the perfusion chamber was flushed with an isotonic D_2O buffer solution (Fig 2c). While the signal measured in the extracellular domain drops with a time constant that is characteristic of the response time of the perfusion chamber, the interior of the cell shows a significantly slower response.

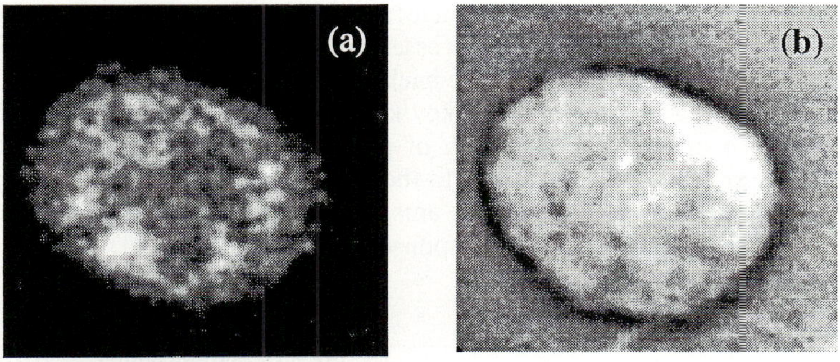

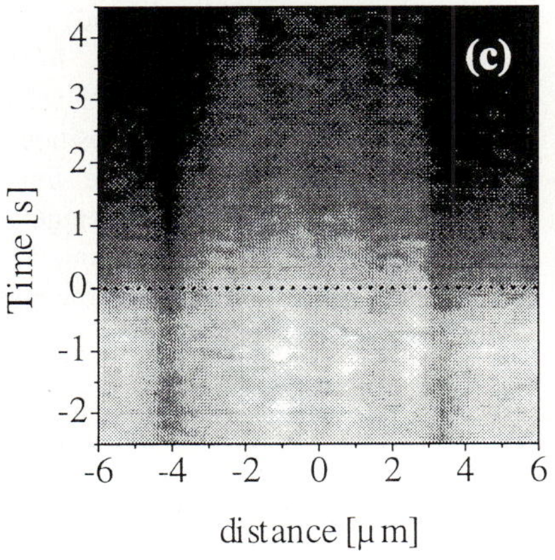

Fig. 2. Simultaneously measured two-photon excited autofluorescence of a Dictyostelium cell (a) and CARS signal (b) using a 0.75 NA objective. Pump beam tuned to 636 nm and Stokes beam tuned to 800 nm. Image size amounts 15µm*15µm. The right image (c) shows the transverse diffusion profile upon the flushing of a single cell by the D_2O buffer at t=0 s (dotted line). Vertically oriented dark regions correspond to the position of the plasma membrane.

Since the efflux rate is characterized by the permeability P_d of the plasma membrane and the effective diffusion coefficient D_w of the intracellular space, examination of the measured spatio-temporal trace may directly eventuate in the determination of both coefficients. As the complete exchange between H_2O and D_2O takes place in a few seconds the observed diffusion coefficients represent an effective measure of the restricted motion that a water molecule experiences on its trajectory to the exterior of the cell. Therefore, the average mobility of the water molecules may be affected by intracellular membrane-bound structures as well. The origin of the restricted motion of water molecules remains to be elucidated to clarify the relative contributions from bound and unbound water in *D. discoideum*.

4. Rotational diffusion of intracellular water

The CARS measurements reveal important information concerning the translational diffusion of water molecules on a second timescale. In order to obtain a deeper understanding of the motional freedom of intracellular water molecules we have studied the water mobility on ultrashort timescales as well. The rotational diffusion of water molecules that reside in the intracellular domain was measured by OKE microspectroscopy. The use of a two color excitation scheme has permitted us to separate the depolarized signal from the pump beam in a collinear beam geometry. Optically heterodyne detection [14] and a polarization sensitive confocal detector [15] were found to be essential in discriminating the weak OKE response of intracellular water from the background contributions. In Figure 3 the Kerr response from African green monkey kidney (COS-1) cells is shown. Based on a comparison of the signal strength of pure water samples, we attribute the intracellular Kerr response exclusively to the water molecules. An exponential fit of the data reveals that the polarization anisotropy inside the cells relaxes almost twofold slower than the extracellular response.

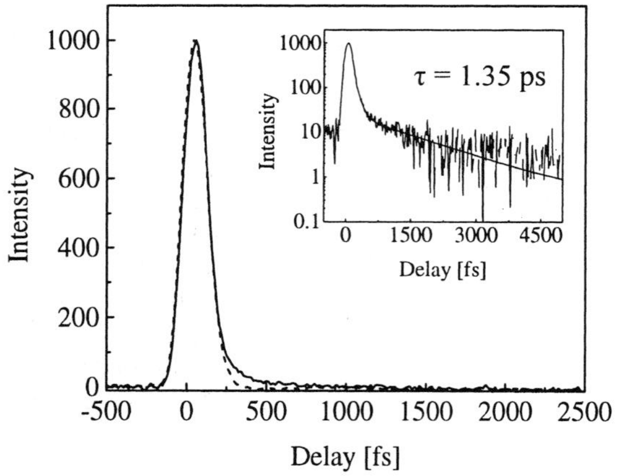

Fig 3. Intracellular OHD-OKE response of COS-1 cells (solid) and corresponding response from coverglass (dotted). Individual cells were probed at 20 different positions of the aqueous domain. The final result was obtained by averaging over the contributions from 5 different cells. Pumped at 800 nm and probed at 620 nm.

5. Conclusions

Advances in femtosecond technology in combination with nonlinear microscopic imaging methods allow detailed spectroscopic mapping of water dynamics in living cells. We have demonstrated that both CARS microscopy and Kerr microspectroscopy are promising tools in cellular research. The application of the CARS method to the study of intracellular hydrodynamics offers a unique approach to the determination of biophysical parameters. Because water molecules can be directly addressed no additional fluorophores are necessary to visualize water distributions in an optical microscope. The high CARS signal yields that are obtained from intracellular water has permitted us to perform dynamical measurements by monitoring the H_2O/D_2O exchange in living cells. From these measurements we were able to extract the membrane permeability and intracellular diffusion coefficients. Excellent morphological contrast is obtained in small cells that measure only 10 μm in diameter and are difficult to study with regular NMR techniques on a single cell level. Whereas CARS microscopy probes the translational water dynamics on a second timescale, the ultrafast rotational diffusion of water molecules is directly addressed in time resolved Kerr microspectroscopy experiments. A multicolor excitation scheme has enabled us to determine the polarization anisotropy decay of intracellular water distributions at high spatial resolution. These experiments illuminate the hidden biophysics of the aqueous cellular domain at the molecular level.

References

1 W. Denk, J.H. Strickler and W.W. Webb, Science **248**, 73, 1990.
2 K. Svoboda, W. Denk, D. Kleinfeld and D.W. Tank, Nature **385**, 161, 1997.
3 P.J. Campagnola, M. Wei, A. Lewis and L.M. Loew, Biophys. J. **77**, 3341, 1999.
4 Y. Barad, H. Eisenberg, M. Horowitz and Y. Silberberg, Appl. Phys. Lett. **70**, 922, 1997.
5 M. Müller, J. Squier, K.R. Wilson and G.J. Brakenhoff, J. Microsc. **191**, 266, 1998.
6 M.D. Duncan, J. Reintjes and T.J. Manuccia, Opt. Lett. **7**, 350, 1982.
7 A. Zumbusch, G.R. Holtom and X.S. Xie, Phys. Rev. Lett. **82**, 4142, 1999.
8 M. Müller, J. Squier, C.A. de Lange and G.J. Brakenhoff, J. Microsc. **197**, 150, 2000.
9 F. Crick, "Diffusion in embryogenesis", Nature **225**, 420-422 (1970).
10 D. Huster, A.J. Jin, K. Arnold, K. Gawrisch, Biophys. J. **73**, 855, 1997.
11 A. Reginald Waldeck, M. Hossein Nouri-Sorkhabi, D.R. Sullivan and P.W. Huchel, Biophys. Chem. **55**, 197, 1995.
12 V. Gulani, G.A. Iwamoto and P.C. Lauterbur, Mag. Res. Med. **41**, 241, 1999.
13 E.O. Potma, W.P. de Boeij, M.S. Pshenichnikov and D.A. Wiersma, Opt. Lett. **23**, 1763 (1998).
14 D. McMorrow, W.T. Lotshaw and G.A. Kenney-Wallace, IEEE J. Quantum Electron. **24**, 443 (1988).
15 P. Higdon, R. Juškaitis and T. Wilson, J. Micrsc. **187**, 8 (1997).

Ultrafast near-field pump-probe spectroscopy of quasi-one-dimensional transport in a single quantum wire

V. Emiliani,[1] T. Guenther,[1] C. Lienau,[1] T. Elsaesser,[1] R. Nötzel,[2] K. H. Ploog[2]

[1] Max-Born-Institut für Nichtlineare Optik und Kurzzeitspektroskopie, Max-Born-Str. 2a, D-12489 Berlin, Germany, E-mail: lienau@mbi-berlin.de
[2] Paul-Drude-Institut, Hausvogteiplatz 5-7, D-10117 Berlin, Germany

Abstract. Carrier-induced reflectivity changes from a single quantum wire are mapped with femtosecond temporal and subwavelength spatial resolution. Carrier relaxation in a single wire on a 100 fs time scale and quasi-one-dimensional carrier transport are resolved.

During the last years, the non-equilibrium dynamics of excitons and free carriers in quasi-two-dimensional semiconductor nanostructures have been extensively investigated using a variety of far-field ultrafast optical techniques [1]. This led to a detailed understanding of those processes. Much less is known about the carrier dynamics in quasi-one-dimensional (1D) systems, such as quantum wires (QWRs). Experimentally, the observation of those dynamics requires optical techniques with ultrahigh spatial and temporal resolution. Here, a promising candidate is the combination of near-field microscopy and femtosecond pump-probe-spectroscopy.

In this contribution, we combine, for the first time, these two techniques to investigate the carrier dynamics in a single quantum wire (QWR) structure with a spatial resolution of 200 nm and a temporal resolution of 200 fs. We give evidence for the ultrafast relaxation of carriers in a single QWR and a transient nonambipolar transport of electrons along the wire on a picosecond time and 100 nm length scale at room temperature.

QWRs with a lateral width of 50 nm and a thickness of up to 13 nm embedded in a 6 nm GaAs quantum well (QW) were grown on patterned GaAs (311)A substrates. Spatially and spectrally resolved quasi-two-color pump-probe experiments are performed with pulses from a 80 MHz Ti:sapphire oscillator. Two optical configurations are used. In the near-field excitation/near-field probe configuration, both pump and probe pulses are transmitted through the same near-field probe. In the far-field excitation scheme, the pump pulses are focused to 30 mm by a far-field lens. The probe light is collected in the far-field and detected with as photodiode as a function of pump-probe time delay.

First, the carrier-induced optical nonlinearity of a single QWR is spatially and spectrally isolated (Fig. 1). The pump laser creates carriers in high-lying QWR states and in the embedding QW. We observe a local increase in the pump-induced change in reflectivity ΔR(QWR) that is energetically localized to the QWR resonance and reflects a carrier-induced decrease in QWR absorption. The temporal variation of ΔR_{QWR} is shown in Fig. 2. With the probe laser being tuned to the QWR absorption resonance around 1.46 eV, we observe an ultrafast rise

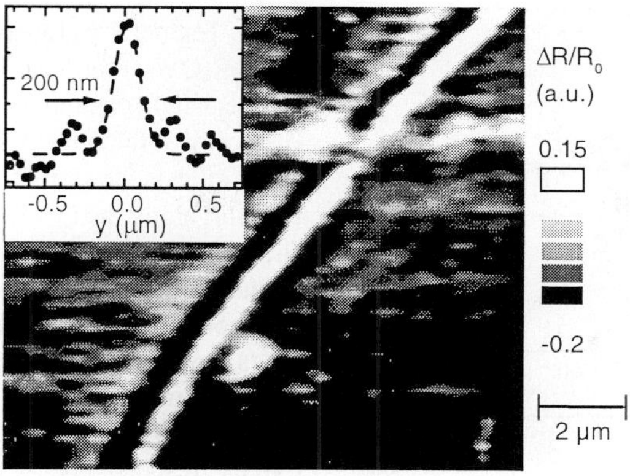

Fig. 1. Spatial map of the pump-induced change in reflectivity at a pump and probe delay time of 10 ps. The pump laser at 1.52 eV excites carriers in high-lying QWR states and in the embedding QW. The probe laser at 1.46 eV is tuned in resonance to the QWR absorption. Inset: line scan perpendicular to the QWR axis

within the time resolution of 200 fs and a constant amplitude up to the maximum delay of 50 ps. The spectral dependence of ΔR(QWR) identifies phase-space filling as the dominant contribution to the excitonic QWR nonlinearity [2]. Therefore ΔR(QWR) probes the **local** carrier concentration within the QWR region. The observation of a reflectivity change for probe energies around 1.46 eV requires carrier redistribution from optically excited high lying QWR and continuum states to the bottom of the QWR. The absence of any slower dynamics on the transient in Fig. 2 suggests that this redistribution process occurs within the first 200 fs. At room temperature and under our excitation conditions, this relaxation involves a complex scattering scenario with contributions from both carrier-LO phonon scattering and carrier-carrier scattering. Wavelength dependent experiments indicate that a thermalized carrier distribution is indeed formed within the first 200 fs, similar to the thermalization dynamics in quantum wells. In our sample, the QWR confinement potentials are markedly different for electrons (65 meV) and heavy holes (15 meV). Therefore, in a thermalized distribution at room temperature, the densities of electrons in the QWR is about 4 times higher than that of confined holes. After the initial thermalization, the temporal variation of ΔR(QWR) thus probes mainly the local dynamics of **electrons** inside the QWR.

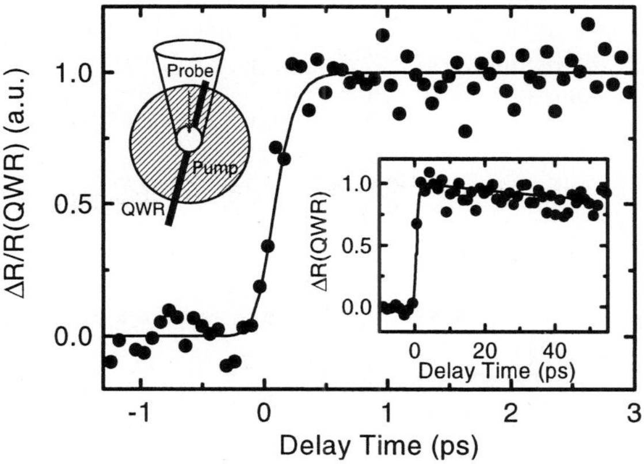

Fig. 2. Temporal variation of the pump-induced change in reflectivity of a single quantum wire, ΔR/R(QWR). The excitation is performed with a far-field spot having a diameter of 30 μm. The probe laser at 1.46 eV probes the local bleaching of the excitonic QWR absorption. Inset: ΔR(QWR) for delay times of up to 50 ps

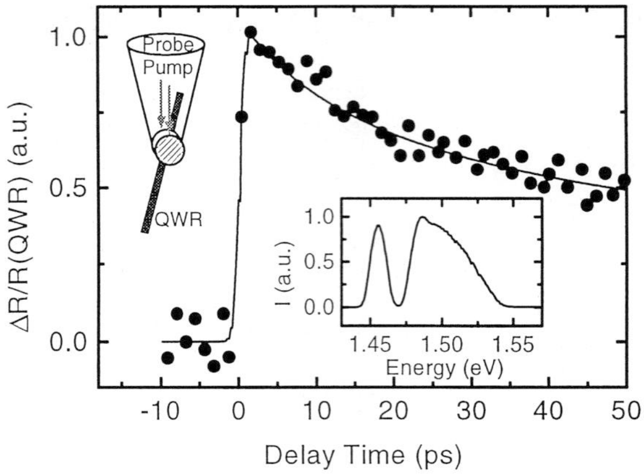

Fig. 3. Temporal variation of ΔR/R(QWR) for localized near-field excitation. The decay of the reflectivity signal reflects the carrier transport along the QWR. The solid line shows a fit to the data within a 1D diffusion model with $D_{QWR} = 80$ cm^2/s Inset: Spectra of pump and probe lasers cemtered at 1.50 eV and 1.46 eV, respectively.

On a time scale of up to 50 ps, the temporal variation of ΔR(QWR) in the two configurations is clearly different [3]. For far-field excitation the signal remains constant up to 55 ps (Fig. 2, inset). For near-field excitation we observe, on the contrary, a clear decay of the reflectivity change on a 10 ps time scale (Fig. 3). Here excitation and probe pulse are spatially overlapping and localized to a small spot of 200 nm diameter. Therefore, the transport of carriers out of the excitation volume results in a decrease of the local carrier concentration and in a reduction of ΔR(QWR). Due to the strong confinement, electrons move predominantly along the QWR. In contrast, the spatial distribution of holes extends well into the QW with negligible influence on the measured reflectivity change. After the fast thermalization process, electron transport is expected to be well described within a one-dimensional diffusion model. In fact, the experimental results are reproduced by assuming a quasi-one-dimensional diffusion coefficient $D_{QWR} = 80 \pm 20$ cm^2/s. This value is more than 5 times larger than the independently measured ambipolar 2D diffusion coefficient $D_{QW} = 12$ cm^2/s [4]. The electron diffusion coefficient, $D_{QWR} = 80$ cm^2/s, corresponds to a mobility of 3000 cm^2/Vs, which is close to the mobility of electrons in GaAs at 300K. This suggests an **unipolar** transport of electrons along the QWR. The strong deviation from convential ambipolar diffusion results from the delocalization of holes over both QWR and QW, which reduces their Coulomb correlation with electrons that are localized in the QWR. This new finding may open up interesting device applications of QWRs, e.g., as contacts for dot-based memory devices.

References

1 J. Shah., *Ultrafast Spectroscopy of semiconductors and semiconductor nanostructures*, Springer, Berlin, 1999.

2 T. Guenther et al., in *Appl. Phys. Lett.* **75**, 3500, 1999.

3 V. Emilinai et al., in *Phys. Rev. B.***61**, R10583, 2000.

4 A. Richter et al., in *Phys. Rev. Lett.* **79**, 2145. 1997.

Femtosecond Near-field Scanning Optical Microscopy of Molecular Thin Films

H. Kawashima[1], M. Furuki[2,3], S. Tatsuura[2], M. Tian[3], Y. Sato[3], L.S. Pu[3], and T. Tani[4]

[1] Electrotechnical Laboratory, Tsukuba, Ibaraki 305-8568, JAPAN
E-mail: e9506@etl.go.jp
[2] FESTA Laboratories, The Femtosecond Technology Research Association, Tsukuba, Ibaraki 300-2635, JAPAN
[3] Corporate Research Laboratory, Fuji Xerox Co. Ltd., Kanagawa 259-0157, JAPAN
[4] Department of Applied Physics, Tokyo University of Agriculture and Technology, Koganei, Tokyo 184-8588, JAPAN

Abstract. We have combined a near-field scanning optical microscope with a two-color time-resolved pump-probe measurement system. The system allowed us to observe an optical gate action of a single domain of molecular thin films.

1. Introduction

A combination of a near-field scanning optical microscope (NSOM) and a femtosecond time-resolved measurement system is a powerful tool for imaging of ultrafast dynamics [1-3]. In this paper, we report on its application to evaluation of molecular thin films. The material of concern is an aggregate of squarylium dye, which is a nanometer-sized crystal [4]. This aggregate reveals an excitonic absorption band like some other aggregates do [5,6]. Nonlinear optical properties of this band have drawn attention with a prospect of optical switching elements. In a recent study on its monomolecular layer films, we have found that dynamic Stark shift occurs [7], and that a two-dimensional sheet of aggregates is composed of small domains, each having an orientational order [8]. In the present study, we focus on its solid film, which is deposited by a spin-coating method. In what follows, we describe development of a measurement system, a response of a single domain, and its spatial distribution.

2. Experimental Methods

We have constructed a two-color measurement system so that the pumping and probing wavelength may be tuned independently [1,3]. The light source is a mode-locked Ti:Sapphire laser, which delivers a 90-MHz train of pulses. The spectrum is centered on 800 nm and is 70-nm broad. The wavelengths are controlled with slits placed in grating and prism pairs for GDD compensation. The two slits are so positioned that the spectral portions around 820 nm and 780 nm may be selected out as the pump and probe pulses, respectively. This choice is essential to make the switching action fast and at the same time to detect the transmittance change sensitively. The probe pulse is coupled to a single-mode fiber. The end of this fiber is Al-coated with a small aperture left at the tip, which serves as a near-field probe. This NSOM tip is used in the illumination mode. The probe light transmitted through the sample is collected by a microscope objective lens and is directed to a Si avalanche photodiode (APD). The pump pulse is introduced with a half-mirror into the same optical path and is focused by the same lens to the sample from the opposite side [1,2]. The optical power of the probe pulse under the NSOM tip was 27 nW and was reduced to 3.2 nW before reaching APD. A dark noise, which was parasitic to a preamp for APD, was 0.053 pW/$\sqrt{Hz}$. The noise equivalent transmittance change $\Delta T/T$ reached 5.0×10^{-5}. The resolution in space and time was estimated to be 200 nm and 100 fs, respectively. We have taken an interferometric approach in order to analyze the temporal profile of weak radiaiton emitted from the fiber tip [9].

The sample was prepared by spin-coating a 1-2 wt% 1,2-dicloroethane solution of the squarylium dyes onto a glass substrate [4]. With a stylus instrument, thickness of the film was measured to be approximately 100 nm.

3. Results and Discussion

A local site under the NSOM tip contains still numerous aggregates. The size varies from one aggregate to another. Although the film is made of a single chemical species, its "composition" can vary in space. By "composition" we mean properties such as how aggregates are distributed over different sizes. The spatial variation of the nonlinear response should be discussed in connection with a linear NSOM image. The upper part of Fig. 1 shows an NSOM transmission image in the absence of the pump beam.

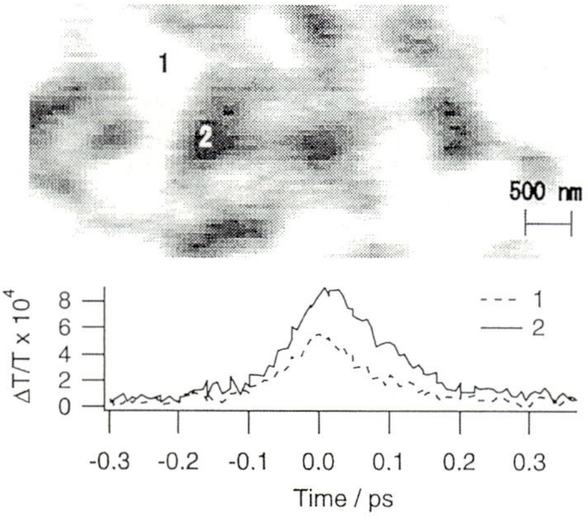

Fig. 1. Linear transmission NSOM image (upper) and induced transmittance at two points in the image (lower)

This looks like a mosaic of bright and dark domains, each with a linear dimension of typically 1 μm or less. The topography was almost flat over this area. This image was taken with linearly polarized light. Upon rotating the polarization direction by 90 degrees, the image altered drastically. This means that the domains are due to dipole orientation as is the case for its Langmuir films [3]. Locating the NSOM tip at several points in the image, we measured transient transmittance. The lower part of Fig. 1 shows the response at two points, "1" and "2". The point "2" in the dark domain gave rise to a greater response. We performed a numerical analysis to eliminate the system response from the observed profiles. The response time proper to the material was found to be about 50 fs. As to the temporal profile, no significant difference was seen among the response at different points. This should suggest that the film composition does not appreciably vary from one domain to another. This result may sound plain, but we should emphasize that verification of spatial homogeneity with a subwavelength resolution, as well as detection of inhomogeneity, is an important task that NSOM only can do. The results of evaluation should serve to improve the process of film deposition.

4. Conclusions

We have put an NSOM apparatus in a two-color pump-probe measurement system. It has a noise-equivalent transmittance change of 5.0×10^{-5} with a probe pulse as weak as 30 nW. The system was used for evaluating molecular thin films that have preserved a domain structure, particularly for observing the response of its single domain. The results include key features to understand an origin of the domains and a spatial range over which the film composition is uniform.

Acknowledgements. This work was supported by New Energy and Industrial Technology Development Organization (NEDO) within the framework of the Femtosecond Technology Project.

References

1 J. B. Stark, U. Mohideen, and R. E. Slusher, QELS'95, OSA Tech. Dig. Ser. Conf. Ed. **16**, 82, 1995.
2 B. A. Nechay, U. Siegner, F. Morier-Genoud, A. Scertel, and U. Keller, Appl. Phys. Lett. **74**, 61, 1999.
3 V. Emiliani, T. Guenther, and Ch. Lienau, Phys. Rev. B **61**, R10583, 2000.
4 S. Tatsuura, M. Furuki, M. Tian, Y. Sato, and L. S. Pu, Mat. Res. Soc. Symp. **567**, 105, 1999.
5 K. Minoshima, M. Taiji, K. Misawa, and T. Kobayashi, Chem. Phys. Lett. **218**, 67 , 1994.
6 D. A. Higgins and P. F. Barbara, J. Phys. Chem. **99**, 3, 1995.
7 M. Furuki, L. S. Pu, F. Sasaki, S. Kobayashi, and T. Tani, Appl. Phys. Lett. **72**, 2648, 1998.
8 M. Furuki, O. Wada, L. S. Pu, Y. Sato, H. Kawashima, and T.Tani, J. Phys. Chem. B **103**, 7607, 1999.
9 H. Kawashima, M. Furuki, and T. Tani, J. Microscopy **194**, 516, 1999.

Background-free THz Imaging using Interferometric Tomography

Jon L. Johnson, Timothy D. Dorney, and Daniel M. Mittleman

Electrical and Computer Engineering Dept., MS-366, Rice University, 6100 Main St., Houston TX 77005 USA
E-mail: daniel@rice.edu

Abstract. We report the implementation of time-domain interferometry in imaging with a terahertz 'T-ray' imaging system. Interferometric tomography provides the possibility of background-free imaging, with enhanced contrast and detectability of sub-wavelength features.

THz imaging is an emerging technology that permits three-dimensional tomographic imaging of non-metallic objects. Numerous applications have been demonstrated where THz imaging could be a valuable complement to existing technologies for non-invasive testing, including the detection of faults or delaminations in packaged integrated circuits and the location of air bubbles or cracks within polymer or ceramic parts [1,2]. In many of these applications, the features one wishes to detect are subtle, in the sense that their interaction with the single-cycle THz pulse impose only a small additional distortion on the waveform. A good example is the detection of a delamination or disbonding between two surfaces. In many practical cases, the gap which opens up between the two surfaces is narrower than the coherence length of the THz pulse, and the waveform is little changed as a result.

Here we report on the use of interferometry in combination with THz tomography, for improving the detectability of such subtle features. This idea has

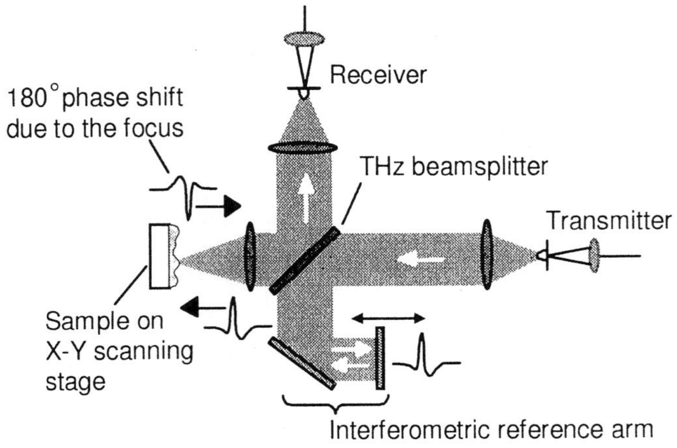

Fig. 1. Schematic of the THz tomographic interferometer.

analogies to optical coherence tomography, in which the signal pulse, reflected off of the sample, is interfered with a reference wave to provide enhanced sensitivity [3]. The experimental layout is shown in figure 1. The collimated THz beam is directed into a Michelson interferometer, in which one arm (signal) contains a 10 cm polyethylene lens. Thus, the THz beam is focused onto the sample at normal incidence. The second arm (reference) contains a planar mirror, mounted on a translation stage for variable delay. The measured signal is the coherent superposition of the electric fields from these two arms. Due to the Gouy phase shift acquired by the signal beam, it is approximately 180° out of phase from the reference beam [4]. As a result, if the sample reflects the THz pulse without distortion, then the delay of the reference arm can be adjusted so that the two pulses destructively interfere at the detector, and almost no signal is detected. Figures 2a and b depict waveforms from the sample and reference arms, respectively. Figure 2c shows the waveform when these two are interfered at the detector, producing a near cancellation of the measured signal. The small residual amplitude of this cancelled waveform can be explained by noting that the Gouy phase shift is not precisely equal to 180° over the entire bandwidth of the THz pulse, due to the fact that the Rayleigh range approaches the focal length of the lens for long wavelengths [4].

The cancellation shown in figure 2c relies sensitively on the delay and distortions acquired by the signal pulse when it interacts with the sample. Any small changes in this waveform produce large fractional changes in the measured signal. This is demonstrated using waveforms reflected off of a mirror with a piece of adhesive tape stuck on the surface. The thickness of this "defect" is

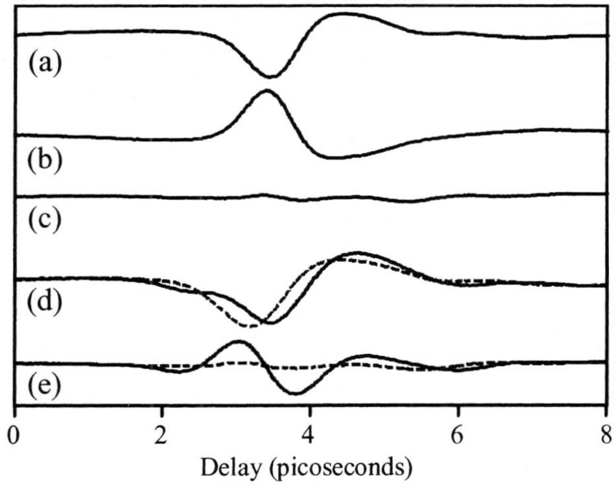

Fig. 2. THz waveforms from (a) the sample and (b) the reference arms, and from (c) the interference between the two arms. In (d) and (e), the solid curves show the waveforms measured with a piece of tape on the sample arm mirror, while the dashed curves show the waveforms without the tape. (d) shows the case without interferometric cancellation, while (e) shows the effect of the interferometry.

approximately 75 microns, or 25% of the coherence length of the pulse. The tape is composed of a transparent, low-index polymer, which has only a weak effect on the THz pulse. The waveforms shown in figures 2d and 2e represent comparisons between waveforms measured with and without the tape in the beam spot, for conventional (2d) and interferometric (2e) imaging. The contrast is evidently enhanced through the use of interferometry.

Figure 3 shows two images of the mirror with thinner pieces of adhesive tape on its surface, with and without the use of interferometric imaging. The weak signal from the tape is substantially enhanced in the interferometric case (fig. 3b), as is evident from these false-gray-scale images. In addition to the enhancement of contrast, the use of a background correction of this type can also suppress the effects of laser amplitude fluctuations or other noise sources. Finally, by canceling out the waveform reflected from a given surface, one can substantially enhance the detectability of a second nearby surface.

(a)

(b)

Fig. 3. Images comparing maximum peak-to-peak amplitude of (a) the normal, non-interferometric imaging and (b) interferometric imaging. The sample consists of two ~50μm thick pieces of adhesive tape on a flat metal surface. In the upper image, only the edges of the two tape strips are faintly discernable. In the lower image, they are much more obvious; also, the tape strip on the right has two air bubbles, only discernable in the interferometric image.

[1] D. Mittleman, S. Hunsche, L. Boivin, and M. C. Nuss, "T-ray tomography," *Opt. Lett.*, **22**(12), 904-906 (1997).

[2] D. Mittleman, R. H. Jacobsen, and M. C. Nuss, "T-ray imaging," *IEEE J. Sel. Top. Quant. Elec.*, **2**(3), 679-692 (1996).

[3] D. Huang, E. Swanson, C. Lin, J. Schuman, W. Stinson, W. Chang, M. Hee, T. Flotte, K. Gregory, C. Puliafito, and J. Fujimoto, "Optical coherence tomography," *Science*, **254**, 1178-1181 (1991).

[4] A. B. Ruffin, J. V. Rudd, J. F. Whitaker, S. Feng, and H. G. Winful, "Direct observation of the Gouy phase shift with single-cycle terahertz pulses," *Phys. Rev. Lett.*, **83**(17), 3410-3413 (1999).

Diagnosis of Dental Cavity and Osteoporosis Using Terahertz Transmission Images

Jaeyoung Ryu[1], Yuchul Jung[1], Seungyong Baek[1], Jongjoo Lee[1], Sera Kim[1], Soontae Kwon[2] and Joungho Kim[1]

[1] Department of Electrical Engineering, Korea Advanced Institute of Science and Technology, 373-1 Kusong-dong, Yusong-gu, Taejon 305-701 Korea
E-mail: forchief@eeinfo.kaist.ac.kr
[2] Department of Diagnostic Radiology, Chungnam National University Hospital, 640 Daesa-dong, Jung-gu, Taejon, 301-721 Korea

Abstract. We demonstrate potential medical applications of terahertz imaging system for diagnosis of dental cavity and osteoporosis. Electrical parameters of tooth and bone are acquired using the transmitted terahertz pulse measurement and genetic algorithm.

1. Introduction

Currently, x-ray is mostly used for diagnosis of dental cavity and osteoporosis[1]-[4]. Although the amount of the irradiated x-ray to human body is relatively small for the clinical diagnosis, the exposition of the x-ray to the human body should be minimized as much as possible, since the x-ray is an ionizing radiation. To investigate other possible diagnosis systems replacing x-ray, ultrasonic imaging and MRI(Magneto-Resonance-Imaging) systems were studied. Unfortunately, an effective and safe diagnosis tool for detecting the dental cavity and the osteoporosis is not developed yet. So we present potential medical applications of THz transmission imaging system for the diagnosis of the dental cavity and the osteoporosis. In the proposed THz transmission image system, clinical information was successfully obtained including density of dentin, electrical properties of dentin, composition of tooth, and locations and size of the cavities[5]. Internal pulp cavity and external cavities were clearly observed in the THz images. Also, this technique has been demonstrated as potential medical applications, monitoring bone fracture, bone diseases such as, osteoporosis and arthritis, and diseases that originate in or affect bone marrow. Density distributions of the marrow were acquired. In addition, we have extracted electrical parameters(permittivity and conductivity) of a sliced tooth and a sliced bone up to 1THz.

2. Diagnosis of Dental Cavity

Figure 1(a) shows a THz transmission image of a dental pulp cavity. For comparison, photograph and x-ray image are shown together. Generally, clinical quality of a dental pulp cavity contains information for health and growth of a tooth. The dental pulp cavity is clearly shown as indicated by a circle. In the THz image, the cavity is detected as a brighter part, implying that the THz pulse is more transparent to the cavity. The THz pulse is absorbed by conductive loss of dentin. Figure 1(b) shows vertical image of a decayed tooth. Both amalgam coverage over the cavity and cavity can be observed in the THz image. The amalgam is a kind of compound metal. The amalgam and cavity is found to be quite contrastive in the THz image comparable to x-ray image. The spatial resolution of the THz image is about 500μm. Figure 1(c) shows the extracted electrical parameters of a sliced tooth whose thickness is 1.1mm in the dentin part. The conductance of the dentin was measured in the THz frequency spectrum region. The parameters can be further used for estimating the density of the dentin from the measured THz waveforms. During the parameter extraction, genetic algorithm was used for time-domain data fitting. Extracted model parameters were well matched to the THz measurement. The auto-correlation coefficient between the measured THz pulse waveform and the reconstructed THz pulse waveform is 0.97, which means that the extracted electrical parameters are accurate.

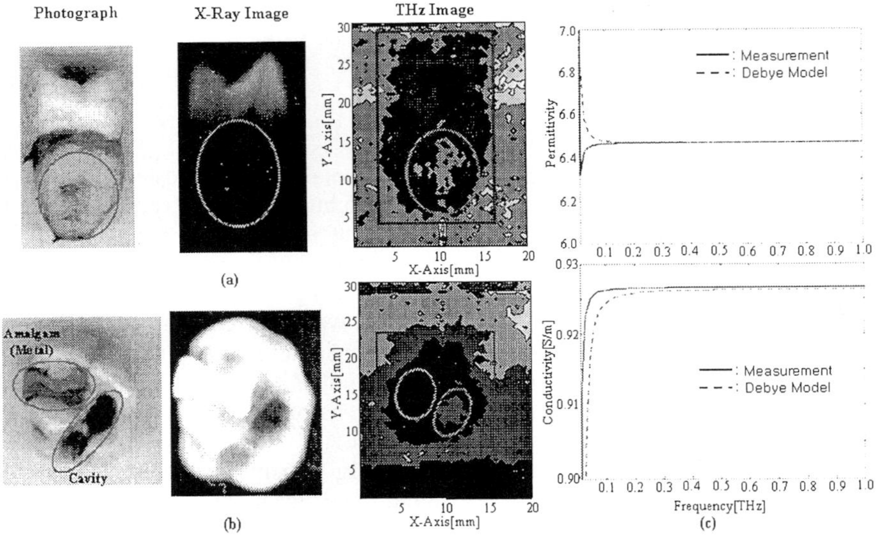

Fig. 1. (a) THz dental image of internal pulp cavity. (b) THz dental image of external cavities. Photographs and x-ray images are also shown for comparison. (c) Extracted electrical parameters of a sliced tooth in dentin part.

3. Diagnosis of Osteoporosis

Figure 2 shows a THz transmission image of head part of a right radius bone and extracted electrical parameters of a sliced cortical part of bone. In the figure 2(a), a part of low marrow density region is shown in a circle. The photograph is showing only surface marrow density. On the other hand, the THz transmission image is showing density distribution inside the bone. Figure 2(b) shows the extracted electrical parameters of a sliced cortical part of the bone whose thickness is 0.85mm.

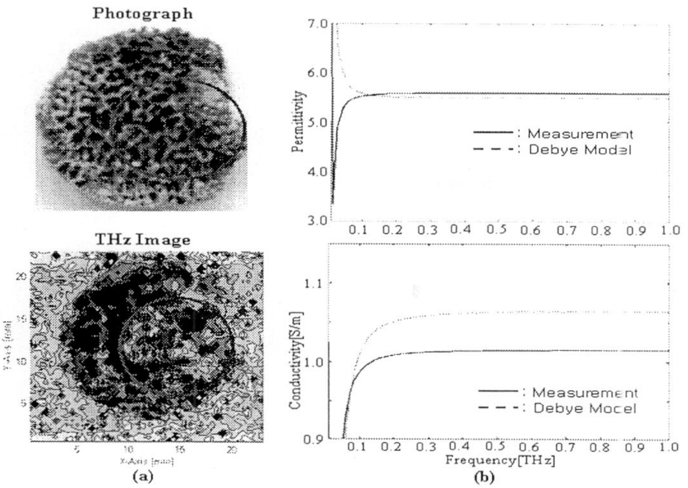

Fig. 2. (a) Cross-sectional THz image of a head part of human right radius for diagnosis of osteoporosis. Photograph is also shown for comparison. (b) Extracted electrical parameters of a sliced cortical part of human rib.

4. Conclusions

We have successfully demonstrated the possibility of the THz imaging system for medical applications including the medical diagnosis for the dental cavity and the osteoporosis. In the THz image, the spatial resolution is about 500μm. Also high-frequency electrical parameters of the tooth and the bone were extracted using the THz time-domain measurement and genetic algorithm. The extracted model parameters were well matched to the THz measurement.

References

1 J. T. Ryaby, "Overview of osteoporosis and current diagnostic techniques", *Proc. IEEE Engineering in Medicine and Biology*, pp. 2122-2123, 1996.

2 M. Benjamin and A. Macovski, "Stimulated positron emission for 3-D tomographic imaging and bone studies-Part I : Method feasibility and system consideration", *IEEE Transactions on Medical Imaging*, vol. 8, no. 2, pp. 113-124, 1989.

3 T. E. Southard and K. A. Southard, "Detection of simulated osteoporosis in maxillae using radiographic texture analysis, *IEEE Transactions on Biomedical Engineering*, vol. 43, no. 2, pp.123-132, 1996.

4 P. I. Saparin, W. Gowin, J. Kurths, D. Felsenberg, "Measures of complexity and processing of vertebral CT-images, *Proc. IEEE/EMBS*, pp.526-529, 1997.

5 Jaeyoung Ryu, Woopoung Kim, Heeseok Lee, Jongjoo Lee and Joungho Kim, "Compact terahertz imaging system using near-field probe", *Proc. of Terahertz Electronics* pp. 81-84, 1999.

Part VI

X-Ray Generation and Diffraction, Higher Harmonics, High Intensity Physics

Femtosecond X-ray Pulses from a Synchrotron

R.W. Schoenlein[1], H.H.W. Chong[2], T.E. Glover[3], P.A. Heimann[3], C.V. Shank[1], A.A. Zholents[4], and M.S. Zolotorev[4]

[1] Materials Sciences Division, Ernest Orlando Lawrence Berkeley National Laboratory, 1 Cyclotron Rd. MS: 2-300, Berkeley CA 94720, USA.
E-mail: rwschoenlein@lbl.gov
[2] Applied Science and Technology Graduate Group, University of California Berkeley, Berkeley CA 94720, USA.
[3] Advanced Light Source Division, Lawrence Berkeley National Laboratory
[4] Accelerator and Fusion Research Division, Lawrence Berkeley National Laboratory

Abstract. An important frontier in ultrafast science is the application of femtosecond x-ray pulses to the study of structural dynamics in condensed matter. We show that femtosecond laser pulses can be used to generate high-brightness, tunable, femtosecond x-ray pulses from a synchrotron. Performance of existing and proposed femtosecond x-ray beamlines at the Advanced Light Source synchrotron are discussed.

1. Introduction

An important new area of scientific research in chemistry, physics, and biology is the investigation of ultrafast structural dynamics and atomic motion in condensed matter using femtosecond x-ray pulses. X-rays are powerful probes of atomic structure since they interact with core electronic levels, and can therefore provide direct information about relative atomic positions and coordination. Synchrotrons generating high flux (photons/sec/0.1% BW), and high-brightness (photons/sec/mm^2/mrad^2/0.1% BW) x-ray beams have proven to be powerful tools for probing the "static" structure of matter. Techniques such as x-ray diffraction, extended x-ray absorption fine structure (EXAFS) and many others are widely used at such facilities to obtain structural information with a resolution of much less than 1 Å. However, the time resolution provided by synchrotrons is limited to >30 ps, as determined by the length of the stored electron bunches. This is orders of magnitude longer than the fundamental time scale for atomic motion, dictated by the period of an atomic vibration, ~100 fs. We demonstrate a novel scheme for generating ultrashort pulses of synchrotron radiation[1]. Femtosecond synchrotron pulses are directly measured from a bend-magnet beamline at the Advanced Light Source (ALS)[2, 3]. A femtosecond x-ray beamline based on a bend-magnet is presently being commissioned at the ALS, and a femtosecond undulator beamline providing higher flux and brightness is proposed.

2. Experimental Methods

The time structure of synchrotron radiation is determined by the duration of the stored electron bunches, typically longer than 30 ps. Storage of ultrashort relativistic electron bunches is problematic due to instabilities arising from bunch-induced wakefields and coherent synchrotron radiation. However, femtosecond-duration electron bunches can exist in a storage ring for a limited time. Our approach is to create femtosecond time-structure on a long electron bunch by using a femtosecond laser pulse to modulate the energy of an ultrashort slice of the bunch.

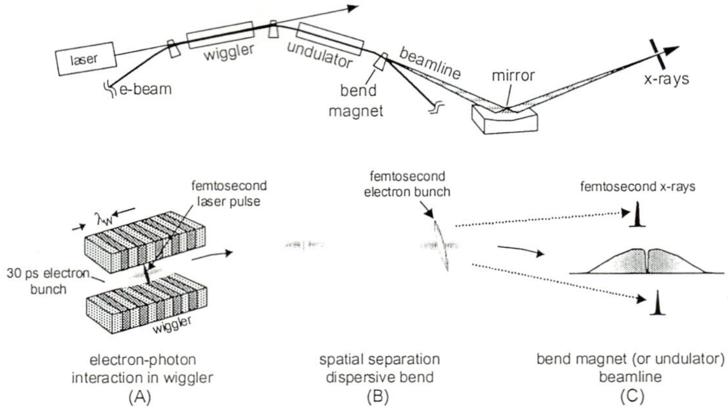

Fig. 1. Schematic of method for generating femtosecond synchrotron pulses. (A) Femtosecond laser pulse interaction with the electron bunch in a resonantly-tuned wiggler. (B) Transverse separation of modulated electrons in a dispersive bend of the storage ring. (C) Generation of synchrotron radiation from a bend-magnet (or undulator) and separation of the femtosecond synchrotron radiation at the beamline image plane.

Figure 1 illustrates the modulation and generation scheme. A femtosecond laser pulse co-propagates with the stored electron bunch through a wiggler (Fig. 1A). The high electric field of the laser pulse ($\sim 10^9$ V/m) modulates the energy of the underlying electrons as they traverse the wiggler (electrons are accelerated or decelerated depending on the optical phase, ϕ, seen by each electron at the entrance of the wiggler). The optimal interaction occurs when the central wavelength of the spontaneous emission from an electron passing through the wiggler, $\lambda_S=\lambda_W(1+K^2/2)/2\gamma^2$, satisfies the resonance condition $\lambda_S=\lambda_L$ where λ_L is the laser wavelength, λ_W is the wiggler period, γ is the Lorentz factor, and the deflection parameter $K=eB_o\lambda_W/2\pi mc$ (where B_o is the peak magnetic field of the wiggler). In addition, the transverse mode of the laser beam must match the transverse mode of the spontaneous radiation from the electron passing through the wiggler, and the laser spectral bandwidth must match the spectrum of the fundamental wiggler radiation averaged over the transverse mode. Under these conditions, the electron energy modulation, ΔE, is given by

$$\Delta E = 2\left(A_L A_W \frac{M_W}{\sqrt{2}M_L} \eta_{emit} \right)^{1/2} \cos\phi \qquad (1)$$

where A_L is the laser pulse energy, M_W is the number of wiggler periods, M_L is the laser pulse length in optical cycles (FWHM), $A_W \cong 4.1\alpha hc/\lambda_L K^2/(2+K^2)$ is the energy spontaneously radiated by a single electron passing through the wiggler, and $\eta_{emit} \approx 0.7$ accounts for the non-zero electron beam size. For example: a 25 fs laser pulse with a photon energy of 1.55 eV, and a pulse energy $A_L=100$ μJ will produce an energy modulation amplitude $\Delta E \approx 9$ MeV using a wiggler of 19 periods.

By creating an energy modulation that is significantly larger than the beam energy spread, the transverse dispersion of the storage ring will cause a spatial displacement of the modulated electrons from the rest of the electron bunch (Fig. 1B). Finally, by imaging the synchrotron radiation from the displaced electrons to the experimental area, femtosecond x-rays can be separated from the long-pulse using an aperture (Fig. 1C). The duration of the synchrotron radiation produced by the laser-modulated electrons will be approximately the same as the laser pulse duration. Furthermore, the extraction of an ultrashort slice of electrons leaves behind an ultrashort hole or "dark pulse" in the core of the electron bunch. This dark pulse will appear in the generated x-rays, and can also be used for time-resolved spectroscopy.

Our experiments are conducted at the ALS storage ring operating at E=1.5 GeV (rms beam energy spread σ_E=1.2 MeV) using a wiggler with M_W=19 periods, λ_W=16 cm, with the gap adjusted for K≅13. Femtosecond pulses (τ_L=100 fs, A_L=400 μJ, λ_L=800 nm, f_L=1 kHz,) from a Ti:sapphire laser system are synchronized to the storage ring RF using phase-locking techniques, and are directed into the main vacuum chamber and co-propagate with the electron beam through the wiggler. Following the interaction region, a mirror directs the fundamental spontaneous wiggler emission and the laser beam out of the storage ring to enable direct measurements of the temporal and spectral overlap, and the spatial mode matching between the laser pulses and the wiggler radiation. The efficiency of the interaction between the laser and electron beams is monitored by measuring the gain in the laser pulse energy. This gain ($\sim 10^{-3}$) is equivalent to the single-pass gain of a free-electron laser, and is optimum under the same mode-matching conditions described above[3].

3. Results and Discussion

The time structure of the temporally incoherent synchrotron radiation is directly determined by the time structure of the electron bunch and is invariant over the entire spectrum of the synchrotron emission, from infrared to x-ray wavelengths. We directly measure the time-structure of the visible synchrotron pulses (at ~2 eV) via cross-correlation (in BBO) with a delayed 50 fs pulse from the laser system. An adjustable knife-edge located in the beamline at an intermediate image plane provides a means to select synchrotron radiation originating from different transverse regions of the electron beam.

Figure 2 shows two cross-correlation measurements made at various knife-edge positions, in units of the rms horizontal beam size, σ_x. Measurements of the central core of the synchrotron beam (Fig. 2A) reveal the femtosecond hole or dark pulse that is created due to the acceleration (and deceleration) of electrons by the laser pulse, and their subsequent transverse spatial separation from the central core of the bunch. The bright pulse is measured with the knife-edge at the $3\sigma_x$ position (Fig. 2B) such that the central core of the beam is blocked. The solid lines in Fig. 2 are predicted pulse shapes based on a model calculation of the

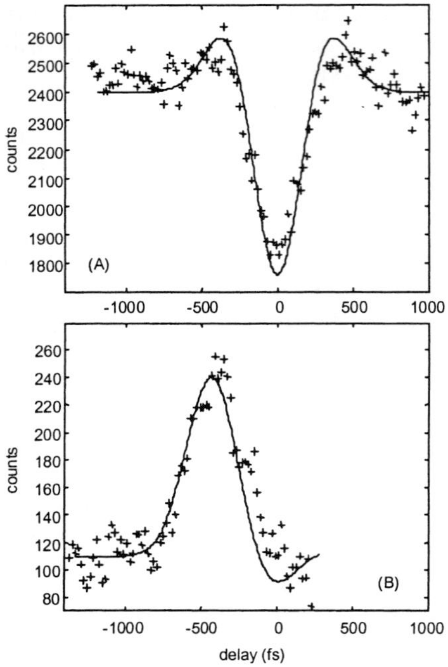

Fig. 2. Cross-correlation measurements between a delayed laser pulse and synchrotron radiation originating from an energy-modulated electron bunch. In (A), synchrotron radiation from the central core ($\pm 3\sigma_x$) of the electron bunch is selected. In (B), synchrotron radiation from the horizontal wings ($+3\sigma_x$ to $+8\sigma_x$) of the electron bunch is selected. Solid lines are from a model calculation of the spatial and temporal distribution of the energy-modulated electron bunch following propagation through 1.5 arc-sectors at the ALS[3].

electron bunch distribution (using known parameters of the electron beam and the storage-ring lattice) [3].

The location of the bend-magnet beamline used in these demonstration experiments is less than optimum due to the distance from the wiggler (in which the laser/electron interaction occurs) to the radiating bend magnet. Stretching of the laser-modulated electron bunch along this storage-ring arc accounts for the observed ~300 fs pulse duration. For a bend-magnet located immediately following the wiggler, the stretching of the electron bunch will be reduced by a factor of three, resulting in an x-ray pulse duration of ~100 fs.

The average flux and brightness of the femtosecond x-rays can be determined from the full flux and brightness from an ALS bend-magnet (or undulator) beamline[4] scaled by three factors: $\eta_1=\tau_L/\tau_e$, $\eta_2=f_L/f_B$, and $\eta_3\approx 0.2$ where τ_L and τ_e are the laser pulse and electron bunch durations, f_L and $/f_B$ are the laser and electron-bunch repetition rates, and η_3 accounts for the fraction of electrons that are in the proper phase of the laser pulse to get the maximum energy exchange suitable for creating the large transverse separation. Synchrotron radiation damping provides for recovery of the electron beam between interactions. With a typical fill pattern of ~300 bunches in the storage ring, femtosecond x-rays can be generated at repetition rates as high as 100 kHz without adversely affecting the other beamlines at the ALS.

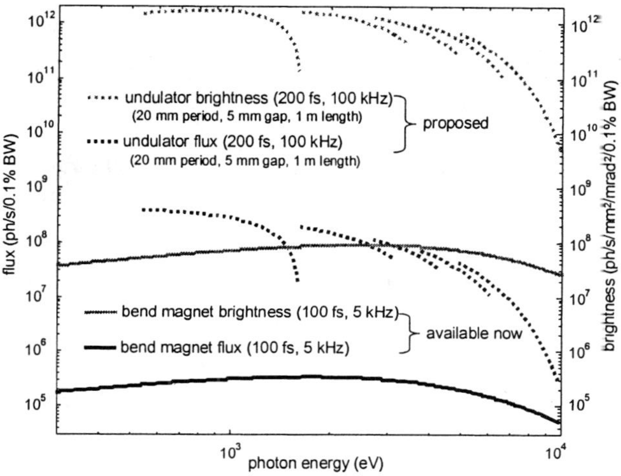

Fig. 3. Flux and brightness for two femtosecond x-ray beamlines based on a bend-magnet and on an undulator. The undulator spectra is the locus of narrow spectral peaks (tuned by adjusting the undulator gap) and represents the envelope of harmonics 1, 3, 5, 7, and 9.

4. Conclusions

Figure 3 shows the calculated flux and brightness from two different femtosecond x-ray beamlines assuming a beam energy of 1.9 GeV, (400 mA average current, 30 ps bunch duration, 500 MHz repetition rate), and rms source size of 200 μm (H) × 20 μm (V) in the straight sections, and 100 μm (H) × 9 μm (V) in the bend sections. The first beamline (presently being commissioned at the ALS) is based on a bend magnet with a field of 1.27 T and an x-ray optic collecting 3 mrad (H) × 0.3 mrad (V) of the broadband bend-magnet emission. The second beamline (proposed for ALS straight sector 6) is based on a small-gap undulator with 50 × 2 cm periods and a maximum deflection parameter of 2.02 (~5 mm min. gap). Undulators offer substantially higher x-ray brightness and flux (per fractional bandwidth) than bend-magnets. These two beamlines are optimally located (immediately following the wiggler) and will generate 100fs and 200 fs x-ray pulses respectively.

Acknowledgement. Supported by the Director, Office of Science, Office of Basic Energy Sciences, of the U.S. Department of Energy under Contract No. DE-AC03-76SF00098.

References

1 A.A. Zholents, M.S. Zolotorev, *Phys. Rev. Lett.* **76**, 912, 1996.
2 R.W. Schoenlein, S. Chattopadhyay, H.H.W. Chong, T.E. Glover, P.A. Heimann, C.V. Shank, A. Zholents, and M. Zolotorev, *Science*, **287**, 2237, 2000.
3 R.W. Schoenlein, S. Chattopadhyay, H.H.W. Chong, T.E. Glover, P.A. Heimann, C.V. Shank, A. Zholents, and M. Zolotorev, *Appl. Phys. B*, (in press), 2000.
4 "1-2 GeV Synchrotron Radiation Source - Conceptual Design Report," Lawrence Berkeley Laboratory LBL PUB-5172 Rev., July 1986.

Direct Observation of Ultrafast Non-thermal Melting by Ultrafast X-ray Diffraction

C.W. Siders[1,2], A. Cavalleri[1], K. Sokolowski-Tinten[3], Cs. Tóth[4], T. Guo[1,5], M. Kammler[6], M. Horn von Hoegen[3], K.R. Wilson[1], D. von der Linde[3] and C.P.J. Barty[7]

[1] Department of Chemistry and Biochemistry, The University of California San Diego, La Jolla CA 92093-0339, USA
E-mail: craig_siders@attglobal.net
[2] School of Optics/CREOL, University of Central Florida, Orlando FL 32816, USA
[3] Institute for Laser- and Plasmaphysics, University of Essen, 45117 Essen, Germany
[4] Institute for Nonlinear Science, The University of California San Diego, La Jolla CA 92093-0339, USA
[5] Department of Chemistry and Biochemistry, The University of California Davis, Davis CA 95616, USA
[6] Institute for Solid State Physics, University of Hannover, 30167 Hannover, Germany
[7] Department of Applied Mechanics and Engineering Sciences, The University of California San Diego, La Jolla CA 92093-0339, USA

Abstract. Ultrafast nonthermal melting of femtosecond laser-irradiated germanium was observed directly for the first time using sub-picosecond, 1.54-angstrom x-ray diffraction. Such measurements allow the study of transient atomic motion fundamental to many processes in physics, chemistry, and biology.

Many fundamental processes in nature, such as chemical and biochemical reactions and phase transitions, involve changes in the structure of matter, i.e. the rearrangement of the constituent atoms and molecules, occurring transiently on time-scales comparable with the natural oscillation periods of atoms and molecules, that is femtoseconds to picoseconds. Ultrashort pulse lasers with such pulse durations have been used for more than two decades to optically pump and dynamically probe a wide array of atomic and molecular, solid-state, and plasma systems, including extreme states of matter normally found only in stellar or planetary interiors and experimentally accessible only by the rapid heating and inertial confinement made possible with ultrashort-pulse irradiation [1,2]. Indeed, with the awarding of the 1999 Nobel Prize in Chemistry to Ahmed Zewail for "showing that it is possible with rapid laser technique to see how atoms in a molecule move during a chemical reaction," the power and utility of ultrafast, optical-pump optical-probe spectroscopy are clear. In such experiments, though, the visible light used inherently cannot resolve atomic-scale features, but rather interacts with valence and free electrons. The deeper lying core electrons and nuclei that most directly delineate structure are not directly probed. Angstrom-scale x-ray radiation, with wavelengths comparable to inter-atomic distances, is well suited to measure structure and atomic rearrangement, and can measure structural dynamics in the interior of samples that are not transparent to ordinary light. Recently, sub-picosecond sources of hard x-rays have been developed [3-5], making possible new classes of experiments in physics, chemistry and biology [4,

6-10]. Visible-pump, x-ray-diffraction-probe experiments which measure the generation and propagation of coherent acoustic pulses in bulk gallium arsenide crystals have been reported [6], with milliangstrom changes in lattice spacing measured with picosecond temporal resolution. Here, we report the first direct observation, using time-resolved x-ray diffraction, of ultrafast non-thermal melting of short-pulse irradiated germanium [7].

In general, if a solid is heated to, or above, the melting temperature, nucleation of the liquid phase occurs around crystal defects or inclusions, usually at the surface itself, and the rate of phase change depends upon the degree of superheating of the solid. In the case of laser irradiation of an absorbing crystalline semiconductor, incident optical energy is initially coupled to the carriers, which undergo band-to-band transitions. Intra-band relaxation and non-radiative recombination cause delayed heating of the lattice [11], which exceeds the melting temperature within several picoseconds. For irradiation with pulses longer than several picoseconds and for near-threshold femtosecond pulses, following nucleation of the liquid phase at the surface, a thin liquid layer grows into the bulk of the material at a velocity limited by the speed of sound, but typically at a rate on the order of angstroms per picoseconds. In contrast, for fluences twice the melting threshold or greater [12-15] a significantly faster, sub-picosecond change of the linear optical properties at the surface is seen, with reflectivities reaching values equal to that of the conventional liquid phase in a few hundred femtoseconds [15]. In other experiments, a reduction in the second-harmonic-in-reflection signal, suggesting a sub-picosecond loss of crystalline order at the very surface, was also seen [12,14]. These observations indirectly indicate that an ultrafast solid-to-liquid phase transition occurs at the surface of the semi-conductor on a time-scale faster than carrier-lattice equilibration times. The possibility of making atom-scale structure-sensitive measurements on an ultrafast timescale to directly observe nonthermal melting has been a goal of ultrafast spectroscopy for decades.

In order to directly observe the non-thermal loss of crystalline order both at the surface and in the bulk of the semiconductor lattice, we have performed time-resolved, optical-pump ultrafast-x-ray-diffraction-probe experiments on laser-irradiated germanium [7]. Our experimental setup consisted of a visible pump, x-ray probe apparatus, where the 8-keV (1.54-Å) bursts of Cu-K_α line radiation were generated by focusing 80-mJ, 30-fs pulses from a multi-terawatt laser [16] at relativistic intensities onto a moving copper wire. X-ray pulses emitted from the plasma point source were diffracted by a photo-excited (111)-Ge crystal in a symmetric Bragg configuration and recorded with a photon-counting area detector. The optical pump pulse, split off from the x-ray generating laser pulse, was independently adjustable in both energy and pulsewidth. The pump fluence was first empirically set, using a silicon wafer target, to the melting fluence of silicon, as evidenced by post-irradiation examination, and then increased to 0.5 J/cm^2, more than twice the known melting-threshold for germanium. To avoid plasma formation in air or at the surface of the sample, the pump-pulse duration was increased to 100-fs, where no effects of plasma formation were observed. Three samples, nominally 160-nm thick, were used, consisting of single crystal germanium films grown using a novel surfactant mediated growth technique [17] over large areas of single crystal silicon substrates.

Diffraction images were measured at different pump-probe time delays (Fig. 1). The $K_{\alpha 1}$ and $K_{\alpha 2}$ spin-orbit-split lines of Cu are clearly resolved with bulk germanium (not shown) with linewidth-limited resolution, but with the 160-nm films, we measure the line-width-convolved rocking curve of the thin film and do not resolve the individual $K_{\alpha 1}$ and $K_{\alpha 2}$ lines. The diffraction image taken at +6-ps delay (Fig. 1B) shows an area of reduced diffraction, indicating disordering in the center of the pumped spot. At later time delays (Fig. 1C) the area widens over which the reduced diffraction (and hence disordering) takes place. A slower disordering process is observed in the outer areas of the pumped spot, where the fluence is closer to the melting threshold. The image taken at +107-ps delay (Fig. 1D) shows a broad disordered area, as well as a shift of the reduced-intensity Bragg line, in the pumped region, toward lower diffraction angles, suggesting that the solid germanium layer beneath the melt is strained due to thermal expansion [6]. Finally, a non-rastered diffraction image (Fig. 1E), taken several seconds after irradiation on a vertical sequence of single-shot damage spots, shows nearly complete recovery on a semi-infinite timescale.

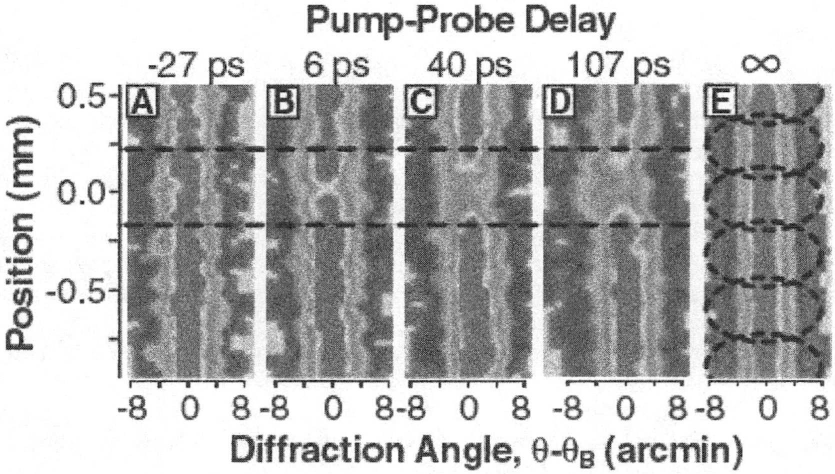

Fig. 1. Images obtained from photon-counting x-ray area detector for five pump-probe time delays. Horizontal axis corresponds to diffraction angle, shown relative to the Bragg angle for germanium. Vertical axis corresponds to position on the semiconductor wafer. The optical pump photo-excites only a portion (indicated by dotted lines in (**A**) through (**D**)) of the entire x-ray probed area. The image at infinite time delay (**E**), including six single-shot damage regions (indicated by dotted circles), was taken non-rastered and with the optical pump blocked during the exposure.

In the plot of the integrated diffraction from the center (solid line) and edge (dashed line) of the pumped spot as a function of visible pump, x-ray probe delay (Fig. 2), the signal from the central region shows a sharp drop of ~20 % within the first seven picoseconds, limited by the time-delay between measurement steps, while the edge region retains the reflectivity of the unpumped crystal. A slower

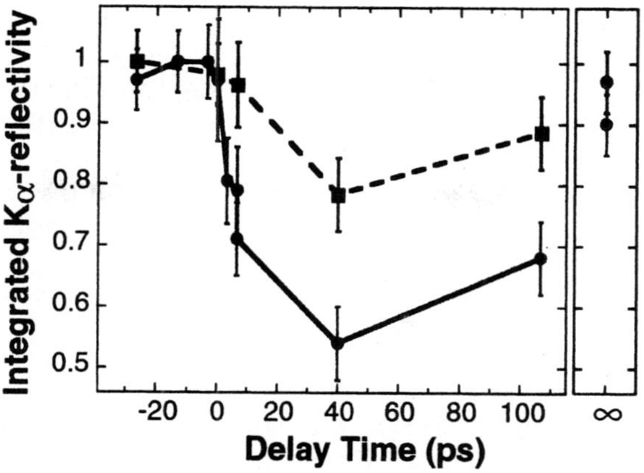

Fig. 2. Time-resolved K_α x-ray reflectivity from 160-nm Ge(111) film, integrated over the central pumped region (solid line) and over a region vertically displaced ~0.2-mm from the center (dashed line).

decay in integrated diffraction is seen in parallel in both regions on a longer time-scale, and a drop in integrated diffraction of ~50% in the central region (~20% on the edge) is observed at 40-ps delay. At 107-ps delay, an increase in the diffraction signal is seen, possibly due to the fact that the expanded solid germanium has higher integrated diffraction efficiency than the unstrained crystal [7]. At infinite time-delay, the diffraction signal recovers, for both regions, to ~90% the initial value, showing that (111)-crystalline order in the molten region is reestablished on the long time-scale, with the incomplete return most likely due to partial ablation at the surface, where the fluence is high enough to generate a super-heated liquid, or possibly amorphous recrystallization of the molten germanium. This clearly rules out the possibility of a solid-gas or solid-plasma phase transition and shows that we're seeing a solid-to-liquid phase transition, followed by recrystallization.

These results, as well as those from similar experiments performed more recently [18], demonstrate the ability of present-day ultrashort-laser driven x-ray sources to study transiently generated extreme states of matter. While further studies of important outstanding problems related to the structure of superheated solids and transient liquid phases [1,2] are natural extensions of this work, expected improvements in table-top laser-driven plasma sources should extend the reach of ultrafast x-ray diffraction to the dynamic study of many other ultrafast processes in physics, chemistry, and biology, including the ultrafast atomic and molecular dynamics by which other solid-state processes and chemical and biochemical reactions take place.

Acknowledgements. K.S.T. gratefully acknowledges support by the Deutsche Forschungsgemeinschaft.

References

1 N. Bloembergen, "First light on fluid carbon", Nature **356**, 110-111 (1992).
2 G. Galli, et al., "Melting of diamond at high pressure", Science **250**, 1547-9 (1990).
3 R. W. Schoenlein, et al., "Femtosecond X-ray pulses at 0.4 Å generated by 90 degrees Thomson scattering: a tool for probing the structural dynamics of materials", Science **274**, 236-8 (1996).
4 C. Rischel et al., "Femtosecond time-resolved x-ray diffraction on laser-heated organic films", Nature **390**, 490-2 (1997).
5 M. M. Murnane, et al., "Ultrafast X-ray pulses from laser-produced plasmas", Science **251**, 531-6 (1991).
6 C. Rose-Petruck, et al., "Picosecond-milliangstrom lattice dynamics measured by ultrafast x-ray diffraction", Nature **398**, 310-12 (1999).
7 C.W. Siders et al., "Detection of nonthermal melting by ultrafast x-ray diffraction", Science **286**, 1340-2 (1999).
8 K. Moffat, "Ultrafast time-resolved crystallography", Nature Structural Biology **5**, 641-643 (1998).
9 A.H. Chin et al., "Ultrafast structural dynamics in InSb probed by time-resolved x-ray diffraction", Phys. Rev. Lett. **83**, 336 (1999).
10 A.M. Lindenberg et al., "Time-resolved x-ray diffraction from coherent phonons during a laser-induced phase transition", Phys. Rev. Lett. **84**, 111-3 (2000).
11 M. C. Downer, C. V. Shank, "Ultrafast heating of silicon on sapphire by femtosecond optical pulses", Phys. Rev. Lett. **56**, 761-4 (1986).
12 C. V. Shank, et al., "Time-resolved reflectivity measurements of femtosecond-optical-pulse-induced phase transitions in silicon", Phys. Rev. Lett. **50**, 454-7 (1983).
13 P. Saeta, et al., "Ultrafast electronic disordering during femtosecond laser melting of GaAs", Phys. Rev. Lett. **67**, 1023-6 (1991).
14 K. Sokolowski-Tinten, et al., "Ultrafast laser-induced order-disorder transitions in semiconductors", Phys. Rev. B **51**, 14186-98 (1995).
15 K. Sokolowski-Tinten, et al., "Thermal and nonthermal melting of gallium arsenide after femtosecond laser excitation", Phys. Rev. B **58**, R11805-8 (1998).
16 C. P. J. Barty, et al., "Generation of 18-fs, multiterawatt pulses by regenerative pulse shaping and chirped-pulse amplification", Opt. Lett. **21**, 668-70 (1996).
17 M. Horn-von-Hoegen, "Surfactants: Perfect heteroepitaxy of Ge on Si(111)", Appl. Phys. A 59, 503 (1994).
18 A. Rousse, et al., Nature (submitted); R. Sauerbrey, et al., private communication.

Time-resolved x-ray diffraction study of ultrafast acoustic phonon dynamics in Ge/Si-hetero-structures

K. Sokolowski-Tinten [1], A. Cavalleri[2], C.W. Siders[2,6], F.L.H. Brown[2], D.M. Leitner[2,7], C. Toth[3], M. Kammler[8], M. Horn von Hoegen[1], D. von der Linde[1], J.A. Squier[4], C.P.J. Barty[5], and K.R. Wilson[2]

[1] Institute for Laser- and Plasmaphysics, University of Essen, 45117 Essen, Germany
E-mail: kst@ilp.physik.uni-essen.de
[2] Department of Chemistry and Biochemistry, [3] Instiute for Nonlinear Science,
[4] Department of Electrical Engineering, [5] Department of Applied Mechanics and Engineering Sciences, The University of California San Diego, La Jolla, CA-92093-0339, USA
[6] School of Optics/CREOL, University of Central Florida, Orlando, FL-32816
[7] Department of Chemistry, University of Nevada, Reno, NV-89557
[8] Institute for Solid State Physics, University of Hannover, 30167 Hannover, Germany

Abstract. Using time-resolved x-ray diffraction the ultafast strain dynamics in fs-laserexcited Ge/Si-heterostructures has been studied. A fluence dependent, anharmonic damping of the impulsively generated acoustic phonons and vibrational transport across the buried Ge/Si-interface are observed.

Short pulse optical excitation of semiconductors creates highly non-equilibrium states of the irradiated material. To re-establish equilibrium a complex sequence of relaxation processes is necessary, starting with intra-band thermalization of the hot carriers, carrier relaxation via phonon-emission and recombination, vibrational acoustic transport into the bulk, and eventual complete thermalization within the phonon-subsystem. The processes involving the electronic sub-system and Raman-active optical phonons have been extensively characterized with time-resolved optical techniques. On the other hand, access to the ultrafast acoustic phonon dynamics has been indirectly achieved only at surfaces.

By combining the temporal resolution of ultrafast laser spectroscopy with the structural sensitivity of x-ray scattering, direct *quantitative* studies of ultrafast atomic motion deep inside the bulk of matter have recently become possible [1-6]. In this contribution, we wish to report on ultrafast x-ray measurements of strain oscillations in fs-laserexcited Ge/Si-heterostructures. Excitation-dependent damping times and vibrational transport across a buried interface are measured simultaneously, thereby revealing different acoustic phonon decay mechanisms.

Single crystalline Ge thin films with 111-surface orientation, grown on 111-oriented Si substrates by surfactant-mediated heteroepitaxy [7], have been optically excited with 30fs laser pulses at 800nm. Evolution of the impulsively generated transient lattice strain is observed by diffraction of short bursts of Cu K_α–line radiation at 8 keV from a laser-produced microplasma in a symmetric Bragg-configuration. Due to the different lattice constants of the two diamond-like materials the Cu K_α-radiation is diffracted at two distinct Bragg-angles. Therefore, we could separate diffraction from the film and the substrate and measure the strain dynamics in *both* components of the heterostructure .

In contrast to bulk crystalline samples [3], where a splitting of the Bragg-line is observed, the rocking curve of the thin film is shifted as a whole towards smaller diffraction angles, without changes of its shape. This indicates homogenous expansion of the film over its entire thickness. In Fig. 1 the time-dependent shift of the centroid of the rocking curve of the Ge-film (left) and the Si-substrate (right) are plotted for four different excitation fluences (15-40mJ/cm^2).

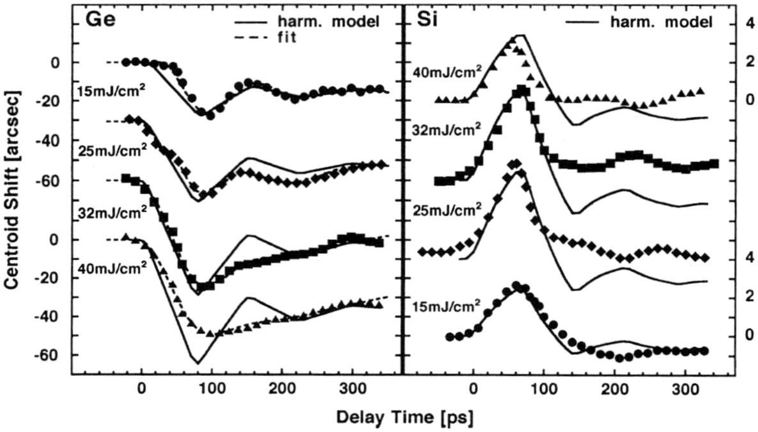

Fig. 1. Time dependent shift of the centroid of the measured rocking curves in the Ge-overlayer (left) and in the Si-substrate (right) for different fluences. Solid lines: harmonic model; dashed lines: anharmonic fit.

From maximum *negative* shifts of the Bragg-angle in the Ge-film between 25-50arcsec, values for the peak strain of 0.05-0.1%, corresponding to an increase in the lattice spacing of about 150-300fm, are obtained. Maximum expansion is reached 80-100ps after excitation, followed by damped oscillations at later times, indicating periodic expansion and compression of the film. Damping of these oscillations is apparently fluence dependent, becoming stronger for higher excitation. In the Si-substrate mainly *positive* line shifts, indicating compression of the material, are observed. The peak of the measured compression corresponds to a 20-40fm change in the average lattice spacing (a few nuclear diameters!).

Interpretation of the experimental data proceeds along the following lines. Irradiation of the sample leads to the generation of a dense e-h-plasma ($N \approx 10^{21}cm^{-3}$) in the Ge-overlayer, but negligible photo-excitation of the Si-substrate. Before complete energy relaxation between the electrons and the lattice can take place (via direct electron-phonon scattering and delayed Auger-heating), fast carrier diffusion [8] distributes the energy over the entire thickness of the Ge-film. X-ray measurements with samples of different film-thickness and on a bulk Ge-crystal further confirm the importance of fast, diffusion-mediated heat transfer in Ge. On the other hand, electronic energy transport into the Si-substrate is inhibited due to the 0.43eV difference of the energy gap in both materials.

Impulsive and homogenous heating produces an initially highly stressed state of the Ge-film. Acoustic relaxation of the stress leads then to vibration of the film with a period of $2d/c_L \approx 140ps$ (d=400nm±20nm: film thickness; c_L=5560m/s: sound velocity). In harmonic approximation, damping of the film vibrations and the compression of the Si-substrate can be attributed to acoustic transmission across the Ge/Si-interface. However, in a real crystal other processes, (defect-, surface- and phonon-phonon-scattering), lead to additional contributions to the decay of the oscillations. While damping due to acoustic transmission and defect- and surface-mediated scattering should be independent of the degree of excitation, the anharmonic phonon-phonon interaction scales with the population of the individual modes and thus with temperature. Our data allow us to distinguish between these mechanisms: the observed fluence dependent damping is clear indication of lattice anharmonicity.

For a quantitative estimate we first compare the data to a fully harmonic description. The initial stress distribution was obtained from a numerical solution of the coupled transport equations for the e-h-plasma and the lattice taking into account Auger recombination and carrier diffusion. The one-dimensional elastic equation was then used to calculate the time-dependent strain in the two-layer system. Finally, dynamic diffraction theory yields the transient x-ray diffraction patterns. The solid curves in Fig. 1 represent the results of these calculations. As expected, the model predicts a fluence-independent damping, originating solely from acoustic transmission across the Ge/Si-interface. For the lowest fluence the calculated curves follow the measured data very closely, but with increasing fluences the harmonic calculations progressively deviate from the measurement, as would be expected for an anharmonic, temperature dependent damping process. To describe lattice anharmonicity we fit the measured Ge-data to a phenomenological response function (dashed curves), which includes the fluence-independent damping due to acoustic transmission and an additional fluence-dependent contribution. The obtained anharmonic damping times range from 500ps at low fluences/temperatures down to 50-70ps at high fluences/temperatures and are consistent with estimated four-body, elastic dephasing times (T_2) for 7-GHz LA-phonons in Ge [9]

Acknowledgements. K.S.T. gratefully acknowledges support by the Deutsche Forschungsgemeinschaft.

References

1 C. Rischel et al., Nature **390**, 490 (1997).
2 J. Larsson et al., Appl. Phys. A **66**, 587 (1998).
3 C. Rose-Petruck et al., Nature **398**, 310 (1999).
4 A.H. Chin et al., Phys. Rev. Lett. **83**, 336 (1999).
5 C.W. Siders et al., Science **286**, 1340 (1999).
6 A.M. Lindenberg et al., ", Phys. Rev. Lett. **84**, 111 (2000).
7 M. Horn-von-Hoegen, Appl. Phys. A **59**, 503 (1994).
8 J.F. Young et al., Phys. Rev. B **26**. 2147 (1982).
9 A. Cavalleri et al., Phys. Rev. Lett. **85**, 586 (2000).

Picosecond time-resolved X-ray diffraction from a silicon crystal under laser-induced breakdown

Kazutaka G. Nakamura, Yoichiro Hironaka, Akio Yazaki, Fumikazu Saito, and Ken-ichi Kondo

Materials and Structures Laboratory, Tokyo Institute of Technology, 4259 Nagatsuta, Midori-ku, Yokohama 226-8503, Japan
E-mail: nakamur2@rlem.titech.ac.jp

Abstract. Picosecond time-resolved X-ray diffraction is performed for a Si(111) under laser-induced breakdown using picosecond pulsed X-rays. Transient change in strain profiles due to the shock compression generated by laser-induced breakdown is observed. The propagating strain profiles inside the Si(111) are determined.

1. Introduction

Time resolved X-ray diffraction using the ultrashort X-ray pulses makes it possible to study structural changes in transient phenomena [1]. The ultrashort pulsed hard X-rays can be generated when femtosecond laser beams are focused onto a metal target at high power density (>10^{16}W/cm^2). Here we report the picosecond time-resolved X-ray diffraction from a Si(111) under pulsed laser irradiation at power density above damage threshold using a laser-pump and X-ray probe technique. The laser-induced breakdown caused ablation and successive shock compression inside the Si(111).

2. Experimental Methods

The used laser system was a T^3 (table-top tera-watt) laser system [2,3]. Before pulse compression, a laser pulse (300 ps) was generated with energy of 400 mJ/pulse at 10 Hz. The 300 ps pulsed beam was divided into two beams by a beam splitter (50 %). One beam was compressed to about 50 fs and used for X-ray generation. The other was used for irradiation to Si through an optical delay line. The femtosecond pulse was focused on an iron target using a parabolic mirror (a focal length of 150 mm) with p-polarization and an incident angle of 60° to the surface normal in a vacuum camber (10^{-6} Torr) at power density of about 10^{17} W/cm^2. The pulse width of the X-rays was measured using an X-ray streak camera to be about 9 ps. Characteristic Kα line of Fe was used for diffraction on the Si(111). The 300 ps laser pulse was irradiated to the Si(111) at a spot size with diameter of 4mm and energy of 180 mJ/pulse (about 1.4 J/cm^2) in air for generating shock wave. The used laser power density is over the damage threshold and can induce breakdown [4]. The Si(111) sample was continuously moved during the experiment in order to reduce the effect of cumulative damage by the optical pump pulse. The X-ray pulse was irradiated at the center of the laser focal spot on the Si(111) and the diffracted X-rays were detected with an X-ray linear sensor and accumulated for 600 shots.

3. Results and Discussion

Time resolved X-ray diffraction from the Si(111) under laser irradiation at above damage threshold was performed using the laser-pump & X-ray probe technique with a time step of 60 ps. Figure 1 shows the measured diffraction profiles as a function of angle (rocking curves) and of pump-probe time delay.

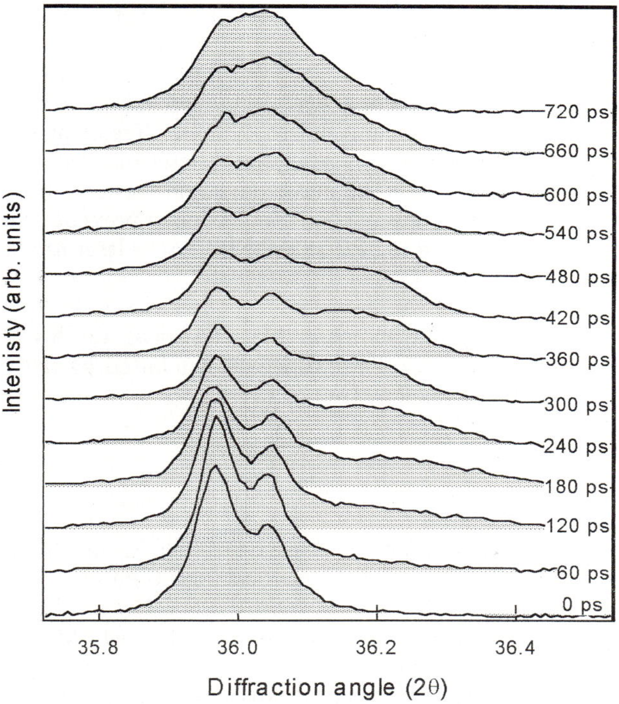

Fig. 1. The diffraction profiles as a function of angle (rocking curves) and of pump-probe time delay measured at time step of 60 ps.

The virgin sample gives the both diffraction lines of Kα1 (1.9360 Å) and Kα2 (1.9399 Å). When the pump laser beam was irradiated, the intensities of these lines decreased and the signals at higher angles appeared. The shift and the intensity of the higher angle diffraction are gradually decreasing after the end of pump beam irradiation. The shift of diffracted signals towards the higher angle corresponds to the decrease of crystal lattice spacing and means that the silicon lattice is compressed by irradiation of the pump laser beam. The strain ($\Delta 2d/2d$) is estimated by the equation: $\Delta 2d/2d = -\Delta\theta \cot\theta$, where d the lattice spacing and the θ is the Bragg angle. The maximum strain was obtained to be -1.05 % (compression) of the lattice spacing and the maximum pressure induced in the present experiment was estimated to be 2.18 GPa. Transient change in strain profiles due to the shock compression, which is generated by the laser-induced breakdown, is observed. The decrease of the shift and the intensity of the higher angle diffraction were explained as that the compressed states propagate (shock wave propagation) inside the silicon deeper than the X-ray penetration depth and the compression near surface is released. The propagating shock profiles were determined using the dynamical diffraction theory and the shock velocity was obtained to be 9.4 km/s.

4. Conclusions

Picosecond time-resolved X-ray diffraction was performed for a Si(111) under laser-induced breakdown using picosecond pulsed X-rays. Transient change in strain profiles due to the shock compression generated by laser-induced breakdown was observed at a time step of 60 ps. The maximum strain was obtained to be -1.05 % (compression) of the lattice spacing and the maximum pressure induced in the present experiment was estimated to be 2.18 GPa for 300 ps laser irradiation at about 1.4 J/cm^2.

Acknowledgements. We would like to thank Mr. Hasegawa for his help for constructing the experimental set-up. This work was supported by the CREST program of Japan Science and Technology Co..

References

1 C.W. Siders, A. Cavalleri, K. Sokolowski-Tinten, Cs. Toth, T. Guo, M. Kammler, M. Horn von Hoegen, K.R. Wilson, D. von der Linde, and C.P.J. Barty, in *Science* Vol. 286, 1340, 1999.

2 M. Yoshida, Y. Fujimoto, Y. Hironaka, K.G. Nakamura, K. Kondo, M. Ohtani, and H. Tsunemi, in *Appl. Phys. Lett.* Vol. 73, 2393, 1998.

3 Y. Fujimoto, Y. Hironaka, K.G. Nakamura, K. Kondo, M. Yoshida, M. Ohtani, and H. Tsunemi, *in Jpn. J. Appl. Phys.* Vol. 38, 6754, 1999.

4 P.P. Pronko, P.A. VanRompay, C. Horvath, F. Loesel, T. Juhasz, X. Liu, and G. Mourou, in *Phys.Rev.B* Vol. 58, 2387, 1998.

Time-resolved x-ray measurements of polaron dynamics of charge-ordered $Nd_{1/2}Sr_{1/2}MnO_3$

I. Kang[1], S. Johnson[1], A. Lindenberg[1], R. Falcone[1], Th. Missalla[2], P. Heimann[2], K. H. Kim[3], T. Katsufuji[4], S-W. Cheong[3,4]

[1] Department of Physics, University of California, Berkeley, CA 94720, USA
E-mail: ikang@grace.lbl.gov
[2] Lawrence Berkeley National Laboratory, Berkeley, CA 94720, USA
[3] Rutgers University, Piscataway, NJ 08854, USA
[4] Bell Laboratories, Lucent Technologies, Murray Hill, NJ 07974

Abstract. We report the first direct observation of polaron dynamics in a charge-ordered manganese oxide material. The excitation and relaxation of lattice distortion coupled to charge-transfer transitions are observed using time-resolved x-ray diffraction.

The hole-doped perovskite manganese oxide materials have recently received much attention due to many fascinating phenomena such as colossal magnetoresistance (CMR). We seek to better understand the strong correlation physics of the CMR materials by time-resolving the dynamic response of electron, lattice, and spin. We specifically investigate the role of the dynamical structural distortion in the CMR materials. It has been proposed that the coupling of electrons to the lattice distortion is crucial for understanding the CMR materials [1]. However, the dynamics of the lattice distortions and their coupling to electronic transitions have not been addressed experimentally to date.

We present time-resolved x-ray measurements of the polaron dynamics of charge-ordered $Nd_{1/2}Sr_{1/2}MnO_3$. This system is chosen for the strong charge-lattice-spin coupling, which is manifested as the charge-ordered, antiferromagnetic insulating phase at low temperature ($< T_{co} \sim 160K$) [2]. The material goes through insulator - metal, structural, and magnetic (antiferro - ferromagnetic) phase transitions at T_{CO} [2]. It exhibits another phase transition (ferro - paramagnetic) at $T_c \sim 230K$ [2]. Below T_{CO}, the Jahn-Teller (JT) polarons around Mn^{3+} ions possess a long-range spatial coherence due to the charge ordering. Thus, the lattice structure of the charge-ordered state can be viewed as a solid composed of static polarons [2]. The lattice distortion due to the polarons gets relieved in the ferromagnetic phase but exists in paramagnetic phase with fluctuating short-range spatial coherence [2].

We monitor how the lattice structure evolves after the material is irradiated with 150-fs, 800-nm laser pulses between 20 and 300K. The laser fluence is maintained less than $5mJ/cm^2$. The pump-laser wavelength overlaps with the broad Mn^{3+} to Mn^{4+} charge-transfer transition [3]. The lattice dynamics is probed with the time-resolved x-ray Bragg diffraction from the orthorhombic *Pbnm* (112) plane of a single crystal $Nd_{1/2}Sr_{1/2}MnO_3$. The experiment was performed at beamline 7.3.3 of the Advanced Light Source using a Ge(111) double crystal monochromator. The diffracted x-ray was collected by a Si-avalanche photodiode and a picosecond x-ray streak camera. The time resolution of the experiment is ~ 2 ps set by the temporal jitter of the streak camera.

The x-ray rocking curve measured 50 ps after the arrival of the laser pulse is shown in the inset of Fig. 1 with the unperturbed rocking curve. The increase of the diffracted intensity at the wings and the reduction of the intensity near the peak are observed. The observed dynamics of the diffracted intensity at different points on the rocking curve are illustrated in the left panel of Fig. 1. The increase in the shorter (longer) wavelength side of the x-ray rocking curve signifies the development of compressive (expansive) strain in the crystal. The experimental data were obtained at 300K and we did not observe a significant dependence of the dynamics on temperature between 20 and 300K.

The observed laser-induced modification of the x-ray Bragg diffraction is a direct result of the strong coupling in this system between the JT lattice distortion and the charge-transfer transition. Impulsive (faster than the lattice can respond) charge-transfer from ions to ions would create the JT lattice distortion. The JT modes which couple to the charge transfer are Q_1, Q_2, and Q_3–type distortions of the Mn-O octahedra (Fig. 2). The laser-induced lattice distortions are anisotropic unlike the simple thermal expansion as both the compressive and expansive components are observed in the data. We model the observed data assuming that the charge-transfer transition creates the JT-lattice modes.

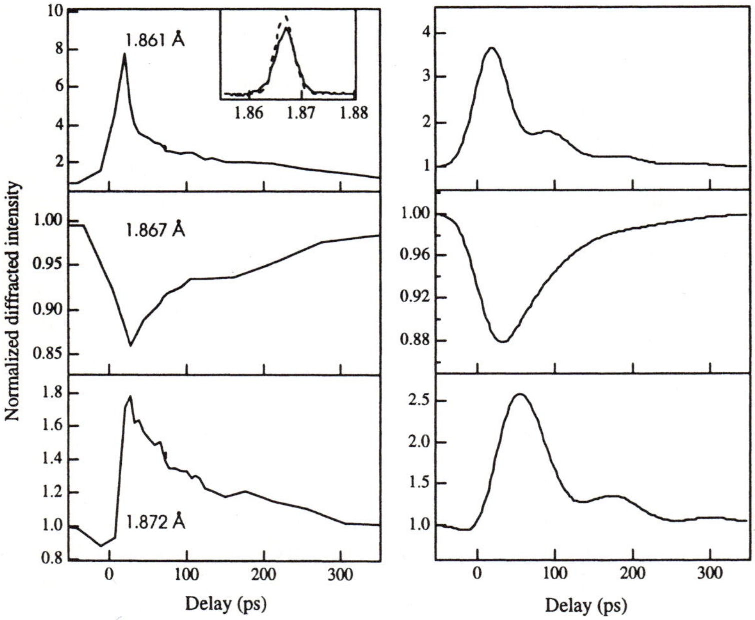

Fig. 1. Left panel: the observed dynamics of the diffracted x-ray intensity at different points of the rocking curve. Right panel: theoretical modeling of the data in the left panel assuming laser-generated transient Jahn-Teller lattice distortion in the crystal. The inset in the left panel shows the rocking curve measured 50 ps after the laser pulse (solid curve) and the unperturbed rocking curve (dashed curve). In the figure, the unperturbed x-ray intensity for each point of the rocking curve is set to unity.

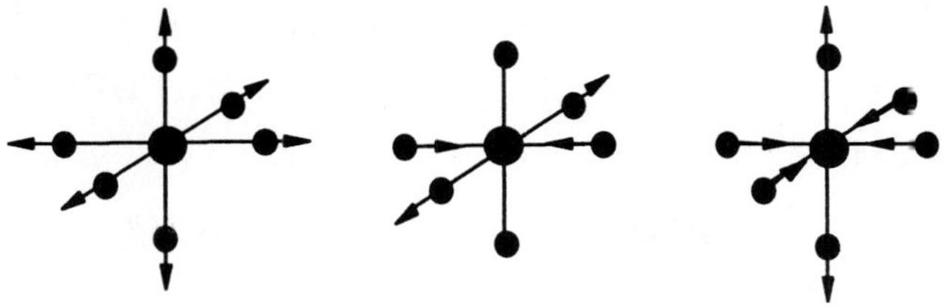

Fig. 2. The Jahn-Teller distortion of the oxygen octahedra around Mn ions that strongly interact with the optical charge-transfer transition.

Relative to the lattice plane, which corresponds to (101) of the cubic perovskite, the distortions amount to longitudinal expansion and transverse compression. The longitudinal and transverse strain associated with the JT distortions propagate into the crystal with the relevant speed of sound. Modifications of the diffracted intensity due to the strain waves are calculated and shown in the right panel of Fig. 1. From the calculation, we estimate that the JT distortions decay in ~ 50 ps and the JT interaction energy is ~ 1 eV independent of temperature. The fact that we observe similar dynamics of the lattice distortion between 20 and 300 K suggests that the JT interaction strength is much larger than the bandwidth of the band derived from the itinerant e_g electrons.

It should be pointed out that the observed effects are not caused by crystal heating. The effect of the heating on the crystal is generation of expansive strain through acoustic phonon creation. We measured that this effect developed over a very long time scale ~ 100 ns). This small rate of acoustic phonon generation is another evidence that the JT interaction is the major channel of energy relaxation of the carriers generated by the charge transfer transitions.

In conclusion, we report the time-resolved x-ray measurements of polaron dynamics of charge-ordered $Nd_{1/2}Sr_{1/2}MnO_3$, which clearly illustrates the strong coupling between charge and lattice in this system.

References

1 A. J. Millis, "Electron-lattice coupling in "colossal" magnetoresistance rare earth manganites," J. Appl. Phys. **81**, 5502-5503 (1997).

2 H. Kuwahara et al, "Spin-charge-lattice coupled phase transitions in bandwidth controlled systems $(Nd,Sm)_{1/2}Mn_{1/2}O_3$," Phys. Rev. B **56**, 9386-9396 (1997).

3 M. Quijada et al, "Optical conductivity of manganites: Crossover from Jahn-Teller small polaron to coherent transport in the ferromagnetic state," Phys. Rev. B **58**, 16093-16102 (1998).

High-Order Harmonic Generation in the Few-Optical-Cycle Regime: Spectral Characteristics and Modeling

Sandro De Silvestri[1], Giulio Cerullo[1], Mauro Nisoli[1], Enrico Priori[1], Salvatore Stagira[1], Orazio Svelto[1], Luca Poletto[2], Giuseppe Tondello[2], Paolo Villoresi[2], Carlo Altucci[3], Riccardo Bruzzese[4] and Corrado de Lisio[4]

[1] INFM, Dipartimento di Fisica, Politecnico di Milano,
Piazza L. da Vinci 32, 20133 Milano, Italy
E-mail: mauro.nisoli@fisi.polimi.it
[2] INFM, Università di Padova, Padova, Italy
[3] INFM, Università della Basilicata, Potenza, Italy
[4] INFM, Università di Napoli "Federico II", Napoli, Italy

Abstract. Few-optical-cycle laser driven high-order harmonic spectral characteristics have been investigated both experimentally and theoretically. The measurements evidence peculiar features typical of quasi-single-cycle regime. A nonadiabatic three-dimensional numerical model is reported, in good agreement with experiments.

1. Introduction

High-order Harmonic Generation (HHG) in noble gases provides a unique source of high brightness, coherent extreme ultraviolet radiation [1-2]. The introduction of high-peak power few-optical-cycle laser pulses has extended the scenario of harmonic generation process, giving rise to a number of open questions to be addressed both from an experimental and a theoretical point of view.

In this work we investigate the spectral properties of harmonic radiation generated by sub-10-fs laser pulses. We also present a three-dimensional numerical model, which takes into account nonadiabatic effects, typical of the few-optical-cycle regime.

2. Results and Discussion

The excitation source used in the experiments is a Ti:sapphire laser providing 30-fs, 0.8-mJ pulses at 1-kHz repetition rate. Sub-10-fs pulses are generated by the hollow-fiber compression technique [3-4]. Typical pulse duration ranges from 5 to 7 fs. Harmonic radiation is analyzed using a high-resolution x-ray spectrometer.

Upper panel of Fig. 1 shows a typical harmonic spectrum obtained in neon using 7-fs excitation pulses; the estimated laser peak intensity in the interaction region was $\sim 9\times 10^{14}$ W/cm^2. The spectrum displays discrete harmonic peaks. We have analyzed the wavelength dependence of the harmonic linewidths. As shown

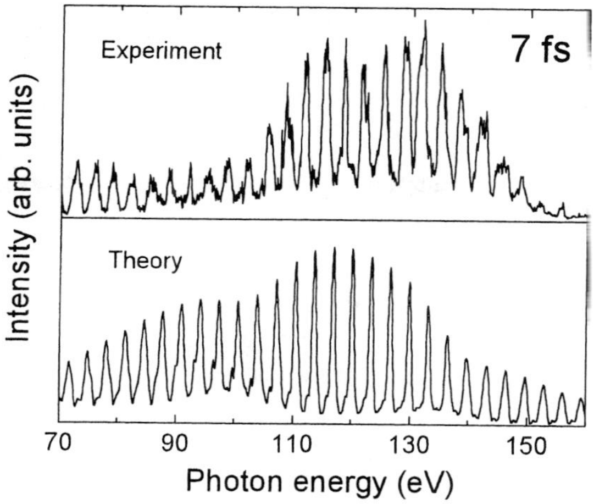

Fig. 1. Comparison between experimental (upper panel) and calculated (lower panel) HHG spectra generated in neon by 7-fs driver pulses.

by filled circles in Fig. 2, the linewidth remains roughly constant in the reported spectral region. In order to understand the role of pulse duration, we have repeated the same measurements using 30-fs excitation pulses.

The experimental results are depicted by the empty triangles in Fig. 2 and show that the linewidth decreases upon increasing harmonic wavelength. In the case of 30-fs driving pulses, this behavior can be at least partially understood in terms of the different emission times of the harmonics: those well in the plateau are emitted over several optical cycles, while those approaching the cutoff are emitted only over a few cycles, close to the peak intensity of the laser pulse. For the 7-fs pulses, the linewidth remains approximately constant with harmonic order, since all the harmonics are emitted over a few cycles of the driving pulse.

The experimental results have been analyzed in terms of a three-dimensional nonadiabatic numerical propagation model. Previous calculations have either used a three-dimensional treatment in the adiabatic approximation [5] or a nonadiabatic approach in a one-dimensional model [6]. For few-optical cycle pulses the adiabatic approach is no longer valid. On the other hand three-dimensional effects are essential to achieve a good agreement with the experimental data. The single-atom response is calculated using the Lewenstein's model, generalized to account for nonadiabatic effects, and it is inserted in a three-dimensional propagation model. We find that, using such short pulses, single-atom emission spectra appear noisy with no recognizable harmonics, while after propagation they are considerably cleaned up and recover a discrete structure. This dramatic change in the spectral features is attributed to phase mismatch due to longitudinal and radial variations of both intensity and phase of the driving pulses: these effects can be considered correctly only in a three-dimensional geometry.

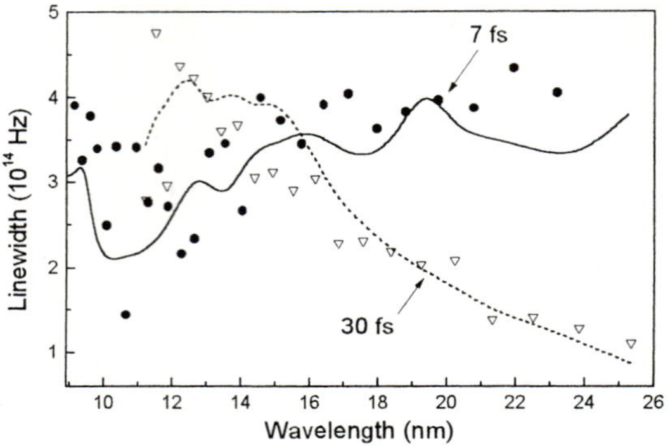

Fig. 2. Linewidths of the harmonics generated by 7- and 30-fs. Points: experimental data; curves: computer simulations.

Calculated HHG spectra for a 1-mm neon jet irradiated by 7-fs pulses are shown in the lower panels of Fig. 1. The computer simulations are in good agreement with the experimental result, in terms of the cutoff position, and of the widths and relative amplitudes of the harmonic peaks. The numerical model reproduces quite well also the wavelength dependence of the harmonic peak linewidths, as shown by the curves in Fig. 2.

3. Conclusions

We have investigated, both experimentally and theoretically, the spectral characteristics of harmonic radiation generated by 7-fs laser pulses. The harmonic spectra show discrete and well resolved peaks, whose linewidths remain almost constant as a function of the harmonic order. A numerical model employing a nonadiabatic three-dimensional treatment, has been developed. The results of the numerical simulations are in remarkable agreement with the experimental data.

References

1 Z. Chang, A. Rundquist, H. Wang, M: M: Murnane, and H. C. Kapteyn, Phys. Rev. Lett. **79**, 2967, 1997.

2 Ch. Spielmann, N.H. Burnett, S. Sartania, R. Koppitsch, M. Schnürer, C. Kan, M. Lenzner, P. Wobrauschek, F. Krausz: Science **278**, 661, 1997.

3 M. Nisoli, S. De Silvestri, and O. Svelto, Appl. Phys. Lett. **68**, 2793, 1996.

4 M. Nisoli, S. De Silvestri, and O. Svelto, R. Szipocs, K. Ferencz, Ch. Spielmann, S. Sartania, and F. Krausz, Opt. Lett. **22**, 522, 1997.

5 P. Salières, A. L'Huillier, and M. Lewenstein, Phys. Rev. Lett. **74**, 3776, 1995.

6 G. Tempea, M. Geissler, and T. Brabec, J. Opt. Soc. Am. B **16**, 669, 1999.

Nonadiabatic Self-Phase Matching of High Harmonic Generation

G. Tempea, M. Geissler, M. Schnürer, F. Krausz, and T. Brabec

Institut für Photonik, Technische Universität Wien,
Gußhausstr. 27/387, A-1040 Vienna, Austria
E-mail: tempea@tuwien.ac.at; brabec@tuwien.ac.at

Abstract. : The nonadiabatic self modulation experienced by a few-cycle optical pulse propagating in a tunnel-ionizing gas can result in phase-matched high harmonic generation. Our calculations predict that the mechanism, which has been termed nonadiabatic self phase-matching, could make the generation of above keV x-ray harmonics possible for the first time.

1. Introduction

Recently, phase-matched high harmonic generation (HHG) using 20fs laser pulses was demonstrated in the wavelength range around 20nm [1]. This mechanism works only as long as the gas is weakly ionized, thus setting a limit to the shortest wavelengths for which phase-matching can be achieved. Our calculations reveal the existence of a different phase-matching mechanism which we have termed "nonadiabatic self-phase-matching" (NSPM) [2,3] that works predominantly for few-cycle, high intensity laser pulses and for wavelengths in the xuv and x-ray regime. In this so-called "nonadiabatic" parameter regime the ionization rate becomes comparable to the center frequency of the laser pulse, leading to a significant change of the free electron density over one optical cycle. The rapidly varying ionization front causes sub-cycle changes of the laser electric field, which results in NSPM. Our analysis predicts that NSPM enhances HHG above 1 keV by 4 orders of magnitude which would for the first time make HHG in the soft x-ray regime possible.

2. Results and Discussion

In Fig.1 we have plotted the calculated harmonic spectrum generated in He at a pressure of 500 Torr by a 5-fs Ti:sapphire pulse (center wavelength 800 nm) after an interaction length of 125μm. The full and the dotted lines denote the spectral intensity of the harmonic signal obtained for laser peak intensities $I_0 = 4x10^{15}$ W/cm^2 and $I_0=1.5x10^{16}$ W/cm^2, respectively. A detailed description of the model is given in Ref. [3]. For $I_0 = 4x10^{15}$ W/cm^2 the harmonic spectrum reveals a strong wavelength dependence (2 orders of magnitude difference between the 300-th and 200-th harmonic) in agreement with previous investigations [3]. The wavelength

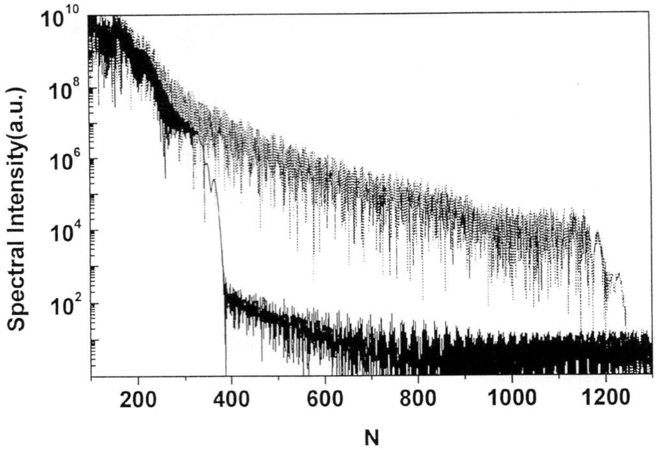

Fig. 1. Spectral intensity versus harmonic order; for parameters see the text.

dependence comes from the phase-mismatch induced by the free electrons, which determines the coherence length, i.e. the length over which the harmonic can grow coherently. Higher harmonics are generated at higher electron densities and experience a stronger phase-mismatch. Extrapolating this trend to higher harmonics one would anticipate that the harmonic signal between N = 300 and N = 600 drops by 6 orders of magnitude, i.e. that it will be impossible to generate measurable keV harmonic radiation by merely increasing the laser peak intensity.

Our calculation reveals a completely unexpected behavior. For I_0=1.5X10^{16} W/cm^2 ionization saturates at half of the peak intensity, where ≈98% of the atoms are ionized. The wavelength dependence of the harmonic yield is much weaker than expected. According to Fig.1 harmonics between 1 and 2 keV can be generated with a spectral intensity which is only 1-2 orders of magnitude lower than the spectral intensity around the 300-th harmonic for I_0=4x10^{15}W/cm^2. As the generation of the 300-th harmonic was already demonstrated experimentally [3], our analysis predicts the possibility to generate and measure coherent (1-2) keV X-ray radiation.

In Fig. 2 we plotted the growth of the harmonic energy in the spectral range between N=950 and 1000 for the same parameters as in Fig. 1. The growth over the first micrometer in the inset is in agreement with conventional theory [3] which predicts a coherence length of ≈0.1μm. For propagation distances much longer as the coherence length NSPM sets in and the harmonic signal exhibits a step like increase which enhances the energy converted into harmonic radiation by four orders of magnitude. During the first step between 4 and 7μm the harmonic signal grows quadratically indicating the coherent nature of the emitted radiation. Finally, our calculations show that NSPM occurs mainly for sub-10 fs pulses. This

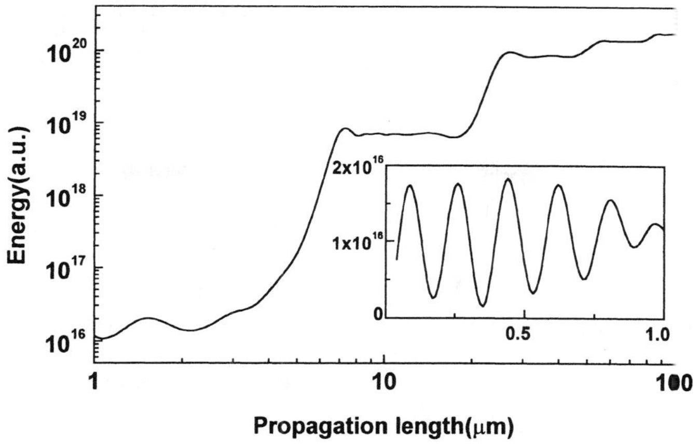

is because, in the so-called nonadiabatic regime realized by few-cycle laser pulses the ionization rate becomes comparable to the center frequency of the laser pulse leading to a significant change of the free electron density over one optical cycle. The rapidly varying ionization front causes sub-cycle changes of the laser electric field, which lead to NSPM. A detailed description of the NSPM mechanism, is given in Ref. [2].

4. Conclusions

We have numerically identified a phase matching mechanism that works predominantly in the limit of high harmonic generation with few-cycle laser pulses. This mechanism, termed nonadiabatic phase matching, could make the generation of above keV x-ray harmonics possible. A harmonic source in the few-keV photon energy range would open the way towards revolutionary applications, such as time resolved x-ray diffraction experiments on a sub-10fs time scale.

Acknowledgements. This work was supported by the Austrian Science Fund, Y142-TPH and by the special research program F016.

References

1 A. Rundquist et al., Science **280**, 1412 (1998).
2 G. Tempea, M. Geissler, M. Schnürer, and T. Brabec, Phys. Rev. Lett. **84**, 4329 (2000).
3 T. Brabec and F. Krausz, Rev. Mod. Phys **72**, 545 (2000).

Diffraction Limited Emission of High Order Harmonics from Solid Surfaces

A.Tarasevitch, C. Dietrich, A. Orisch, and D. von der Linde

Institut fuer Laser- und Plasmaphysik, University of Essen, D-45117 Essen, Germany
E-mail: tar@ilp.physik.uni-essen.de

Abstract. High order harmonics are generated from solid surfaces using femtosecond pulses. For high-contrast-ratio laser pulses the harmonic emission is highly directional. Influence of the plasma scale length on harmonic efficiency is investigated. At high pump intensity the decay of the plasma mirror is observed.

1. Introduction

Generation of optical harmonics of high order is one of the most promising tools for generating short wavelength coherent radiation. In noble gases harmonics up to about the 300th order have been produced [1]. A different approach is high order harmonic generation (HOHG) in reflection from solid targets [2-5]. According to the theoretical predictions at relativistic intensities ($I \geq 10^{18}$W/cm2) the number of harmonics and the conversion efficiency are strongly enhanced [2]. For HOHG in reflection a step-like plasma-vacuum interface is essential. The use of femtosecond pump pulses leads quite naturally to the formation of a steep gradient of the plasma density. However, laser pulses of very high contrast are required. A prepulse or a slowly rising leading front of the pump pulse lead to premature ionization of the target and destruction of the initial density profile. In that case, the interaction with the tail of underdense plasma and rippling of the plasma surface lead to substantial spectral broadening and a very wide angular distribution[3]. The importance of a clean rising edge of the laser pulses in HOHG was demonstrated in [4]. So far, however, the direct observations of the influence of the plasma scale length on harmonic efficiency, as well as detailed measurements of the angular distribution demonstrating the directionality of the harmonics, are lacking.

The aim of the present work was (i) to determine the divergence of the harmonics, (ii) to carry out direct measurements of the influence of the plasma scale length on efficiency and coherence of HOHG, and (iii) to detect the possible decay of the plasma-vacuum interface at high intensities.

2. Experimental

In our experiments we used a titanium sapphire system with a pulse duration of 120 fs and an energy up to 200 mJ. The contrast ratio was nearly 10^{-5} at 1 ps from the pulse maximum. The incident laser pulses at the wavelength of 800 nm were focused onto the target with the help of an off-axis parabolic mirror. The spectrum of the reflected light was imaged onto a phosphor screen with the help of a toroidal grating, and the phosphorescent light was recorded with a CCD camera. The targets were optically polished glass substrates, which were raster-scanned to provide a fresh surface for each laser pulse. The pressure in the experimental chamber was 10^{-3} Torr. At the pump intensity of 0.2×10^{18} W/cm^2 harmonics up to 20th order were observed.

Measurements of harmonic divergence and efficiency. In order to study the angular divergence, the harmonic beams were scanned across entrance aperture of the spectrometer by rotating the target. The measured angular dependence of the harmonic energy shows that the divergence of the harmonics is smaller than that of the fundamental. Assuming a power law for the harmonic intensity and a Gaussian beam profile we find that the emitted harmonics are nearly diffraction limited. The measured conversion efficiency for 10th and 15th harmonics exceeded 10^{-7} and 10^{-8} respectively.

Dependence of the harmonic efficiency on plasma scale length. Pump-probe experiments were carried out in order to investigate the dependence of HOHG efficiency on the plasma scale length. A plasma was created on the surface of the target by a first, relatively weak laser pulse (0.5×10^{15} W/cm^2). A second pulse (approximately 0.2×10^{18} W/cm^2) with an adjustable time delay τ with respect to the first pulse was used to produce harmonics. The measured dependence of the harmonic energy on τ for several harmonic orders is depicted in Fig. 1. The decay rate of the harmonic signals is approximately the same for all harmonic orders. The expansion velocity of the plasma is about 0.075 λ/ps and the data can be fitted by $\exp(-L/L_s)$ with $L_s = 0.015\ \lambda$.

Decay of the plasma mirror. Completely different angular distributions were observed when the pump intensity exceeded 10^{18} W/cm^2. The reflected light is no

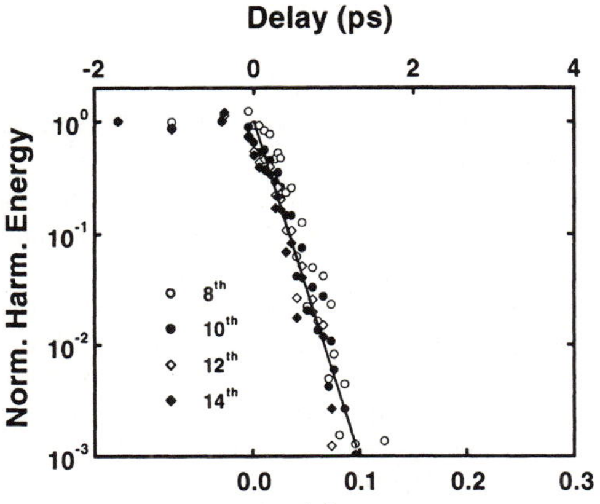

Fig. 1. Energy of different harmonics as a function of the time delay (upper horizontal scale). Lower horizontal scale: Estimated plasma scale length. Solid line: $\exp(-L/0.015\lambda)$

longer collimated, but spread over certain solid angle around the specular direction. We have developed a femtosecond interference microscope to study the deformation of the plasma mirror with high spatial resolution. The deformation of the interference fringes at the pump intensity of 2×10^{18} W/cm^2 demonstrate the distortions of the plasma mirror.

4. Conclusions

Nearly diffraction limited high efficient emission of high order harmonic from solid targets is demonstrated. With a help of an interference microscope we observe a spatially resolved picture of the distortions of the plasma mirror at the intensities exceeding 10^{18} W/cm^2.

We do not observe a certain optimum scale length for HOHG during the expansion of the plasma, in contradiction with PIC simulations. The efficiency of HOHG drops more rapidly with plasma scale length than predicted by these simulations. Plasma surface distortions, which destroy the specular reflected beam and are not taken into account in one-dimensional PIC calculations, could explain our observations.

Because a substantial increase of the efficiency of harmonic generation is expected for laser intensities exceeding 10^{18} W/cm^2, it is a great challenge to overcome the present limits. The threshold of the plasma instabilities should strongly depend on the plasma scale length. Therefore, more detailed pump-probe experiments with the diagnostics of plasma surface and of the spatial and angular distribution of the harmonic are important.

Acknowledgements. The support by the European Union in the framework of "Training and Mobility by Research" and Deutsche Forschungsgemeinschaft is gratefully acknowledged.

References

1 Z. Chang, A. Rundquist, H. Wang, M. M. Murnane, and H. C. Kapteyn, Phys. Rev. Lett. **79**, 2967-2970 (1997); Ch. Spielmann, N. H. Burnett, S. Sartania, et al., Science **278**, 661-664 (1997).

2 R. Lichters, J. Meyer-ter-Vehn, and A. Pukhov, Phys. Plasmas **3**, 3425-3437 (1996); P. Gibbon, Phys. Rev. Lett. **76**, 50-53 (1996); D. von der Linde and K. Rzazewski, Appl. Phys. B 63, 499-506 (1996).

3 P. A. Norreys, M. Zepf, S. Moustaizis, et al., Phys. Rev. Lett. **76**, 1832-1835 (1996); J. Zhang, M. Zepf, P. A. Norreys, et al., Phys. Rev. A **54**, 1597-1603 (1996).

4 R. Lichters, J. Meyer-ter-Vehn, Inst. Phys. Conf. Ser. **154**, 221-230 (1997); M. Zepf, G. D. Tsakiris, G. Pretzler, et al., Phys. Rev. A **58**, R5253-R5256 (1998).

5 S. Kohlweyer, G. D. Tsakiris, C. G. Wahlstrom, et al., Opt. Commun. **117**, 431-438 (1995); D. von der Linde, T. Engers, G. Jenke, et al., Phys. Rev. A **52**, R25-R27 (1995).

Femtosecond Time-Resolved Luminescence Spectroscopy of Inner-Shell Excitation by High-Order Harmonics

Taro Sekikawa, Tomohiro Yamazaki, Satoshi Miura, Yasuo Nabekawa, and Shuntaro Watanabe

Institute for Solid State Physics, University of Tokyo, 5-1-5 Kashiwanoha, Kashiwa 277-8581, Japan
E-mail: sekikawa@issp.u-tokyo.ac.jp

Abstract. High-order harmonics of a Ti:sapphire laser were applied to femtosecond luminescence spectroscopy for the observation of the relaxation dynamics of inner-shell excitation in an alkali-halide crystal for the first time.

Development of the femtosecond pulse lasers has revealed many relaxation processes of excited or non-equilibrium electronic states, which are mainly generated by the photo-excitation of the valence electrons, in the visible, infrared, and hard x-ray regions. The relaxation processes by using femtosecond pulse lasers have been investigated so far in terms of the interactions between valence electrons and their environments, such as phonons, other electrons, solvent, and so on. On the other hand, the relaxation dynamics of inner-shell excitations in the femtosecond region have been rarely researched because of the lack of the intense femtosecond light sources in the extreme ultraviolet (XUV) and soft x-ray regions.

Recent development of the high-peak-power Ti:sapphire (TiS) laser makes it possible to overcome such a difficulty by generating high-order harmonics (HH). HH is a promising tool for the investigation of ultrafast and nonlinear phenomena in the soft x-ray and XUV regions, where inner-shell excitations are expected to play an important role in the nonlinear optical processes. In this paper, we demonstrate the first application of HH to the femtosecond time-resolved luminescence spectroscopy of inner-shell excitation in an alkali-halide crystal, CsCl, to investigate the relaxation processes related to inner-shell excited states. Figure 1 shows the energy diagram of CsCl. In the nanosecond time scale, the inner-shell excitation with the energy of 14.1 eV is relaxed by the radiative recombination between the valence-band electron and the inner-shell hole. In the femtosecond time region, the time dependence of the luminescence intensity and spectrum is expected to reflect the relaxation processes, such as the Auger

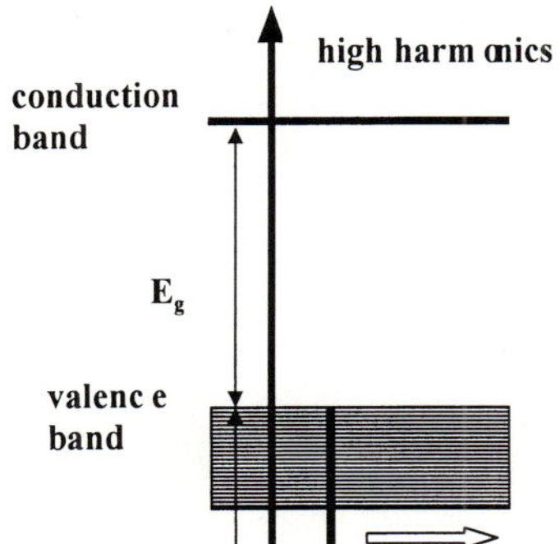

Fig. 1: Energy diagram of CsCl.

decay, electron-electron correlation, electron-phonon interaction. Especially, the Auger decay is the characteristic process of inner-shell excitation and its rate has never been measured in the time domain.

The time dependence of the luminescence in the femtosecond region was measured by the down-conversion method, in which the intensity dependence of the difference frequency between the luminescence and the gate pulse on the gate timing corresponds to the time evolution of the luminescence. Figure 2 shows the experimental setup for the measurement. For the HH generation, 10-mJ TiS laser pulses with a center wavelength of 800 nm were focused into the Kr-gas filled target at the repetition rate of 1 kHz in the vacuum chamber. [1] The generated HH were separated from the fundamental laser by a beam splitter and focused on the sample by a troidal mirror after passing through an Al filter cutting off the harmonics lower than the 9th. Consequently, the harmonics between the 11th (=16.6 eV) and the 25th (=37.8 eV) excited the sample. The transmitted fundamental pulses were used as the gate pulses. The luminescence was collimated and focused into the BBO crystal by parabola mirrors. The generated difference-frequency photons were detected by a photomultiplier after passing through a monochromator. The timing between the gate pulse and luminescence was adjusted by the computer-controlled optical delay. It is difficult to measure the cross-correlation between the exciting HH and the gate pulses, which corresponds to the time resolution of the system. Thus, we measured the width of the cross-correlation between the third harmonic and fundamental pulses as the upper limit of the time resolution, which was 140 fs.

Figure 3 shows the luminescence spectrum excited by HH. [2] Figures 4 (a) and (b) show the time dependence of the luminescence at the peaks of 240 and 266 nm, respectively, in the femtosecond region. At 240 nm, the luminescence rose within the time resolution and decayed with the long lifetime (1.3 ns) determined by the

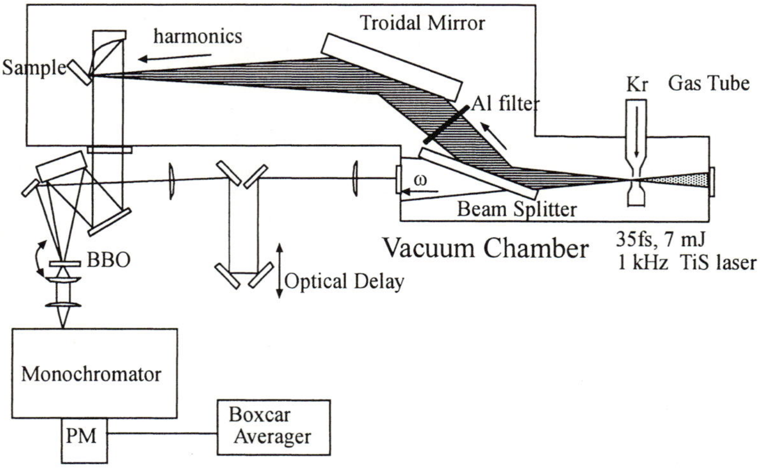

Fig. 2: Experimental set-up for measuring the femtosecond time-resolved luminescence by using high-order harmonics.

radiative recombination rate between the valence-band electron and the inner-shell hole. On the other hand, the luminescence at 266 nm has the rise time of 300 fs, indicating that the origins of these peaks are different. The luminescence at 266 nm was suggested to be due to the lattice relaxation related to the phonon of 14 meV from the temperature dependence of the luminescence spectrum. [3] Taking account of this, the rise time corresponds to the lattice relaxation time due to the inner-shell hole and phonon interaction.

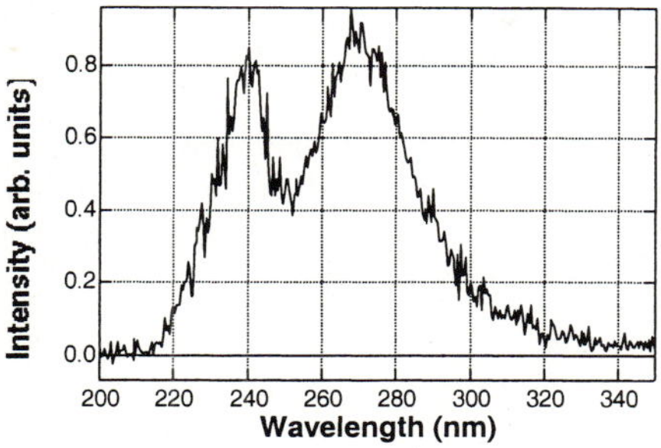

Fig.3 Luminescence spectrum of CsCl excited by high-order harmonics

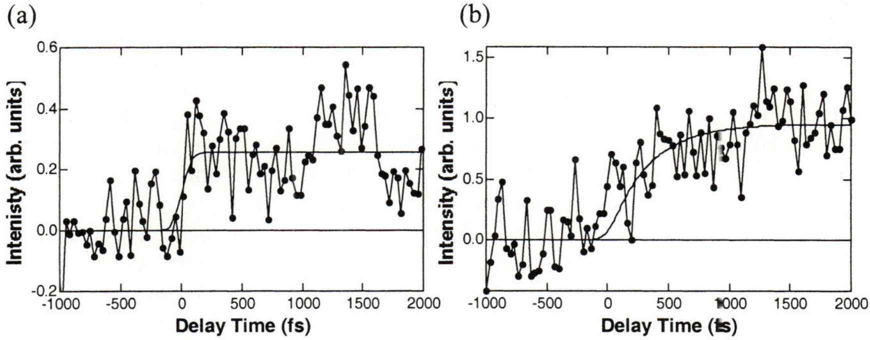

Fig.4: Dots show the time dependence of the luminescence intensities at the peaks of (a) 240 and (b) 266 nm, respectively. The solid lines are fitting results, taking account of a rise time.

References

1 Y. Nabekawa, Y. Kuramoto, T. Togashi, T. Sekikawa, and S. Watanabe, Opt. Lett. **23**, 1384, 1998.

2 T. Sekikawa, T. Ohno, Y. Nabekawa, and S. Watanabe, J. Lumin. **87-89**, 827, 2000.

3 M. Itoh, M. Kamada, and N. Ohno, J. Phys. Soc. Jpn. **66**, 2502, 1997.

Femtosecond x-ray fluorescence from light elements excited by laser harmonics

Michael Hentschel[1], Matthias Schnürer[1], Reinhard Kienberger[1], Christian Spielmann[1], Ferenc Krausz[1], Christina Streli[2], and Peter Wobrauschek[2]

1 Institut für Photonik, Technische Universität Wien, Gusshausstr. 27 / 387, A-1040 Wien, Austria; E-mail: michael.hentschel@tuwien.ac.at, ferenc.krausz@tuwien.ac.at
2 Atominstitut der Österreichischen Universitäten, A-1020 Wien, Austria

Abstract. Using few-cycle-driven coherent laser harmonic radiation, K-shell vacancies have been created in light elements (boron and carbon) on a time scale of a few femtoseconds for the first time. Our experiments constitute the first application of a coherent laboratory source of water-window radiation.

1. Introduction

XUV radiation sources based on high harmonic generation have evolved rapidly in recent years. The cut-off has been extended to the water window by using few-cycle drivers [1-3] and the conversion efficiency has been optimized to reach the XUV absorption limit by phase-matched generation [4] for various wavelength regions extending from 50 nm [5] to as short as 10 nm [6].

Theoretical investigations predict that the XUV emission may be temporally confined to a fraction of the oscillation period when high harmonic radiation is produced by few-cycle pump pulses [7]. The feasibility of generating isolated XUV/soft-X-ray pulses with durations of a few femtoseconds or less holds promise for probing ultrafast processes at short wavelengths.

In this paper we report on the detection of X-ray fluorescence emission following the creation of K-shell vacancies in light elements such as boron and carbon within less than 5 fs by few-cycle-driven soft-X-ray harmonics. The capability of producing measurable X-ray fluorescence excited by few-femtosecond radiation paves the way to probing the evolution of the microscopic environment of selected atoms in chemical and biochemical reactions on previously inaccessible time scales (< 100 fs) by tracing the temporal evolution of the "chemical shift" of peaks associated with inner-shell electronic transitions in time-resolved X-ray fluorescence, photoelectron or absorption spectra.

2. Experimental Methods

In our experiment we investigate the fluorescence signal originating from the recombination of K-shell vacancies in light elements. The vacancies are created via photoionization within a few femtoseconds. Filling the vacancy by relaxation of an L-shell electron releases energy, which can be either transferred to another electron (Auger relaxation) or emitted in form of a fluorescence photon. For low-

Z elements Auger relaxation has a 2-3 orders of magnitude higher probability than fluorescence emission. Because of the low X-ray fluorescence yield extraction of the fluorescence signal from a strong background originating from scattering of the incident radiation is a challenging task. A carefully designed excitation and detection geometry as well as appropriate reference measurements are indispensable for extracting the K-shell fluorescence signal of boron and carbon from scatter background.

The exciting radiation was generated in a quasi-static helium gas cell (Fig. 1) pumped by few-cycle optical pulses (τ = 7 fs, E = 0.7 mJ) delivered by a Ti:sapphire-amplifier/pulse-compressor system at a 1 kHz repetition rate [8]. The well-collimated XUV beam propagated down a miniature beam line passing a 200 nm thick silver filter to block low-oder harmonics and the fundamental light. In the detection chamber the XUV photons impinged on the samples near 45 degree ($\approx$ Brewster's angle) and the detector was placed with its axis perpendicular to the incoming beam to reduce the background caused by scattering. The X-ray detector consisted of an cryogenically-cooled, energy-dispersive silicon-PIN-diode, a preamplifier and a multichannel pulse height analyzer. The detector window was covered with a 200 nm thick Ti foil for additional spectral filtering.

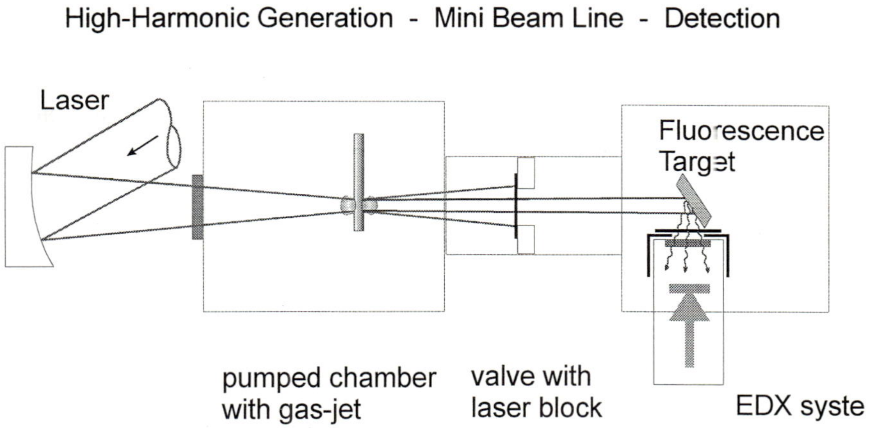

Fig. 1. Experimental setup.

3. Results and Discussion

Because of the low fluorescence yield for low-Z elements, e.g. 0.17% in boron, a strong background caused by coherent scattering of the exciting radiation was superimposed on the fluorescence photons. We managed to significantly suppress undesirable background with the above described excitation geometry and by using p-polarized XUV light. Residual scattering background was eliminated by performing reference measurements [9], yielding the pure fluorescence signal from the boron and carbon sample as shown in Fig. 2 (measurement time: 1500 s) and Fig. 3 (measurement time: 5000 s), respectively. Gaussian fits (solid lines)

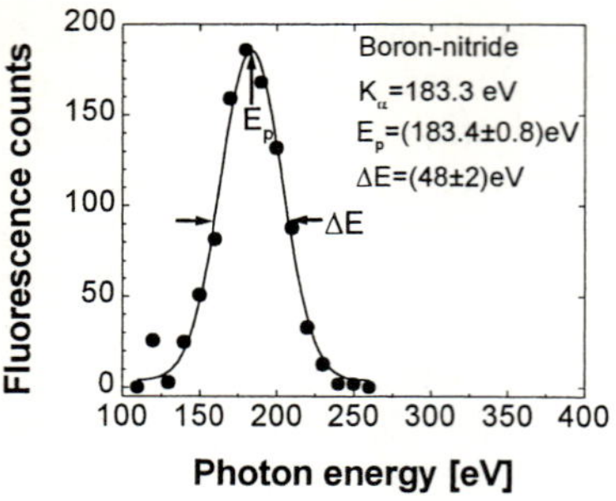

Fig. 2. Fluorescence signal from boron-nitride.

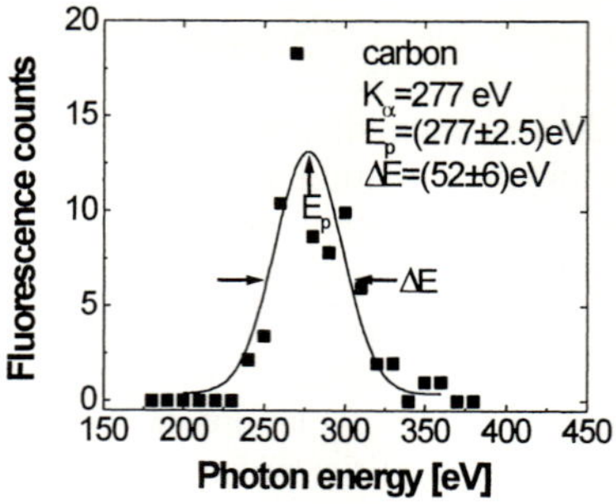

Fig. 3. Fluorescence signal from carbon.

resulted in fluorescence peaks (E_p) close to the predicted values (K) for boron and carbon.

4. Conclusions

We demonstrated the feasibility of detecting fluorescence signals from light elements excited with a few-femtosecond laser-driven XUV source. Our experiments demonstrate the potential of tracing molecular dynamics and inner-shell relaxation processes with few-cycle-driven X-ray harmonics on never-before-accessed time scales.

Acknowledgements. This work has been sponsored by the Austrian Science Fund, Grant Y44-PHY and within the framework of the Special Research Program F016. C.S. acknowledges financial support from the Austrian Academy of Science (APART fellowship).

References

1 Spielmann *et al.*, Science **278**, 661 (1997).
2 Z. Chang *et al.*, Phys. Rev. Lett. **79**, 2967 (1997).
3 M. Schnürer *et al.*, Phys. Rev. Lett. **80**, 3236 (1998).
4 A. Rundquist *et al.*, Science **280**, 1412 (1998).
5 E. Constant *et al.*, Phys. Rev. Lett. **82**, 1422 (1999).
6 M. Schnürer *et al.*, Phys. Rev. Lett. **83**, 722 (1999).
7 See e.g. T. Brabec and F. Krausz, Rev. Mod. Phys. **72**, 545 (2000).
8 S. Sartania *et al.*, Opt. Lett. **22**, 1562 (1997).
9 M. Schnürer *et al.*, "Femtoseocnd X-ray fluorescence," Phys. Rev. Lett. (to be published).

Emission of highly–charged ions of heavy elements from a picosecond laser–produced plasma

J. Badziak, J. Makowski, P. Parys, J. Wolowski, E. Woryna, A.B. Vankov

Institute of Plasma Physics and Laser Microfusion, 23 Hery St., P.O. Box 49, 00-908 Warsaw, Poland
E-mail: badziak@ifpilm.waw.pl, fax: +48 (022) 666 83 72

Abstract. The results of investigation of ion emission from picosecond laser-produced plasmas of various atomic number are presented. The emission of Ta^{38+}, Au^{33+} as well as Al^{13+} ions is demonstrated. Some characteristics of the picosecond plasma are also determined.

Ion streams emitted from a laser–produced plasma are a useful source of information on properties and parameters of the plasma [1] as well as they may be found interesting for some practical application, particularly for ion implantation technology [2] and heavy ion accelerators [3]. In these applications, and especially for heavy ion accelerators highly stripped ions are needed. A possibility of producing intense streams of highly–charged heavy ions (z_{max} ~ 50 for Ta, Au, Pb in a far expansion zone) has been demonstrated recently with the use of high–energy (tens of joules), high–intensity (~10^{15} W/cm^2) subnanosecond laser pulses irradiated a solid target [4, 5]. Pico– and femtosecond lasers may also be potential drivers for laser ion sources. Unlike large nanosecond laser installations, ultrashort–pulse laser make it possible to obtain high radiation intensities and, as a result, high plasma temperatures at low energies of a laser pulse (< 1 J). This in turn, enables the laser to work with a considerable repetition rate (much higher than attainable for high-energy nanosecond lasers), which is indispensable for the practical applications of laser–driven ion sources. On the other hand, a short time of the target ionisation and rapid cooling of ultrashort–pulse–produced plasma, are the factors making the attainment of considerable amounts of highly–stripped ions at a long distance from a target difficult.

Properties of ion streams emitted from a picosecond or femtosecond laser-produced plasma were investigated in few experiments carried out with solid targets of low (Z < 15) [6 – 9] or medium (Z = 29, Cu) [10] atomic number. Maximum charge state z_{max} = 13 (for Cu) was registered in a far expansion zone [10]. Our paper presents the results of investigations of ion streams emitted from solid targets of various Z–number (up to Z = 79) irradiated with a high–intensity 1–ps laser pulse. The emission of highly–charged heavy ions (Ta^{38+}, Au^{33+}) is demonstrated for the first time for a plasma produced with an ultrashort (≤ 1 ps) light pulse.

In the experiment, terawatt chirped-pulse-amplification Nd:glass laser [11], generating 1–ps pulse of short-time scale (<1 ns) contrast ratio ≥$5\cdot 10^3$, was used.

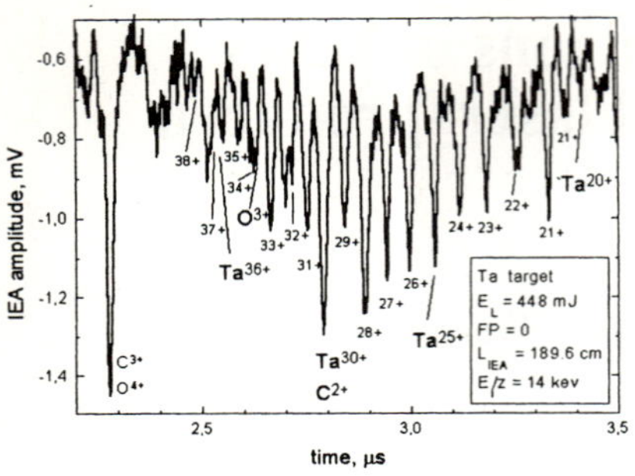

Fig. 1. Ion charge spectrum of Ta ions measured by the electrostatic ion-energy analyser

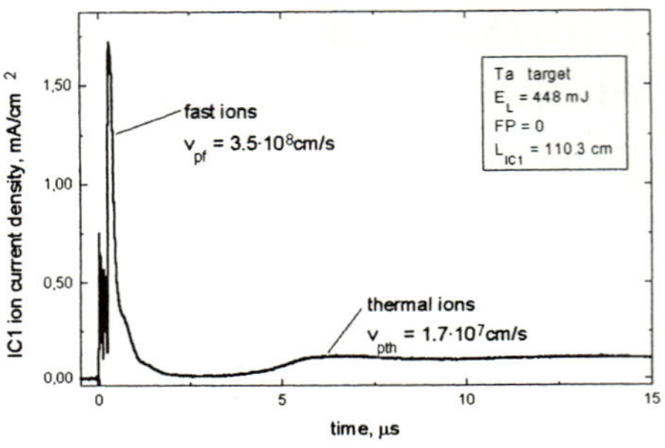

Fig. 2. Ion collector signal of Ta plasma registered in the direction normal to the target surface

The linearly polarised laser beam was focused onto flat polished targets with the use of a parabolic mirror of the focal length f = 27 cm and with a hole in the mirror center. The laser beam's axis was perpendicular to the target surface. The measurements were based on the time-of-flight method [1] and were performed with the use of ion collectors (ICs) and electrostatic ion-energy analyser (IEA).

The IEA and one of the ion collectors (IC1) registered the ion current through the hole in the mirror in the direction normal to the target surface. The paths of

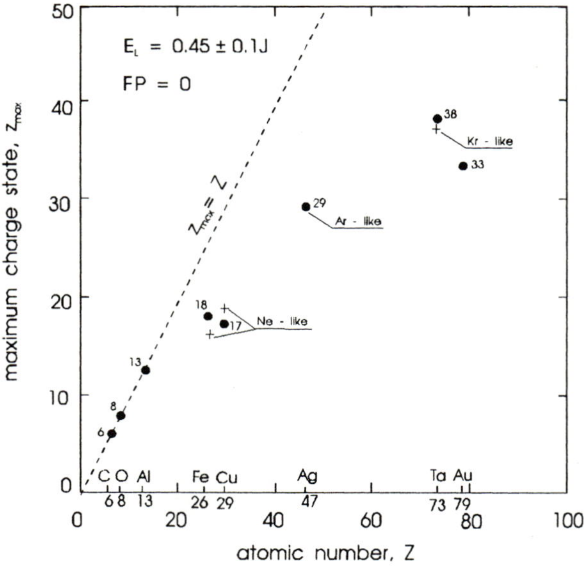

Fig. 3. Maximum charge state of ions measured by the IEA as a function of Z-number of a target.

flight from the target to the IEA and the IC1 were 189.6 cm and 110.3 cm, respectively. Most of the measurements were performed with a laser pulse energy $E_L \approx 0.45 \pm 0.1$ J and light intensity on the target near 5 x 10^{16} W/cm^2.

Fig. 1 presents a part of ion charge spectrum (measured by the IEA) demonstrating the existence of highly–charged Ta ions. Ta^{38+} ions visible in the spectrum have the energy of 530 keV for presented case (registration of the spectrum was performed at E_i/z = 14 keV). The spectrum also contains the peaks of ions of light elements (H, C, O) originating from contaminates (H^+ ions are not visible in the part of the spectrum shown in the figure).

Fig. 2 illustrates a typical structure of the signal from the ion collector registering the ion current in the direction normal to the target surface. Two ion groups can be seen: a thermal and fast ones. The thermal ions propagate with relatively slow velocity of the value v_{pth} = 1.7x10^7cm/s in the peak of the distribution. For this velocity the energies of corresponding ions are as follows: $E^H{}_{pth}$= 0.15 keV, $E^C{}_{pth}$= 1.8 keV, $E^O{}_{pth}$= 2.4 keV, $E^{Ta}{}_{pth}$= 27 keV. The velocity of fast ions reaches the value v_{pf} = 3.5 x 10^8cm/s in the peak. The energies of ions corresponding to this velocity are equal, respectively: $E^H{}_{pf}$ = 63 keV, $E^C{}_{pf}$= 760 keV, $E^O{}_{pf}$=1.02 MeV, $E^{Ta}{}_{pf}$= 11.5 MeV. However, it should be noted that fast ions group is probably dominated by light ions from contaminants, so it is not clear how many Ta ions propagate with such high energies. This issue needs a more detailed study.

The diagram showing maximum charge state of ions measured by the IEA as a function of Z–number of a target, is presented in Fig.3. Full circles correspond to the results of the measurements and crosses indicate position of Ne–like ions (for Fe and Cu) and Kr–like ions (for Ta) in the diagram. Fully stripped ions up to Z = 13 and highly-charged ions for heavier elements are visible.

In conclusion, the results of investigation of ion emission from solid targets of low, medium and high Z–number, irradiated with intense picosecond laser pulse are presented. The emission of fully stripped Al ions, Ar–like Ag ions and Kr–like Ta ions is demonstrated. Some properties of ion streams emitted from a picosecond laser-produced plasma are determined.

References:

1 E. Woryna, P. Parys, J. Wolowski et al., *Laser Part. Beams*, **14**, 293, 1996, and references therein.
2 F.P. Boody, R. Hopfl, and H. Hora, *Laser Part. Beams*, **14**, 443, 1996.
3 H. Haseroth and C.E. Hill, *Rev. Sci. Instrum.*, **67**, 1328, 1996.
4 E. Woryna, P. Parys, J. Wolowski et al., *Appl. Phys. Lett.*, **69**, 1547, 1996.
5 L. Laska, J. Krasa, K. Masek et al., *Rev. Sci. Instrum.*, **69**, 1072, 1998.
6 D.D. Meyerhofer, H. Chen, J.A. Delletrez et al., *Phys. Fluids,* **B5**, 2583, 1993.
7 A.P. Fews, P.A. Norreys, F.N. Beg et al., *Phys. Rev. Lett.*, **73**, 1801, 1994.
8 G. Guethlein, M.E. Ford, and D. Price, *Phys. Rev. Lett.*, **77**, 1055, 1996.
9 P.A. VanRompay, M. Nantel, and P.P. Pronko, *Appl. Surf. Sci.*, **127**, 1023, 1998.
10 J. Badziak, A.A. Kozlov, J. Makowski et al., *Laser Part. Beams*, **47**, 323, 1999.
11 J. Badziak, S.A. Chizhov, A.A. Kozlov et al., *Opt. Commun.*, **134**, 495, 1997.

Evanescent-Wave Acceleration of Ultrashort Electron Pulses

Justyna Zawadzka, Dino A. Jaroszynski, John J. Carey, Klaas Wynne

University of Strathclyde, Dept. of Physics and Applied Physics, Glasgow G4 0NG, UK
URL: http://dutch.phys.strath.ac.uk/

Abstract. High-power femtosecond laser pulses have been used to excite surface plasmons in metal films. Non-linear excitation results in the emission of electron bunches through multi-photon excitation at low power and laser-induced field emission at high power. The electron beam is directional and perpendicular to the prism surface. A time-of-flight set-up has been used to measure the kinetic-energy distribution of the photoelectrons, which is found to extend to energies as high as 0.4 keV. Calculations show that these high energies are due to acceleration in an evanescent field extending from the metal film out into the vacuum. The results suggest that femtosecond electron pulses with relativistic energies can be generated using stronger laser fields or longer wavelengths.

A number of schemes based on plasma waves and wake fields have been developed to accelerate electrons using high-power lasers with the objective of building the next generation of high-gradient accelerators. [1] Because laser light is a transverse electromagnetic wave, the acceleration is also transverse unless complicated schemes are devised to produce a light gradient field. Here, we will present a new scheme in which electrons are accelerated by an *evanescent* electromagnetic field.

In the experiments, a 1-kHz regenerative amplifier producing 0.8-mJ, 100-150 fs pulses has been used with a centre wavelength of ~800 nm. Approximately 500-Å thick films of gold and silver were vacuum deposited (at 10^{-5} Torr) on the hypotenuse face of BK7 prisms. With a prism mounted on a vacuum chamber at a pressure of 10^{-7} Torr, surface plasmons (SPs) were excited in the metal film in the Kretschmann geometry. [2] Maximum electron emission was observed for TM-polarisation and an internal angle of incidence of $41.3^0/42.4^0$ in silver/gold. No electron emission was detectable for angles away from the SP-resonance angle or for TE-polarised laser light. The laser intensity was adjusted using a half-wave retardation plate and a polariser.

The energy distributions of the electrons have been calculated from time-of-flight (TOF) data, which have been recorded with a multi-channel array plate (MCP) placed at a distance of 20 cm. In experiments measuring the average photocurrent, lock-in detection has been used to measure the current collected by copper collector plates with the potential difference between the prism and the collector set between zero and a few tens of volts.

Solutions for the Maxwell equations for TM-polarised light at the glass/metal/vacuum interfaces show that due to SP excitation, the electromagnetic field on the metal surface is enhanced with respect to the incoming laser beam by a factor of 8 to 20 depending on the metal and prism material used. The samples have been examined using an AFM and features approximately 10 nm high and

20 nm wide were found suggesting that the overall field enhancement might be magnified by the localisation of SPs. [3] The resulting field enhancement is strong enough to cause non-linear excitation of electrons. Because of the boundary conditions, the SP electric field must be perpendicular to the film and its wave-vector in the plane of the film. When SPs are formed, they will acquire momentum directed in the plane of the film.

Several experiments have been carried out to determine the spatial distribution of the emitted electron beam. These show that in the optical plane of incidence, the majority of photoelectrons are emitted perpendicularly to the metal surface whereas the out-of-plane distribution is random. The inset of FIG 1 shows that at zero bias voltage between prism and collector, the current increases approximately as the fourth power of the incident laser intensity before saturating at the highest intensities. When a decelerating bias voltage is applied to the metal layer, the slope of curves increases up to ninth order. FIG 1 shows the measured kinetic-energy distribution of photoelectrons emitted from a silver film. For the lowest power density it is found that the distribution has a width of ~0.2 eV. With increasing laser power density, the peak broadens with a maximum at about 0.5 eV and there appears to be a cut-off energy, which is proportional to the incident laser power and is about 50 eV above the work function of the metal (corresponding to 33 photons) for 16.1 GW/cm^2. At the highest laser intensities used, self-phase modulation in the prism material caused white-light continuum generation and laser ablation resulted in degradation of the metal film.

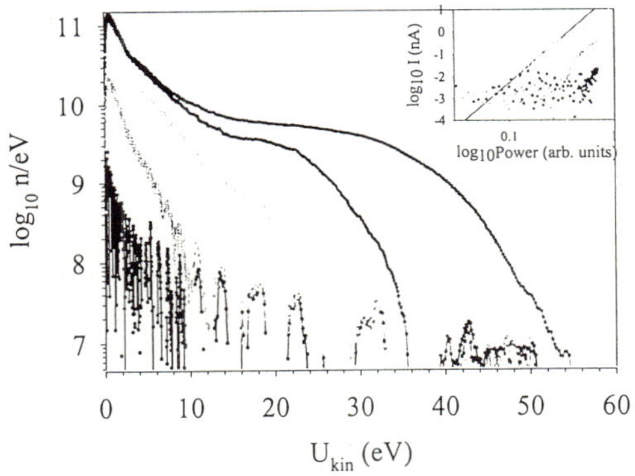

FIG 1. Kinetic-energy distribution of photoelectrons from a silver film measured as a function of incident intensity. The incident power densities varied from 0.5 to 12.5 GW/cm^2. The inset shows the dependence of the measured photocurrent from gold on incident laser power. The data were measured with a DC bias between the prism and the detector of 0, -10 and -30 V, and with incident power density ranging from ~5 to ~50 GW/cm^2. The lines show fourth, seventh and ninth-order power dependencies.

The high kinetic energies observed in the experiments (at peak current densities of about 1-100 A/cm^2) are not consistent with an origin in space-charge effects. The 800-nm laser impinging on the metal film in the Kretschmann geometry will produce an evanescent field in the vacuum. Using our laser parameters one calculates a field strength of more than ~3 10^{11} V/m including the field-enhancement effects of SPs.

If the field strength on the prism surface is $E_0(t)$, the field as a function of position z away from the surface is $E(z,t)=E_0(t)\cos(\omega t)\exp(-z/\lambda)$ where λ is the evanescent-wave decay length calculated to be ~1 μm. The Lorentz force on the electrons results in a non-linear equation of motion, which may be converted to an equation for motion on a ponderomotive potential:

$$U(z)=U_{final}\exp(-2z/\lambda),\ U_{final}=e^2E_0^{\ 2}/\left(4m_e\omega^2\right) \qquad . \qquad (1)$$

The ponderomotive potential can be thought of as being caused by an effective electric field, which on the surface has magnitude $E_{effective}=eE_0/(2\lambda m\omega)$. At the highest laser intensities used, one calculates an effective surface field strength of $4\ 10^9$ V/m.

Outside the metal film, the potential falls off exponentially to $\varphi-U_{final}$ (where ϕ is the work function of the metal), which ranges from 0 to ~2 keV in our experiments. At the high intensities ($U_{final}\cong 2keV$), electrons may tunnel through the potential in a laser-induced field emission process. [3] At lower intensities ($U_{final}\cong 100eV$), the tunnel barrier at the Fermi level is thickened (~20 nm) and field emission stops. Absorption of one photon at 1.5 eV increases the probability of field emission by 10^7. At the lowest intensities ($U_{final}<\varphi$), field emission is strictly impossible and the multi-photon photoelectric effect dominates. In practice, the laser pulse is bell-shaped and all three regimes of photoemission may occur for each laser pulse.

When an electron is ejected from the metal layer, the evanescent optical field pushes it away from the surface and it would gain energy U_{final} if the field remained on infinitely long. Numerical simulations were performed using Eq. (1) for a square laser pulse with a 100 fs duration. Solutions reproduce the observed cut-off energy. For low laser intensity, the cut-off energy scales quadratically with the laser intensity. If the laser intensity or the pulse duration is increased, the kinetic energy will saturate and the final kinetic energy will scale as U_{final}.

The experiments and theory presented here show that evanescent fields produced by amplified femtosecond laser pulses can be used very effectively to accelerate electrons. The relatively long pulse and short wavelength produced by our laser system limited the maximum observed kinetic energy to ~0.4 keV. However, little technical advancement would be required to be able to accelerate electrons to relativistic energies. Such extremely short electron bunches would be of immense value to, for example, time resolved electron diffraction studies. [4]

Acknowledgements. We gratefully acknowledge financial support from the Royal Society and EPSRC, and technical support from Andrew Hughes and Sebastian Lesniewski. JZ gratefully acknowledges financial support from an Overseas Research Students Award Scheme.

References

[1] C. Gahn, et al., Phys. Rev. Lett. **83**, 4772 (1999).
[2] T. Tsang, T. Srinivasan-Rao, and J. Fischer, Phys. Rev. B **43**, 8870 (1991).
[3] C. A. Brau, Nucl. Instrum. Meth. Phys. Res. A **407**, 1 (1998).
[4] J. Cao, H. Ihee, and A. H. Zewail, Proc. Natl. Acad. Sci. USA **96**, 338 (1999).

Pulse Compression in Plasma: Generation of Femtosecond Pulses Without CPA

G. Shvets[1], N.J. Fisch[1], A. Pukhov[2], and J. Meyer-ter-Vehn[2]

[1] Princeton Plasma Physics Laboratory, Princeton NJ 08543, USA
[2] Max-Planck-Institut für Quantenoptik, D-85748 Garching, Germany

Abstract. Laser pulses can be efficiently compressed to femto-second duration when a smaller-frequency short pulse collides with high frequency long pulse in a tenuous plasma, absorbing most of its energy. The mechanism of the short pulse amplification is the nonlinear superradiance.

Chirped Pulse Amplification [1] (CPA) is the standard approach to generating high-intensity ultra-short laser pulses. Among the limitations of CPA are (i) limited pulse energy due to the thermal damage to the compression gratings, and (ii) requirement for an amplifying medium which is both broad-band *and* has a high saturation fluence. In this paper we introduce a novel approach to pulse compression which does not require pulse stretching/compression. It involves the parametric conversion in the plasma of a long pumping beam (PB) into an ultra-short *counter-propagating* laser pulse [2,3]. The advantages are that (i) there is no thermal damage threshold – fresh plasma can be used for each shot, and (ii) the pumping beam can be produced in a high-fluence narrow-bandwidth medium (e.g. Nd:Yag) while the short seed pulse – in a low-fluence broad-bandwidth medium (e.g. Ti:Sapph).

The counter-propagating geometry is chosen for two reasons. First, it enables the amplified pulse (AP) to sweep through the entire PB, absorbing most of its energy and averaging over all the imperfections of the pumping signal. Second, the cold electron plasma exhibits remarkable properties as a parametric medium: its χ_3 coefficient has a very strong dependence on the wavenumber difference Δk between the two interacting laser pulses. In particular, for counter and co-propagating lasers ($\uparrow\downarrow$ and $\uparrow\uparrow$, respectively) the ratio of the χ_3 coefficients $\chi_3(\uparrow\downarrow)/\chi_3(\uparrow\uparrow) \approx 4\omega_0^2/\omega_p^2$ [4], where ω_0 and ω_1 are the frequencies of the AP and the PB, $\omega_p = (4\pi e^2 n_0/m)^{1/2}$ is the plasma frequency, n_0, $-e$, and m are the plasma density, electron charge, and mass, respectively. In a tenuos plasma this ratio can be several hundred. This distinguishes parametric amplification in cold electron plasma from that in gasses, liquids, and fibers. Therefore, a short pulse can be rapidly amplified by *backscattering* a counter-propagating pump without suffering from forward-propagating instabilities (such as filamentation) [3]. Physically, the electron density perturbation is ponderomotively driven by the interference between the PB and AP. This density perturbation then serves as a grating which scatters the pump.

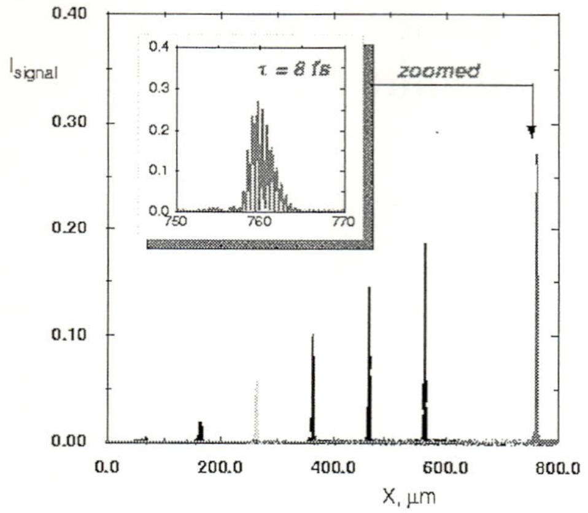

Fig. 1. Amplification of 10 fs pulse with initial intensity $I_0 = 3.5 \times 10^{15}$ W/cm^2 by a counter-propagating pump with the same intensity $I_1 = I_0$ inside the plasma with density $n_0 = 1.5 \times 10^{19}$ cm^{-3}. Amplification 60× after 700 μm

There are two regimes of pulse compression in plasma. Accessing the first (Compton) regime requires an ultra-short $\tau_L \sim 2\omega_p^{-1}$ and sufficiently intense initial pulse: $4\omega_0^2 a_0 a_1 \geq \omega_p^2$, where $a_{0,1} = eA_{0,1}/mc^2$ are the normalized vector potentials of the AB and PB, respectively. When this condition is satisfied, electron motion is determined by the ponderomotive force and not by the space-charge electric field of the plasma wave. The resonance condition for the plasma wave excitation $\Delta\omega = -\omega_p$ doesn't need to be satisfied precisely, making the amplification process robust to the plasma inhomogeneity.

Numerical simulation of the pulse amplification in this regime using particle-in-cell (PIC) numerical code is shown in Fig. 1, where the amplification of a short (initially $\tau_L = 10fs$) $\lambda_0 = 1\mu$m laser pulse is demonstrated as it propagates through 700μm of the plasma while interacting with a long counter-propagating pulse whose frequency is 3% higher. Simulation parameters are given in the caption. As Fig. 1 indicates, the short pulse intensity is amplified by a factor 60 while its duration shrinks to about 8 fs. The energy conversion efficiency from the long to short pulse is 40% in this example.

Superradiant amplification is a strongly nonlinear process. The dependence of the short pulse intensity and duration on the propagation distance z can be qualitatively estimated as

$$|a_0|^2 \sim \frac{|a_1|^2 \omega_p^4 z^2}{16\omega_0^2 c^2} \qquad \tau_L \approx \frac{2}{\omega_p a_1}\left(\frac{c}{\omega_0 z}\right)^{1/2} \tag{1}$$

Quadratic dependence of the short pulse intensity on the propagation distance (and, therefore, on the number of electrons $\mathcal{N}$ encountered by the pulse) gives this amplification process its name [5]. The dependence of τ_L has a somewhat unusual scaling $1/\sqrt{\mathcal{N}}$. This is because the pulse narrowing is caused by the electron recoil and not by the pump depletion. In particular, the characteristic bounce frequency of a plasma electron inside the ponderomotive potential created by the interference of the pump and the pulse is given by $\omega_B = 2\omega_0\sqrt{a_0 a_1}$. In the Compton regime frequency replaces the Rabi fre-

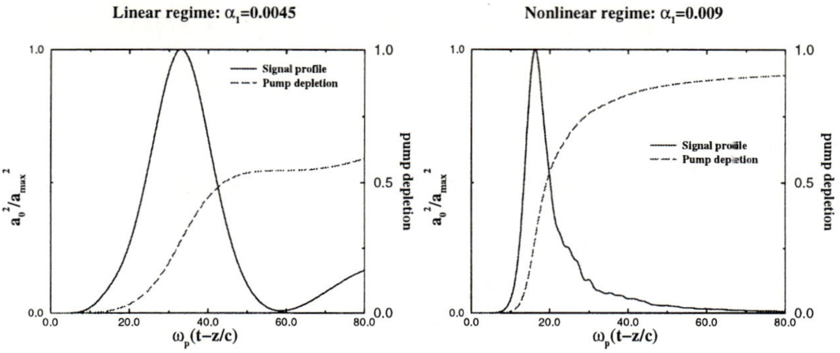

Fig. 2. Normalized intensity profiles for the amplified pulse for two pump amplitudes: below (left) and slightly above (right) the wavebreaking amplitude $a_{\rm wb} = 0.008$. In this example $n_0 = 10^{19}$ cm^{-3} and the amplification distance $L = 4$ mm

quency which, in atomic systems, scales $\sim a_0$. In the absense of significant pump depletion pulse amplification is described by the self-similar solution which maintains $\omega_B \tau_L \approx \pi$ throughout the amplification process. In the second (Raman) regime the motion of the plasma electrons remains linear, and the dominant nonlinearity is the pump depletion. This regime is described by the standard set of equations for three-wave interaction. Frequency detuning between the pump and the pulse has to be precisely equal to the plasma frequency: $\omega_1 - \omega_0 = \omega_p$, imposing stringent requirements on plasma homogeneity. The important figure of merit that separates the two regimes is the ratio of the pump amplitude a_1 to the critical $a_{\rm wb} = 0.25(\omega_p/\omega_0)^{3/2}$: for $a_1 > a_{\rm wb}$ plasma wave breaks before the pump is depleted, and Compton regime is accessed. In the opposite low-intensity limit of $a_1 < a_{\rm wb}$, the three-wave Raman description is appropriate.

Raman regime is characterized by the "Burnham-Chiao ringing" [6]: periodic energy exchange between the pulse and the pump due to the long-lived coherent plasma wave. This ringing can be suppressed by operating near the wavebreaking limit $a_1 \approx a_{\rm wb}$. In this case the plasma wave breaks and thermalizes near the intensity maximum of the pulse. The suppression of the ringing is demonstrated in Fig. 3: only 50% of the pump energy flows into the leading spike in the linear case, and about 80% in the nonlinear case.

References

1. D. Strickland and G. Mourou, Opt. Commun. **56**, 219 (1985).
2. G. Shvets, N.J. Fisch, A. Pukhov, and J. Meyer-ter-Vehn, Phys. Rev. Lett. **81,** 4879 (1998).
3. V.M. Malkin, G. Shvets, and N.J. Fisch, Phys. Rev. Lett. **82,** 4448 (1999).
4. G. Shvets and A. Pukhov, Phys. Rev. E **59,** 1033 (1999).
5. R. Dicke, Phys. Rev. **93,** 99 (1954).
6. D.C. Burnham and R.Y. Chiao, Phys. Rev. **188,** 667 (1969).

Dynamic Stark Effect in Rydberg NO Induced by Intense Laser Pulses

R.B. López-Martens, T.W. Schmidt and G. Roberts

University of Cambridge, Department of Chemistry, Lensfield Road, Cambridge CB2 1EW, England
E-mail: rl212,ts237,gr17@cam.ac.uk

Abstract. The dynamic Stark effect in Rydberg NO $A^2\Sigma^+$ induced by ultrafast laser pulses is quantitatively determined at different laser intensities: dynamic Stark shifts of 0.36 eV are reported and assessed in relation to the electron ponderomotive energy.

1. Introduction

The dynamic distortion of atomic and molecular eigenstates by intense light is of concern in the application of femtosecond laser pulses to problems in intense laser physics. In general, the polarizability induced by an applied optical field shifts all atomic and molecular energy levels to some degree, either through Thomson scattering or the AC (dynamic) Stark effect [1]. For a free electron exposed to an oscillating electric field $E = E_0 cos(\omega t)$ at frequency ω, the energy $U_p = e^2E^2/4m\omega^2$ gained at the expense of the field is equivalent to the classical, time-averaged ponderomotive energy [2]. Electronic states of atoms and molecules that lie close to the ionization potential are Stark-shifted relative to the ground state by an amount approximately equal to U_p when irradiated by laser light. States located at energies intermediate between the ground state and ionization potential are Stark shifted by amounts smaller than U_p, the magnitude and sign of which depends upon the details of their couplings to neighboring levels induced by the optical field. In this report, we discuss a quantitative determination of the AC Stark effect in the $A^2\Sigma^+$ Rydberg state of NO [3].

2. Experiment

Measurements were made by monitoring the intensity of NO $A^2\Sigma^+ \rightarrow X^2\Pi_r$ fluorescence, induced by frequency-doubled sub-100 femtosecond pulses from an amplified Ti:Sapphire laser focussed into gaseous NO, as a function of peak incident intensity up to $I_p = 6 \times 10^{13}$ W cm^{-2} and at photon energies scanned from non-resonantly below to non-resonantly above the NO $A^2\Sigma^+$ $n' = 2 \leftarrow X^2\Pi_r$ $n'' = 0$ two-photon transition in the weak-field limit (λ = 395-415 nm). Temporal characterization of the pulses was performed by cross-correlating fundamental and second-harmonic beams in a BBO crystal. Spatial characterization was performed using a CCD camera to record images of the transverse profile and from knowledge of the confocal parameter along the propagation direction

3. Results and Discussion

Figure 1 shows graphs of the intensity S_F of $A^2\Sigma^+$ n′ = 2 → $X^2\Pi_r$ n″ = 0 fluorescence at λ = 205 nm plotted as a function of I_p at three different wavelengths, with experimental points displayed as dark circles. The result for the zero-field two-photon resonant wavelength of 410 nm, figure 1(a), may be interpreted as an AC Stark shift of the $A^2\Sigma^+$ n' = 2 level with I_p such that, at $I_p > 1 \times 10^{13}$ W cm^{-2}, the two-photon energy becomes non-resonant and further increases in intensity do not bring about additional population of the n′ = 2 level. At higher photon energies, further increases in I_p are required before the $A^2\Sigma^+$ n′ = 2 level is tuned into resonance, as shown in figures 1(b) and (c). This interpretation is supported by the dispersed fluorescence spectrum recorded at $I_p > 6 \times 10^{13}$ W cm^{-2} and λ = 400 nm, which exhibits lines due to $A^2\Sigma^+ \rightarrow X^2\Pi_r$ emission from the n′ = 1 and 0 levels [3]. The point of inflexion (POI) of each of the curves of figure 1 lies very near the value at which I_p of the laser pulse has shifted the $A^2\Sigma^+$ n′ = 2 level

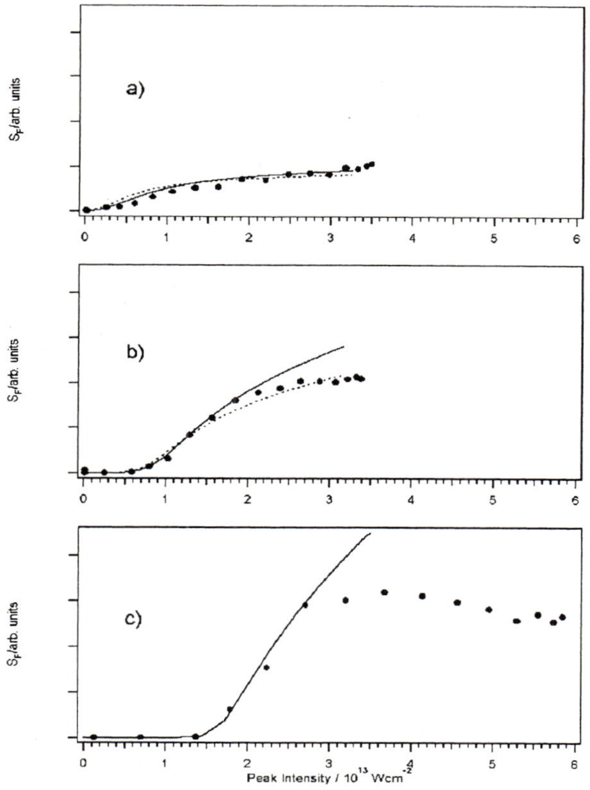

Fig. 1. Graphs of $A^2\Sigma^+$ n′ = 2 → $X^2\Pi_r$ n″ = 0 fluorescence intensity S_F versus peak incident laser intensity at different two-photon $A^2\Sigma^+ \leftarrow X^2\Pi_r$ transition wavelengths: (a) 410 nm; (b) 405 nm; (c) 400 nm.

into resonance with the two-photon energy. (The wavelengths which are resonant or nearly resonant at low intensities are obviously an exception.

The observations of figure 1 can be accounted for quantitatively by a kinetic model which simulates fluorescence $A^2\Sigma^+ \rightarrow X^2\Pi_r$ emitted by NO molecules subject to spatiotemporally varying laser field; details are presented elsewhere [3]. The model yields a final population in the n′ = 2 level of the $A^2\Sigma^+$ state, integrated over spacetime mapped out by the laser pulse, which is taken to be directly proportional to the fluorescence signal S_F. Stark-tuning of the $A^2\Sigma^+$ vibrational levels into resonance is accounted for by rendering the two-photon $A^2\Sigma^+ \leftarrow X^2\Pi_r$ absorption cross section intensity-dependent, whereby the 'resonant' transition energy shifts linearly with intensity. This is described by a parameter χ which scales the magnitude of the AC Stark shift of the 3sσ Rydberg electron as a fraction of U_p, where U_p is itself a function of $I_p(x,y,z;t)$. For each wavelength studies, χ is expected to be the same. Optimum nonlinear least-squares fits to the variation of S_F with I_p, shown as solid lines, were obtained with values of $\chi = 0.26$, 0.32 and 0.40 for figures 1(a), (b) and (c) respectively. Fitting the data for 410 and 405 nm excitation with a common value of $\chi = 0.40$ gives the dashed lines in figures 1(a) and (b), consistent with the best fit to the 400 nm data shown in figure 1(c). The POI predicted by the kinetic model is its most robust feature, being resilient to large artificial changes in two-photon ionization cross sections and laser spectral width.

The observed dependence of S_F on laser intensity could not be reproduced by the kinetic model for $I_p \geq 2.5 \times 10^{13}$ W cm^{-2}, at which point there is complete depletion of ground-state molecules at the center of the laser focal volume. We suspect that resonantly enhanced ionization through high-lying Rydberg states, facilitated by Stark shifting during the leading edge of the intense laser pulses, is at least partially responsible for the breakdown in the kinetic model at $I_p > 2.5 \times 10^{13}$ W cm^{-2}. Fluorescence observed above this intensity (figure 1) most likely originates from molecules located around the periphery of the laser profile, where the electric field is lower and where Stark-induced population of the $A^2\Sigma^+$ n′ = 2 level can still occur.

In summary, the data of figure 1 point to a dynamic Stark effect of 0.4 times Up for the 3sσ Rydberg electron of NO, equivalent to 0.36 eV at an incident intensity of 6×10^{13} W cm^{-2} at 400 nm. It may be conjectured that such a result is a manifestation of the still-dominant Coulombic interaction experienced by an electron in the $A^2\Sigma^+$ state.

Acknowledgements. We thank EPSRC, the Isaac Newton Trust, the Royal Society of London and Churchill College, Cambridge, for support of this work.

References

1 M.H. Mittleman, *Theory of Laser-Atom Interactions*, Plenum, New York, 1982.
2 R.R. Freeman and P.H. Bucksbaum, J. Phys. B **24**, 325 (1991).
3 R.B. López-Martens, T.W. Schmidt and G. Roberts, Phys. Rev. A **62**, 013414 (2000).

Ultrafast transient-grating study of molecules after high intensity excitation

Matthew Comstock, Igor Pastirk and Marcos Dantus

Department of Chemistry and Center for Fundamental Materials Research, Michigan State University, East Lansing MI 48824,
E-mail: dantus@msu.edu

Abstract: We report a method based on transient-grating spectroscopy to assess the degree of alignment and electronic state mixing caused by intense femtosecond laser pulses. Molecular deformations are determined by analysis of the filed-free rotational recurrences.

1. Introduction

High-intensity off-resonance laser fields from ultrafast laser pulses can cause very large electric-field gradients. Molecules experience a large torque along the polarization vector of the field due to the anisotropic polarizability. With long laser pulses the torque is enough to cause adiabatic alignment [1, 2], however, with ultrafast pulses the torque provides an 'instantaneous' kick towards alignment [3]. For high enough fields, electronic state mixing can occur. When the electronic states have different geometry, this process leads to molecular deformation. For our measurements, we use two high-intensity laser pulses followed by a weak probe pulse. The signal corresponds to the diffraction of the probe laser from the transient grating formed by the two intense pulses [4]. The degree of molecular alignment and deformation are measured from changes in the rotational recurrences that are observed after field-free evolution of the molecular sample.

2. Experimental

The experiments were carried out with three laser beams in the forward-box geometry. The first two laser pulses where overlapped in time while the third pulse was delayed by a computer controlled actuator. The intensity of the first two laser pulses was varied between 10^{10} to 10^{13} W/cm^2 using a set of matched neutral density filters, all with the same thickness. The third pulse was maintained at a fixed 10^{10} W/cm^2 intensity. The signal was collected background-free at the phase matching geometry. The sample consisted of a quartz cell containing CS_2. A cold finger in the cell was kept in ice to reduce the vapor pressure to 100 Torr. The 60 fs laser pulses were centered at 620 nm.

3. Results

The full rotational recurrence of CS_2 is expected at 76.5 ps [4, 5]. When the laser pulses are weak, this feature can be observed and an accurate rotational constant can be determined for the linear molecule [4, 5]. In Fig. 1a, we show experimental data for the case when all laser pulses were weak. The simulation of the data with no adjustable parameters fits well the experimental results (see Fig. 1a). When the laser intensity of the first two pulses is increased by an order of magnitude, the full rotational recurrence observed increases in intensity. The data can still be fitted by a room temperature distribution of linear CS_2 molecules (see Fig. 1b).

When the intensity is increased above 10^{12} W/cm^2, the full rotational recurrence increases in intensity and changes shape (see Fig. 1c). Most markedly, the first feature increases in intensity by a factor of two. The features observed at long time delays, >77.5 ps, increase in intensity and new oscillations are observed. This region of the transient is amplified by a factor of five in the figure.

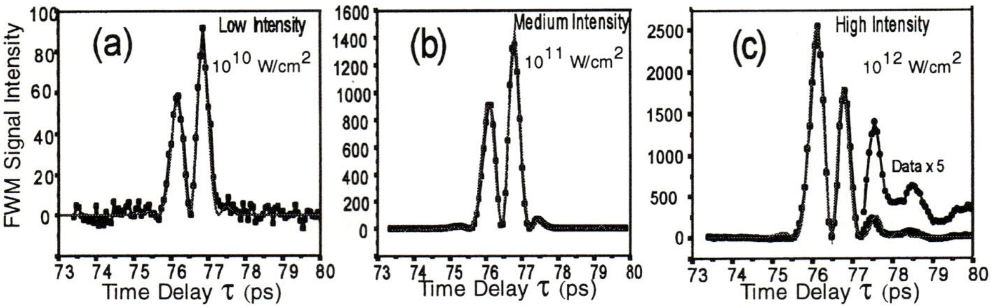

Figure 1. Experimental data and simulations of transient-grating measurement on gas phase CS_2. The first two laser pulses are overlapped in time and their peak intensity is regulated from (a) weak, to (b) medium, to (c) high. The effect of the high-intensity laser field on the molecules is measured after a long time delay by the probe pulse, here we only show the time delay where the first full-rotational recurrence takes place. Note that for weak and medium intensity the data can be fit by a simulation without adjustable parameters. However, for high intensity fields the shape of the recurrence changes. Simulation of the high intensity transient includes bent CS_2 molecules.

Simulation of the data for high intensity excitation shown in Fig. 1c, requires the following modifications caused by the high intensity fields. It is possible that the molecules experience multiple Raman transitions during the first two laser pulses, changing the population of the various rotational levels. Because of centrifugal distortion, the molecules stretch and the rotational revival is delayed in time. Bending can also result from mixing with excited electronic states. Both singlet and triplet excited states are bent for this molecule. The simulation in Fig. 1c is based on a combination of bending and stretching induced by the electric field. A bond angle of about 160° is required for this simulation, and the C-S bond lengths have been extended about ten percent in our model. These parameters are consistent with the $^1B_2$ state.

4. Discussion

We explore the alignment and deformation induced by high-intensity off-resonance laser pulses on gas phase CS_2 molecules. When the laser intensity is ~ 10^{10} W/cm^2, the data can be fit without adjustable parameters. For intermediate intensities, a large increase in the signal is observed, this enhancement results from the initial torque towards alignment that the molecules experience. Beyond 10^{12} W/cm^2, the structure of the rotational recurrence changes. The data could only be simulated in this case by including bent molecules. In conclusion, we consider the ultrafast transient-grating technique to be ideal to measure the changes that molecules experience in the presence of high-intensity laser pulses.

Acknowledgements

This work was supported by the National Science Foundation Grant CHE-9812584. Additional funding comes from a Packard Science and Engineering Fellowship, an Alfred P. Sloan Research Fellowship and a Camille Dreyfus Teacher-Scholar award. We are grateful to Tamar Seideman for stimulating discussions on high-filed molecular alignment and rotational revivals.

References

1 B. Friedrich, and D. Herschbach, Phys. Rev. Letters **74**, 4623, 1995.

2 W. Kim and P. M. Felker, J. Chem. Phys. **104**, 1147, 1996.

3 C. M. Dion , A. Keller, O. Atabek and A. D. Bandrauk, Phys. Rev. A **59**, 1382, 1999.

4 E. J. Brown, Q. Zhang, and M. Dantus, J. Chem. Phys. **110**, 5772, 1999.

5 J. P. Heritage, T. K. Gustafson, and C. H. Lin, Phys. Rev. Letters **34**, 1299, 1975.

Part VII

Ultrafast Dynamics of Atoms, Molecules and Clusters in the Gas Phase

Stepwise Electron Emission from Autoionizing Magnesium Stark States

J.B.M. Warntjes[1], C. Wesdorp [1], F Robicheaux[1,2] and L.D. Noordam[1].

[1]FOM-Institute for Atomic and Molecular Physics (AMOLF), Kruislaan 407, 1098 SJ Amsterdam, The Netherlands. (http://www.amolf.nl/research/femto_physics). E-mail: warntjes@amolf.nl
[2]Department of Physics, Auburn University, Auburn, AL 36849. Tel. 334-844-4366.

Abstract. Magnesium is excited to a doubly excited state that is energetically allowed to autoionize. Electron emission from autoionization is usually exponential. Excitation with a short (ps) pulse, however, results in a stepwise electron emission.

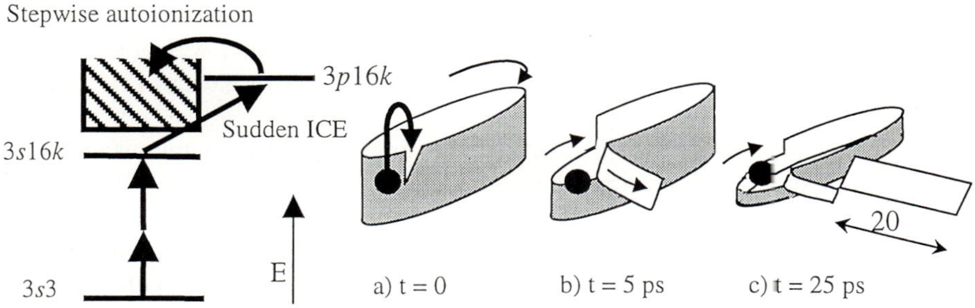

Fig. 1. (left). Potential energy scheme for magnesium. We prepare an initial, high Rydberg state by a two-photon transition from the $3s^2$ ground state. The isolated core is subsequently excited with a short laser pulse, leading to a stepwise autoionization. **(right).** Intuitive picture for stepwise decay of an autoionizing two-electron atom. In the initial state one electron is in a high Rydberg orbit. The electron density of this Rydberg orbit is depicted as the gray ribbon around the black core. At t = 0 a short laser pulse excites the inner electron creating an exit for the outer electron. The autoionization probability (50% here) is constant and the electron emission has a constant intensity during a complete roundtrip period (20 ps in this example). After one roundtrip time there is less electron density to autoionize and the outgoing flux will suddenly drop and remain constant again during the second roundtrip period.

Decay of an autoionizing two-electron atom involves a doubly excited state coupled to a state in which one electron is ejected while the second electron relaxes to a deeper bound state. If one of the excited electrons is in a Rydberg state, only the fraction of the Rydberg wavefunction close to the nucleus can exchange energy with the excited core electron. The emitted electron flux of such an autoionization process is most often *exponential* in nature.The key element for *stepwise* decay is that the second electron is excited with a laser pulse of short duration with respect to the classical period of the spectating Rydberg electron. An intuitive picture of stepwise decay is given in Fig. 1. Most of the initial Rydberg wavefunction is far away from the nucleus. The electron density is depicted as the gray ribbon in Fig. 1a. Sudden isolated core excitation [1] causes a shockwave [2] that travels through the Rydberg wavefunction. Simultaneously the fraction of the initial Rydberg wavefunction that happens to be near the core can directly autoionize which results in electron emission. The probability for the Rydberg electron to autoionize during one collision with the core electron is 50% in Fig. 1. The outgoing electron flux due to autoionization remains constant during one oscillation period since the Rydberg density far from the nucleus is unaffected by the short excitation (Fig. 1b). After one oscillation period the shockwave returns to the core. At that time the all incoming electron density is reduced and the autoionization flux suddenly decreases to $50 \times 50 = 25\%$. This reduced electron emission remains constant for a second Rydberg oscillation period and so on (Fig. 1c). The result, a series of steps in the outgoing electron flux, was predicted in 1991 [2]. Here we report on the first observation of this stepwise decay [3].

In our experiment we investigated the time dependent autoionization of doubly excited magnesium atoms in a strong electric field of typically 1000 V/cm. In a field the angular momentum l of an electron is not conserved and it will oscillate between $l = n\text{-}1$ and $l = 0$ where n is the principle quantum number. A narrow band laser excites one electron towards a stationary $3snk$ Rydberg-Stark state, where n is typically around 16. The radial oscillation time of a $n = 16$ Rydberg electron is much faster (2 ps) than its angular oscillation time (10 ps in a field of 1000 V/cm). Only an electron with low angular momentum penetrates the core region and is able to exchange energy with the isolated core electron leading to autoionization. As a result, the angular oscillation period is the relevant time scale in our experiment.

The remaining isolated core electron is excited with a short (2 ps) laser pulse ($3s16k \rightarrow 3p16k'$). The low angular momentum fraction of the Stark electron can immediately autoionize. In the electric field all angular momenta oscillate up to $l = n\text{-}1$ and back to $l = 0$. Therefore, during an angular oscillation period, the fraction of low angular momentum is constant and so is the emitted electron flux. After one angular oscillation period, however, there is less electron density to autoionize and the electron emission suddenly drops to remain constant again for a second period, etc. The autoionizing electron yield shows a stepwise decay which is detected by a streak camera [4], with picosecond resolution, as shown in figure 2. The observed step-period of about 9 ps is the angular oscillation time of the initial Rydberg-

Stark electron. The drop in emission of about 60% per step is the probability of the electron to autoionize per core passage.

The dotted line in the figure is a full quantum calculation, based on MQDT, which is in good agreement with the experimental results. To our knowledge this is the first match between theory and experiment of ICE in an electric field. Deviation between the curves is partially due to a 'timeblurring' effect in the camera due to the large excess kinetic energy of the autoionizing electrons (4.3 eV) [5].

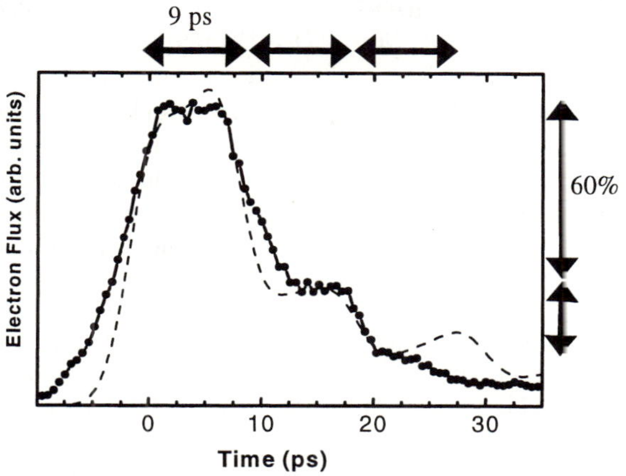

Fig. 2. Electron emission yield after isolated core excitation with a 2 ps pulse, starting from an initial, stationary Stark state. The emission has a stepwise appearance with a period of about 9 ps, corresponding to the oscillation time of the Rydberg-Stark electron. The decrease in stepheight is 60%, corresponding to the probability of electron emission upon core passage. The dotted line is the MQDT calculation.

The work described in this paper is part of the research program of the FOM (Foundation for Fundamental Research on Matter) and was made possible by the financial support from NWO (Netherlands Organization for the Advancement of Research). FR. is also supported by the NSF.

References

1 W.E. Cooke *et al*, Phys. Rev. Lett. **41**, 178 (1978).

2 Wang and W.E. Cooke, Phys. Rev. Lett. **67**, 976 (1991); Phys. Rev. A, **46**, 4347 (1991); Phys. Rev. A, **46**, R2201 (1991).

3 The time dependent ICE spectra are published in J.B.M. Warntjes, C. Wesdorp, F. Robicheaux and L.D. Noordam, Phys. Rev. Lett. **83** 512 (1999).

4 G. M. Lankhuijzen and L. D. Noordam; Opt. Commun. **129**, 361 (1996); Phys. Rev. Lett. **76**, 1784 (1996).

5 Recently, narrowband ICE spectra were measured and fitted with high accuracy. The manuscript has been accepted for publication by Phys. Rev. A.

Centrifugal Dissociation of a Molecule Using the Optical Centrifuge

D. M. Villeneuve[1], S. A. Aseyev[1], P. Dietrich[2], M. Spanner[1], M. Yu. Ivanov[1] and P. B. Corkum[1]

[1] National Research Council of Canada, 100 Sussex Drive, Ottawa, ON K1A 0R6, Canada. E-mail: david.villeneuve@nrc.ca
[2] Freie Universität Berlin, Arnimallee 14, D-14195 Berlin, Germany.

Abstract. We will show that we are able to use a molecule's nonresonant polarizability to cause it to rotate in a controlled manner. We can accelerate the rate of rotation from 0 to 6 THz in 50 ps, a rate so high that chlorine molecules dissociate due to the centrifugal force. The ensemble of molecules spins coherently, making possible a time-varying nonisotropic optical medium. This experiment is based on the theoretical work described by Karczmarek et al.

It has recently been predicted that diatomic molecules can be rotationally accelerated from rest up to the point where the molecular bond is torn apart by the centrifugal force [1], in less than one ground state rotational period. In this process, a molecule has to undergo several hundred successive Raman transitions up the ladder of rotational states, each in a fraction of a picosecond and with nearly 100% probability (see Fig. 1). To achieve such high efficiency, the laser intensity must be high enough to force the transition, and the frequency content of the laser pulse must constantly change to match the changing Raman transition energy. This can be accomplished using femtosecond laser technology which gives precise control over the amplitude, phase, frequency, and polarization of the laser radiation.

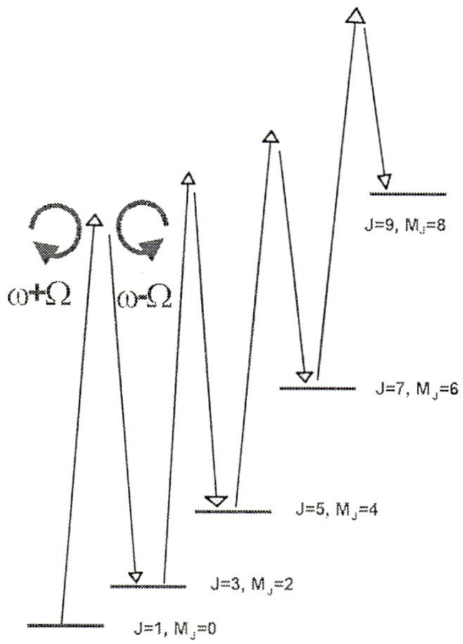

Fig. 1. The optical centrifuge requires over 200 Raman transitions, each with ΔJ=2 to reach a high rotational state.

We will demonstrate that we can rotationally accelerate Cl molecules up to the point where the molecular bond is broken by the centrifugal force.

We use either a rotationally cold or nearly room-temperature ensemble. In the experiment, angular traps with depths of 50–100 meV, exceeding room temperature, were achieved for Cl with minimal ionization during the 50 ps pulse.

The intention of the experiment is to demonstrate centrifugal dissociation. If instead we want to demonstrate efficient spinning of molecules, then we would choose a lighter molecule that better suits the energy capability of our laser system. Although spinning anisotropic molecules using the nonresonant dipole force is general, Cl is an ideal candidate for a first experiment. Near the equilibrium internuclear separation, single photon absorption is not possible [see Fig. 2(a)] with an 800 nm photon—the frequency region where the most advanced femtosecond sources operate. In this region we have both exquisite control over the optical pulse and access to high intensities and, hence, very deep angular traps.

Figure 2(b) is a plot of the Cl ground state potential for different angular momenta J. As J increases, the bond starts to stretch until it is lost near J~420. Bond breaking requires a rotational frequency of almost 5 THz, well within

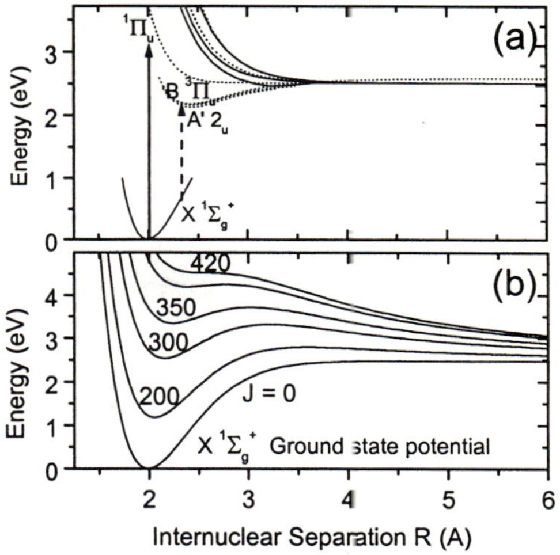

Fig. 2. (a) Lower surfaces of chlorine. A 400 nm photon (shown) can dissociate it. (b) Potential energy surface of the ground state of chlorine with the centrifugal term. When J>420 the bond disappears.

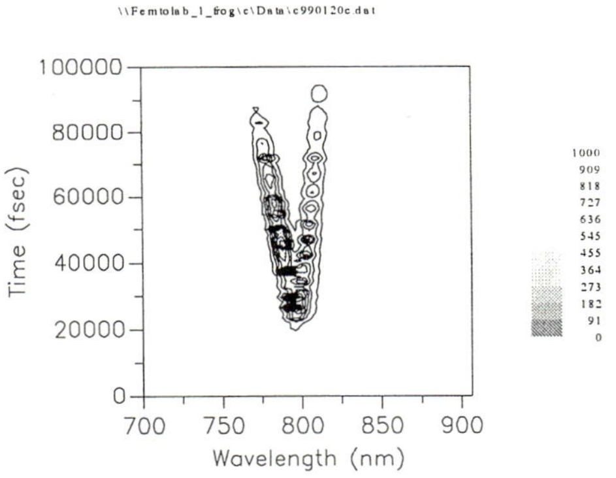

Fig. 3. Frequency-resolved cross-correlations of the two oppositely circularly polarized chirped pulses that form the optical centrifuge.

the bandwidth of our Ti:sapphire laser pulse.

Up to four laser beams were used, all split from a 5 mJ pulse produced by a 10 Hz Ti:sapphire laser system. The FWHM bandwidth of the pulse was over 30 nm with a central wavelength of 810 nm, compressible to 50 fs, and stretched to a duration of 50 ps before amplification; the stretching resulted in a positive frequency chirp.

Two beams were used to make an optical centrifuge [1,2], a pulse with rotating linear polarization that spins the molecules. Such rotation, at frequency Ω, is accomplished by combining two counterrotating circularly polarized fields with frequencies ω_L-Ω and ω_L+Ω. To increase Ω with time we chirp the two circularly polarized pulses in opposite frequency directions.

To accomplish this in practice, the stretched pulse was incident onto a stretcher composed of three 1200 line/mm gold gratings and three 150 mm focal length achromatic doublet lenses. Half of the spectrum was reflected by a dielectric mirror placed in the Fourier plane of one grating-lens pair, and the other half was transmitted to the other grating-lens pair. Figure 3 shows the frequency resolved cross-correlation measurement of the optical centrifuge pulse.

An ionizing pulse (800 nm, 15 μJ , 50 fs) came another 300 ps later and served to ionize any atoms or molecules so they could be detected in the time-of-flight system.

Figures 4(a)–4(c) show a portion of the mass spectrum around the arrival time for Cl^{+1}, with only the spinning pulse and the 50 fs

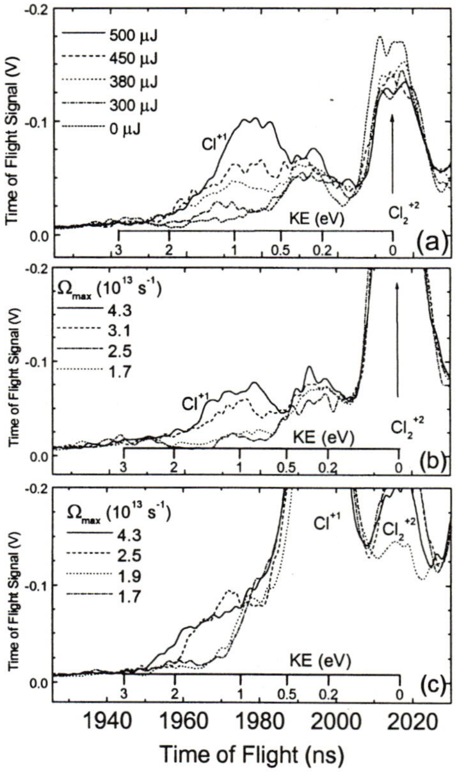

Fig. 4. Time of flight data showing the kinetic energy of chlorine atomic fragments. The atoms are produced by dissociation of the molecule by the optical centrifuge. (a) shows power dependence of 1 eV fragments. (b) shows how the reduction of Ω_{max} inhibits dissociation. (c) shows the dissociation induced by a 400 nm pulse.

ionizing pulse (to ionize the fragments). The most prominent peak on the right of Figs. 4(a)–4(c) is produced by zero kinetic energy Cl_2^{2+}. This ion is quite stable since double ionization of Cl_2 by an 800 nm pulse removes antibonding electrons. We show only the region to the left of the $_{35,35}Cl_2^{2+}$ molecular isotope because on the right of the spectrum the atomic fragment peaks are masked by the $_{35,37}Cl_2^{2+}$ molecular isotope. A kinetic energy scale for the Cl^{+1} atoms is superimposed.

According to Fig. 2(b), rotational dissociation of Cl_2 should yield atomic fragments with kinetic energy of about 1 eV. Figure 4(a) shows that chlorine atom fragments around 1 eV are associated with the spinning pulse. The inset shows that this peak increases nonlinearly with intensity. These results are consistent with chlorine atoms being produced by the rotational dissociation of Cl: as the spinning angular potential well gets deeper, more and more molecules are spun to dissociation. We have obtained similar results without the buffer gas in the jet expansion, i.e., with rotationally hot molecules, although the spinning efficiency was somewhat reduced.

We made several tests to verify that the 1 eV peak in Fig. 4(a) is due to molecules being spun by the centrifuge until they dissociate. (1) When one of the two beams that make up the optical centrifuge was blocked or when the two beams were linearly polarized, no atomic fragments were seen. (2) We could reduce the maximum spinning frequency, Ω_{max} by truncating in frequency one of the two optical centrifuge beams. As expected, the atomic fragments around 1 eV disappeared for $\Omega_{max} < 2.6\times10^{13}$ s^{-1} (4.2 THz), i.e., when Ω_{max} was insufficient to reach $J = 420$ [see Fig. 4(b)]. (3) The rotating angular trap could be turned on at higher initial spinning frequency, Ω_{init} , by delaying one of the two chirped circularly polarized pulses of the optical centrifuge.

More insight can be gained by using an additional 2 ps 400 nm probe pulse, delayed by 200 ps with respect to the centrifuge pulse. According to Fig. 2(a), a single 400 nm photon (shown with the long arrow) can dissociate Cl_2 by exciting it to the dense manifold of electronically excited states. Hence this 400 nm pulse gives us a method to detect spinning molecules by "cutting the bond" and converting the rotational energy into translational.

Figure 4(c) shows the time-of-flight spectrum obtained using all four pulses—pulses for spinning, dissociation, and ionization. Molecules with no rotational energy lead to fragments with kinetic energy of about 0.3 eV, equal to the photon energy of 3.1 minus 2.5 eV dissociation energy for the two atoms. These are seen in the figure as the offscale Cl^{+1} fragments. With the spinning pulse present, fragments with kinetic energies of up to 2 eV were seen, while without the spinning pulse no energetic atoms are seen.

Every measurement that we have performed is consistent with spinning Cl_2 in an optical centrifuge. We have also performed preliminary experiments with O_2 and CS_2. Our observations are consistent with successful spinning of these molecules, although the maximum spinning frequency achieved with our laser was not sufficient for rotational dissociation of the ground state.

These results have important implications for both chemistry and physics. From a gas phase chemistry perspective, we open a new pathway for photochemistry with molecules in extreme angular momentum states. This would be especially interesting for polyatomic molecules, where the centrifugal force modifies the much more complex potential energy surface. Huge changes in the electronic states are possible (see Fig. 2).

From a surface chemistry perspective, we can produce molecules with high J and M_J and direct them towards a surface. The rotational energy of such a molecule can be in the multi-eV range, sufficient to stimulate chemistry on the surface. When J control is combined with molecular focusing, a new experimental pathway for precision etching and surface studies will be opened.

From a physical perspective, this experiment shows that molecules are devices that can be manipulated with precision. All aspects of the spinning pulse are under our control, allowing us to control the time dependent molecular alignment. If the gas density is high enough, spinning molecules can serve as a precisely controlled high-frequency modulator with the modulation synchronizable with femtosecond precision. If the molecules have a dipole moment, they could provide an intense source of tunable multi-THz radiation.

References

[1] J. Karczmarek *et al.,* Phys. Rev. Lett. **82,** 3420 (1999).
[2] D. M. Villeneuve et al., Phys. Rev. Lett. **85**, xx (2000).

State-selective Rotational Coherence Spectroscopy by Two-color Pump-Probe Schemes

Andreas Weichert, Christoph Riehn, and Bernhard Brutschy

Johann Wolfgang Goethe-Universität Frankfurt/M., Institut für Physikalische und Theoretische Chemie, Marie-Curie-Str. 11, D-60439 Frankfurt/M., Germany
E-mail: riehn@chemie.uni-frankfurt.de

Abstract. Rotational constants of *para*-cyclohexylaniline in the electronic ground and excited state were measured in the gas phase by time-resolved Raman spectroscopy and stimulated emission pumping. A new mass-selective scheme by pump-probe photoionization is presented.

1. Introduction

Rotational coherence spectroscopy (RCS) serves as a powerful method to analyze the molecular structures of large molecules and clusters in the gas phase by determination of their rotational constants and alignment of their transition dipole moment [1,2]. In this Boltzmann averaged quantum beat spectroscopy recurrences of a rotational wavepacket, created by a linearly polarized picosecond pump pulse, are detected by a polarization sensitive probe process and retraced to structural parameters of the system under study. For instance, a near-prolate symmetric rotor with a perpendicular transition dipole moment, like *para*-cyclohexylaniline (*p*CHA), can exhibit recurrences of *J*-type at $t_J = n/[2(B+C)]$ and of *K*-type at $t_K = n/[4A-2(B+C)]$, $n = 1,2...$, with A, B, and C being the rotational constants of *p*CHA. The intention of this contribution is to investigate and compare different two-color pump-probe schemes for state-selective RCS. The methods are applied to the large molecule *p*CHA, which was the subject of several studies before [3-5].

2. Experimental

A solid state two-color laser system based on a chirped pulse amplifier (CPA) has been employed. The second harmonic of the CPA output was used to pump two optical parametric generators / optical parametric amplifiers (TOPAS) to achieve wavelength tunability for both the pump and the probe pulse. The output parameters are: wavelength range 250-400 nm, repetition rate 500 Hz, pulse energy 30 μJ, autocorrelation pulse width 3 ps, bandwidth $\approx 10\ cm^{-1}$. The measurements were performed in a vacuum apparatus on a supersonic expansion by laser induced fluorescence or on a molecular beam by photoionization and

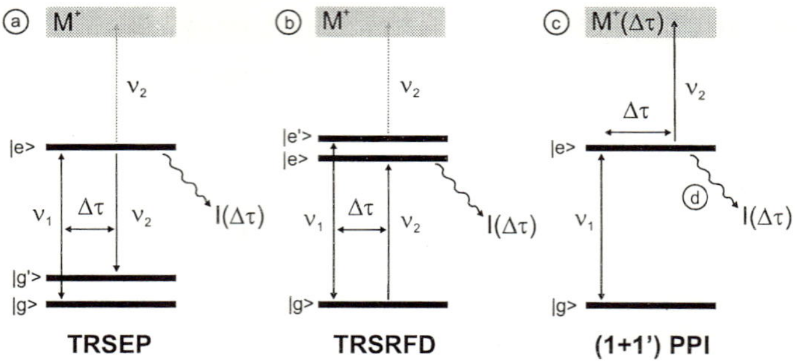

Fig. 1. Excitation schemes for: a) Time-resolved stimulated emission pumping (TRSEP). b) Time-resolved stimulated Raman induced fluorescence depletion (TRSRFD). c) (1 + 1') pump-probe ionization (PPI). d) (1 + 1') time-resolved fluorescence depletion (TRFD).

subsequent time-of-flight mass spectrometry. Further details of the setup are given in the literature [5,6]. The pump-probe excitation schemes for the applied spectroscopic methods are depicted in Fig. 1.

3. Results and Discussion

The majority of RCS experiments has been performed by one-color excitation schemes with fluorescence detection [1]. As a starting point of our investigation the one-color time-resolved fluorescence depletion (TRFD) RCS spectrum of *p*CHA is depicted in Fig. 2.

A splitting of the *K*-type transients was readily analyzed as a contribution of both the electronic ground and the excited state. The particular assignment was achieved by analysis of RCS transients up to 8 ns delay upon consideration of the fluorescence lifetime of *p*CHA (6.5 ns). However, the *J*-type transients in Fig. 2 exhibit no obvious splitting and the few recurrences are disturbed due to

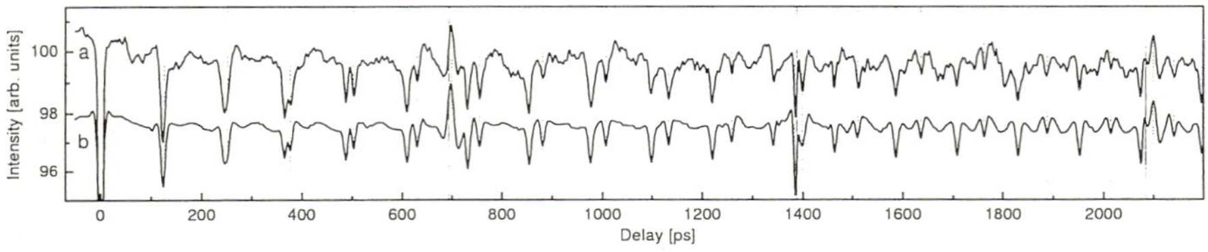

Fig. 2. Time-resolved fluorescence depletion (TRFD) RCS spectrum for *p*CHA [5]. a) Experimental data. b) Numerical simulated RCS spectrum based on the fitted rotational constants for the electronic ground state and the excited state.

the overlap with K-type transients. Thus, state-selective measurements, where either the electronic ground or the excited state are detected, have been performed in this study.

In order to analyze the ground state time-resolved stimulated Raman fluorescence depletion spectroscopy (TRSRFD) was employed as depicted schematically in Fig. 1b. The RCS spectrum obtained is given in Fig. 3a. In accordance with our previous assignment only one set of K-type transients is left over. The numerical simulation (Fig. 3b) agrees very well with the experimental trace. For the analysis of the excited state we performed time-resolved stimulated emission pumping (TRSEP). The experimental spectrum is depicted in Fig. 3c and the simulation given below (Fig. 3d). Here, the other set of K-transients is observed. Moreover, the J-type transients of the excited state could be measured and analyzed separately. The result of these investigations is summarized in Table 1.

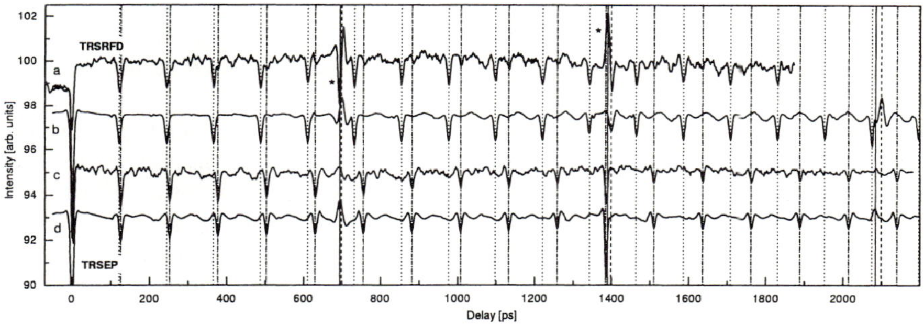

Fig. 3. State-selective RCS spectra of *para*-cyclohexylaniline. a) TRSRFD experimental data. Transients due to (1 + 1') PPI are marked by *. b) Simulation of the ground state. c) TRSEP experimental data. d) Simulation of the excited state.

Table 1. Rotational constants of pCHA of S_0 and S_1 as a result of RCS

S_0	[MHz]	S_1	[MHz]
A''	2406.4 ± 0.6	A'	2346.3 ± 1.3
B''	358.3 ± 0.3	B'	362.9 ± 1.8
C''	356.5 ± 0.3	C'	356.4 ± 1.8
$B'' + C''$	714.8 ± 0.4	$B' + C'$	719.3 ± 2.1

New transients of opposite polarity have been detected in the TRSRFD spectra (marked with an asterisk *) and assigned to a transition to higher lying states above the ionization threshold. This assumption has been checked by detecting the ion current as a function of time delay as depicted in Fig. 4b. The transient positions coincide with the corresponding peaks in the fluorescence detected spectrum of Fig. 4a. Moreover, the position of the J-type transients agrees well with the simulated trace for the excited state (Fig. 4c). Thus the RCS spectrum in Fig. 4b is the result of a (1 + 1') pump probe photoionization (PPI)

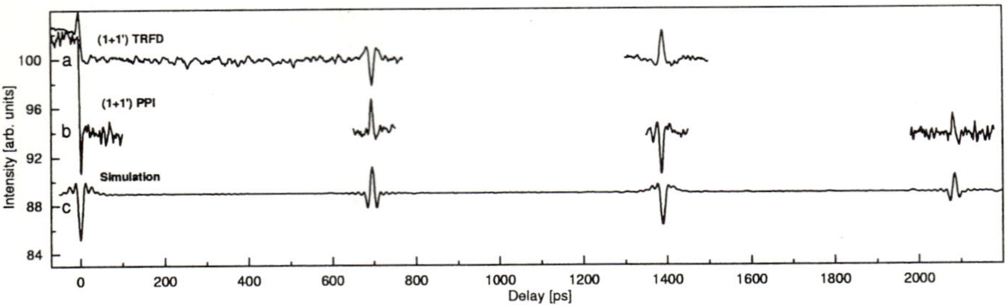

Fig. 4. RCS spectra of *para*-cyclohexylaniline obtained by: a) (1 + 1') pump-probe photoionization. b) (1 + 1') time-resolved fluorescence depletion. c) Simulation.

process. This is remarkable, since it was claimed in the literature that (1 + 1') PPI is not suitable for the detection of rotational coherences due to the undefined state of the molecular ion and relaxation of selection rules for angular momentum [7]. To the best of our knowledge, no experimental results for RCS by (1 + 1') PPI have been reported so far.

In summary, a systematic investigation of two-color RCS schemes revealed the subtle interplay between time-resolved stimulated emission, stimulated Raman and pump-probe photoionization processes and demonstrated how they affect the detected RCS spectrum. As a result state-selective rotational constants of high precision for *p*CHA are obtained. Moreover, these studies led to a new, fairly simple, mass-selective RCS scheme, (1 + 1') PPI, which has great potential for future investigations of molecular clusters.

Acknowledgements. We appreciate the collaboration with D. Podenas (University of Vilnius, Lithuania) and P.M. Felker (UCLA). We acknowledge gratefully financial support by the Herrmann-Willkomm-Stiftung and the Deutsche Forschungsgemeinschaft (DFG).

References

1 A.H. Zewail, in *Femtochemistry*, J. Manz, L. Wöste, ed., (VCH, Weinheim, 1995), vol. I., ch. 2.
2 P.M. Felker, J. Phys. Chem. **96**, 7844, 1992.
3 P.G. Smith, J.D. McDonald, J. Chem. Phys. **92**, 3991, 1990.
4 P.G. Smith, T. Troxler, and M.R. Topp, J. Phys. Chem. **97**, 6983, 1993.
5 C. Riehn, A. Weichert, U. Lommatzsch, M. Zimmermann, and B. Brutschy, J. Chem. Phys. **112**, 3650, 2000.
6 C. Riehn, A. Weichert, M. Zimmermann, and B. Brutschy, Chem. Phys. Lett. **299**, 103, 1999.
7 N.F. Scherer, L.R. Khundkar, T.S. Rose, and A.H. Zewail, J. Phys. Chem. **91**, 6478, 1987.

Vibrational revivals in the Iodine B-state from femtosecond fluorescence interference noise

Ch. Warmuth[1], A. Tortschanoff[1], F. Milota[1], M. Leibscher[2], I. Sh. Averbukh[2], W. Jakubetz[3], and H. F. Kauffmann[1]

[1] Institut für Physikalische Chemie der Universität Wien, A-1090 Wien, Austria
E-mail: kauffm@ap.univie.ac.at
[2] Department of Chemical Physics, Weizmann Institut of Science, Rehovot, 76100 Israel
[3] Institut für Theoretische Chemie und Strahlenchemie der Universität Wien, A-1090 Wien, Austria

Abstract. Observing the fluctuations of femtosecond correlated fluorescence generated by pairs of pulses with randomized relative phase, fractional revivals (order 1/2 and 1/4) of vibrational wavepackets have been found in the B-state of molecular Iodine.

1. Introduction

In small molecules with shallow potentials, such as iodine I_2, electronic excitation through pulses <100fs prepares a superposition of several vibronics that is not an eigenstate of the system, but oscillates as a wavepacket (WP) along the interatomic axis [1]. Basic work on the photophysics of WPs was experimentally done by two-color pump-probe (PP) with laser-induced fluorescence [2], via ionization techniques using two-photon excitation probing pulses [3], as well as with WP-fluorescence interferometry with phase-locked pulses (PL-FI) [4,5].

An alternative technique is based on the use of phase-randomized excitation pulses (Coherence Observation of Interference Noise, COIN [6]). In this semicoherent method one measures the fluctuations about the average F-signal, thereby avoiding the problem of controlling the phase of pulses or the necessity to resolve fast optical oscillations. The technique has been applied to the measurement of coherent transients and static energy-splitting in an atomic gas [6] and in a molecular, organic solid [7]. Very recently [8], COIN has been shown to be a valuable concept to probing WP-dynamics in the I_2 B-state. Since the fluctuations in the COIN measurement directly yield the "slowly" varying amplitude of WP-oscillation without resort to resolving the pulse central frequency this interferometric technique is very robust, which offers a challenge to the exploration of the quantum-classical border in vibrational WP-motion, i.e. the transition from the classical, short-time region to the quantum-mechanical long-time scale. The long-time property of the B-state WP in I_2 was already addressed in the pioneering PL-FI [5]. This type of nonlinear vibronic WP-motion was experimentally probed by high-resolution PP-photoionization [9].

In this work the novel aspect is to demonstrate that the quantum-mechanical revival phenomenon, namely the typical patterns of both a fractional and a full revival in the I_2 B-state can be probed by COIN-interferometry.

2. Experimental Methods

The experimental setup is explained in more detail elsewhere [8]. Briefly, the Iodine sample at vapour pressure at room temperature was exited by 100fs pulses from a Ti/Sa-laser Rega/OPA system. The fluorescence was collected in a 90° geometry, spectrally resolved by a monochromator, and detected by a photomultiplier tube and standard photon counting electronics.

3. Results and Discussion

In Fig.1 the variance of the fluorescence fluctuation var F is plotted over the pulse-to-pulse time delay. In the recurrence-regime (below 2.5ps) 3 peaks with a period of T_1=661fs are seen. The range between 4-7ps is too noisy to be identified in terms of revivals of the order >4. The time base from 7-10ps is the quarter-revival of the WP with a period of 350fs, which was previously observed in I_2 [1] and in Na_2 [10]. The half revival is found between 15-19ps. Here the WP almost recovers its original shape. The two revival periods (661 and 350fs, respectively) are in good accordance with the results of an unbiased Fourier transformation.

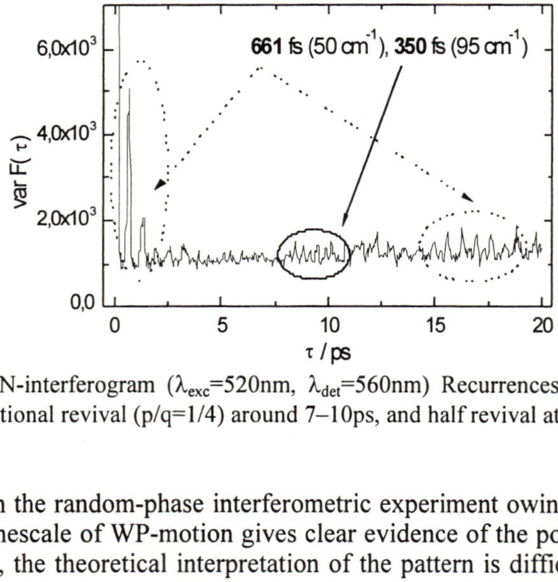

Fig. 1. COIN-interferogram (λ_{exc}=520nm, λ_{det}=560nm) Recurrences at <3ps (T_1≈661fs, 50cm^{-1}), fractional revival (p/q=1/4) around 7–10ps, and half revival at 15<τ<19ps.

Although the random-phase interferometric experiment owing to its stability on the long-timescale of WP-motion gives clear evidence of the post-classical revival phenomena, the theoretical interpretation of the pattern is difficult. Since the system investigated is a thermal ensemble, higher, vibrational ground-state modes that give rise to large Franck-Condon factors in the WP-initialization have to be considered. This leads to a superposition of WPs with slightly different round-trip frequencies that strongly interfere with each other. Another complex effect is concerned with the efficient damping process [8] at the early stage of the WP free evolution. This phenomenon is due to the coupling of a rotational manifold to the vibrational modes involved in the excitation process. Quite obviously, the coherent COIN-experiment, similar to PL-FI [5] is more sensitive to rotational multi-line dephasing than the PP-experiment. This is caused, first of all, by the difference of the rotational constants in the X- and B-state, respectively, which is an intrinsic feature of the interferometric double-pulse excitation in one and the same excited-state level. Note that, therefore, our COIN-pattern does not correspond to the data obtained by PP [2], but rather to a profile that is governed by the autocorrelation of population amplitudes and the differences in the rotational details of the B-and X-state, respectively. The polarization is not damped via some dissipative relaxation channels, but rather is preserved in this inhomogeneous dephasing process as a delocalized packet over a time-span of several ps, and the WP, finally, nearly recovers its shape after ~17ps (Fig.1) because of the discreteness of the WP-spectrum. An analytic theory that takes into account these effects is in satisfactory agreement with the experimental data and provides a sound basis in the interpretation of the various time-regimes of WP-motion [11].

4. Conclusions

The measurement presented here in this preliminary long-time WP-study and done under less favorable thermal averaging conditions is a first demonstration of a quantum-mechanical, coherent observable brought about by the stochastic interplay between the optical processes of enhanced absorption and amplified fluorescence from random-phased pulses.

Acknowledgements. This research was supported by the Austrian Science Fund within the framework of the special research program F016 and project P12566-PHY.

References

1 M. Gruebele and A.H. Zewail, J. Chem. Phys. **98**, 883, 1993.
2 R.M. Bowman, M. Dantus, and A.H. Zewail, Chem. Phys. Lett. **161**, 297, 1989.
3 T. Baumert, M. Grosser, R. Thalweiser, and G. Gerber, Phys. Rev. Lett. **67**, 3753, 1991.
4 N. F. Scherer, A. Ruggiero, M. Du, and G. R. Fleming, J. Chem. Phys. **93**, 856, 1990; N.F. Scherer, A. Matro, L.D. Ziegler, M. Du, R.J. Carlson, J.A. Cina, and G.R. Fleming, J. Chem. Phys. **96**, 4180, 1992.
5 N.F. Scherer, R.J. Carlson, A. Matro, M. Du, A.J. Ruggerio, V. Romero-Rochin, J.A. Cina, G.R. Fleming, and S.A. Rice, J. Chem. Phys. **95**, 1487, 1991.
6 O. Kinrot, I.Sh. Averbukh, and Y. Prior, Phys. Rev. Lett. **75**, 8822, 1995.
7 A. Tortschanoff, K. Brunner, Ch. Warmuth, and H.F. Kauffmann, J. Phys. Chem. A **103**, 2907, 1999.
8 Ch. Warmuth, A. Tortschanoff, F. Milota, M. Shapiro, Y. Prior, I.Sh. Averbukh, W. Schleich, W. Jakubetz, and H.F. Kauffmann, J. Chem. Phys. **112**, 5060, 2000.
9 M.J.J. Vrakking, D.M. Villeneuve, and A. Stolow, Phys. Rev. A **54**, R37, 1996.
10 T. Baumert, V. Engel, C.R. Röttgermann, W.T. Strunz, and G. Gerber, Chem. Phys. Lett. **191**, 639, 1992.
11 Ch. Warmuth, A. Tortschanoff, F. Milota, M. Leibscher, Y. Prior I.Sh. Averbukh, W. Jakubetz, and H.F. Kauffmann, to be submitted.

Untangling π-π*/n-π* Orbital Interactions Via Ultrafast Time-Resolved Photoelectron Spectroscopy.

James P. Shaffer, Thomas Schultz, Michael Schmitt, Jonathan G. Underwood and Albert Stolow
The Steacie Institute for Molecular Sciences, The National Research Council, Ottawa, ON, K1A 0R6, Canada
e-mail: albert.stolow@nrc.ca

Abstract: The ultrafast photochemical dynamics of π-π*/n-π* orbital interactions are determined using time-resolved photoelectron spectroscopy. We compare the spectra of several linear α,β-enones and 1,3-butadiene. We find two different ultrafast decay paths for the α,β-enones, both predicted by *ab initio* calculations.

1. Introduction

Photodissociations and photoisomerizations are fundamental reactions in chemical physics that are amenable to study due to the rapid development of ultrafast laser sources. Time-resolved spectroscopy is advantageous for studying complicated dynamics where many electronic states interact and overlap with one another. This is usually the case in intermediate and large sized molecules where the interaction between π-π* and n-π* molecular orbitals is an important example. In these types of systems it is difficult, if not impossible, to use continuous wave techniques to assign the primary photophysical processes. Photoelectron spectroscopy is particularly sensitive to electronic configurations. It is well known from Koopmans' theorem that specific neutral electronic state configurations correlate upon photoionization more or less strongly with certain ion electronic state configurations. It is expected that the continuum electronic structure should be sensitive to the electronic population dynamics of the intermediate prepared state in a pump-probe experiment. This sensitivity was demonstrated to follow electronic population dynamics during intersystem crossing[1]. Recently, it was demonstrated that Koopmans' type correlations may be used to disentangle vibrational from electronic population dynamics in ultrafast processes[2]. In this work we use time resolved photoelectron spectroscopy to determine complicated ultrafast decay paths due to π-π*/n-π* orbital interactions using linear α,β-enones as a representative model system. These molecules are photochemically reactive and play a role in atmospheric chemistry. Much is known about the photodissociation products of these compounds[3] but very little is known about the ultrafast dynamics which lead to these product channels. By determining the

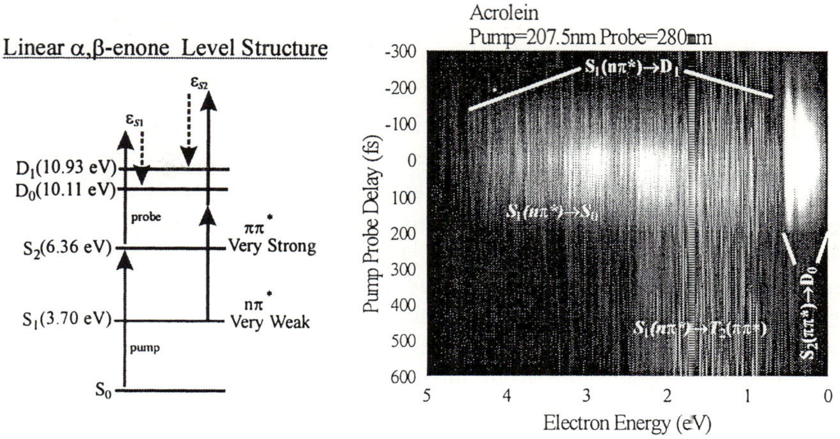

Fig. 1. The upper part of the figure shows the singlet electronic structure and the ion continua of acrolein. The lower figure shows the time-resolved photoelectron spectra of acrolein. The horizontal axis is the electron kinetic energy. The vertical axis is the pump-probe delay. Positive delays correspond to the probe pulse following the pump pulse. The italic labels indicate the features which correspond to the nonradiative transitions from $S_1(n\pi^*)$. The plain text indicates the ionization processes responsible for each region of the photoelectron spectrum.

ultrafast dynamics of α,β-enones we provide a valuable testing ground for extending cutting edge quantum chemical calculations to larger molecules.

2. Results and Discussion

In Figure 1 we show the level scheme and the time-resolved photoelectron spectrum of acrolein. The experiment is initiated when a pump pulse centered at 207.5nm prepares a wavepacket in $S_2(\pi\pi^*)$. A time-delayed pulse centered at 280nm probes the wavepacket by ionization. The level diagram shows $S_1(n\pi^*)$ is strongly coupled to the D_1 continua (1+2' process) while $S_2(\pi\pi^*)$ is strongly coupled to the D_0 continua (1+1' process). At photoelectron kinetic energies <0.5eV, we see a short-lived signal due to ionization of $S_2(\pi\pi^*)$. The $S_2(\pi\pi^*)$ population is depleted by internal conversion to $S_1(n\pi^*)$. The broad photoelectron signal at electron energies >1eV is due to the ionization of population in $S_1(n\pi^*)$ by the 1+2' mechanism shown in Figure 1. The delay in the onset of this photoelectron feature is the time for the $S_2(\pi\pi^*)\rightarrow S_1(n\pi^*)$ internal conversion. The $S_1(n\pi^*)$ band broadens in time as the molecule relaxes to its electronic minimum (IVR). $S_1(n\pi^*)$ decays via two separate channels[3]. Internal conversion to the ground state (<50fs) dominates the photoelectron signal for electron kinetic energies >2eV. Near the electronic minimum of $S_1(n\pi^*)$ fast intersystem crossing (~1.8-2.1ps)[4] between $S_1(n\pi^*)$ and $T_2(\pi\pi^*)$ competes with internal conversion (860fs) to the ground state. This is observed as the longer decay around 1.5eV. The intersystem crossing is fast because it is spin-orbit allowed, the spin-orbit interaction is relatively large, the energy splitting between $S_1(n\pi^*)$ and $T_2(\pi\pi^*)$ is small, and the transition state (considering spin-orbit interaction) geometry[5] is efficient for coupling these two states. Our data on methyl vinyl ketone and crotonaldehyde shows the same qualitative behavior as that described for acrolein. The substitution effects of the methyl groups observed in methyl vinyl ketone and crotonaldehyde allow us to isolate the state to state couplings. The overall behavior is largely due to the strong interaction between the C=C and C=O bonds along the backbone of these compounds. In stark contrast to the complicated behavior of the α,β-enones, 1,3-butadiene which has a conjugated C=C backbone internally converts to the ground state in <50fs. Our findings agree well with CASSCF *ab initio* calculations[4].

3. Conclusion

This work provides a more complete picture of complex non-adiabatic coupling of π-π* and n-π* molecular orbitals using the ultrafast photochemistry of linear α,β-enones as an example. We find two primary decay pathways in this class of compounds. The photochemistry is found to be dominated by interactions between the C=C and C=O bonds along the backbone of the molecule. These results help to confirm our conjecture that time-resolved photoelectron spectroscopy, being sensitive to both electronic configurations and vibrational dynamics, is well suited to study the complex dynamics of non-adiabatic processes in polyatomic molecules.

References

[1]B. Kim, C.P. Schick, and P.M. Weber, "Time-Delayed Two Color Photoelectron Spectra of Aniline, 2-Aminopyridine and 3-Aminopyridine: Snapshots of the Nonadiabatic Curve Crossing," J. Chem. Phys., **103**, 6903-6913(1995).

[2]V. Blanchet, M.Z. Zgierski, T. Seideman, and A. Stolow, "Discerning Vibronic Dynamics Using Time-Resolved Photoelectron Spectroscopy," Nature, **401**, 52-54(1999).

[3]W.H. Fang, "A CASSCF Study on the Photodissociation of Acrolein in the Gas Phase," J. Am. Chem. Soc., **121**, 8376-8384(1999).

[4]K.W. Paulisse, T.O. Friday, M.L. Graske, and W.F. Polik, "Vibronic Spectroscopy and lifetime of S_1 Acrolein," J. Chem. Phys, **113**, 184-191(2000).

[5]M. Regeuro, M. Olivucci, F. Bernardi, and M.A. Robb, "Excited State Potential Surface Crossings in Acrolein: A Model for Understanding the Photochemistry and Photophysics of α, β-Enones," J. Am. Chem. Soc., **116**, 2103-2114(1994).

Change of ultrafast ammonia cluster dynamics by a femtosecond infrared control pulse

P. Farmanara, V. Stert, H.-H. Ritze and W. Radloff

Max-Born-Institut für Nichtlineare Optik und Kurzzeitspektroskopie
Max-Born-Strasse 2a, D-12489 Berlin, Germany
E-mail: radloff@mbi-berlin.de

Abstract: Using femtosecond time-resolved photoelectron spectroscopy we have analyzed the modification of the dynamics and fragmentation of small ammonia clusters which firstly were transferred to the H-transfer state and subsequently excited by an additional infrared laser pulse.

1. Introduction

As femtosecond time-resolved photoelectron spectroscopy has shown [1] small ammonia clusters are transferred to the H-transfer state (e.g. NH_4NH_2 for the dimer) about 200fs after the primary excitation of their electronic Ã state. The H-transfer state cannot be directly reached from the electronic ground state due to Franck-Condon reasons and it exists on the ps time scale. The properties of this state determined experimentally were nicely confirmed by ab initio calculations of the corresponding potential energy surfaces. The energetic scheme for the ammonia dimer resulting from this calculation is shown in Fig.1. For the excitation with 6.0eV photons and probing with 4.5eV photons the electron spectrum should peak at E_{el}=0.6eV which, indeed, was found in the experiment [1].

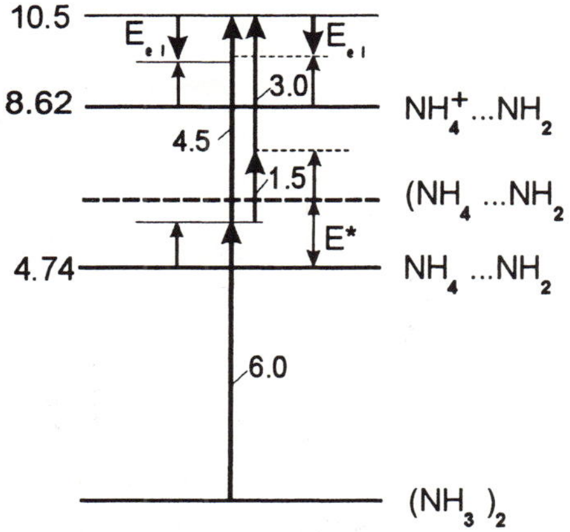

Fig. 1. Energy scheme (in eV) of the ammonia dimer in the H-transfer state.

In a new type of experiments the ammonia dimer in the H-transfer state was excited to the next higher electronic state $(NH_4NH_2)^*$ by an additional, time-delayed, femtosecond control pulse. Thus, the vibrational energy and, hence, the dynamics and the fragmentation rates of the cluster states will be changed. We started the experiments with infrared control photons of 1.5eV energy because in an earlier theoretical study [2] an electronic excitation energy E^*=1.8eV for the H-transfer state of the dimer has been estimated.

2. Experimental Results and Discussion

In the experiment the ammonia clusters are excited to the electronic $\tilde{A}$ state by pump photons ($h\nu_1$=6.0eV) produced by quadrupling the 100fs pulses of a Ti:sapphire laser. The second harmonic of the same laser ($h\nu_2$=3.0eV) is used for probing in the pump-probe experiment. Both laser beams are crossed with the cluster beam in the ionization region of a combined time-of-flight mass spectrometer and a time-of-flight electron spectrometer. By applying the photoion-photoelectron-coincidence detection in combination with the pump-probe technique we were able to study the time-resolved photoelectron spectra for each cluster size [1]. At the delay time τ=0 between pump and probe pulse no ion and electron signals are observable because the sum energy $h\nu_1$+ $h\nu_2$ is below the vertical ionization potentials of the small ammonia clusters (cf. Fig. 2). Irradiating now the clusters with a third pulse ($h\nu_3$=1.5eV) delayed by 2ps with respect to the pump pulse we excite the H-transfer state populated secondarily after this delay time from which the ionization is possible. A lifetime of about 230fs for the

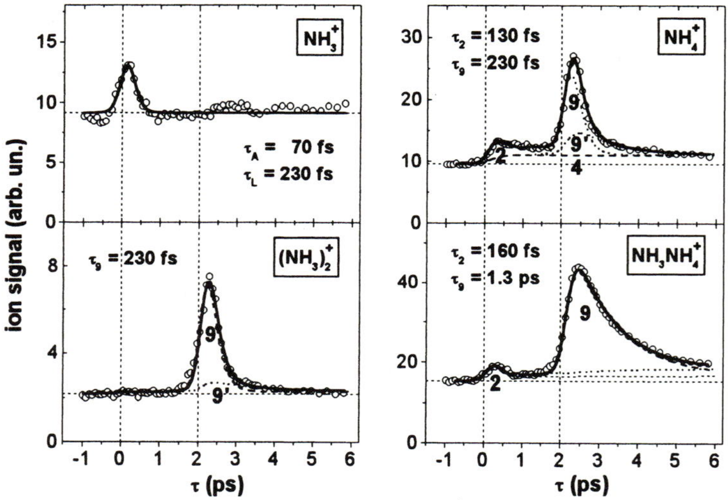

Fig. 2. Ion signals vs. delay time τ between pump (208nm) and probe (416nm) pulse and a control pulse (832nm) delayed by 2ps with respect to the pump pulse.

excited H-transfer state of the dimer has been derived from the time-dependent ion signals (Fig.2) which is about one order of magnitude shorter than that of the ground H-transfer state [1]. In contrast, the excited H-transfer state of the trimer decays much slower and its lifetime is even increased by a factor of about 30 after deuteration of the clusters, whereas the dimer signal hardly changes by deuteration.

Treating the corresponding electron spectra (Fig. 3) we obtained direct insight in the energetic changes caused by the control pulse. For the electron spectrum of $(NH_3)_2^+$ with the control pulse irradiated at 2ps the maximum electron signal is

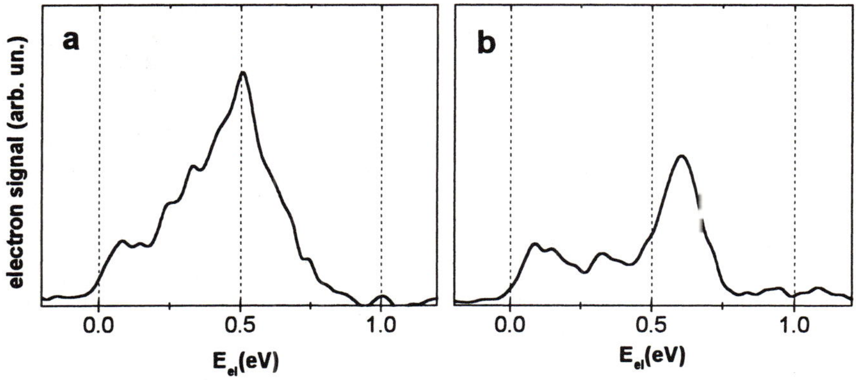

Fig. 3 Electron spectra of $(NH_3)_2^+$ at parameters given in Fig. 2
a) with control pulse, background subtracted, b) without control pulse.

found at E_{el}=0.5eV (Fig. 3a) whereas the corresponding electron signal without excitation of the H-transfer state peaks at 0.6eV (Fig. 3b). This result is excellently confirmed by our ab initio calculations of the potential energy surface for the excited H-transfer state where an excitation energy $E^*=(1.4\pm0.1)$eV has been obtained in contrast to 1.8eV cited above from [2]. Furthermore, the measured lifetimes of the excited H-transfer state for the small ammonia clusters can be explained by crossing of the corresponding potential curve with that of the next higher electronic state.

Further experiments were carried out with control photons of lower energy (e.g. 1.03eV) in which the vibrational energy in the neutral and ionic cluster states and, thus, their fragmentation rates, are strongly reduced (published elsewhere).

References

[1] P. Farmanara, W. Radloff, V. Stert, H.-H. Ritze, and I.V. Hertel, J. Chem. Phys. 111, 633 , 1999.

[2] J. K. Park and S. Iwata, J. Phys. Chem. 101, 3613, 1997.

Observation of the Lorentz Local Field in a Dense Atomic Vapor Using Transient Spectroscopy

Steven T. Cundiff, Justin M. Shacklette and Emily A. Gibson

JILA, University of Colorado and National Institute of Standards and Technology, Boulder, CO, 80309-0440, USA
E-mail: cundiffs@jila.colorado.edu

Abstract. The Lorentz local-field modifies the properties of a dense medium, compared to isolated atoms. It results in the Lorentz-Lorenz shift, and a negative delay signal in transient-four-wave mixing, which we have observed in potassium vapor.

The electric field experienced by an atom in a dense ensemble is not just the externally applied electric field, but also includes a contribution due to the polarization induced in all the other atoms. This is known as the Lorentz local field and results in the Lorentz-Lorenz resonance shift [1]. It has proven difficult to experimentally isolate the Lorentz-Lorenz shift and it has been accomplished only recently [2,3]. These experiments observe an extra density dependent shift, which is attributed to window collisions [3].

Local fields are also important in direct gap semiconductors, where time domain techniques are preferred because of the very short dephasing times. A common time domain spectroscopic technique is transient-four-wave-mixing (TFWM). TFWM is very sensitive to many-body interactions, including local field effects. The Lorentz local field causes a signal for "negative time delays" in TFWM, where there would be no signal for an isolated atom [4]. However, in semiconductors the contribution due to Lorentz local field cannot be separated because there are several many-body phenomena that give rise to negative delay signals. In an atomic vapor, the other many-body effects are absent, allowing clean observation of the Lorentz local field contribution.

We performed TFWM in a dense potassium vapor ($N \sim 10^{18}$ cm^{-3}). We clearly observe negative delay signals due to the Lorentz local field. To the best of our knowledge, this has not previously been observed in an atomic vapor.

Fig. 1 is a schematic of the experiment showing what is meant by a "negative time delay." In a TFWM experiment, two incident pulses have wave vectors $\mathbf{k}_1$ and $\mathbf{k}_2$. For a two-level system, without local fields, a signal is produced in the direction $2\mathbf{k}_2 - \mathbf{k}_1$ if the incident pulse in direction $\mathbf{k}_1$ precedes that in direction $\mathbf{k}_2$. For the opposite time ordering, the signal is emitted in direction $2\mathbf{k}_1 - \mathbf{k}_2$. These directions are shown in Fig. 1(a). Because the vapor is optically thick, these signals are not transmitted. However, they also occur in reflection, as shown in Fig. 1(b), where the superscript *R* denotes reflections. When local fields are included, the time ordering requirements on the incident pulses are relaxed and a signal is emitted in both directions for either time ordering. This means that a

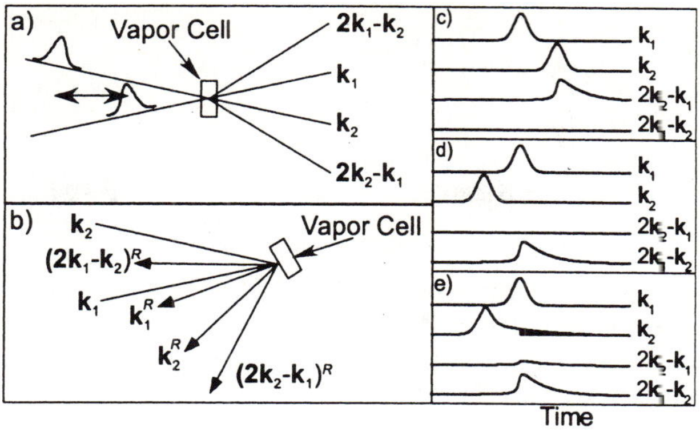

Fig. 1. Schematic of a TFWM experiment in (a) forward and (b) reflection geometries. Parts (c) and (d) show the connection between geometry and time ordering without local fields, in (e) local fields are included, allowing signals in both directions.

signal can be emitted in the direction $2\mathbf{k}_2 - \mathbf{k}_1$ when pulse $\mathbf{k}_2$ precedes pulse $\mathbf{k}_1$, which we call negative delay. This occurs because the polarization in direction $\mathbf{k}_2$ continues to radiate until after pulse $\mathbf{k}_1$ arrives (shown by gray shading in Fig. 1(e)).

The incident 70 fs laser pulses are produced by a Kerr-lens-modelocked Ti:sapphire laser. The laser cavity is stretched so that the repetition rate is 25 MHz, yielding a pulse separation of 40 ns, which is longer than the potassium excited state lifetime of 10-20 ns. The center wavelength is tuned to 768 nm, which is between the potassium $4^2S_{1/2}$? $4^2P_{3/2}$ transition at 766.49 nm and the $4^2S_{1/2}$? $4^2P_{1/2}$ transition at 769.87 nm. The pulses have a spectral width of 9 nm and hence resonantly excite both transitions. A slow detector spectrally and temporally integrates the signal. The TFWM signal is measured as a function of the delay, τ, between the pulses. The vapor cell is constructed from poly-crystalline alumina and a sapphire window. Potassium is loaded under an argon atmosphere. The cell is evacuated to a background pressure of less than 10^{-6} torr before the hot valve is closed. It is then heated to 600 to 800 C, yielding number densities between 10^{18} and 10^{19} cm^{-3}. At these densities, the on resonance absorption length is a few wavelengths, necessitating the use of the reflection geometry. The window is held 50 C above the reservoir to prevent condensation of the potassium.

Fig. 2(a) shows the TFWM signal for several different densities, the quantum beats occur because two transitions are excited. A signal is clearly observable for negative delays. An autocorrelation is also plotted to verify that the negative time signal is not an artifact. The autocorrelation has been corrected to compensate for it being a 2nd order measurement while TFWM is 3rd order.

Perturbation theory can be used to calculate the signal from the Optical Bloch equations for a two level system, including the Lorentz local field. Using delta-

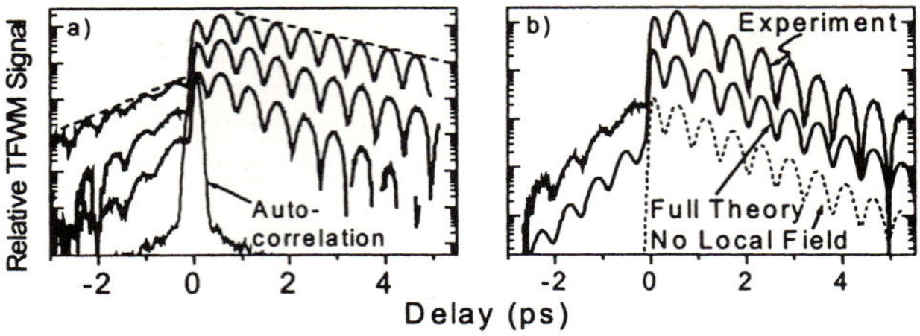

Fig. 2. (a) TFWM mixing signal at densities of 2.3 x 10^{18}, 3.7 x 10^{18} and 5.7 x 10^{18} cm^{-3} (top to bottom, curves offset for clarity). Dashed lines are perturbation theory results. (b) Comparison of numerical calculations with experimental results. Theory is shown with and without local field (curves offset for clarity).

function pulses, perturbation theory predicts that for positive times the signal is given by $I_+(\tau) = Ae^{-2\gamma\tau}$, while for negative delays it is $I_-(\tau) = Be^{4\gamma\tau}$, where γ is the dephasing rate (linewidth). The dashed lines in Fig. 2 plot these functions, verifying the factor of two in slopes for $I_+(\tau)$ and $I_-(\tau)$. For weak local fields, perturbation theory also yields $B/A = [(N\mu^2/3\varepsilon_0\hbar)/\gamma]^2$ where μ is the dipole moment. The Lorentz-Lorenz shift is given by $N\mu^2/3\varepsilon_0\hbar$, hence it can be extracted from the time-domain data.

To include finite pulse widths, a numerical calculation has been performed. The calculation is based on performing a spatial Fourier expansion of the incident field and the elements of the density matrix. To match the experiment, a three level system is used. Fig. 2(b) shows a comparison between the numerical calculation and experimental results. The results agree very closely with each other. The calculation has also been performed without the local field. It is clear that the negative delay signal completely vanishes in the absence of the local field, confirming the origin of the signal.

These time-domain measurements show a clear, unambiguous signature of the Lorentz local field in a dense atomic vapor. The effect of the local field is more distinct than in previous frequency domain measurements of the Lorentz-Lorenz shift.

References

1 H.A. Lorentz, *The Theory of Electrons*, 2nd ed. (Dover, New York, 1952).
2 J.J. Maki, M.S. Malcuit, J.E. Sipe and R.W. Boyd, Phys. Rev. Lett. **67**, 972-975 (1991)..
3 J. Guo, A. Gallagher and J. Cooper, Opt. Commun. **131**, 219-222 (1996); P. Wang, A. Gallagher, J. Cooper, Phys. Rev. A **56**, 1598-1606 (1997).
4 M. Wegener, D.S. Chemla, S. Schmitt-Rink and W. Schäfer, Phys. Rev. A **42**, 5675-5683 (1990).

Part VIII

Ultrafast Processes in Semiconductors

Resonant Rayleigh Scattering from Semiconductor Nanostructures: Ultrafast Spectral Interferometry

Dan Birkedal[1,2], Andrei Shchegrov[3] and Jagdeep Shah[1]

[1] Bell Laboratories, Lucent Technologies, Holmdel, NJ 07733, USA
[2] present address: Research Center COM, TU Denmark, DK-2800 Lyngby, Denmark
[3] U. of Rochester, Rochester, NY 14627, USA; now at: KLA-Tencor, San Jose, CA 95134

Abstract. Resonant secondary emission (RSE) from semiconductor nanostructures consists of an incoherent component (resonant luminescence), and a coherent component (resonant Rayleigh scattering (RRS)) that originates from disorder. Ultrafast spectral interferometry unambiguously isolates this coherent component, and demonstrates the presence of speckles, contrary to earlier expectations. It further allows the investigation of the complex amplitude and phase dynamics of the coherent emission, and provides new insights into the physics of RRS and disorder. We review recent developments in the field.

1. Introduction

Resonant excitation of excitons in semiconductors by a femtosecond pulse creates a coherent ensemble of dipoles whose radiation interferes destructively in all directions other than the specular (or primary) directions (transmission and reflection). In an ideal quantum well without any disorder, radiation in secondary (i.e. non-specular) directions becomes possible only if the momentum of the excitons parallel to the quantum well interfaces is altered by collisions. The resultant radiation, resonant luminescence of excitons, is incoherent. Disorder in quantum wells localizes excitons, with spatially varying exciton energies (inhomogeneous broadening). The phases of excitons with different energies evolve at different rates so that the excitons no longer remain in phase with each other as time progresses. Therefore, even in the absence of any collisions, the destructive interference in the secondary directions is reduced and coherent radiation in the secondary directions becomes possible. Another way of looking at this is to say that disorder effectively lifts the momentum conservation in the parallel directions and allows elastic scattering of light in the secondary directions. This coherent radiation is the well-known resonant Rayleigh scattering (RRS) [1].

RRS from semiconductor quantum wells was first investigated by Hegarty *et al.* [2,3] nearly 20 years ago using continuous-wave lasers. The interest in the field was revived in mid-1990s when it became possible to study the dynamics of RSE with femtosecond time resolution [4]. Wang *et al.* [4] observed that Resonant Secondary Emission (RSE) is delayed with respect to the exciting pulse and the delay decreases with increasing density of excitons. This was attributed to increased momentum scattering rate due to increased exciton-exciton scattering, and led to the conclusion that resonant luminescence dominated under these conditions. Haacke *et al.* [5] extended these measurements to lower densities and, by comparing the results with the theoretical predictions [6-8], concluded that RSE was dominated by RRS, the coherent component, at low densities. Gurioli *et al.* [9] analyzed the coherent and incoherent components of the emission using two phase-locked excitation pulses.

Birkedal and Shah [10] argued that the truly distinguishing feature of RRS is its coherence with the exciting electromagnetic field. They unambiguously isolated this coherent component [10] and determined its dynamics using the technique of spectral interferometry [11,12]. These studies also demonstrated the presence of speckles in this coherent component. These results show that RRS can not be properly described by a theory using ensemble averaging, but that the results must be analyzed in terms of a single realization of disorder. This was accomplished by means of a Monte Carlo simulation by Shchegrov *et al.* [13]. The subject has continued to be an active field of research. Woerner *et al.* [14] time-resolved the emission excited by two phase-locked pulses and found a novel component due to "quantum coherence". Langbein *et al.* [15] showed that statistical analysis of the speckle pattern yields important information. Savona and Zimmermann [16] theoretically investigated the influence of level repulsion (quantum effects) on RRS, and Savona *et al.* [17] presented supporting experimental results. Birkedal *et al.* [18] showed that *intrinsic* coherent secondary emission is possible from two-exciton states (biexcitons) in semiconductors. RRS from semiconductor *microcavities* has been investigated experimentally using time resolved studies of emitted intensity [19] and using spectral interferometry [20], and theoretically [21,22]. RRS from semiconductor Bragg structures has also been investigated very recently using spectral interferometry by Prineas *et al.* [23]. After this brief review, we now focus on the results obtained using spectral interferometry.

2. Experimental Methods

The spectral interferometry measurements [10] were performed using a Ti:Sapphire laser with 100 fs pulses at 82 MHz repetition rate. GaAs/AlGaAs single and multiple quantum well samples with different well widths were investigated between 10 K and 70 K. Here we discuss the results obtained on a 20-period quantum well structure with 13 nm GaAs quantum wells and 15 nm $Al_{0.3}Ga_{0.7}As$ barriers at 10 K. The laser is tuned close to the excitonic resonance and focussed on the sample with a spot diameter of $\approx$ 40 μm. The resonant secondary emission is collimated by a lens and combined collinearly with a suitably attenuated and delayed reference laser pulse. An aperture in the combined beam optimizes the overlap and assures investigation of a single speckle (diffraction-limited performance). Time-integrated spectra are obtained using a CCD detector (spectral resolution of 0.11 meV). *Spectral interferogram* is obtained by subtracting the spectrum of the reference laser and that of the secondary radiation from the combined spectrum. Since only the coherent component of the RSE can interfere with the reference laser, the spectral interferogram unambiguously isolates RRS and allows investigation of its properties.

3. Results and Discussion

The total RSE spectrum of excitons in a 13 nm GaAs quantum well at 10 K exhibits a strong peak at the heavy-hole (HH) exciton and a weaker peak at the light-hole (LH) exciton (Fig. 1a). The corresponding spectral interferogram (Fig. 1b) shows strong and high contrast fringes at the same energies, demonstrating that RRS signal originates from both these excitons. In addition, laser light non-resonantly scattered from the sample surface exhibits fringes over the spectrum of the laser pulse. Since the fringe spacing is inversely proportional to the delay between the reference laser pulse and RRS, the smaller fringe spacing for the excitons demonstrates that RRS is delayed with respect to the non-resonantly scattered laser (marking the time t=0). The two coherent signals also have different phases.

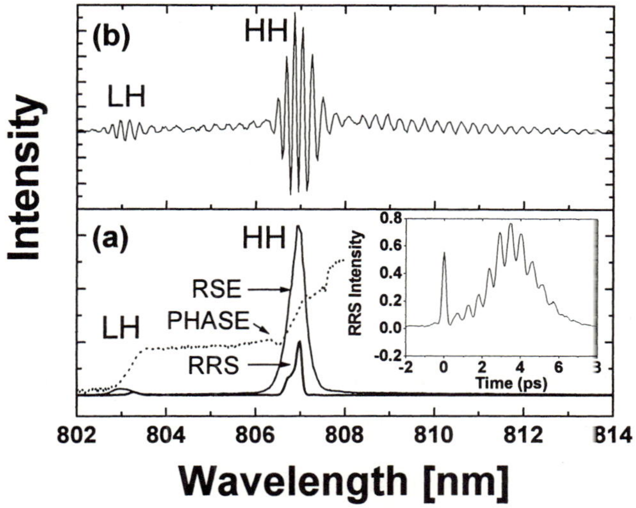

Fig. 1. 13-nm GaAs quantum well at 10 K; 4×10^8 excitons/cm^2-quantum well ; (a) RSE spectrum, coherent (RRS) spectrum (heavy line) and phase spectrum (dots); the inset shows the dynamics of RRS obtained from the spectral interferogram in (b).

The dynamics of RRS, determined from the spectral interferogram using Fourier transforms [11], is depicted in the inset of Fig. 1a. The surface scattered laser light leads to the peak at t=0, whereas the RRS signal is delayed and exhibits a peak near 4 ps. The beats observed in the RRS signal correspond to beats between HH and LH excitons, which are simultaneously excited by the spectrally broad 100 fs pulses. Fig. 1a also depicts the coherent spectrum and the phase spectrum obtained by using inverse Fourier transform of the dynamics in the inset. The phase spectrum shows jumps by 2π at the exciton resonances.

These results unambiguously isolate RRS, and determine its dynamics and spectrum for the first time. Dynamics of the phase can be obtained in a similar fashion. Furthermore, the results show that RRS forms speckles much like those formed in non-resonant scattering of a He-Ne laser from a rough surface. The speckles in the case of RRS, however, exhibit extremely complex temporal dynamics. Each speckle, in fact, exhibits its own complex dynamics. The dynamics of different speckles and different realizations of disorder (i.e. different spots on the sample) have been investigated by Birkedal *et al.* [10,24,25].

The variations in the speckle dynamics for different speckles and different spots on the sample underscores that the behavior of RRS can not be properly predicted or explained on the basis of a theory that relies on ensemble averaging. Shchegrov *et al.* [13] have calculated the dynamics of RRS for a single realization of disorder using Monte Carlo simulations. This approach provides a good agreement with the data on RRS dynamics depicted in the inset of Fig. 1a.

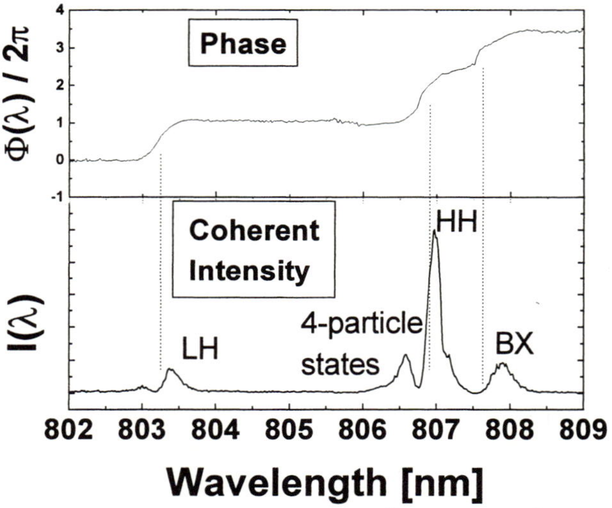

Fig. 2. 13 nm GaAs quantum well at 10 K; 5×10^{10} excitons/cm^2-quantum well RRS and phase spectra; Bound and unbound states of biexcitons make important contributions.

Finally, Birkedal *et al.* [18] have shown recently that increasing the intensity of excitation dramatically changes the nature of RRS from semiconductor quantum wells. The results show that biexcitonic effects begin to make a strong contribution to RRS. This contribution does *not* depend on the presence of disorder and hence is intrinsic in nature. The spectral phase and amplitude of the RRS at higher excitation density are depicted in Fig. 2. The biexcitonic coherent signal is present not only at the bound biexciton energy (BX) but also at the energies of the unbound biexcitons (4-particle states). Understanding four-particle (2 electrons and 2 holes) correlations is one of the outstanding problems of condensed matter physics at this time [26], and spectral interferometry of RRS from quantum wells is potentially a powerful means of investigating such correlations.

4. Conclusions

Spectral interferometry provides a unique perspective on the nature of resonant Rayleigh scattering and the nature of disorder in semiconductor quantum wells.

Acknowledgements. We thank L. N. Pfeiffer and K. W. West for the high quality quantum well samples, and M. Woerner for illuminating discussions.

References

1. J. W. Strutt (Lord Rayleigh), Phil. Mag. 47, 375 (1899).
2. J. Hegarty, M. D. Sturge, C. Weisbuch, A. C. Gossard and W. Wiegmann, Phys. Rev. Lett. 49, 930 (1982).
3. J. Hegarty, L. Goldner and M. D. Sturge, Phys. Rev. B 30, 7346 (1984).
4. H. Wang, J. Shah, T. C. Damen and L. N. Pfeiffer, Phys. Rev. Lett. 74, 3061 (1995).
5. S. Haacke, R. A. Taylor, R. Zimmermann, I. Bar-Joseph and B. Deveaud, Phys. Rev. Lett. 78, 2228 (1997).
6. H. Stolz, Time-Resolved Light Scattering from Excitons, Springer-Verlag, Berlin, (1994).
7. R. Zimmermann, Il Nuovo Cimento 17D, 1801 (1995).
8. D. S. Citrin, Phys. Rev. B 54, 14572 (1996).
9. M. Gurioli, F. Bogani, S. Ceccherini and M. Colocci, Phys. Rev. Lett. 78, 3205 (1997).
10. D. Birkedal and J. Shah, Phys. Rev. Lett. 81, 2372 (1998).
11. L. Lepetit, G. Cheriaux and M. Joffre, J. Opt. Soc. Am. B 12, 2467 (1995).
12. D. N. Fittinghoff, J. L. Bowie, J. N. Sweetser, et al. Opt. Lett. 21, 884 (1996).
13. A. V. Shchegrov, D. Birkedal and J. Shah, Phys. Rev. Lett. 83, 1391 (1999).
14. M. Woerner and J. Shah, Phys. Rev. Lett. 81, 4208 (1998).
15. W. Langbein, J. M. Hvam and R. Zimmermann, Phys. Rev. Lett. 82, 1040 (1999).
16. V. Savona and R. Zimmermann, Phys. Rev. B 60, 4928 (1999).
17. V. Savona, S. Haacke and B. Deveaud, Phys. Rev. Lett. 84, 183 (2000).
18. D. Birkedal, J. Shah and L. N. Pfeiffer, Phys. Rev. B 60, 15585 (1999).
19. G. R. Hayes, S. Haacke, M. Kauer, et al. Phys. Rev. B 58, R10175 (1998).
20. J. Bloch, D. Birkedal, J. Shah, and V. Thierry-Meig, Postdeadline paper presented at QELS99; p.QPD6-1
21. D. M. Whittaker, Phys. Rev. B 61, R2433 (2000).
22. A. V. Shchegrov, J. Bloch, D. Birkedal and J. Shah, Phys. Rev. Lett. 84, 3478 (2000).
23. J. P. Prineas, J. Shah, B. Grote, C. Ell, G. Khitrova, H. M. Gibbs and S. W. Koch, unpublished.
24. D. Birkedal, J. Shah, A. V. Shchegrov and L. N. Pfeiffer, Physica Status Solidi (a) 178, 5 (2000).
25. D. Birkedal and J. Shah, Adv. Sol. St. Phys., 39, 495 (1999).
26. D. S. Chemla and J. Shah, Proc. of the National Academy of Sciences, **97**, 2437 (2000).

Radiative interwell coupling of hh-, lh-excitons, and correlated electron-hole pairs in multiple quantum well Bragg samples

D. Ammerlahn[1], J. Kuhl[1], B. Grote[2], S.W. Koch[2], G. Khitrova[3], and H. Gibbs[3]

[1] Max-Planck-Institut für Festkörperforschung, D-70569 Stuttgart, Germany
E-mail: kuhl@servix.mpi-stuttgart.mpg.de
[2] Physics Department, Philipps University, D-35032 Marburg, Germany
[3] Optical Sciences Center, University of Arizona, Tucson, AZ 85721, USA

Abstract. Spectrally and temporally resolved linear reflectivity measurements with fs optical pulses on (In,Ga)As/GaAs Bragg structures show pronounced coherent interwell polarization coupling on light-hole excitons and for correlated electron-hole pairs close to the band edge.

1. Introduction

Exciton-photon coupling in multiple quantum well (MQW) Bragg structures leads to drastic changes of the optical features in dependence on the number of quantum wells (QW's) N and the mismatch between the QW separation d and multiple integers of $\lambda/2$ (λ is the wavelength of the exciton transition) [1]. Variations with sample geometry have been reported for the radiative contribution to optical dephasing [2], the homogeneous linewidth, absorption [3] and reflectivity [4], photoluminescence efficiency [5] or the 3rd order nonlinear susceptibility [6]. All these effects are explained by the formation of a superradiant exciton-polariton mode propagating through the sample.

So far all demonstrations of coherent interwell coupling phenomena have been restricted to the lowest heavy hole (hh) exciton transition which is distinguished by a high oscillator strength and a comparatively small phase relaxation rate which both favor the exciton-photon coupling. Here we present spectrally and temporally resolved linear reflectivity measurements with 100 fs laser pulses on Bragg samples with a large number of QW's which provide clear evidence for radiative coupling phenomena and the formation of exciton-polaritons for light hole (lh) excitons and correlated electron-hole states in the continuum [7].

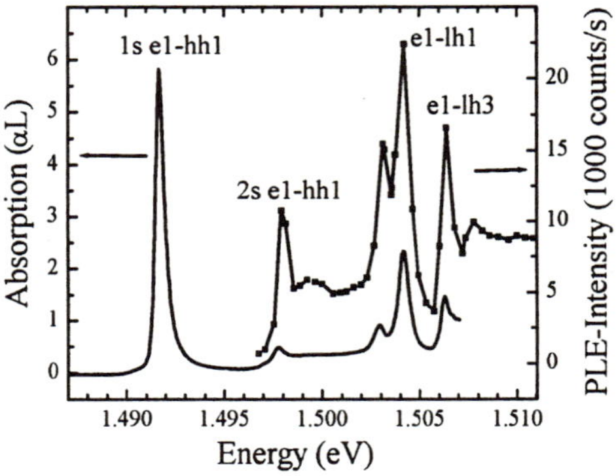

Fig. 1. Absorption and PLE spectrum of the 30 QW (In,Ga)As Bragg sample at $d = 0.4\lambda$.

2. Experiment

We used two samples with $N = 30$ and 60 (In,Ga)As/GaAs QW's grown on GaAs substrates with a well width of 8.5 nm and an In content of 4 %. The barrier thickness of these MQW's decreases from the center to the edge of the wafer by approximately 30 %. Thus it is possible to tune different excitonic transitions into Bragg resonance by moving the laser spot on the sample. For measurements of the linear reflectivity, we applied low intensity 100 fs pulses from a mode-locked Ti:sapphire laser. The 15 meV broad spectrum was centered at 1 505 eV and covered the whole excitonic region including the band gap (Fig. 1). The spectra of the reflected beams were analyzed by a 0.75 m double monochromator.

3. Results and Discussion

Figures 2(a) and 2(b) show measured reflectivity spectra for two different positions of the laser spot on the wedge-shaped sample which correspond to resonances between the hh1- or lh1-exciton transition, respectively, and the structure as demonstrated by the high amplitude and broad linewidth of the reflectivity. Figure 2(c) displays a pronounced peak at the high energy end of the spectrum which shifts with decreasing barrier width to higher energies. Since the absorption and photoluminescence excitation spectra display no corresponding peak in this region which overlaps with the interband continuum transitions, this observation provides clear evidence for radiative coupling between correlated electron-hole pairs in the continuum close to the band edge. All important spectral features are qualitatively well reproduced by solutions of the semiconductor Maxwell-Bloch equations (SMBE) shown in Figs. 2(d) -2(f).

The high reflectivity at resonance can be interpreted as the formation of a photonic bandgap pushing any electromagnetic field which is resonant to the spatial frequency of the Bragg structure out of the crystal. This behavior is confirmed by time-resolved reflectivity measurements on the hh-exciton resonance of the $N = 60$ sample for different detunings (Fig. 3). These reflection signals are composed of two contributions: an almost constant peak at delay 0 due to the residual reflectivity of the AR coated sample surface and a second peak at later times strongly varying in intensity with sample position and shifting to later times with decreasing barrier width. It reflects the coherent

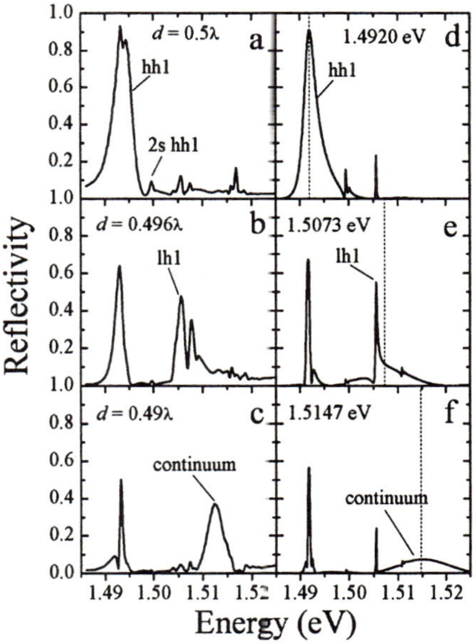

Fig. 2. (a-c) Measured and (d-f) calculated reflection spectra at different detunings corresponding to Bragg resonance with the hh-, lh-exciton and continuum transitions.

reemission of the coherent polarization created on the hh exciton into the reflection direction. The strong variation of the time-integrated coherent reemission with detuning (inset of Fig. 3) and the changes in the temporal distribution of the reemitted light prove the sensitivity of the radiative contribution to dephasing on the QW separation in a MQW structure.

Finally we have measured the influence of increasing nonradiative contributions to the homogeneous linewidth onto amplitude and width of the hh-exciton reflectivity at Bragg resonance [7]. The nonradiative linewidth has been continuously varied by either changing the lattice temperature (increased exciton-phonon scattering) or higher laser pulse intensity (increased exciton-exciton-scattering). Comparison of reflectivity data measured at Bragg resonance and for large detuning reveals the presence of strong radiative coupling phenomena at temperatures up to 150 K and laser pulse intensities up to 5 kW/cm^2.

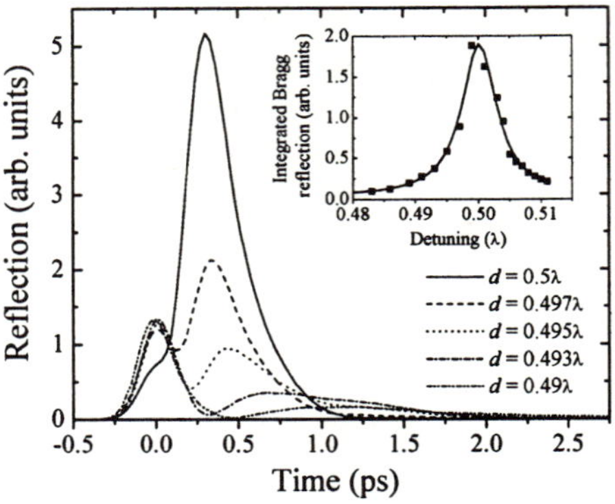

Fig. 3. Time-resolved reflectivity of a 60 QW (In,Ga)As Bragg sample. Inset: time-integrated coherent reemission vs. detuning.

4. Conclusions

Spectrally and time-resolved reflectivity studies on (In,Ga)As/GaAs MQW-Bragg samples provided clear evidence for the presence of substantial radiative interwell coupling of heavy- and light-hole excitons as well as of correlated electron-hole pairs in the continuum. The experimental results are well described by solutions of the SMBE.

Acknowledgements. We would like to thank the Deutsche Forschungsgemeinschaft, the Sonderforschungsbereich 383, the Leibniz prize, and NSF AMOP for financial support of different parts of this project.

References

1 T. Stroucken, A. Knorr, and S.W. Koch, Phys. Rev. B **53**, 2026 (1996).

2 M. Hübner, J. Kuhl, T. Stroucken et al., Phys. Rev. Lett. **76**, 4199 (1996).

3 G.R. Hayes, J.L. Staehli, U. Oesterle et al., Phys. Rev. Lett. **83**, 2837 (1999).

4 J.P. Prineas, C. Ell, E.S. Lee et al., Phys. Rev. B **61**, 13863 (2000).

5 M. Hübner, J.P. Prineas, C. Ell et al., Phys. Rev. Lett. **83**, 2841 (1999).

6 J. Kuhl, M. Hübner, D. Ammerlahn et al., in *Festkörperprobleme/Advances in Solid State Physics* **38**, ed. by B. Kramer, Vieweg, Braunschweig (1998).

7 D. Ammerlahn, J. Kuhl, B. Grote, S.W. Koch, G. Khitrova, and H. Gibbs, accepted for publication in Phys. Rev. B (2000).

Highly Energetic Excitons in GaAs Studied via Femtosecond Transmission Spectroscopy

Alfred Leitenstorfer[1], Gernot Göger[1], Markus Betz[1], Max Bichler[2], Werner Wegscheider[3], and Gerhard Abstreiter[2]

[1] Physik-Department E11, TU-München, James-Franck-Str., D-85748 Garching, Germany
E-mail: aleitens@ph.tum.de

[2] Walter-Schottky-Institut, TU-München, D-85748 Garching, Germany

[3] Institut für Angewandte und Experimentelle Physik, Universität Regensburg, D-93040 Regensburg

Abstract. The first optical investigation of excitons with large wave vectors is presented, based on violated momentum conservation at interfaces. We study the ultrafast dynamics of these „dark" excitations via scattering with free carriers and phonons.

It is usually assumed that first-order interband transitions in crystalline solids occur almost vertically in the band structure. Especially, optical excitation and probing of free excitons in direct semiconductors is thought to be possible only near the center of the Brillouin zone. We demonstrate that this restriction can be overcome by taking advantage of the symmetry break at any semiconductor interface. In time and energy resolved femtosecond transmission experiments, we find spectral beats between photon-like and weak exciton-like modes.

The impact of exciton-light coupling on light propagation in the frequency regime close to the fundamental resonance has been known for many years [1]. For the first time, the high sensitivity of our femtosecond technique allows the observation of related effects high in the absorption continuum and gives spectroscopic access to excitons with large momenta [2]. We perform pump-probe experiments using a two-color femtosecond Ti:sapphire laser with two independently tunable synchronized pulse trains. Our MBE grown samples consist of 500 nm, 200 nm and 50 nm thick GaAs layers oriented along <100>. They are kept at low lattice temperature $T_L = 20$ K. Nonequilibrium free carriers are injected by a pump pulse with a pulse duration of 80 fs. A time delayed 17 fs probe pulse is spectrally analyzed in a double monochromator after transmission through the sample. In Fig. 1(a) the pump induced transmission changes in the 500 nm specimen are shown versus probe photon energy, measured at a delay time of $t_D =$ 10 ps. At probe photon energies in the absorption continuum nonequidistant oscillations appear in the differential spectra. This phenomenon is explained as follows: When probe photons of energy $h\nu_{Probe}$ (see inset of Fig. 2) are incident on the sample surface, the break of the translational symmetry of the crystal allows excitation of excitons with wave vectors much larger than that of the electromagnetic wave. The excitonic polarization couples coherently to the light field resulting in new quasiparticles, called exciton polaritons. For excess energies of a few meV above the band gap, the photon-like polariton branch P approaches the usual linear dispersion relation, while the hh and lh exciton-like polaritons (X_{hh}

and X_{lh} in the inset of Fig. 2) recover their parabolic exciton center-of-mass dispersion. Frequency components of the incident pulse with energy in the absorption continuum are converted predominantly into photon-like polaritons. Only a small fraction propagates on the exciton-like branches. At the other sample surface the three components are reconverted into photons. Interference of the electric field amplitudes results in weak fringes superimposed on the spectrum of the transmitted probe pulse. The efficient damping of exciton-like polarizations via the scattering with free carriers injected by the pump pulse modifies the fringe contrast in the probe beam. Therefore, the nonlinear femtosecond experiments are very sensitive to polariton effects.

Two distinct components are identified in Fig. 1: One fast beating mode corresponding to the 1s hh exciton polariton, whose amplitude and period exhibit a sharp discontinuity at a probe photon energy of the fundamental 1s exciton plus one LO phonon (X+LO, LO phonon energy in GaAs: 36 meV). Slower oscillations without step-like behavior at X+LO are superimposed originating from the lh exciton polariton. Apparently, lh excitons conserve their polarization over large distances high above the LO threshold energy (see Fig. 1 (c) and (d)). In contrast, the coherence of hh excitons is destroyed rapidly as soon as emission of LO phonons becomes energetically allowed.

In order to analyze our experimental findings, we have performed extensive numerical simulations based on the Pekar-Hopfield additional boundary conditions [3] and Fröhlich scattering of hot excitons with

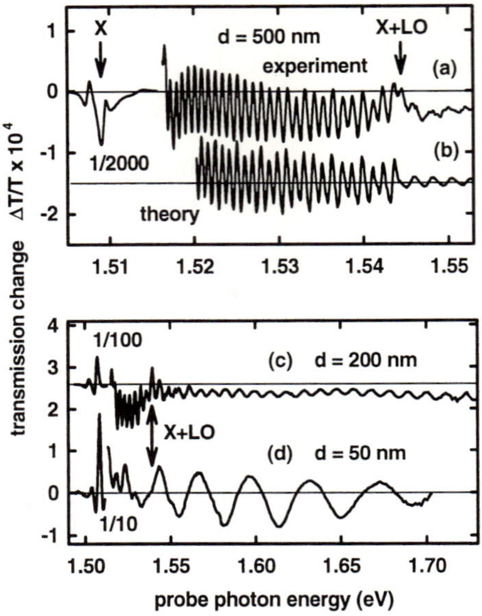

Fig. 1. Spectrally resolved transmission changes measured 10 ps after hot carrier injection in GaAs films of different thicknesses d at $T_L = 20$ K (a, c, d). Free carrier densities are $N_{Pump} = 3 \times 10^{15}$ cm^{-3} (a), 5×10^{15} cm^{-3} (c) and 3×10^{16} cm^{-3} (d), respectively. (b) Transmission changes as calculated theoretically for d = 500 nm.

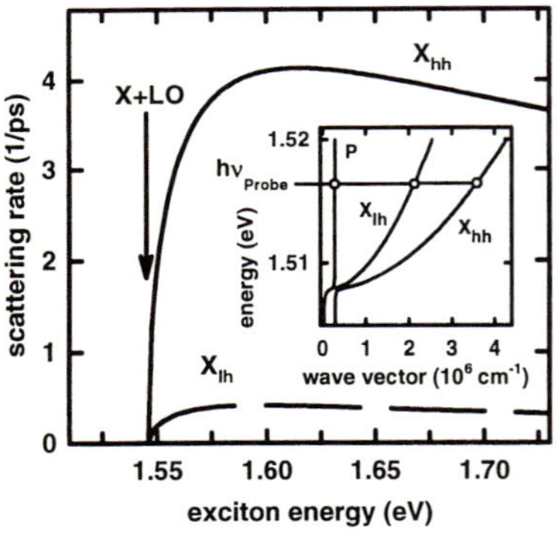

Fig. 2. Calculated LO phonon emission rates of 1s hh (X_{hh}, full line) and lh (X_{lh}, dashed line) excitons in GaAs at $T_L = 0$ K, assuming parabolic bands. Inset: Exciton polariton dispersion consisting of one photon-like branch (P) and two exciton-like branches (X_{lh}, X_{hh}).

LO phonons [4]. The spontaneous LO emission rate calculated for GaAs is shown in Fig. 2: hh excitons couple strongly to an LO polarization field resulting in scattering rates of approximately 4 ps^{-1}. On the other hand, we obtain a strongly reduced LO phonon emission rate for the lh excitons caused by destructive quantum interference in the matrix element for particles composed of oppositely charged carriers with similar effective masses. Our model reproduces the experimental data well: As shown in Fig. 1 (b), both hh and lh excitons contribute to the spectral beats below an energy of X+LO. Above this threshold energy, the hh component is strongly reduced and the lh beats remain. From the data in Fig. 1 (a) and (c), we determine effective hh and lh exciton masses of 0.52 m_0 and 0.19 m_0, respectively. Another indication for the splitting of an ultrashort probe pulse into three components is obtained by examining the wave packet dynamics at various negative delay times. In Fig. 3 the pump induced transmission changes in the 500 nm sample are plotted versus the probe photon energy. At positive time delays, both hh and lh probe polaritons suffer efficient damping. Since they propagate at a lower group velocity than the photon-like component of the pump pulse, the exciton-like polaritons scatter with pump generated free electrons and holes along the entire crystal thickness. If the probe pulse enters the sample at t_D = -2 ps, the spectrum contains exclusively one component (hh excitons), ending at a photon energy of X+LO. The faster mode (lh excitons) escapes the sample before carrier excitation by the pump. For a delay time of t_D= -3.5 ps, the hh excitons are damped only toward the end of their propagation through the film and the differential beat amplitude is strongly reduced.

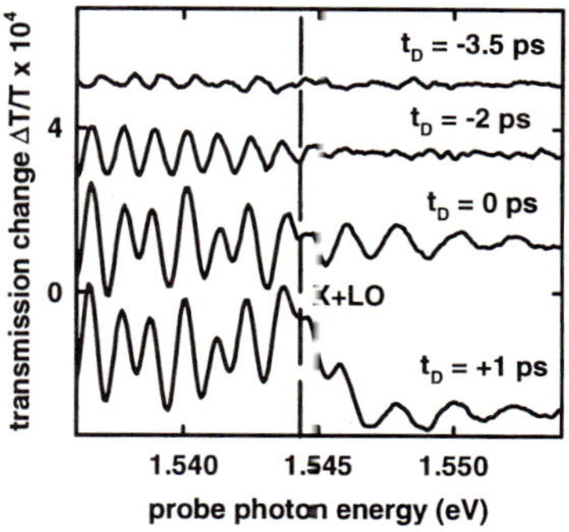

Fig. 3. Differential transmission spectra of a 500 nm GaAs film for different delay times t_D.

In conclusion, we have investigated the ultrafast dynamics, dispersion and interface properties of highly energetic excitons in GaAs. A wealth of surprising results is obtained on these fundamental excitations which have up till now been inaccessible to experiments. As an example, lh excitons turn out to be almost inert against LO phonon scattering due to destructive quantum interference in the polaron coupling.

References

1 For an overview see: *Excitons in Confined Systems*, Proceedings of the International Meeting, Rome, 1987, Edited by R. Del Sole, A. D'Andrea, and A. Lapiccirella, 1988.

2 G. Göger, M. Betz, A. Leitenstorfer, M. Bichler, W. Wegscheider, and G. Abstreiter, Phys. Rev. Lett. **84**, 5812, 2000.

3 S. I. Pekar, Soviet Physics JETP **34**, 1176, 1958.

4 Y. Toyozawa, Prog. Theor. Phys. **12**, 111, 1959.

Ultrafast Electro-Absorption at the Transition between Classical and Quantum Response

Alan H. Chin[1] and Junichiro Kono[2]

[1] W. W. Hansen Experimental Physics Laboratory, Stanford University, Stanford, CA 94305, USA
E-mail: chinah@stanford.edu
[2] Present address: Department of Electrical and Computer Engineering, Rice University, P.O. Box 1892, Houston TX 77251, USA
E-mail: kono@rice.edu

Abstract. We have made the first observation of induced absorption extending ~ 1 eV below the band edge in semiconductors strongly driven by intense ultrafast mid-infrared laser pulses, which we attribute to the dynamical Franz-Keldysh effect.

We have observed for the first time, to our knowledge, ultrafast induced absorption below the band edge in semiconductors strongly driven by intense mid-infrared (MIR) pulses, which we attribute to the dynamical Franz-Keldysh effect (DFKE). This ultrafast electro-absorption has the largest extent (~ 1 eV) below the band edge ever observed, to our knowledge. Phenomena of this type occur in matter when the applied electromagnetic field has a ponderomotive potential, $U_p = \left(2\pi e^2/mc\right)\left(I/\omega^2\right)$, comparable to or greater than the photon energy, $\hbar\omega$ [1-4]. Under such conditions, the behavior of matter is at the transition between classical and quantum response, i.e. photon effects of the applied field cannot be ignored.

To be able to strongly drive semiconductors, i.e. generate conditions in semiconductors where $U_p \approx \hbar\omega$ (accounting for the effective mass), while avoiding multi-photon absorption and sample damage, we used MIR laser pulses. We used tunable intense MIR pulses (~ 1 ps or ~ 200 fs pulse duration, 3 μm to 10 μm) from an optical parametric amplifier system with difference frequency mixing. Using either 3.5 μm or 6.2 μm MIR pump pulses, we studied the DFKE by measuring the transmission spectrum around the band edge of a continuum probe after transmission through a strongly driven sample.

We observed a dramatic *decrease* in transmission below the band edge, due to induced absorption, in strongly driven GaAs (350 μm thick, (100) orientation). As shown in Fig. 1, this induced absorption using ~ 1 ps driving pulses at 3.5 μm (~ 2×10^{10} W/cm^2, $U_p \approx \hbar\omega$) is significant (~ 20% absolute transmission change) and extends more than 0.06 eV below the band edge. In addition, the effect persists only during the presence of the intense MIR pump, demonstrating the virtual nature of the effect (no carriers or lattice heating). The effect is not specific to GaAs, as we also observed the DFKE in ZnSe and ZnTe. Fig. 2 shows the effect using the same driving field as in Fig. 1, but in ZnSe (3 mm thick,

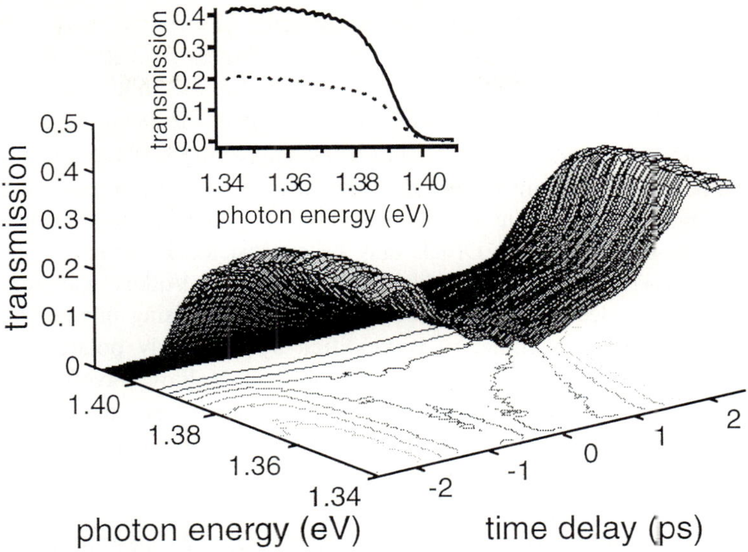

Fig. 1. Induced absorption below the band edge of GaAs strongly driven by a 3.5 μm MIR pump pulse as a function of the probe photon energy and time delay between the pump and probe (negative time delays correspond to pump pulses arriving before probe pulses). The inset shows the transmission at 2.5 ps delay (solid curve) and 0 ps delay (dashed curve) between the MIR and NIR continuum pulses.

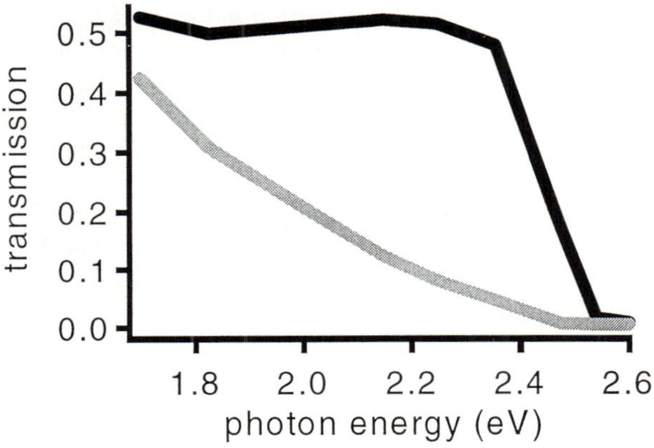

Fig. 2. Transmission below the band edge of ZnSe (3 mm thick) with (gray) and without (black) an intense 3.5 μm MIR driving field with $U_p \approx \hbar\omega$.

polycrystalline). We observed even more dramatic absorption using ~ 200 fs driving pulses at 3.5 μm (~ 2×10^{11} W/cm^2) due to the higher field strength.

We attribute this ultrafast electro-absorption to the high-frequency or dynamical Franz-Keldysh effect (DFKE) [1, 5]. Since the (static-field) Franz-Keldysh effect ($U_p >> \hbar\omega$) can be thought of as a tunneling assisted transition [6, 7], the DFKE ($U_p \sim \hbar\omega$) may be intuitively thought of as the point where the tunneling time is comparable to the period of the applied electric field. A quantitative description of the DFKE may be constructed in terms of effective mass Bloch-Volkov wavefunctions [1, 3], in analogy to Volkov wavefunctions that represent free electrons in an electromagnetic field. Using this perspective, the conduction band wavefunction is modified by a linearly polarized strong driving field to be given by the effective mass Bloch-Volkov wavefunction [3]. Using a model based on these Bloch-Volkov wavefunctions and the theory of x-ray absorption in gases in the presence of a strong laser field [8], we are able to simulate the data in GaAs driven by a ~ 1 ps, 3.5 μm MIR driving field with $U_p \approx \hbar\omega$.

Besides the scientific interest, our measurements may have an impact on future technological development. Due to the large electro-optic modulation of NIR light using MIR pulses, one may use the DFKE using MIR lasers as a basis for an ultrafast optical modulator. Alternatively, electronics may be developed that are able to apply strong fields (~ 10^4 V/cm) to electro-optic devices at THz frequencies. In this case, effects at the transition between classical and quantum behavior will play a role, due to the non-negligible photon energy.

In conclusion, we have made the first observation of ultrafast electro-absorption below the band edge in semiconductors due to the dynamical Franz-Keldysh effect. The observed ultrafast absorption induced by intense MIR pulses has an extent below the band gap that is the largest ever observed, to our knowledge. The DFKE occurs at the transition between classical and quantum behavior, i.e., when the ponderomotive potential is comparable to the photon energy of the strong driving field and photon effects cannot be ignored.

Acknowledgements. We gratefully acknowledge support from NSF Grant DMR-9970962, ONR N00014-94-1024, and the NEDO International Joint Research Grant Program. J. K. is supported in part by the Japan Science and Technology Corporation PRESTO Program.

References

1 Y. Yacoby, *Phys. Rev.* **169**, 610 (1968).
2 L. V. Keldysh, *Sov. Phys. JETP* **20**, 1307 (1965).
3 H. D. Jones and H. R. Reiss, *Phys. Rev. B* **16**, 2466 (1977).
4 Y. I. Balkarei and E. M. Epshtein, *Sov. Phys. Solid State* **15**, 641 (1973).
5 K. B. Nordstrom, K. Johnsen, S. J. Allen, A.-P. Jauho, B. Birnir, J. Kono, T. Noda, H. Akiyama, and H. Sakaki, *Phys. Rev. Lett.* **81**, 457 (1998).
6 D. A. B. Miller, D. S. Chemla, and S. Schmitt-Rink, *Phys. Rev. B* **33**, 6976 (1986).
7 L. V. Keldysh, *Sov. Phys. JETP* **34**, 788 (1958).
8 M. Jain and N. Tzoar, *Phys. Rev. A* **15**, 147 (1977).

Sub-Picosecond Dynamics of the Photo-Induced Magneto-Optical Kerr Effect in CdTe at Room Temperature

A. V. Kimel[1], F. Bentivegna[2], V. N. Gridnev[1], V. V. Pavlov[1], R. V. Pisarev[1], and Th. Rasing[2]

[1] A. F. Ioffe Physical Technical Institute of the Russian Academy of Sciences, Politekhnicheskaya 26, 194021 St. Petersburg, Russia
E-mail: kimel@pop.ioffe.rssi.ru
[2] Research Institute for Materials, University of Nijmegen, Toernooiveld 1, 6525 ED Nijmegen, The Netherlands
E-mail: fb@sci.kun.nl

Abstract. A time-resolved, pump and probe technique based on the photo-induced magneto-optical Kerr effect is applied to the spectral study of the room temperature spin dynamics in CdTe between 1.44 and 1.63 eV. Two relaxation processes with respective decay times of up to 200 fs and 2.5 ps are distinguished. A change of the spectral dependence regime of the Kerr effect during the faster relaxation is observed and discussed.

1. Introduction

As it is well known, circularly-polarized light produces a spin polarization in a medium [1]. During the past few years time-resolved magneto-optical techniques were successfully used to the studies of ultrafast spin dynamics in semiconductors [2, 3]. The intense pump light pulse induces a spin imbalance and its decay as a result of relaxation processes provokes the decay of the photo-induced magneto-optical Faraday or Kerr effect experienced by a time-delayed, linearly polarized probe pulse. The transient magneto-optical Kerr effect (MOKE) affecting the reflected probe pulses is then due to the pump-induced difference between the optical coefficients for right-handed (σ^+) and left-handed (σ^-) circularly polarized light. This paper presents results on dynamics of the photo-induced MOKE in CdTe at room temperature. We have found that this effect shows a `diamagnetic'-like (DM) spectral behavior [4], i.e. a sign reversal of the Kerr rotation at the gap energy, that can be explained by the splitting of spin sub-levels. However, the optical pumping of CdTe leads to the selective excitation of one particular transition, in which the probe pulse itself will be less likely to engage. Thus the optical coefficients of CdTe will be different for probe photons with two circular polarizations of opposite helicities and an additional contribution to the photo-induced MOKE will appear [5]. This contribution features a spectral behavior named `paramagnetic (PM) with a maximum of the Kerr rotation at the interband transition energy.

2. Experimental Procedure and Results

A pulsed Ti:sapphire laser was used (pulse duration of ~ 100 fs in the spectral range from 1.44 to 1.63 eV). Coherent pump and probe pulses (intensity ratio of 10:1, pump fluence of about 10 $\mu J/cm^2$) were focused on the sample. The Kerr rotation θ_K and ellipticity ε_K experienced by the reflected probe pulses were measured as a function of the time delay Δt between the pump and probe pulses with a temporal resolution of 6.6 fs. The circular polarization of the pump pulses was modulated at a frequency of 50 kHz between σ^+ and σ^-, and the reflected probe beam was splitted by a Wollaston prism into two spatially separated beams with orthogonal polarizations. The intensities of the beams were measured with two equivalent photodiodes, followed by a differential amplifier, the output of which, proportional to θ_K (or to ε_K in case of a quarter-wave plate placed in front of the prism), was detected by a lock-in amplifier.

A linear dependence of the photo-induced Kerr effect with respect to the pump intensity was found. Typical relaxation dynamics of θ_K are shown in Fig. 1 for optical excitation below (a) and above (b) the room temperature band-gap E_g = 1.52 eV. Clearly, the initial peak of θ_K does not exactly follow the shape of the cross-correlation function of pump and probe pulses (dashed lines), but shows a fast relaxation component up to Δt • 600 fs, after which a slower relaxation takes place within a few picoseconds. All over the photon energy range the decay times were found between 100 and 200 fs for the faster contribution, and between 1 and 2.5 ps for the slower one.

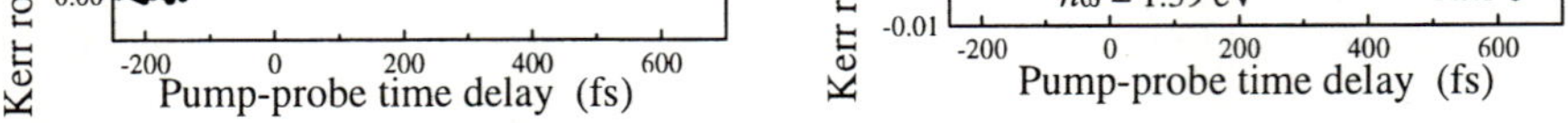

Fig. 1. Relaxation dynamics of θ_K in CdTe below (a) and above (b) the band-gap.

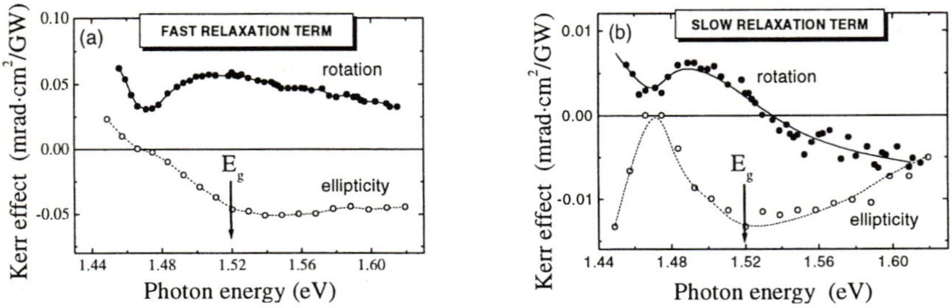

Fig. 2. Spectra of θ_K and ε_K in CdTe for the fast (a) and the slow relaxation term (b).

Rotation θ_K and ellipticity ε_K as a function of photon energy are shown in Fig. 2, both for the fast (a) and the slow relaxation term (b), for respective pump-probe delays of 0 and 600 fs. The spectral dependence of θ_K in the fast relaxation term shows a maximum near the interband transition E_g at 1.52 eV, hence a PM behavior. The increase towards a second maximum when the photon energy decreases below 1.44 eV is likely due to a transition at an energy E_i between the top of the valence band and a bound impurity level within the band-gap [7]. According to Kramers-Kronig relations ε_K ought to experience a sign reversal at E_g, but actually changes sign slightly below 1.52 eV (see Fig. 2b), which may indicate that the spectral weight of the transition at E_i is larger than that of the interband and excitonic transitions. In contrast, θ_K in the slow relaxation term changes sign at the interband transition energy while ε_K shows an extremum at this point, i.e. they exhibit a DM spectral behavior. Thus, an evolution of the MOKE from a PM to a DM spectral dependence has taken place during the fast relaxation process.

According to the qualitative mechanisms described above, the PM behavior observed in the fast relaxation term can be attributed to the difference in populations of excited states with different magnetic quantum numbers m_j. Therefore the decay of the PM behavior is assigned to the electron relaxation to the bottom of conduction band and exciton ionization. The predominant DM contribution to the slow relaxation term of the spectra is related to the splitting of spin sub-levels and caused by the repulsive interaction among carriers due to Pauli exclusion principle [8]. Therefore, the relaxation of this contribution is assigned to the spin scattering of electrons in the conduction band. The measured time of the spin relaxation is in reasonable agreement with that reported elsewhere [3, 9]

Acknowledgements. The authors thank B. Koopmans and A. V. Petukhov for fruitful discussions. This work was supported by NWO, INTAS, RFBR, the EU and the Robert Havemann Foundation.

References

1 *Optical Orientation*, Edited by F. Meyer and B. P. Zakharchenya (North-Holland, Amsterdam, 1984).
2 J. M. Kikkawa and D. D. Awschalom, Nature **397**, 139 (1999); I. Malajovich, J. M. Kikkawa, D. D. Awschalom, J. J. Berry, and N. Samarth, Phys. Rev. Lett. **84**, 1015 (2000).
3 B. Koopmans and W. J. M. de Jonge, Appl. Phys. B **68**, 525 (1999).
4 A. K. Zvezdin and V. A. Kotov, in *Modern Magnetooptics and Magnetooptical Materials* (IoP Publishing, Bristol, 1997).
5 B. Montegu, A. Laugier, and R. Triboulet, J. Appl. Phys. **56**, 3061 (1984).
6 W. H. Knox, R. L. Fork, M. C. Downer, D. A. B. Miller, D. S. Chemla, C. V. Shank, A. C. Gossard, and W. Wiegmann, Phys. Rev. Lett. **54**, 1306 (1985).
7 A. Tackeuchi, S. Muto, T. Inata, and T. Fujii, Appl. Phys. Lett. **56** 2213 (1990).
8 L. Viña, L. Muñoz, E. Pérez, J. Fernández-Rossier, C. Tejedor and K. Ploog, Phys. Rev. B **54**, R8317 (1996).
9 S. Bar-Ad and I. Bar-Joseph, Phys. Rev. Lett. **68**, 349 (1992).

Femtosecond Optical Time-of-Flight Measurement of the Electron Velocity in GaN

M. Wraback[1], H. Shen[1], J.C. Carrano[2], T. Li[3], and J.C. Campbell[3]

1 Sensors and Electron Devices Directorate, U.S. Army Research Laboratory, 2800 Powder Mill Rd., Adelphi, MD 20783, USA
2 Photonics Research Center, U.S. Military Academy, West Point, NY 10996, USA
3 Microelectronics Research Center, Department of Electrical and Computer Engineering, The University of Texas at Austin, Austin, TX 78712

Abstract. A femtosecond optically-detected time-of-flight technique has been used to determine the electron transit time and steady-state velocity-field characteristic in a GaN *p-i-n* diode. The peak electron velocity of 1.9×10^7 cm/sec is attained at 225 kV/cm.

Theoretical calculations [1,2] of the electron velocity-field characteristic in GaN predict a peak steady-state velocity of $2\text{-}3\times10^7$ cm/sec, implying that this material may be an important candidate for high frequency devices. Nevertheless, no direct measurements of carrier velocities in III-N materials have been reported. In this paper we present the first measurement of the electron velocity at high electric fields in GaN. An optically-detected time-of-flight technique [3-5] with femtosecond resolution that monitors the change in the electroabsorption due to charge transport in a GaN *p-i-n* diode has been used to determine the electron velocity-field characteristic at room temperature.

The nominal thicknesses of the *p*-layer, intrinsic (ν-type) region, and *n*-layer in the sample used in this study are 0.2 μm, 0.5 μm, and 3.4 μm, respectively. Visible ultrashort pulses derived from the signal beam of a 250 kHz regenerative amplifier-pumped optical parametric amplifier were compressed to less than 60 fs and frequency doubled to obtain a tunable source of ultraviolet pulses for pump-probe measurements. The pulses were incident on the *p*-layer of the sample. The spectrum of the pump pulse was centered at 3.41 eV, a photon energy corresponding to the peak photocurrent of the device. The diode was biased using 200 ns voltage pulses synchronized with the optical pulses. The change in transmission ΔT of the probe was monitored at a sub-bandgap photon energy chosen to maximize the sensitivity of this quantity to a change in the broadening of the absorption edge under applied electric field (Franz-Keldysh effect). Pump-generated electrons and holes moving under the applied plus built-in electric field E are swept toward the *n*- and *p*-layers, respectively. As the carriers drift apart, a space-charge field of opposite sign creates a potential difference that grows in time until transport is completed. The pump-induced charge density is kept low enough that this screening field ΔE is much smaller than E. The space-charge field

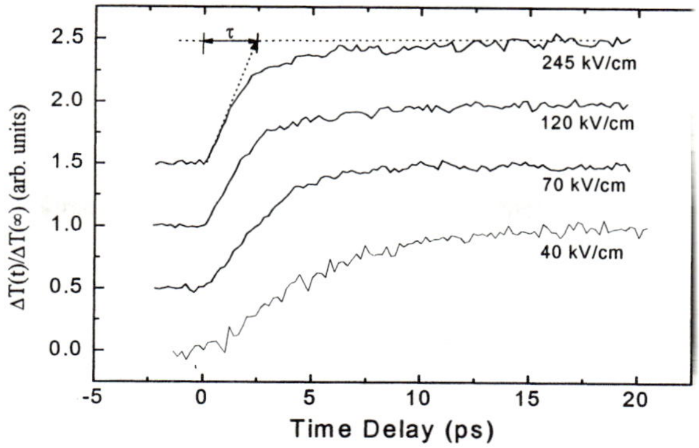

Fig. 1. Normalized photoinduced bleaching of the p-i-n diode electroabsorption as a function of time delay for four electric fields. The method of obtaining the transit time from the initial slope and the plateau of the transient is illustrated.

causes a reduction in the broadening of the absorption edge and concomitant induced bleaching of the relative probe transmission.

Figure 1 shows the normalized bleaching $\Delta T(t)/\Delta T(\infty)$ as a function of time delay for four electric fields. The normalization factor $\Delta T(\infty)$ refers to the change in transmission after all the carriers have traversed the sample. All of these curves are characterized by a quasi-linear increase in this quantity that begins at t=0 and eventually reaches a plateau at $\Delta T(t)/\Delta T(\infty)$ =1. The slope of the bleaching curves near t=0 is given by [3,4]

$$\frac{\partial}{\partial t}\frac{\Delta T(t)}{\Delta T(\infty)} \equiv \frac{1}{\tau} \cong \frac{v_e + v_h}{d}, \tag{1}$$

where v_e and v_h, the electron and hole velocities, respectively, are assumed to be constant, and τ is defined as the transit time across the sample, as depicted in figure 1. Since the carriers are generated at the interface of the *p*- and *i*-layers, the electrons drift much further than the holes, and the majority of the signal is due to electron transport. One may also notice that the slope changes more with increasing electric field at low fields than it does at high fields. This phenomenon implies that the electron transit time and velocity become 'saturated' – i.e., relatively independent of electric field – at high fields.

The experimentally determined electron velocity shown in figure 2 was obtained using equation 1. The calculated v_h [6] was included in equation 1 to achieve a more accurate estimate of the electron velocity. In the intermediate field regime (50-100kV), the electron velocity v_e possesses a weak, quasi-linear dependence on E. This behavior may be associated with electrons attaining

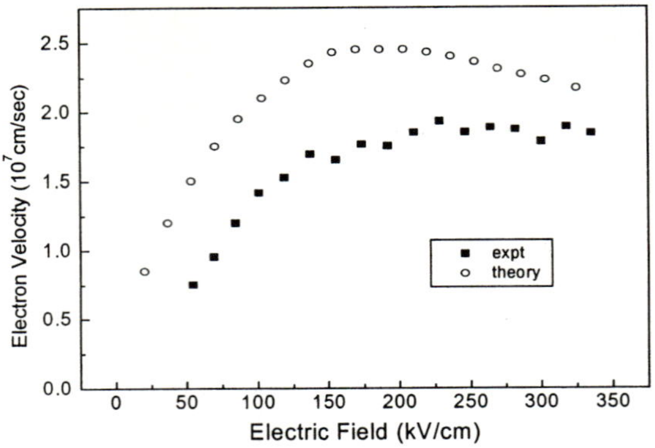

Fig. 2. Steady-state electron velocity as a function of electric field. Squares: experiment; Circles: theoretical calculation using a full zone band structure (from ref. 1).

enough excess energy to emit polar optical phonons [2]. In the high field regime, the electron velocity gradually becomes independent of electric field, reaching a peak of 1.9×10^7 cm/sec at ~ 225 kV/cm that corresponds to a transit time of ~2.5 ps. The steady-state velocity-field characteristic derived from an EMC calculation including a full Brillouin zone band structure [1] is shown in figure 2 for comparison with the measurements. The theoretical results are in qualitative agreement with the data. The discrepancies between the experimental and theoretical curves may be associated with poorly known deformation potential scattering parameters, as well as high defect density and polaron effects not accounted for by theory.

References

1. J. Kolnik, I.H. Oğuzman, K.F. Brennan, R. Wang, P.P. Ruden, and Y. Wang, J. Appl. Phys. **78**, 1033, 1995.
2. U.V. Bhapkar and M.S. Shur, J. Appl. Phys. **82**, 1649, 1997.
3. M. Wraback, H. Shen, J.C. Carrano, T. Li, J.C. Campbell, M.J. Schurman, and I.T. Ferguson, Appl. Phys. Lett. **76**, 1155, 2000.
4. C.V. Shank, R.L. Fork, B.I. Greene, F.K. Reinhart, and R.A. Logan, Appl. Phys. Lett. **38**, 104, 1981.
5. E.A. Schiff, R.I. Devlen, H.T. Grahn, J. Tauc, and S. Guha, Appl. Phys. Lett. **54**, 1911, 1989.
6. I.H. Oğuzman, J. Kolnik, K.F. Brennan, R. Wang, T.-N. Fang, and P.P. Ruden, J. Appl. Phys. **80**, 4429, 1996.

Femtosecond intersubband dynamics of holes in p-type $Si_{1-x}Ge_x$/Si multiple quantum wells

R. A. Kaindl[1], M. Wurm[1], K. Reimann[1], M. Woerner[1], T. Elsaesser[1], C. Miesner[2], K. Brunner[2], and G. Abstreiter[2]

[1] Max-Born-Institut für Nichtlineare Optik und Kurzzeitspektroskopie, Max-Born-Straße 2A, D-12489 Berlin, Germany, E-mail: woerner@mbi-berlin.de
[2] Walter-Schottky-Institut, Technische Universität München, Am Coulombwall, D-85748 Garching, Germany

Abstract: Ultrafast hole dynamics in $Si_{0.5}Ge_{0.5}$/Si multiple quantum wells is studied with resonant excitation of the HH1 → HH2 heavy-hole intersubband transition. Pump-probe experiments with 150-fs mid-infrared pulses demonstrate intersubband scattering and thermalization on a subpicosecond timescale.

Optical intersubband excitations in semiconductor nanostructures provide direct insight into the ultrafast dynamics of coherent optical polarizations and nonequilibrium carriers and play a key role for novel devices such as quantum cascade lasers. So far, most experiments have concentrated on intersubband excitations of quasi-two-dimensional electron plasmas. Their coherent and incoherent dynamics including many-body effects has been analyzed in quite some detail [1]. In contrast, intersubband transitions of quasi-two-dimensional holes, which are highly relevant for device applications [2], have been much less explored. Both the stationary optical spectra and the dynamics of holes after intersubband excitation are not well understood.

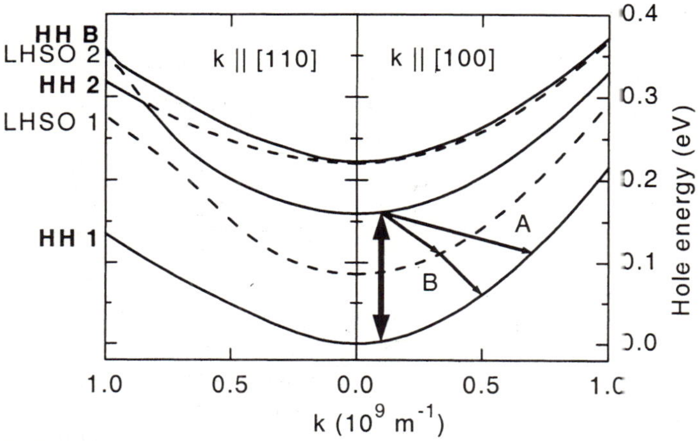

Fig. 1. Calculated in-plane dispersion of the valence band in a strained, modulation-doped $Si_{0.5}Ge_{0.5}$/Si multiple quantum well structure with a well width of d = 4.4 nm. In the femtosecond experiments, holes are excited from the HH1 to the HH2 subband (vertical arrow). Subsequent intersubband relaxation occurs predominantly via deformation potential scattering with optical phonons (diagonal arrows: A, B).

For instance, theoretical predictions for hole scattering rates between subbands separated by more than the energy of an optical phonon differ by orders of magnitude [2, 3]. Investigation of hole dynamics in real time is required to clarify these issues. Here, we present the first time-resolved study of hole plasmas in quantum wells using direct intersubband excitation and probing of heavy holes. We demonstrate intersubband scattering and thermalization of holes in p-type modulation-doped $Si_{0.5}Ge_{0.5}$/Si multiple quantum wells on a subpicosecond time scale. Emission of optical phonons via the deformation potential represents the dominant mechanism of intersubband relaxation whereas intraband optical phonon and Coulomb scattering of holes are important for thermalization. Thermalization results in a transient heating of the hole distribution which cools down to lattice temperature on a ≈ 25-ps timescale.

We investigate high-quality $Si_{1-x}Ge_x$/Si multiple quantum wells grown by solid-source molecular beam epitaxy [4]. The sample consists of ten-periods of p-type modulation-doped ($p = 1.2 \times 10^{12} cm^{-2}$) 4.4-nm wide $Si_{0.5}Ge_{0.5}$ quantum wells separated by 18-nm Si barriers.

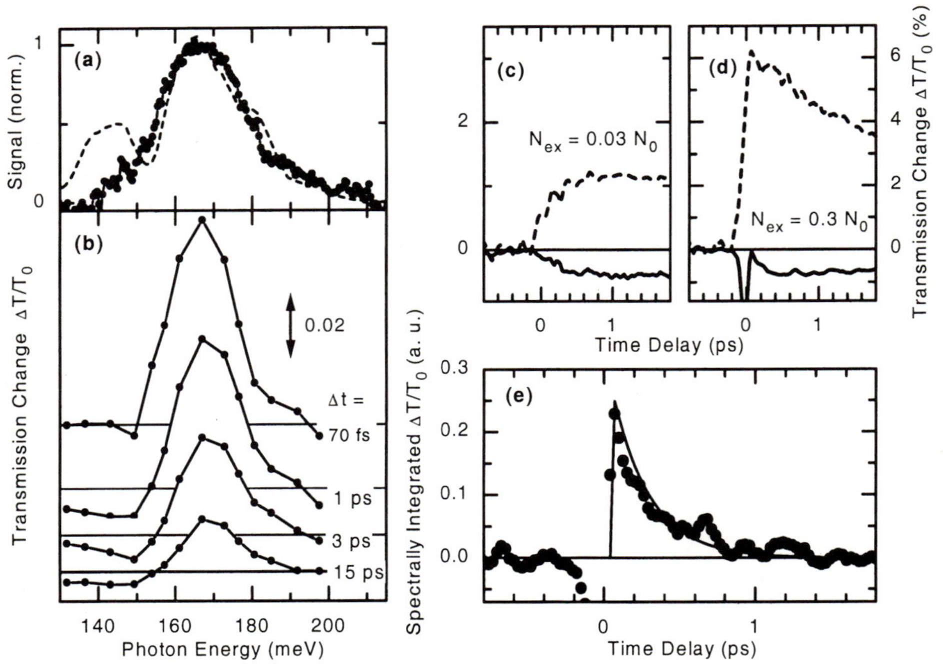

Fig. 2. (a) Normalized stationary intersubband absorption at T = 15 K (circles) and spectrum of the 150-fs pump and probe pulses (dashed line). (b) Transient spectra of intersubband absorption at several delay times Δt after femtosecond HH1 → HH2 excitation. (c, d) Time evolution of the transmission changes for different excitation densities N_{ex} and probe photon energies. (c) Transient bleaching (dashed line) at 167 meV and enhanced absorption (solid line) at 143 meV for $N_{ex} = 0.03\ N_0$. (d) Transient bleaching at 167 meV and enhanced absorption at 140 meV for $N_{ex} = 0.3\ N_0$. (e) Fast bleaching signal reflecting intersubband relaxation (symbols) as extracted from the transient spectra in (b). Solid line: Single exponential decay with a time constant of 250 fs.

Results of an eight-band k·p band structure calculation are shown in Fig. 1. Two-dimensional confinement and the $Si_{0.5}Ge_{0.5}/Si$ built-in strain induce a splitting between heavy-hole and light-hole bands and determine the sequence of subbands. These subbands can be classified at k = 0 as heavy-hole (HH n) or mixed light-hole split-off (LHSO n) subbands. At low temperatures, the carriers are confined to the HH1 subband. Polarized absorption spectra [Fig. 2(a): circles] reveal an intersubband line due to HH1 $\rightarrow$ HH2 transitions, centered at 167 meV in accordance with the calculated subband energy difference. The coupling to the intersubband dipole is maximized using a prism geometry. Mid-infrared pump and probe pulses of $\approx$ 150 fs duration with a spectrum shown as the dashed line in Fig. 2(a) are derived from the output of a 1-kHz optical parametric amplifier by difference frequency mixing in GaSe. Broad pulse spectra allow to probe across the whole intersubband line by spectral selection after the sample

Figure 2(b) shows transient spectra of time-resolved transmission changes $\Delta T/T_0$ for various delay times after femtosecond excitation of $\approx$ 30% of the holes initially present in the HH1 subband to the HH2 subband. The transients (solid circles) at the spectral positions of the largest bleaching (E = 167 meV) and of the largest induced absorption (E $\approx$ 141 meV) are shown in Figs. 2(c) and 2(d) for two excitation densities. Note that the fast contribution at early delay times is absent in the transients at E = 167 meV for 3% excitation density.

The transmission changes reflect the transient hole distributions: because of the different dispersions of the two subbands (Fig. 1) the optical intersubband transition energies are correlated with specific HH1 and HH2 energies. Transmission changes at photon energies E $\approx$ 170 meV are related to a non-equilibrium population in the HH2 band and to the depletion of HH1 states below the initial Fermi level E_{F0}. Theoretical calculations show a markedly different intensity dependence of these two contributions. Induced absorption at lower photon energies (E < 150 meV) results exclusively from increased HH1 carrier populations above E_{F0}.

At low excitation densities, the transmission changes are dominated by the holes in the HH1 band. The $\approx$ 1 ps rise and subsequent slow decay on a $\approx$ 25-ps timescale corresponds to heating and cooling, respectively. In contrast, the fast component observed at high excitation densities [solid circles in Fig. 2(d)] is due to a non-equilibrium population in the HH2 band. We have obtained this component by subtracting the heating dynamics in the HH1 band (overall induced absorption) from the transmission increase in Fig. 2(d). The resulting transient is shown in Fig. 2(e) and essentially represents the population dynamics of the HH2 band. We deduce an intersubband relaxation time of $\tau \approx 250$ fs. This is in excellent agreement with our calculation for the depopulation of the HH2 band by the emission of phonons via the optical deformation potential, which gives 230 fs.

References

1. R. A. Kaindl et al., Phys. Rev. Lett. **80**, 3575, 1998, and references therein.
2. L. Friedman, R. A. Soref, and G. Sun, J. Appl. Phys. **83**, 3480, 1998.
3. P. Lautenschlager, P. B. Allen, and M. Cardona, Phys. Rev. **B 33**, 5501, 1986.
4. T. Fromherz et al., Phys. Rev. B **50**, 5073, 1994.

Femtosecond mid-infrared study of electron dynamics in self-organized InAs quantum wires on InAlAs/InP (001)

E. Péronne, T. Polack, J. F. Lampin*, and A. Alexandrou
Laboratoire d'Optique Appliquée, ENSTA - Ecole Polytechnique - CNRS UMR 7639, F-91761 Palaiseau, France
E-mail: alexand@enstay.ensta.fr

F. Fossard and F. H. Julien
Institut d'Electronique Fondamentale, Université Paris-XI, CNRS URA 22, F-91405 Orsay, France

J. Brault and M. Gendry
Laboratoire d'Electronique-LEOM, Ecole Centrale de Lyon, CNRS UMR 5512, F-69131 Ecully, France

Abstract: We performed pump-mid-infrared-probe measurements in InAs/InAlAs self-organized quantum wires and observed a fast (3-ps) *electron* capture time from the barrier to the ground state and, at higher densities, a contribution of Coulomb-collision processes.

1. Introduction

Promising quantum-nanostructure devices such as lasers and infrared photodetectors have lead to an intensive study of the optical properties of quantum wires and dots. The physics of carrier dynamics in nanostructures is of particular importance for devices especially because a drastic slow-down of the relaxation is expected due to the discrete nature of the electronic energy states in quantum dots (phonon bottleneck effect). Nevertheless, short capture and relaxation times have been measured [1] and explained in terms of Coulomb collisions between electrons or electrons and holes [2]. In quantum wires, a slow-down of the relaxation through LO-phonon emission is also expected because of the reduced density of states with respect to the case of quantum wells.

2. Visible-pump mid-infrared-probe technique

We here present femtosecond pump-probe experiments in the mid-infrared range where, in contrast to previous experiments, the electron dynamics is clearly separated from that of holes. Electron-hole pairs are injected into the barrier by a 800-nm pump and the probe is tuned to a transition in the conduction band. Thus, the differential transmission signal reflects the electron population difference between the two conduction band states. The tunable mid-infrared pulses (3-18 μm) are obtained with a Ti:Sapphire oscillator followed by a 200-kHz regenerative amplifier, optical parametric amplification and difference frequency mixing [3].

3. Self-organized nanostructures

The samples were grown on an InP (001) substrate by solid-source molecular-beam epitaxy. Depending on the growth conditions, atomic-force microscopy images show the formation of either quantum-wire-like structures along [-110] or quantum dots elongated along [-110] [4,5]. The results presented below were obtained in ten periods of 3 monolayers InAs separated by 50-nm thick $In_{0.52}Al_{0.48}As$ spacers leading to wire-like InAs nanostructures with a wire width of about 22 nm at the base and a height of about 2.7 nm [6]. The samples were studied by means of photoluminescence, photoluminescence excitation spectroscopy and photo-induced infrared absorption and exhibit strong [110]-polarized mid-infrared electronic absorption around 89 meV at normal incidence with an oscillator strength comparable to that of quantum wells [5]. We attribute this absorption to the intraband transition between the first and second wire subband in the conduction band. The polarization and the energy of the transition are in agreement with this attribution.

4. Results and discussion

The differential transmission signal for two different probe polarizations is shown in Fig. 1. For a polarization along [-110], the signal is a step-like function as in the case of an InP substrate (see inset) and is due to the free-carrier absorption. In the [110] direction, the signal is mainly due to the transition between the ground electronic subband and the first excited subband. The rise time of this signal is 3 ps and corresponds to an effective capture time, i.e. capture of the electron in the wire plus subsequent relaxation to the ground state. Capture directly to the first subband is expected to be less probable.

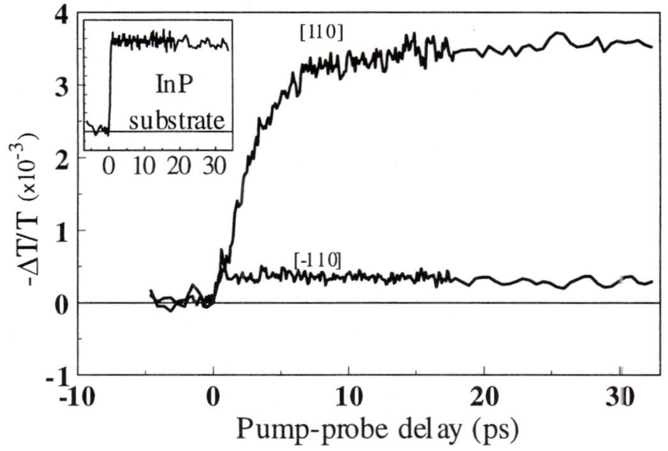

Fig. 1. Differential transmission of the quantum wire sample for probe polarizations along [110] and [-110]. The pump pulse energy was 0.1 nJ corresponding to an estimated carrier density of $7x10^4$ cm^{-1}. The inset shows the differential transmission of an InP substrate.

No negative signal is observed which means that the electrons rapidly scatter from the second to the first subband and that there is no electron accumulation in the second subband. To our knowledge, this is the first measurement of *electron* dynamics in quantum wires. We measured similar capture times in n- and p-doped quantum wires. This indicates that the capture and relaxation process is due to LO-phonon emission.

The differential transmission results as a function of pump energy are shown in Fig. 2 together with mono-exponential fits. In order to subtract away the polarization-independent free-carrier absorption and isolate the contribution of the quantum-wire absorption, we show the difference between the [110]- and [-110]-polarized signal. We measured a constant 3-ps rise time up to an excitation energy of 0.5 nJ/pulse, followed by a decrease of the capture time down to 1.5 ps for an excitation energy of 24 nJ/pulse. This behavior indicates a contribution of Coulomb-collision processes at high densities.

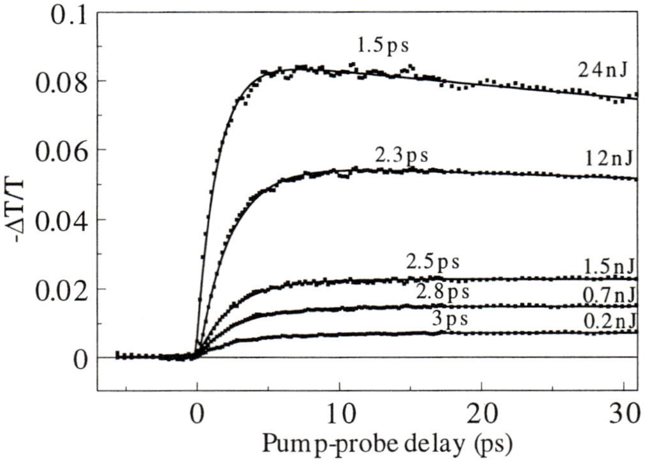

Fig. 2. Differential transmission for different pump pulse energies. The solid lines are exponential fits with rise-time constants ranging from 3 to 1.5 ps. For the two highest pump pulse energies, a second exponential was used in order to describe the slow decay time which we ascribe to electron-hole recombination.

References

* Present address: IEMN-Dpt.ISEN, BP 69, 59652 Villeneuve d'Ascq Cedex, France

1 T. S. Sosnowski, T. B. Norris, H. Jiang, J. Singh, K. Kamath, and P. Bhattacharya, Phys. Rev. B **57**, R9423 (1998).

2 A. V. Uskov, F. Adler, H. Schweizer, and M. H. Pilkuhn, J. Appl. Phys. **81**, 7895 (1997);R. Ferreira and G. Bastard, Appl. Phys. Lett. **74**, 2818 (1999).

3 M. K. Reed and M. K. Steiner Shepard, IEEE J. Quantum Electron. **32**, 1273 (1996).

4 J. Brault M. Gendry, G. Grenet, G. Hollinger, Y. Desières, and T. Benyattou, Appl. Phys. Lett. **73**, 2932 (1998).

5 A. Weber, O. Gauthier-Lafaye, F. H. Julien, J. Brault, M. Gendry, Y. Desières, and T. Benyattou, Appl. Phys. Lett. **74**, 413 (1999).

6 J. Brault, M. Gendry, O. Marty, M. Pitaval, J. Olivares, G. Grenet, G. Hollinger, Applied Surface Sciences (to be published).

Size and Surface Dependence of Ultrafast Carrier Dynamics in CdSe Quantum Dots Determined by Fluorescence Upconversion Spectroscopy

David F. Underwood, Tadd Kippeny, Rachel Ward, Andreas V. Kadavanich, Jason Taylor, and Sandra J. Rosenthal

Department of Chemistry, Vanderbilt University, Nashville TN 37235
E-mail: sjr@femto.cas.vanderbilt.edu

Abstract. Dynamics for band edge and deep trap emission in CdSe nanocrystals (NCs) were determined for several sizes ranging from 27Å to 72Å using the femtosecond fluorescence upconversion technique. The band edge rise and decay timescales are directly correlated to NC size and surface passivation. A new model is presented which implicates surface selenium dangling bond electrons in the ultrafast appearance of the deep trap emission and incorporates the previously suggested formation of triplet states for the long-lived components of the fluorescence decay.

Semiconducting quantum dots have gained tremendous attention because of their unique electronic and optical properties. Due to their ability to harvest much of the visible solar spectrum and their robust nature, cadmium selenide (CdSe) nanocrystals (NC's) are particularly suited for photovoltaic applications. The dynamics of the photo-induced charge carriers (electrons and holes) and their relaxation pathways are crucial in developing the most efficient solar cell device. Fluorescence upconversion spectroscopy is unique in its ability to monitor only the emitting state (the exciton) and is not obfuscated by interference from other dynamical processes such as excited state absorption or ground state recovery.

CdSe Nanocrystals. A series of narrow size-distributed (±5%) CdSe nanocrystals were prepared in the strict absence of oxygen using the method of Murray, Norris, and Bawendi [1], as modified by the Alivisatos group [2]. The nanocrystals were characterized by static absorption and emission spectroscopy, powder XRD, and high-resolution transmission electron microscopy. The stoichiometry and surface composition of the nanocrystals was determined by RBS [3]. When synthesized in absence of oxygen nanocrystals larger than ~45Å display no static emission at either the band edge or red-shifted surface emission wavelengths. Upon photo-irradiation with a CW argon ion laser in presence of air both emission features began to grow in, and with time the intensity of both features increased until the emission stabilized at its maximum. The absorption spectrum of the sample did not change and the sample remained optically clear, indicating the photo-oxidation treatment of the surface did not remove TOPO capping ligands or degrade the nanocrystals. The extent of photo-oxidation is being determined by NRA using ^{15}N.

Fluorescence Upconversion Spectroscopy. The fluorescence upconversion apparatus employs an elliptical reflector for sample fluorescence focusing and a 1mm $LiIO_3$ nonlinear mixing crystal, and will be described elsewhere[4]. Tunable excitation throughout the visible was obtained from a Coherent® titanium sapphire oscillator-regenerative amplifier-optical parametric amplifier laser system. 800 nm light leftover from the optical parametric amplification was used to gate the fluorescence. Prism compression on the visible and 800 nm beam paths lead to an instrument response function of ~200 fs.

The excitation wavelength is tuned on average to 20 nm above the first exciton peak and the excitation is < 4 nJ per pulse at 250 kHz. The nanocrystals are dissolved in anhydrous toluene (concentration ~50 μM) and flowed through a 1mm cell under argon. This sample concentration and excitation energy corresponds to ~1 excitation photon per nanocrystal. Data was collected with a 1 second dwell time per step and typically 5-10 scans were added together to make a data set. The static absorption and emission spectra of the samples did not change during the experiment.

Decay profiles for several CdSe nanocrystals ranging in sizes from 27 to 72 Å were collected and fit to a sum of three exponentials. Independent of size, each band edge trace contains an ultrafast rise component of less than 300 fs, followed by a sub-3 ps decay and finally a long-lived nanosecond component. For a given size NC, the band edge dynamics shows no wavelength dependence for emission monitored at 15 nm to the blue and red side of the peak of the static fluorescence spectrum. The band edge kinetics shows no dependence on excitation polarization, with an initial anisotropy value of 0.15 – 0.20.

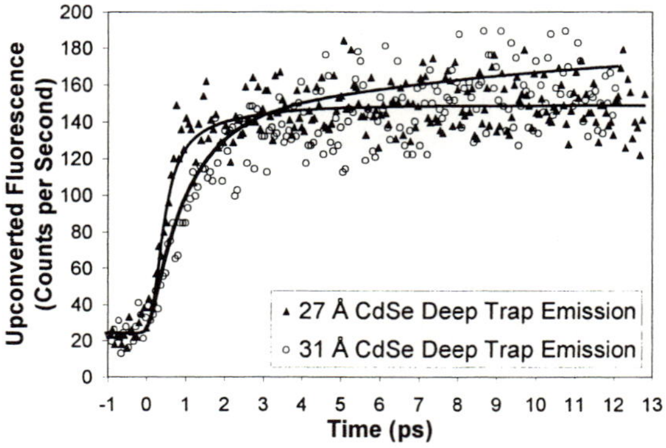

Figure 1. Fluorescence upconversion of the 885 nm deep trap emission for 27 and 31 Å CdSe nanocrystals in toluene.

To the best of our knowledge, we have obtained the first data on the "deep trap" dynamics from surface states with fluorescence upconversion (Fig. 1). Two sizes of NC's were studied with radii of 27 and 31 Å. The 27 Å NC's had rise times of 318 fs, and 1.61 ps, whereas the 31 Å sample had longer rise times of 929 fs and

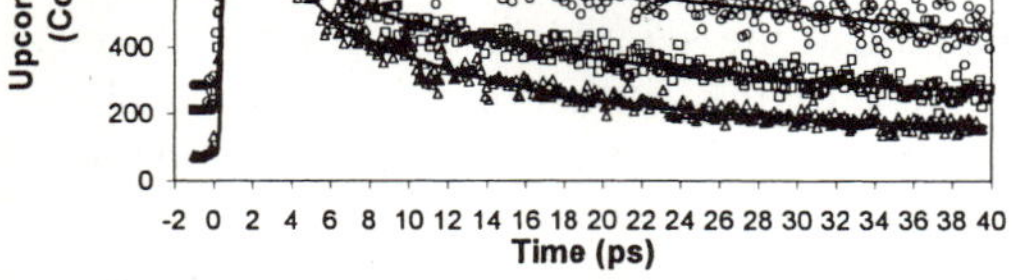

Figure 2. Size dependence of the decay dynamics for 30, 52, and 72 Å CdSe NCs

9.7 ps; both samples contained a long-lived nanosecond decay component.

The size dependence of the band edge emission is illustrated by the dynamics of 30 Å, 52Å, and 72 Å CdSe (Fig. 2). The smaller NC's fit to a sum of three exponentials with time constants of 155 fs (rise), 1.7 ps (decay), and 19 ps (decay); the larger NC's fit to time constants of 327 fs (rise), 3.2 ps (decay), and 35.5 ps (decay). The size dependence of the deep trap emission is evident from the upconversion of the 27 Å and 31 Å NC fluorescence at 885 nm. Fig. 3 shows an increase in the band edge rise time is seen upon increasing excitation energy, which is attributed to phonon-mediated relaxation to the lowest emitting state.

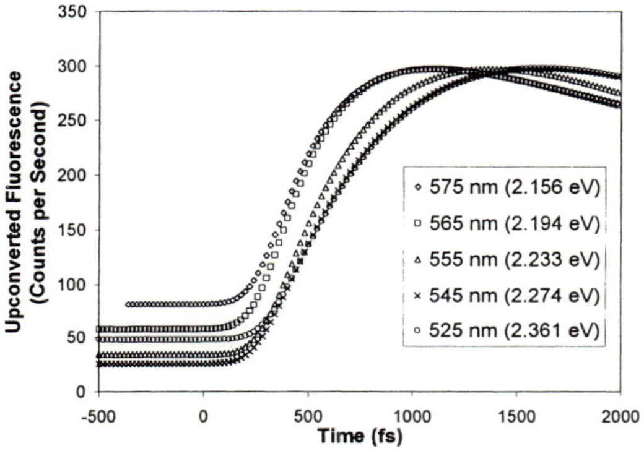

Figure 3. Excitation dependence on the rise time of 52 Å CdSe NCs.

A new model is presented which accounts for all the data we have thus far collected on CdSe NCs. The deep trap emission that appears within 2 ps is attributed to ultrafast relaxation of a surface selenium dangling bond electron to the valence band where it combines radiatively with the initial photo-generated hole. The long lifetime of the band-edge emission is due to its origin in a triplet state, with an energy lying just below the lowest electronic level in the conduction band. The extended deep trap emission arises from the relaxation of the excited state conduction band electron to a surface hole or vice-versa.

The size dependence of the ultrafast band edge and surface emission from CdSe nanocrystals has been studied by femtosecond fluorescence upconversion spectroscopy. For a given size the dynamics of the band edge emission did not depend on emission wavelength, but the band edge kinetics for CdSe nanocrystals ranging in size from 30 to 72 Å show appreciable size dependence. A new model is presented which describes the relaxation of excitons in CdSe nanocrystals.

This work was supported by the National Science Foundation under Career Award 9875875 and by the National Renewable Energy Laboratory under RFP #RAD-8-18688.

[1] C. B. Murray, D. B. Norris, and M. G. Bawendi, *J. Am. Chem. Soc.*, vol. 115, pp. 8706, 1993.

[2] J. E. B. Katari, V. L. Colvin, and A. P. Alivisatos, *J. Phys. Chem.* 98, pp. 4109-4117, 1994.

[3] J. Taylor, T. Kippeny, J. A. Bennett, M. Huang, L. C. Feldman, and S. J. Rosenthal, *Proceedings of the Materials Research Society, Microcrystalline and Nanocrystalline Semiconductors*, vol. 536, pp. 413, 1999.

[4] D. F. Underwood, T. Kippeny, and S. J. Rosenthal, *J. Phys. Chem.*, Submitted.

Investigation of phase change in $Ge_2Sb_2Te_5$ films by coherent phonon spectroscopy

M. Först[1], T. Dekorsy[1], M. Laurenzis[1], C. Trappe[1], H. Kurz[1], and B. Béchevet[2]

[1] Institut für Halbleitertechnik II, RWTH Aachen, Sommerfeldstr. 24, 52074 Aachen, Germany
E-mail: foerst@iht-ii.rwth-aachen.de

[2] Laboratoire d'Electronique de Technologie et d'Instrumentation, 17 Rue de Martyrs, 38054 Grenoble, France

Abstract. Thermally induced phase transitions of $Ge_2Sb_2Te_5$ are studied by coherent phonon spectroscopy. Changes of the coherent phonon spectra related to the crystallization of the sputter-deposited amorphous film and a further modification of the crystallographic structure are observed. The phonon modes detected can be assigned by a comparison with coherent phonon spectra of the binary Sb_2Te_3 and GeTe constituents.

1. Introduction

The development of future reversible optical storage media depends entirely on the fundamental knowledge of the microscopic nature of the phase changes involved. One of the most favoured material for optical data storage is the pseudo-binary GeTe-Sb_2Te_3 system whose reflectivity difference between the amorphous and crystalline state is exploited in DVD discs. Dependent on the composition, Ge-Sb-Te films crystallize from the as-deposited so-called "amorphous" state to a cubic NaCl-type lattice at temperatures between 120 °C and 150 °C. A second change of the crystallographic structure, i.e. the transition from the cubic to a hexagonal crystal lattice, occurs at temperatures above 200 °C [1].

Here, the thermally induced crystallization of "amorphous" $Ge_2Sb_2Te_5$ is investigated by coherent phonon spectroscopy. In a sputter-deposited $Ge_2Sb_2Te_5$ film, the crystallization of the "amorphous" phase to the cubic crystal lattice and the second phase transition to the hexagonal lattice are observed by changes in coherent phonon spectra detected in optical pump-probe experiments. Solely the crystallization of the as-deposited phase is accompanied by a huge increase of the static reflectivity.

2. Experiment

The sample investigated is a 200 nm thick $Ge_2Sb_2Te_5$ film sputter-deposited on a Si substrate. Coherent phonons are studied in a standard optical pump-probe setup in reflection geometry. Femtosecond laser pulses are derived from a Ti.sapphire oscillator at a central wavelength of 800 nm. Pump and probe beam are focussed to a spot size of approximately 40 μm in diameter on the sample with average intensities of 40 mW and 4 mW, respectively. The sample is mounted on a heating stage to increase the sample temperature from room temperature up to 300 °C.

2. Results and Discussion

Fig.1 shows time-resolved reflectivity changes detected at different sample temperatures. The reflectivity changes are modulated by coherent phonons that are launched by the femtosecond laser pulses [2]. The phonon signatures, i.e. their frequencies and amplitudes, strongly depend on the sample temperatures at which certain phases are stabilized.

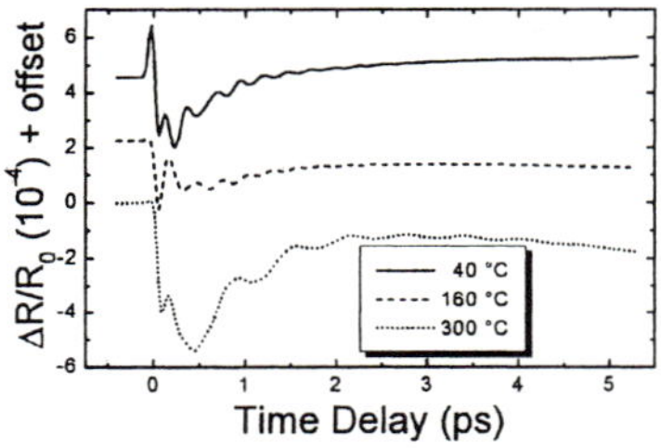

Fig. 1. Time-resolved reflectivity changes in a $Ge_2Sb_2Te_5$ film at different temperatures.

Fig.2 depicts coherent phonon spectra obtained by Fourier transforms of the phonon induced signal contributions and the static reflectivity of the film at temperatures between 100 °C and 300 °C. Here, the distinct phase changes and the transition temperatures are visualized. The as-deposited film crystallizes to the cubic lattice around 140 °C monitored by a modification in the coherent phonon spectra and an increase of the static reflectivity. A second strong change in the coherent phonon spectra occurs above 210 °C and is ascribed to the transition from the cubic to the hexagonal lattice structure. Contrary to the first phase transition, the static reflectivity does not change significantly in our measurements.

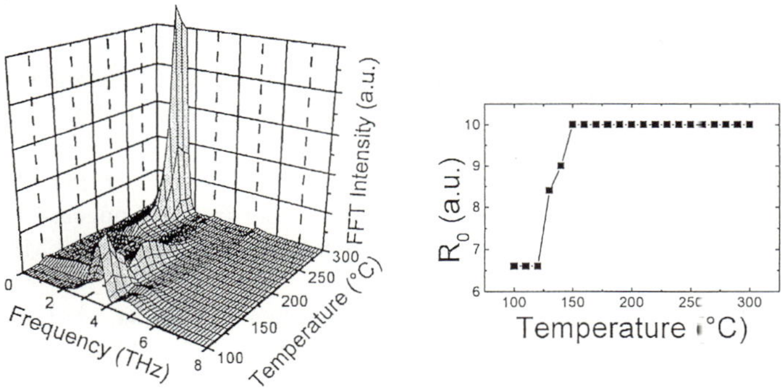

Fig. 2. Temperature-dependent coherent phonon spectra and static reflectivity of the $Ge_2Sb_2Te_5$ film.

The identification of the phonon modes observed is possible by comparative investigations of the phonon spectra of $Ge_2Sb_2Te_5$ and its counterpart of the binary GeTe and Sb_2Te_3 constituents (not shown). A comparison between "amorphous" $Ge_2Sb_2Te_5$ and "amorphous" GeTe reveals the phonon modes at 3.7 THz and 5.1 THz as symmetric vibrations of $GeTe_4$ tetrahedra and disordered Te chains, respectively [3]. Both modes are excited via the DECP mechanism (displacive excitation of coherent phonons) that selectively excites symmetric vibrations [4] .

The phonon modes at 2.0 THz, 3.5 THz, and 5.1 THz detected in the cubic lattice of $Ge_2Sb_2Te_5$ are observed in crystalline Sb_2Te_3, too. Thus, they are assigned to Raman-active modes of Sb_2Te_3 packets present in the randomly occupied NaCl-type lattice [5]. Fractions of ß-crystalline GeTe are likewise present in the cubic lattice. The associated degenerate phonon mode is only infrared-active and, hence, does not modulate the time-resolved reflectivity changes due to its vanishing Raman tensor [6].

In the hexagonal phase, the phonon spectra are dominated by a strong line at 1.5 THz that is attributed to symmetric vibrations of atomic layers stacked along the c-axis of the unit cell [1].

4. Conclusion

In conclusion, thermally induced phase changes in $Ge_2Sb_2Te_5$ films are investigated by coherent phonon spectroscopy. The crystallization of the sputter-deposited "amorphous" film to the cubic crystal lattice around 140 °C and the further transition to the hexagonal crystal lattice around 230 °C are clearly observed by modifications of the coherent phonon spectra. A comparison to the binary Sb_2Te_3 and GeTe constituents enables the identification of the phonon modes observed.

Acknowledgements. This work was financially supported by the *Deutsche Forschungsgemeinschaft* and the *Bundesministrium für Bildung und Forschung*.

References

1 N. Yamada, E. Ohno, K. Nishiuchi, N. Akahira, and M. Takao, J. Appl. Phys. **69**, 2849 (1990).

2 for a review: T. Dekorsy, G.C. Cho, and H. Kurz, in *Light Scattering in Solids VIII*, ed. by M. Cardona and G. Güntherodt (Springer, Berlin, 1999), and references therein.

3 G.B. Fisher, J. Tauc, and Y. Verhelle, in *Proc. 5th Intern. Conf. Liquid and Amorphous Semiconductors, Garmisch 1973* (Taylor and Francis, London, 1974), 1259.

4 H.J. Zeiger, J. Vidal, T.K. Cheng, E.P. Ippen. G. Dresselhaus, and M.S. Dresselhaus, Phys. Rev. B **45**, 768 (1992).

5 W. Richter, H. Köhler, and C.R. Becker, Phys. Stat. Sol. (b) **84**, 619 (1977).

6 E.F. Steigmeier and G. Harbeke, Sol. Stat. Comm. **8**, 1275 (1970).

Generation of Coherent LO Phonons in GaAs/AlGaAs MQW's by the Impulsive Stimulated Raman Scattering

K. J. Yee[1], Y. S. Lim[2], and D. S. Kim[1]

[1] Department of Physics, Seoul National University, Seoul 151-747, Korea
E-mail: kedroi@phya.snu.ac.kr
[2] Department of Applied Physics, Konkuk University, Chungju, Chungbook 380-701, Korea
E-mail: yslim@kku.edu

Abstract. We report on generation mechanisms of coherent LO phonons in GaAs/AlGaAs multiple quantum wells at resonant conditions. Both the forbidden and the allowed impulsive Raman scatterings are shown to generate coherent phonons, in stark contrast to bulk GaAs where the dominant generation mechanism is screening of surface fields.

1. Introduction

After the invention of femtosecond pulse lasers, generating and detecting coherent optical phonons in various materials became possible [1-5]. One intriguing problem in this area is what the generation mechanisms are. In general impulsive stimulated scattering is known to play an important role in transparent materials [4], and displacive excitation mechanism such as the DECP was proposed in absorbing materials [2]. In bulk GaAs, coherent LO phonons are known to be generated by the ultrafast screening of the surface space-charge fields [3]. However, little is known about the generation mechanisms of coherent phonons in GaAs quantum structures.

In this paper, we report on coherent LO phonon generation by the impulsive stimulated Raman scattering (ISRS) in absorbing region of a quasi-two dimensional GaAs/AlGaAs MQW. This is in sharp contrast with bulk GaAs, where ISRS plays only a minor role, if any.

2. Results and Discussion

We used the reflective electro-optic sampling technique (REOS) on a GaAs/$Al_{0.36}Ga_{0.64}As$ multiple quantum well (MQW) structure and a bulk GaAs at the sample temperature of 12 K. The MQW sample has the well width of 15 nm, and the barrier width of 5 nm. With the probe beam polarization fixed parallel to one of principal axes ([100] or [010]), the pump beam polarization was rotated to characterize the polarization dependence of the coherent LO phonon oscillations.

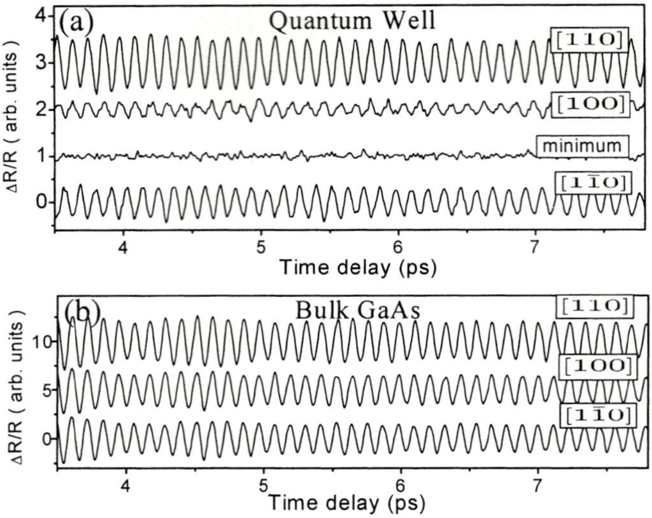

Fig. 1. The REOS signals in time domain obtained for a GaAs/AlGaAs MQW(a), and a bulk GaAs(b) at pump beam polarizations of [110], [100], [1 1 0] and at 'minimum' at which the phonon oscillation is minimum.

Figure 1 shows the coherent phonon oscillations at different pump-beam polarizations in our MQW sample (Fig. 1(a)) and a bulk GaAs (Fig. 1(b)) where 'minimum' corresponds to the angle of minimum phonon oscillation. Drastic changes in both the amplitude and the phase of the oscillations are observed in the MQW sample. In bulk GaAs sample, however, no polarization dependence is observed. In addition, noticeable asymmetry in the oscillation amplitude is observed between the [110] and the $[1\bar{1}0]$ directions for the MQW sample. More detailed angle dependence of the Fourier transformed intensity shows this more clearly in Fig. 2. While the overall angle dependence is sinusoidal, there is almost a factor of 2 difference between the two peaks.

All of our observations can be understood if we introduce both the allowed and forbidden Raman scattering tensors as well as their interference [6,7]. The "allowed" deformation potential Raman tensor is given as follows: $R_D = \begin{pmatrix} 0 & a \\ a & 0 \end{pmatrix}$, whereas the "forbidden", Fröhlich Raman tensor reads: $R_F = \begin{pmatrix} b & 0 \\ 0 & b \end{pmatrix}$. The angle dependence of the coherent phonon oscillation intensity is then: $|b + a\cos(2\varphi)|^2$. This equation fits Fig. 2 rather well, as shown in the dotted lines using b/a $\cong 0.25$.

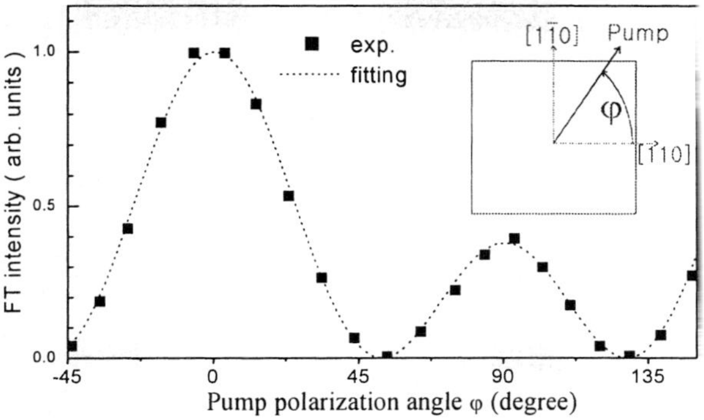

Fig. 2. FFT intensity of phonon oscillations as a function of the pump polarization angle φ which is defined in inset, the dotted line is the fitting to the data.

3. Conclusions

We have shown that coherent phonon generation in GaAs MQW is dominated by the ISRS mechanism, in sharp contrast to bulk GaAs where the screening mechanism is much more dominant.

References

1 G. C. Cho, W. Kütt, and H. Kurz, Phys. Rev. Lett. **65**, 764 (1990).

2 T. K. Cheng *et al.*, Appl. Phys. Lett., **57**, 1004 (1991).

3 T. Pfeifer, T. Dekorsy, W. Kütt, and H. Kurz, Appl. Phys. A: Solids Surf. **55**, 482 (1992).

4 Y. Liu, A. Frenkel, G. A. Garrett, J. F. Whitaker, S. Fahy, C. Uher, and R. Merlin, Phys. Rev. Lett., **75**, 334 (1995).

5 K. J. Yee, D. S. Yee, D. S. Kim, T. Dekorsy, G. C. Cho, and Y. S. Lim, Phys. Rev. B **60**, R8513 (1999).

6 C. Y. Chen. Phys. Rev. B 27, 1436 (1983).

7 J. Menendez, and M. Cardona, Phys. Rev. Lett. **51**, 1297 (1983)

Observation of coherent confined LO phonons in $^{70}Ge/^{74}Ge$ isotope superlattices

M. Nakajima[1], H. Harima[1], K. Morita[2], K. M. Itoh[3], K. Mizoguchi[4], E. E. Haller[5], and M. Hangyo[6]

[1] Department of Applied Physics, Osaka University, 2-1 Yamadaoka, Suita 565-0871, Japan
E-mail: nakajima@ap.eng.osaka-u.ac.jp
[2] Department of Applied Physics And Physico-Informatics, Keio University, 3-14-1 Hiyoshi, Kohoku-ku, Yokohama 223-8522, Japan
[3] Department of Applied Physics And Physico-Informatics, Keio University and PRESTO-JST, 3-14-1 Hiyoshi, Kohoku-ku, Yokohama 223-8522, Japan
[4] Department of Applied Physics, Osaka City University, 3-3-138 Sugimoto, Sumiyoshi-ku, Osaka, 558-8585, Japan
[5] UC Berkeley and Lawrence Berkeley National Laboratory, Berkeley, California 94720, USA
[6] Research Center for Superconducting Materials and Electronics, Osaka University, 2-1 Yamadaoka, Suita 565-0871, Japan

Abstract. Coherent higher-order confined LO phonons are reported for the first time by a pump-probe technique in $^{70}Ge/^{74}Ge$ isotope superlattices. The Fourier-transformed spectra are well explained by a linear chain model and a bond polarizability approach.

1. Introduction

Investigations of coherent phonons have been conducted with various materials including bulk semiconductors and semiconductor superlattices (SLs) [1-3]. However, some basic issues are still unclear including the generation and detection mechanism of coherent phonons in light-absorbing materials. For example, it is not easy to predict the observable phonon modes in GaAs-based superlattices. As for the coherent confined LO phonons, only the first-order GaAs-like mode has been observed in GaAs/AlAs [3], and no higher-order modes have been reported.

$^{70}Ge/^{74}Ge$ isotope superlattices (abbreviated hereafter as Ge I-SLs) present an excellent model to study vibrational properties of SLs, because the atomic structure is subject to a periodic isotopic mass modulation while the electronic structure is the same as that of a bulk crystal. The Raman spectra have been investigated by Cardona and co-workers. [4].

2. Experimental Methods

The samples were Ge I-SLs with a periodic structure of $^{70}Ge_n/^{74}Ge_n$ grown by molecular-beam-epitaxy (MBE). Here, n (= 8, 16, 32) is the number of monolayers in each constituent layer. High purity Ge (001) wafers of natural isotopic composition have been used as substrates. ^{70}Ge buffer layers of 20 nm thickness were grown prior to the SL formation. All samples had a total thickness of 50nm, thus the total number of periods of each SL varied with *n*.

A reflection-type pump-probe measurement was performed at liquid N_2 temperature by using a mode-locked Ti:sapphire laser. The laser produces optical pulses with a duration of 20 fs and a wavelength of 820 nm.

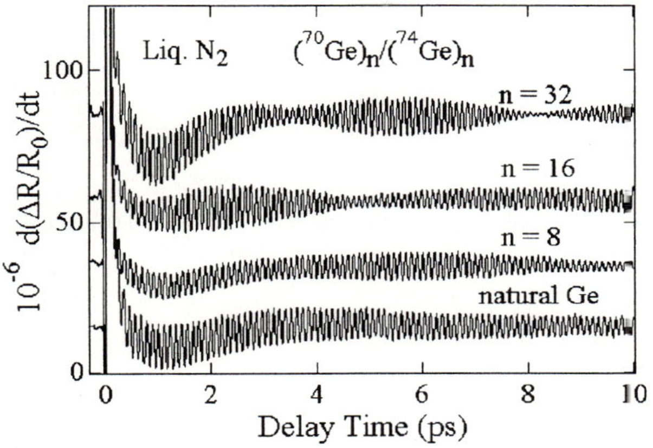

Fig. 1. Oscillatory component of the time-domain signals in $^{70}Ge_n/^{74}Ge_n$ isotope superlattices with various layer thicknesses and natural Ge.

3. Results and Discussion

Figure 1 shows the time-derivative traces of reflectivity changes for Ge I-SLs with various layer thicknesses and natural Ge. Beatings are observed for all Ge I-SL samples, indicating that multiple phonon modes are generated. In order to analyze the observed coherent phonon oscillations, Fourier transformation of the time-domain signals was performed as depicted in Fig. 2(a). Calculated spectra are shown in Fig. 2(b) for comparison with the observed spectra. The calculation is based on a linear chain model and a bond polarizability approach with appropriate frequencies and masses for the Ge isotopes [5-7]. We assumed that the bond polarizability is uniform for all the bonds. The calculated spectra consist of LO phonon modes confined in each constituent layer and reproduce well the FT spectra in Fig. 2 (a). From comparison with the calculated spectra, the modes observed by the pump-probe technique are assigned to confined SL phonons as denoted Fig. 2 (a). An assignment (3, 70) means the 3-rd order confined mode in ^{70}Ge layers. Peaks for bulk ^{70}Ge from the buffer layer (9.26 THz) and natural Ge from the substrate (9.12 THz) appear clearly for all Ge I-SLs as marked by vertical dashed lines. As *n* increases, the first-order mode confined in ^{74}Ge layer (open triangle) is enhanced relative to that in the ^{70}Ge layers, and the counterpart confined in the ^{70}Ge layers (filled triangle) shifts to higher frequency. Higher order modes confined in ^{70}Ge layers (asterisk) are observed for *n* =16 and 32.

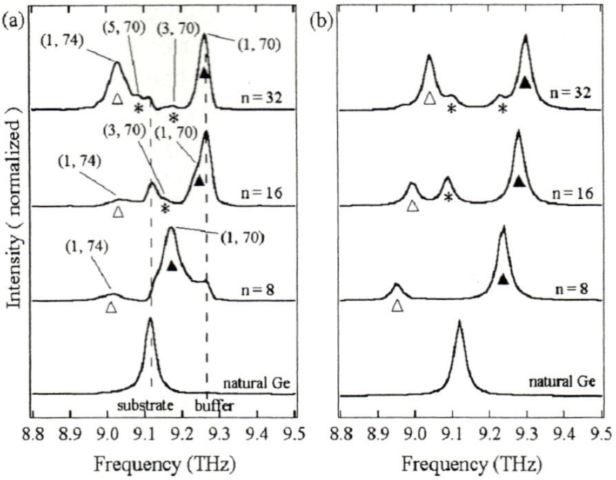

Fig. 2. (a) Fourier transformed spectra of time-domain signals, and (b) calculated spectra.

4. Conclusions

We have observed coherent higher-order confined LO phonon oscillations in $^{70}Ge/^{74}Ge$ isotope superlattices using a pump-probe technique. The Fourier-transformed spectra of the time-domain signal agree well with results of a calculation based on a linear chain model and a bond polarizability approach.

References

1 J. J. Baumberg, D. A. Williams, and K. Kohler, Phys. Rev. Lett. **78**. 3358, 1997.
2 K. Mizoguchi, M. Hase, S. Nakashima, and M. Nakayama, Phys. Rev. **B60**, 8262, 1999.
3 T. Dekorsy, A. M. T. Kim, G. C. Cho, H. Kurz, A. V. Kuznetsov and A. Förster, Phys. Rev. **B53**, 1531, 1996.
4 J. Spitzer, T. Ruf, M. Cardona, W. Dondl, R. Schoner, G. Abstreiter, and E. E. Haller, Phys. Rev. Lett. **72**, 1565, 1994.
5 B. Jusserand and M. Cardona, in: *Light Scattering in Solids V*, ed M. Cardona and G. Güntherodt (Springer, 1989) pp. 49.
6 H. D. Fuchs, C. H. Grein, M. Cardona, W. L. Hansen, K. Itoh, and E. E. Haller, Solid State Commun. **82**, 225, 1992.
7 H. D. Fuchs, C. H. Grein, C. Thomsen, M. Cardona, W. L. Hansen, E. E. Haller, and K. Itoh, Phys. Rev. **B43**, 4835, 1991.

Effect of lattice defects on LO phonon-plasmon coupled modes in n-GaAs

Muneaki Hase[1], Kunie Ishioka[1], Masahiro Kitajima[1], and Kiminori Ushida[2]

[1] National Research Institute for Metals, 1-2-1 Sengen, Tsukuba 305-0047, Japan
E-mail: hasedon@nrim.go.jp
[2] RIKEN (The Institute of Physical and Chemical Research), 2-1 Hirosawa, Wako 351-0198, Japan

Abstract. We have studied effect of point defects on generation and relaxation of coherent LO phonon-plasmon coupled modes in He-ion irradiated n-GaAs. The results suggest that defect-induced carrier trapping dominates annihilation of the coherent coupled modes.

1. Introduction

GaAs is a key material for ultrafast switching and THz radiation because the carrier lifetime of low-temperature (LT) grown GaAs is very short (sub-picosecond). The dynamics of photo-excited carriers and coherent phonons have been investigated in as-grown and annealed LT-GaAs [1]. In doped GaAs, the plasmon and the LO phonon form coupled modes through Coulomb interactions, and the frequencies of the LO phonon-plasmon coupled (LOPC) modes depend on the carrier density. Ultrafast relaxation of the coherent LOPC modes in n-doped GaAs has been studied using pump-probe techniques, and indicated that electron-hole scattering plays a dominant role in dephasing of the LOPC modes [2,3]. For highly n-doped GaAs, relaxation time of the LOPC modes did not depend on photo-excited carrier density and the dephasing of the coupled modes was thought to be caused by impurity or defect scattering when doping level was higher than photo-excitation [3]. Detailed study for the generation and relaxation of the coherent LOPC modes under the existence of active defects is required for understanding those natures. Here we report the effect of lattice defects on coherent LOPC modes in He^+ irradiated n-GaAs for different doses.

2. Experimental Methods

The samples used were Si doped n-GaAs with the carrier density of $n_{dop}=1.4\times10^{18}$ cm^{-3}. In order to examine the effect of point defects, 5 keV He^+ ions were irradiated to n-GaAs samples at various doses in a UHV chamber with base pressure of 3×10^{-9} Torr. The femtosecond pump-probe measurements were carried out at room temperature using a mode-locked Ti:sapphire laser whose pulse width was 20 fs. The pump- and probe-beam powers were fixed to 60 mW and 6 mW, respectively. Isotropic reflectivity changes ($\Delta R/R$) were measured to obtain relaxation dynamics of non-equilibrium carriers, and anisotropic reflectivity changes ($\Delta R_{eo}/R$) were measured using an EO sampling technique to obtain coherent oscillations as a function of the time delay by scanning the optical path length of the probe beam.

3. Results and Discussion

Figure 1 shows the time derivatives of $\Delta R_{eo}/R$ signal with the Fourier transformed (FT) spectra. The beating pattern in the time-domain data is due to co-existence of the strong LO mode and the L_- mode as shown in its FT spectrum. Disappearance of the mode-beating pattern at high ion dose suggests that the pump pulse does not generate the L_- mode. The decay time of the LO phonon estimated from fitting with damped harmonic oscillations slightly increases from 2.1 to 3.0 ps as increasing the ion dose. However, the decay time of the L_- mode decreases from 970 ± 40 to 340 ± 40 fs with increasing ion dose. These results suggest the decrease in effective carrier density that is due to point defects introduced by He^+ irradiation. As a result, electron-phonon interaction becomes less efficient, which would lead to a longer lifetime of the LO phonon, and the L_- mode changes from a phonon-like mode with a long dephasing time to a plasmon-like mode with a short dephasing time [2,3].

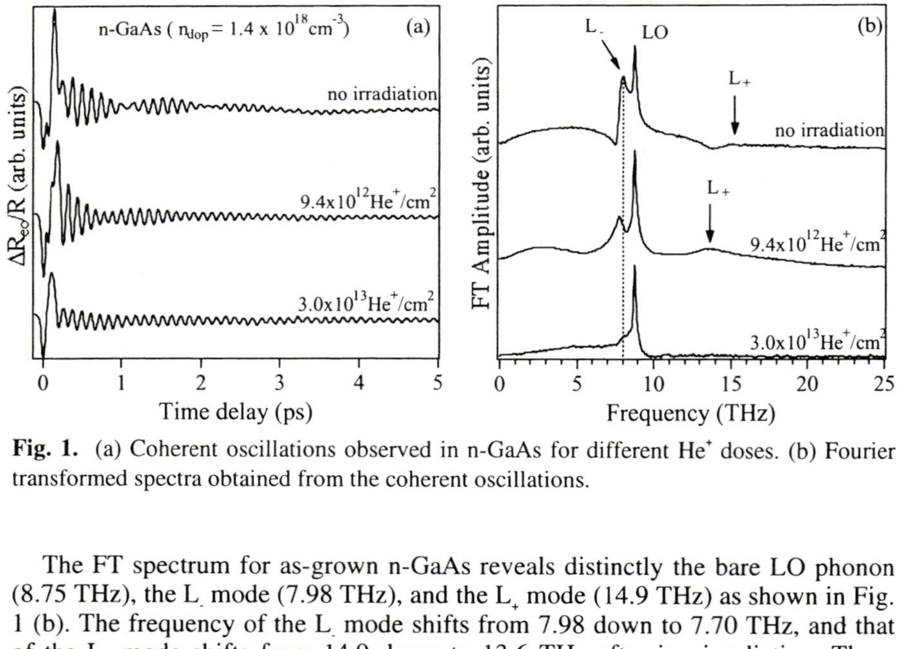

Fig. 1. (a) Coherent oscillations observed in n-GaAs for different He^+ doses. (b) Fourier transformed spectra obtained from the coherent oscillations.

The FT spectrum for as-grown n-GaAs reveals distinctly the bare LO phonon (8.75 THz), the L_- mode (7.98 THz), and the L_+ mode (14.9 THz) as shown in Fig. 1 (b). The frequency of the L_- mode shifts from 7.98 down to 7.70 THz, and that of the L_+ mode shifts from 14.9 down to 13.6 THz after ion irradiation. These frequency-shifts of the coupled modes are thought to be resultant of the change in the effective carrier density. The decay time of the L_+ mode obtained by a time partitioning Fourier transform (TPFT) [3] increases from 100 ± 40 fs (for as-grown) to 150 ± 40 fs (for $9.4\text{x}10^{12}\,He^+/cm^2$). This increase can be explained by the assumption that effective carriers in ion-irradiated samples are less than those in the as-grown sample and the fact that dephasing time of the L_+ mode was longer for low doped n-GaAs than high doped n-GaAs [3].

As shown in Fig. 2, the isotropic reflectivity change consists of an initial fast relaxation (sub-picosecond time scale) and a following slow relaxation. The fast one corresponds to carrier-carrier and carrier-phonon scattering, and the slow one corresponds to recombination of carriers. The relaxation time of the fast and slow components is deduced by fitting the time-domain data to a double exponential decay. The obtained fast scattering time does not depend on ion irradiation, being 150 ± 40 fs for all the samples. However, the slow relaxation time strongly depends on ion irradiation. They are over 15.0, 11.1, and 2.4 ps for the sample labeled (a), (b), and (c), respectively. These results suggest that point defects involve deep levels (or deep centers) below the conduction bands, and as a result, photo-excited carriers decay into the deep levels after carrier-carrier and carrier-phonon scattering. Thus an existence of defect-induced deep centers, at which majority carriers are trapped, is observed for ion-irradiated n-GaAs as the decrease in recombination time of photo-excited carriers.

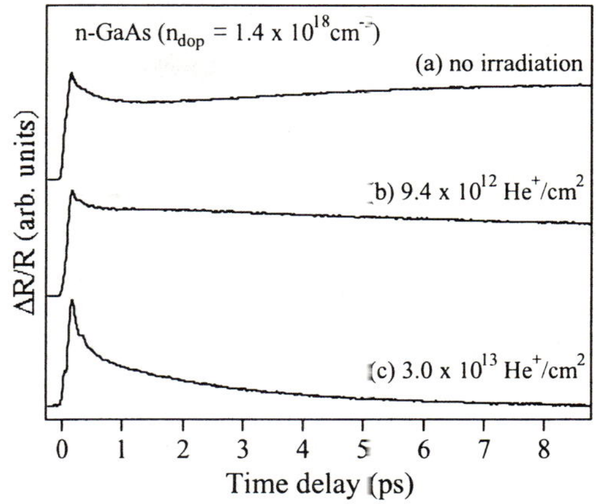

Fig. 2. Traces of the isotropic reflectivity change obtained for the as-grown n-GaAs (a) and the He^+ irradiated n-GaAs (b and c).

4. Conclusions

The effect of point defects on the ultrafast dynamics of the coherent LO phonon-plasmon coupled modes in ion-irradiated n-GaAs has been studied by using a femtosecond pump-probe technique. The annihilation of the coherent LOPC modes is observed with He-ion irradiation, which is explained by trapping of majority carriers at deep levels below the conduction bands.

References

1 T. Dekorsy, H. Kurz, X. Q. Zhou, and K. Ploog, Appl. Phys. Lett. **63**, 2899 (1993).

2 G. C. Cho, T. Dekorsy, H. J. Bakker, R. Hövel, and H. Kurz, Phys. Rev. Lett. **77**, 4062 (1996).

3 M. Hase, S. Nakashima, K. Mizoguchi, H. Harima, and K. Sakai, Phys. Rev. B **60**, 16526 (1999).

Part IX

Isolators, Metals and Superconductors

Ultrafast dynamics of excitonic self-trapping: The role of the electron-phonon interaction

A. D. Van Pelt and S. L. Dexheimer*

Department of Physics, Washington State University, Pullman, WA 99164-2814, USA
E-mail: dexheimer@wsu.edu

Abstract. The ultrafast electronic and vibrational dynamics of the exciton self-trapping process in quasi-one-dimensional systems are investigated as a function of the strength of the electron – phonon coupling using femtosecond vibrationally impulsive excitation techniques.

1. Introduction

In a wide range of ultrafast processes, including ultrafast electron transfer, localization of electronic excitations, and photoinduced structural transformations, the electron – lattice interaction directly drives the dynamics, largely determining the structure, formation, and evolution of the photoexcited system. In this work, we directly address the role of the electron-phonon coupling by studying the dynamics of photoexcitations in materials in which the strength of the electron – phonon coupling can be systematically varied, providing unique insight into the ultrafast dynamics.

In our initial work [1], we directly time-resolved the self-trapping of an exciton, a photoinduced structural change in which an electronic excitation becomes localized as a result of its interaction with the lattice. This process was detected using femtosecond vibrationally impulsive excitation of a quasi-one-dimensional material, the mixed-valence metal-halide linear chain (or MX) complex $[Pt(en)_2][Pt(en)_2Br_2]\cdot(PF_6)_4$, (en = ethylenediamine, $C_2H_8N_2$), also abbreviated as PtBr(en). In these experiments, the formation of the self-trapped exciton is detected by monitoring the appearance of its characteristic red-shifted absorbance, and the associated vibrational dynamics are detected as a large amplitude, strongly damped wavepacket response that accompanies the formation of the self-trapped exciton absorbance. We found that the self-trapped exciton forms on a time scale comparable to a single vibrational period of the lattice, consistent with the idea that the vibrational motion directly drives the dynamics of the self-trapping process.

The MX complexes are ideal for studying the role of the electron – phonon coupling in the self-trapping process because of their structural tunability, and we have extended these studies by investigating the dynamics of a series of complexes with varying electron – phonon coupling strengths. The quasi-one-dimensional structure along the chain axis can be represented schematically as:

$$\ldots \text{-------} X^- - Pt^{+(3+\delta)} - X^- \text{-------} Pt^{+(3-\delta)} \text{-------} X^- - Pt^{+(3+\delta)} - X^- \text{-------} \ldots$$

which shows both the periodic charge disproportionation (or mixed-valence character) and the periodic bond length distortion (or Peierls distortion) characteristic of the charge density wave ground state. The strength of the electron-phonon interaction varies dramatically with chemical composition, and ranges from very strong coupling in the highly distorted, valence-localized PtCl system to weaker coupling in the more valence-delocalized PtI system. These effects can be seen clearly in the ratio ρ of the short to the long Pt-X bond lengths: ρ = 0.75, 0.83, and 0.92 for the PtCl(en), PtBr(en), and PtI(en) complexes, respectively [2].

In the linear chain complexes, as in other quasi-one-dimensional systems, the electron – lattice interaction can lead to the formation of localized excitations including self-trapped excitons, solitons, and polarons. In the case of the MX complexes, excitation of the optical intervalence charge-transfer transition, which effectively transfers charge between the inequivalent metal sites, creates a highly nonequilibrium lattice configuration, and subsequent lattice relaxation and charge transfer processes lead to the evolution the excitations. Theoretical work based on Peierls-Hubbard models has predicted the formation of a self-trapped exciton state following photoexcitation and its subsequent decay into solitons, polarons, and bipolarons [3].

2. Experimental Methods

To resolve the self-trapping dynamics, we have carried out femtosecond transient absorption measurements in the vibrationally impulsive limit. By exciting the system with a pulse short compared to the period of the characteristic lattice motion, we generate nonstationary vibrational wavepackets, and we can follow the vibrational dynamics by measuring the resulting modulation of the optical response. In parallel with the vibrational wavepacket motion, the dynamics of the formation of the self-trapped exciton can be identified by the appearance of its characteristic red-shifted absorbance [4].

Transient absorption measurements were carried out using an amplified Ti:S laser system operating at a repetition rate of 1 kHz, thereby minimizing potential artifacts due to multiple excitation effects. For the PtBr(en) complex, the material was excited using pulses 35 fs in duration centered at 800 nm, near the onset of its intervalence charge transfer band, and probe pulses were generated using a compressed white light continuum. The PtCl(en) complex was excited near the onset of its intervalence charge transfer band using pulses 25 fs in duration centered at 600 nm that were generated by spectral filtering and compression of a hollow fiber continuum. Wavelength resolution in the differential transmittance measurements was achieved by spectrally filtering the probe beam after it had passed through the sample. The measurements were carried out on single crystal samples of good optical quality. Recent advances in the synthesis of the metal-halide complexes have produced materials that are largely free of defects and that are resistant to photoinduced damage [5].

3. Results and Discussion

A representative measurement on the strongly-coupled PtCl(en) complex is shown in Fig. 1 [6]. The time-resolved differential transmittance shows the formation of an induced absorbance, corresponding to the characteristic absorption of the self-trapped exciton state, on a time scale of ~ 200 fs. The induced absorbance is strongly modulated by wavepacket oscillations, and the frequency components are evident in the Fourier power spectrum shown in Fig. 2. A strong Fourier component appears at ~ 315 cm^{-1}, which is the frequency of the symmetric stretch chain axis mode of the complex observed by resonance Raman spectroscopy [7], and is consistent with resonantly-enhanced impulsive stimulated Raman excitation of this fully symmetric ground state mode. A second component, not present in the ground state Raman spectrum, appears at ~ 180 cm^{-1} in the Fourier spectrum.

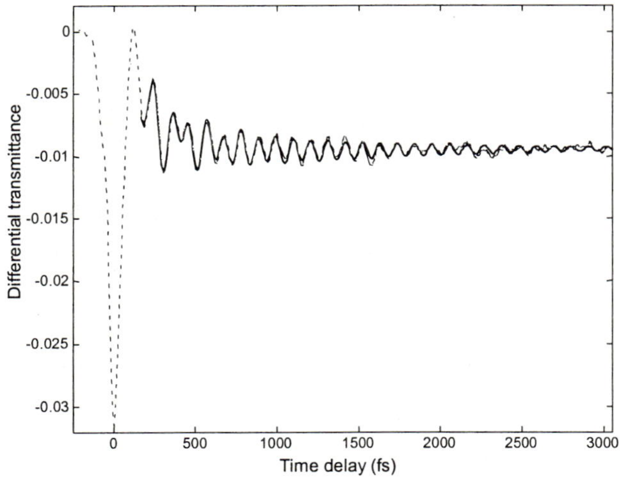

Fig. 1. Time-resolved differential transmittance of PtCl(en) following excitation of the intervalence charge transfer transition with pulses 25 fs in duration centered at 600 nm, probed at a detection wavelength of 800 nm. The feature near t = 0 includes contributions from cross-phase modulation that occurs during the temporal overlap of the pump and probe pulses. The solid line corresponds to the linear prediction fit.

The time-resolved measurements were analyzed using the linear prediction / singular value decomposition method, which effectively fits the response to a sum of decaying exponentials and exponentially damped cosine waves. The linear prediction fit reveals a ~ 200 fs exponential formation time for the underlying induced absorbance signal, providing the formation time of the self-trapped exciton state. The oscillatory component at ~ 180 cm^{-1} (period of 185 fs) is seen to be heavily damped with a time constant that matches the formation time of the self-trapped exciton induced absorbance signal, and has an amplitude that gives rise to ~ 100% modulation of the induced absorbance. In contrast, the ~ 315 cm^{-1} oscillation damps on a picosecond time scale, consistent with its assignment to coherent vibrational excitation in the ground electronic state.

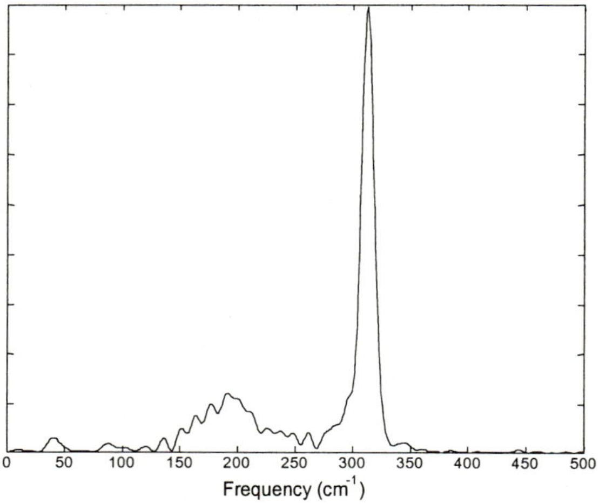

Fig. 2. Fourier power spectrum of the oscillatory part of the PtCl(en) differential transmittance response displayed in Fig. 1.

The features observed in the time resolved response of PtCl(en), specifically a low-frequency oscillatory modulation that damps in concert with the formation of the induced absorbance, together with impulsive Raman ground state oscillations, essentially parallel the features observed in the somewhat slower dynamics of the less strongly coupled PtBr(en) complex. Detailed measurements of PtBr(en) at a series of detection wavelengths throughout the absorption band of its self-trapped exciton state provide a clear picture of the excited-state dynamics in which the observed low frequency oscillation (which appears at ~ 110 cm^{-1} in PtBr(en)) is consistent with a wavepacket corresponding to the lattice motion that carries the excited system to the self-trapped state [1]. In both PtCl(en) and PtBr(en), the exciton formation time (and correspondingly, the wavepacket dephasing time) is comparable to a single oscillation period of the excited-state wavepacket, consistent with a barrierless process. It is interesting to note, however, that in both systems, the excited state wavepacket evolution exhibits clear recurrences. Both PtCl(en) and PtBr(en) also display a significant softening in the excited state vibrational modes: in each case, the excited state wavepacket frequency is only ~

60 % of the ground state symmetric stretch vibrational frequency. Despite the similarities in the responses in PtCl(en) and PtBr(en), comparison of the overall speed of the self-trapping processes shows significantly more rapid dynamics resulting from the stronger electron-phonon coupling in PtCl(en). In addition, measurements in a structurally distorted PtI complex show comparatively slower dynamics [6], providing a useful comparison with the linear PtCl(en) and PtBr(en) systems.

4. Conclusions

We have time-resolved the coupled electronic and vibrational dynamics of the exciton self-trapping process in a class of quasi-one-dimensional molecular solids. The associated vibrational dynamics are evident as a strongly damped, low frequency wavepacket modulation corresponding to the lattice motion that carries the photoexcited system to the self-trapped state. Similar physical mechanisms are observed in both PtBr(en) and PtCl(en), complexes that differ only in the bridging halide ion, and a clear increase in the speed of the self-trapping dynamics is observed for increased electron-phonon coupling strength.

Acknowledgements. This work was supported by the donors of The Petroleum Research Fund, administered by the American Chemical Society, by the National Science Foundation, and by Washington State University. We gratefully acknowledge J. A. Brozik, W. Buschmann, and B. I. Swanson for providing us with samples of the MX materials.

References

1 S. L. Dexheimer, *et al.*, Phys. Rev. Lett. **84**, 4425, 2000.
2 B. Scott, *et al.*, J. Mol. Struct. **356**, 207, 1995.
3 J. T. Gammel, *et al.*, Phys. Rev. B **45**, 6408, 1992; M. Suzuki and K. Nasu, Phys. Rev. B **45**, 1605, 1992; A. Mishima and K. Nasu, Phys. Rev. B **39**, 5758, 1989.
4 For example, H. Ooi, *et al.*, Solid State Comm. **86**, 789, 1993.
5 J. A. Brozik, *et al.*, Inorg. Chem. Acta **294**, 275, 1999.
6 S. L. Dexheimer, *et al.*, to be published.
7 R. J. H. Clark, Adv. Infrared Raman Spectr. **11**, 95 (1984); S. P. Love, *et al.*, Phys. Rev. B **47**, 11107 (1993).

Femtosecond and Attosecond Giant Optical Responses and Fluctuations in Disordered Clusters, Nanocomposites, and Rough Surfaces

Mark I Stockman

Department of Physics and Astronomy, Georgia State University, Atlanta, GA 30340, USA
E-mail: mstockman@gsu.edu, web site: http://www.phy-astr.gsu.edu/stockman

Abstract. We predict that excitation of the disordered nanostructured systems with ultrashort pulses causes transient concentration of excitation energy in nanometer size spatial regions on the time scale of femtoseconds. Concurrently, giant attosecond fluctuations of local fields occur within the optical period. We discuss manifestations of these effects and possible experiments.

1. Introduction

In this paper we consider ultrafast optical responses of strongly disordered nanostructured systems (fractal clusters, rough surfaces, and random composites) whose size is mesoscopic, i.e., much larger than the atomic size but much smaller than the light wavelength. We predict from our computations that femtosecond linear responses of such systems show what we call *"The Ninth Wave Effect"* [1]. Namely, in the course of evolution of the system induced by an ultrashort (femtosecond) laser pulse, dissipation does not have time to run its course, and the excitation energy is accumulated in the system. The giant fluctuations of local fields lead to accumulation of this energy initially spread over the extended system in a narrow (on a nanometer scale) region, where a local field develops that can exceed the average and exciting fields by orders of magnitude. The spatio-temporal behavior of the local fields on a femtosecond-nanometer scale is deterministically chaotic, i.e., random, but fully reproducible for the given system and the exciting pulse. In another related effect, the local fields *within* the optical period experience giant fluctuations on the *attosecond* temporal and nanometer spatial scales [2].

2. Methods

Quantitatively, consider a system consisting of N particles (monomers) positioned at coordinates $\mathbf{r}_i$, $i=1,\cdots,N$. The electric field of the exciting pulse at an ith monomer at time t, denoted $\mathbf{E}_i^{(0)}(t)$, is assumed to be non-saturating and known. We define a local field at this monomer $\mathbf{E}_i(t)$ in terms of the corresponding induced dipole moment $\mathbf{d}_i(\omega)=\alpha_0(\omega)\mathbf{E}_i(\omega)$, where $\alpha_0(\omega)$ is the dipole polarizability of an isolated monomer at a frequency ω. In specific computations, we consider the monomers as spheres of radius R_m, for which

$\alpha_0(\omega) = R_m^3 [\varepsilon(\omega)-1]/[\varepsilon(\omega)+2]$, where $\varepsilon(\omega)$ is the relative dielectric function of the monomer material. The local field $\mathbf{E}_i(t)$ at a time t at an ith monomer is given by a retarded Green's function of the system G^r,

$$E_{i\beta} = \sum_{j=1}^{N} \int_{-\infty}^{t} G^r_{i\beta,j\gamma}(t-t')E^{(0)}_{j\gamma}(t')dt' . \tag{1}$$

Here and below, Greek subscripts denote Cartesian components with summation over recurring indices implied. We have computed the Green's function. (1) in the form of a spectral expansion using two well-tested approximations: the quasistatic and dipolar approximations [1].

3. Results and Discussion

Results of computation of femtosecond dynamics of local fields for a cluster-cluster aggregate (CCA) cluster are shown in Fig.1. Importantly, the local field at the site of its global maximum is enhanced by more than two orders of magnitude reaching its maximum ("The Ninth Wave") by the end of the exciting pulse The averaged fields are not significantly enhanced (data not shown). The dynamics of local fields shows pronounced beats (recurrence of the coherence) after the end of the exciting pulse, i.e., at the free induction stage.

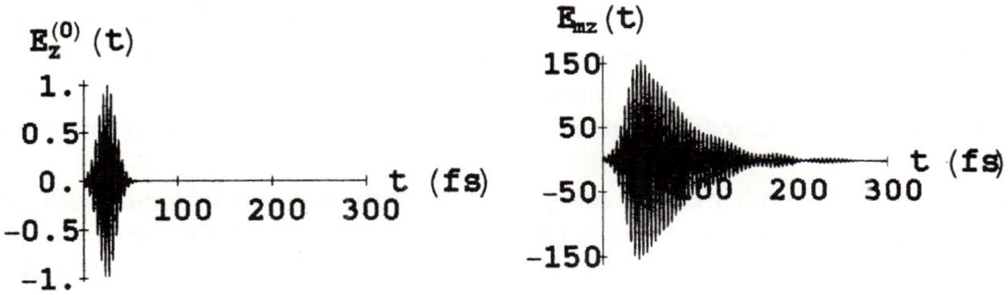

Fig. 1. For a 3D cluster-cluster aggregate, dependencies on time of the exciting field $E_z^{(0)}(t)$ and the local field at the site of the maximum field $E_{mz}(t)$ are displayed. The carrier frequency of the exciting pulse is $\hbar\omega_0 = 1$ eV, and the pulse length is $T = 25$ fs.

To illustrate "The Ninth Wave" effect, two snapshots of local field distribution for a CCA cluster consisting of silver nanospheres are shown in Fig. 2. The shortest time shown ($t = 6.4$ fs) is already sufficient for the giant fluctuations to develop, which is manifested in sharp irregular spikes of the local fields. By the end of the exciting pulse ($t = 40$ fs), "The Ninth Wave" forms: there is a significant concentration of the conserved (accumulated) excitation energy at a narrow spatial region, where the local field exceeds the exciting field by more than two orders of magnitude. At a comparatively long time ($t \geq 200$ fs, data not shown), the dissipation takes place, resulting in the decreased peak intensity and some delocalization of the excitation.

The spatial distributions shown in Fig. 2 are dramatically different from distributions for very long excited pulses ($T >> 100$ fs) that is virtually identical with the distribution for steady state excitation (data not shown). A very similar behavior has been found for other disordered clusters (CCA in two dimensions) and Maxwell Garnett nanocomposites (suspensions of nanospheres in dielectric hosts). This supports universality of "The Ninth Wave" effect that is due to the long range of the dipole interaction that causes giant fluctuations in the systems.

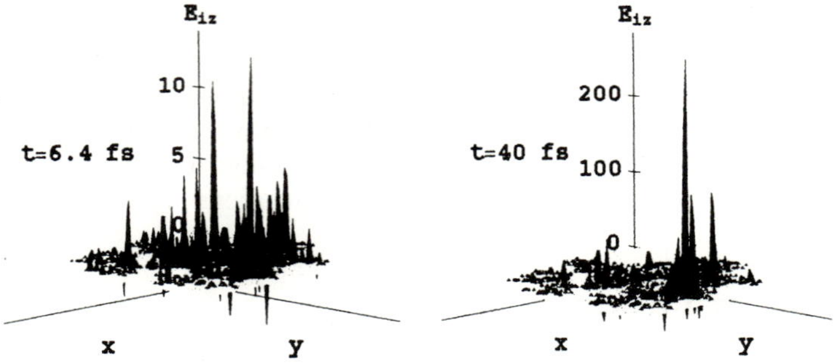

Fig. 2. Local fields $E_{iz}(t)$ for a 3D CCA cluster as functions of the spatial coordinates (x, y) for the moments of time (after the beginning of the exciting pulse) indicated.

"The Ninth Wave" effect described above is linear, similar to quantum chaos in that respect. However, it will influence a wealth of nonlinear phenomena such as generation of high harmonics [3], production of x-rays [4,5], and energetic particles and photonuclear reactions [6,7], because nonlinear phenomena are known to be enhanced by high local fields. This effect can also be observed directly if a scanning near-filed optical microscopy with femtosecond resolution is developed.

We also predict another effect, *attosecond giant fluctuations of local fields* [2], which is one of the fastest phenomena possible in optics. It follows from our computations that in disordered nanostructured systems the spatial distribution of the local optical fields changes dramatically and chaotically (resembling strong turbulence) on an *attosecond* temporal scale (i.e., *within* the optical period) and on a nanometer spatial scale (data not shown). The underlying physical cause of these attosecond-nanometer giant fluctuations is chaos of the polar eigenmodes of the system that leads to random magnitudes and phases of the local fields in the vicinity of different nanoparticles (monomers) of the system.

References

1. M. I. Stockman, Phys. Rev. Lett. **84**, 1011 (2000).
2. M.I.Stockman, Phys. Rev. B (2000) (in press).
3. Z. Chang et al., Phys. Rev. Lett. **79**, 2967 (1997).
4. M. M. Murnane et al., Appl. Phys. Lett. **62**, 1068 (1993).
5 T. Ditmire et al., Phys. Rev. Lett. **75**, 3122 (1995).
6. K. W. D. Ledingham et al., Phys. Rev. Lett. **84**, 899 (2000).
7. T. E. Cowan et al., Phys. Rev. Lett. **84**, 903 (2000).

Electron Energy Relaxation Dependent on Size and Matrix in Gold Nanocrystal-Dielectric Composites

Y. Hamanaka[1], J. Kuwabata[1], A. Nakamura[1], I. Tanahashi[2], and S. Omi[3]

[1] Center for Integrated Research in Science and Engineering and Department of Crystalline Materials Science, Nagoya University, Chikusa-ku, Nagoya 464-8603, Japan and CREST, JST, Japan
E-mail: hamanaka@cirse.nagoya-u.ac.jp
[2] R&D Planning Office, Matsushita Electric Industrial Co., Ltd., Moriguchi, Osaka 670-8501, Japan
[3] R&D Center, HOYA Corporation, Akishima, Tokyo 196-8510, Japan

Abstract. We represent size- and matrix-dependent relaxation behaviors of hot electrons in gold nanocrystals. The observed results have been well explained by a model taking into account an electron-breathing vibration interaction in addition to an electron-bulk phonon interaction and thermal diffusion through the interface as electron relaxation pathways.

1. Introduction

Metal nanocrystals embedded in dielectric matrix have attracted great interest because of the large third-order susceptibility and fast response time [1-6]. Fast response time of a few picoseconds reflects relaxation dynamics of non-equilibrium electrons created by femtosecond optical pulses [3-6]. The energy relaxation of the photoexcited electron system has been interpreted in terms of the two-step mechanism: cooling of hot electrons via electron-phonon interaction to be in equilibrium with the lattice and thermal diffusion from nanocrystals to the matrix. In this study, we have observed diameter- and matrix-dependent relaxation, and performed numerical calculations taking into account both the cooling of hot electrons to the lattice via interactions of bulk and surface modes and the rapid thermal diffusion to the matrix. We point out that the electron-breathing mode interaction plays an important role in the nonequilibrium electron relaxation in metal nanocrystal-dielectric composites.

2. Experimental Methods

Gold nanocrystal-dielectric composites have been prepared by magnetron sputtering method [7]. We used three different materials, SiO_2 glass, TiO_2 and Al_2O_3 with thermal conductivities of 1.38, 8.4 and 36 $Wm^{-1}K^{-1}$ as a matrix. Average diameters of gold nanocrystals were varied between 3.0 and 17.5 nm. Femtosecond pump and probe measurements have been carried out with 150 fs laser pulses from an amplified Ti:sapphire laser. The photon energy of the pump pulse was set to 3.1 eV corresponding to the band to band transition associated with 5*d* electrons of gold nanocrystals.

3. Results and Discussion

Figure 1 shows the time evolution of differential absorption detected at the surface plasmon resonance energy for gold nanocrystals embedded in SiO_2 glass (Au/SiO_2) and Al_2O_3 (Au/Al_2O_3). Pump intensity was adjusted so that the absorbed pump energy by unit volume of nanocrystals was kept constant for all the measurements. The observed decay kinetics clearly indicates size dependent behaviors. The relaxation times, τ_1, obtained by fitting the data to an exponential function are plotted as a function of diameter in Fig. 2. The relaxation becomes fast with decreasing nanocrystal diameter. The decay curve for Au/Al_2O_3 with a diameter of 4.2 nm is also shown in Fig. 1. The decay is faster than for Au/SiO_2, and τ_1 is as short as 1.5 ps. These results suggest that the thermal conductivity is one of the parameters governing the relaxation dynamics, since the thermal conductivity is about 26 times larger in Al_2O_3 than in SiO_2 glass.

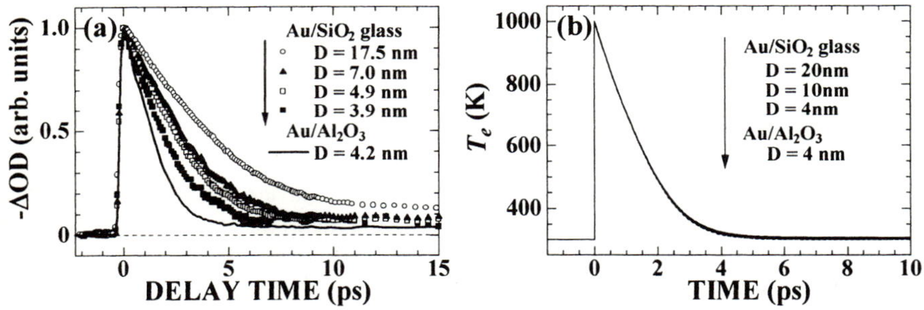

Fig. 1. (a) Time evolutions of differential absorption for Au/SiO_2 (3.9, 4.9, 7.0 and 17.5 nm) and Au/Al_2O_3 (4.2 nm). (b) Calculated time evolutions of electron temperature, T_e, for Au/SiO_2 (solid curves) and Au/Al_2O_3 (dashed curve). Diameters of gold nanocrystals are 4, 10 and 20 nm.

In order to discuss size- and matrix-dependent decay behaviors, we calculated energy relaxation rates by using an extended two-temperature model which takes into account both the hot electron relaxation to the lattice via electron-phonon interaction [3] and the thermal diffusion from nanocrystals to the matrix [2]. Since the relaxation behavior depends on the matrix as well as the size, the size-dependence does not originate from the electron-bulk phonon interaction in gold nanocrystals. Therefore, we assume that the electron-phonon coupling strength is constant in the two-temperature model. We have numerically calculated an electron-temperature T_e from simultaneous differential equations for electron, lattice and matrix temperatures. Time evolutions of T_e for gold nanocrystals with different diameters embedded in SiO_2 glass and Al_2O_3 are shown in Fig. 1 (b). Electrons which are initially in equilibrium with the lattice at 300K are heated up to 1000 K by photo-excitations, and then cooled down to 300 K. The calculated dependence of T_e on size and matrix is much smaller than the observed one. This result suggests that there exists another relaxation pathway giving rise to the size- and matrix-dependence. Here, we consider the relaxation of nonequilibrium electrons via breathing vibration of nanocrystals as an additional pathway. For Au/SiO_2 with a diameter of 17.5 nm, we observed the oscillatory behavior of differential absorption due to excitation and damping of a breathing mode by the pump pulse with relatively weak intensity [8]. The total relaxation rate is calculated from a sum of the relaxation rate via bulk modes and that via breathing modes coherently excited by the optical pulse. The calculated results reproduce well the observed dependence for Au/SiO_2 (solid curve in Fig. 2), Au/Al_2O_3 and Au/TiO_2. The relaxation due to the breathing mode damping gives rise to the size- and matrix-dependence of the relaxation time.

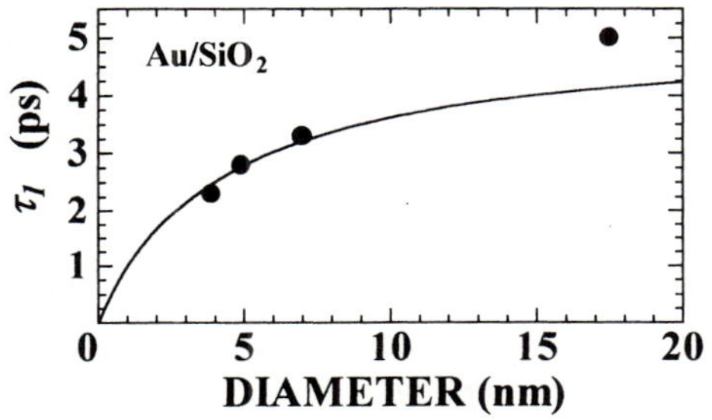

Fig. 2. Diameter dependence of relaxation times for Au/SiO_2. Closed circles and a solid curve indicate experimental results and calculated one, respectively.

4. Conclusion

We have observed electron relaxation dynamics of gold nanocrystal-dielectric composites by femtosecond pump-probe experiments. The size- and matrix-dependent relaxation behaviors are explained by the model taking into account the relaxation via electron-bulk phonon interaction, damping via electron-breathing mode interaction and thermal diffusion through the interface.

References

1 F. Hache et al., J. Opt. Soc. Am. B **3**, 1647 (1986).
2 M. J. Bloemer et al., J. Opt. Soc. Am. B **7**, 790 (1990).
3 T. Tokizaki et al., Appl. Phys. Lett. **65**, 941 (1994).
4 M. Perner et al., Phys. Rev. Lett. **78**, 2192 (1997).
5 Y. Hamanaka et al., Appl. Phys. Lett. **75**, 1712 (1999).
6 V. Halté et al., Appl. Phys. Lett., **75**, 3799 (1999).
7 I. Tanahashi et al., Surf. Rev. Lett. **3**, 1071 (1996).
8 N. Del Fatti et al., J. Chem. Phys. **111**, 8613 (2000).

Femtosecond photoemission studies of the transient electron temperature in supported metal nanoparticles

M. Merschdorf, W. Pfeiffer, A. Thon, S. Voll, and G. Gerber

Physikalisches Institut, Universität Würzburg, Am Hubland, 97074 Würzburg, Germany. E-mail: gerber@physik.uni-wuerzburg.de

Abstract: Double excitation of the surface plasmon with 388 nm (3.2 eV) femtosecond radiation allows to distinguish photoemission from silver nanoparticles and substrate. We apply time-resolved photoemission to measure the electron temperature in the nanoparticles subsequent to excitation with a 776 nm (1.6 eV) femtosecond pulse.

1. Introduction

Metal nanoparticles are candidates for new optoelectronic materials [1] and can be used as a model catalyst [2]. Usually, electron cooling in nanoparticles is studied by transient absorption [3-6]. Assuming a linear dependence of the dielectric function and the electron temperature in the nanoparticles on the time-dependent transmission, cooling times can be measured. However, so far no experiment allowed to measure the electron temperature in metal nanoparticles quantitatively. Femtosecond photoemission spectroscopy has been used to determine electron temperatures at surfaces [7]. Recently we demonstrated that time-resolved two-photon photoemission spectroscopy can be used to study the electron dynamics of supported metal nanoparticles by resonantly exciting the surface plasmon in the nanoparticles [8]. Here we show that it is possible to apply this new method to determine the time-dependent electron temperature in the nanoparticles.

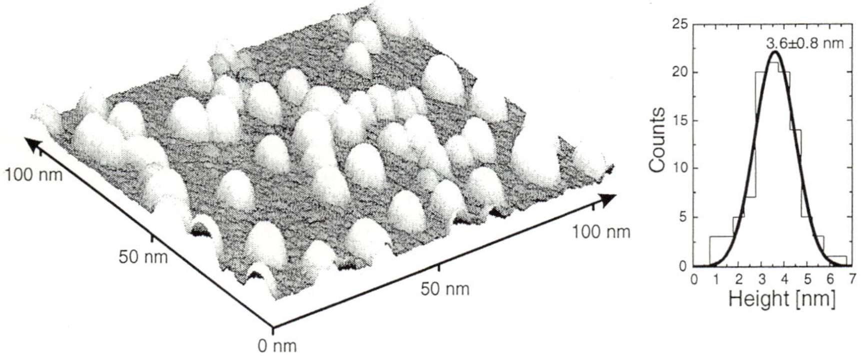

Fig. 1. STM topography of silver nanoparticles on HOPG (U = 300 mV, I = 30 pA)

2. Experiment

The preparation of the nanoparticles follows the procedure described in [9]. Silver is evaporated on graphite (HOPG) and condenses into prestructured nanometer-sized pits forming the particles. *In-situ* STM topographies allow to measure the particle height distribution (see Figure 1). For excitation we use a home-built femtosecond Ti:Sapphire laser (100 kHz, 3 μJ, 60 fs, 776 nm) and subsequent frequency doubling (100 nJ, 50 fs, 388 nm). The photoelectrons are analyzed using a time-of-flight spectrometer [10].

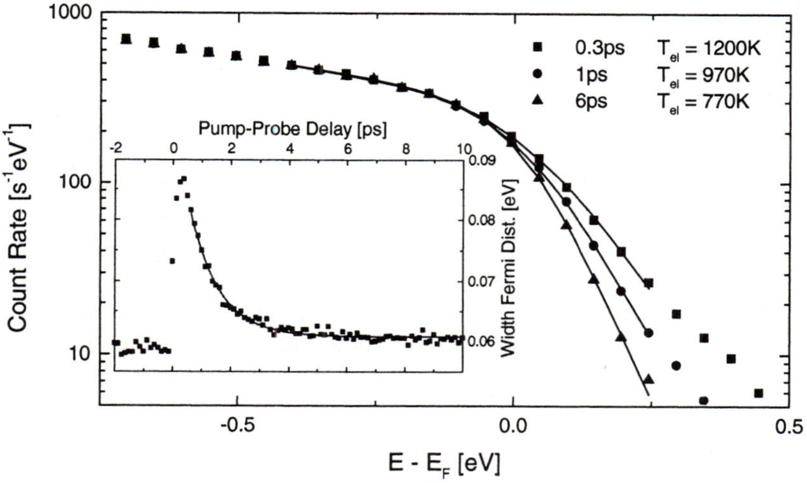

Fig. 2. Electron spectra from silver nanoparticles on graphite shown for the energy range corresponding to initial states close to the Fermi energy, measured with a 388 nm (3.2 eV) femtosecond pulse subsequent to excitation with an 776nm pump pulse. The electron temperatures at different delay times are visible as different slopes for energies above the Fermi energy. The inset shows the time-dependent electron temperature.

3. Results and Discussion

When exciting the surface plasmon in silver nanoparticles using p-polarized 388 nm laser pulses, the photoelectron yield from the nanoparticles is two orders of magnitude higher than that from the substrate and therefore completely dominates the photoemission spectra [8]. Photoemission occurs via the doubly excited surface plasmon in the nanoparticles. As no scattering in an intermediate single-particle state is envolved in the excitation process, the electron distribution in the silver nanoparticles can be monitored directly. We probe the time-dependent electron distribution in the nanoparticles subsequent to excitation with a 776 nm pulse. We fit the electron spectra using an exponential decay multiplied by a Fermi distribution yielding the electron temperature as a fitting parameter (see Fig. 2). The different electron temperatures are clearly visible as different slopes in the electron spectra for energies above the Fermi energy. The inset shows the time dependence of the electron temperature. After a fast rise, it decays on a ps time scale. We measured the decay times of the electron temperature for a number of samples with particle sizes ranging from 1.5 nm to 4.4 nm and found decay times between 1.0 ± 0.1 ps and 1.4 ± 0.1 ps. These times are longer than the decay time of 0.9 ± 0.1 ps which we measured in a silver crystal.

4. Conclusion

Summarizing, we applied femtosecond time-resolved multiphoton photoelectron spectroscopy to study Ag nanoparticles grown on a graphite substrate. The nanoparticles can be probed selectively by tuning the excitation wavelength near the surface plasmon resonance of the nanoparticles. This method was applied to probe the transient electron temperature in silver nanoparticles. We found decay times slightly longer than the bulk value. These experiments represent the first direct measurement of electron temperatures in metal nanoparticles.

References

1 R. Service, Science **271**, 920, 1996.

2 G. Ertl and H. J. Freund, Physics Today **52/1**, 32, 1999.

3 J. Bigot, J. Merle, O. Cregut, and A. Daunois, Phys. Rev. Lett. **75**, 4702, 1995.

4 M. Perner et al., Phys. Rev. Lett. **78**, 2192, 1997.

5 V. Halte, J. Guille, J. Merle, I. Perakis, and J. Bigot, Phys. Rev. B **60**, 11738, 1999.

6 S. Link, C. Burda, Z.L. Wang and M.A. El-Sayed, J. Chem. Phys. **111**, 1255, 1999.

7 W.S. Fann, R. Storz, H.W.K. Tom and J. Bokor, Phys. Rev. Lett. **68**, 2834, 1992.

8 J. Lehmann, M. Merschdorf, W. Pfeiffer, A. Thon, S. Voll and G. Gerber, J. Chem. Phys. **112**, 5428, 2000

9 H. Hövel et al., J. Appl Phys. **81**, 154, 1997.

10 J. Lehmann, M. Merschdorf, A. Thon, S. Voll, and W. Pfeiffer, Phys. Rev. B **60,** 17037, 1999.

Acoustic vibrations of silver nanoparticles

S. Grésillon[1], M. Perner[1], G. von Plessen[1], J. Feldmann[1], J. Porstendorfer[2], K.-J. Berg[2], and G. Berg[2]

[1] Photonics and Optoelectronics Group, Department of Physik and CeNS
University of Munich, Amalienstraße 54, D-80799 München, GERMANY.
Email : samuel.gresillon@physik.uni-muenchen.de

[2] Fachbereich Physik, Universität Halle-Wittenberg, Friedemann-Bach-Platz 6, D-06108 Halle, GERMANY

Abstract. Using time-resolved pump-and-probe experiments we show that the acoustic vibrations of silver nanoparticles excited by ultrashort laser pulses are triggered by two mechanisms: the lattice anharmonicity and the thermal pressure of the hot electrons.

The small size of metal nanoparticles leads to vibrational properties different from those of bulk metals. These different properties are observable in phenomena like the melting point depression of small clusters [1] and a modified electron-phonon coupling strength [2]. An important consequence of the small particle size is the existence of confined acoustic vibrational modes with discrete frequencies. In the time domain, they are observable as periodic signal modulations in ultrafast optical pump-probe experiments on spherical metal nanoparticles. The vibrations in these experiments have been argued to be excited by the thermal expansion of the laser-heated particles, which results from the lattice anharmonicity [3-5]. So far only the lowest (n = 1) vibrational breathing (ℓ = 0) mode of the spherical particles has been observed.

Here we investigate the acoustic vibrations of ellipsoidal and spherical silver nanoparticles. We perform time-resolved optical pump-probe experiments which show that the n = 1 vibrational mode splits into two non-degenerate modes in ellipsoidal particles. In addition, a small contribution from higher-frequency modes is

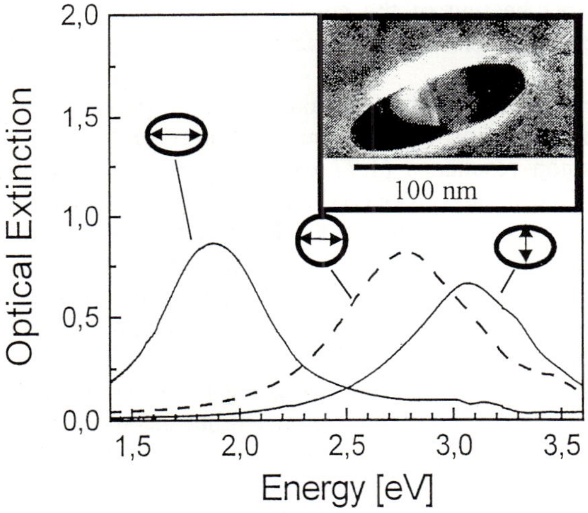

Fig. 1: Optical extinction spectra of spherical (dashed line) and ellipsoidal (solid lines) silver nanoparticles. The light polarizations are indicated by arrows. Inset: Transmission electron micrograph of an ellipsoidal nanoparticle embedded in glass.

observed in large spherical particles. An analysis of the time dependence of the vibrations shows that the lattice anharmonicity is not the only mechanism driving the thermal expansion of the particles; there is an additional contribution from the thermal pressure of the hot conduction electrons.

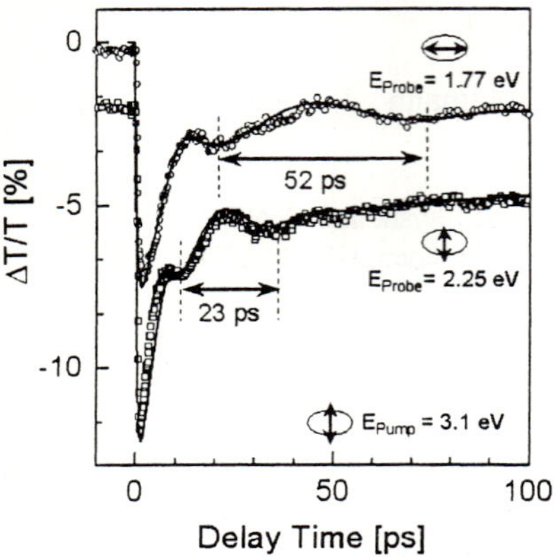

Fig. 2: Differential transmission transients (vertically offset for clarity). The light polarizations and photon energies of the pump and probe pulses are indicated.

The silver nanoparticles studied here were grown via Ag+/Na+ ion exchange in a glass matrix. The spherical particles have a mean diameter of 60 nm. The ellipsoidal particles were made by subjecting spherical particles to tensile deformation; they have mean axis lengths of a=b≈40 nm and c≈100 nm, respectively, and a uniform orientation. Upon deformation, the dipolar surface-plasmon (SP) resonance of the spherical particles splits up into two spectrally separate surface-plasmon resonances that can be excited with light polarised parallel to the short and long axes, respectively, as shown by the extinction spectra in Fig. 1. Time-resolved optical pump and probe experiments are performed with 150 fs laser pulses generated by a regenerative amplifier and subsequent frequency doubling and white light generation. The particles are excited with 3.1 eV pump pulses, resulting in a considerable heating of the conduction electrons and the metal lattice. The pump-induced changes of the surface-plasmon line [5] are detected with the probe pulse.

Fig. 2 shows two differential transmission transients taken with the probe pulses centered at the low-energy sides of the short-axis surface-plasmon resonance (2.25 eV) and long-axis surface-plasmon resonance (1.77 eV), respectively. In both cases, an initial rapid increase of the signal is observed during the first picosecond after excitation, followed by a two-step decay with decay times of several picoseconds. In addition, the signals exhibit periodic modulations during the second decay step. The initial signal increase is caused by changes in the dielectric properties of the laser-heated electron gas in the particles [6,7]. The time taken by the initial signal increase to reach its maximum (i.e., ~ 1ps) is the thermalization time of the non-equilibrium electron-hole distribution resulting from the decay of the surface-plasmon resonance excited by the pump pulse. During the subsequent 5-10 ps the hot electrons cool down via phonon emission, thereby heating up the lattice rapidly; this electron cooling results in a reduction of the dielectric signal component. In the second decay step the particle slowly cools to room temperature via heat transfer to the surrounding glass matrix.

An important observation is that the red-shift in the second signal-decay step exhibits periodic modulations (Fig. 2). Similar modulations have been observed in pump-probe experiments on spherical metal clusters, and explained as the

signature of acoustic particle vibrations [3-5]. A remarkable difference is that in our ellipsoidal clusters the time period of the modulation depends on the particle axis probed. A time period of ≈ 23 ps is observed when probing the short-axis surface-plasmon (lower trace in Fig. 2), while a period of ≈ 52 ps is observed at the long-axis surface-plasmon (upper trace). Additional measurements at different probe photon energies (not shown here) demonstrate that the signal modulations are caused by periodic red shifts of the two SP resonances. These red-shifts are explained as an electromagnetic retardation effect resulting from the length increase of the laser-heated particles. These modulations are perfectly reproduced by the results (solid lines) of thermoelastic model calculations, which take into the account the calculated eigenfrequencies of the lowest (n = 1) vibrational modes of the particles along the two axes, and the thermal expansion of the particles. A careful comparison of the temporal phase of the calculated and measured signal modulations demonstrates that the thermal expansion triggering the vibrations can not be explained by the lattice anharmonicity alone [5]; there is a also considerable contribution from the thermal pressure of the hot conduction electrons, which reach an estimated peak temperature of 3000 K under the present optical excitation conditions. This is consistent with theoretical predictions that demonstrate the enhancement of the electron pressure at high electron temperature [8].
Time-resolved pump and probe experiments are also performed on very large spherical silver nanoparticles with the probe pulse centered at a photon energy of 2.25 eV. Modulations are observed which exhibit, in addition to the dominant lowest (n=1) vibration mode, a small contribution from higher-frequency components. Work is presently in progress to investigate these higher vibration components in the large spherical particles.

Reference

1 S.L. Lai, et al., Phys. Rev. Lett. **77**, 99 (1996).
2 A. Stella, et al., Phys. Rev. B **53**, 15497 (1996).
3 M. Nisoli et al., Phys Rev B 55, 13424 (1997)
4 J.H. Hodak et al., J. Chem. Phys **108**, 9210 (1998);
5 N. Del Fatti et al., Physica B **263**, 54 (1999)
6 M. Perner, S. Grésillon, et al., Phys. Rev. Lett. (2000, in press).
7 Y. Hamanaka, N. Hayashi, A. Nakamura, S. Omi, J. Lum. **76-77**, 221 (1998).
8 V.E. Gusev, Opt. Comm. **94**, 76 (1992).

Ultrafast Interferometric Studies of Multiple Scattering of Light in Photonic Structures

Yish-Hann Liau and Norbert F. Scherer

Department of Chemistry and James Franck Institute, University of Chicago, 5735 South Ellis Avenue, Chicago, IL 60637, USA
E-mail: nfschere@uchicago.edu

Abstract. We report the first observation of pronounced recurrent signals in the pump-probe interferometry measurements of structured metallic interfaces. The coherent multiple scattering of surface plasmons from the microsphere overlayer causes light localization in the photonic structure.

Photonic structures have attracted extensive attention in the past decade because of their unprecedented potential for manipulating light.[1] Previous research on photonic structures almost exclusively employed CW light, thereby limiting the studies to static and linear properties.[2] Elucidation of the dynamics of wave packet propagation in the photonic structure requires ultrafast spectroscopic techniques.[3] Further, the use of high peak-intensity femtosecond pulses opens possibilities to exploit the nonlinearity of photonic materials.[4]

Scattering of light plays a critical role in many important phenomena including multiple scattering as a precursor for light localization in a random medium.[5] Interference of scattered light in the photonic structure is responsible for the formation of photonic band-gaps. In our earlier studies, coherent multiple scattering of surface plasmon polaritons on rough metal surfaces was observed in pump-probe measurements[6]. In the present study, surface waves associated with the surface plasmon that propagates on a decorated metal film are used as a 2-D model system. Surface plasmons can be launched by light by satisfying the dispersion relation at the interface.[7] Plasmon scattering can be introduced by spatially modulating the interfacial dielectric properties.

In this paper, we report the first observation of pronounced recurrent interferometric signals resulting from multiple scattering of surface plasmons propagating in photonic structures. The ultrafast laser system and the experimental setup have been described elsewhere.[8] The surface plasmon was launched in the Kreschmann geometry.[7] The pump-probe laser pulses with controlled interferometric delay were sent to a prism coupler and focused onto a thin metallic film sample. A CCD camera was used to acquire the plasmon field distribution. The sample, a 46-nm thickness Ag film, thermally evaporated on a glass slip, was spin-coated with a monolayer of latex spheres (diameter = 2.1μm). Optical contact between the glass slip and the prism hypotenuse was made by index matching oil. The interferometric signal resulting from the surface plasmon created by the pump beam scattering into the direction of the probe beam was detected by the lock-in demodulation detection technique. A 100-μm wide slit was placed before the detector to achieve angular (hence momentum) resolution.

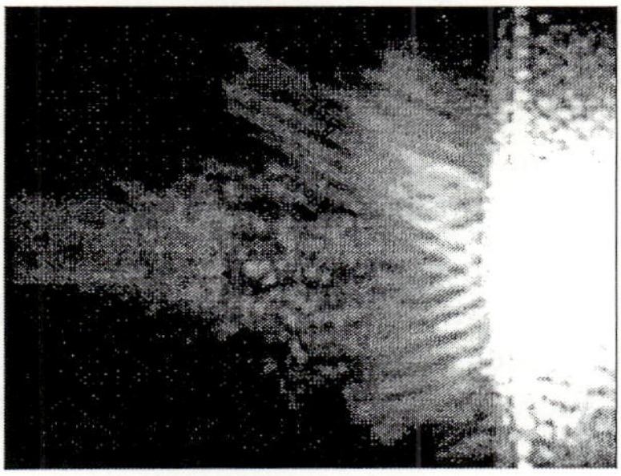

Figure 2. A microscopy image (60 x 44 μm) acquired with a 100x objective showing the directional back scattering of surface plasmons. The propagation direction is from left to right.

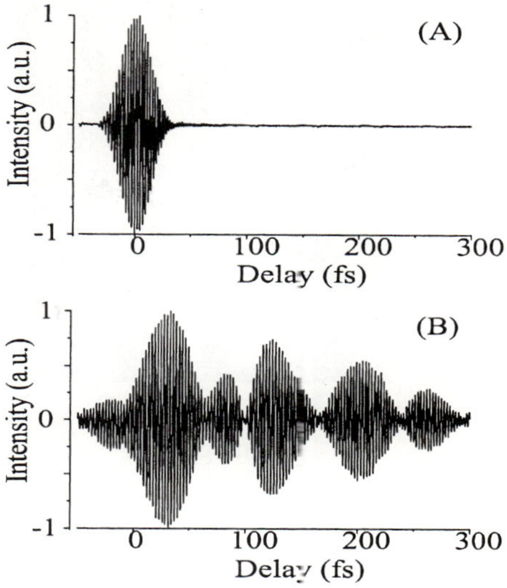

Figure 1. Pump-probe interferometric signals measured at non-ATR-resonance condition (2A) and when the laser beams were directed to the photonic structures (2B).

At the resonance angle attenuated total internal reflection (ATR) was observed indicating that light was converted to surface plasmons. Almost no scattered light is observed from the region of the smooth Ag film. Efficient plasmon scattering occurs when the surface plasmon propagates into the perfect hexagonal lattice region. In particular, Fig. 1 clearly shows the directional backward scatter from the lattice. The forward and backward propagating fields create the standing wave pattern observed.

The dynamics of the plasmon scattered out of the 2-D photonic structure are monitored with interferometric pump-probe measurements. Figure 2A shows the signals measured at a non-ATR-resonance angle representing the instrument response. Fig. 2B is the result acquired as the surface plasmon propagating in the periodically patterned region. Pronounced recurrent interferometric signals lasting as long as 1 ps were observed. The decay of the interferogram indicates a loss of amplitude (and coherence) due to propagation or scattering out of the film. Further, the temporal pattern is strongly altered by the lattice orientation with respect to the optical/plasmon k-vector in the film.

Interpretation of these data is aided by considering the Fourier power spectra. The reference interferogram gives a smooth laser spectrum. The Ag film interferogram (not shown) reflects the resonance absorption that follows from satisfying the single valued dispersion relation of the unstructured Ag film. The Fourier spectrum of Fig. 2B shows two "dips", reflecting the dual valued solution of the dispersion relation, separated by the "stop band", a characteristic of a photonic band-gap structure.

The momentum-energy relation of the surface plasmon propagating in the photonic structure is obtained by Fourier transformation of the angle-resolved pump-probe interferograms displayed in Figure 3. The upper part of the dispersion diagram shows the single-valued dispersion relation of the surface

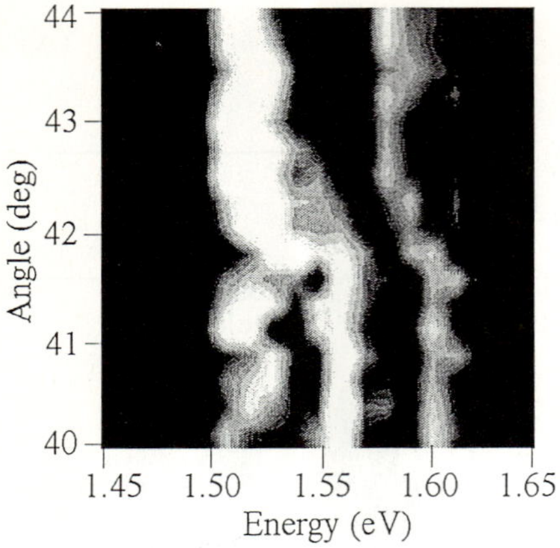

Figure 3. The dispersion diagram of surface plasmons propagating in the photonic structure.

plasmon. Notably, the dispersion of the surface plasmon is split as shown at Fig. 3 with a peak in the center indicating the existence of a photonic bandgap. The dispersion of the surface plasmon is altered by the spatially modulated dielectric function at the interface. The surface plasmon scattered elastically by the dielectric lattice carries a different momentum while maintaining the same energy resulting in the spitting of the dispersion curve. The lattice momentum can be obtained with spatial Fourier transformation of the real-space dielectric lattice pattern. The location of the bandgap is well predicted by taking into account the lattice momentum accessible by the surface plasmon.

Studies of the intensity distribution of the plasmon (field) and the spatially dependent dynamics are being conducted with scanning probe microscopy methods.[9] Other dielectric decorations, including metals and polymers, are being employed to expand the range of perturbations and to facilitate the construction of nonlinear photonic structures. We are also pursuing the fabrication of programmable devices using patterned self-assembled monolayer formation methods with electrochemically reversible attachment chemistry[10,11].

References

1 E. Yablonovitch, Phys. Rev. Lett. **58**, 2059, 1987; S. John, Phys. Rev. Lett. **53**, 2169, 1984; J. D. Joannopoulos, R. D. Meade, J. N. Winn, *Photonic Crystals*, Princeton University Press, Princeton, 1995.

2 Kitson, et al., Phys. Rev. Lett. **77**, 2670, 1996; T. W. Ebbesen, et al., Nature **391**, 667, 1998.

3 C. Spielmann, et al., Phys. Rev. Lett. **73**, 2308, 1994.

4 V. Berger, Phys. Rev. Lett. **81**, 4136, 1998; N. G. R. Broderick, et al, Phys. Rev. Lett. **84**, 4345, 2000.

5 D. S. Wiersma, et al., Nature **390**, 671, 1997

6 W. Wang, et al,, Chem. Phys. Lett. **262**, 573, 1996.

7 H. Raether, *Surface plasmons on smooth and rough surfaces and on gratings*, Springer-Verlag, Berlin, 1988.

8 Y.-H. Liau, et al., Appl. Opt. **38**, 7386, 1999.

9 M. Feldstein, at al., J. Phys. Chem. **100**, 4739, 1996.

10 M. N. Yousaf and M. Mrksich, J. Am. Chem. Soc. **121**, 4286, 1999.

11 Acknowledgement: W. M. Keck Foundation and NSF-MRSEC.

Resolving Ultrafast Dynamics of Electron Thermalization in Gold using Surface SHG

Chunlei Guo, George Rodriguez, and Antoinette J. Taylor

Condensed Matter and Thermal Physics Group, Materials Science and Technology Division, Los Alamos National Laboratory, Los Alamos, NM 87545

Abstract. Using time-resolved surface second harmonic generation (SHG), an electron temperature dependent enhancement in the SHG signal is observed, for the first time in the high temperature regime, resolving the thermalization of nonequilibrium hot electrons.

The study of the dynamics of electron-electron and electron-phonon interactions is of primary importance in understanding the fundamental and technological properties of metallic and semiconductor systems. Even though some limited studies have been performed in the relatively low electron temperature (T_e) regime to try to understand the early stage of the electron thermalization process [1,2], the thermalization dynamics of hot electrons from a nonequilibrium distribution following ultrafast laser excitation has never been resolved in the high T_e regime (e.g. ~1 eV) due to experimental difficulties. It is of premier importance to understand this dynamics in the high T_e regime to gain further information on phenomena related to phase transitions, ablation, and photochemistry.

In this paper, we describe a novel pump-probe optical measurement that reveals the details of hot electron thermalization dynamics by monitoring the time evolution of both the fundamental (1.55 eV) and second harmonic (3.1 eV) generation (SHG) signals from a gold (Au) surface. The experimental system follows the basic setup described in Ref. [3]. The 110-fs laser pulse from an amplified Ti:sapphire laser system is split into an intense pump beam and a weak probe beam with crossed polarization (p-polarization for probe). Surface SHG is generated by the probe beam, and the collinear fundamental and SHG beams are separated by a dichroic beamsplitter and monitored simultaneously. The Au sample used in the experiments is a polycrystalline sample prepared by high vacuum deposition.

By varying the pump pulse intensity up to, but just below, the Au damage threshold, the dynamics of both the fundamental and SHG can be resolved with the pump-probe measurements for T_e up to ~1.1 eV. (Note that temperature is an ill-defined quantity for a nonequilibrium system. When we quote a temperature, we always refer to the final temperature when equilibrium is reached by assuming no energy is lost during the thermalization process). The time-resolved fundamental and SHG signals at a pump intensity of 1.3×10^{12} W/cm^2 for Au is plotted in Fig. 1. The signals at both fundamental and SHG initially decrease in amplitude after pump excitation, reaching minimum values, and then recovering. In contrast to the monotonic recovery in the fundamental reflectivity, the SHG signal, rises up to a maximum value and then drops down again, revealing a peak structure at a pump-probe time delay of approximately 260 fs.

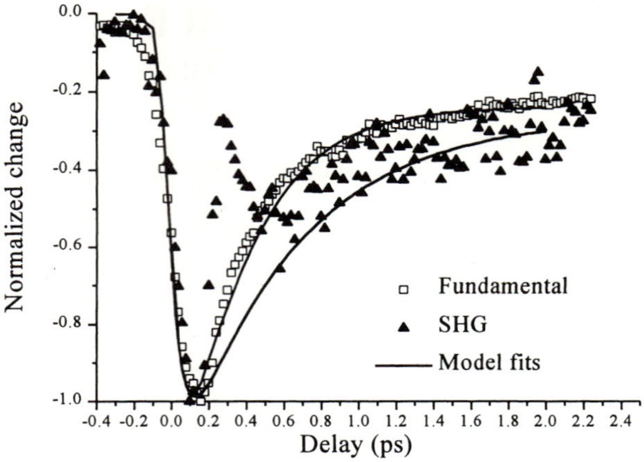

Fig. 1. Time-resolved fundamental and surface SHG signals at an initially heated electron temperature of ~1.03 eV in Au. The solid lines are the model fits. The fit to the SHG data is obtained by considering only the temperature dependent change of Fresnel factors.

To understand the dynamics of the SHG signal following pump excitation, we model the SHG signal based on a phenomenological model from Ref. [4]. Generally speaking, SHG depends on Fresnel factors and the second order nonlinear susceptibility, $\chi^{(2)}(T_e)$. We first assume that $\chi^{(2)}$ is temperature independent and fit the change of SHG only with temperature dependent Fresnel factors. Using the deduced parameters from the two-temperature model [5], the dynamics of the fundamental reflectivity and the main feature in the SHG data can be reproduced except for the local peak enhancement at ~260 fs (Fig.1). The modeled change of SHG will always be a smooth function if we assume $\chi^{(2)}$ is temperature independent. Therefore, the discrepancy between the SHG data and the model fit must come from a temperature effect on $\chi^{(2)}$ due to laser heating.

To determine the origin of this discrepancy, we calculate the temperature dependence of $\chi^{(2)}$ and plot it as a function of initial T_e in Fig. 2. $\chi^{(2)}$ increases with T_e for both equilibrium and nonequilibrium distributions. The values of $\chi^{(2)}$ for both kinds of distributions are similar for electron temperatures below 0.7 eV, but the value of $\chi^{(2)}$ in an equilibrium electron distribution surpasses that of nonequilibrium distribution when T_e is above 0.7 eV. The subtle dependence of $\chi^{(2)}$ on both temperature and thermalization is in fact directly exhibited in our experimental data. The local enhancement around 260 fs after pump in SHG at $T_e = 1.03$ eV (Fig. 1) results from the thermalization of hot electrons from a nonequilibrium to an equilibrium state since the value of $\chi^{(2)}$ is higher in an equilibrium electron distribution for $T_e > 0.7$ eV. The local enhancement in the SHG signal also scales with the pump intensity, as predicted in Fig. 2. Therefore, the deviation of the SHG signal from the model fit reveals the dynamic process of electron-electron thermalization and subsequent electron-phonon coupling.

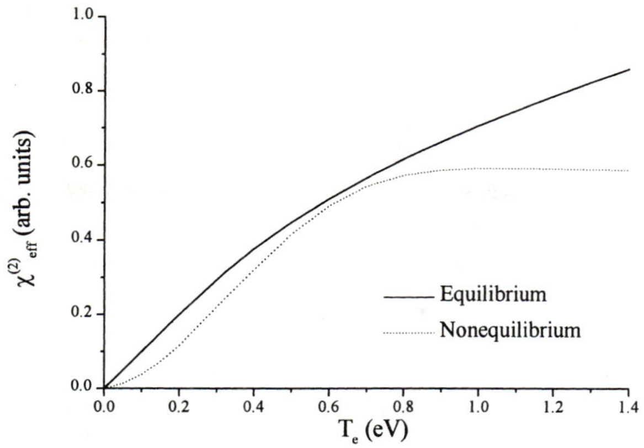

Fig. 2.: Nonlinear susceptibility $\chi_{eff}{}^{(2)}$ versus T_e in both equilibrium and nonequilibrium distributions.

In summary, we monitor the time-resolved SHG from an Au surface following intense 1.55-eV ultrafast laser excitation. For the first time, an electron temperature dependent enhancement in the SHG probe signal is observed and shown to directly reflect the thermalization of hot electrons from nonequilibrium to equilibrium states. Therefore, the relaxation dynamics of the hot electrons are directly resolved by monitoring the time evolution of the SHG signal in the high T_e regime. The resolved hot electron dynamics in this temperature regime will improve our understanding of electronic effects in high temperature phenomena in metals, such as laser-induced structural phase transitions and chemical reactions.

References

1 W.S. Fann, R. Storz, H.W.K. Tom, and G.D. Aumiller, *Phys. Rev. Lett.* **68**, 503 (1992).

2 C-K. Sun, F. Vallee, L. Acioli, E.P. Ippen, and J.G. Fujimoto, *Phys. Rev. B* **48**, 12365 (1993).

3 C. Guo, G. Rodriguez, A. Lobad, and A.J. Taylor, *Phys. Rev. Lett.* **84**, 4493 (2000).

4 J .E. Sipe, D.J. Moss, H.M. van Driel, *Phys. Rev. B*, **35**, 1129 (1987).

5 S.I. Anismov, B. L. Kapeliovich, T.L. Perel'man, *Sov. Phys. JETP* **39**, 375 (1975).

Analysis of Hot Electron Cascades in Copper

Miles J. Weida[1], Susumu Ogawa[1], Hisashi Nagano[1], and Hrvoje Petek[2]

[1]Advanced Research Laboratory, Hitachi, Ltd., Hatoyama, Saitama 350-0395, Japan
E-mail: mweida@harl.hitachi.co.jp
[2]Department of Physics and Astronomy, University of Pittsburgh, Pittsburgh, PA 15260 U.S.A., E-Mail: petek@pitt.edu

Abstract A combination of interferometric time-resolved two-photon photoemission techniques and rigorous kinetic modeling is used to consider the contribution of secondary hot electrons to measured hot electron dynamics in copper.

1. Introduction

The decay times for electrons excited above the Fermi level (i.e., hot electrons) are necessary to test theoretical descriptions of electron dynamics in metals[1]. In time-resolved two-photon photoemission (TR-2PP) measurements of hot electron lifetimes, electrons in the intermediate state (see Fig. 1) can arise from several processes: primary photoexcitation, and secondary processes such as Auger decay and hot electron cascade from higher energy levels. The presence of secondary hot electrons can complicate the analysis of the primary hot electron lifetimes. Therefore, it is important that measurements and models be made to separate the effects of primary and secondary hot electrons in experiments.

In the present work we consider a model for cascade secondary hot electrons that can be used with rigorous kinetic analysis of the data to separate primary and secondary hot electron processes. Copper is chosen for the test system for its theoretical tractability and wealth of experimental information[1]. Low energy (1.55 eV) excitation on cesiated Cu(100) and Cu(111) is used to exclude the more complicated *d*-band Auger processes in interferometric two-photon correlation (I2PC) measurements. The data are modeled based on the optical Bloch equations[2] with a secondary electron source term described below.

2. Theory

In the optical Bloch equation description of population dynamics[2], the population of the intermediate state is governed by a coupled differential equation that includes effects due to photoexcitation and scattering out of the state. To include hot electron cascade processes, an additional source term is added:

$$\frac{\partial P(E_i)}{\partial t} = (\ldots) - \frac{P(E_i)}{\tau_i} + \int_{E_i}^{E_{\text{vac}}} \alpha(E_i, E') \frac{P(E')}{\tau_{E'}} dE'. \tag{1}$$

Here, E_i is the energy of the ith state, $P(E_i)$ is the population, and τ_i is the time scale for decay. The unlisted terms in parenthesis represent coupling to other levels via photoexcitation, while the next term represents depopulation due to scattering. The integral term of Eq. (1) is the secondary cascade electron source term. The source term depends on the population and lifetimes of all states higher in energy. The parameter $\alpha(E_i, E')$ is the weighting coefficient for scattering into the state of interest. Here we are assuming that electron-electron scattering allows transitions to the intermediate state that involve large changes in **k** such that the projected band structure can be ignored in calculating $\alpha(E_i, E')$.

Equation (1) is approximated by a discrete sum. The source term then becomes

$$S(t) = \sum_{j=i+1}^{N} \alpha(E_i, E_j) \frac{P(E_j)}{\tau_j}, \tag{2}$$

where the Nth level is the highest-energy bound level excited by a single photon. For the *sp* bands of copper, the density of states is nearly constant, so Auger and electron scattering processes are identical, and the weighting term is

$$\alpha(E_i, E_j) = \frac{4(E_j - E_i)}{E_j^2} \Delta E, \tag{3}$$

where ΔE is the discrete step used to calculate the sum in Eq. (2).

The source term in Eq. (2) can be calculated for each energy of interest provided that the lifetimes of the higher energy states are already known. The calculated source term can then be used in a least-squares fit of the data, where the fitting parameter is an overall scaling of the complete sum $S(t)$. The total secondary contribution is determined by integrating the time-dependent populations of the intermediate state after one photon excitation with and without the source term. The ratio of the integrated populations provides the secondary contribution.

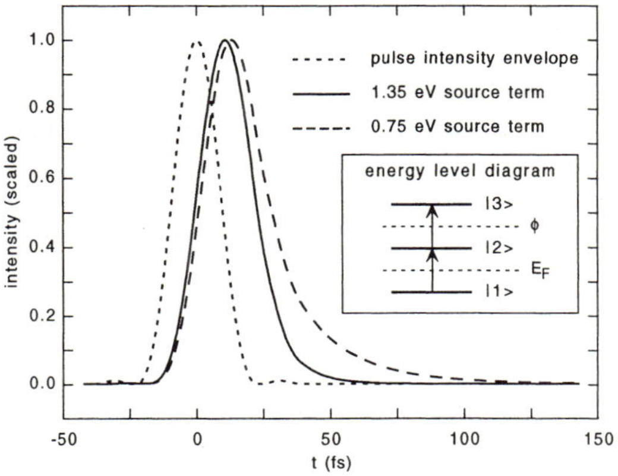

Fig. 1. Sample source terms and energy level diagram.

Consider the calculated source terms in Fig. 1 for two different intermediate state energies after 1.55 eV excitation. A Fermi liquid lifetime scaling is used for the model. As can be seen, for higher energies the electron cascade occurs quite quickly, while for lower energies the cascade occurs through a number of higher energy levels and thus stretches the source term out to longer time scales.

The presence of a secondary source term can affect the ratio of the peak of the I2PC to the constant offset at large delays. For a second-order interferometric autocorrelation, this ratio should be 8:1. Simulations of an I2PC with and without the source term verify that the presence of secondary electrons reduces the ratio to a value less than 8:1. Moreover, this effect is most pronounced when the source term has significant components on the time scale of the laser pulse.

3. Results and Conclusions

I2PC measurements are made on cesiated Cu(100) and Cu(111) surfaces with a 19 fs intensity FWHM 1.55 eV pulsed laser. In rigorous fits of the data to an optical Bloch equation model with a secondary source term, it is found that the contribution to the data from cascade secondaries is less than 1% for all conditions considered[3]. However, these experiments reveal a complication with I2PC measurements and metal systems where phonon-mediated transitions are possible, *i.e.*, multiple contributions to the I2PC due to transitions through either real or virtual intermediate states. When this is taken into account, the fitted I2PC data have a characteristic residual pattern that was initially taken to be a strong indication of secondary electrons. However, it is most likely an artifact of a signal with multiple contributions from different pathways.

The measured I2PC peak to baseline ratios, however, do suggest that there is a measurable presence of cascade secondary hot electrons. For example, for Cu(100) with S polarization, the measured ratios are as low as 6.6:1. Such a low ratio corresponds to a secondary contribution as high as 10%. The other measured ratios are higher, but suggest secondary contributions on the order of 5%. It is likely that more sensitive pump-probe measurements that are able to isolate and exclude transitions through virtual states will be sensitive to such fractions of secondaries, thus providing a better test of the model and the possibility of moving on to consider Auger secondary contributions as well.

Acknowledgments. M.J. Weida thanks the National Science Foundation and the Center for Global Partnership for support (NSF grant INT-9819100).

References

1. H. Petek and S. Ogawa, Prog. Surf. Sci. **56**, 239 (1997).
2. M. J. Weida, S. Ogawa, H. Nagano and H. Petek, J. Opt. Soc. Am. B **17**, (2000).
3. M. J. Weida, S. Ogawa, H. Nagano and H. Petek, (in preparation).

Phonon-Pair combination states driven by second-order Raman scattering in $KTaO_3$

Albrecht Bartels, Thomas Dekorsy, and Heinrich Kurz

Institute of Semiconductor Electronics II, RWTH Aachen, Sommerfeldstr. 24, 52056 Aachen, Germany
E-mail: bartels2@iht-ii.rwth-aachen.de

Abstract. Excitation of $KTaO_3$ with ultrashort laser pulses drives phonon-pair combination states via impulsive second-order Raman scattering. A model explains the resulting covariance-oscillations of the displacements of the two correlated phonon modes that are observed in a spectrally and temporally resolved pump-probe experiment.

1. Introduction

Coherent phonon spectroscopy has been a field of extensive efforts since the first observation of coherent phonons in bulk semiconductors in 1990. Several excitation mechanisms have been identified in solids among which the most prominent is impulsive first-order Raman scattering [1]. Coherent states exhibit an oscillating phonon displacement but the statistic properties of the lattice, i.e. displacement-variances and -covariances, remain static. Dynamics of these can be driven by second-order Raman scattering (SOR). States with a noise modulation, i.e. displacement-variances oscillating at twice the phonon frequency, have first been reported by Garret et al. [2]. We report here for the first time on phonon-pair combination states driven by ultrashort light pulses. Two harmonic oscillators, that are decoupled in the absence of perturbation (neglecting anharmonicities) become correlated through impulsive SOR and exhibit oscillating displacement-covariances as demonstrated in $KTaO_3$ [3].

2. Model

The SOR excitation of the lattice is modelled by two harmonic oscillators coupled for a very short time by a light pulse via the second-order polarizability. Phonon-pair combination states out of the entire Brillouinzone are driven under wavevector selection rules. They result in oscillations of the displacement-covariance at the sum- and difference-frequency of the two involved phonons. The detection in a transmission pump-probe experiment can be described within the framework given by Yan et al. [4]. Normalized transmission changes after impulsive excitation at certain optical detection frequencies ω in a spectrally resolved detection scheme are described by [3]:

$$\frac{\Delta T}{T_0} \propto \sum_{q,m,n} \frac{I_0}{\sqrt{\omega_{q,m}\omega_{-q,n}}} \times$$

$$\left[A^{+}_{q,m,n} \cos(\omega_{q,m} + \omega_{-q,n})\tau + A^{-}_{q,m,n} \cos(\omega_{q,m} - \omega_{-q,n})\tau\right] \textit{with}$$

$$A^{+}_{q,m,n} \propto 1 + \langle n(\omega_{q,m})\rangle + \langle n(\omega_{-q,n})\rangle \; \textit{and} \; A^{-}_{q,m,n} \propto \langle n(\omega_{q,m})\rangle - \langle n(\omega_{-q,n})\rangle$$

(1)

Here are I_0 the integrated intensity of the pulse and the $\omega_{q,m}$ are phonon frequencies for wavevector **q** and band m. <n> denotes the Bose phonon occupation number expectation value. Dominant contributions will arise from mode-pairs having a maximum in their combined density of states $N_{q,m}N_{-q,n}$. Due to the <n> dependence the amplitudes exhibit a pronounced variation with temperature: The difference frequency contribution $\propto A^-$ vanishes at low temperatures while the sum frequency $\propto A^+$ contribution approaches a constant non-zero value. Also a linear dependence of the signal amplitude on the optical detection frequency with a sign reversal at the laser's central frequency is expected.

3. Experiment, Results and Discussion

A 25-fs Ti:sapphire pulse laser with a repetition frequency of 2 GHz and a central wavelength of 800 nm is used. A pump-pulse is focused to a spot of 25 μm into the $KTaO_3$-sample and a spectrally filtered (1 nm bandwidth) probe-pulse is used to monitor transmission changes as a function of time delay between the pulses and the optical detection frequency. Average pump and probe powers are 200 mW and 40 mW, respectively.

Fig. 1a shows the transient transmission changes after pulsed excitation at 290 K and a fixed detection wavelength of 835 nm together with the corresponding Fourier transform. Besides the thermal noise modulation at twice the zone-boundary TA-phonon frequency, combinations of phonon-modes of the TA, TO_1, TO_2 (the last two are close in frequency and appear as one peak in the spectrum) and TO_4-phonon-branches as indicated in the figure appear. The peaks correspond to maxima in the combined density of states, i.e. they arise from combinations close to the zone-boundary. Normalized amplitudes of the combination modes are shown in Fig. 1b as function of temperature in comparison to the model amplitudes A^- and A^+. The expected main features, i.e. vanishing difference frequency contribution at T→0 K and rising amplitudes with temperature, are reproduced. There are, however, significant deviations for temperatures above 100 K partly due to phonon-phonon interaction that increases the decay of the covariance oscillations at elevated temperatures.

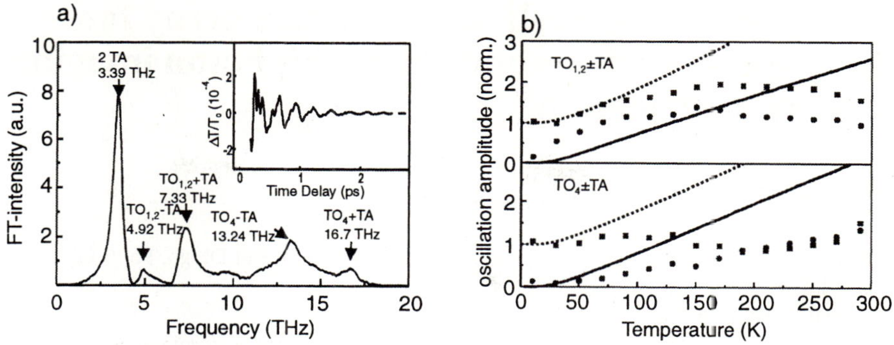

Fig. 1: a) Fourier intensity of the transient transmission change data shown in the inset. The data were taken at room temperature at a fixed detection wavelength of 835 nm. Excited phonon-mode combinations are indicated in the figure. b) Normalized temperature dependent amplitudes of the $TO_{1,2}$±TA and TO_4±TA combinations. Squares (circles) correspond to sum frequencies (difference frequencies). Solid and dashed lines are calculated amplitudes of the sum frequency (dotted line) and difference frequency (solid line) contribution.

Finally the amplitude of the displacement-covariance oscillations depends linearly on the optical detection frequency. A sign reversal at the laser's central frequency as predicted by the model confirms the validity of eq. (1) [3].

4. Conclusions

In conclusion, combination states of phonon-pairs out of the whole Brillouinzone of $KTaO_3$ are driven by a fs laser pulse via SOR. Oscillators that are decoupled in the absence of perturbation become coupled for the period of excitation through SOR and remain statistically correlated resulting in oscillations of their displacement-covariance. Specific and essential experimental details such as increasing sum and difference frequency contributions with temperature and vanishing difference frequency combinations for $T \rightarrow 0$ are explained by a quantum mechanical model satisfactorily.

Acknowledgements. We thank A. Stahl, M. Artoni and W. Xu for helpful and stimulating disussions.

References

[1] for a review see T. Dekorsy et al., in *Light Scattering in Solids VIII*, ed M. Cardona and G. Güntherodt (Springer, Berlin, 1999) and references therein.
[2] G.A. Garret et al., Science **275**, 1638 (1997).
[3] A. Bartels et. al, Phys. Rev. Lett. **84**, 2981 (2000).
[4] Y.X. Yan, J. Chem. Phys. **83,** 5391 (1985).

Temporally decorrelated multifocal array for high-speed, high-resolution, multiphoton imaging and micromachining

David N. Fittinghoff[1], and Jeffrey A. Squier[2]

[1] Institute for Nonlinear Science, University of California at San Diego, 9500 Gilman Dr., La Jolla CA 92093-0339, USA
E-mail: dnf@ucsd.edu
[2] Electrical Engineering and Computer Science, University of California at San Diego, 9500 Gilman Dr., La Jolla CA 92093-0339, USA
E-mail: jsquier@ucsd.edu

Abstract. We demonstrate, for the first time, a temporally decorrelated, multifocal multiphoton microscope for imaging and micromachining. This new technique allows multiphoton processes using high numerical aperture optics at spatial foci densities that were previously prohibitive.

1. Introduction

Multiphoton microscopy has recently benefited from the introduction of multifocal arrays, which have made it possible to produce images at rates exceeding 30 frames per second[1-3]. To date, these foci have been produced by placing a microlens array in the beam to produce a series of beamlets. Scanning is achieved with galvanometric mirrors[3] or by rotating the lenslet array[1]. and the multiphoton signal is imaged on a CCD array. A limitation of this method is that, as the foci are placed closer together, interference between the overlapping, out-of-focus, tails of the beamlets can generate significant amounts of multiphoton signal. For wavelengths near 800 nm and numerical apertures (NA) near 1.3, this effect greatly degrades the sectioning for separations between the foci that are less than ~7-10 μm [3]. Assuming sufficient laser power, this effect limits the speed of acquisition for a given scan area by reducing the total number of usable foci. In addition, the multifocal arrays have not produced diffraction limited focusing for pulse durations below 100 fs, due to chromatic effects. We have demonstrated, for the first time, a temporally decorrelated, multifocal multiphoton microscope that does not have these limitations. Using an etalon, we create a 1-D array of beamlets in which the pulses arrive at a plane perpendicular to the propagation direction at different times. We then use a system of lenses to overlap the collimated beams with slightly different angles at the entrance aperture of the objective and produce a 1-D array of foci that are temporally decorrelated. By measuring the two-photon absorption (TPA) point spread function (PSF) of these foci, we show that the temporal decorrelation eliminates any interference between the foci and allows multifocal multiphoton imaging with diffraction limited focusing, even for pulses as short as 20-fs. This new method of imaging will make it possible to completely overlap the foci and eliminate the need for scanning. This makes efficient use of the power available from typical ultrafast lasers, increasing the frame rate in microscopy and the throughput in micromachining applications.

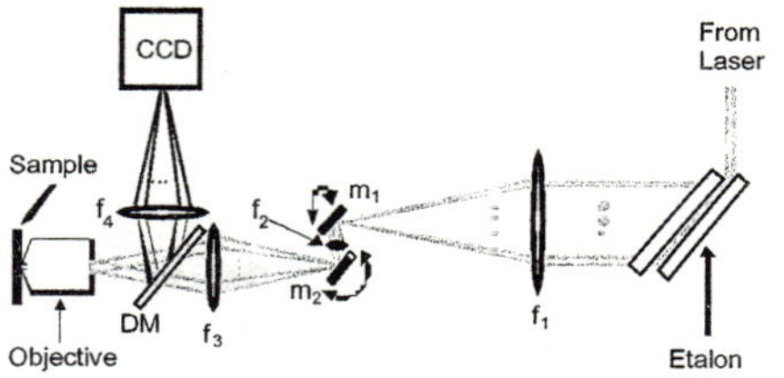

Fig. 1 Temporally decorrelated multiphoton microscope. An etalon divides the beam into beamlets, which are then manipulated to produce temporally decorrelated foci in the sample. f_1, f_2, f_3 and f_4 are the focal lengths of the lenses. *DM* is a dichroic mirror, and m_1 and m_2 are scan mirrors.

2. Experimental Configuration

In these experiments, we use a passively mode-locked $Ti{:}Al_2O_3$ oscillator, which produces a 150-mW, 76-MHz train of ~20 fs pulses at 800 nm. The temporally decorrelated microscope itself is shown in Fig. 1. We use the leakage from an etalon to produce a series of spatially parallel beamlets in which the pulses are temporally separated by the transit times of the air paths in the etalon. The beamlets are then collimated at the correct diameter to fill the entrance aperture of the objective. Since the incidence angle of each beamlet is slightly different at the entrance aperture of the objective, the beamlets are mapped to an evenly spaced 1-D array of temporally decorrelated foci. For imaging or machining the entire line of foci is scanned.

The etalon is made from two high-reflectivity (R > 99%), 45° s-polarized mirrors. The high reflectivity guarantees that the beams will have near uniform intensity, but it does reduce the available power per beamlet. To obtain a greater percentage of the incident light in each beamlet, a mirror with spatially varying reflectivity could be used. The mirror separation in the etalon, d, and incidence angle of the beam, θ, control the physical separation of the beamlets, $S = d/\cos\theta$. In the configuration used, S is also the maximum clear aperture for the initial beam entering the etalon, so the minimum beamlet separation is the minimum clear aperture required for the input beam. While the material paths, excluding the air gap between the etalons, are identical for all the beams, the pulse delay within the etalon increases by $\tau_d = 2S/c$, with each successive beam. Here c is the speed of light in air. In this work, $d = 3.0$ to 4.0 mm and $\theta = 45°$, so there are 28 to 38 ps delays between the foci.

Fig. 2 Line of foci produced by the time-decorrelated multifocal microscope in cell containing coumerin dye.

3. Results

Fig. 2 shows a TPA image of a row of 8 foci produced in a thin cell containing $2x10^{-3}$M coumerin dye in ethanol. The cell was made from a coverslip and slide since the Zeiss 100x/1.25-NA Achroplan oil objective is designed for optimal correction when imaging through the coverslip. The foci are 2.2 μm apart, which is much closer together than the 7-10 μm limit set by the interference between the beams that was mentioned in previous work [3]. Fig. 3 shows a measurement of the PSF with only with 8 pJ per pulse per focus, which we measured using the CCD camera, integrating over the image. To eliminate any residual fundamental light, we filtered the signal with 2-mm of BG-39 glass. This measurement is a dramatic demonstration of the temporal decorrelation. The figure shows the theoretical best result for a single beam and the data. The agreement is excellent with the only deviation occurring deeper in the dye cell where the fall off of the signal is expected due to spherical aberration, which are not accounted for in the theory. It is important to note that without the temporal decorrelation a 40% background is expected with these spatial beam densities.

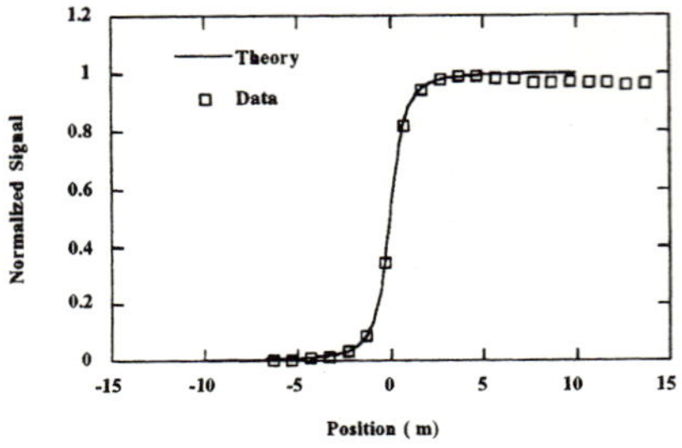

Fig. 3 Measurement of the point spread function for 8 foci produced using a Ti:sapphire oscillator and comparison to theory. The theory assumes a 1.25 NA Gaussian beam focus.

4. Summary

We have demonstrated for the first time a temporally decorrelated, multifocal multiphoton microscope and showed that the decorrelation eliminates interference between the foci and, allows multifocal multiphoton imaging with diffraction limited focusing for pulses as short as 20 fs. Future work on multifocal imaging using oscillators will increase the image acquisition speed to video frame rates by producing completely overlapping foci, and increasing the power per beamlet by optimizing the etalon. Using more energetic pulses, micromachining is possible over very large fields with high resolution. By using liquid crystal arrays, for instance, the amplitude of the foci can be modulated making it possible to write complex structures easily. This technique can also be extended to both dimensions completely eliminating the need for scanning, and subsequently scan errors.

References

1. J. Bewersdorf, R. Pick, and S. W. Hell, Optics Letters **23**, 655 (1998).
2. M. Straub and S. W. Hell, Applied Physics Letters **73**, 1769 (1998).
3. A. H. Buist, M. Muller, J. Squier, and G. J. Brakenhoff, Journal of Microscopy **192**, 217 (1998).

Femtosecond laser-induced ablation of graphite

K. Sokolowski-Tinten[1], S. Kudryashov[1,2], V. Temnov[1], J. Bialkowski[1], D. von der Linde[1], A. Cavalleri[3], H. O. Jeschke[4], M. E. Garcia[4], K. H. Bennemann[4]

[1] Institute for Laser-and Plasmaphysics, University of Essen, 45117 Essen, Germany
E-mail: kst@ilp.physik.uni-essen.de

[2] Laser Chemistry Division, Dep. of Chemistry, Moscow State University, 119899 Moscow, Russia
E-mail: serge@laser.chem.msu.ru

[3] Dep. of Chemistry and Biochemistry, The University of California San Diego, La Jolla, CA 92093-0339, USA
E-mail: axc@ucsd.edu

[4] Institute for Theoretical Physics, Freie Universität Berlin, 14195 Berlin, Germany
E-mail: garcia@physik.fu-berlin.de

Abstract. The dynamics of fs-laser ablation of graphite has been investigated experimentally and theoretically. The experimental observation of two different ablation mechanisms is supported by molecular dynamics calculations, which incorporate the changes of the interatomic potentials due to electronic excitation.

Femtosecond laser ablation of solids has attracted increasing interest in the past few years, both from a fundamental physics point of view, and with respect to technical applications of ultrashort laser pulses for material processing. With such short pulses transient extreme states of matter can be reached, opening unique paths for structural modifications and material removal. Recently we have shown [1] that near-threshold ablation of absorbing solids with femtosecond and picosecond pulses exhibits a material-independent behavior, which is characterized by a distinct transient interference phenomenon in the nanosecond time range. Here we present a combined experimental and theoretical study of ultrashort laser pulse ablation of graphite, emphasizing how the specific, but still not well characterized features of the carbon phase diagram modify the observed *universal* ablation behavior.

Highly oriented pyrolytic graphite (HOPG) was excited with single 100fs laser pulses. Using time resolved microscopy [2] the reflectivity of the irradiated surface was monitored with 100fs temporal and micrometer spatial resolution. Fig. 1 shows a sequence of micrographs covering a time span from the excitation up to the appearance of the final modifications of the surface.

For early times the surface exhibits an increase in reflectivity as has been observed previously [3] and interpreted as ultrafast melting of the material. Compared to other group IV and III-V semiconductors [4,5] the reflectivity increase upon melting is less pronounced. As already pointed out in [3] and supported by recent theoretical calculations [6] this may be explained by a metallic liquid state with less than four electrons per atom, which is resistively saturated. The initial state of increased reflectivity, indicative of a high-density molten phase of carbon, *survives* only for a few ps. For fluences below the ablation threshold (F_{abl} = 185mJ/cm^2) a reflectivity value between the level of the initial liquid and that of solid graphite is reached within 10-20ps. It must be noted that the melting threshold F_m = 165mJ/cm^2, as determined from a *post-mortem* analysis of the final surface modifications, is very close to the ablation threshold.

The onset of ablation is marked by a drastic decrease of reflectivity to values well below the level of the solid (Δt=50ps). This zone of decreased reflectivity subsequently transforms into the interference pattern typical for the near-threshold ablation of absorbing solids [1] (Δt=300ps - 2.4ns). The interference indicates the existence of an optically steep ablation front towards vacuum. Its formation can be explained by a *thermal* ablation model, which describes material removal as hydrodynamic expansion of the highly pressurized fluid layer through the liquid-gas coexistence regime of the phase-diagram [1,7].

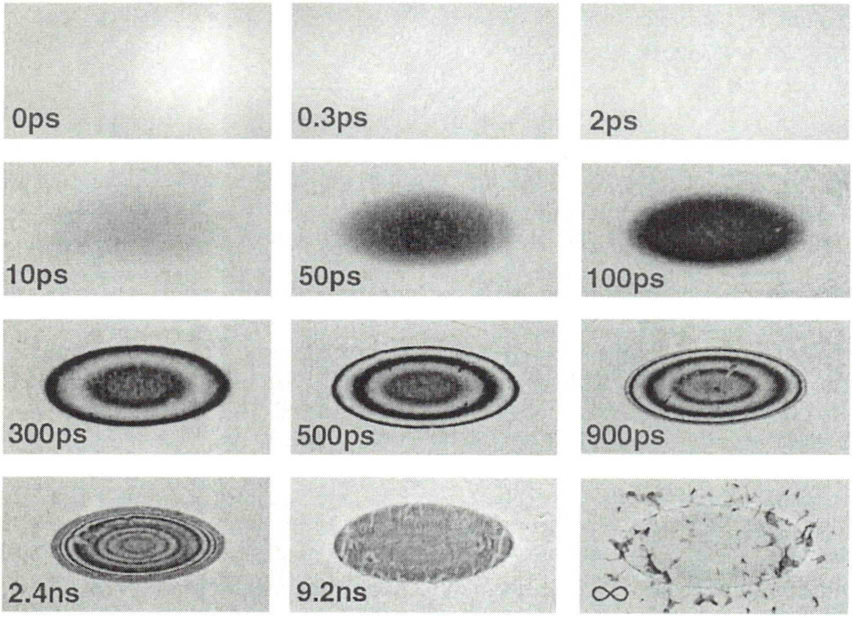

Fig. 1. Sequence of micrographs of a HOPG surface excited with a 100fs, 240mJ/cm^2 laser pulse for different delay times between pump and probe pulse.

In contrast to all the other materials investigated so far, the interference observed on graphite does not just vanish due to the progressive transformation of expanded liquid material to the gas phase. Instead, the area covered by interference fringes appears to be the source of μm-sized fragments which are ejected at later times (Δt=9.2ns) and finally deposited in and around the ablation crater (Δt=∞). We relate this behavior to the peculiar and still not well understood properties of the liquid state of carbon. The high temperatures and pressures required to observe l-C and the possibility of a liquid-liquid phase transition [8-10] will surely influence the ablation dynamics. Therefore, formation and ejection of the fragments may indicate solidification of the liquid component of the ablation plume before a complete transformation into the gas-phase can occur.

Above a threshold fluence of $F_{ab2} \approx 250 mJ/cm^2$ we neither observe the interference phenomenon, nor the formation of fragments. According to our understanding of femtosecond ablation [1], the absence of interference implies that hydrodynamic expansion of the laser-generated fluid does *not* include the passage through the liquid-gas co-existence regime of the phase diagram, but proceeds most likely *above* the critical point. In this frame of interpretation the ablating material can be completely transformed into a volatile phase (atomic gas, clusters) and will be permanently removed from the surface, explaining the absence of fragments in this case.

The ablation process has been also theoretically studied using molecular dynamics simulations, which allow a separate treatment of the electronic and ionic degrees of freedom [11]. Although a direct comparison of simulation and experiment is difficult due to the fact that the 100ps- to ns-time-scale is not accessible for these calculations, the numerical results show also two distinct thresholds corresponding to different microscopic processes. For lower intensities optical excitation leads to drifting and sliding of the graphite layers. This process may be associated with melting of the material. At higher intensities a second threshold indicates bond breaking within individual graphite layers. As a consequence single atoms and small clusters are emitted.

In conclusion, we have studied the dynamics of melting and ablation of graphite after fs-laser-excitation, using time- and space-resolved measurements of the optical reflectivity. Compared to other covalently bonded materials, the experimental data on graphite reveal significant differences, which we relate to the unique features of the carbon phase diagram. Supported by theoretical simulations, two different pathways of ablation are identified leading to different final states of the removed material.

References

1 K. Sokolowski-Tinten et al., Phys. Rev. Lett. **81**, 224-227 (1998)
2 M.C. Downer et al, J. Opt. Soc. Am. **B 4**, 595-599 (1985).
3 D.H. Reitze et al., Phys. Rev. B **45**, 2677 (1992).
4 C.V. Shank et al., Phys. Rev. Lett. **50**, 454-457 (1983).
5 K. Sokolowski-Tinten et al., Phys. Rev. B **51**, 14186-14198 (1995).
6 P.L. Silvestrelli et al., J. of Appl. Phys. **83**, 2478 (1998).
7 N.A. Inogamov et al., JETP Lett. **69**, 310 (1999).
8 M. van Thiel et al., Phys. Rev. B **48**, 3591 (1993).
9 M. Togaya, Phys. Rev. Lett. **79**, 2474 (1997).
10 J.N. Glosli, Phys. Rev. Lett. **82**, 4659 (1999).
11 H.O. Jeschke et al., Phys. Rev. B **60**, R3701 (1999).

Femtosecond mid-infrared study of the high-T_C superconductor $YBa_2Cu_3O_{7-\delta}$

M. Woerner[1], R. A. Kaindl[1], T. Elsaesser[1], D. C. Smith[2], J. F. Ryan[2], G. A. Farnan[3], M. P. McCurry[3], and D. G. Walmsley[3]

[1] Max-Born-Institut für Nichtlineare Optik und Kurzzeitspektroskopie, Max-Born-Straße 2A, D-12489 Berlin, Germany
E-mail: woerner@mbi-berlin.de
[2] Department of Physics, Clarendon Laboratory, Oxford University, Oxford OX1 3PU, UK
[3] Department of Pure and Applied Physics, Queen's University of Belfast, Belfast BT7 1NN, UK

Abstract. Ultrafast near-infrared excitation of optimally-doped and underdoped $YBa_2Cu_3O_{7-\delta}$ leads to an ultrafast fill-in of the (ab)-plane mid-infrared conductivity gap. Underdoped material shows two gap constituents in the transients with different recovery dynamics and saturation behaviour.

Since the discovery of high transition temperature superconductivity more than a decade ago a controversial debate on the underlying pairing mechanism is not yet resolved. A key parameter is the carrier concentration in the conducting copper-oxide (a,b)-planes which in the case of $YBa_2Cu_3O_{7-\delta}$ (YBCO) is varied by altering the oxygen content from the underdoped, via the optimally-doped (i.e. highest T_C) to the overdoped region. Below T_C in optimally-doped materials, the largest changes of the (ab)-plane reflectivity $R(\omega,T)$ appear in the mid-infrared spectral range around $\hbar\omega \approx 100$ meV which are directly connected to the opening of a gap for electronic transitions near the Fermi level. In underdoped cuprates, these features ("pseudogap") already occur at temperatures T* above T_C [1].

Here, we present experiments [2] probing, for the first time, this mid-infrared range in optimally ($T_C = 88$ K) and underdoped ($T_C = 68$ K) YBCO on ultrafast time scales. The superconductor is excited with 13-fs pulses at 1.6 eV from a 2-MHz cavity-dumped Ti:sapphire laser. Mid-infrared probe pulses tunable from 60 - 180 meV with $\approx$150 fs duration are derived from the output of the Ti:Sapphire laser by difference frequency mixing [3].

Fig. 1 (a) shows (ab)-plane reflectivity transients at $E_{probe} = 90$ meV for the optimally-doped sample. For temperatures below T_C (solid line) we observe a fast sub-ps increase and a longer decay on a $\approx$5 ps timescale. With increasing temperature the signal amplitude decreases and completely changes its dynamics around T_C, above which much faster transients of opposite sign are observed (dashed line). The nonlinear response is determined by carrier heating similar to metals with flat spectral response [diamonds, Fig. 1(c)] for $T > T_C$ [4] and by depletion and recovery dynamics of the superconducting condensate for $T < T_C$ [5]. This fill-in and recovery of the stationary conductivity gap is underlined by a comparison of the femtosecond spectra [circles in Fig. 1(b)] with the stationary reflectivity difference below and above T_C [solid line, Fig. 1(b)]. As we vary the intensity of the excitation pulses [Fig. 1(c)] for optimally-doped YBCO, we observe a saturation behaviour that is different for below and above T_C. While the amplitude of the heating related signal for $T > T_C$ (diamonds) follows almost linearly the excitation density, the condensate related signal for $T < T_C$ (circles) saturates at a certain level due to the finite number of pairs in the superconducting condensate prior to excitation.

A direct association of the conductivity gap, however, to the superconducting gap has been unclear since in underdoped cuprates it already appears below the higher pseudogap temperature T* [1]. In Ref. [2] we have shown that the mid-infrared response of underdoped material shows two gap constituents in the transients. Nonlinear reflectivity changes above T_C show at various spectral positions almost the same transient (i.e. only a single component) whereas below T_C the transients are described by two separate components, with a fast (700 fs) and a slow (≈ 5 ps) decay, respectively. For temperatures above T_C only the fast component is observed.

Here, we discuss the excitation intensity dependence of the slow and fast components observed in the underdoped sample at T = 14 K. In Fig. 2(a) we show $\Delta R/R_0$ transients for various excitation intensities at the prominent probe frequency of $\hbar\omega = 146$ meV. At low excitation intensities the amplitudes of the slow and the fast component are almost equal but have opposite sign. We found that all transients at various excitation intensities nicely fit to a linear combination of the normalized slow and fast components.

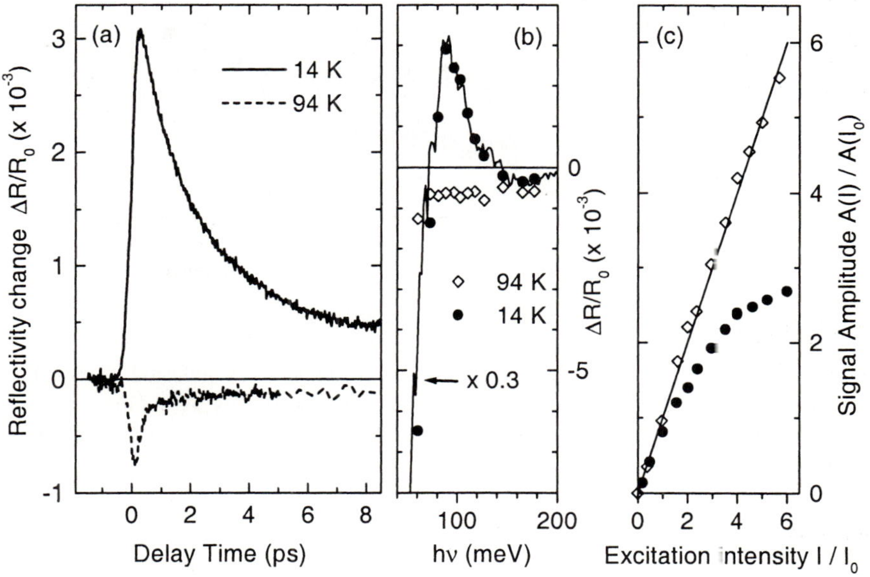

Fig. 1. Optimally-doped YBCO ($T_C = 88$ K): (a) Transient reflectivity change $\Delta R/R_0 = (R(t) - R_0)/R_0$ as a function of pump-probe delay at a probe photon energy for temperatures as indicated. (b) Transient spectra at the extrema of time curves. Solid line: difference between stationary reflectivity spectra $\Delta R = R(95K) - R(20K)$ multiplied by 0.3. (c) Symbols: excitation intensity dependence of the signal amplitudes (extremum of $\Delta R(t)/R_0$) normalized to their values at $I = I_0$. Solid line: linear fit for T = 94 K.

The decomposed amplitudes are shown in Fig. 2(b). Similar to the optimally doped case, the amplitude of condensate related component saturates at a certain excitation density. In contrast, the fast decaying component which is also related to the conductivity gap around $\hbar\omega = 100$ meV shows no saturation at all for the excitation intensities applied in our experiments [triangles, Fig. 2(b)].

Thus, the different saturation behaviour of the two reflectivity components additionally supports the picture that two different mechanisms govern the mid-infrared response of underdoped cuprates. Ultrafast excitation of the electronic system leads to a breakup of both condensed pairs and a second type of correlated carriers. The dominant slow (≈5 ps) component follows the gap related to condensed pairs and can be fully destroyed by strong enough optical excitation as in optimally doped samples. In contrast, the fast (≈700 fs) component which exists up to the pseudogap temperature T* cannot be saturated by optical excitation even at intensities where the superconducting condensate is completely destroyed.

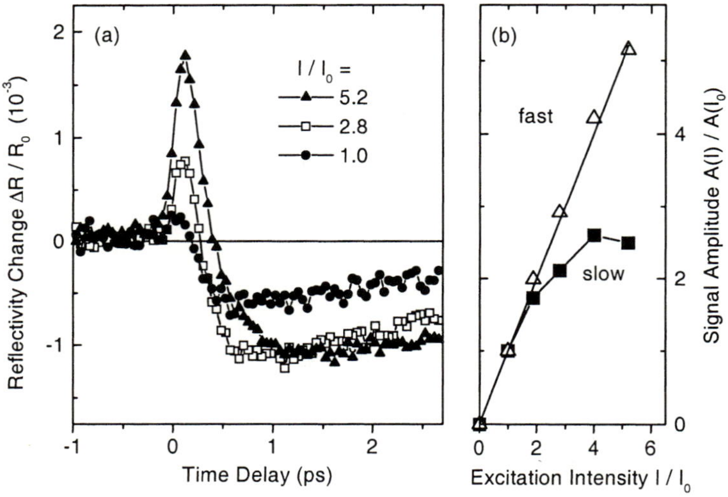

Fig. 2. Excitation intensity dependence of the signal amplitudes of the slow and fast components in underdoped YBCO at T = 14 K. (a) Measured $\Delta R/R_0$ transients at the probe frequency $\hbar\omega = 146$ meV for three different excitation intensities as indicated. (b) Symbols: Decomposed signal amplitudes of the slow and fast components as a function of the excitation intensity normalized to their values at $I = I_0$. Solid line: linear fit for the fast component.

References

1 Z. Schlesinger et al., Physica **C 235-240**, 49 1994.
2 R. A. Kaindl et al., Science **287**, 470 2000.
3 R. A. Kaindl et al., Appl. Phys. Lett. **75,** 1060, 1999.
4 S. D. Brorson et al., Phys. Rev. Lett. **64**, 2172, 1990.
5 J. L. W. Siders et al. in: *Ultrafast Phenomena XI,* Edited by Elsaesser, Fujimoto, Wiersma, Zinth (Springer) 365, 1998.

Observation of the Josephson Plasma Resonance in $Tl_2Ba_2CaCu_2O_8$ using THz Spectroscopy

V. K. Thorsmølle, R. D. Averitt, M. P. Maley, L. N. Bulaevskii and A. J. Taylor
Los Alamos National Lab, Los Alamos, NM 87545
E-mail: vthorsmolle@lanl.gov, ttaylor@lanl.gov

Abstract. Employing terahertz time-domain spectroscopy in transmission, we have unambiguously observed the Josephson Plasma Resonance in $Tl_2Ba_2CaCu_2O_8$ high-T_c thin films in zero magnetic field as a function of temperature.

1. Introduction

The c-axis Josephson Plasma Resonance (JPR) in highly anisotropic layered cuprate superconductors originates from the interlayer tunneling of Cooper pairs. These Josephson-coupled systems have received much interest in recent years due to their electrodynamic properties. The JPR, related to the c-axis penetration length, $\omega_{pc} = c / \lambda_c (\varepsilon_\infty^c)^{1/2}$, is a fundamental probe of the superconducting state and an excellent tool to study these highly anisotropic systems. For T << T_c, the spectral shift of the JPR with temperature is related to the symmetry of the order parameter. For T close to T_c, the appearance of the JPR probes the onset of interlayer phase coherence. Furthermore, the JPR spectral width is a measure of the quasiparticle scattering rate. In addition, the JPR probes the tunneling mechanism and provides information about the validity of the interlayer-tunneling model [1]. In a magnetic field, the JPR probes the correlation of pancake vortices along the c-axis and is a tool to study the B-T phase diagram [2,3]. The advantage of using terahertz time-domain spectroscopy (TTDS) to map out the phase diagram is that this technique allows measurements to be made over a broad frequency range with fixed magnetic fields and temperatures. Importantly, the JPR and the quasiparticle scattering rates of high-T_c superconductors with extreme anisotropy such as the BSCCO, thallium, or mercury based compounds lie in the THz frequency range.

We have measured the JPR in $Tl_2Ba_2CaCu_2O_8$ (Tl-2212) thin films in zero applied field as a function of temperature employing terahertz spectroscopy in transmission. The JPR frequency peak is 692 GHz at 10 K decreasing to 386 GHz at 90K (T_c = 98.8 K). We also observe an increase in the JPR spectral width with increasing temperature, indicative of an increase in the quasiparticle scattering rate.

2. Results and Discussion

The experiments were carried out in transmission using TTDS. The experimental setup is shown and explained in [4]. The Tl-2212 film (~700 nm thick) was grown on an MgO substrate using a two-step process. First an amorphous TBCCO film was deposited by laser ablation. Next it was annealed at high temperature in oxygen to form the 2212 phase. The measured T_c for this film was 98.8 K.

In order to excite the c-axis JPR, a component of the terahertz field has to be along the c-axis of the sample. The a-b plane plasma frequency is on the order of 1.5 eV, while the JPR along the c-axis lies in the terahertz range. Therefore, tilting the sample at an angle Θ will transmit the incident p-polarized electric field component parallel to the c-axis at frequencies above the plasma edge while reflecting the component that is parallel to the a-b plane. A more detailed analysis shows that for frequencies $\omega << \omega_{ab}$, and taking into account that the system consists of discrete layers, that electromagnetic waves can only propagate along the c-axis in a narrow frequency window above the plasma edge.

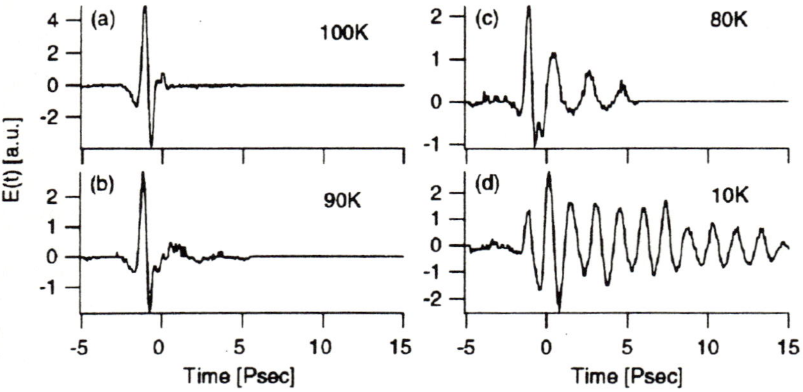

Fig 1. (a)-(d)) Transmission of incident THz pulse through Tl-2212 thin film. (a) The THz pulse transmitted through the film above the onset of superconductivity. (b) Just below T_c, there is a slight oscillation after the main pulse due to excitation of the c-axis JPR – the oscillation is highly damped. (c) At 80K, the JPR oscillation is more pronounced since the quasiparticle density is reduced resulting in a decrease in the scattering. (d) At 10K, the JPR oscillations persist at least to 15 ps – note also the increased oscillation frequency in comparison to (c).

Figure 1(a)-(d) shows the transmission of a THz pulse through a Tl-2212 thin film as a function of temperature. Fig. 1(a) shows the THz pulse transmitted through the film above the onset of superconductivity. At 90K (Fig. 1(b)), the onset of c-axis coherent Josephson tunneling is just perceptible as a slight oscillation after the main THz pulse. The JPR is highly damped due to quasiparticle scattering. At lower temperatures (Fig. 1(c)-(d)), the JPR is more pronounced due to a reduction in

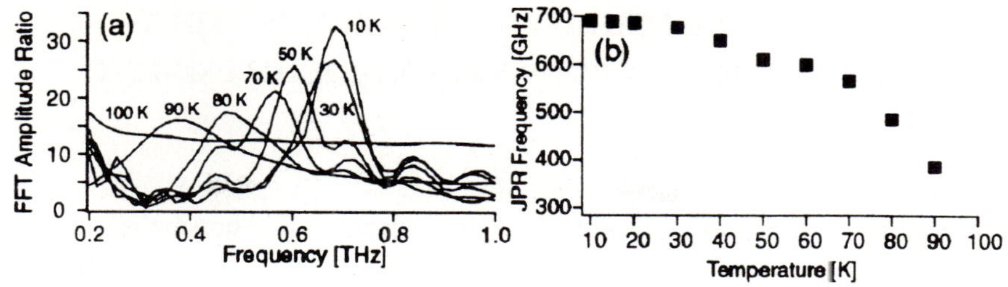

Fig 2. (a) The JPR as a function of frequency at various temperatures. With increasing temperature, the peak shifts to lower frequencies and broadens. (b) The JPR peak maximum as a function of temperature.

scattering. The time domain data also indicates that the JPR shifts to higher frequencies as the temperature is lowered as indicated by the increase in oscillation frequency.

Figure 2 shows the amplitude ratio of the Fast Fourier Transform (FFT) of the time-domain data in Fig. 1 divided by a reference scan on bare MgO. Fig. 2(a) clearly shows that the JPR shifts to higher frequencies and narrows as the temperature is decreased. The JPR frequency shift vs. temperature in Fig. 2(b) agrees well with previous reflectivity measurements on Tl-2212 [1] and we have been able to extend the measurements to temperatures approaching T_c. Future experiments aim to utilize this new capability of measuring the JPR in transmission to probe the B-T vortex phase diagram of anisotropic high-T_c superconductors.

Acknowledgements. We would like to thank Superconductor Technologies Inc. for providing the Tl-2212 films used in these experiments. This work was funded in part by CLC.

References

1 A.A. Tsvetkov, et al., Nature **395**, 360 (1998).
2 L.N. Bulaevskii, M.P. Maley, M. Tachiki, Phys. Rev. Lett. **268**, 309 (1995).
3 A.E. Koshelev, Phys. Rev. Lett. **77**, 3901 (1996).
4 R.D. Averitt, G. Rodriguez, J.L.W. Siders, S.A. Trugman, A.J. Taylor, J. Opt. Soc. Am. B. **17**, 327 (2000).

Ultrafast THz conductivity dynamics: spin-lattice relaxation in colossal magnetoresistive oxides

R. D. Averitt[1], A. I. Lobad[1], C. Kwon[2], S. A. Trugman[1], V. K. Thorsmølle[1], and A. J. Taylor[1]

[1] Los Alamos National Lab, Los Alamos, NM 87545
E-mail: raveritt@lanl.gov, ttaylor@lanl.gov
[2] Department of Physics, California State University-Long Beach, Long Beach, CA 90840
E-mail: ckwon@csulb.edu

Abstract. Ultrafast measurements of optically induced changes in the conductivity (0.4-1.0 THz) of $La_{0.7}M_{0.3}MnO_3$ films (M = Ca, Sr) from ~10K to ~$0.9T_c$ reveal a spin-lattice relaxation that follows the temperature dependence of the spin specific heat.

1. Introduction

Transition metal oxides are an important class of materials that display an enormous range of phenomena due primarily to the interplay between the charge, spin, lattice, and orbital degrees of freedom. An important subclass of these materials is the mixed-valence manganite perovskites $R_{1-x}D_xMnO_3$ where R is a trivalent rare earth (e.g. La, Nd) and D is a divalent alkali (e.g. Ca, Sr). The observation of "colossal" negative magnetoresistance (CMR) in the hole-doped manganites demonstrates the sensitivity of the electronic conduction to the underlying magnetic structure in these materials [1]. Experimental and theoretical studies have also revealed the importance of the lattice degrees of freedom in determining the electronic properties of CMR materials above and below T_c [2,3]. Despite these important advances, the understanding of transport in CMR materials below T_c is incomplete. Thus, novel probes of the conducting ferromagnetic state of CMR materials can provide complementary data in helping to elucidate the dependence of electrical transport on the spin and lattice structure.

We have employed time-resolved terahertz spectroscopy (TRTS) to measure dynamic conductivity changes in hole-doped $La_{0.7}Ca_{0.3}MnO_3$ (LCMO) and $La_{0.7}Sr_{0.3}MnO_3$ (LSMO) thin films. The induced conductivity changes occur on two time scales. The slower component, related to spin-lattice relaxation, has a lifetime that increases upon approaching T_c from below in accordance with an increasing spin specific heat.

2. Results and Discussion

The TRTS experiments were performed on LCMO and LSMO epitaxial thin films (~100 nm) grown on $LaAlO_3$ substrates using pulsed laser deposition [7]. Magnetization measurements yield T_c = 250, 360 for the LCMO and LSMO films respectively. The experiments utilized a commercial-based regeneratively amplified $Ti{:}Al_2O_3$ system operating at 1 KHz and producing nominally 1.0-mJ, 150-fs pulses at 1.5 eV. The THz pulses were generated and detected using electrooptic techniques. A He cryostat permitted temperature dependent measurements from 4-400 K. Further details of the film growth and TRTS experiments are described elsewhere [4,5].

Measurements of σ (T) without optical excitation can be fitted using the following equation: $\sigma(T) = \sigma_0 \exp(M(T)/M_o)$ where $M(T) \propto (1 - T/T_c)^{\beta}$ ($\beta \sim 0.33$) is the magnetization and T_c is the transition temperature. This demonstrates, in agreement with DC resistivity measurements, that the conductivity primarily follows the magnetization below T_c [6]. However, phonons also play an important role in determining σ (T) as revealed by our dynamic conductivity measurements.

Upon optical excitation, a decrease in the conductivity occurs due to energy exchange between the photoexcited electrons and the lattice and spin systems. Figure 1 (A and B) shows the temporal evolution of the conductivity at various

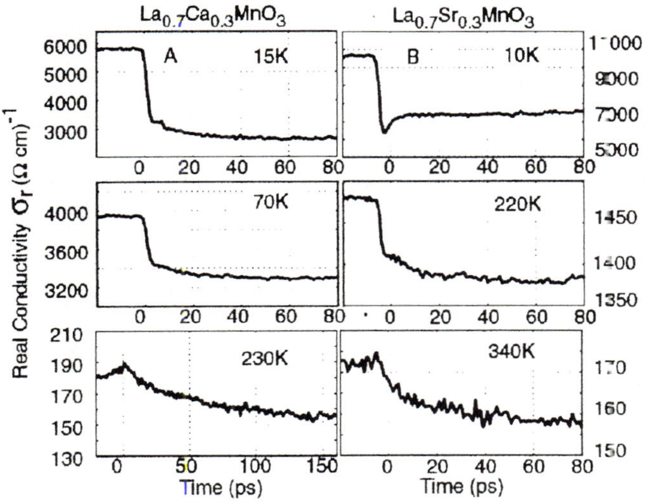

Fig 1. (A and B) Transient real conductivity vs. time for LCMO and LSMO films. The fast ~2 ps decrease observed for both films decreases to zero upon approaching the transition temperature. The lifetime of the slower component increases with increasing temperature.

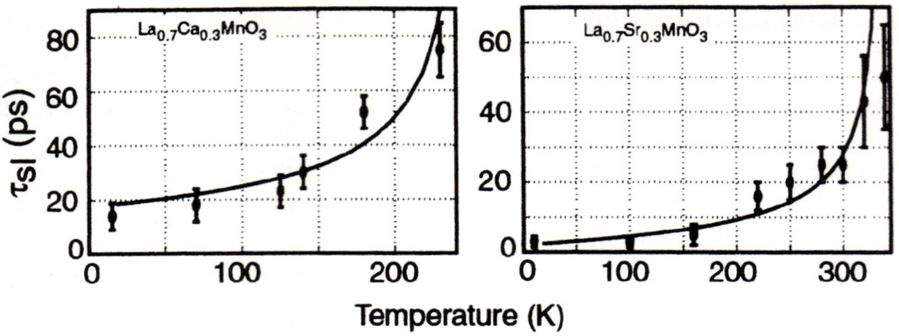

Fig 2. Temperature dependence of the spin-lattice relaxation for LCMO and LSMO respectively. The solid lines are results of a two-temperature calculation using approximate fits to the specific heat data for LCMO [9] and LSMO [10]. The lifetime follows the temperature dependence of the spin component of the specific heat.

temperatures for LCMO and LSMO films. There is a fast component (~2 ps) whose magnitude (relative to the slow component) decreases with increasing temperature. This ultrafast conductivity decrease is due to an optically induced modification of the effective phonon temperature [11]. Also evident is a slower component whose lifetime increases with increasing temperature. Figure 2 shows the lifetime of the slow component as a function of temperature for LCMO and LSMO. This increase in the lifetime follows the increase in the spin specific heat indicating that the dynamics are associated with spin-lattice equilibration. These conductivity dynamics agree with all-optical pump-probe studies of the dynamic spectral weight transfer in CMR films [7,8] providing a consistent picture of the ultrafast spin-lattice dynamics in LCMO over a broad energy range.

References

1 For an introduction to the manganites see *Colossal magnetoresistive oxides*, edited by Y. Tokura (Gordon and Breach, Amsterdam, 2000).

2 A. J. Millis, Nature **392**, 147 (1998).

3 C. H. Booth et al., Phys. Rev. Lett. **80**, 853 (1998).

4 C. Kwon, Q. X. Jia, Y. Fan, M. F. Hundley, D. W. Reagor, J. Y. Coulter, D. E. Peterson, Appl. Phys. Lett. **72**, 486 (1998).

5 R. D. Averitt, G. Rodriguez, J. L. W. Siders, S. A. Trugman, A. J. Taylor, J. Opt. Soc. Am. B **17**, 327 (2000).

6 M. F. Hundley et al., Appl. Phys. Lett. **67**, 860 (1995).

7 A. I. Lobad, A. J. Taylor, C. Kwon, S. A. Trugman, T. R. Gosnell, Chem. Phys. **251**, 227 (2000).

8 A. I. Lobad, R. D. Averitt, C. Kwon, A. J. Taylor, submitted.

9 A. P. Ramirez, et al., Phys. Rev. Lett. **76**, 3188 (1996).

10 M. N. Khlopkin, et al., Phys. Sol. State **42**, 114 (2000).

11 R. D. Averitt, A. I. Lobad, C. Kwon, S. A. Trugman, V. K. Thorsmølle, A. J. Taylor, submitted.

Part X

Electron Dynamics and Charge Transfer at Interfaces and in Liquids

Femtosecond Electron Dynamics at Surfaces and Interfaces.

C.B. Harris, A.D. Miller, S.H. Liu, and K.J. Gaffney

Department of Chemistry, University of California, Berkeley, California 94720 and Chemical Sciences Division, E.O. Lawrence Berkeley National Laboratory, Berkeley, California 94720

Abstract. Time and angle resolved two photon photoemision has been used to study dynamic electron localization and solvation at methanol and n-heptane covered metal surfaces. Coverage dependant localization is observed for methylthiolate on Ag(111).

1. Introduction

Electrons at interfaces provide a unique opportunity to study the influence of reduced dimensionality on electron carrier dynamics. The dynamics and band structure of interfacial electrons can be measured with femtosecond time and angle resolved two photon photoemission (TPPE). In addition, the changes in the interfacial electronic structure that result from the adsorption of atoms and molecules on a metal substrate can be monitored directly with angle resolved TPPE. Among the excited states observed with TPPE, image potential states have proven to be a sensitive probe of the influence of the adsorbate layer. The spatial extent of the image potential electrons parallel to the interface proves to be particularly interesting.

Localized interfacial electrons can result from either the static potential of the adsorbate, or a dynamically induced potential within which the electron can self trap. Small polaron formation in overlayers of n-heptane [1,2] and electron solvation in a methanol adsorbate both display dynamic localization from an interaction of the electron with the adsorbed layer. Localized electrons at low coverages of a chemisorbed thiol layer result from the static layer potential rather than a dynamic interaction between the electron and the adsorbate. The spatial extent of an electron is determined by a competition between localizing and delocalizing factors. Electronic localization in a crystal depends on the dimensionality of the system [3]. Localization of electrons in one dimensional systems is energetically favorable, while three dimensional systems tend to support delocalized electrons. In two dimensional systems, the competition between the kinetic energy cost and the potential energy benefit of localization is fairly balanced, and the spatial extent of the electron is uncertain. Angle resolved TPPE can be used to probe the spatial extent of two dimensional electrons at surfaces and at interfaces. Electrons whose behavior is free electron like parallel to the surface display a parabolic dispersion with the electronic energy proportional to the square of k-parallel. The curvature of the parabola is fitted by an effective mass parameter, m*, which is normalized to the mass of a free electron. An electron with m*=1 has a free electron like dispersion parallel to the interface, while a less dispersive electron has a larger effective mass. The effective mass of the electron can be used as a measure of its interaction with the surrounding media, as well as a measure of the spatial extent of the electron.

2. Adsorbed n-Heptane Overlayers

The electron affinity of the adsorbate layer has a large effect on the electronic properties of the image states. Image states at the n-heptane/Ag(111) interface demonstrate the effect of a negative electron affinity. The n=1 image state lifetime for a monolayer of n-heptane is 150 fs. With an increase in layer thickness to a trilayer, the lifetime becomes 17,600 fs. The negative electron affinity of the n-heptane overlayer excludes the electron from the layer, forcing it to reside at the n-heptane/vacuum boundary. This decouples the electron from the substrate and the adsorbate layer and increases the lifetime of the electron. The layer becomes a barrier through which the electron must tunnel in order to decay into the metal, resulting in electronic lifetimes that increase exponentially with the thickness of the layer [4]. The longer image state lifetimes for multilayer n-heptane allow the nuclei of the layer enough time to respond to the electron. The electron induced changes in the overlayer result in localization via small polaron formation [1,2]. This process competes with the decay of the electron into the metal, and is thus mostly unobserved for a monoayer of n-heptane. The image state electrons for the monolayer return to the metal before they can induce a localizing potential in the layer.

Angle resolved lifetime measurements were used to determine mechanism for localization. The localized polaron was observed to grow in with a 360 fs rise time and decay with a 1600 fs time constant. These values were found to be independent of angle. Time constants for the delocalized n=1 image state were found to vary with angle, with decay constants decreasing from 810 fs at 0° to 200 fs at the highest angle measured. These angle dependent lifetimes can be explained with a transition state theory analogous to Marcus electron transfer theory [5,6]. Electronic states with higher k-parallel (i.e. higher angle) have a smaller barrier to localization, which results in faster decay. A path integral technique is used to account for coupling between the adiabatic and non-adiabatic limits [7]. Oscillations in the log of the decay rate as a function of the reaction exothermicity and the non-Arrhenius temperature dependence are explained via the use of a single quantum mode approximation.

3. Methanol Adsorbates

Two dimensional electron solvation and localization by a monolayer and a bilayer of methanol adsorbed onto a Ag(111) surface have been studied with angle and femtosecond time-resolved TPPE. The time dependent stabilization of the electron binding energy is used to monitor electron solvation by the methanol molecules. The electron is initially excited from Ag substrate into an image state. The image state electron then interacts with methanol molecules and induces structural changes in the layer. Conversely, structural changes in the methanol layer can affect the binding energy of the electron. The time dependent binding energies for an electron in a bilayer of methanol are shown in Figure 1. Experimental results reveal that the solvation process occurs on multiple time scales. The initial fast solvation takes place in the first 500 femtoseconds, followed by a slower solvation that takes place for several picoseconds. Although the solvation time scales observed in two dimensional electron solvation show a strong resemblance to those observed in molecular dynamics simulations of electron solvation in liquid methanol [8], the initial fast Gaussian component of relaxation observed in the non-adiabatic simulation is absent from the experiment [9]. This Gaussian component of the relaxation is due to the rotational motion of the OH group in the solvation process. The stabilization of the electron binding energy can be attributed to the structural changes in the methanol layers to reduce the interaction energy between the electron and the solute.

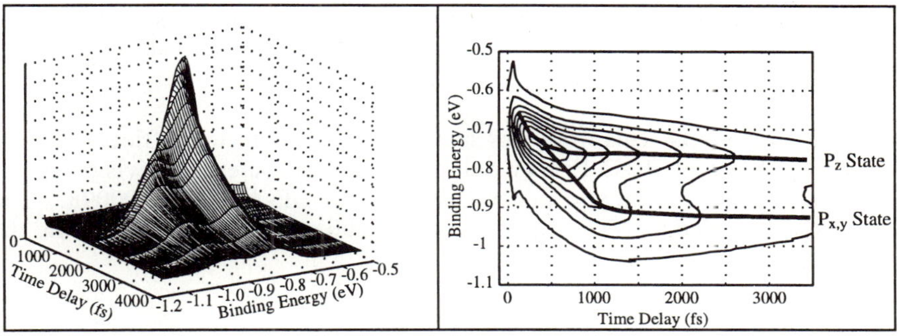

Fig. 1. Time dependent stabilization of electron binding energy by methanol. 3-D plot (left) and contour plot (right)

Angle resolved TPPE measurements have been used to determine the spatial extent of the electron parallel to the surface. Initially delocalized image state electrons are observed to localize dynamically within the first 500 femtoseconds of solvation. As the electron interacts with the solvent layer, the methanol molecules reorient to form a cavity around the electron, resulting in a localized electron. The dynamics of electron localization demonstrate the time scale of cavity formation in methanol.

Two different solvated states are observed due to the reduced dimensionality of the system. Excited P-type electrons oriented parallel and perpendicular to the surface experience different environments due to the symmetry imposed on the system by the surface geometry. These differences in geometry result in distinct states with different binding energies. Polarized light can be used to distinguish between these states, with S polarized light preferentially photoemitting electrons oriented parallel to the surface, and P polarized light preferentially photoemitting electrons oriented perpendicular to the surface. Results from polarization studies show that the $P_{x,y}$ state (parallel to the surface) has a lower energy compared to the P_z state (perpendicular to the surface).

4. Localization in Other Systems

Dynamical localization, as discussed above, results from electron induced distortion of the nuclear degrees of freedom of the layer. Electrons have also been observed to localize as a function of the static layer geometry. These electrons do not induce a potential within which they then trap, but rather trap into an already existing potential. Statically localized electrons are observed for low coverages of alkylthiol molecules chemisorbed onto a Ag(111) surface. The sulfur head of the thiol molecule interacts strongly with the silver, creating a chemical bond. Two non-image potential interfacial states are observed for this system, one below the Fermi level with an energy consistent with a chemical bond. The other state is initially unoccupied. Both states are delocalized for a fully covered surface, but do not disperse at lower coverages. These states are not dynamically localized, but rather localize as a result of the structural inhomogeneity of the adsorbed layer.

Acknowledgements. This work was supported by the Director, Office of Energy Research, Office of Basic Energy Science, Chemical Sciences Division of the U.S. Department of Energy, under Contract No. DE-AC0376SF00098. The authors acknowledge NSF support for specialized equipment used in the experiments described herein.

References

1 N.-H. Ge, C.M. Wong, R.L. Lingle, Jr., J.D. McNeill, K.J. Gaffney and C.B. Harris, "Femtosecond Dynamics of Electron Localization at Interfaces" Science, **279**, 202-205 (1998).

2 N.-H. Ge, C.M. Wong, and C.B. Harris, "Femtosecond Studies of Electron Dynamics at Interfaces" Acc. Chem. Res., **33**, 111-118, (2000).

3 Y. Toyozawa, 'Electron Holes and Excitons in Deformable Lattice" in *Relaxation of Elementary Excitations*, R. Kubo, E. Hanamura, ed. (Springer-Verlag, Berlin, 1980).

4 R.L. Lingle, Jr., N.-H. Ge, R.E. Jordan, J.D. McNeill, and C.B. Harris, 'Femtosecond Studies of Electron Tunneling at Metal-Dielectric Interfaces' Chem. Phys., **205**, 191-203 (1996).

5 T. Holstein, "Studies of Polaron Motion, Part II. The 'Small' Polaron" Ann. Phys., **8**, 343-389 (1959).

6 R.A. Marcus, N. Sutin, "Electron Transfers in Chemistry and Biology" Biochim. Acta., **811**, 265 (1985).

7 A.A. Stuchenbrukhov, X.Y. Song, "Quantum Effects in Electron Transfer Reactions with Strong Electronic Coupling" J. Chem. Phys., **101**, 9354-9365 (1994).

8 A.A. Mosyak, O.V. Prezhdo, P.J. Rossky, "Solvation Dynamics of an Excess Electron in Methanol and Water" J. Chem. Phys., **109**, 6390-6395 (1998).
9 P. Minary, L Turi, P.J. Rossky, 'Nonadiabatic Molecular Dynamics Simulation of Photoexcitation Experiments for the Solvated Electron in Methanol" J. Chem. Phys., **110**, 10953-10962 (1999).

Spectral Sensitization, Supersensitization and Ultrafast Exciton Migration on AgBr Semiconductor Surface

Igor V. Rubtsov[1,2], Koujirou Ebina[1], Keitaro Yoshihara[1], Jasper Knoester[3] Takeshi Suzumoto[4] and Tadaaki Tani[4]

[1] School of Materials Science, Japan Advanced Institute of Science and Technology, Tatsunokuchi, Ishikawa 923-1292, Japan; e-mail: ivr@jaist.ac.jp; yosihara@jaist.ac.jp
[2] Present address: Department of Chemistry, University of Pennsylvania, Philadelphia, PA 19104-6323, USA; e-mail: irubtsov@sas.upenn.edu
[3] Theoretical Physics, University of Groningen, Nijenborgh 4, 9747 AG Groningen, The Netherlands
[4] Asahigara Research Laboratories, Fuji Photo Film Co., Ltd., Minamiashigara, Kanagawa 250-01, Japan

Abstract. The fast electron photo-injection rate from the adsorbed dye to AgBr was found to be dependent on the type of AgBr surface, size of grains, and *pAg* of emulsion. Ultrafast exciton migration is discussed in terms of dimensionality of J-aggregates.

1. Introduction

Sensitization of semiconductors by dye molecules is a process, which has applications in different areas, photography, solar cells, data storage, sensors etc. Dye molecules can be adsorbed on a semiconductor surface up to a monolayer level forming aggregates. Variety of aggregate structures of the dye on the surface affects the electron injection rate. Surface defects, in turn, influence energy levels of the adsorbed sensitized and vary electron photo-injection rate.

In this work spectral sensitization and supersensitization of the AgBr microcrystals were investigated by the femtosecond fluorescence up-conversion. The electron injection from J-aggregates of cyanine dye, adsorbed on the surface of AgBr grains, into the conduction band of AgBr semiconductor has been studied. Two different types of surfaces, [100] and [111], of the AgBr grains of different sizes were examined. The dependence of the injection rate upon *pAg* (-log[Ag^+]) of the solution is reported. The effect of supersensitizer (SS), a second dye which is co-adsorbed on AgBr surface to improve the quantum yield of spectral sensitization is studied in detail.

2. Experimental Methods

Sample Preparation. The emulsions used in this work were composed of the suspensions of AgBr microcrystals in aqueous gelatin solution. Two types of crystal were studied, octahedral ([111] surface) and cubic (mostly [100] surface) with edge lengths of 0.04, 0.2, 0.4, and 0.9 μm. The cyanine dye (3,3'-disulfopropyl-5,5'-dichloro-9-ethylthiacarbocyanine, Dye 1, see inset of Fig. 1 was used as a sensitizer. The preparation procedure of the emulsions was similar to that of ref. [1]. *pAg* of the emulsion was about 6.8, except for the experiment where it was varied; *pH* was about 6.6.

Fluorescence up-conversion setup. A second harmonic (SH) of the chromium-forsterite femtosecond laser (610 - 660 nm) was used to excite the sample [2]. To avoid exciton-exciton annihilation the excitation flux was carefully controlled to be about $(0.5–1)\times10^{12}$ photons cm^{-2} $pulse^{-1}$. The full width at half maximum of the instrument response was 98 ± 2 fs. Fluorescence signals were detected at wavelengths from 672 to 729 nm. During experiments the sample was circulated in a 0.5 mm thick sample cell at temperature of 50 °C.

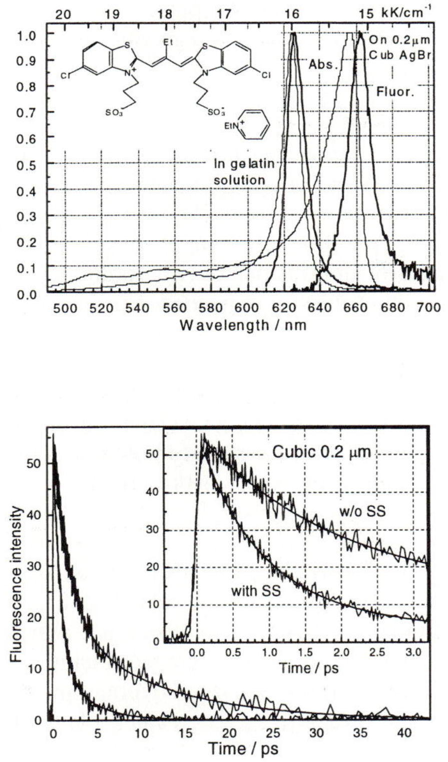

Fig. 1. Upper panel: steady-state absorption and fluorescence spectra of J-aggregates of Dye 1 on different surfaces; Lower panel: effect of supersensitizer on [100] surface.

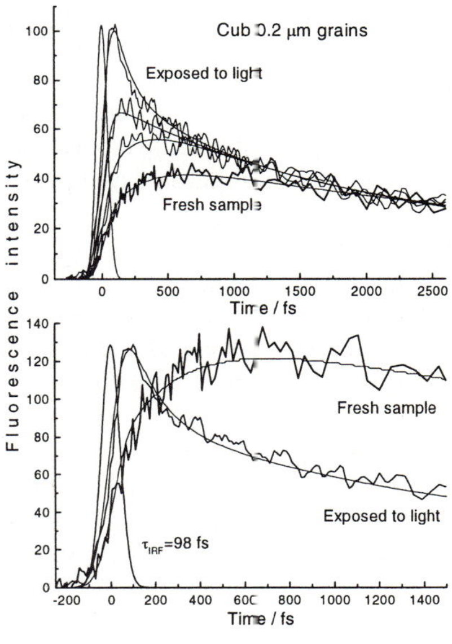

Fig. 2. The initial part of the fluorescence signal changes from rising to decay, while the sample is exposed to light. The fast decay is attributed to accumulation of the oxidized dyes.

3. Results and Discussion

3.1 Sensitization and surface habit

Fluorescence from J-aggregates adsorbed on the surface of AgBr grains has been measured with a sub-hundred femtosecond time resolution. The fluorescence decay is determined by the electron injection process from the excited dye to the conduction band of AgBr as well as by radiative and nonradiative decay processes. The fluorescence decay is strongly nonexponential, which is discussed in terms of distribution of surface sites. The electron injection rate on the octahedral grains was found to be about 8 times faster than that on cubic grains.

3.2 Supersensitization

In the presence of SS the excited state lifetime was found to decrease up to 4 times (Fig. 1), with the corresponding increase of the injection efficiency. Different supersensitization schemes were considered. It is concluded that the "hole-trapping" mechanism (an "Auger-type") is dominant in the present case. The "hole-trapping" rate (electron transfer from SS to the excited J-aggregate) is found to be similar for the cubic and octahedral grains.

3.3 Grain Size Effect

The electron injection rate was found to increase about twice with the increase of the grain size of octahedral grains from 0.2 to 0.9 μm. The aggregate size and *pAg* of the emulsion were kept constant in this experiment. The *pAg* was about 6.8, which is higher than *pAg* corresponding to the point of zero charge (p.z.c.), equal to 5.4. A shift of the surface potential, caused by an influence of two double layers on each other, is suggested as a possible reason of the size dependence.

3.4 *pAg* Effect

Significant dependence of the injection rate on *pAg* of the emulsion was found for cubic grains (change of about 40 times), and smaller effect for octahedral grains (about twice). Data ere explained by the dependence of the surface potential upon *pAg* of the solution. Comparison for two surfaces of two types is presented.

3.5 Exciton Migration

The rising component in the fluorescence signal was observed with the rise time of about 200 fs, which is much faster than the electron injection rate (Fig. 2). At the same time fast anisotropy decay in similar time scale was observed, which suggests exciton migration in direction other than along the aggregate chain. By comparison of the anisotropy decay at different values of surface coverage by the dye it is concluded that the two-dimensional structure of aggregates plays an important role in the exciton migration and trapping.

References

1. T. Suzumoto, T. Tani, J. Im. Sci. Technol. **40**, 56 (1996).
2. I. V. Rubtsov, H. Shirota, K. Yoshihara, J. Phys. Chem. A **103**, 1801 (1999).

Ultrafast interfacial charge separation from the singlet and triplet MLCT states of $Ru(bpy)_2(dcbpy)$ adsorbed on nanocrystalline SnO_2 under applied bias

S. Iwai, K, Hara, R. Katoh, S. Murata, H. Sugihara, and H. Arakawa

National Institute of Materials and Chemical Research, 1-1 Higashi, Tsukuba, JAPAN
E-mail: iwai@nimc.go.jp

Abstract. The lifetime of stumulated emission from the singlet MLCT of Ru complex adsorbed on SnO_2 nanocrystal was measured under applied biases in order to estimate the interfacial charge separation rate. Unnegligible numuber of excited electrons are injected directly from the singlet MLCT state before they relax to triplet state.

1. Introduction

Among many processes involved in dye sensitized solar cells (Graetzel cell) [1], the ealiest events are the relaxation within the dye molecules and the dye-semiconductor interfacial charge separation (CS) [2-4]. In the most famous sensitizer : $[Ru(bpy)_3]^{2+}$ complex based dye, ultrafast (<300 fs) intersystem crossing (ISC) from the excited singlet (^{1}MLCT) state to the thermalized excited triplet (^{3}MLCT) state has been observed in solution under resonant excitation to the excited singlet state (^{1}MLCT) [5]. So, one of the most fundamental problems in the primary processes in $Ru(dcbpy)_2(NCS)_2$ or similar Ru-complex dye sensitized systems is how the CS competes with the ISC of the dye molecules, in other words, how high the CS efficiencies from the ^{1}MLCT state is.

In this study, we measured the stimulated emission decay from the ^{1}MLCT state of $Ru(bpy)_2(dcbpy)$ adsorbed on SnO_2 microcrystal, while changing the energy gap between the conduction band edge of the semiconductor and LUMO of the dye by appling external bias. In general, the lifetime of the dye excited state does not always reflect the CS rate, since other quenching pathways may exist. However, the interfacial CS rate can be changed,by applying an external bias on the dye-semiconductor interface [6]: it is suppressed under negative biases. On the other hand, relaxation processes other than the interfacial CS are not likely to be affected by the applied bias. Therefore, by analyzing the change in the lifetime of the excited state caused by the applied bias, we can obtain information about the interfacial CS rate from the excited state and the internal relaxation rate.

2. Experimental Methods

The three-electrode cell for photoelectrochemical measurements used in this study is similar to the one reported by Kamat et al.[6]. An OTE (=Optically Transparent Electrode) /SnO_2/Ru(bpy)$_2$(dcbpy) working electrode (WE) is set in this cell containing an electrolyte solution (0.1 mol dm^{-3} (Bu_4NClO_4) in acetonitrile. A reference electrode (RE) of Ag/Ag^+ and a counter electrode (CE) of Pt wire are inserted through the side arms of the cell. The maximum IPCE (incident photon-to-current conversion efficiency) was about 10% at positive biases (vs. Ag/Ag^+). Below -0.7 V, the observed IPCE was almost zero. This means that the CS is suppressed completely at -0.7 V. To measure the stimulated emission decay from ^{1}MLCT of the dye, two-color femtosecond pump-probe (transient absorption) method was employed. A Ti: Al_2O_3 amplifier system operating at 1KHz (Spectra Physics Sub-50 fs Super Spitfire system) with dual OPA systems (Spectra Physics OPA800) was used as a light source. The response function of the measurement system has a FWHM of about 60 fs. The pump intensity was limited below 0.5 mJ/cm^{-2} to avoid various undesierable effects caused by the high-density excitation. The reproducibility was confirmed for all the results. All the measurements were made at 296 K.

3. Results and Discussion

Fig. 1 shows the transient absorption spectrum of Ru(bpy)$_2$(dcbpy) adsorbed on SnO_2 film. Steady state absorption spectrum is indicated in the top frame. Closed circles show the absorption change (in ΔOD) observed at several wavelengths. The time profile at each wavelength can be analyzed using a two-component exponential function. The observed spectra was deconvoluted into the fast (open triangles: lifetime = 70 fs) and the slow (open squares: lifetime >ns) components. The spectrum of the slow component is well coincident with that of the steady state absorption. We can attribute this long-lived component to the bleaching due to the ground state depletion of the dye as reported earlier [6]. On the other hand, the spectrum of the fast component has a maximum at longer wavelength side (520 nm) of the absorption peak. So, the fast decay component is considered to be due to the stimulated emission from the excited singlet (^{1}MLCT) state of the Ru complex dye.

Fig. 2 shows the time evolution of the transient absorption of Ru(bpy)$_2$(dcbpy) adsorbed on SnO_2 film measured at 480nm under applied biases between 0 and -0.7 V. As the bias becomes more negative, the decay of the stimulated emission (= fast component) becomes slower. The lifetimes obtained at applied biases of 0, -0.3, -0.5 and -0.7 V are 70, 90, 190, 320 fs, respectively. The continuous change of the ^{1}MLCT lifetime in one hundred femtosecond time region is considered to be due to the change in the CS rate, since the rates of other relaxation processes from ^{1}MLCT are supposed not to change. The decay profile of the stimulated emission does not change in Ru(bpy)$_2$(dcbpy) on SiO_2 where the interfacial CS does not occur. This confirms our interpretation.

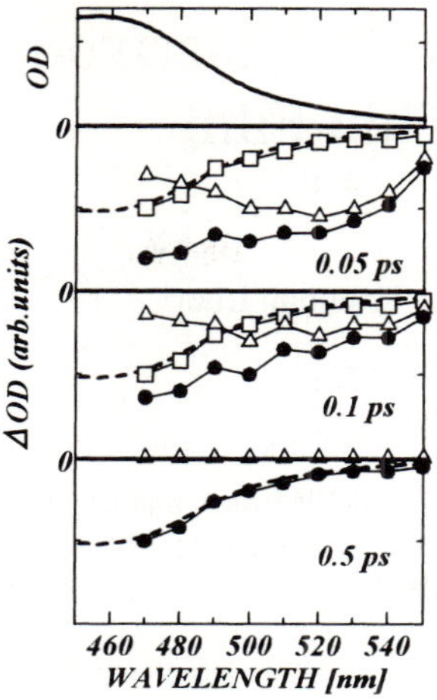

Fig.1 Transient absorption spectrum of Ru(bpy)(dcbpy) on SnO_2 (closed circles). Deconvoluted fast and slow components are shown by open triangles and squares.

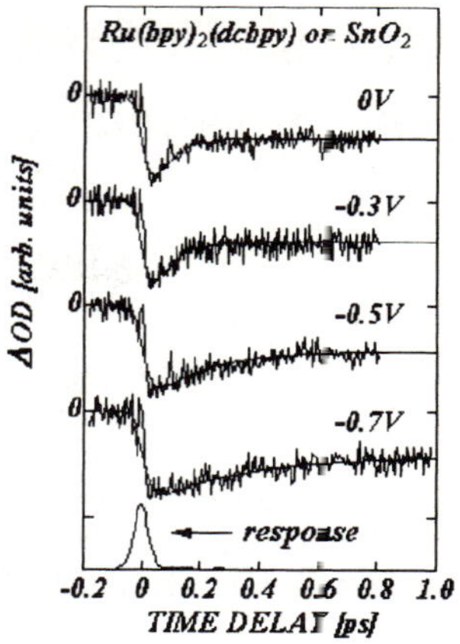

Fig.2 Time evolution of the transient absorption of Ru(bpy)(dcbpy) under applied biases between 0 and -0.7 V. The fast component shows the stimulated emission from ^{1}MLCT

Observed decay of the stimulated emission from ^{1}MLCT state (τ_1) is considered to be expressed by $1/\tau_1 = 1/\tau_{CS} + 1/\tau_{ISC} + 1/\tau_{nr}$ where $1/\tau_{CS}$, $1/\tau_{ISC}$, $1/\tau_{nr}$ are the rate of interfacial CS, intersystem crossing and other non-radiative relaxation processes. From the stimulated emission decay observed in $Ru(bpy)_2(dcbpy)$ adsorbed on SnO_2 at the bias of -0.7 V (320 fs) and in $Ru(bpy)_2(dcbpy)$ adsorbed on SiO_2 (350 fs) , the lifetime of ^{1}MLCT in the case that the interfacial CS is completely suppressed is about 330 fs The value of $(1/\tau_{ISC} + 1/\tau_{nr})$ and is estimated to be 1/(330 fs). Therefore, the interfacial CS rate is obtained to be 1.1×10^{13} /s. This result shows that the unnegligible number of excited electrons (about 70 %) is injected directly from ^{1}MLCT to SnO_2 nanocrystals before the internal relaxation of the dye is completed.

References

1 A. Hagfeldt, M. Gratzel, Chem. Rev.**95**, 49-68 (1995).

2 Y. Tachibana, J. E. Moser, M. Gratzel, D. R. Klug, J. R. Durrant, J. Phys. Chem. **100**, 20056-20062 (1996).

3 T. Hannappel, B. Burfeindt, W. Stork, F. Willig, J. Phys. Chem. B **101**, 6799-6802 (1997).

4 R. J. Ellingson, J. B. Asbury, S. Ferrere, H. N. Ghosh, J. R. Sprague, T. Lian, A. J. Nozik, J. Phys. Chem. B **102**, 6455-6458 (1998).

5 N. H. Damrauer, G. Cerullo, A. Yeh, T. R. Boussie, C. V. Shank, J. K. McCusker, Science, **275**, 54-57 (1997).

6 P.V. Kamat, I. Bedja, S. Hotchandani, L. K. Patterson, J. Phys. Chem. **100**, 4900-4908 (1996).

Femtosecond IR Study of Ultrafast Electron Injection in Nanocrystalline Thin Film Electrodes

John B. Asbury, Encai Hao, Yongqiang Wang, and Tianquan Lian*

Department of Chemistry, Emory University, Atlanta, GA 30322
Phone: (404) 727-6649, Fax: (404) 727-6586, tlian@emory.edu

Abstract: Ultrafast electron transfer dynamics from $Re(CO)_3Cl(dcbpy)$ and derivatives on TiO_2 nanocrystalline thin films were studied by simultaneously measuring the injected electron absorption and the transient absorption of the sensitizer by femtosecond IR spectroscopy.

1. Introduction

Interfacial electron transfer (ET) between semiconductor nanoparticles and molecular adsorbates has been the subject of intense research interest in recent years[1]. Recently, Gratzel and co-workers reported that a solar cell based on a TiO_2 nanocrystalline photo-electrode sensitized with $Ru(NCS)_2(dcbpy)_2$ (dcbpy = 4,4'-dicarboxy-2,2'-bipyridine) (referred to as Ru N3) achieved 10% solar to electrical conversion efficiency[2]. This report has focused much of the interest in interfacial electron transfer on the electron transfer properties of Ru N3 and related complexes on TiO_2[3-5]. The high efficiency of the solar cell based on Ru N3 has been attributed to efficient charge separation by fast electron injection and slow back electron transfer.

The fast electron injection time reported for Ru N3 to TiO_2 is on the same time scale or faster than vibrational relaxation of the ^{1}MLCT excited state[3,6] and may occur in the adiabatic limit. As a result, it is unclear whether electron injection occurs from the lowest ^{3}MLCT state or from some vibrationally hot excited state. In this work, we examine the transient mid-IR spectra of a series of Re polypyridyl complexes in order to investigate the competition between electron injection and excited state relaxation and to understand the electronic coupling dependence of the electron injection rate by varrying the length of CH_2 bridging spacers.

2. Results and Discussion

The transient absorption spectra of the three Re dyes on TiO_2 film are shown in Figure 1. Each spectrum consists of several delay times in the specified time range averaged together in order to improve the quality of the spectra. Figure 1A shows the transient spectra of ReC0A sensitized TiO_2 film after 400 nm excitation. The transient spectra of ReC1A and ReC3A, shown in Figure 1B and 1C respectively, are averaged spectra at the same delay times as the ReC0A spectra and are coded in the same manor . The transient absorption spectra consist of the following components with band positions indicated in parenthesis, ground state(2030-2040 cm^{-1}), excited state(2055-2065 cm^{-1}), and oxidized state(~2100 cm^{-1}) of the complexes superimposed on a broad electron absorption signal.

In Figure 1A, the transient absorption spectra of ReC0A on TiO_2 shows the simultaneous formation of electrons injected into TiO_2 and formation of the oxidized form of the dye before the spectral evolution of the excited state peak at 2065 cm^{-1} was complete. In our earlier paper, we assigned the spectral evolution at 2065 cm^{-1} to vibrational cooling of the excited state[7]. The energy of the vibrationally relaxed ^{3}MLCT state of ReC0A is estimated to be –0.35 V from the ground state Re(II/I) reduction potential and the 0-0 transition energy[8]. This energy lies below the conduction band edge of TiO_2, which is estimated to be around –0.5 V[1]. The electrons injected into the TiO_2 film from ReC0A must therefore have been injected before the excited state relaxed to the lowest ^{3}MLCT state. The vibrational cooling dynamics in the excited state peak are consistent with hot excited state electron injection predicted from the energetics.

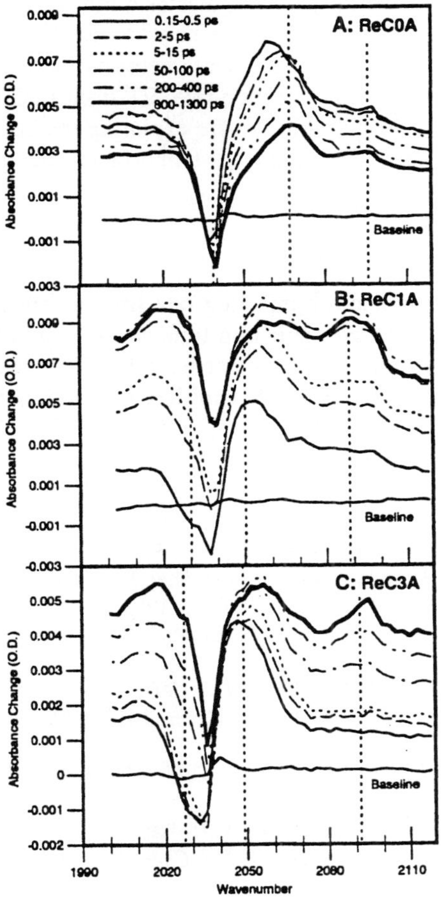

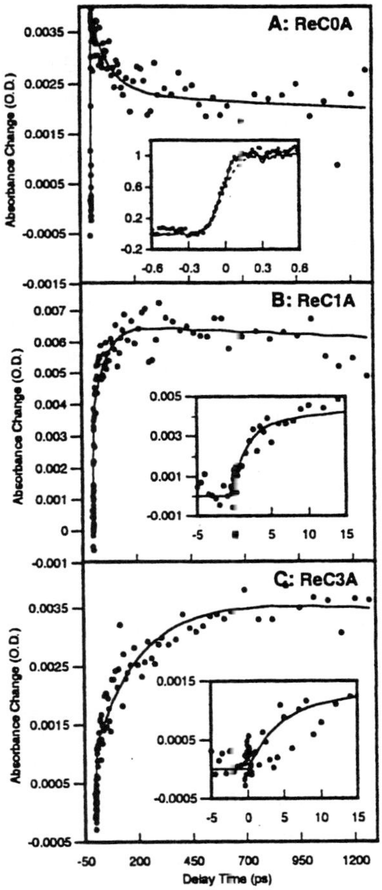

Fig 1. Comparison of transient IR spectra of Re dye sensitized TiO_2.

Fig 2. Comparison of electron absorption kinetics of Re dye sensitized TiO_2 at 2020 cm^{-1}.

The kinetics of ReC0A, measured at 2120 cm^{-1}, are shown in Figure 2A. The inset shows the rise time of the signal, which occurs on the <50 fs time scale. For comparison, the instrument response function measured in a thin silicon wafer and a curve reflecting a 50 fs exponential rise are also shown in the inset. The absorption rise kinetics of ReC1A and ReC3A were fit with a stretched exponential rise that was convoluted with a 10 ns single exponential electron cross section decay function. The fit functions had the following lifetimes and distribution parameters in parentheses, 19 ps (0.45) for ReC1A and 260 ps (0.45) for ReC3A. The insets in Figure 2B and 2C show the same data plotted on a shorter time scale.

The ReC1A and ReC3A complexes, whose ^{3}MLCT states are at –0.65 V[8], undergo slow injection into TiO_2. The electron injection rate changed by a factor of 13.7 for ReC1A and ReC3A. Assuming an exponential dependence of ET rate on coupling, this change corresponds to a β value of 1.3/ per C-C bond, which is consistent with the value reported in a recent work[9]. This indicates a non-adiabatic Et process. However, the change in the electron injection rate between ReC0A and ReC1A was >200, indicating a different type of injection process between these two complexes. In light of the measured ultrafast electron injection rate, ET from ReC0A, may occur in the adiabatic limit. On going studies in our lab are exploring the nature of donor and acceptor states involved in this type of ultrafast interfacial electron transfer process.

3. Conclusion

In conclusion, we have directly measured the electron injection rate from a series of Re polypyridyl dyes to TiO_2. We found the electron injection time was < 100 fs for ReC0A, occurring probably in the adiabatic limit. Much slower ET rates that depend exponentially on bridge length were found for ReC1A and ReC3A, indicating nonadiabatic ET processes.

References

(1) Hagfeldt, A.; Gratzel, M. *Chemical Reviews* **95**, 49-68, 1995.

(2) Nazeeruddin, M. K.; Kay, A.; Rodicio, I.; Humphrybaker, R.; Muller, E.; Liska, P.; Vlachopoulos, N.; Gratzel, M. *J. Am. Chem. Soc.* **115**, 6382-6390, 1993.

(3) Asbury, J. B.; Ellingson, R. J.; Ghosh, H. N.; Ferrere, S.; Nozik, A. J.; Lian, T. *J. Phys. Chem. B* **103**, 3110-3119, 1999.

(4) Heimer, T.; Heilweil, E. J. *J. Phys. Chem. B* **101**, 10990-10993, 1997.

(5) Hannappel, T.; Burfeindt, B.; Storck, W.; Willig, F. *J. Phys. Chem. B* **101**, 6799-6802, 1997.

(6) Damrauer, N. H.; Cerullo, G.; Yeh, A.; Boussie, T. R.; Shank, C. V.; McCusker, J. K. *Science* **275**, 54, 1997.

(7) Wang, Y. Q.; Asbury, J. B.; Lian, T. *J. Phys. Chem. A* **104**, 4291-4299, 2000.

(8) Worl, L. A.; Duesing, R.; Chen, P.; Ciana, L. D.; Meyer, T. J. *J. Chem. Soc. Dalton Trans.* , 849-858, 1991.

(9) Smalley, J. F.; Feldberg, S. W.; Chidsey, C. E. D.; Linford, M. R.; Newton, M. D.; Liu, Y. P. *J. Phys. Chem.* **99**, 13141, 1995.

Unique curve crossing and vibrational wavepacket behavior of femtosecond heterogeneous electron transfer

C. Zimmermann, F. Willig, S. Ramakrishna, B. Pettinger[§], B. Burfeindt, N. Biswas, R. Eichberger
Hahn-Meitner-Institut, Department SE4, 14109 Berlin, Germany, willig@hmi.de
[§]*Fritz-Haber-Institut der MPG, Department PC, Faradayweg 4-6, 14195 Berlin, Germany*

Abstract: Femtosecond transient absorption signals of heterogeneous electron transfer show that the latter occurs with identical probability irrespective of the specific position of the vibrational wavepacket. These dynamics are very different from the molecular donor-acceptor pair.

Femtosecond electron transfer from a molecular donor (chromophore) to a wide conduction band of empty electronic levels in a solid (electrode) is a fundamental step in several photo-electrochemical processes of practical importance [1]. It controls also the important lifetime parameter of the photo-reduced adsorbed molecule in photo-desorption scenarios. The dynamics of femtosecond heterogeneous electron transfer has several unique distinguishing features compared to femtosecond electron transfer between a molecular donor and a molecular acceptor. In the wide band limit the heterogeneous electron transfer rate is controlled only by the electronic interaction and is independent of Franck-Condon factors [1-3]. Thus, in contrast to the case of a molecular donor-acceptor pair heterogeneous electron transfer in the wide band limit does not occur only at a specific curve crossing point. Rather, it occurs with identical probability irrespective of the specific vibrational wavefunction in the donor molecule [1] and correspondingly also irrespective of the specific position of a vibrational wavepacket in the donor molecule [2]. Accordingly, the probability for the occupancy of both, reactant and product state, is expected to show a smooth decay or a smooth rise without oscillations. In the wide band limit recurrences of the injected electron are negligible. However, the motion of a vibrational wavepacket in the potential curve for the reactant or product state can be monitored as oscillations in the respective transient absorption signals recorded at a fixed observation wavelength [3]. The above picture of the dynamics is suggested already by the short electron transfer time of several 10 femtoseconds reported in our earlier investigations of heterogeneous electron transfer [4]. However, direct experimental evidence for this unique dynamic behavior is presented here for the first time.

It turns out that vibrational wavepacket motion, probed as oscillations in the transient absorption signal of the product state of heterogeneous electron transfer, is continuing long after the electron has been completely transferred to the solid. The Fourier Transform of these oscillations yields the vibrational frequencies of the modes that are contributing to the vibrational wavepacket in the ionized chromophore. Steps or oscillations in the rate of electron transfer are not observed, in contrast to the behavior expected for an electronic two level system, e.g. a molecular donor-acceptor pair.

The following four conditions have to be met in the experimental system in order to observe unambiguously the above unique heterogeneous electron transfer dynamics. Firstly, the energy of the donor orbital of the molecule must be located well below the ionisation threshhold but also well above the lower edge of the empty electronic states, i.e. it should be roughly in the center of a sufficiently wide conduction band in the electrode material. Secondly, the edge of the donor chromophore must be kept a certain minimum distance from the surface atoms of the solid, e.g. via insertion of a non-aromatic molecular anchor group, such that electronic coupling is neither too weak nor too strong, i.e. it should roughly correspond to an electron transfer time in the range of 100 femtosconds. Thirdly, the chromophore must have vibrational modes of sufficiently low energy, e.g smaller than 1000 cm^{-1}, such that a vibrational wavepacket is easily generated with the impinging laser pulse in the excited electronic state of the donor molecule. Fourthly, the laser pulse must be short enough, i.e. 30 fs or shorter, to provide a sufficiently wide energy window for coherent excitation of the respective vibrational modes in the donor chromophore.

These requirements are fulfilled in an experimental system where the large aromatic chromophore perylene injects an electron from its first excited singlet state into the empty conduction band of the wide band gap semiconductor TiO_2 [4]. Bulky side groups attached to the perylene chromophore prevent the formation of dimers, aggregates, and of a 2-dimensional molecular solid. A molecular anchor group, i.e. carboxylic or phosphonic acid, is attached to the chromophore and forms a strong chemical bond with the surface atoms of the semiconductor. Thus, the edge of the chromophore is separated by the length of this anchor group from the surface atoms of the solid, and the electronic coupling is reduced to yield the desired injection time around 100 fs. Complications are known to arise with the stability of such a heterogeneous electron transfer system when immersed into a solvent environment. They are prevented if the measurements are carried out in ultra-high-vacuum [4]. Fig.1 shows the transient absorption signal of the product state of heterogenous electron transfer, i.e. the ionized perylene chromophore, on the TiO_2 electrode in ultra-high-vacuum. The time-dependent behavior of the signal is controlled by the strength of the electronic coupling and by vibrational wavepacket motion [2,3]. There are neither oscillations nor steps in the rise of the signal that could be linked to an enhanced probability for electron transfer at a specific position of a vibrational wavepacket. Rather, there is a smooth increase in the electron transfer rate without any crossing point. This is characteristic of heterogeneous electron transfer in the wide band limit [1,2]. There are oscillations but their amplitude is a constant fraction of the magnitude of the signal. The maximum oscillation amplitude is reached when transfer of the electron is nearly completed. These oscillations are caused by periodic changes in the optical transition probability due to wavepacket motion in the ionized chromophore, i.e. in

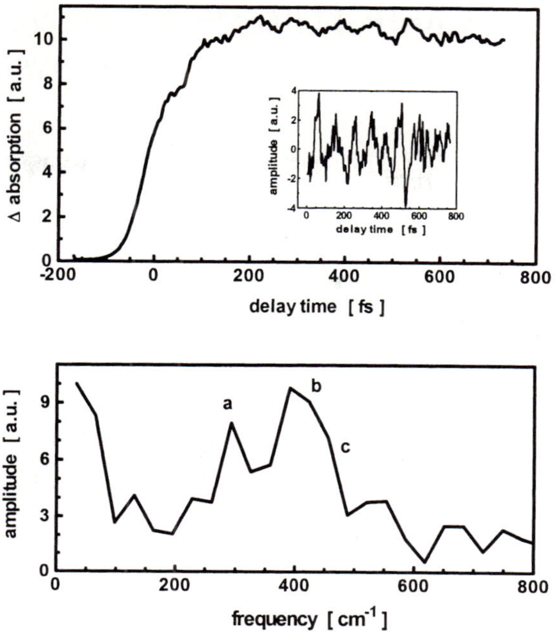

Fig. 2. Upper panel: Absorption of the product state of heterogenous electron transfer, i.e. of the ionized perylene chromophore with anchor group and side groups on the TiO_2 electrode. Inset: Oscillatory component obtained from a smooth fit to the curve. Lower panel: Fourier Transform of the oscillatory component in the upper panel.

the product state of the electron transfer reaction. The lifetime of these oscillations is at least 10 times longer than the electron transfer time of 70 fs.
Thus, a nice agreement is found in the behavior of the experimental signals with the dynamics of femtosecond heterogeneous electron transfer in the wide band limit.

[1] R.J.D. Miller, G.L. McLendon, A.J. Nozik, W. Schmickler, F. Willig, *Surface Electron Transfer Processes* (VCH Publishers, New York, 1995), Chapter 5
[2] S. Ramakrishna, F. Willig, V. May, "Photo-induced ultrafast electron injection from a surface attached molecule: Control of electronic and vibronic distributions via vibrational wavepackets", to be published
[3] S. Ramakrishna and F. Willig, "Pump-probe spectroscopy of ultrafast electron injection from the excited state of an anchored chromophore to a semiconductor surface in uhv: a theoretical model", J. Phys. Chem. B **104**, 68-77 (2000)
[4] B. Burfeindt, C. Zimmermann, S. Ramakrishna, T. Hannappel, B. Meißner, W. Storck, and F. Willig, "Femtosecond electron transfer from the excited state of chemically anchored chromophores into the empty conduction band of nano-crystalline sponge-like TiO_2 films", Z. physik. Chem, **212**, 67-75 (1999)

Surface States Control Ultrafast Electron Injection in Dye/Semiconductor Colloidal Systems

Robert Huber[1], Sebastian Spörlein[1], Jacques E. Moser[2], Michael Grätzel[2] and Josef Wachtveitl[1*]

[1] Lehrstuhl für BioMolekulare Optik, Sektion Physik, Ludwig-Maximilians-Universität München, Oettingenstr. 67, 80538 München, Germany
E-mail: Josef.Wachtveitl@physik.uni-muenchen.de

[2] Laboratory for Photonics & Interfaces, Department of Chemistry, Ecole Polytechnique Fédérale de Lausanne, CH-1015 Lausanne, Switzerland

Abstract. Electron injection from photoexcited dye molecules into semiconductor colloidal particles (TiO_2 and ZrO_2) is investigated by femtosecond (fs) spectroscopy. Ultrafast reaction dynamics in the dye / ZrO_2-system with an inaccessible conduction band demonstrates efficient electron transfer to surface defect states.

1. Introduction

Electron injection from organic dye molecules coupled onto semiconductor colloids has received much interest during the last years. Applications like the conversion of solar energy to electricity or the decontamination of water are important reasons. Electron injection in such systems takes place on time scales from µs down to several fs [1-3]. The underlying process is an electron transfer from a system with discrete energy levels to a continuum of acceptor states (usually, the conduction band of the semiconductor colloid). Apart from a few recent studies [4], the influence of energetically low lying surface states, a consequence of the disrupted crystal structure on the colloidal surface, has not been in the focus of detailed studies yet. Here we show that also for systems like dye / ZrO_2 ultrafast electron transfer dynamics occurs. Since the conduction band of ZrO_2 lies energetically far above the S_1 state of the investigated dye, the observed dynamics must be caused by low lying surface defect states.

2. Experimental Methods

We performed visible pump / white light probe experiments using a non-collinear optical parametric amplifier (NOPA) [5,6] for tunable excitation on (i) free alizarin, (ii) alizarin adsorbed onto TiO_2 colloids ($r \approx 8$ nm, E_{gap}=3.2eV) as injecting system and (iii) alizarin coupled onto ZrO_2 colloids as "reference" ($r \approx 8$nm, E_{gap}= 4.5eV). Adsorbed alizarin exhibits strong charge transfer (CT) character (Fig.1). A very high electronic matrix element of ≈ 3100 cm^{-1} (coupling a ground state to a CT excited state) can be estimated. Therefore, regarding classical electron transfer theory by Marcus, electron injection should occur on a fs time scale.

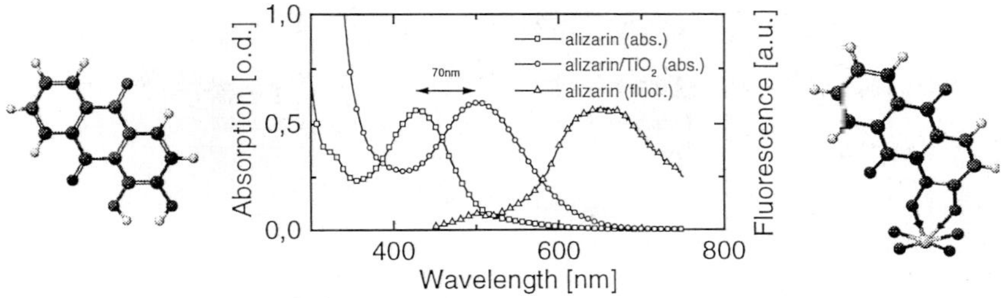

Fig. 1. The coupling of alizarin (left) via the two hydroxyl groups to the titanium / zirconium atom (right) results in an equally strong shift of the absorption band for both systems (center). This indicates a large electronic coupling between alizarin and the TiO_2 / ZrO_2 surface and a strong ligand-to-metal charge transfer character of the band. The fluorescence of free alizarin (-Δ-) is completely quenched in alizarin / TiO_2.

3. Results and Discussion

The reaction dynamics of coupled alizarin (Fig. 2) is completely different compared to the 63 ps excited state decay observed in the free dye. For alizarin / TiO_2 (Fig. 2, top) an additional fast decrease ($\tau = 430$ fs) of the initial signal amplitude is observed throughout the entire investigated spectral range. Probing with a white light continuum helps to identify the reaction products of the injection mechanism. For the system alizarin / TiO_2 we could observe the population density of the ground state (S_0, $\lambda_{probe} = 500$ nm), the excited state (S^*, $\lambda_{probe} < 600$ nm) and the cation of alizarin (S^+, $\lambda_{probe} = 500$ nm - 600 nm) in addition with the related cooling processes, as well as the injected electron in the conduction band of the colloid ($\lambda_{probe} > 700$ nm). The recorded transients allow us to suggest an upper limit for the charge injection time of ≈100 fs and reveal a highly non-exponential character of the ground state recovery reaction. The long term characteristics of the spectra (t_D = 1 ns) correlate well with the reported absorption of the charge separated state [7] and confirm the injection of the electron into the TiO_2 conduction band.

In ZrO_2 the dye undergoes the same coupling process as in TiO_2 but any electron injection into the conduction band is inhibited due to the significantly higher band gap of ZrO_2. Nevertheless, we observe a complex, ultrafast reaction dynamics in the alizarin / ZrO_2 system (Fig. 2 left, middle and bottom), indicative for electron injection. The reaction mechanism (Fig. 2, right) can be derived from global data analysis: after excitation of the alizarin / ZrO_2 system an electron is injected into surface defect states and cools therein with $\tau < 100$ fs. From the traps one part of the electrons is reinjected in the lower vibrational levels of the alizarin S_1 state, which is repopulated within 450 fs. The subsequent $S_1 \rightarrow S_0$ transition occurs in about 160 ps. The second reaction pathway leads to a direct ground state recombination (without S_1 repopulation) with a dominant time constant of 2.4 ps. A few of the electrons become localized in traps with a weak electronic coupling to the alizarin and therefore recombine very slow ($\tau_D > 1$ ns). The heterogeneous nature of colloidal suspensions may be primarily responsible for the dispersive

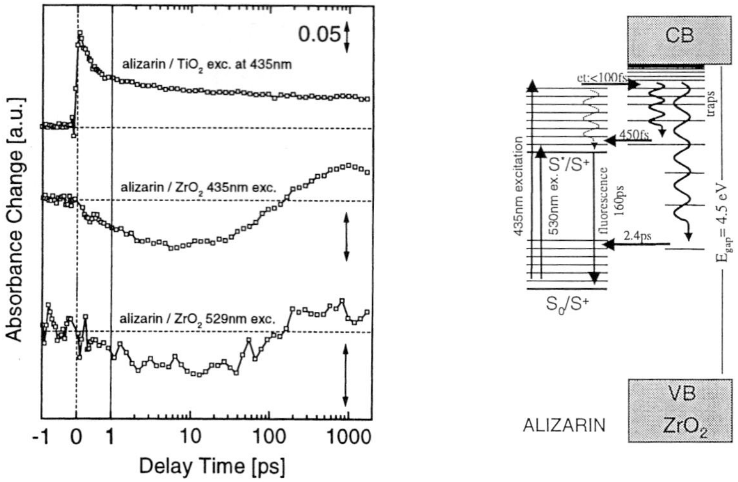

Fig. 2. (left): Transient absorbance changes for alizarin / TiO_2 and alizarin / ZrO_2, excited with 435 nm and 529 nm at a probing wavelength of 720nm. (right): energy scheme and assignment of the various time constants observed for alizarin / ZrO_2.

character of this reaction dynamics. The spectra of the residual signal at $\tau_D > 1$ ns correlate almost perfectly with the data of alizarin / TiO_2. Obviously, also in the ZrO_2-system a part of the excited molecules ends up in a long lived charge separated state. Excitation at 529nm (Fig. 2 left, bottom) demonstrates that the reaction dynamics is independent of the vibrational excess energy, i.e. effective injection into trap states does not require a "hot" electron.

In conclusion, Alizarin / ZrO_2 represents a system where only the primary reaction step, the electron injection into trap states is observed. The ultrafast time scale of this transition ($\tau < 100$ fs) underlines the importance of defect states in the injection mechanism of electrons from photoexcited dye molecules to semiconductor surfaces. A defined implementation or manipulation of surface defects and trap states to influence and optimize the charge injection dynamics can be envisioned as an attractive application in the design of mesoscopic electron transfer systems.

References

1 Y. Tachibana, J. E. Moser, M. Grätzel, D. R. Klug and J. R. Durrant, Chem. Phys. Lett. **272**, 489, 1997.

2 T. Hannappel, B. Burfeindt and W. Storck, J. Phys. Chem. **101**, 6799, 1997.

3 P. V. Kamat, Prog. React. Kinet. **19**, 277, 1994.

4 Y. Weng, Y. Wang, J. Asbury, H. Gosh and T. Lian, J. Phys. Chem., **104**, 93, 2000.

5 H. Huber, M. Meyer, T. Nägele, I. Hartl, H. Scheer, W. Zinth and J. Wachtveitl, Chem. Phys. **197**, 297, 1995.

6 T. Wilhelm, J. Piel and E. Riedle, Opt. Lett. **22**, 1474, 1997.

7 J. E. Moser, M. Grätzel, Chem. Phys. **176**, 493, 1993.

THz Pump-Probe Measurements of Electrons in Non-Polar Liquids

E. Knoesel[1], J. Shan[1], M. Bonn[2], M. Wolf[3], T.F. Heinz[1]

[1] Departments of Physics and Electrical Engineering, Columbia University, 538 W 120 Street, New York, NY 10027, USA.
tony.heinz@columbia.edu
[2] Leiden Institute of Chemistry, P.O. Box 9502, 2300RA Leiden, The Netherlands
[3] Department of Physical Chemistry, Fritz-Haber-Institut der MPG, Faradayweg 4-6, 14195 Berlin, Germany

Abstract: Terahertz time-domain spectroscopy is applied to examine the nature and dynamics of photo-generated carriers in non-polar liquids. The experimental results yield the complex conductivity of the material and its temporal evolution. From these data, carrier concentrations, scattering times, and dynamics are inferred.

1. Introduction

Terahertz time-domain spectroscopy (THz TDS) has emerged as a powerful probe of charge carriers in condensed matter. The method provides direct information on the complex dielectric constant or, equivalently, the complex conductivity, over a broad range of frequencies extending up to a few THz. In this spectral region, the free carrier response is generally strong and readily interpretable. Of particular note is the possibility of performing such measurements under nonequilibrium conditions subsequent to photo-excitation of carriers. This approach permits the study of systems in which charge carriers are absent under equilibrium conditions. It also provides a means of examining the ultrafast dynamics of the carriers subsequent to their generation. To date relatively few THz TDS measurements have been performed in the pump-probe experimental scheme [1,2]. In this paper, pump-probe THz TDS is applied to investigate carrier dynamics in the non-polar liquid hexane.

The nature of solvated electrons in liquids is a classic problem in condensed-phase chemistry, and one that has been addressed by a variety of techniques [3]. The present method offers, however, certain compelling features. It may be distinguished from radiolysis measurements [4] by its excellent time resolution. On the other hand, in contrast to optical pump-probe experiments [5], the THz response may be correlated directly with the response and properties of the quasi-free electrons, which are responsible for the conductivity, rather than with the bound states that typically contribute to the optical signature.

2. Experimental Methods

Fig. 1 shows the experimental scheme employed in these measurements. The laser source is an amplified, modelocked Ti:sapphire laser, which provides pulses of approximately 1-mJ energy and 100-fs duration. In THz TDS, we produce a controlled ultrafast electromagnetic transient from an ultrafast laser pulse and record the corresponding electric field directly in the time domain. In our implementation, we made use of optical rectification for the generation process and electro-optic sampling to record the transient. Both generation and detection of the transient were accomplished in ZnTe crystals. The equilibrium response of the material was measured by comparing the transmission of the THz beam in the presence and absence of the sample. The photo-induced change in the THz response of the materials was obtained correspondingly by recording the THz transient passing through the sample with and without photo-excitation. To optimize the signal-to-noise ratio, we exploited balanced detection in the electro-optic sampling scheme, as well as appropriate lock-in techniques. The production of carriers was accomplished through a single- or multiphoton excitation process induced by the third-harmonic of the Ti:sapphire laser.

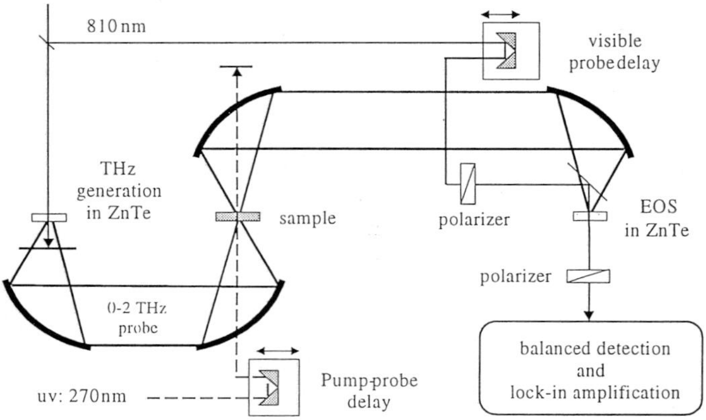

Fig. 1. Experimental set-up for pump-probe THz TDS of carriers in various materials. The THz TDS measurements were performed using a combination of optical rectification and electro-optic sampling. An ultrafast UV pulse at 270 nm served as the pump for the photo-generation of the charge carriers.

3. Results and Discussion

Figure 2 displays THz waveforms obtained for a sample of the non-polar liquid n-hexane. Corresponding data have also been measured for cyclo-hexane. The dotted line represents the transmitted THz electric field E(t), while the solid line corresponds to the change in the transmitted THz electric field ΔE(t) induced by the injection of free electrons. In this case, the free electrons are produced by the 270-nm pump pulse through a two-photon ionization process in the neat liquid. As can be seen in the Figure, the induced changes in the THz waveform are relatively small (0.5%). Still, the use of lock-in techniques permits these waveforms to be recorded with considerable precision. From such waveforms, we may extract the real and imaginary parts of the THz dielectric response through the conventional analysis, taking into account the effect of reflections [2], as required.

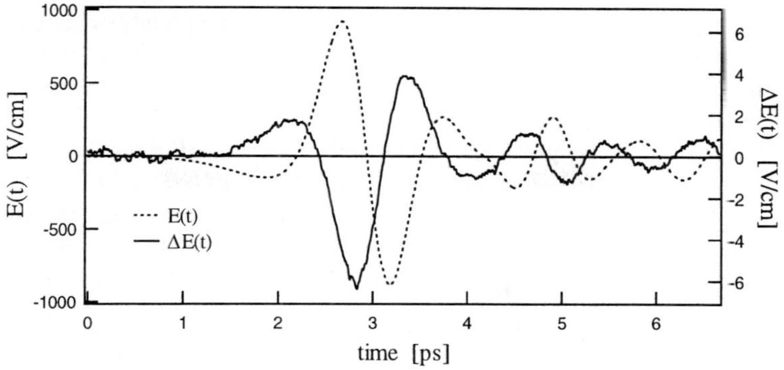

Fig. 2. The transmitted THz electric field E(t) (dotted) and its change ΔE(t) (solid) induced by electron injection in n-hexane. The pump-probe delay was 67 ps.

The solvated electrons in non-polar liquids are generally described within a two-state model [4] in which the electrons are in equilibrium between the quasi-free state and a bound state formed by a favorable local configuration of the solvent. In contrast to optical techniques, THz radiation is predominantly absorbed by the free electrons, the source of the observed ΔE signal. Their contribution to the dielectric response could be represented satisfactorily by the Drude model. From such fits we have extracted the basic parameters for the transient conductivity, namely, the carrier density and electron scattering time (~ 250 fs).

By varying the delay time between the UV pump pulse and the THz probe pulse we can record the dynamics of the photo-generated electrons (Fig. 3). In the limit of low pump fluence, we see little change in ΔE with delay time. We infer a lifetime for the free electrons in excess of 1 ns. At higher pump fluences ($f_o = 0.3$ J/cm^2), however, a significantly reduced decay time of 360 ps is observed. This behavior is indicative of a decay mechanism in which the electron capture rate depends on the density of the available ions, i.e., a non-geminate recombination

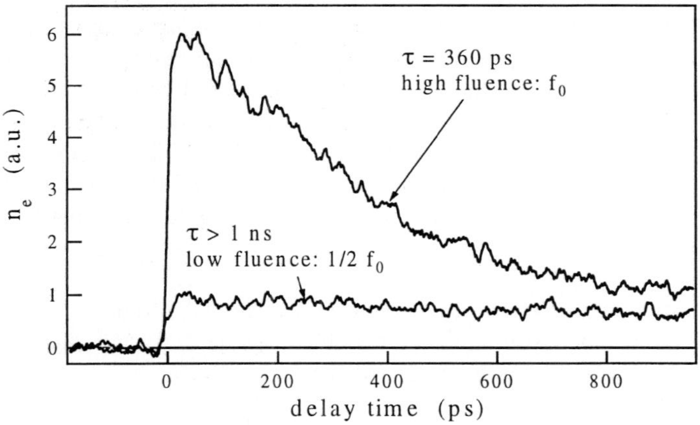

Fig. 3. Time-resolved measurements of the density of electrons in n-hexane. Data are shown for two pump fluences.

process. Analysis of the electron scattering rate indicates no appreciable change as a function of pump fluence or of delay time. This observation shows that the scattering of the free electrons is dominated by their interactions with the neutral molecules in the liquid, rather than with the cations.

THz TDS also permits insight to be gained into the energetics of the bound state where the vast majority of electrons (~99%) reside. This is accomplished by measuring the temperature dependence of the quasi-free electron concentration. From these data, a binding energy of 150 meV was deduced for n-hexane, comparable to that inferred previously [4]. These results, it should be noted, also verify that the induced THz response is dominated by the contribution of the free electrons.

In an additional series of measurements, we have applied the same experimental techniques to examine carrier scattering and dynamics in wide band-gap solids. We have investigated the photo-induced THz response in sapphire, previously identified in measurements of Haran *et al.* [1], as a prototype of a wide-gap (8.7 eV) crystalline solid. The scheme of interpretation of these data is generally similar to that for liquids, except that the role of holes needs to be considered in a more symmetrical fashion with that of electrons: unlike in the case of liquids where the positive charge carriers are ionic species, in solids the hole mobility may be significant. The experimental results reveal a rapid decay (< 100 ps) in the density of mobile charge carriers. Since this time scale is too short to be governed by radiative electron-hole recombination, the data suggest that the decrease in carrier density results from localization at trapping sites, such as impurities, or the formation of excitons [6]. A decrease in the decay rate of the carrier concentration with increasing pump fluence is observed in sapphire. We attribute this behavior to the depletion of available trap sites with increasing carrier density.

4. Conclusion

THz time-domain spectroscopy permits one to probe directly the transient conductivity associated with photo-generated electrons in liquids and wide band-gap materials. Information may be obtained on the carrier density and scattering rate. By introducing the THz probe pulse at variable delay with respect to the pump beam, the time evolution of the carriers may also be followed. For the case of photo-excited n-hexane presented in this paper, a non-geminate recombination process was observed at high ion concentration. The method of THz TDS applied in a pump-probe scheme should be suitable for a broad class of materials in which short carrier lifetimes may preclude conventional approaches to the measurement of charge transport.

Acknowledgments. This work was supported by the US National Science Foundation, the Alexander von Humboldt Foundation (E.K.), and the Royal Netherlands Academy of Arts and Sciences (M.B.).

References

1 See, for example: P.N. Saeta *et al.*, Appl. Phys. Lett. **60**, 1477 (1992); R.H.M. Groeneveld and D. Grischkowsky, J. Opt. Soc. Am. B **11**, 2502 (1994); J. Zielbauer, M. Wegener, Appl. Phys. Lett. **68**, 1223 (1996); G. Haran, W. Sun, K. Wynne and R.M. Hochstrasser, Chem. Phys. Lett. **274**, 365 (1997); R. McElroy and K. Wynne, Phys. Rev. Lett. **79**, 3078 (1997); S.S. Prabhu, S.E. Ralph, M.R. Melloch and E.S. Harmon, Appl. Phys. Lett. **70**, 2419 (1997) R.D. Averett, G. Rodriguez, J.L.W. Siders, S.A. Trugman and A.J. Taylor, J. Opt. Soc. Am. B, **17**, 327 (2000).

2 M. Schall and P.U. Jepsen, Opt. Lett. **25**, 13 (2000).

3 R.A. Holroyd and W. F. Schmidt, Annu. Rev. Phys. Chem. **40**, 439 (1989).

4 N. Gee, P.C. Senanayake and G. R. Freeman, J. Chem. Phys. **89**, 3710 (1988).

5 L.D.A. Siebbeles, U. Emmerichs, A. Hummel and H. J. Bakker, J. Chem. Phys. **107**, 9339 (1997).

6 M. Kirm, G. Zimmerer, E. Feldbach, A. Lushcik, Ch. Lushcik and F. Savikhin, Phys. Rev. B **60**, 502 (1999).

Photon-Echo Spectroscopy of the Hydrated Electron with 5-fs Pulses

Maxim S. Pshenichnikov, Andrius Baltuška, Michel F. Emde, and Douwe A. Wiersma

Ultrafast Laser and Spectroscopy Laboratory, University of Groningen, Nijenborgh 4, 9747 AG Groningen, The Netherlands
E-mail: M.S.Pshenichnikov@chem.rug.nl

Abstract. Employing photon-echo techniques with extremely short laser pulses that consist of only few optical cycles, we investigate the dynamics of the equilibrated hydrated electron. The pure dephasing time of the hydrated electron deduced from the measurements is ~1.6 fs. The shape of the absorption spectrum is described excellently by an extended Lorentzian contour the spectral width of which is fully determined by the pure dephasing time.

Excess electrons in condensed-phase media play a crucial role in the dynamics of important chemical processes. Among those are solution photochemistry, non-radiative electronic transitions, and charge transfer reactions. Hydrated electrons, i.e. electrons captured in potential wells formed by the neighboring water molecules, are of special interest. They can be viewed as an exceptional instrument for extracting information about the solvation process in water that plays an outstanding role in nature. Another motivation for a detailed study of the hydrated electron stems from the unique possibility to confront the predictions of mixed classical-quantum mechanical molecular dynamics simulations with experimental evidence. This presents a direct way of verifying the basic *a priori* assumptions that radically influence the outcome of computer modeling [1].

The issue of homogeneous *vs.* inhomogeneous broadening in the optical absorption spectrum of the excess electron in water and other fluids, as well as the explanation of its asymmetry and extraordinary spectral width, has remained a standing problem for over three decades. Numerous attempts have been made to fit the experimental data by various line shapes and/or superposition of lines. In this Contribution, photon-echo spectroscopy on the hydrated electron is performed with an unprecedented time resolution offered by 5-fs pulses [2].

Experiments with 5-fs pulses are a daunting task. Besides the trivial experimental nuisances such as the pulse lengthening during its propagation before and inside the sample because of group velocity dispersion, there are also more fundamental problems to be addressed. The conventionally employed slowly varying envelope approximation [3], implying that the change of the pulse amplitude on the duration of an optical cycle is negligible compared to the magnitude of the amplitude itself, can no longer be maintained [4]. Furthermore, the phase-matching bandwidth that is limited due to dispersion in the nonlinear medium rapidly gains importance with the broadening of the pulse spectrum. Another point of serious concern is the frequency-dependent variation in the sensitivity of the photodetector employed to register the signal generated in the nonlinear process. On top of that, artificial lengthening of the observed time dependencies is a direct consequence of the noncollinear geometry employed in experiments. In our analysis of these issues, we derived the complete expressions for the non-linear signal valid even for single-cycle-pulse applications [5]. We also assert that the influence of geometrical delay smearing does not introduce a significant distortion of the observed traces provided the geometry of the experiment is carefully optimized.

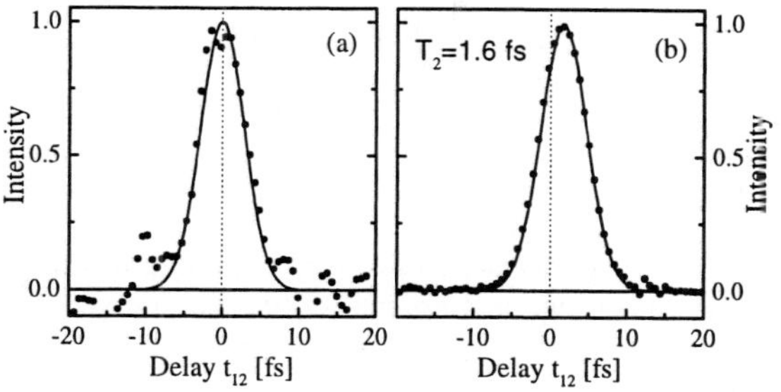

Fig.1. Results of two-pulse photon echo experiments on neat water (a) and hydrated electrons (b). Circles represent experimental data points and solid curves show fits to the data.

Figure 1 presents the two-pulse photon echo signals obtained from the water and hydrated electrons, respectively. A very minute difference in the widths of these two traces suggests that the electronic dephasing of the hydrated electrons is extremely fast. The finite population lifetime of the electrons in the excited state causes the delay of the echo trace in Fig.2b with respect to zero. The best fit to the experimental data yields the dephasing time of T_2= 1.6 fs. The addition of any appreciable amount of inhomogeneous broadening immediately results in pulling the echo maximum away from zero delay time and in the appearance of a noticeable asymmetry of the trace.

Evidently, such a 1.6-fs dephasing time should manifest itself in the Lorentzian–shaped absorption contour [3]. However, the latter is centro-symmetric while the absorption line of the hydrated electron shows a strong asymmetry (Fig.2a, solid circles). On the other hand, the use of a standard Lorentzian line shape is not warranted for the spectra with the widths comparable to the central frequencies. Instead, a more general relation should be employed:

$$\sigma_A(\omega) \propto \omega \,\mathrm{Im}\left[\frac{1}{(\omega_{eg}-\omega)-iT_2^{-1}}+\frac{1}{(\omega_{eg}+\omega)+iT_2^{-1}}\right]=\frac{4\omega^2 T_2^{-2}}{(\omega_{eg}^2-\omega^2)^2+4\omega^2 T_2^{-2}} \qquad (1)$$

In Eq.3 σ_A is the absorption cross-section and ω_{eg} is the transition frequency. In the case of a narrow absorption band, i.e. $T_2^{-1} << \omega_{eg}$, one can make use of the approximation $\omega \cong \omega_{eg}$ and recover the conventional Lorentzian contour:

$$\sigma_A(\omega) \propto \frac{T_2^{-2}}{(\omega_{eg}-\omega)^2+T_2^{-2}} \qquad (2)$$

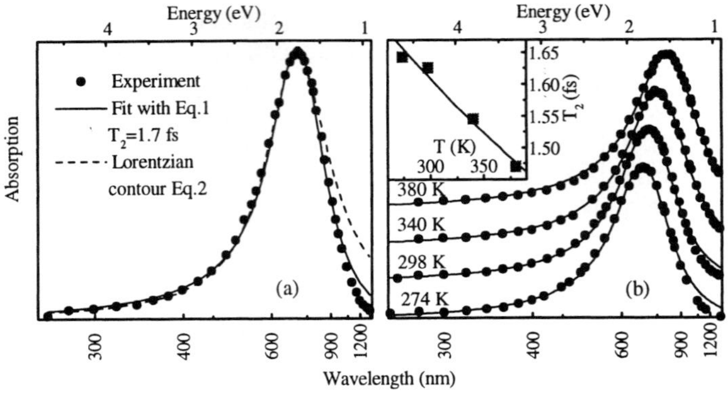

Fig.2. Absorption spectra of hydrated electrons. The circles show measured absorption of hydrated electrons produced by electron beams in bulk water [6]. Solid curves depict fits by homogeneously broadened line-shape calculated according to Eq.1. Dotted line in (a) shows for a comparison a conventional Lorentzian line-shape given by Eq.2. Numbers in (b) indicate the relevant temperatures while the inset demonstrates as the dephasing time changes with temperature.

Employing the more general relation for a homogeneously broadened line-shape given by Eq.1, we obtained fits to absorption spectra of hydrated electrons at different temperatures (solid curves in Fig.2). Evidently, the whole absorption contour can be excellently reproduced by a single homogeneously broadened line-shape. The dephasing time of T_2=1.7 fs, deduced from the fit of the spectrum at 298 K, perfectly agrees with the value obtained from the photon-echo experiment. This leaves no room for doubts about the homogeneous nature of spectral broadening of the hydrated electron absorption band.

In fact, the rejection of the second term of the sum in Eq.1 amounts to the application of the so-called rotating-wave approximation (RWA) [3]. The need to account for the experimentally observed asymmetry precluded the use of a single Lorentzian line-shape in the past attempts to model the absorption spectrum of hydrated electrons. In our explanation, the asymmetry is a natural consequence of the breakdown of the RWA.

References

1. B.J. Schwartz, P. J. Rossky, J. Chem. Phys. **101**, 6902 (1994).
2. A. Baltuška, M.S. Pshenichnikov, D.A. Wiersma, Opt. Lett. **23**, 1474 (1998).
3. S. Mukamel, *Principles of Nonlinear Optical Spectroscopy* (Oxford University Press, New York, 1995).
4. T. Brabec, F. Krausz, Phys. Rev. Lett. **78**, 3282 (1997)
5. A. Baltuška, M.F. Emde, M.S. Pshenichnikov, D.A. Wiersma, J. Phys. Chem. A **103**, 10065 (1999)
6. F.-Y. Jou, G.R. Freeman, J. Phys. Chem. **83**, 2383 (1979).

Ultrafast thermalization dynamics of hot photoelectrons injected into water

V. H. Vilchiz, J. A. Kloepfer, A. C. Germaine, V. A. Lenchenkov and S. E. Bradforth

Department of Chemistry, University of Southern California, Los Angeles, CA 90089
E-mail: bradfort@chem1.usc.edu

Abstract. We present results on the mechanism of solvated electron production by photoionization of water and photodetachment of iodide with 50fs deep-UV pulses. The relaxation of photoejected electrons from the two systems differs markedly, consistent with the suggestion that iodide detachment leads to an electron:iodine caged pair.

1. Introduction

Recent advances in ultrashort UV pulse sources has allowed for the investigation of the early time dynamics of photoejected electrons in the condensed phase. This in turn allows detailed comparisons with, and calibration of, molecular dynamics simulations for quantum solutes in water. Because of the simplicity of the electron transfer system, comparisons may also be made with experiments in gas-phase water clusters.

In this paper, we present broadband studies of the nascent electron following photoexcitation of dilute aqueous I^- and H_2O itself. Electrons are produced either by photoionizing water via two-photon absorption (9.7 eV) or, subsequent to excitation into the iodide ion charge-transfer-to-solvent (CTTS) state with a single 4.9 eV photon, by break up of the short-lived CTTS state. It has been suggested that electrons produced by two-photon excitation of liquid water into its conduction band have enough energy to populate the excited *p*-state of the trapped hydrated electron.[1] However, electrons produced through the solvent induced destruction of the CTTS state should have insufficient energy to reach the excited electron state.[2] Furthermore, the electrons generated from water ionization, having greater kinetic energy, are ejected several solvent shells away from the parent molecule.[3] For excitation into the iodide CTTS state, quantum simulations suggest that adiabatic detachment leads to a ground s-state electron and an iodine radical arranged as a contact pair.[2,4] Evidence based on picosecond geminate recombination dynamics from our lab suggests these contrasting ejection scenarios are indeed occurring.[5] In this paper, we present our early-time, multicolor probe data for both photoejection systems. Global fitting of both datasets with a spectral evolution model as suggested by Jay-Gerin [1] allows for quantitative comparison. We suggest that the caged pair formed slows the relaxation dynamics of the photodetached electron.

2. Experimental Methods

The pump-probe experiments are performed using a 200 kHz Ti:Sapphire regeneratively amplified system and OPA. The UV pump pulses were obtained by frequency doubling 510 nm from the OPA output using a Type-I BBO crystal. A deliberate positive chirp on the 510 nm pulse is optimized, using a pair of fused silica prisms in the beam path prior to the doubling crystal, to yield maximum UV bandwidth and conversion efficiency. The resulting beam passes through a pair of calcium fluoride prisms for dispersion compensation of the deep-UV pulses. This process results in 255 nm pump pulses of ~50 fs.[5] The probe pulse is obtained by using a white light continuum and selecting the color of interest with either an interference filter or by using a monochromator after the entire continuum has been transmitted through the sample. Data is recorded for probes wavelengths every 50 nm from 400 – 1000 nm. The sample is a free flowing jet of either an aqueous 60 mM KI solution or neat deionized water at room temperature. The instrument response is measured for each probe wavelength by DFG cross-correlation; it is ~80 fs over the entire probe continuum.

3. Results and Discussion

Iodide ions are resonantly pumped into the CTTS state by a 255 nm laser pulse. The experiment monitors the appearance of the detached electron produced upon break up of this initial state. Fig. 2(a) shows the results obtained by detecting the ejected electrons at different probe wavelengths. Absorption appears first at the red edge; an electron appearance time of ~200 fs is observed. To gain further

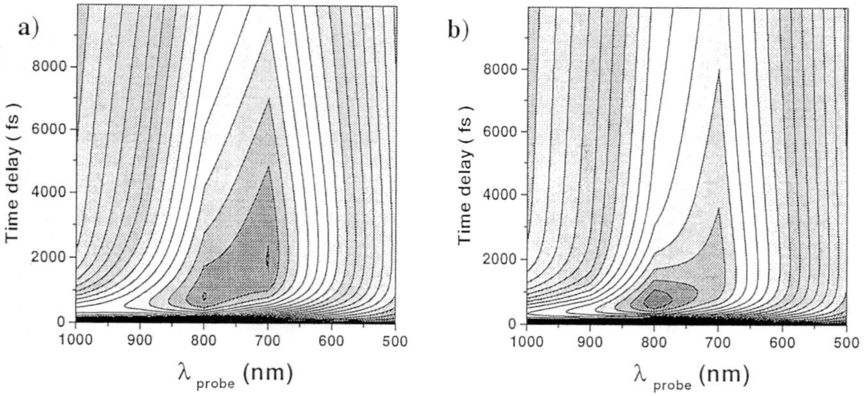

Fig. 1. (a) Transient absorption data for photodetached electrons generated by resonantly exciting iodide ions into the lowest CTTS-state. (b) Transient absorption data for ejected electrons produced by 2-photon ionization of neat deionized water. Each contour plot is generated by fitting high S/N experimental data for each probe wavelength to an unbiased functional form and then normalizing each absorption signal at 10 ps to the literature equilibrium solvated electron spectrum.[6]

insight into the results for anion photodetachment, we performed complementary experiments and analysis on a previously investigated system, the photoionization of water. Fig. 2(b) shows our results for pure water at higher pump intensity. The appearance time of the solvent ionized electron at the reddest wavelength measured is ~170 fs. The principal observation is that photoelectrons produced by both processes undergo substantial spectral shifting, but the iodide-detached electrons manifest appreciably slower relaxation. This spectral shifting is primarily due to solvation of the electron due to rearrangement of its hydration shell. The electron relaxation phase is followed by geminate recombination which takes place over ~20 ps. The recombination kinetics are qualitatively different for the two systems.[5] Recombination after two-photon water ionization can best be described by a diffusion-limited random walk of the electron from a 15 Å ejection radius. In the case of anion CTTS detachment, recombination is better modeled by competitive kinetics of recombination or escape of the geminate partners from a rapidly-formed caged pair.[4]

First, we have compared the electron relaxation results for pure water to the comprehensive photoionization study by Jay-Gerin and co-workers. A spectral evolution model was proposed for the early time dynamics of the nascent electron.[1] The model takes into account population relaxation and solvation dynamics of both excited *p*- and ground *s*-state electrons. In contrast, we are able to globally fit all of our data with a simple one state (ground state) spectral evolution model with only four adjustable parameters. We note that the *p*-state transient absorption is centered at 1250 nm, slightly beyond the range of our current experiments. The extent of ground state shifting is 0.46 eV, and a similar spectral correlation time is found (450 fs) to that of Jay Gerin. We then applied the same model to the relaxation of photodetached electrons. Global fitting of Figure 2(a) yields a longer spectral correlation timescale of 850 fs and a smaller overall band shift of 0.35 eV.

If the electron is adiabatically detached from the CTTS state to form a contact pair, as suggested by quantum simulations [2,4] and supported by the recombination kinetics, one might expect the electron spectrum and its evolution due to solvent rearrangement to be distorted by the presence of the iodine within the same solvent shell. The current results indicating a factor-of-two slower solvation for photodetached electrons attest to a rather different ejection mechanism in operation. This work is supported by NSF, the Donors of the Petroleum Research Fund and the Research Corporation.

References

1. Pepin, C.;Goulet, T.;Houde, D.;Jay-Gerin J.-P. J. Phys. Chem. A 1997, 101, 4351
2. Sheu, W.-S.; Rossky, P. J. Phys. Chem. 1996, 100, 1295
3. Crowell, R.A.; Bartels, D.M. J. Phys. Chem. 1996, 100, 17940
4. Staib, A.; Borgis, D. J. Chem. Phys. 1996, 104, 9027
5. Kloepfer, J. A.; Vilchiz, V.H., Lenchenkov, V.A.; Bradforth, S.E J. Chem. Phys. submitted; Kloepfer, J.A.; Vilchiz, V.H.; Lenchenkov, V.A.; Bradforth, S.E.Chem. Phys. Lett. 1998, 298, 120
6. Jou, F.-Y.; Freeman, G.R. J. Phys. Chem. 1979, 83, 2383

Mid-IR Detection of a Precursor to the Prehydrated Electron

Neil A. Anderson, Kun Hang, John B. Asbury, and Tianquan Lian

Department of Chemistry, Emory University, Atlanta, GA 30322, USA
Phone: (404)727-6649; E-mail: tlian@emory.edu

Abstract. Ultrafast transient absorption spectroscopy probing in the mid-IR is used to investigate electron hydration following ejection from $Fe(CN)_6^{4-}$. An intense, broad absorption signal shows an instrument-limited appearance and 170±20 fs decay. The signal is assigned to the direct precursor to the presolvated electron.

1. Introduction

Study of the electron hydration process has been facilitated by use of ultrafast spectroscopy. Electrons can be generated either through excitation of an anion with an accessible charge-transfer-to-solvent (CTTS) state, or through multiphoton ionization of water. The probe for the previous experiments has been the visible and near-IR. For a overview of much of the work in this area, see reference [1]. Absorption signals from the fully solvated species (peak at 710 nm) and its direct precursor, the presolvated electron (peak near-IR) have been observed. However, the presolvated electron is a localized species, and its absorption appears on a 160-180 fs timescale.[1,2] Prior to localization in a trap site, the ejected electron is expected to be quasifree (delocalized). However, no signal arising from the earliest steps in solvation has been observed.

A model of the quasifree electron's spectrum predicted that the species will show a strong absorption in the mid-IR. [3] Since previous experiments focused on the visible and near-IR, this may explain the failure to observe the initial steps in electron solvation. To explore this possibility, this report presents results from a pump-probe transient absorption experiment of $Fe(CN)_6^{4-}$ (ferrocyanide) in D_2O and H_2O solutions exciting at 266 nm and probing from 3-7 μm. This excitation wavelength allows one photon access to the ferrocyanide CTTS state, which has been reported to be very short lived [4], so comparable timescales to direct water ionization are anticipated. The details of the femtosecond mid-IR spectrometer have been described elsewhere. [5]

2. Results and Discussion

An advantage to using ferrocyanide as the electron donor is the simultaneous ability to monitor IR CN stretching peaks in the reactant (2034 cm^{-1}) and product (2114 cm^{-1}). The spectral evolution in this region following excitation of ferrocyanide/D_2O is shown in Figure 1. These spectra show an intense, broad absorption at early time, with overlying narrower features assignable to the iron complexes. At the earliest times, a significant absorption increase at 2034 cm^{-1} is seen, which quickly decays to a bleach signal. The early peak likely arises from an excited state of the ferrocyanide complex. The $Fe(CN)_6^{3-}$ (ferricyanide) absorption peak at 2114 cm^{-1} is present at all time delays, although its size is comparable to the noise level at the earliest times. This peak confirms rapid electron ejection. After the broad signal has decayed, only the ferrocyanide bleach and ferricyanide absorption are present. The ratio between the peaks is consistent with the static difference between the species. [6] The narrow peaks persist without distinct decay for >1 ns.

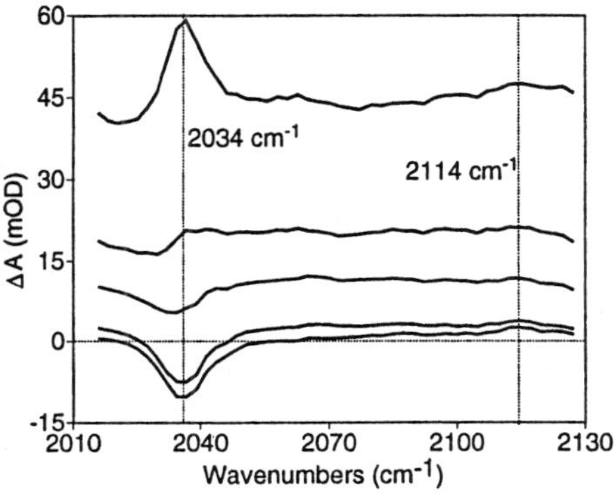

Figure 1: Spectral evolution in the region of the ferrocyanide CN stretch at 2034 cm^{-1} and the ferricyanide CN stretch at 2114 cm^{-1}, indicated on the figure. An intense, broad absorption underlies the narrow iron complex peaks at early times.

A broader spectrum of the signal at 160 fs is shown in figure 2. Over this range, the intensity steadily rises for higher energy probes. Solvated electron species have quite broad absorption spectra. Since the signal is much too broad to assign to the iron complexes, it must arise from the ejected electron. As a result, comparison of the long-time ferrocyanide bleach to the amplitude of the electron signal nearby gives an estimate for the extinction coefficient. Using this method, the extinction coefficient at 2100 cm^{-1} is > 40 000 $M^{-1}cm^{-1}$.

A kinetic trace probing at 1960 cm^{-1} is shown in figure 3. This trace is typical of mid-IR traces at probe wavelengths from 3-7 μm, with an instrument-limited rise was followed by a 170±20 fs wavelength-independent decay. The rise and decay timescales were identical within error for both D_2O and H_2O solutions.

The mid-IR electron signal is not consistent with assignment to either the presolvated or fully solvated electron species. The presolvated electron has been shown to appear in 160-180 fs and decay in 410-540 fs.[1,2] Clearly, both the rise and decay of the mid-IR signal are too fast to support assignment to the presolvated electron. Additionally, the decay of the mid-IR signal occurs on a 170 fs timescale, which agrees with the risetime of the presolvated electron. Therefore, the signal observed here is assigned to the direct precursor to the presolvated electron. For confirmation, a kinetic trace probing at 650 nm is shown in Figure 3. At this wavelength, the absorption is dominated by the fully solvated electron population dynamics. This data fits well with a 480 fs risetime, in excellent agreement with a previous result.[3] Since this signal basically corresponds to the appearance of solvated electron, it further confirms that the mid-IR signal does not arise from presolvated electron.

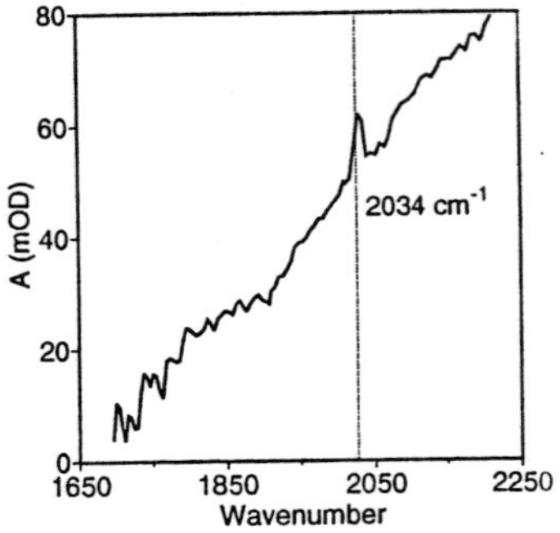

Figure 2: A spectrum at 160 fs covering a broader spectral region shows a monotonic increase toward higher energy probes.

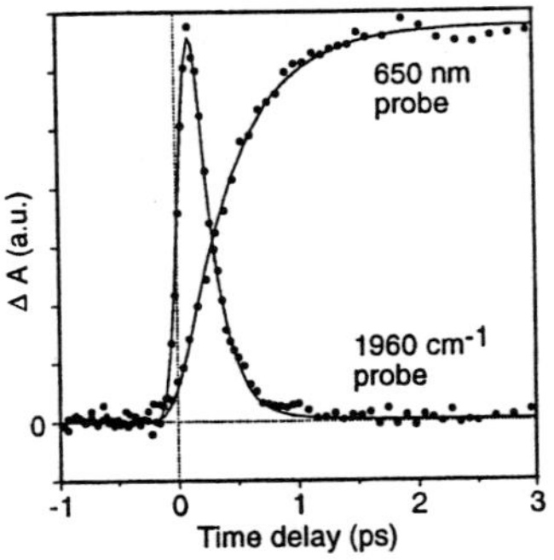

Figure 3: Kinetic traces probing in the mid-IR fit to an instrument-limited rise and 170 fs decay. At 650 nm, the fully solvated electron appears on a 480 fs timescale.

3. Conclusions

Electron ejection from $Fe(CN)_6^{4-}$ following 266 nm excitation was confirmed by monitoring the CN stretching frequencies of ferro- and ferricyanide. In addition to the narrow IR peaks, a broad mid-IR signal is observed that is consistent with assignment to the direct precursor to the presolvated electron. This absorption may arise from the initially formed quasifree electron, or from a high-lying localized state.

References

1. C. Pepin, T. Goulet, D. Houde, J.-P. Jay-Gerin, J. Phys. Chem. A, **101**, 4351, 1997.
2. A. Reuther, A. Laubereau, D.N. Nikogosyan, J. Phys. Chem., **100**, 16794, 1996.
3. E. Keszei, T. H. Murphrey, P. J. Rossky, J. Phys. Chem. **99**, 22-28, 1995.
4. S. Pommeret, R. Naskrecki, P. van der Meulen, M. Menard, G. Vigneron, T. Gustavsson, Chem. Phys. Lett. **288**, 833-840, 1998.
5. Y. Q. Wang, J. B. Asbury, T. Lian, J. Phys. Chem. A, **104**, 4291, 2000.
6. N.A. Anderson, K. Hang, J.B. Asbury, T. Lian, Chem. Phys. Lett. submitted.

Novel Fs-Precursors of Solvated Electrons in Water in the MIR: A Charge Transfer Process

R. Laenen, T. Roth, and A. Laubereau

Physik-Department E11, Technische Universität München,
James-Franck-Str., D-85748 Garching, Germany
e-mail: Robert.Laenen@ph.tum.de

Abstract. Two intermediates with peak absorptions at 2.9 μm and 1.6 μm are observed for the first time during the generation of hydrated electrons. A charge transfer process is proposed involving an OH:e^--complex and partially solvated e^-:$(H_2O)_n$.

The solvated electron in water has been of special interest due to the importance of water to physics, chemistry and biology. The first ultrashort time-resolved investigations of the generation of solvated electrons in water have been performed more then 10 years ago [1, 2]. One of the questions which are up to now controversely debated concerns the nature of one of the intermediates, the so called "wet electron", seen during the generation process of the solvated electron. In the literature an assignment of the "wet electron" to a p-state can be found as well as to a hot ground state of the solvated electron. A shortcoming of all the earlier measurements on the generation process is the accessible probing wavelength range, commonly restricted to the visible and near IR.

In this study we present the first data on the generation of solvated electrons in neat water which extends to probing wavelengths of 5.5 μm in the mid IR. The electrons are generated via two photon absorption of the third harmonic (4.5 eV, 150 fs) of an amplified Ti:sapphire laser pulse. The tunable probe pulse is derived via various parametric processes and allows for investigation of the dynamics in the range of 0.4–5.5 μm with pulse duration of $\simeq$ 150 fs. The sample is tridistilled neat water and to avoid for thermal effects we have utilized a jet with thickness of $\simeq$ 75 μm. Delay time zero is determined independently via two photon absorption in a thin diamond specimen [3].

Experimental results on the induced absorption of the water sample are depicted in Fig. 1. Here we have summarized time-dependent measurements on the induced absorption of the water sample taken at different wavelengths between 1.4 and 2.3 μm for the first 200 fs. A strong shift of the maximum induced absorption is noticed from the MIR to the NIR from the data which has already finished after 200 fs. Secondly a transient isosbestic point is seen for the time intervall 75 fs to 150 fs around 1.85 μm. This indicates that the

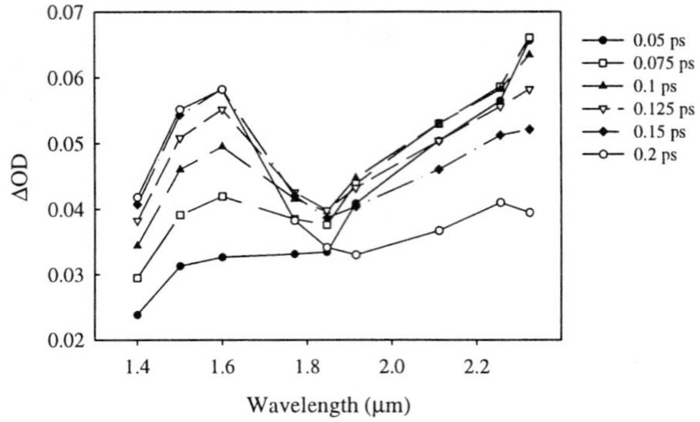

Fig. 1. Spectra of the generation of solvated electrons in neat water; notice the strong shift of the maximum from the MIR to the NIR within the first 200 fs

initial dynamics can be described by a two level scheme with subsequent relaxation to further energy levels. Our data are fitted with the help of a 6 level scheme which extends earlier approaches [4] to account for important details of the generation process in water. Two-photon absorption of a water molecule (0) results in an excited water molecule H_2O^* (1), which relaxes within our temporal resolution to a first intermediate (2) with absorption in the MIR. Subsequently relaxation proceeds within 110 fs to a second species (3), which corresponds to the induced absorption with peak at 1.6 μm. The next intermediate (4) which contributes to the transient spectrum is the already well known "wet electron", which builds up with a time constant of 200 fs. The corresponding absorption is red-shifted in comparison to the equilibrated hydrated electron (5). The lifetime of the "wet electron" amounts to 540 fs in well agreement to earlier published results [2, 4]. With an efficiency of 64% long-lived hydrated electrons (5) are generated while the remainder relaxes back to the ground state of water due to geminate recombination with a time constant of 2.5 ps.

In order to interpret the new two intermediates (2,3) seen from our data, we have plotted their respective molar extinction coefficients as derived from fitting of the data with the relaxation scheme in Fig. 2, together with the conventional IR-absorption of the 75 μm thin water sample. The conventional infrared absorption of the water sample (r.h.s., dotted line) is dominated by a strong band with peak at 2.9 μm which is related to the OH-stretching modes of water. The other peaks corresponds to absorption of combination modes and are of minor importance. The first identified intermediate (2, l.h.s., hollow circles, dashed line) shows a strong induced absorption centered at the peak absorption of the OH-stretch, but notably wider. For the second intermediate (3, l.h.s., filled circles, solid line) we find a peak absorption at 1.6 μm with a weaker absorption extending to the long wavelength edge of our probing range. Taking this into account we derive the following interpretation of the measured dynamics:

The increased and spectrally broadened OH-stretch absorption of the first intermediate (2) is proposed to represent an $OH{:}e^-$ complex, as supported

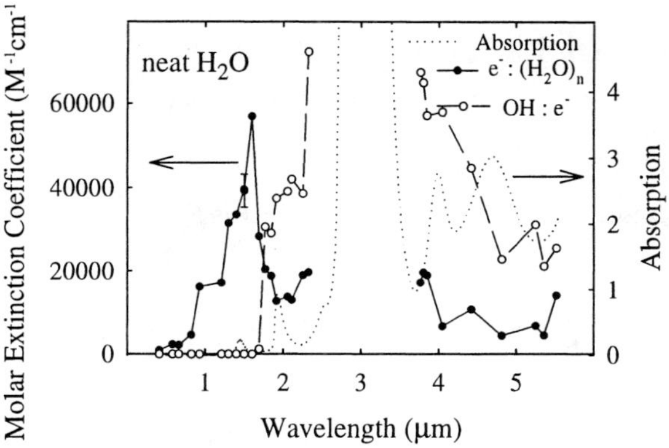

Fig. 2. The molar extinction coefficient for the first two identified intermediates during the generation of solvated electrons (l.h.s.) and the conventional absorption of the water sample (r.h.s., dotted line) are shown here; line as a guide for the eye

by theoretical investigations of electrons in water clusters in the gas phase [5] and further arguments [3].

The generation process of hydrated electrons investigated here is consequently explained by a proton transfer along the H-bond of the excited water molecule resulting in an H_3O^+ and the mentioned OH:e^- complex. The excitation energy of 9 eV is obviously too low for the generation of electrons in the conduction band of water. This is in accordance with recent investigations [6].

The second intermediate (3) is assigned to an electron in a trap consisting of fewer water molecules in comparison to e_{solv}. This is supported by gas phase investigations of the absorption of electrons in water clusters with known size [7]. Accordingly we interprete the second intermediate (3) as a mix up of hot ground-state electrons within traps consisting of different numbers of water molecules.

Our study clearly indicates that the "wet electron" (4) is a hydrated electron in its electronic ground state in a fully developed trap of hot water molecules.

References

1. A. Migus, Y. Gauduel, J.L. Martin, and A. Antonetti, Phys. Rev. Lett. **15**, 1559 (1987)
2. F.H. Long, H. Lu, and K.B. Eisenthal, Phys. Rev. Lett. **64**, 1469 (1990)
3. R. Laenen, T. Roth, and A. Laubereau, submitted for publication
4. A. Reuther, A. Laubereau, and D.N. Nikogosyan, J. Phys. Chem. **100**, 16795 (1996)
5. R.N. Barnett, U. Landman, A. Nitzan, J. Chem. Phys. **89**, 2242 (1988)
6. C.L. Thomsen, D. Madsen, S.R. Keiding, and J. Thøgersen, J. Chem. Phys. **110**, 3453 (1999)
7. P. Ayotte and M.A. Johnson, J. Chem. Phys. **106**, 811 (1997)

Femtosecond Study of Electron Photodetachment from Complex Anions: $Fe(CN)_6^{4-}$ and $CuBr_2^-$ in H_2O

Victor Lenchenkov, Victor Vilchiz, Jeremiah Kloepfer and Stephen Bradforth

Department of Chemistry, University of Southern California, Los Angeles, CA 90089
E-mail: lenchenk@aludra.usc.edu

Abstract. The ultrafast dynamics of photoelectron detachment from the molecular anions, $Fe(CN)_6^{4-}$ and, for the first time $CuBr_2^-$ are studied in detail by means of pump-multicolor probe spectroscopy and compared to the prototypical I^- system.

1. Introduction

The development of sub-50 fs UV laser sources allows one to investigate the influence of the solvent environment on very simple chemical reactions in the aqueous phase [1]. Photolysis of the ferrocyanide system has been used as an efficient source of solvated electrons in several recent experiments [2], however the mechanism of photodetachment in complex anions is not well understood. The initial process of the generation of the solvated electron was first observed by Rentzepis with ~2 ps resolution [3], which, however, was not sufficient to observe the electron formation time. In this paper, we present a detailed pump-probe study using a broadband visible/IR probe and explore the effect of excitation into different electronic states of the ferrocyanide ion, as well as the first ultrafast study of another molecular anion, $CuBr_2$.

$$Fe(CN)_6^{4-} + h\nu \rightarrow Fe(CN)_6^{4-}* \rightarrow Fe(CN)_6^{3-} + e^-(aq) \qquad (1)$$

$$CuBr_2^- + h\nu \rightarrow CuBr_2^-* \rightarrow CuBr_2 + e^-(aq) \qquad (2)$$

In each case, the initial excited state has been assigned as charge-transfer-to-solvent (CTTS) in analogy to the bands seen in halide ions [4, 5]. However, the existence of overlapping intramolecular electronic bands for these ions may lead to a more involved mechanism for electron detachment. A fascinating possibility exists for reaction (2) that on ultrafast removal of the electron, we will be able to track the neutral $CuBr_2$ fragment as it undergoes dissociation or solvolysis – to lead to independent solvated ions.

2. Experimental Methods

A 200 kHz Ti: sapphire regenerative amplifier producing pulses of 45-50 fs in duration drives a visible-pumped optical parametric amplifier (OPA). Additional 780 nm and 390 nm driven supercontinua are used in our experiment to probe in the 500-1000 and 350-500 nm spectral range respectively. A 255-300 nm photodetachment pump pulse is generated by temporal-walkoff-controlled second harmonic generation of the OPA signal beam, and is dispersion compensated with pairs of calcium fluoride prisms. At 255 nm, the pump pulse is ~50 fs – the pump-probe instrument response throughout is ~ 100 fs. The pump and probe beams are focused into a free flowing jet of the room temperature aqueous solutions containing either 0.04 M $Fe(CN)_6^{4-}$ or 0.005 M $CuBr_2^-$ ions. In the case of $Fe(CN)_6^{4-}$, the time resolved signal for each probe color has been fitted to an arbitrary multi-exponential function to give an unbiased representation of the data, Figure 1. The transient absorption at 3 ps delay for each probe is normalized to the literature equilibrated hydrated electron spectrum, which is assumed as the only long-time signal carrier over the spectral region displayed. Transient absorption spectra for several pump-probe delays are shown in Figure 2.

3. Results and Discussion

The appearance and subsequent time evolution of the electron absorption band subsequent to photodetachment is observed in Figure 1, where it is compared with data for the atomic iodide anion studied in our laboratory [1]. Rise times are

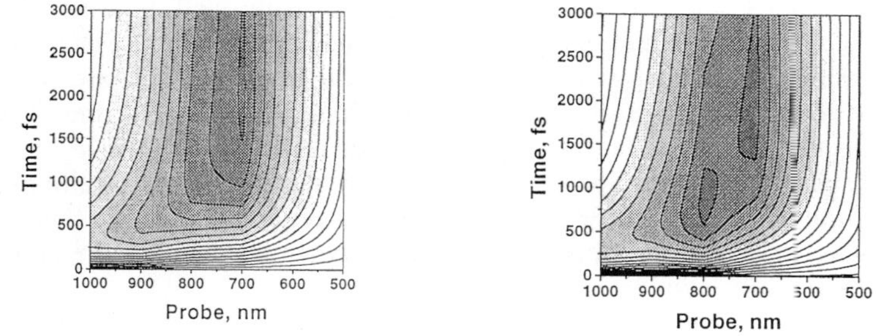

Fig. 1. 255 nm pump – multicolor probe experimental data for (left) $\mathbf{Fe(CN)_6^{4-}}$ in H_2O, (right) $\mathbf{I^-}$ in H_2O. The thermalization of the ejected electron is faster for $Fe(CN)_6^{4-}$.

fastest at 1000nm (180 fs, compared to 600 fs at 500 nm), and the band is almost fully developed within 900 fs after resonant photo-oxidation of the Fe^{II} center, Figure 2. The relaxation of the electron is apparently more rapid than in the case of iodide detachment. As we probe further blue than shown in Figure 1, the absorption rise time becomes faster than in the visible (~ 370-350 fs at 387 and 400 nm). Both $Fe(CN)_6^{3-}$ and the solvated electron absorb in this region. Therefore, results of UV and blue probe experiments, as reported by others [6], are more complex to analyze without separating out the component due solely to the evolution of the ejected electron.

Geminate recombination of the products of reaction (1) is observed over a 100 ps timescale. We find that the recombination yield although small, increases with increasing solution ionic strength over a range of 9 to 20 %. We attribute this to the fact that $Fe(CN)_6^{4-}$ does not exist as a totally dissociated ion in aqueous solution; the fraction of ion pairing with the K^+ counter ion and thus the effective charge on the geminate recombination target varies with overall solute concentration.

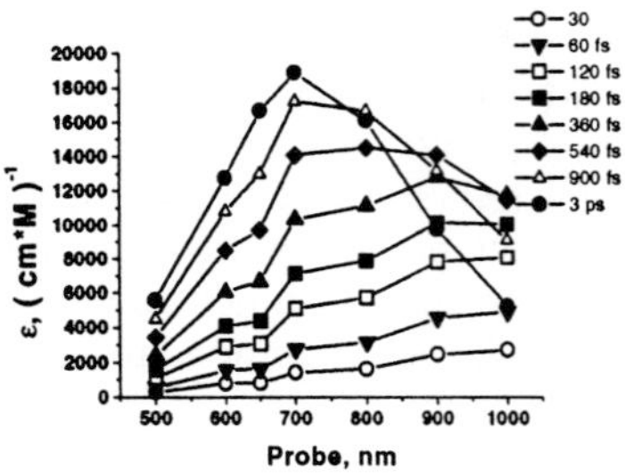

Fig. 2. Transient absorption spectra at several delays; 255 nm pump -multi color probe $Fe(CN)_6^{4-}$ in H_2O.

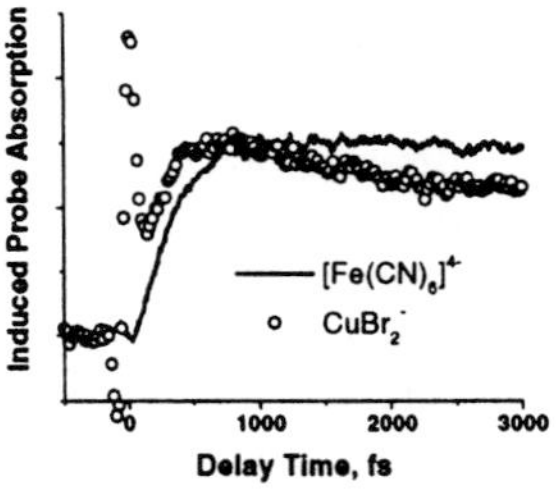

Fig. 3. 255 nm Photodetachment of $CuBr_2^-$ (circles) and $Fe(CN)_6^{4-}$ (solid line) with an 800 nm probe, rise times are 140 fs and 350 fs respectively.

The photodetachment of the Fe^{II} center occurs over a wide excitation range with widely varying quantum yield [4]; this is not well understood. To explore the possibility of contrasting ejection pathways on pumping into an intramolecular rather than CTTS state, we have photodetached with well separated pump energies. Surprisingly, pump-probe signals obtained for 255 nm, assigned to a CTTS transition [4], and for 300 nm, an iron *d-d* transition [7], are perfectly overlapped for all probe wavelengths recorded. One explanation for the identical electron ejection dynamics is that the poorly resolved CTTS band is in fact very broad and substantially overlaps with the intramolecular electronic transitions.

Finally, the data obtained for electron photodetachment from $CuBr_2^-$, Figure 3, indicate considerably faster electron detachment dynamics than for the multiply-charged ferrocyanide complex ion.

References

1 J.A. Kloepfer, V.H. Vilchiz, V.A. Lenchenkov, S.E. Bradforth, Chem. Phys. Lett. **298**, 120, 1998; J.A. Kloepfer, V.H. Vilchiz, V.A. Lenchenkov, A.C. Germaine and S.E. Bradforth, *J. Chem. Phys.* submitted

2 K. Yokoyama, C. Silva, D. H. Son, P.K. Walhout, P.F. Barbara, J. Phys. Chem. A **102**, 6957, 1998; A. Kummrow, M.F. Emde, A. Baltuska, M.S. Phenichnikov, and D.A. Wiersma, J. Phys. Chem. A **102,** 4172, 1998.

3 P.M. Rentzepis, R.P. Jones, J. Jortner, Chem. Phys. Lett. **15**, 480, 1972.

4 M. Shirom and G. Stein, J. Chem. Phys. **55**, 3372, 1971.

5 O. Horvath, K.L. Stevenson, Charge Transfer Photochemistry of coordination Compounds, (VCH Publishers, Inc., New York, NY, 1993), Chap. 4.

6 S. Pommeret, R. Naskrecki, P. van der Meulen, M. Menard, G. Vigneron, T. Gustavsson, Chem. Phys. Lett. **288,** 833, 1998.

7 H. B. Gray and N. A. Beach, J. Am. Chem. Soc. **85**, 2922, 1963.

Femtosecond Studies of the Charge-Transfer-to-Solvent Transition of the Sodium Anion in THF

Erik R. Barthel, Ignacio B. Martini, and Benjamin J. Schwartz

Department of Chemistry and Biochemistry, University of California, Los Angeles, Los Angeles, CA 90095-1569 USA. E-mail: schwartz@chem.ucla.edu

Abstract. Following excitation of the CTTS band of the sodium anion in THF, recombination of the solvated electrons with the parent sodium atom takes place on two time scales: ~2 ps, and ~200 ps. The fraction of electrons that recombine in ~2 ps depends on the initial excitation energy; the data suggest that a stable atom:electron contact pair is formed following CTTS. All the results show that Na^- provides a unique model system for studying charge transfer: only electronic degrees of freedom are involved, so that the spectroscopy is well-defined for all the species in the reaction.

1. Introduction

The detailed dynamics underlying charge transfer reactions is of great interest in chemical physics. The simplest example of such reactions is that of the charge-transfer-to-solvent (CTTS) transitions of atomic anions; excitation of the CTTS band produces a neutral atom and a solvated electron. While most of the attention has focused on aqueous halides, the halides are challenging to study because the time scales for charge transfer are exceedingly fast (≤ 200 fs) and because the relevant spectroscopic transitions are deep in the UV [1].

Recently, we have explored the CTTS dynamics following excitation of Na^- in THF [2]. Our choice was predicated by spectroscopic convenience: both the Na^- CTTS ground and excited states, and the absorptions of both the solvated electron and the Na atom, are easily and independently accessible in the visible or IR [2].

In this paper, we explore the dynamics of recombination between the ejected electron and the neutral Na atom following CTTS excitation. We find that this back electron transfer takes place on two very different time scales, suggesting that the CTTS electron is bound in a contact pair with the neutral atom product.

2. Experimental Methods

The pump-probe experiments utilize a regeneratively amplified Ti:Sapphire laser that produces 1 mJ, ~100 fs pulses near 800 nm at a 1 kHz repetition rate. These pulses pump a dual-pass OPA to produce light at ~2 μm for probing solvated electrons in THF. The visible pump light is produced by sum-frequency mixing the residual 800 nm fundamental with either the OPA signal or idler. Details on the set-up of the pump-probe spectrometer are described elsewhere [2].

Samples of Na^- were prepared following a modification [2] of the procedure of Dye [3]. The basic idea is to shift the equilibrium 2 $Na_{(s)} \leftrightarrow Na^+ + Na^-$ to the right by complexing the sodium cation with the addition of a crown ether.

3. Results and Discussion

Figure 1 summarizes the spectroscopy of the Na^- CTTS system. The heavy solid curve is the UV-visible absorption of Na^- in THF: excitation of this band produces a Na atom and a solvated electron. Note that the acronym CTTS is somewhat of a misnomer: the CTTS excited state is *localized,* bound not only by the nuclear attraction of the electron to the Na core but also by the polarization of the solvent [4]. It is solvent motion subsequent to the excitation that causes the electron to detach; the Franck-Condon excitation does not directly produce a solvated electron.

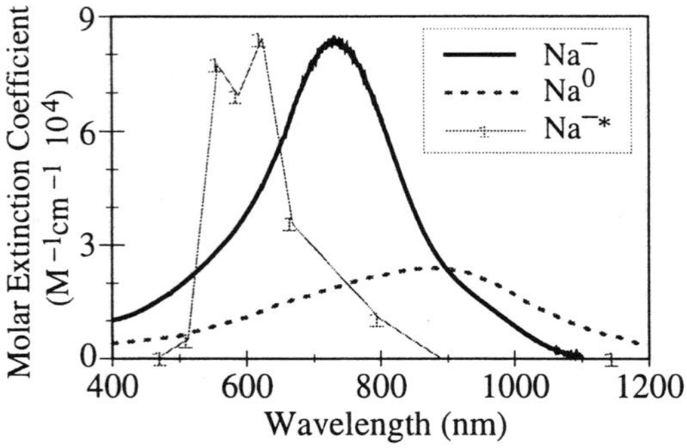

Fig. 1: Absorption of species involved in the Na^- CTTS reaction.

The dashed curve in Fig. 1 shows the spectrum of the solvated Na atom in THF [5]. This species could be described as a Na^+/solvated electron contact pair [5]; however, a species with this absorption spectrum is detected immediately following CTTS electron detachment [2]. As will be shown below, a significant fraction of geminate recombination occurs within 2 ps of the CTTS excitation. Since it is unlikely that a 3-body recombination event can take place on this time scale, we believe that the species whose absorption is represented by the dashed curve in Fig. 1 is better described as a solvated neutral Na atom.

Finally, the symbols in Figure 1 show the absorption spectrum of the CTTS excited state prior to electron detachment, as determined by fitting a variety of pump-probe transients to a kinetic model [2]. The strong absorption near 590 nm is reminiscent of the gas phase absorption spectrum of neutral sodium. The error bars on the data (not shown) are larger than the apparent structure near 550 nm.

Figure 2 shows the results of femtosecond pump-probe experiments which monitor the ~2 μm absorption of the THF-solvated electron following excitation at a variety of visible wavelengths. The left panel shows the dynamics in the first 8 ps following excitation, with the data normalized to the same maximum change in absorbance. The right panel shows the data on a much longer time scale; here, the data have been normalized to the same change in absorbance at 8 ps delay.

The ~700 fs rise of the absorption in Fig. 2(a) corresponds to the time needed for solvent motions to cause the electron to detach from the CTTS excited state [2]. The rapid decay on the ~2 ps time scale is independent of probe wavelength, verifying that it results from geminate recombination, or back electron transfer.

The amount of back electron transfer changes drastically with excitation wavelength but the time scale does not. This suggests multiple pathways for charge transfer. Following CTTS excitation, electrons are ejected to a nearby

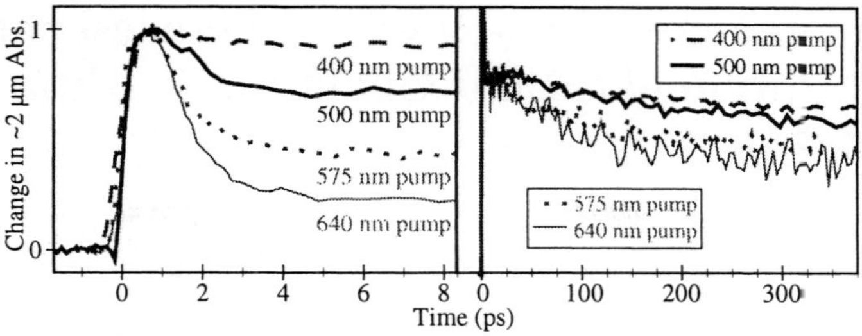

Fig. 2: 2 μm pump-probe transients for Na^- at different pump wavelengths

solvent cavity. Since the electronic wavefunction is highly delocalized in non-polar solvents like THF, there is some probability of overlap with the Na atom parent, leading to the possibility of recombination via direct non-adiabatic transition. At blue excitation wavelengths, the electron is ejected further into the solvent, reducing the overlap. The right panel of Fig. 2 makes it clear that those electrons which do not recombine in ~2 ps do not recombine for at least hundreds of ps. This long time recombination cannot be diffusive. Fits to standard diffusive recombination models give a thermalization distance ≥ 45 Å; a visible photon cannot eject electrons this far (especially given that some fraction of them are close enough to recombine directly within 2 ps). Thus, an electron ejected by CTTS forms a contact pair with the parent atom. If the electronic wavefunction overlaps the parent, recombination occurs quickly. If the electron localizes just a small distance further away (solvent-separated pair), there is no overlap and recombination cannot take place until a significant solvent fluctuation occurs. Diffusion does not appear to play any important role in the recombination process.

4. Conclusions

Overall, if the solvated electron's wavefunction has overlap with the neutral partner, then recombination can take place via a direct non-adiabatic transition on a time scale ≤ 2 ps. If there is no such overlap, on the other hand, recombination is inhibited (taking ≥ 200 ps) because there is a large free energy barrier to break up the local solvent structure around the contact pair: in other words, the back electron transfer can be thought of as occurring in the Marcus inverted regime.

Acknowledgements. We thank the NSF (CHE-9733218). BJS is a Cottrell Scholar of Research Corporation and an Alfred P. Sloan Foundation Fellow.

References

1. J. A. Kloepfer, V. H. Vilchiz, V. A. Lenchenkov and S. E. Bradforth, *Chem Phys. Lett.* **298**, 120, 1998 and references therein.
2. E. R. Barthel, I. B. Martini and B. J. Schwartz, *J. Chem. Phys.* **112**, 9433, 2000; I. B. Martini, E. R. Barthel and B. J. Schwartz, submitted.
3. J. L. Dye, *J. Phys. Chem.* **84**, 1084, 1980.
4. W.-S. Sheu and P. J. Rossky, *J. Phys. Chem.* **100**, 1295, 1996.
5. P. Piotrowiak and J. R. Miller, *J. Am. Chem. Soc.* **113**, 5086, 1991.

(Sub)picosecond Time-Resolved Fluorescence Depolarization of OLED Compounds Alq_3, Gaq_3, and Inq_3

E. van Veldhoven, H. Zhang and M. Glasbeek

Laboratory for Physical Chemistry, University of Amsterdam, Nieuwe Achtergracht 129, 1018 WS Amsterdam, The Netherlands
E-mail: glasbeek@fys.chem.uva.nl

Abstract. Femtosecond fluorescence depolarization upconversion measurements were performed for the OLED compounds Alq_3, Gaq_3 and Inq_3. For these compounds in solution it is shown that intraligand electron transfer in the emissive state is controlled by solvation.

1. Introduction

Alq_3 (tris(8-hydroxyquinoline) aluminum) (Fig. 1) is the most widely used electroluminescent molecule in organic light-emitting devices (OLED). Alq_3 gives rise to a bright fluorescence after either electron injection or photoexcitation. The lowest excited state is localized at one of the quinolinolate ligands. At this ligand site, the electronic charge has been partially transferred from the phenoxide to the pyridyl ring (intraligand charge transfer) [1]. Recently we have performed solvation dynamics studies using fluorescence upconversion methods for Alq_3 and Gaq_3 in liquid solution [2]. It has been suggested that concomitant with the solvation, an excited-state adiabatic electron transfer occurs. The increasing charge transfer character of the emissive state as solvation progresses, suggests that strong fluorescence depolarization effects on a (sub) picosecond time scale may be observed upon excitation of the metal complexes with polarized laser pulses (< 100 fs). We report here on time-resolved fluorescence upconversion depolarization measurements for Alq_3, Gaq_3 and Inq_3 in a few polar solvents and in a $Al(acac)_3$ host crystal. The depolarization measurements reveal (i) ultrafast internal conversion (< 150 fs) of the initially excited state (ii) solvation (1-10 ps) of the emissive charge transfer state in small polar solvents, and (iii) rotational diffusion of the probe molecules (80 ps – 1.3 ns, depending on the solvent viscosity).

2. Experimental

Alq_3, 8-hydroxyquinoline and $Al(acetylacetonate)_3$ were purchased from Aldrich and used without further purification. The fluorescence of Alq_3 was measured for the probe molecule both in liquid solution and doped (1:1000) in a host crystal of

Al(acetylacetonate)$_3$. Details of the fluorescence upconversion set-up (time response ~150 fs) and the time-correlated single-photon counting picosecond fluorescence set-up (time response ~ 16 ps) were as described in [2]. In the time-resolved depolarization measurements, a linear polarizer was placed in the detection pathway, for measuring the fluorescence intensity polarized parallel ($I_{||}$) and perpendicular ($I_{\perp}$) to the polarization direction of the exciting laser pulses.

3. Results and Discussion

Typical polarized upconversion transients measured for Alq$_3$, dissolved in DMF, when excitation is at 360 nm, are given in Fig. 1. The figure shows the fluorescence upconversion transients (in a 10 ps time window) for $I_{||}(t)$ and $I_{\perp}(t)$, detected at 505 nm (a) and at 600 nm (b). The temporal dependence of the anisotropy (r), where $r = (I_{||} - I_{\perp})/(I_{||} + 2 I_{\perp})$, as obtained from the experimental transients is plotted (dots) in the panels (c) and (d), respectively. The time dependence of the data points, representative of the depolarization, could be fitted to a biexponential decay function convoluted with the system response function (drawn lines in Figs. 1c and 1d). As is well-known, for a fluorophore in liquid solution the maximum possible value for r is 0.4; this value is attained when the absorption and emission dipoles are colinear. In fact, from Fig. 1c it is seen that the initial value of r for Alq$_3$ in DMF is 0.2 when detection is at 505 nm. This value (0.2) indicates that the absorption and emission dipoles in Alq$_3$ are oriented at an angle of $\alpha = 35^0$ with respect to each other. For excitation wavelengths

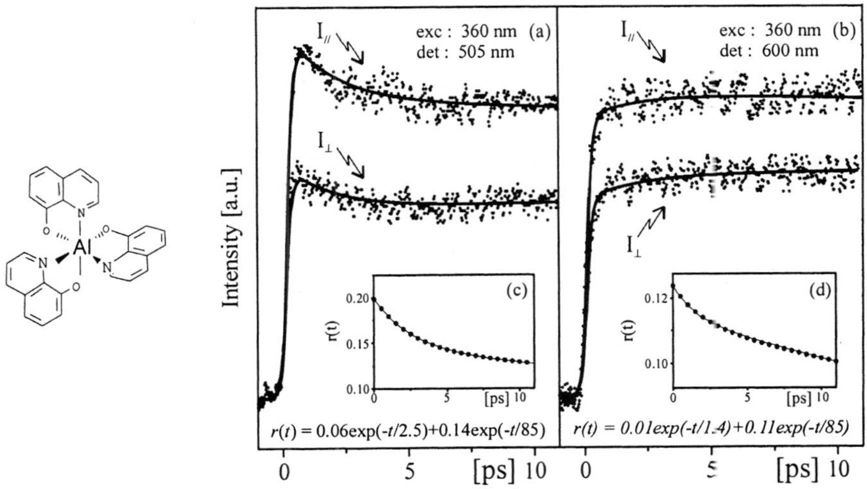

Fig. 1. Polarized fluorescence upconversion transients of Alq$_3$ in DMF detected at 505 nm (a) and 600 nm (b). Corresponding anisotropy decays are given in panels (c) and (d), respectively. Best fit expressions for the anisotropy, $r(t)$, are inserted.

shorter than about 330 nm, however, *no* anisotropy could be resolved within the time window of 10 ps. This absence of fluorescence polarization, when excitation is at higher energies, shows that $r(0)$, i.e., the value of the anisotropy immediately after the pulsed excitation, has dropped to zero within the system response time of about 150 fs. Since Alq_3 has a number of close lying excited ligand-localized electronic states, relaxation of the initially excited upper state(s) to the emissive state involves internal conversion to these states. Apparently during this internal conversion the initial anisotropy is averaged out. It is thus found that *internal conversion from the higher excited electronic levels to the emissive electronic state occurs within 150 fs*. This result is compatible with the ultrafast interligand relaxation previously reported for another metal complex, namely $Ru(bpy)_3^{2+}$ [3]. Measurement of the fluorescence upconversion transients under magic angle conditions, yields transients analogous to those in Figs. 1a and 1b. It has recently been verified that the ps fluorescence components can be attributed to the excited state solvation process [2]. The time constants characteristic of the biexponential decay of the fluorescence anisotropy of Alq_3 in DMF (Figs. 1c and 1d), are 2.0 ps and 85 ps, respectively. It is remarkable that solvation, as manifested by the faster (2.0 ps) component, can effect the fluorescence depolarization. Normally, the electronic state is not changed during solvation and thus the direction of the electronic transition dipole remains unaffected. Very recently, however, we conjectured for Alq_3, that an adiabatic mixing of nearby excited electronic states of different radiative character occurs during the solvation process [2]. It is implicit to this conjecture that the solvation process may affect the polarization of the fluorescence from Alq_3. We consider the 2.0 ps component in the anisotropy decay as *independent support for an intraligand charge separation in photoexcited* Alq_3 *that adiabatically follows the solvation process.*

4. Conclusions

For the metal complexes, Alq_3, Gaq_3 and Inq_3, the excited-state dynamics could be studied by means of the fluorescence upconversion technique. Three excited-state dynamical processes were discerned. The fastest of these concerns internal conversion from higher initially excited electronic states to the fluorescent state, within 150 fs. The second process involves adiabatic charge separation, in the fluorescent state, in a few picoseconds. Its dynamics is controlled by solvation dynamics. Finally, the slowest fluorescence anisotropy decay component is due to the rotational diffusion motions of the solute molecules in the liquid solution.

References

1. A. Curioni, M. Boero, and W. Andreoni, Chem. Phys. Lett. 294 (1998) 263.
2. W. Humbs, E. van Veldhoven, H. Zhang, and M. Glasbeek, Chem.Phys.Lett. 304 (1999) 10.
3. N. H. Damrauer, G. Cerullo, A. Yeh, T.R. Boussie, C. V. Shank, J. K. McCusker, Science 275 (1997) 54.

Investigation of the Ultrafast Dynamics of Charge Recombination of Ion Pairs Using Multiplex Transient Grating Spectroscopy

Eric Vauthey

Institut de chimie-physique de l'Université de Fribourg, Pérolles, CH-1700 Fribourg, Switzerland
E-mail: eric.vauthey@unifr.ch

Abstract. The dynamics of charge recombination of ion pairs formed by electron transfer quenching of excited cyano-anthracenes has been investigated using multiplex transient grating spectroscopy. The directly measured rate constants of charge recombination are considerably larger than those deduced indirectly.

1. Introduction

Over the last decades, the dynamics of charge recombination (CR) within geminate ion pairs (GIP) generated by bimolecular electron transfer (ET) reactions has been intensively investigated. In almost all studies, the rate constant of CR, $k_{CR,}$ was determined indirectly [1]. However, it has been shown that this procedure does not give reliable rate constants [2-4].

We report here on direct measurements of CR dynamics of GIPs and on its free energy dependence using the multiplex transient grating (TG) spectroscopy [5]. The GIPs were generated by ET quenching of excited electron acceptors (cyano-anthracene derivatives) by electron donors (methoxy-substituted benzenes (MSB) and aromatic amines (AA)) at high concentration (1 M) in ACN.

2. Experimental procedure

Excitation was carried out with two time-coincident pulses at 400 nm crossing the sample with an angle of 1.2°. These pulses were produced by frequency doubling the output of an amplified Ti:S system. The resulting gratings were probed with white light pulses produced by self phase modulation. The ensuing TG spectra contain contributions from absorbance, ΔA, and refractive index changes, Δn:

$$I_{TG} \propto \Delta A^2 + \left(\Delta n_r + \Delta n_{nr}\right)^2 \qquad (1)$$

where Δn_r is the resonant contribution to Δn, connected to ΔA through the Kramers-Kronig relationship, and Δn_{nr} is a non-resonant contribution, which arises from the optical Kerr effect (OKE) in the sample solution. Figure 1A shows the time profile of the square root of the TG intensity at 712 nm measured with a

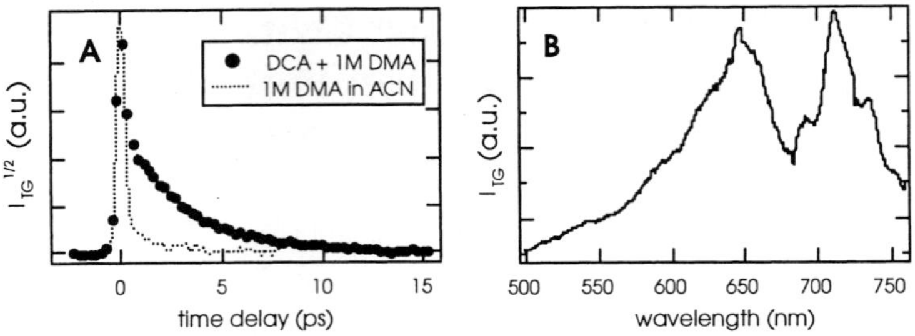

Fig. 1. A) Time profile of the square root of the TG intensity measured at 712 nm with a solution of DCA/DMA and non resonant response of the sample measured with DMA in ACN. B) TG spectrum measured with DCA/DMA, 1 ps after excitation

solution of N,N-dimethylaniline (DMA) in ACN. The initial spike is due to the electronic OKE, whose contribution vanishes after about 300 fs, while the small remaining tail arises from nuclear OKE. Figure 1A also shows the TG intensity measured after excitation of 9,10-dicyanoanthracene (DCA) with DMA in ACN. After time delays longer than the pulse duration, the non-resonant contribution to the signal is negligible compared to the resonant one. Figure 1B, shows the TG spectrum measured 1 ps after excitation of the same sample. All the TG bands can be assigned to $DCA^{\cdot-}$, formed upon ET quenching of $^1DCA^*$ by DMA. Thus, the time profile shown in Figure 1A corresponds to the decay $DCA^{\cdot-}$ and reflects the CR dynamics of the GIP.

3. Results and discussions

The rate constant of BET was measured with more than 40 different acceptor/donor pairs, allowing the free energy of CR, ΔG_{CR}, to be tuned from -3 eV to -0.6 eV. The measured k_{CR} were found to vary over four orders of magnitude, the fastest CR taking place in less than 1 ps.

A clear correlation between k_{CR} and ΔG_{CR} can only be obtained if the GIPs are sorted according to the electron donor. Figure 2 shows a plot of k_{CR} vs. ΔG_{CR} for

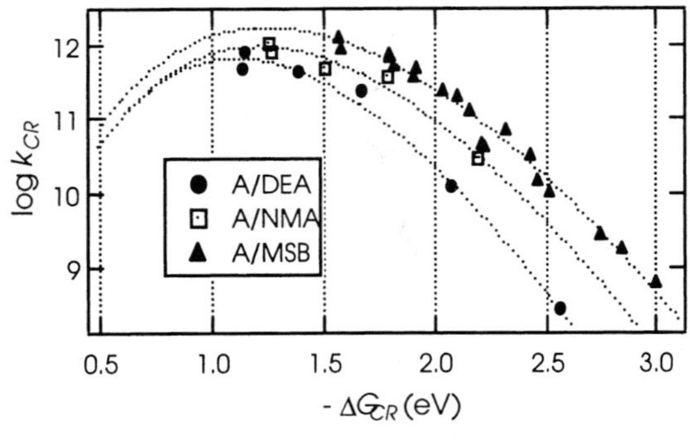

Fig. 2. Free energy dependence of k_{CR} within GIPs containing DEA, NMA and MSB and best fits of the semi-classical expression of non adiabatic ET

GIPs containing different donors. The dotted curves are the best fits of the semi-classical expression for non-adiabatic ET [6]. CR processes take place in the inverted and the barrierless free energy regions. The best fit parameters indicate that the solvent reorganization energy, λ_s, is almost constant. For AAs, both the electron coupling matrix element V, and the vibrational reorganization energy, λ_v, decrease with increasing substitution on the N atom. These two effects can be rationalized by considering the geometry change of the donor upon ionization and the delocalization of the hole with increasing the size of the substituents. On the other hand, increasing the substitution of MSBs is predicted to lead to a decrease of V and to an increase of λ_v, the effect of these changes on k_{CR} canceling each other. Therefore, the single free energy dependence observed with pairs containing MSB is fortuitous.

The normal region is not clearly observed even with pairs containing the strong acceptor TMPD (N,N,N',N'-tetramethyl-1,4-phenylenediamine). This effect might be due to the fact that GIPs containing TMPD are formed in a highly exergonic ET. Figure 3 shows intensity normalized TG spectra measured with the pair tetracyanoanthracene(TCA)/TMPD. The free energy of formation of this GIP amounts to -2.1 eV. The spectral changes of the $TMPD^+$ band observed with increasing time delays can be assigned to vibrational cooling. For this GIP, CR takes place in less than 2 ps, i.e. while the vibrational temperature of the GIP is still much higher than the equilibrium temperature. Consequently the measured CR rate constant might be substantially larger than the rate constant at 300 K.

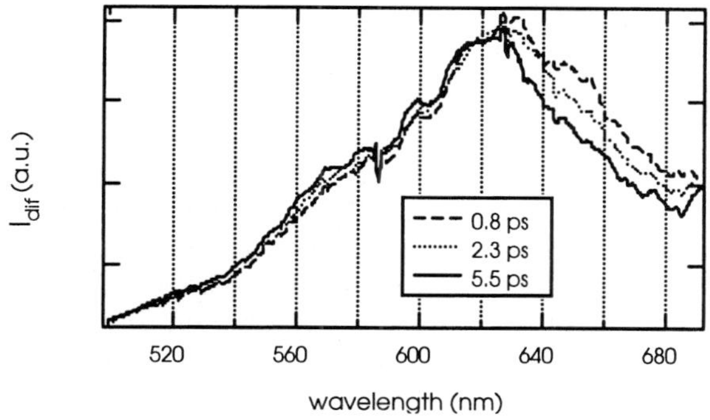

Fig. 3. Intensity normalized spectra measured with the pair TCA/TMPD at various time delays after excitation

References

1 I. R. Gould, D. Ege, J. E. Moser and S. Farid, J. Am. Chem. Soc. **112**, 4290, 1990.
2 N. Mataga, T. Asahi, Y. Kanda, T. Okada and T. Kakitani, Chem. Phys. **127**, 249, 1988.
3 E. Vauthey, C. Högemann and X. Allonas, J. Phys. Chem. A, **102**, 7362, 1998.
4 E. Vauthey, J. Phys. Chem. A, **104**, 1804, 2000. **16**, 409, 1978.
5 C. Högemann and E. Vauthey, Rev. Sci. Instrum. **67**, 3449, 1996.
6 R. A. Marcus and N. Sutin, Biochim. Biophys. Acta, **811**, 265, 1985.

Ultrafast adiabatic elctron transfer cooperated with exciplex formation induced by strong interaction between donor and acceptor at short distances

S. Iwai, S. Murata, and M. Tachiya

National Institute of Materials and Chemical Research,1-1 Higashi, Tsukuba, JAPAN
E-mail: iwai@nimc.go.jp

Abstract. A relation between free energy change vs. electron transfer (ET) rate between donor and acceptor in contact was investigated by time-resolved fluorescence study. Non-Marcus type ET was found to occur on the adiabatic barrierless potential surface which can be realized by the strong interaction between donor and acceptor molecules.

1. Introduction

In recent 10 years, several new aspects of intermolecular electron transfer (ET) in organic molecular systems have been revealed by femtosecond time-resolved spectroscopy. These results have been explained by the dynamic solvent effects cooperating with the molecular vibrations [1]. Basically, the results of these previous studies have been explained on the basis of the non-adiabatic ET theory by Marcus [2] which assumes weak interaction between the donor (D) and acceptor (A). On the other hand, when D and A are close enough to each other, they can interact more strongly. ET in strongly interacting systems is of particular interest, since such ET is considered to occur in molecular complexes and aggregates including CT complexes and exciplexes. They are expected to be much faster than the predictions by the theory of weak interaction. In this study, we have investigated ET between D and A at short distances by measuring the fluorescence from A molecules (used as fluorescer) dissolved in solvents containing high concentration of D (used as quencher).

2. Experimental

9-Cyanoanthracene (CA), 9,10-dicyanoanthracene (DCA) and 1,2,9,10-tetracyanoanthracene (TCA) were used as acceptors and *N,N*-dimethylaniline (DMA) and aniline (ANL) as donors. The concentration of the acceptors was 2×10^{-3} M in 0.2 mm cell. The donor concentration was changed between 0.15M to 8M(neat). Acetonitorile (ACN) was used as solvent. Hereafter, for simplicity an ACN solution containing CA and DMA will be represented as "CA-DMA in ACN". Neat solution will be represented as "CA-DMA". Measurement was made by fluorescence up-conversion with the time resolution of 70 fs.

3. Results and Discussion

Fig. 1 shows the decay curves of DCA fluorescence quenched by DMA in ACN solution at several quencher concentrations (dots). The decays calculated using the diffusion equation and the Marcus equation are shown by solid curves. In the calculation, the parameters determined by the experimental results in the ns [3] and 10-100 ps [4] time regions were used. At concentrations higher than 0.6 M, an additional decay component appears which cannot be reproduced by the analysis based on Marcus equation. This component has a very fast decay of about 200 fs. The fast decay component becomes more evident at higher concentrations. In the donor (DMA and ANL) solvents, only this fast component remains as shown by open circles in Fig. 1(b). Its lifetime is 210 fs for CA-DMA. The spectra of this short-lived fluorescence observed in 450-500 nm were reconstructed from the decay curves measured at various wavelengths. It was found that the spectra are close to those of the corresponding acceptors. So, the observed decays are considered to be due to the acceptor fluorescence. Another fluorescence with a maximum at 620nm is observed for CA-DMA. This is due to the exciplex formed between CA and DMA. The rise time of the exciplex fluorescence observed at 550 nm is equal to the decay time of the above-mentioned short-lived fluorescence. The coincidence of the decay of short-lived fluorescence with the rise shows that the exciplex is formed as a result of very fast ET.

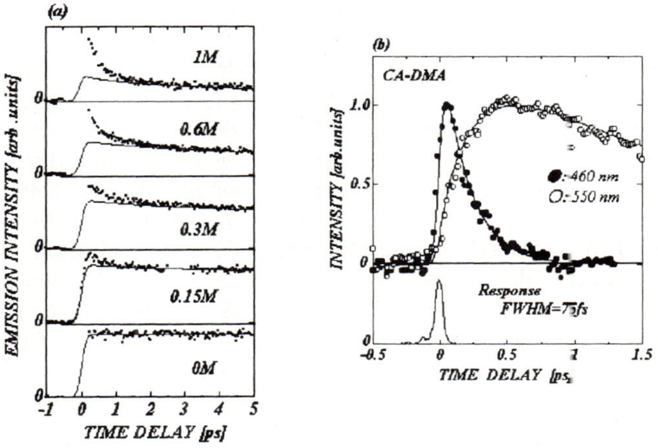

Fig.1 (a) Fluorescence decay curves of DCA quenched by DMA in ACN solution in ps time region. The fluorescences were observed at 457 nm. (b) Fluorescence decay of CA dissolved in donor (DMA) solvent. Closed and open cirrcles show the results observed at 460 nm and at 550 nm, respectively

In the following we consider how fast this rate can be. The reciprocal of the lifetime of the short-lived fluorescence gives the rate constant of formation of the exciplex or the charge separation rate constant k_{CS} from the excited acceptor. Fig. 2 (a) shows k_{CS} vs. ΔE obtained in the present study together with the data reported by Mataga et al. [5] for the charge recombination rate in charge transfer complexes. ΔE is the energy difference between the excited state of the free acceptor and the exciplex state estimated from the absorption and fluorescence

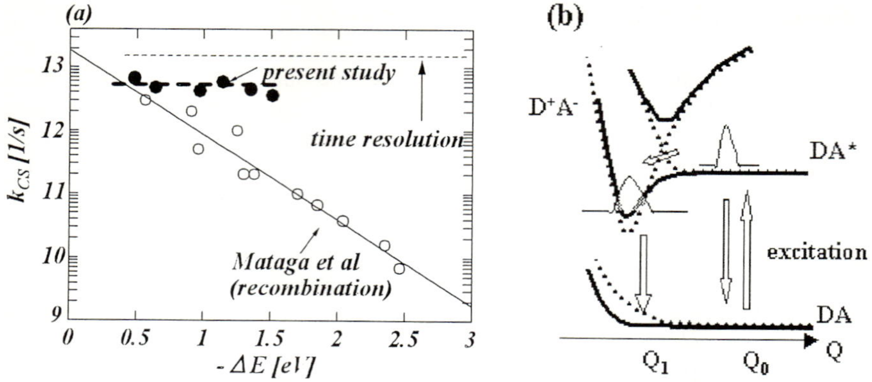

Fig. 2 (a) Charge separation rate constant k_{CS} (closed circles) found in this study and charge recombination rate (open circles) found by Mataga et al. [6] . (b) Potential energy diagram as a function of intermolecular relative coordinate including distance and angle (Q) illustarting the charge separation.

spectra and redox potentials. The result of Mataga et al. shows an exponential dependence of k_{CS} on ΔE and has been explained using a theory of intramolecular radiationless transitions. The result obtained in the present study is quite different from theirs: k_{CS} is almost independent of ΔE.

Fig. 2(b) shows the potential energy curves as a function of intermolecular distance. In Fig. 2(b), we take into account the level mixing of the zeroth-order states of DA* and D^+A^-(indicated by dotted lines) at shorter distances. As a result of the mixing, the potential energy curves change to those indicated by solid lines. At Q_0, D and A are in contact but do not form a CT complex. Excitation of A at Q_0 produces the excited state of free A. D and A* at Q_0, however, strongly interact and are converted to the exciplex whose equilibrium distance is Q_1. The rate of conversion can be very fast, since the potential barrier between Q_0 and Q_1 would disappear as a result of the strong interaction at these distances. The rate would be of the order of the intermolecular stretching vibration. If the potential barrier disappears, the rate of conversion will not depend on the energy gap ΔE between the excited state of the acceptor and the exciplex state. This is what we found in Fig. 2(a). The results presented here show that the strong interaction between the donor and acceptor at short distances can cause an ultrafast charge separation and that the rate of charge separation does not depend on the energy gap involved in the charge separation process.

References

1 Y. Nagasawa, A. P. Yarsev, K. Tominaga, A. E. Johnson, K. Yoshihara J. Chem. Phys. **101**, 5717-5726 (1995).
2 H. Sumi, R. A. Marcus, J. Chem. Phys. **84**, 4894-4914 (1986).
3 S. Murata and M. Tachiya, J. Phys. Chem. **100**, 4046-4070 (1996)
4 S. Iwai, S. Murata, and M. Tachiya, J. Chem. Phys. **109**, 5963-5970 (1998).
5 T. Asahi and N. Mataga, J. Phys. Chem.**95**, 1956-1963 (1991).

Femtosecond multicolor pump-probe investigation of ultrafast electron transfer of $(NH_3)_5Ru^{III}NCRu^{II}(CN)_5^-$ in solution

Dong Hee Son, Patanjali Kambhampati, Tak W. Kee and Paul F. Barbara

Department of Chemistry and Biochemistry, The University of Texas, Austin TX 78712, USA
E-mail: p.barbara@mail.utexas.edu

Abstract. Ultrafast measurements on $(NH_3)_5Ru^{III}NCRu^{II}(CN)_5^-$ in solution reveal an ultrafast spectral component due to vibrationally unrelaxed stimulated emission, allowing for new insights on the interplay of solvent and solute degrees of freedom in an electron transfer reaction.

1. Introduction

Ultrafast pump-probe experiments on the mixed valence compound ($(NH_3)_5RuNCRu(CN)_5^-$; RuRu) has provided insight on the role of solute and solvent degrees of freedom in electron transfer reactions [1-3].

$$(NH_3)_5Ru^{III}NCRu^{II}(CN)_5^- \underset{ET}{\overset{h\nu}{\rightleftarrows}} (NH_3)_5Ru^{II}NCRu^{III}(CN)_5^-$$

In detail, it is known that the timescale for reverse electron transfer (τ_{ET}) is much shorter than the slowest components of solvation dynamics. It is also known τ_{ET} is comparable to the vibrational relaxation times of key vibrational modes in the ground state, as evidenced by red-shifted "hot" features in the ultrafast spectroscopy. In this paper we report for the first time information on the timescale for relaxation of vibrational degrees of freedom of RuRu in the excited state.

2. Experimental Methods

The experiments were performed with a Ti:sapphire based multipass amplifier system producing 300 μJ, transform limited 35 fs pulses centered at 800 nm at 1 kHz. Sample solutions in water, ethylene glycol or glycerol were pumped by 10 μJ pulses from either the fundamental output from the amplifier or by tunable output from an optical parametric amplifier generating 25 fs pulses at 615 nm and probed with white light continuum generated in a sapphire crystal producing cross correlations of 40 fs (FWHM).

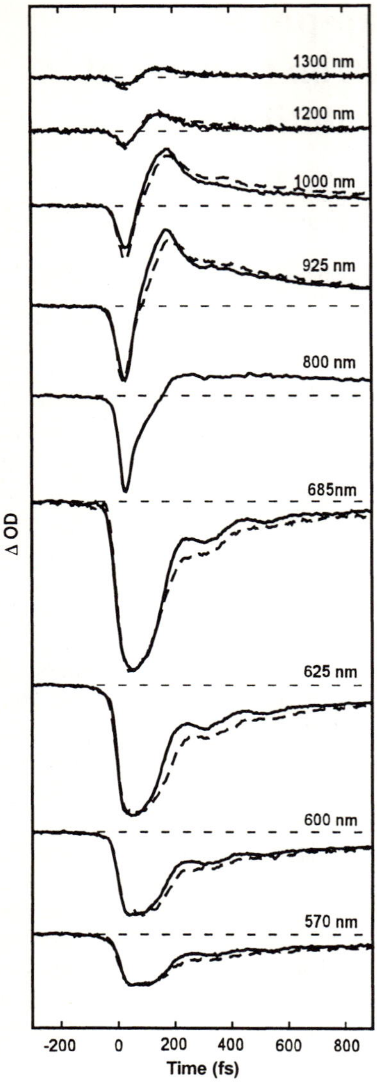

Fig. 1. Pump-probe data of RuRu in H_2O (solid line) and D_2O (dashed line) obtained with 800 nm pump and variable wavelength probe. Probe wavelengths are indicated next to each pump-probe data.

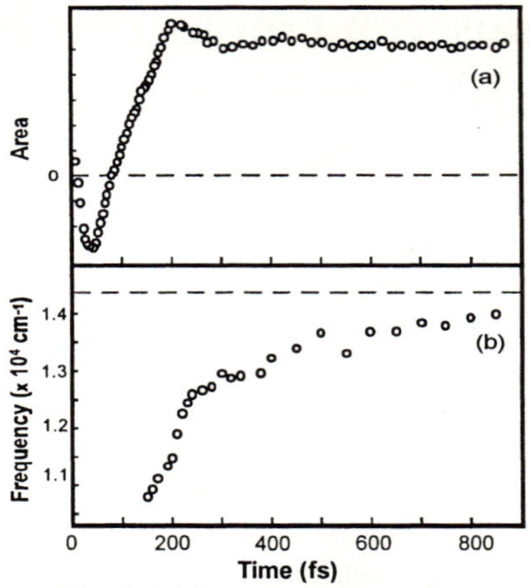

Fig. 2. (a) Integrated absorption strength of RuRu in H_2O as a function of time. (b) Average transition frequency of absorption recovery.

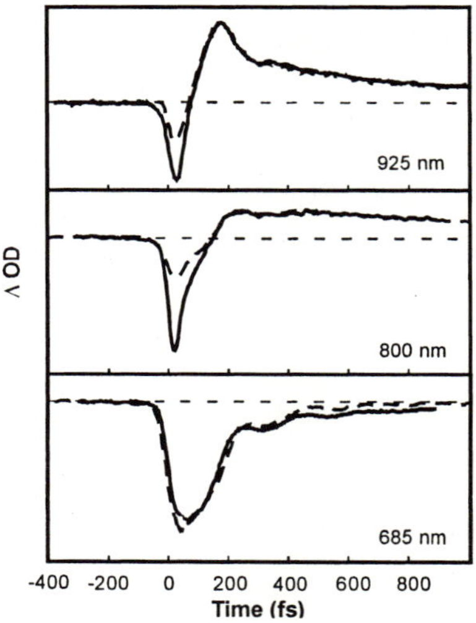

Fig. 3. Comparison of the pump-probe data of RuRu in H_2O obtained with 800 nm (solid line) and 615 nm pump(dashed line).

3. Results and Discussion

The femtosecond pump-probe spectroscopy on RuRu in aqueous solution with variable pump and probe wavelengths is summarized in Figs 1-3 The data reveal (Fig. 1): (i) a pulse-limited bleach due to photoinduced ET; (ii) a rapid recovery on the 80-100 fs timescale due to the reverse ET (ground state recovery) and (iii) various features, especially on the red and blue edge due to the disposal of energy into the solvent and vibrational degrees of freedom from the fast ET process and subsequent equilibration of these modes. The dynamic absorption spectrum constructed from the pump-probe data enables measurement of the ET time (τ_{ET}) and characterization of the role of solvent motions during and after the ET process. The integrated absorption strength as a function of time of the absorption recovery (Fig. 2a) allows for an estimate of τ_{ET} of 80-100 fs in H_2O, which shows an isotope effect upon deuteration of the solvent, indicating the participation of librational motion of the solvent. In alcohols, τ_{ET} is ~220 fs and does not show solvent dependence, which indicates that the intramolecular vibrational modes are playing a major role in promoting the ET. The evolution of the average transition frequency of the absorption recovery (Fig. 2b) occurs over a range of timescales reflecting the different relaxation times of the various vibrational and solvent modes that are coupled to the ET process. Pump-probe experiments (Fig. 3) using a variable wavelength pump pulse reveal the presence of contributions from stimulated emission in the pump-probe data. The pump wavelength dependent feature near zero time is assigned to rapidly evolving, stimulated emission from coherently prepared non-equilibrium populations of excited vibronic states. The observation that the stimulated emission does not equilibrate on the excited state (ET) timescale implies that the ET timescale (80-100 fs) is much shorter than the timescale for vibrational wavepacket motion, intramolecular vibrational energy redistribution, and vibrational relaxation of the most strongly coupled modes.

4. Conclusions

Electron transfer of RuRu in solution is promoted by intramolecular vibrational modes and inertial solvation, and occurs before the equilibration of the coupled nuclear degrees of freedom. The excess energy from ET is deposited in both solvent and vibrational coordinates, which subsequently undergo relaxation.

Acknowledgements. This work was supported by grants from the DOE-BES and the Robert A. Welch Foundation.

References

1 P. J. Reid, C. Silva, P. F. Barbara, L. Karki and J. T. Hupp, J. Phys. Chem., 99, 2609, 1995.

2 G. C. Walker, E. Åkesson, A. E. Johnson, N. E. Levinger and P. F. Barbara, J. Phys. Chem. 96, 3728, 1992.

3 C. Wang, B. K. Mohney, B. B. Akhremitchev and G. C. Walker, J. Phys. Chem. A, 104, 4314, 1999.

Ultrafast Solvent-Induced Charge Localization in Tris-(2,2'-Bipyridine) Ruthenium(II)

Alvin T. Yeh[a,b], Charles V. Shank[a,b] and James K. McCusker[a]

[a]Department of Chemistry, University of California, Berkeley, CA 94720
[b]Materials Sciences Division, Lawrence Berkeley National Laboratory, Berkeley, CA 94720.

Abstract. Two distinct excited state processes are resolved in tris-(2,2'-bipyridine) ruthenium(II). Time resolved anisotropy measurements reveal a solvent dependent depolarization indicative of charge localization from an initially delocalized Franck-Condon state. Transient absorption measurements resolve wavepacket motion on the excited state surface(s) independent of solvent.

Tris-(2,2'-bipyridine) ruthenium(II) or $[Ru(bpy)_3]^{2+}$ is the prototype to study the photophysical properties of transition metal charge transfer complexes[1]. Previous work from this lab focused on time-resolving the evolution of the Franck-Condon state of this molecule to its low energy ^{3}MLCT excited state using transient absorption pump-probe measurements[2]. The excited state absorption spectrum for $\Delta t \geq 300$ fs matched spectra acquired on the nanosecond time scale. This was an indication that the formation of the ^{3}MLCT state occurs with a time constant of $\tau \sim 100$ fs. Although this study helped to define the overall time scale of excited-state evolution in this system, it left unanswered a number of important questions concerning localization/delocalization of the excited state as well as the role of solvent in the formation of the thermalized charge-transfer state. These issues are addressed by examining the time dependence of the anisotropic absorption of the molecule[3].

Time dependent anisotropy data for $[Ru(bpy)_3]^{2+}$ in a nitrile solvent series ($CH_3(CH_2)_nCN$, n=0,1,2) are shown in Figure 1. These measurements were performed at 480 nm, which is at the red edge of the ^{1}MLCT absorption and corresponds to a (x,y)-polarized transition from the D_3 symmetry ground state. The initial anisotropy in all three solvents is $r \sim 0.55$ and depolarizes to a value of $r \sim 0.4$. The direct interpretation of the time-dependent anisotropy values at 480 nm is complicated by contributions from excited state absorption[4]. The total observed anisotropy with possible contributions from excited state absorption and stimulated emission can be calculated as a weighted sum of the anisotropies of the contributing transition dipole moments[5]

$$r = [\alpha/(\alpha+\beta+\gamma)]r_\alpha + [\beta/(\alpha+\beta+\gamma)]r_\beta + [\gamma/(\alpha+\beta+\gamma)]r_\gamma, \qquad (1)$$

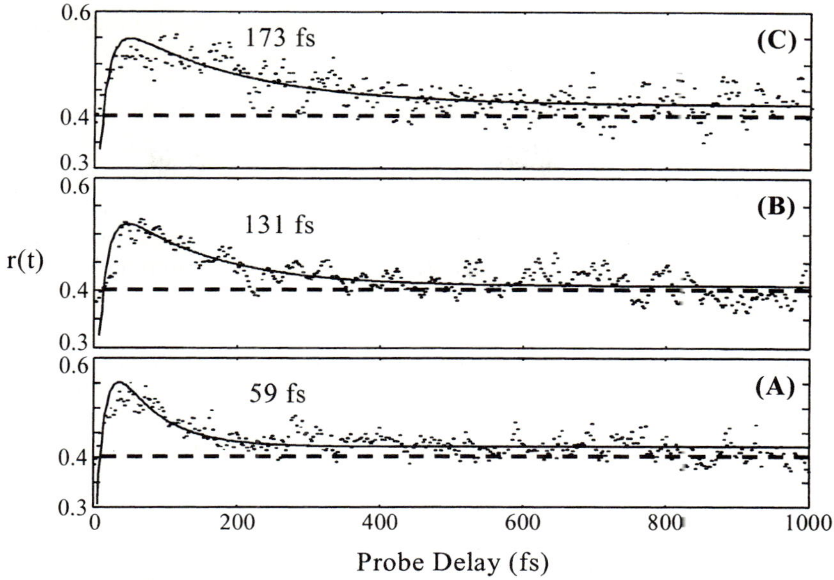

Figure 1 Solvent dependence of time-resolved anisotropic absorption measurements of $[Ru(bpy)_3]^{2+}$ in CH_3CN (A), CH_3CH_2CN (B) and $CH_3CH_2CH_2CN$ (C).

where r_i are the anisotropies and α, β and γ are weighting factors for ground state bleach, excited state absorption and stimulated emission, respectively. A localized model, assuming random selection of one of the three bpy ligands, would be expected to yield a total anisotropy of r = 0.45[3]. This is consistent with the long time values seen in Figure 1. The observed time dependence of r(t) is indicative of a change in symmetry of the molecule subsequent to excitation. The data suggest an initial observation of a partially dephased two-fold degenerate transition dipole moment of a delocalized electronic state evolving to a non-degenerate, charge localized state.

As seen in Figure 1, the rate of charge localization is observed to be dependent upon solvent. The nitrile solvents are polar and prior to excitation will solvate the dicationic $[Ru(bpy)_3]^{2+}$ in such a way as to screen charge. Thus, the negative end of the solvent dipole will be oriented toward the solute in the ground state. This electrostatic condition is changed dramatically upon photoexcitation with the introduction of additional negative charge density on the periphery of the chromophore. The initial response to this new charge distribution will most likely be dictated by non-diffusive inertial motion of the solvent. The solvent dependence of the time for charge localization scales with the inertial moment of the solvent molecules. The driving force for charge localization is postulated to be the minimization of the number of unfavorable solvent-solute interactions in the excited state.

Magic angle transient absorption measurements were made to assess the correlation between population dynamics and the time scale for charge localization[3]. No difference in the rate of spectral evolution was observed between $[Ru(bpy)_3]^{2+}$ in CH_3CN and $CH_3CH_2CH_2CN$. This suggests that the wavepacket motion associated with population dynamics is largely governed by intramolecular factors and occurs relatively independent of environmental forces.

Acknowledgements. This research was supported by the Director, Office of Science, Office of Basic Energy Sciences of the U.S. Department of Energy Contract No. DE-AC03-76F00098 (C.V.S.) and Grant No. DE-FG03-96ER14665 (J.K.M.).

References

1. V. Balzani, A. Juris, M. Venturi, S. Campagna, S. Serroni, *Chem. Rev.* **96**, 759 (1996).
2. N. H. Damrauer, G. Cerullo, A. Yeh, T.R. Boussie, C.V. Shank, J.K. McCusker, *Science* **275**, 54 (1997).
3. For a more detailed account see: A. T. Yeh, C. V. Shank, J. K. McCusker, *Science* (in press).
4. N. Sutin, C. Creutz, *Adv. Chem. Ser.* **168**, 1 (1978).
5. D. M. Jonas, M. J. Lang, Y. Nagasawa, T. Joo, G. R. Fleming, *J. Phys. Chem.* **100**, 12660 (1996).

Part XI

Two-Dimensional Spectroscopy

Structural Dynamics of Small Peptides by Time Dependent 2D-IR Spectroscopy

Sander Woutersen, Peter Hamm

Max-Born-Institut für nichtlinear Optik und Kurzzeitphysik, D-12489 Berlin, Germany
E-mail: hamm@mbi-berlin.de

Abstract. Using nonlinear 2D-IR spectroscopy on the amide I states of trialanine, its backbone structure is uniquely determined. Time-dependent 2D-IR spectroscopy furthermore allows to observe ultrafast structural motion which gives rise to exciton spectral diffusion and inter-exciton cross-relaxation (exciton migration).

1. Introduction

Protein and peptides are highly flexible and it is known from numerous examples that the dynamics of protein structures is as crucial for their functionality as the structures themselves. Our knowledge of protein dynamics, however, is mostly based on computer simulations, with little or rather indirect feedback from experimentally obtainable observables. Solvation of optical chromophors embedded into a protein, measured by dynamic hole burning, dynamic Stokes shift or stimulated photon echoes, is predominantly driven by long ranged Coulomb interaction between the fluctuating surrounding (i.e. the protein) and the test chromphore. While this approach, which averages over all details of the interaction, appears to be appropriate to describe liquid dynamics, it fails to account for the heterogeneous complexity of proteins. Here, we describe our effort in developing novel spectroscopic tools to study fast protein motion in a potentially site-selective manner.

2. Static 2D-IR Spectroscopy

Nonlinear 2D-IR spectroscopy [1-3] is the infrared spectroscopic analogue of 2D-NMR spectroscopy (COSY), where the nonlinear response of the sample is recorded as a function of probe frequency and as a function of the center frequency of a tunable narrow band (ca. 10 cm^{-1}) pump pulse. The various amide I modes of a peptide (which involves mostly the C=O stretching motion of the peptide unit) are coupled by through-space electrostatic Coulomb interaction [2,3]. As a consequence, cross peaks arise in the off-diagonal region of the 2D-IR spectra, the strength of which is given by the secondary structure of the peptide which determines the relative positions and orientations of the peptide groups. As a first step, we investigate to what extent the structure of a peptide can be derived from its 2D vibrational spectrum. We have chosen the simplest possible peptide

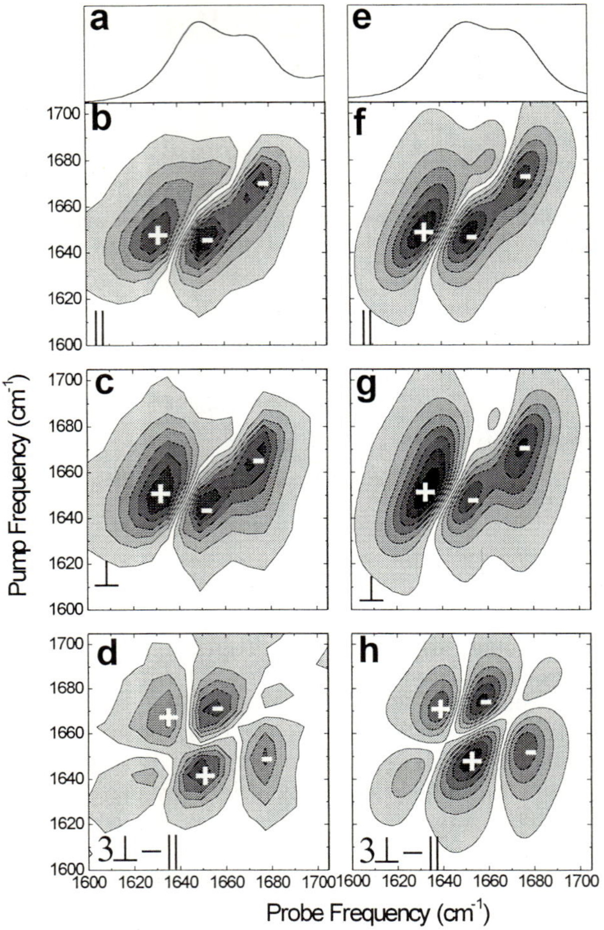

Fig. 1. Linear absorption spectrum (a) and 2D-IR spectrum of trialanine in D_2O with parallel (b) and perpendicular (c) polarization of pump and probe pulse. (d) Scaled difference, revealing cross peaks. Negative (bleach and stimulated emission) and positive bands (excited state absorption) are indicated. (e-h) Fit of the signal, rendering for the structure-related parameters $\beta = 6$ cm^{-1} and $\theta = 106°$.

accessible for 2D studies, tri-L-alanine in D_2O (pD=1) [3]. Since trialanine has only two peptide groups, the only structure-related parameters are the coupling strength β between the two amide I oscillators and the angle θ between their transition dipoles, both of which determined by the dihedral angles ϕ and ψ (see Fig. 2a). Vibrational circular dichroism studies have suggested that trialanine adopts a single conformation in aqueous solution [4].

Figs. 1bc show the two-dimensional pump-probe signal at a delay of 1.5 ps for parallel and perpendicularly polarized pump and probe pulses, respectively. In

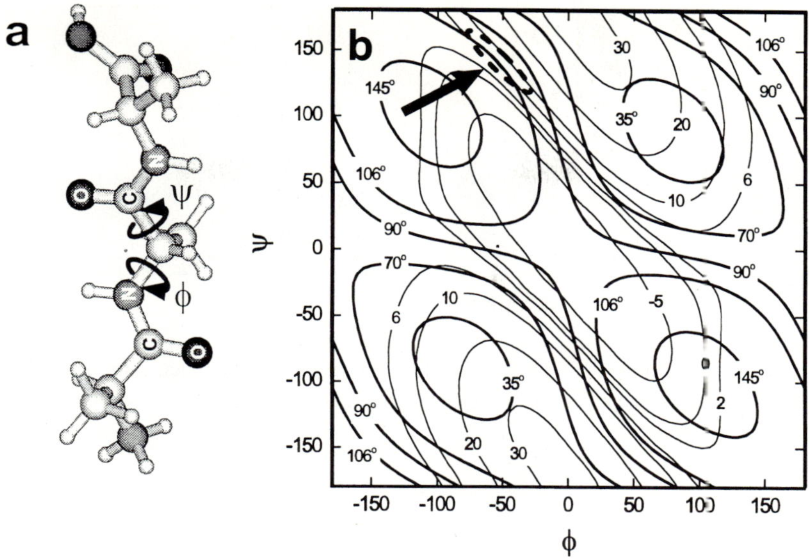

Fig. 2. (a) Structure of trialanine with the dihedral angles ϕ and ψ indicated. (b) Calculated angle θ between the amide I dipole moments (thick lines) and coupling strength β (thin lines) as a function of the dihedral angles ϕ and ψ. The latter was kindly provided to us by the authors of Ref. [5]. The region, which is predominantly populated by trialanine, is marked (arrow).

order to reveal the cross peaks, which overlap with the strong and broad diagonal peaks, we exploit their polarization dependence. The pumped and probed transition dipoles giving rise to the cross-peak signal will in general not be parallel, which will render the anisotropy of the cross peaks different from that of the diagonal peaks (which is ideally 2/5). Hence, by subtracting the parallel from the perpendicular spectra (both scaled to the maximum occurring on their respective diagonals) the diagonal peaks are eliminated, and the cross peaks are clearly revealed (Fig.1d). From a least-squares fit to the experimental results (Fig. 1e-h), applying explicit diagonalization of the one- and two-exciton Hamiltonian and simple models for line broadening mechanisms [3], we obtain values for the coupling strength $\beta = 6\ \mathrm{cm}^{-1}$ between the two amide I modes and the angle $\theta = 106°$ between their transition dipoles. The coupling model is verified on a ^{13}C isotope-edited sample, which leaves the backbone structure, and hence the coupling parameters β and θ, unchanged. The cross peak intensity, however, is now considerably smaller since both amide I states are shifted out of resonance (see Fig 3a, cross peaks marked by arrows). A model calculation, using the same parameters as those obtained from the fit in Fig. 1e-h and merely taking into account the known isotope shift, reproduces the experimental data very well (see Fig. 3b).

The dependence of the angle θ between the transition dipoles on both dihedral angles ϕ and ψ is given by elementary geometry and is depicted in the Ramachandran plot shown in Fig. 2b (thick lines). The coupling strength β, on the

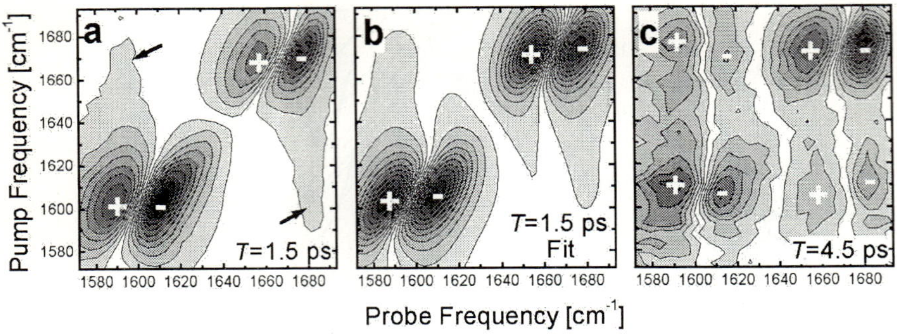

Fig. 3. (a) 2D-IR spectrum of ^{13}C labeled trialanine and (b) a fit of the data, using the same parameters as in Fig. 2 except for the isotope shift. (c) The delay between pump and probe pulse increased to 4.5 ps.

other hand, has been computed using *ab initio* molecular orbital calculations [5] and is also shown in Fig. 2 (thin lines). The structure of trialanine is given by that region in the Ramachandran plot, where the $\beta = 6$ cm^{-1} and the $\theta = 106°$ contour lines intersect. There are two such regions, $\phi,\psi \approx -60°,140°$ (see arrow) and $\phi,\psi \approx 60°,-140°$, situated center-symmetrically in the ϕ,ψ configuration space. Because of sterical hindrance by the residual methyl group of L-alanine, only the former conformation is physically realizable. This conformation coincides with one of the two well-known local minima of the ϕ,ψ free energy surface, which corresponds to secondary structures such as β-sheets and Poly(Gly) II. We conclude that significantly less, if any, molecules populate the second local minimum, which is related to α-helical structures.

3. Time-Dependent 2D-IR Spectroscopy

After having established the equilibrium conformation of trialanine in water, we now address its dynamics. These studies were performed on the ^{13}C labeled sample in order to facilitate distinction of the various phenomena. Fig. 3c shows the 2D-IR spectrum with the delay time between pump and probe pulse increased to 4.5 ps. Besides relaxation into the vibrational ground state, which diminishes the total signal size and which has been divided out in the representation of Fig. 3, two inter-exciton relaxation processes can be identified: (a) Inter-exciton cross relaxation (or exciton migration) enhances the cross peak intensity relative to the diagonal peak intensity. This phenomenon represents an incoherent population transfer between the initially pumped state and dynamically coupled states and gives rise to additional cross-peak intensity superimposed to the coherent contribution described above. (b) Exciton spectral diffusion: The diagonal peaks are initially elongated along the diagonal of the 2D-IR graph, indicating that the amide I states are inhomogeneously broadened (i.e. the hole position follows the pump frequency). As delay time increases, the diagonal peaks broaden and rotate so that they are now orientated parallel to the pump frequency axis, suggesting a homogeneous band (i.e. the hole position does not depednent on pump frequency).

In the early 2D-IR experiment, the pump pulse labels a certain inhomogeneous distribution which is probed instantaneously. As delay time increases, structural motion modifies the backbone configuration, blurring the initial inhomogeneity until it vanishes completely at some large enough time. Since the excitonic coupling depends on the secondary structure of the peptide, the dynamics of both are intimately connected. Fluctuations of the peptide-solvent complex may affect the effective coupling Hamiltonian $H(t)$ in two ways: (a) Since the distance and orientation of the peptide units determine the coupling strength, the fluctuating backbone will translate directly into fluctuating off-diagonal elements of the coupling Hamiltonian. (b) Since the strength of hydrogen bonds changes the C=O vibrational frequencies, the diagonal terms of the Hamiltonian will experience the surrounding solvent. The fluctuating term $H'(t)$ around the time-averaged Hamiltonian $H^{(0)}$ will steadily remix the zero-order (or time-averaged) excitonic states $|v_k\rangle$ [6]:

$$H(t) = H^{(0)} + H'(t) = \sum_k \varepsilon_{kk}^{(0)} |v_k\rangle\langle v_k| + \sum_k q_{kk}(t) |v_k\rangle\langle v_k| + \sum_{k\neq l} q_{kl}(t) |v_k\rangle\langle v_l|$$

The significance of the Hamiltonian for nonlinear spectroscopic experiments has been worked out theoretically in great detail [6]. The fluctuating diagonal terms $q_{kk}(t)$ cause dephasing and exciton spectral diffusion, while the off-diagonal terms $q_{kl}(t)$, $k\neq l$, couple different states and give rise to cross-relaxation (exciton migration). These are the very phenomena that have been identified in the time dependent 2D-IR spectra. The timescale of exciton spectral diffusion (i.e. a few ps) directly represents that of structural motion of the peptide-solvent complex. The exciton migration rates, on the other hand, are determined by the spectral densities of the fluctuating term $q_{kl}(t)$ [6]. Since time dependent 2D-IR spectroscopy in principle allows to determine the complete 2D set of inter-exciton relaxation rates, the inversion of the problem appears to be feasible so that fluctuations of small peptides may be assessed in a site specific way. A dynamic modeling is underway.

Acknowledgements. We gratefully acknowledge H. Torii and M. Tasumi for providing us with the results of their calculations [5]. This work was supported by the Deutsche Forschungsgemeinschaft (DFG).

References

1 P. Hamm, M. Lim and R. M. Hochstrasser, J. Phys. Chem. B **102**, 6123, (1998).
2 P. Hamm, M. Lim, W. F. DeGrado, and R. M. Hochstrasser, Proc. Natl. Acad. Sci. USA **96**, 2036 (1999).
3 S. Woutersen, P. Hamm, J. Phys. Chem B. submitted.
4 A. Kocak, R. Luque, and M. Diem, Biopolymers **46**, 455, (1998).
5 H. Torii and M. Tasumi, J. Raman Spectrosc. **29**, 81, (1998).
6 W. M. Zhang, T. Meier, V. Chernyak and S. Mukamel, J. Chem. Phys. **108**, 7763, (1998).

Frequency Resolved and Heterodyned Femtosecond Infrared Echoes of Peptides; Multiple Pulse Coherent Vibrational Analogues of NMR

Martin T. Zanni[1], Matthew C. Asplund[1], Sean M. Decatur[2], and Robin M. Hochstrasser[1]

[1]Department of Chemistry, University of Pennsylvania, Philadelphia, PA 19104-6323, USA
[2]Department of Chemistry, Mount Holyoke College, S. Hadley, MA 01075, USA

Abstract. Two-dimensional (2D) infrared (IR) spectra of a dipeptide and a ^{13}C substituted 20 unit peptide are obtained from heterodyned photon echoes (PE). Off-diagonal peaks reflect the coupling between amide groups. The two amide groups of acetylproline-NH_2 show different vibrational dynamics.

1. Introduction

Nonlinear IR spectroscopy holds great promise for determining the vibrational dynamics and conformations of proteins and peptides in solution. We recently demonstrated the utility of spectrally resolved PE's to separately monitor the fundamental and anharmonically shifted contributions to the echo signal in HbCO [1], and in a separate paper reported 2D IR spectra from heterodyned fs PE's that exhibit off-diagonal peaks which reflect the coupling between the amide groups in peptides [2]. Here we use these new techniques to separately measure the PE of two acetylproline-NH_2 vibrational bands, and to determine the real part of their PE response. The real part of the 2D IR spectrum enhances the frequency resolution and permits the coupling between these bands to be accurately determined.

Although the two amide bands are well separated in acetylproline-NH_2, this is not usually the case for α-helical and other peptides or small proteins where individual amides have similar environments. We show that ^{13}C isotope substitution can rectify this problem in 2D IR spectroscopy for a 20 amino acid α-helical peptide [3] by shifting selected amide frequencies by ~35 cm^{-1}, allowing their coupling to the unsubstituted amides to be measured.

2. Experimental Methods

Femtosecond IR pulses (120 fs, 1 μJ energy, 1650 cm^{-1}) were generated by mixing the signal and idler beams of an 800 nm pumped OPA in an $AgGaS_2$ crystal. The output was split into three equal beams (wavevectors k_1, k_2, and k_3; each 300 nJ) and a local oscillator beam (k_{LO}; 3 nJ). For the heterodyne experiments, k_{LO} was combined with the signal field in the phase-matched direction $k_s = -k_1 + k_2 + k_3$, and focused into a monochromator. The spectrum was either dispersed onto a 32 element HgCdTe detector or the zero order diffracted beam was collected at an exit slit and measured with a single channel HgCdTe detector. The time interval between k_1 and k_2 is τ, T is the interval between k_2 and k_3, and t_{LO} is the interval between k_3 and the local oscillator.

3. Spectrally Resolved Photon Echo of Acetylproline-NH_2

Shown in Fig. 1a are the spectrally resolved PE's for the 1620 and 1680 cm^{-1} bands in acetylproline-NH_2 at τ=200 fs and T=0 measured as a function of τ and T delay times; their integrated areas are plotted in Fig. 1b with T=0. These curves reflect the vibrational lifetimes and delayed signals resulting from the inhomogeneous distributions of the bands, which clearly differ. The evolution of the areas of the (1590-1625 cm^{-1}) and (1625-1650 cm^{-1}) halves of the 1620 cm^{-1} band, which correspond mainly to v=1→v=2 and v=0→v=1 transitions of the underlying conformations, also differ. The frequency dependence of the dynamics is a result of differing peptide conformations and solvent interactions occurring at the C- and N-termini. The N-terminus amide transition (1680 cm^{-1}) is considerably more inhomogeneous on the ps time scale.

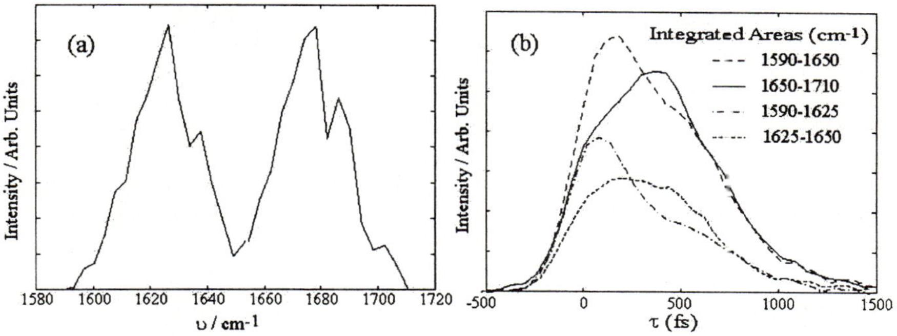

Fig. 1. (a) Spectrally resolved PE, and (b) integrated areas of PE as a function of τ for acetylproline-NH_2.

4. Two-dimensional Spectra of Acetylproline-NH_2

By overlapping the PE with a weak LO oscillator pulse, we measured the echo field as a function of τ. The double Fourier transform with respect to τ and t_{LO} gives the 2D IR spectrum, the real part of which is shown in Fig. 2. Positive peaks below the diagonal, e.g. $(\omega_\tau,\omega_{LO})$=(1620,1630), and negative peaks above the diagonal, e.g. (1630,1620), are observed, corresponding to the two positive and one negative Feynman diagrams for a 3-level vibrational system. The real part of the spectrum is much better resolved than the magnitude spectrum presented previously [2], and off-diagonal peaks are observed at e.g. (1660,1640), (1640,1680), and (1630,1580). The multiple off-diagonal peaks suggest that acetylproline-NH_2 has several conformations in chloroform. The inhomogeneous distribution of frequencies and slow spectral diffusion is clearly evidenced by the elongation of the peaks along the diagonal. This asymmetry is not seen with the more homogeneous transition of N-methylacetamide [2].

5. Two-dimensional Spectrum of a ^{13}C Substituted α-helix

We substituted the first 4 N-terminus alanines in a 20 amino acid, α-helical peptide [3]: Ac-AAAAKAAAAKAAAAKAAAAK-NH_2, for which the 2D IR spectrum at 5° C is plotted in Fig. 3 (magnitude of the echo response). The diagonal peaks at 1630 and 1600 cm^{-1} are due to the ^{12}C and ^{13}C carbonyls. An off-diagonal peak is present at (1620,1600), and a weaker one at (1600,1630), which are due to the coupling between the ^{12}C and ^{13}C carbonyls. The intrinsic asymmetry is caused by the cross-peaks constructively and destructively interfering with the diagonal peaks, as was proven by simulations.

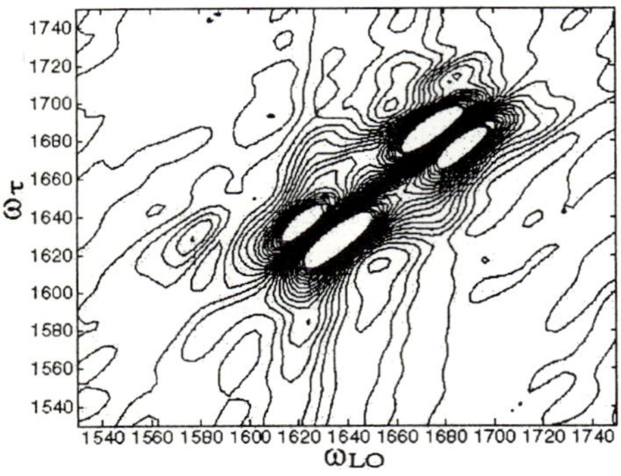

Fig. 2. Real part of the 2D IR response of acetylproline-NH_2 in $CHCl_3$.

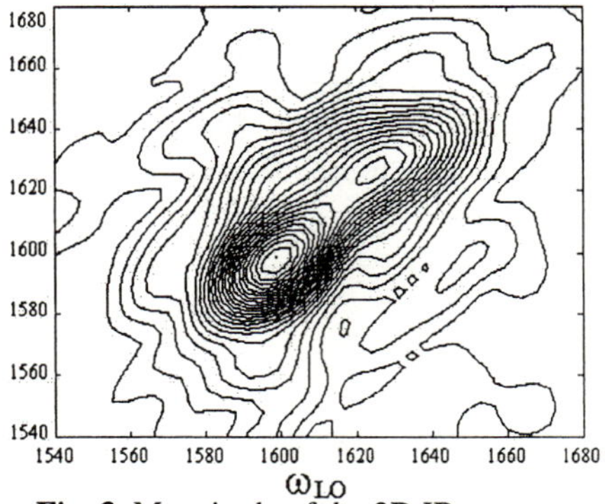

Fig. 3. Magnitude of the 2D IR response of a ^{13}C substituted α-helix in D_2O.

6. Conclusion

The 2D IR method presented above is the direct IR analogue of multiple pulse coherent 2D NMR. The off-diagonal peaks directly measure the couplings between amide groups. Spectrally resolved PE measurements compliment 2D IR by measuring the frequency dependence of the vibrational dynamics. Since the timescales of these experiments are less than a few ps, structural changes slower than this appear static. Each structure in the distribution shows cross peaks centered at different values of (ω_τ,ω_{LO}).

References

1 M. C. Asplund, M. Lim, and R. M. Hochstrasser, Chemical Physics Letters, **323**, 269, 2000.

2 M. C. Asplund, M. T. Zanni, and R. M. Hochstrasser, Proceedings of the National Academy of Sciences, **97**, 8219, 2000.

3 S. M. Decatur, Biopolymers, **54**, 180, 2000.

Nonlinear two–dimensional IR spectroscopy of unfolding processes in β–peptides

Christoph Scheurer, Andrei Piryatinski, and Shaul Mukamel

Department of Chemistry, University of Rochester, Rochester, NY 14627–0216, USA

Abstract. We present simulations of two–dimensional IR Photon Echo spectra for the unfolding process of a helical β–peptide following a temperature jump. A detailed microscopic model of the vibrational Hamiltonian is developed that includes disorder effects and allows to perform realistic simulations of a large number of two– and three–dimensional IR experiments. This should allow the development of new pulse sequences for studying fast protein dynamics.

1. Introduction

Ultrafast nonlinear multidimensional IR spectroscopy [1] can directly probe the time–dependent fast dynamics involved in protein folding. In order to judge the versatility of possible IR multidimensional techniques for following fast structural changes in peptides and to develop new pulse sequences that will reveal more detailed information about the dynamics, it is crucial to construct a realistic microscopic model system of vibrational dynamics during peptide folding. In the present study we have chosen the helical β–hepta–peptide shown in Fig. 1 [2, 3].

2. Model Calculations

Molecular Dynamics Simulation. The peptide was prepared in its solution NMR structure [2] (see Fig. 1**A**) and embedded in a cubic methanol solvent box. The system was equilibrated at 300 K and a 2 ns MD trajectory at 300 K was generated [4]. The folded NMR solution structure is stable under these conditions as can be seen from the time–dependent backbone (residues 2–6) root mean square deviation (RMSD) shown in Fig. 1. After 2 ns, the temperature was raised to 340 K to simulate a temperature jump (T–jump) experiment. A subset of 3300 structures with a timestep of 3 ps, which is longer than the amide–I vibrational correlation time [5], was taken from the trajectory. A clustering procedure based on the backbone RMSD as distance measure yields mainly two ensembles of structures representing the partially unfolded state **B** and the folded state **A**, respectively [4].

Vibrational Exciton Model. The amide–I band of the peptide is modelled as six interacting three–level systems representing the peptide bonds. The exciton Hamiltonian which describes the optical response of the model was given in Ref. [6]. The amide–I frequencies are uniformly chosen as $\Omega_i = 1655\,\mathrm{cm}^{-1}$ and can be red–shifted by hydrogen bonding to neighboring NH groups [5]. The magnitude of the

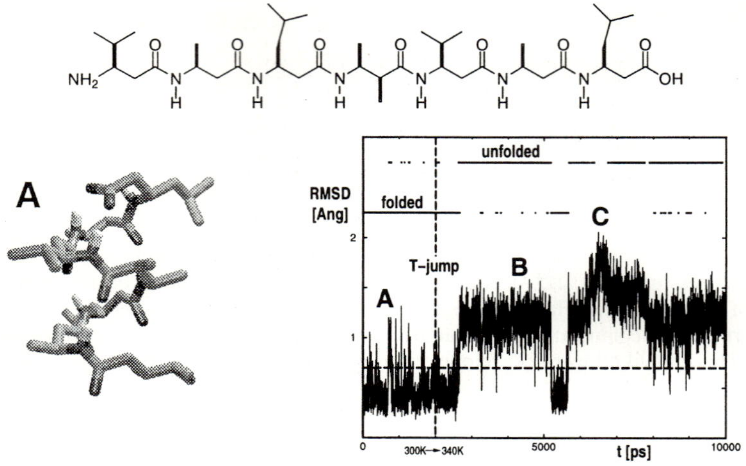

Fig. 1. Time dependent backbone RMSD with respect to the solution NMR structure **A** for the β–hepta–peptide with the given chemical structure (**B** is partially unfolded).

transition dipole moments and the direction and effective location of the individual point transition dipoles with respect to the peptide group were obtained from Ref. [7]. The bath Hamiltonian is treated through relaxation kernels which are described by two homogeneous linewidth parameters Γ and $\gamma^{(2)}$, representing the $0 \rightarrow 1$ and $0 \rightarrow 2$ dephasing rates [1, 8].

NEE calculations of 2 D Photon Echo Spectra. Closed expressions for the third–order nonlinear response function were obtained in Ref. [8] by solving the Nonlinear Exciton Equations (NEE). For each MD structure, the 2 D PE signal $S(\Omega_2, \Omega_1)$ was calculated based on the NEE [6]. The signals were then averaged over the two clusters of MD structures to obtain the inhomogeneously broadened spectra for the folded and unfolded state. Different values of Γ (with $\gamma^{(2)} = 2\Gamma$) were chosen to account for varying homogeneous broadening. The resulting absolute PE signals $|S(\Omega_2, \Omega_1)|$ for the unfolded and folded state are shown in Fig. 2 for $\Gamma = 5\,\text{cm}^{-1}$.

We have found clear signatures for the unfolding of a β–hepta–peptide in 2 D PE spectra. The average couplings and the inhomogeneous broadening, which are derived from the MD geometries, are in agreement with experimental estimates for α–peptides [5, 4]. The inhomogeneously broadened peak around $1640\,\text{cm}^{-1}$ (a, b) is indicative for a hydrogen bonded helical structure, while the appearance of a peak near $1660\,\text{cm}^{-1}$ (c) for the unfolded ensemble is due to the partial unfolding at the N–terminus. As expected for PE spectra, the peaks are inhomogeneously broadened along the diagonal ($\Omega_2 = -\Omega_1$) while the anti–diagonal is only homogeneously broadened. For the current dephasing rate, which yields the experimentally observed broadening, it is only possible to resolve cross peaks (d) for the unfolded spectrum. In the folded spectrum the effect of couplings and anharmonicity can only be seen as a shoulder (a') of the $1640\,\text{cm}^{-1}$ peak in Ω_2–direction.

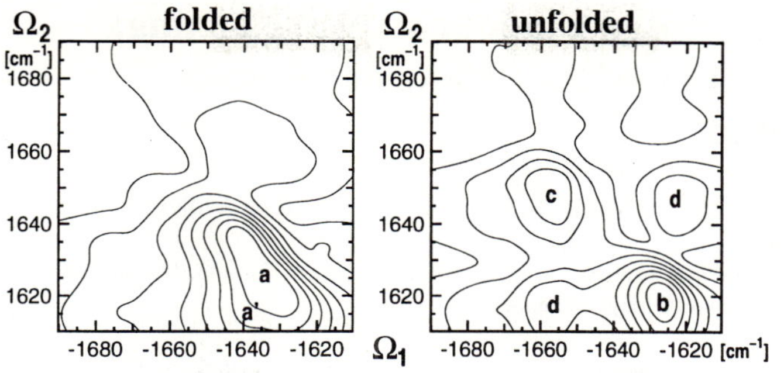

Fig. 2. Absolute value of the total 2 D Photon Echo signal $|S(\Omega_2, \Omega_1)|$ for the folded and unfolded state of the β–heptapeptide shown in Fig. 1 ($\Gamma = 5\,\mathrm{cm}^{-1}$).

Conclusions. The current Hamiltonian neglects contributions to the diagonal disorder caused by local deformations of the peptide plane or by hydrogen bonding to solvent molecules. It is also known that the dipole approximation for the off–diagonal coupling elements is not adequate for a quantitative description of all spectroscopical features. These approximations may be avoided by direct quantum chemical calculations for all parameters of the vibrational exciton Hamiltonian for each individual MD structure which will also include couplings due to nonlocal anharmonicities of the molecular potential energy surface.

The calculation of the full third order response based on our microscopic model will allow us to explore all possible three–pulse experiments including relaxation and exciton transport effects. This should allow to devise a set of 3 D experiments that will yield all accessible information about the third order response which might not be easily extracted from a single 2 D experiment due to limited resolution and line broadening effects.

Acknowledgements. The support of the National Science Foundation is gratefully acknowledged. C.S. thanks the Alexander von Humboldt Foundation for a Feodor Lynen fellowship.

References

1. S. Mukamel, A. Piryatinski, and V. Chernyak, *Acc. Chem. Res.* **32**, 145, 1999. A. Piryatinski, S. Tretiak, V. Chernyak, and S. Mukamel, *J. Raman Spectrosc.* **31**, 125, 2000. S. Mukamel, *Annu. Rev. Phys. Chem.* **51**, 691, 2000.
2. D. Seebach, P. E. Ciceri, M. Overhand, B. Jaun, D. Rigo, L. Oberer, U. Hommel, R. Amstutz and H. Widmer, *Helv. Chim. Acta* **79**, 2043, 1996.
3. X. Daura, K. Gademann, B. Jaun, D. Seebach, W. F. van Gunsteren, and A. E. Mark, *Angew. Chem. Int. Ed.* **38**, 236, 1999.
4. C. Scheurer, A. Piryatinski, and S. Mukamel, in preparation
5. P. Hamm, M. Lim, and R. M. Hochstrasser, *J. Phys. Chem. B*, **102**, 6123, 1998. P. Hamm, M. Lim, W. F. DeGrado, and R. M. Hochstrasser, *J. Phys. Chem. A*, **103**, 10049, 1999.
6. W. M. Zhang, V. Chernyak, and S. Mukamel, *J. Chem. Phys.* **110**, 5011, 1999.
7. H. Torii, and M. Tasumi, *J. Chem. Phys.* **96**, 3379, 1992.
8. V. Chernyak, W. M. Zhang, and S. Mukamel, *J. Chem. Phys.* **109**, 9587, 1998.

Diffractive optics based 2-colour six wave mixing: heterodyne detection of the fifth-order Raman response of liquids

V. Astinov, K.J. Kubarych, C.J. Milne, S. Lin, and R.J. Dwayne Miller*

Departments of Chemistry and Physics, 60 St. George Street, University of Toronto, Toronto, Ont. Canada, M5S-1A7. E-mail: dmiller@lphys.chem.utoronto.ca

Abstract: $\chi^{(5)}$ of CS_2 is observed using phase contrast provided by a diffractive optic approach to generate two distinct phase matching geometries with a reference for heterodyne detection and passive phase stabilization of all six fields.

Fifth-order two-dimensional Raman spectroscopy has been the subject of great experimental and theoretical interest.[1] This interest is largely driven by the unique information this spectroscopy contains with respect to intra- and intermolecular anharmonic mode-coupling terms and nonlinear polarizabilities. Unfortunately the direct fifth order susceptibility ($\chi^{(5)}$) of interest is several orders of magnitude smaller than the third order susceptibility ($\chi^{(3)}$) and is very difficult to isolate from parasitic cascaded processes.[2] Heterodyne detection can significantly amplify the signal to offset some of the problem. Most importantly, the $\chi^{(5)}$ response is 90° phase shifted from the lower order cascaded processes[1,3], making it possible to explicitly use the phase dependence of the signal in relation to a reference field to isolate the different signal contributions and provide a contrast agent for the higher

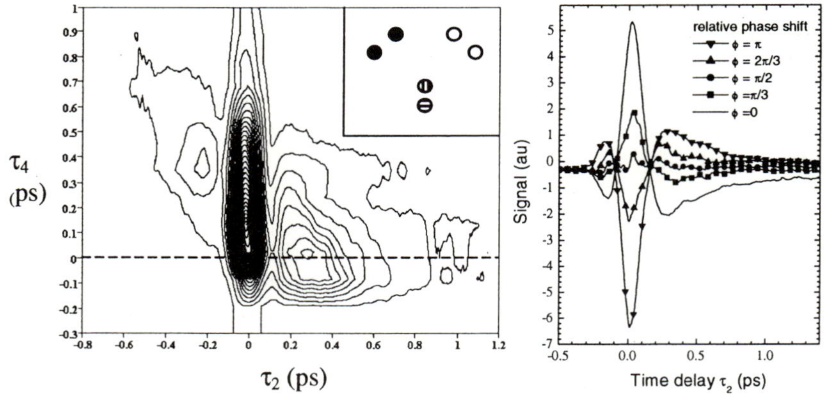

Fig 1. Fifth-order response of liquid CS_2. (left) homodyne-detected two-dimensional plot, where the dashed horizontal line corresponds to the slice in the right panel. (right) Heterodyne-detected slices of the fifth-order signal at different phase shifts between the signal and reference (ϕ). (inset) Phase-matching geometry—solid and empty circles denotes pump pairs, vertical filling denotes probe, and the horizontal filling the signal/reference. (300 μm pathlength)

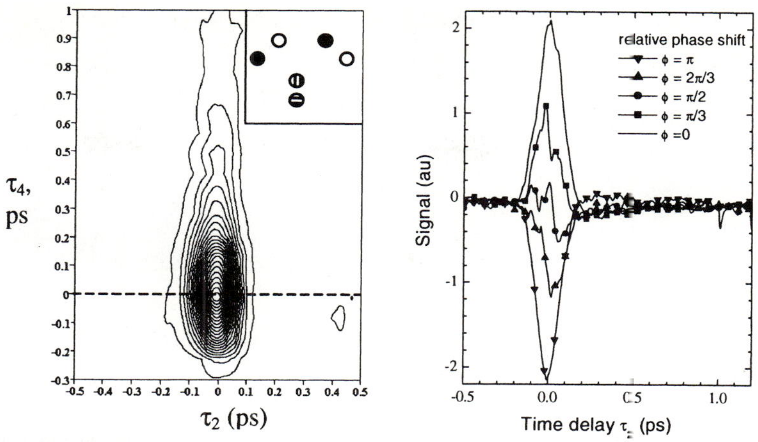

Fig 2. Fifth-order response of liquid CS_2 in the crossed-beam geometry (inset). (left) Homodyne-detected two-dimensional plot; (right) Heterodyne-detected slices of the fifth-order signal. Key: Fig. 1.

order response function. The present work uses a specifically designed diffractive optic (DO) to provide the complex beam geometry for six wave mixing and a reference beam to exploit heterodyne detection for maximum signal contrast. In total, spectral (two color), temporal signatures, wave-vector mismatch, signal phase and path length are explored to provide maximum contrast against the parasitic cascaded $\chi^{(3)}$ processes. The salient experimental details have been published elsewhere.[3] The second-harmonic (400nm) of an amplified Ti:Sapphire laser (70 fs) was used for the pump beams, while the fundamental was used for the probe and reference. Spectroscopic grade liquid CS_2 was flowed though the sample cell with a path length of either 0.1mm or 0.3mm.

Fig. 1 (left) shows the homodyne-detected 2D fifth-order response of CS_2 with the "baseball diamond" phase-matching geometry. The homodyne detected signal is in relatively good agreement with previous reports,[2] albeit with longer path lengths. Since there is far more information contained in a heterodyne-detected signal, the clear advantage of the DO is in its facilitation of heterodyne detection, both geometrically and temporally. Fig. 1 (right) depicts heterodyne fifth-order traces as slices along τ_2 when $\tau_4 = 0$ in the baseball diamond geometry. The polarization states are: all parallel for the excitation pulses; 45° for the probe/reference; and the analyzer was oriented 90° relative to the probe for greater reduction in scatter. Tilting one of the cover slips in the reference beam changes the relative phase between the signal and reference. The distinctive phase components can be seen in the right panel of Fig. 1. The total fifth-order signal at $\tau_2 \approx 0$ changes from its positive value, through minimal signal and ends with negative signal when the phase shift in the reference beam has been changed by 180°. These components cannot be separated by direct detection (homodyne). Fig. 1. clearly shows that the initial component near $\tau_2 = 0$ fs and the nuclear peak at $\tau_2 = 200$ fs are opposite in sign.

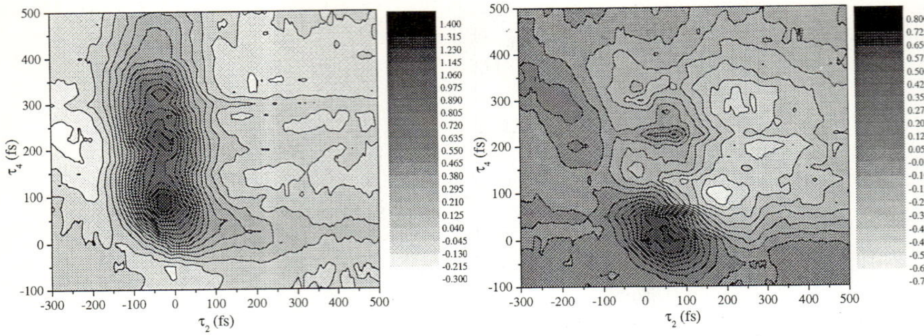

Fig 3. Fifth-order response of liquid CS_2 via the $\chi^{(5)}{}_{121121}$ tensor element in the baseball diamond geometry. (left) In-phase signal. (right) $\pi/2$ phase shifted signal.

To further discriminate against cascaded $\chi^{(3)}$ processes, a modified (crossed-beam) phase-matching geometry was used to reduce the contribution from the parallel processes by increasing the wave-vector mismatch by an order of magnitude. This beam geometry can be readily adopted by simply changing spatial filters such that the effect of phase matching discrimination can be observed under identical conditions (Fig. 2). Due to the increased phase-mismatch for the cascaded third-order processes, the shoulder at $\tau_2 = 200$ fs cannot be seen even in the heterodyne slices. Comparison of Figs. 1 and 2 illustrates that there are two signal components in Fig. 1 with different phases; whereas Fig. 2 is dominated by a component distinct from cascaded $\chi^{(3)}$ processes. The characteristic signature of the cascaded (both parallel and sequential) is a ridge along τ_2. The absence of this contribution with greater discrimination reveals the underlying direct $\chi^{(5)}$ response.

To explore the effect of pathlength in further discriminating against cascades (grow as l^2), Fig. 3 shows 2D heterodyne detected contour plots for the $\chi^{(5)}{}_{121121}$ tensor element using a 100 μm cell. Two relative phase settings are shown; one for maximum signal along τ_4, and 90° from this setting. The nuclear response is more discernable in the latter configuration. Using a synchronized $\tau_2 = \tau_4$ scan, the peak in the nuclear response is observed at 150 fs.

The response shown in Figs. 2 and 3 is assigned to the direct $\chi^{(5)}$ and is in qualitatively in accord with theoretical studies.[4] The absence of any appreciable rephasing indicates that CS_2 lacks significant inhomogenous broadening.

1. S. Mukamel, A. Piryatinski, and V. Chernyak, *Acc. Chem. Res.* **32**, 145 (1999) and references contained therein.
2. D. A. Blank, L. J. Kaufman and G. R. Fleming, *J. Chem. Phys.* **111**, 3105 (1999).
3. V. Astinov, K. J. Kubarych, C. J. Milne, and R. J. D. Miller, *Opt. Lett.* **25**, 853 (2000); *Chem. Phys. Lett.* (in press).
4. Saito and I. Ohmine, *J. Chem. Phys.* **108**, 240 (1997); A. Ma and R. M. Stratt, *Phys. Rev. Lett.* in press (2000).

We acknowledge NSERC and Photonics Research Ontario for funding.

Polarized and magic angle fifth-order 2D electronically non-resonant Raman scattering from CS_2

David A. Blank, Laura J. Kaufman and Graham R. Fleming

Department of Chemistry, University of California, and Physical Biosciences Division, Ernest Orlando Lawrence Berkeley Lab, Berkeley, CA, 94720, USA
E-mail: GRFleming@lbl.gov

Abstract. We have successfully used phase matching considerations to provide significant discrimination against cascaded lower-order responses and measured the direct fifth-order response from the intermolecular motions in liquid CS_2. Here we report the measured polarized and magic angle responses and discuss comparisons with simulations.

The extension of time domain electronically non-resonant Raman spectroscopy to two dimensions has received considerable attention due to its potential to provide microscopic information about the coupling between both inter- and intra-molecular nuclear motions.[1]. Meauring the fifth-order response as a function of the two adjustable time variables allows one to measure the coherent transfer of one vibrational coherence to another. Generation of the signal requires nonlinearity either in the interaction of the system with the field, nonlinearity in the polarizability operator, or in the form of anharmonicity in the vibrational potential. The requirement of at least one of these two forms on nonlinearity in the system provides the potential for the measurement to reveal the coupling between different nuclear motions as well as the potential for the rephasing of nuclear coherences that can deconvolute homogeneous and inhomogeneous contributions to the vibrational lineshape.[1]

Experimental attempts to measure the fifth-order non-resonant response have proven to be particularly challenging. The greatest difficulty has come from the presence of cascading third-order responses with much larger signal intensities than the desired direct fifth-order response. These interfering signals were originally suggested by Wright and coworkers[2] as the dominant signal in frequency domain experiments and were latter implicated as the the origin of the signals in experiments by Albrect and coworkers.[3] We recently demonstrated experimentally that all of the previous time domain fifth-order measurements on the intermolecular modes of CS_2 are dominated by these cascading third-order signals.[4]

In order to measure the direct fifth-order response we have exploited phase matching considerations to provide significant discrimination against the cascaded signals. Contour plots for the measured 2D polarized, R_{zzzzzz}, and magic angle, R_{zzmmzz} m=54.7^o, tensoral responses are shown in figure 1. The phase matching geometry implemented in these experiments provides a factor of $> 10^3$ discrimination against the cascaded signals while still maintaining excellent phasematching for the overall fifth-order process.[5] Comparison of the measured responses in figure 1 to simulations of third-order cascaded responses demonstrated that the measured responses are not the result of third-order cascades (not shown).

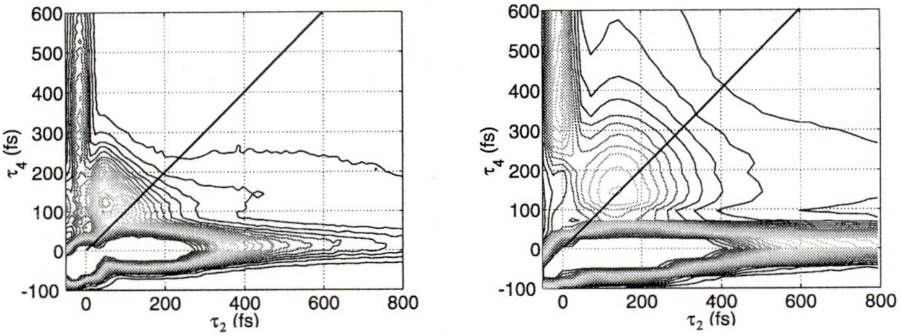

Fig. 1. Contour plots of the measured 2D Raman a)polarized, R_{zzzzzz}, and b) magic angle, R_{zzmmzz}, tensoral responses.

Of course, one of the greatest challenges lies in the iterpretation of the measured responses and acquisition of the microscopic information discussed in the introduction of this summary. Initially one might consider using the third-order response as a starting point. Unfortunately, this approach cannot be employed when discussing the continuous, unresolved, set of intermolecular motions. Any tractable fitting of the 1D response requires collection of the thousands of intermolecular modes into a few generalized timescales, for example Brownian oscillators. Use of the collected modes to fit the fifth-order response must then assume that the nonlinearity inherent to the 2D response, and [essentially] absent from the 1D response, is the same for all of the actual modes in each collected mode. There appears to be no justification for this assumption. In fact, Murry and Fourkas have demonstrated using instantaneous normal mode simulations, INM, that this assumption cannot be made for CS_2 when the nonlinearity is in the polarizability operator.[6]

One cannot overemphasize the fact that lower-order measurements cannot be used to interpret the fifth-order responses from intermolecular motions. However, we can make comparisons with simulations. Using INM and quenched normal mode, QNM, methods Saito and Ohmine have simulated the fifth-order response from CS_2.[7] Their simulations of the polarized response agree quite well with figure 1a. However, when they only consider nonlinearity in the polarizability operator as a source of mode coupling the simulation has a small diagonal ridge that does not appear in our measured signal. The response along the diagonal is a measure of the rephasing ability of the system, and the residual ridge in the simulation demonstrates some degree of inhomogeneity on the timescale of the response. When Saito and Ohmine included vibrational relaxation, which can effectively be considered mode coupling from anharmonicity in the vibrational potential, the diagonal ridge is no longer apparent in the simulation. This indicates that our measured signal is not only the result of mode coupling induced by the field-matter interactions, via nonlinearity in the polarizabiliy operator, but also reflects the anharmonicity of the vibrational potential.

Although they have not simulated the full 2D response, Murry and Fourkas have considered the tensoral nature of the fifth-order response originating from nonlinearity in the polarizability operator.[6] Our measured polarized signal shows a more rapid decay than the magic angle signal, which is qualitatively consistent with the prediction of Murry and Fourkas that the magic angle response shows slightly greater weighting towards lower frequency motions. However, we find that the relative intensity between the polarized and magic angle responses differ by more than an order of magnitude, while the simulations predict very similar relative intensities.

Another prediction from the tensoral analysis of Murry and Fourkas is that the rephasing ability of the system, reflected in the response along the time diagonal, should be more apparent in the magic angle signal. This is due to the fact that rephasing events only occur via interaction pathways where the nonlinear field-matter interaction happens in the second step. While the polarized spectrum will weigh all of the possible interaction pathways equally, the magic angle response will weigh the rephasing pathways more heavily. Qualitative inspection of figure 1 shows that the polarized reponse is flat along the anti-diagonal ($\tau_2 + \tau_4$ = constant) but the magic angle response appears to have some echo-like character with a peak at $\tau_2 = \tau_4$ along the antidiagonal for $\tau_2 + \tau_4 < 600$ fs. Along the diagonal the magic angle signal extends to longer time than the polarized response giving good agreement with the analysis of Murry and Fourkas and the idea that the rephasing ability of the system is reflected more strongly in the magic angle response.

The diagonal in figure 1 also shows that any rephasing ability of the intermolecular motions is lost within 400 fs. This rapid loss of rephasing occurs on a timescale comparable to or faster than one period of the intermolecular frequencies being considered, 0-200 cm^{-1}. As a result, we conclude that the intermolecular motions in CS_2 at room temperature are effectively homogeneous. As discussed above, based on comparisons with the simulations of Saito and Ohmine, the coupling that leads these motions to appear homogeneous on a 400 fs timescale is a combination of both the nonlinearity in the polarizability (field-matter interaction) and anharmonicity in the mechanical potential.

References

1. S. Mukamel, A. Piryatinski, and V. Chernyak, Acc. Chem. Res., **32**, 145-154 (1999), *and references therein.*
2. J. E. Ivanecky and J. C. Wright, Chem. Phys. Lett., **206**, 437-444 (1993).
3. D. J. Ulness, J. C. Kirkwood, and A. C. Albrecht, J. Chem. Phys., **108**, 3897-3902 (1998).
4. D. A. Blank, L. J. Kaufman, and G. R. Fleming J. Chem. Phys., **11**, 3105-3114 (1999).
5. D. A. Blank, L. J. Kaufman, and G. R. Fleming, J. Chem. Phys , **113:1**, (2000).
6. R. L. Murry and J. T. Fourkas, J. Chem. Phy., **109**, 2814-2825 (1998); *ibid* 7913-7922.
7. S. Saito and I. Ohmine, J. Chem. Phys. **108**, 240-251 (1998).

Separating the direct and cascaded contributions to two-dimensional Raman signals using phase-sensitive detection

O. Golonzka, N. Demirdöven, and A. Tokmakoff

Department of Chemistry, Massachusetts Institute of Technology, Cambridge, MA 02139
Email: tokmakof@mit.edu

Abstract: We demonstrate a method of phase-sensitive detection for separating the direct and the cascaded contributions to fifth-order 2-D Raman spectroscopy. The phase of an intrinsic local oscillator is cycled to distinguish the two components that are radiated in quadrature to each other.

Introduction. 2-D fifth-order Raman experiment, originally proposed by Tanimura and Mukamel [1], carries information pertaining to the coupling between vibrational coordinates of liquids [2]. Of considerable concern to the development of this method is the observation that the earlier 2-D Raman data of the low frequency intermolecular motions of CS_2 are dominated by cascaded signals [3]. The cascaded response can be exactly modeled as a product of two 1-D third-order responses and, consequently, does not carry any 2-D information. It is indisputable that the utility of fifth-order 2-D Raman spectroscopy hinges upon the experimental ability to separate the direct fifth-order and cascaded contributions.

Blank, et al. have demonstrated a way to suppress the cascaded contributions through selective wave-vector matching [4]. We implement a different experimental approach that exploits the phase shift between the electric field due to cascades and the direct fifth-order signal. Since the generation of the cascaded signal involves an additional radiation event, the electric field due to the direct fifth–order polarization is $\pi/2$ phase shifted with respect to the cascades [5]. Therefore, heterodyne detection using a local oscillator (LO) in phase with the direct fifth-order signal will suppress cascaded contributions.

Experimental. Phase-sensitive detection of fifth-order signals is possible with an intrinsic heterodyne-detection geometry (Fig.1a) in which a third-order signal field acts as the LO [6]. A third-order signal $\mathbf{E}^{(3)}{}_{LO}(\tau_4)$, ($\mathbf{k}_{lo}$=$\mathbf{k}_3$-$\mathbf{k}_4$-$\mathbf{k}_5$) is generated by three fields incident in a "box car" geometry. An additional pump beam generates the fifth-order signal $\mathbf{E}^{(5)}(\tau_2, \tau_4)$ collinear with the third-order signal ($\mathbf{k}_{sig}$=$\mathbf{k}_1$-$\mathbf{k}_2$+$\mathbf{k}_3$-$\mathbf{k}_4$+$\mathbf{k}_5$). The polarizations of $\mathbf{E}_3$, $\mathbf{E}_4$ and $\mathbf{E}_5$, as well as the LO $\mathbf{E}^{(3)}{}_{LO}(\tau_4)$, are parallel (0°), while the polarizations of $\mathbf{E}_1$ and $\mathbf{E}_2$ are set at 45°. An analyzing polarizer is set to extinguish the third order signal (90°) and is slightly tilted to introduce a third-order LO for the heterodyne measurement ($\mathbf{E}^{(3)}{}_{LO}(\tau_4)$= $\mathbf{E}^{(3)}{}_{ZZZZ}(\tau_4)$). The fifth-order tensor elements measured are $\boldsymbol{R}^{(5)}{}_{YZZZYZ}$ and $\boldsymbol{R}^{(5)}{}_{YZZZZY}$. For a purely non-resonant experiment in which the third-order response is purely birefringent, the LO field $\mathbf{E}^{(3)}{}_{LO}(\tau_4)$, as well as the direct fifth-order field $\mathbf{E}^{(5)}{}_{dir}(\tau_2, \tau_4)$, are in

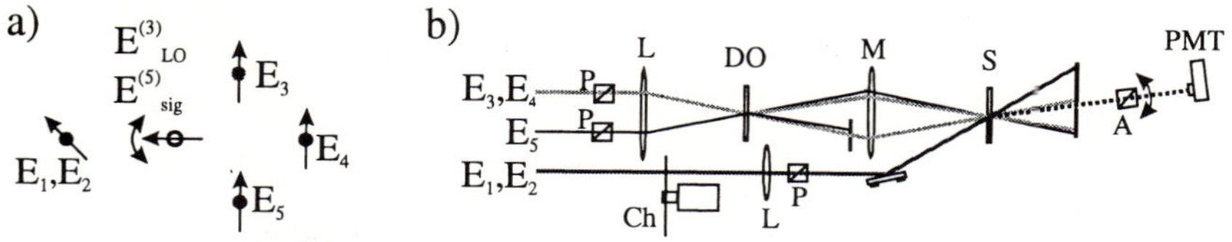

Fig. 1. (a) Wavevector matching and polarizations of the electric fields involved in the experiment. The orientation of the analyzer (empty circle) is slightly varied to control the strength and phase of the LO field. **(b)** Experimental layout. Pump fields E_3, E_4 and the probe field E_5 are generated using a diffractive-optic (DO) in combination with a lens (L) and a curved mirror (M).

quadrature ($\pi/2$ phase shifted) with the probing field and in phase with each other. The cascaded contributions, on the other hand, are in phase with the probe field and in quadrature with the $\mathbf{E}^{(3)}{}_{LO}(\tau_4)$ and $\mathbf{E}^{(5)}{}_{dir}(\tau_2, \tau_4)$. Since tilting the analyzer clockwise and counterclockwise changes the sign of the projection of the third order field onto the analyzer axis, it is equivalent to cycling the phase of the LO between 0° and 180° relative to the fifth order signal.

The experiment (Fig. 1b) uses 35 fsec near-transform-limited pulses (λ=650 nm) derived from doubling the signal output of a 1 kHz Ti:sapphire-pumped near-IR BBO OPA. A diffractive optic is used to generate fields $\mathbf{E}_3$, $\mathbf{E}_4$ and $\mathbf{E}_5$ in a box-car geometry.

Results and discussion. Figure 2a shows homodyne 2-D data of liquid CS_2 detected by extinguishing the third-order LO with the analyzer. The contours are highly asymmetric, with the signal decaying rapidly along τ_4 axis. The observed response is expected to be a mixture of the direct fifth-order signal and the cascaded contributions, with the latter being the dominant term. Our simulations (Fig. 2b) show that the experimental data is successfully reproduced by considering only the parallel cascade. The absence of the sequential cascade is due to the interference of two pathways differing by the sequence of interactions with the third and the fourth incident fields.

The heterodyne data was collected by tilting the analyzer ±2° to introduce the LO with phase shifts of ϕ = 0° and 180° relative to the direct fifth-order signal. The difference of these two data sets, shown in Fig. 2c, eliminates the homodyne and cascaded signals and represents the cross-term:

$$S^{dif} \propto 4E^{(5)}{}_{dir}(\tau_2,\tau_4)E^{(3)}{}_{LO}(\tau_4) \tag{1}$$

The direct 2D response is obtained by normalizing to the third order response. Since the cascaded field is in quadrature with the birefringent part of the third order response, the proposed method suppresses the cascaded contributions by a factor equal to the ratio of birefringent and dichroic third-order responses. The resulting contour map (Fig. 2d) cannot be reproduced from cascaded signals and differs from the cascaded signals observed in other polarization geometries.

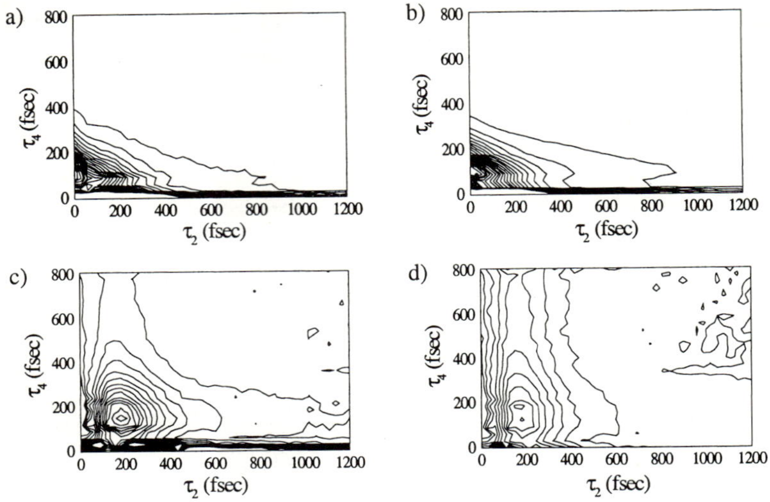

Fig. 2. (a) Experimental homodyne fifth-order signal for CS_2. **(b)** Simulated homodyne cascaded signal calculated using only "parallel" cascades. **(c)** Difference data obtained by subtracting the results of heterodyne measurements with analyzer axis tilted clockwise and counterclockwise from being crossed with the polarizations of $\mathbf{E}_3$, $\mathbf{E}_4$ and $\mathbf{E}_5$. **(d)** Amplitude of the combined fifth-order field $\mathbf{E}^{(5)}(\tau_2,\tau_4)$ in phase with the LO field.

The projection of $\boldsymbol{R}^{(5)}{}_{YZZZZY}$ onto the τ_2 is dominated by the time scale of the inertial response of CS_2. The τ_4 projection, however, shows no pronounced inertial peak and a *psec* decay that maps best onto the time scale of diffusive reorientation in the liquid. The dominant contributions to the fifth-order signal are predicted to be due to the interaction-induced terms in the polarizability. While molecular reorientation should not contribute significantly it may enter through molecular-interaction-induced cross-correlations. These contributions are presumably the reason for observation of long tome dynamics along τ_4 in the experiment.

References

1 Y.Tanimura and S.Mukamel, J. Chem. Phys., **99** ,Vol. 12, 9496, 1993.

2 A. Tokmakoff, M. J. Lang, D. S. Larsen, G. R. Fleming, V. Chernyak, and S. Mukamel, Phys. Rev. Lett., **79**, 2702, 1997.

3 D. A. Blank. L. J. Kaufman, and G. R. Fleming, J. Chem. Phys., **11**, 3105, 1999.

4 D. A. Blank. L. J. Kaufman, and G. R. Fleming, J. Chem. Phys., (to be published).

5 M. Cho, D. A. Blank, J. Sung, K. Park, S. Hahn, and G. R. Fleming, J. Chem. Phys., **112**, 2082, 2000.

6 A. Tokmakoff, M. J. Lang, D. S. Larsen, G. R. Fleming, Chem. Phys. Lett., **272,** 48, 1997.

Time dependent 2D Fourier transform spectra reveal femtosecond solvation dynamics

John D. Hybl, Allison W. Albrecht, and David M. Jonas

Department of Chemistry and Biochemistry, University of Colorado,
Boulder, CO 80309-0215 USA
E-mail: david.jonas@colorado.edu

Abstract: Femtosecond 2D Fourier transform spectra simultaneously provide time resolution and two dimensions (excitation, signal) of frequency resolution limited only by the sample. 2D spectra of IR144 in methanol provide a clear view of ultrafast solvation.

Introduction. In femtosecond spectroscopy, high time resolution usually implies low frequency resolution, necessitating a balance between pulse duration and frequency resolution. In magnetic resonance, "two-dimensional" experiments with high excitation and signal frequency resolution use broadband pulses and obtain frequency resolution by Fourier transformation of the signal field with respect to time intervals between and after pulses [1]. The 2D signals presented here can be viewed as a Fourier separation of the spectrally resolved pump-probe signal [2] according to excitation frequency within the pump pulse. Broadband pulses and a stimulated photon echo pulse sequence provide powerful inhomogeneous line-narrowing advantages, demonstrated here for the dye IR144 in methanol. The IR144 absorption frequency essentially reflects a collective "solvent coordinate" that governs the rate of electron transfer in solution. The solvent coordinate (hence the IR144 absorption frequency) fluctuates on a 100 fs timescale due to the random thermal motion of solvent molecules surrounding the dye. 2D spectra of IR144 in methanol provide pictures of this time dependent solvent motion.

Experiment. Near transform limited 27 fs pulses from a cavity dumped titanium sapphire laser generate the non-collinear three pulse scattering signals used to generate 2D Fourier transform spectra. The first pulse (wave-vector $\boldsymbol{k}_a$) sets electronic dipoles oscillating and the second pulse ($\boldsymbol{k}_b$) enhances or suppresses these oscillations depending on the dipole frequency, delay τ, and position in the sample. This creates a grating with the dipole oscillation frequency stored in the τ dependent grating phase. After a relaxation time T, the third pulse ($\boldsymbol{k}_c$) "scatters" off the grating into a background free direction ($\boldsymbol{k}_s = \boldsymbol{k}_c + \boldsymbol{k}_b - \boldsymbol{k}_a$), transferring the grating phase to the phase of the "scattered" field radiated during t. The sample is in one arm of a small interferometer that combines the signal with a time delayed reference pulse for complete characterization by spectral interferometry [3]. The signal field $E_s(t;\ \tau, T)$ is detected at the sample for a set of τ, and the set is Fourier transformed with respect to t and τ to make a 2D spectrum in ω_t and ω_τ [4]. For the Fourier transformation, delays between excitation pulses are measured to ±0.04 fs by spectral interferometry for each signal. Integrals of the real 2D spectra over the excitation frequency ω_τ match spectrally resolved pump-probe signals at the same T. Fig. 1 shows the nonzero regions of a few 2D spectra.

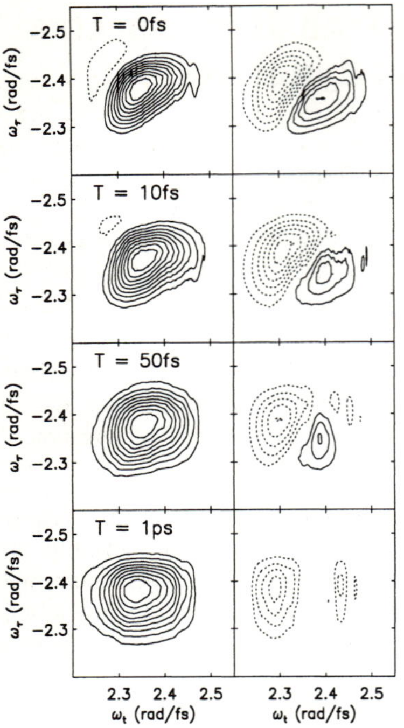

Fig. 1. Experimental real (left) and imaginary (right) 2D spectra for 4 values of T (top to bottom).

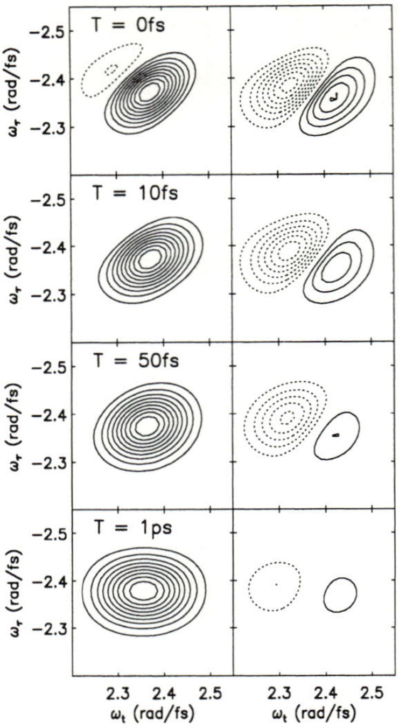

Fig. 2. Calculated real and imaginary 2D spectra for 25 fs Gaussian pulses. Contour intervals are 10% of the real maximum, dotted contours are negative.

Results and Discussion. 2D spectra calculated with a multi-mode Brownian oscillator model [5] are shown in Fig. 2. Positive real 2D amplitude indicates that excitation at ω_τ causes a reduction in absorption at ω_t. The two frequencies have opposite signs because photon echo type rephasing signals have been selected for $\tau > 0$. The imaginary 2D spectrum separates the refractive index change according to the excitation frequency. A near Kramers-Kronig relation with respect to ω_t between real and imaginary slices of the 2D spectrum for each ω_τ is evident at T=0. With increasing T, absorption changes outside the pulse spectrum cause apparent deviations from this 2D Kramers-Kronig relation. The limitation of real 2D amplitude to a region around the diagonal $\omega_t = -\omega_\tau$ at T=0 indicates inhomogeneous broadening: different molecules absorb different frequencies. The width of the 2D spectrum perpendicular to the diagonal is narrower than the vibrational width of the IR144 electronic spectrum by about a factor of 2. The 2D spectrum has frozen out rapidly moving vibrational and solvent coordinates as "inhomogeneities". Calculations verify that the displacement of the T=0 2D spectrum off the diagonal so that $|\omega_t|<|\omega_\tau|$ arises from the dynamic Stokes shift during dipole oscillation. The negative region above the diagonal comes from the signal field decay along the time domain echo slice $t=\tau$. The field along the echo slice decays through the loss of photon echo rephasing as the ensemble average $<\cos[(\omega_t-\omega_\tau)t]>$. Pre-existing thermal vibrational and solvent motions lead to an "inertial" time derivative of the optical frequency $\dot{\omega}$, which dominates the early response, so that $\omega_t = \omega_\tau + \dot{\omega}\tau$. The resulting $[1-(1/2)\dot{\omega}^2 t^4]$ decay along $t=\tau$ begins slowly and then suddenly drops, which produces "ringing" on frequency domain axes parallel to $\omega_t=\omega_\tau$, hence the negative region. To model the 2D spectra, the timescale and energetic stabilization attributed to inertial solvation in prior three pulse echo peak shift studies [6] were increased to match the T=0 2D spectrum (yielding τ_g = 125 fs and $2\lambda_g$ =840 cm^{-1}), the 10 vibrations were left unchanged, and the total diffusive solvation energy was reduced to maintain the total Stokes' shift. In this model, inertial solvation is crucial for predicting the rapid changes in the 2D spectrum observed as T increases from 0 to 100 fs. Subtle, slow changes in the 2D spectrum on the 1-100 ps timescale are attributed to diffusive solvent reorganization. 2D spectra indicate homogeneous broadening at T = 100 ps.

Conclusions. Femtosecond 2D Fourier transform spectroscopy has several remarkable properties: 2D spectra provide frequency resolution within the pulse spectrum, a nearly undistorted window on the delta function 2D spectrum with finite duration pulses, and a narrowing of both vibrational and inhomogeneous broadening. 2D spectra let the sample determine time and frequency resolution, so the shortest pulses available can be used for every femtosecond experiment [7].

Acknowledgements. This work was supported by the National Science Foundation, the David and Lucile Packard Foundation, and the Sloan Foundation. JDH and AWA are Fellows in the NSF Optical Science and Engineering Program.

References

1 R. R. Ernst, G. Bodenhausen, and A. Wokaun, *Principles of nuclear magnetic resonance in one and two dimensions* Oxford University Press, Oxford, 1987.
2 C. H. Brito Cruz, et al., IEEE J. Quantum Elect. **24,** 261, 1988.
3 L. Lepetit, G. Chériaux, and M. Joffre, J. Opt. Soc. Am. B **12,** 2467, 1995.
4 J. D. Hybl, et al., Chem. Phys. Lett. **297,** 307, 1998.
5 S. Mukamel, *Principles of Nonlinear Optical Spectroscopy* Oxford University Press, New York, 1995.
6 T. Joo, et al., J. Chem. Phys. **104,** 6089, 1996.
7 S. M. Gallagher Faeder and D. M. Jonas, J. Phys. Chem. A **103,** 10489, 1999.

Two-dimensional time-resolved THz spectroscopy of solvent response to photoexcitation

M. C. Beard, G. M. Turner, and C. A. Schmuttenmaer

Yale University, Department of Chemistry, 225 Prospect St., P.O. Box 208107, New Haven, CT 06520-8107, USA
E-mail: charles.schmuttenmaer@yale.edu

Abstract. Low-frequency intermolecular solvent dynamics of chloroform have been measured with two-dimensional time-resolved THz spectroscopy.

1. Introduction

Solvation and liquid dynamics play a very important role in essentially all liquid phase processes, such as photoexcitation, photoionization, electron transfer, proton transfer, energy dissipation, and chemical reactions. Therefore, it seems reasonable to directly probe the low frequency solvent modes during one of the above processes in an effort to understand the solvent response to the event.

We report the low frequency intermolecular response of chloroform upon photoexcitation of TBNC (2,11,20,29-tetra-*tert*-butyl-2,3-naphthalocyanine). Unlike typical time-resolved laser studies, the response of the *solvent* can be monitored. TBNC does not have a dipole moment in either its ground or excited state, so the dynamics probed are due to solute-solvent interactions and/or solvent-solvent interactions, and not purely a solute response.

The generation of far-infrared (FIR) pulses with sub-picosecond duration is a relatively new development [1]. The spectral content of these pulses extends from 3 to 100 cm^{-1} (0.09 to 3 THz), which allows the FIR spectrum of a sample to be measured with previously unattainable temporal resolution.

There have been many studies that utilize the broadband FIR coverage of THz pulses, but few that have used them in a time-resolved fashion. Of these, there have been 3 studies of the overall change in THz transmission of a solution upon photoexcitation of a dye molecule [2, 3]. In this study, we extract the time-dependent change in response function for the first time. As described earlier [4, 5], the measured change in THz transmission depends on two time variables, and the only way to obtain the time-dependent response function is to perform a series of scans and build up a two-dimensional data set.

2. Experimental Methods

The experimental apparatus has been described previously [3, 5]. Figure 1 shows a THz reference pulse which has broadened somewhat due to absorption by the 200 μm thick $CHCl_3$ sample.

Concentrations of TBNC in various solvents ranging from 0.003 M to 0.03 M have been investigated, and typically 0.01 M is used. The TBNC molecules do in fact dimerize at high concentrations, but the absorption maximum of the dimers is shifted to shorter wavelengths to such a degree that the 800 nm pump beam is primarily absorbed by monomers. The dye solution is flowed through a 200 μm thick cuvette at a rate such that each molecule experiences roughly 5 laser pulses as it passes through the beam waist. We have verified that our signal is not dependent on flow rate.

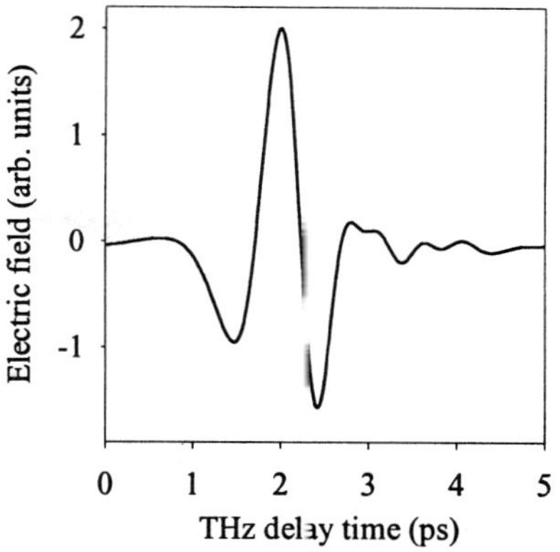

Fig. 1. THz reference pulse.

3. Results and Discussion

Figure 2 shows the 2-D time-domain data set for TBNC/chloroform. We find that this system has a reasonably good signal-to-noise (S/N) ratio because the chloroform solution does not absorb strongly in the FIR (the Naperian power absorption coefficient is roughly 20 per cm), yet chloroform possesses a moderate dipole moment. Reasonable S/N is also obtained in the TBNC/toluene system (not shown) due to a slightly smaller dipole moment and absorption coefficient.

While it is possible to extract the change in response function directly from the data [4], it is well known that deconvolution procedures can be adversely affected by experimental noise. Therefore, we calculate the expected change in THz

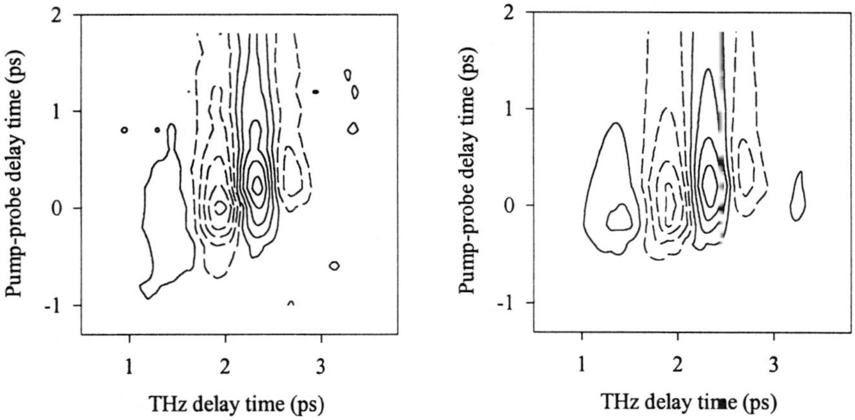

Fig. 2. Contour plot of 2D TRTS difference signal (THz pulse propagated through with laser on minus that with laser off). Comparison of measured (left), with calculated (right) values. Positive values shown with solid lines, negative values with dashed.

Table 1. Brownian oscillator parameters for TBNC/chloroform. The asterisk denotes the photoexcited medium.

Oscillator #	$\mu\,(\mathrm{cm}^{-1})^2$	$\mu^*\,(\mathrm{cm}^{-1})^2$	$\omega_0\,(\mathrm{cm}^{-1})$	$\omega_0^*\,(\mathrm{cm}^{-1})$	$\gamma\,(\mathrm{cm}^{-1})$	$\gamma^*\,(\mathrm{cm}^{-1})$
1	178.2	244	41.8	48.0	30.8	31.0
2	104.2	151.4	6.3	7.6	19.3	19.2
electronic	---	15.18	---	4.7	---	13.6

transmission for a parameterized model of the solvent (based on the Brownian oscillator), and find the best fit to the data using a nonlinear least squares fit.

The Brownian oscillator model describes a damped, driven oscillator. The equation of motion for the "position" of an oscillator as a function of time is $\ddot{x}(t) = -\omega_0^2 x(t) - 2\gamma\dot{x}(t) + (\mu\varepsilon_0)^{1/2} E(t)$, where ω_0 is the frequency of the oscillator, γ is its damping constant, and the coupling to the driving field is $\mu = \omega_0^2\,(\varepsilon_s - \varepsilon_\infty)$, where ε_s and ε_∞ are the static and high frequency dielectric constants, respectively. The mode is overdamped if $\gamma^2 > \omega_0^2$, and underdamped if $\gamma^2 < \omega_0^2$.

Both electronic and nuclear responses can contribute to the signal. It is customary to attribute the response close to time $t = 0$ (when the two pulses are overlapped) to electronic processes, while those that persist for longer times to nuclear or molecular processes. We include both electronic and nuclear responses in our model. The static FIR absorption and index of refraction of chloroform are well fit by a model of two Brownian oscillators, one underdamped and one overdamped. Thus, the sample is described as a mixture of two non-photoexcited oscillators, two photoexcited oscillators whose concentration is time-dependent, and an oscillator representing an electronic response whose contribution tracks the intensity profile of the excitation pulse. The results of the fit agree quite well with the data (Fig. 2). It is seen in Table 1 that both oscillators couple strongly to the electric field, and that their responses to photoexcitation are similar. That is, the damping coefficient barely changes, but the resonant frequency increases by roughly 20%. While the electronic contribution is necessary to obtain a good fit to the data, it is seen to be less important than the molecular modes.

Acknowledgements. We acknowledge the National Science Foundation and the Sloan Foundation for partial support of this work.

References

1 Ch. Fattinger and D. Grischkowsky, Appl. Phys. Lett. **54**, 490, 1989; P. R. Smith, D. H. Auston, and M. C. Nuss, IEEE J. Quant. Electron. **24**, 255, 1988.

2 G. Haran, W. D. Sun, K. Wynne, and R. M. Hochstrasser, Chem. Phys. Lett. **274**, 365, 1997; R. McElroy and K.Wynne, Phys. Rev. Lett. **79**, 3078, 1997.

3 D. S. Venables and C. A. Schmuttenmaer, in Ultrafast Phenomena XII, *Springer Ser. Chem. Phys.* (1998), Vol. 63, pp. 565-567.

4 J. T. Kindt and C. A. Schmuttenmaer, J. Chem. Phys. 110, 8589 (1999).

5 M. C. Beard, G. M. Turner, C. A. Schmuttenmaer, Phys. Rev. B, submitted.

Part XII

Dynamics in Liquids

Ultrafast intermolecular transfer of vibrational excitations in liquid water

Sander Woutersen and Huib J. Bakker

FOM-Institute for Atomic and Molecular Physics,
Kruislaan 407, 1098 SJ Amsterdam, The Netherlands
E-mail: woutersen@mbi-berlin.de, h.bakker@amolf.nl

Abstract. Using femtosecond vibrational pump-probe spectroscopy, we find that extremely fast resonant intermolecular transfer of the OH-stretch excitation occurs in liquid water, with a characteristic transfer time of less than 100 femtoseconds.

With the advent of techniques to generate subpicosecond pulses in the mid-infrared, it has become possible to study the vibrational dynamics of liquid water in a time-resolved manner, which provides information on the microscopic structure and dynamics that cannot be obtained with conventional methods. As yet, most time-resolved vibrational studies [1-3] on liquid water have been concerned with the OH-stretch mode of strongly diluted solutions of HDO in D_2O. However, such a dilute solution of HDO in D_2O can only be regarded as a good model system for 'real' water (liquid H_2O) if there is no resonant intermolecular transfer of the vibrational excitation. Such intermolecular energy transfer is a well-known process for electronic excitations, but is an unlikely process for vibrational transitions because of their small transition dipoles and short lifetimes.

To investigate the occurrence of resonant intermolecular energy transfer in liquid water, we have used polarization-sensitive vibrational pump-probe spectroscopy to study the OH-stretch dynamics in both liquid H_2O and solutions of HDO in D_2O with different HDO concentrations. In our experiment [4], an intense pump pulse tuned to the $v_{OH} = 0 \rightarrow 1$ transition frequency (3400 cm^{-1}) creates an anisotropic orientational distribution of vibrationally excited OH-groups, with a maximum in the direction of the polarization of the pulse. Intermolecular transfer of the OH-stretch excitation will lead to a change of the polarization of the excitation, and hence to a decay of the anisotropy. This decay is monitored by a weak delayed probing pulse tuned to the $v_{OH} = 1 \rightarrow 2$ transition frequency (3000 cm^{-1}).

Fig. 1 shows the decay of the rotational anisotropy as a function of delay observed in pure H_2O and in solutions of HDO in D_2O with three different HDO concentrations. The decay of the rotational anisotropy of the OH-stretch excitation becomes significantly faster with increasing concentration of OH oscillators. This concentration dependence of the decay of the rotational anisotropy clearly shows that resonant intermolecular energy transfer of the OH-stretch excitation occurs in liquid water. The rate of intermolecular energy transfer, and thus of depolarization, strongly increases with decreasing average distance between the OH groups, and therefore increases with the concentration of OH groups. Hence, we find that a high concentration of OH-stretch oscillators leads to a rapid delocalization of the OH-stretch vibrational excitation.

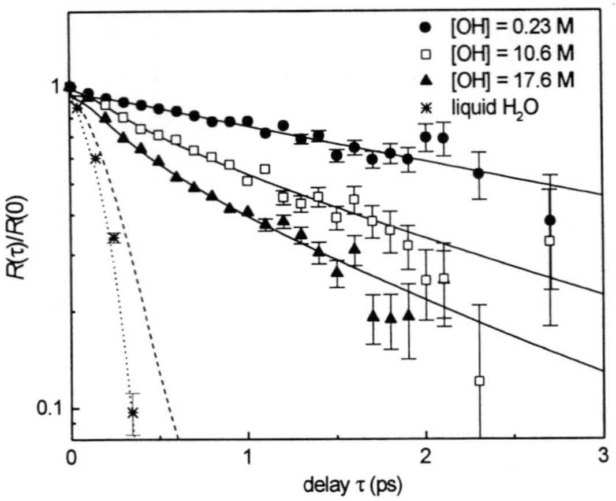

Fig. 1. Rotational anisotropy as a function of delay, normalized with respect to the value at zero delay, as observed in $HDO:D_2O$ solutions with three different HDO concentrations, and in pure H_2O. The solid curves have been calculated assuming a Förster mechanism (Eq. 1) with r_0 = 2.1 Å, and using the appropriate values for the OH concentration. The dashed curve has been calculated using the same value for r_0 and an OH concentration of 111 M. The dotted curve has been calculated assuming an instantaneous decay of the rotational anisotropy.

In order to quantitatively analyze our data, we must take into account that the decay of the rotational anisotropy is caused not only by intermolecular transfer of the OH-stretch excitation, but also by the orientational random motion of the OH-groups. Clearly, the latter contribution to the depolarization will not depend on the concentration. At very low HDO concentration, the intermolecular energy transfer becomes negligible, and the decay of the anisotropy results from orientational diffusion only. We find that at low HDO concentration the decay of the anisotropy can be well described by a single exponential with decay constant τ_{or}.

If the intermolecular energy transfer is described by a dipole-dipole (Förster-type) interaction, the rate of energy transfer between two OH-oscillators is given by $k = T_1^{-1}(r_0/r)^6$, where T_1 is the lifetime of the excited state, r the distance between the oscillators, and r_0 a constant usually referred to as the Förster radius. The Förster radius r_0 is a parameter that characterizes the intermolecular energy transfer. If r_0 is smaller than the average distance between the OH groups, no significant energy transfer will occur during the lifetime T_1 of the excitation. If r_0 is comparable to or larger than the average distance between the OH groups, the energy transfer will be significant for the dynamical behaviour. Using the above expression for k and incorporating the orientational diffusion, an equation for the decay of the rotational anisotropy with increasing delay τ can be derived [5]:

$$\frac{R(\tau)}{R(0)} = \exp\left(-\frac{\tau}{\tau_{or}} - \frac{4\pi^{3/2}}{3}[\mathrm{OH}]\sqrt{\frac{r_0 \tau}{T_1}} \right) \tag{1}$$

where [OH] is the concentration of molecules containing an OH-group. We find that all delay scans recorded in $HDO:D_2O$ solution can be well described with this expression, and from a least-squares fit to the $HDO:D_2O$ data (solid curves), we obtain $\tau_{or} = 4.0\pm0.4$ ps for the decay constant of the orientational diffusion of the OH groups, and a Förster radius $r_0 = 2.10\pm0.05$ Å. The value we find for r_0 is comparable to the intermolecular distance of approximately 2.8 Å in liquid water, which implies that intermolecular energy transfer will be very fast in liquid H_2O.

Using the Förster radius obtained from the fit to the $HDO:D_2O$ data, we have calculated the decay of R predicted for liquid H_2O by the Förster mechanism (dashed curve). We used an OH concentration of 111 M, which is twice the concentration of H_2O molecules in liquid H_2O, since each H_2O molecule has two OH-stretch modes (symmetric and asymmetric). Surprisingly, the experimentally observed decay occurs significantly faster than predicted by the Förster mechanism. In fact, we find that the H_2O data can be well described by assuming an instantaneous decay of the rotational anisotropy (dotted curve). This means that in liquid H_2O the intermolecular energy transfer takes place faster than our time resolution of approximately 100 fs. The rapid decay of R cannot be caused by *intra*molecular redistribution of the OH-stretch excitation between the symmetric and asymmetric OH-stretch modes of the H_2O molecule, since this can at most lead to a decay of the rotational anisotropy to a quarter of its initial value [6]. Hence, the rapid decay of the anisotropy can only be caused by an extremely fast *inter*molecular energy transfer process that leads to a rapid delocalization of the OH-stretch vibration over many water molecules.

The difference between the observed, very fast, energy transfer in liquid H_2O and the behaviour predicted by the dipole-dipole mechanism is probably due to the very small distance between the OH groups in liquid H_2O. At such intermolecular distances, not only dipole-dipole, but also dipole-quadrupole, quadrupole-quadrupole, and higher-order interactions will contribute significantly to the energy transfer. An additional contribution may arise from kinematic coupling between the OH-oscillators on neigbouring H_2O molecules, possibly through the hydrogen bond.

The extremely fast intermolecular energy transfer in liquid H_2O renders its vibrational dynamics essentially different from that observed in diluted $HDO:D_2O$. In liquid H_2O, an OH-stretch excitation will be delocalized over many molecules before being dissipated by energy relaxation. As a consequence, water molecules can efficiently transport vibrational energy, and can act as an efficient sink for the vibrational energy situated on an OH group.

References

1 H. Graener, G. Seifert, and A. Laubereau, Phys. Rev. Lett. **66**, 2092 (1991).

2 S. Woutersen, U. Emmerichs, H.-K. Nienhuys, and H. J. Bakker, Phys. Rev. Lett. **81**, 1106 (1998).

3 G. M. Gale, G. Gallot, F. Hache, N. Lascoux, S. Bratos, and J.-C. Leicknam, Phys. Rev. Lett. **82**, 1068 (1999).

4 S. Woutersen and H. J. Bakker, Nature **402**, 507 (1999).

5 K. B. Eisenthal, Chem. Phys. Lett. **6**, 155 (1970).

6 A. Szabo, J. Chem. Phys. **81**, 150 (1984).

Time and frequency-gated FID: a new approach to study the vibrational dephasing of water

Julie A. Gruetzmacher and Norbert F. Scherer

Department of Chemistry, James Franck Institute, and Institute for Biophysical Dynamics, University of Chicago, Chicago, IL 60637. E-mail: nfschere@uchicago.edu

Abstract. The ultrafast frequency-resolved upconverted vibrational free induction decay (FID) of HDO in D_2O was measured, allowing extraction of the amplitude and phase of the coherent transient. The FID Wigner spectrogram shows evidence of λ-dependent dynamics.

Considerable experimental and theoretical effort has been expended toward understanding the behavior of liquid water at the molecular level, much of it striving to understand the structural and dynamical character of its hydrogen bonding interactions. Investigations of the hydrogen bonding nature of water via vibrational spectroscopy have focused on the OH-stretching band in the 3 μm region. The broad frequency profile of this band and the correlation between stretching frequency and hydrogen bond strength has prompted debate over the possible existence of distinct hydrogen-bonded species and their timescales of interconversion. Recent time-resolved infrared studies of liquid water, primarily in isotopically diluted samples (HDO in D_2O), have shed light on vibrational population [1-3] and rotational [1a, 3] relaxation in the liquid, as well as vibrational Stokes shift [1b] and resonant energy transfer [1c]. These studies do not, however, concur on the nature of the heterogeneity of the OH-stretching band and are limited to resolving dynamics on $\geq$ 150 fs timescales.

The interplay between H-bond strength and OH-stretch transition frequency suggests value in a mixed time-frequency approach to learning about the influence of H-bonds on the vibrational dynamics of water. Such an approach can be realized in time-resolved experiments using ultrashort (broadband) infrared pulses by determination of the electric fields of the signals. Knowledge of the signal amplitude and phase allows translation of the field into the time-frequency domain via Wigner transformation. [4] The Wigner spectrogram provides a means of visualizing frequency-dependent dynamics as it represents the time-dependent spectrum of the signal. Field measurements of four-wave mixing signals have been accomplished in one-color optical experiments by interferometric techniques. [5] In this paper, we demonstrate the first use of cross-correlation frequency-resolved optical gating (XFROG, [6]) as a means of extracting the amplitude and phase of a spectroscopic signal, that of the free induction decay of HDO molecules in liquid D_2O. This measurement, performed with 55 fs infrared pulses, resolves the complex linear polarization and provides a foundation for similar studies of nonlinear responses such as pump-perturbed FIDs.

Femtosecond mid-infrared (MIR) pulses were generated by a type-I optical parametric amplifier (OPA) based on the nonlinear crystal $KNbO_3$ (KN). [7] The OPA is pumped by the fundamental of a home-built Ti:sapphire regenerative amplifier and seeded with the near-infrared portion of a white light continuum generated in a [0001]-cut sapphire plate. The design is similar to that of ref. 8, but is modified for shorter pulse generation. Cross-correlations of the pump pulse with the continuum in the 980-1150 nm range are approximately 60 fs in duration. The pump and continuum are combined collinearly in a 1 mm KN crystal. The center wavelength of the MIR pulses is tunable between 2.9 and 4.2 μm with only minor crystal angle adjustment by varying the temporal overlap of the continuum and pump. Upconversion of MIR radiation for pulse and signal characterization was accomplished by sum-frequency generation in a 300 μm KN crystal using a portion of the 800 nm pump. The gate pulse was characterized by SHG autocorrelation and FROG in BBO. Frequency-resolved signals were detected by a PMT through a monochromator. The sample consisted of a dilute (~1:200) solution of HDO in D_2O in a variable-pathlength cell with CaF_2 windows. The OPA was tuned to ~ 2.94 μm to overlap the OH-stretching transition of HDO, and the pulse spectrum spanned the OH-stretch absorption band. The pulses were shortened at the sample location by precompensating the CaF_2 entrance window's dispersion with a Ge window before the sample. The pulses were then re-optimized at the SFG crystal with germanium after the entire empty sample cell.

The XFROG traces of the (empty cell) MIR reference pulse and the HDO FID are shown in Figures 1A and 1B, respectively. Analysis of the XFROG signal of the MIR reference pulse yields a FWHM pulse duration of 55 fs. The FID measurement was made on a 500 μm thick HDO solution by insertion of the sample in the infrared beam with the same dispersion compensation optics. The

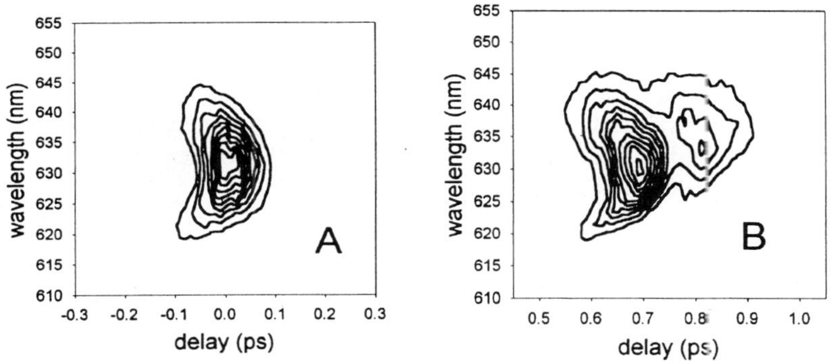

Fig. 1. XFROG data for (A) reference pulse; (B) HDO FID.

XFROG trace of the FID exhibits different temporal behavior on the high and low energy portions of the signal. Analysis of the two-dimensional signals yielded temporal intensity and phase profiles. The phase of the FID of HDO is quadratic in shape for the duration of the free induction decay field.

The amplitudes and phases of the extracted fields were used to construct the Wigner spectrograms (WS) of the MIR reference and FID complex fields as shown in Figures 2A and 2B, respectively. The FID spectrogram exhibits asymmetry along both the time and frequency axes. Simulations show that the general features observed in the FID WS can be reproduced by a superposition of

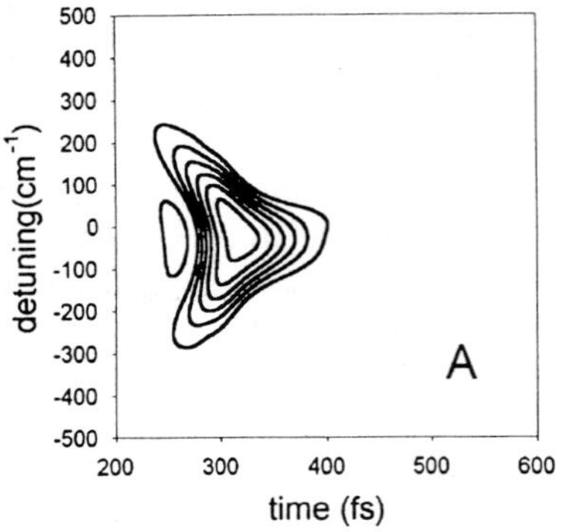

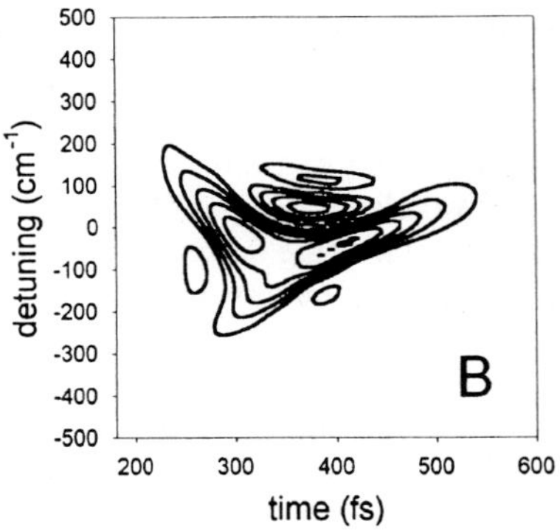

Figure 2. Wigner spectrograms of (A) reference and (B) FID complex fields.

components with different frequencies. Simulations of one- and two-dimensional signals, as well as further experiments with different sample path lengths and concentrations, are in progress to explore the origin(s) of the observed features. These will establish whether the structure in the FID WS is due to true frequency substructure of reflects propagation effects and phase shifts from the dispersion of the index of refraction.

The experiment presented here is a promising approach for the measurement of amplitude and phase dynamics of spectroscopic signals. In the present case we are able to ascertain information from the frequency-resolved upconverted FID that is not accessible from a traditional (frequency-integrated) FID measurement. We are extending this technique to the study of the frequency-dependent dynamics of infrared signal fields of non-equilibrium (pre-pumped) samples, giving access to higher order nonlinear optical information.[9]

References

1 (a) S. Woutersen, U. Emmerichs, and H. J. Bakker, Science **278**, 658 (1997); (b) S. Woutersen and H. J. Bakker, Phys. Rev. Lett. **83**, 2077 (1999); (c) S. Woutersen and H. J. Bakker, Nature **402**, 507 (1999);

2 G. M. Gale, G. Gallot, and N. Lascoux, Chem. Phys. Lett. **311**, 123 (1999).

3 R. Laenen, C. Rauscher, and A. Laubereau, J. Phys. Chem. B **102**, 9304 (1998).

4 S. Mukamel, C. Ciordas-Ciurdariu, and V. Khidekel, IEEE J. Quantum Electron. **32**, 1278 (1996).

5 S. M. Gallagher, A. W. Albrecht, J. D. Hybl, B. L. Landin, B. Rajaram, and D. M. Jonas, J. Opt. Soc. Am. B **15**, 2338 (1998).

6 S. Linden, H. Giessen, and J. Kuhl, Phys. Stat. Sol. (b) **206**, 119 (1998).

7 J. A. Gruetzmacher and N. F. Scherer, manuscript in preparation (2000).

8 J. D. Kafka and M. L. Watts, *Ultrafast Phenomena X*, Springer Series in Chemical Physics Vol. 62 (Springer-Verlag, Berlin, 1996), 38.

9 Acknowledgement: NIH (GM57768) and NSF Predoctoral Fellowship (JAG).

Observation of the Bose Peak in Nanometer-Sized Liquid Water Clusters using Terahertz Spectroscopy

Joel Boyd,[1] Ari Briskman,[1] Vicki Colvin,[1] and Daniel Mittleman[2]

[1] Chemistry Dept., MS-60, Rice University, 6100 Main St., Houston TX 77005 USA
[2] Electrical and Computer Engineering Dept., MS-366, Rice University, 6100 Main St., Houston TX 77005 USA
E-mail: daniel@rice.edu

Abstract. We report the first observation of the disorder-induced Bose peak in nanometer-sized liquid water pools. The size-dependent properties of the peak permit us to rule out Rayleigh scattering as a possible mechanism for the formation of this ubiquitous feature of amorphous materials.

The origin of the spectral feature known as the Bose peak remains an important outstanding question in the study of amorphous materials and the liquid-to-glass transition. The Bose peak is a universal phenomenon in amorphous solids, appearing as an anomalous enhancement in the vibrational density of states of the system in the far infrared spectral range [1]. It has been observed in many different glassy materials by a wide range of experimental techniques. These excess vibrational modes play a role in many aspects of the glassy state. For example, they are thought to be related to the well-known low-temperature specific heat anomaly of glassy materials. They are also correlated with the fragility (i.e., the behavior of the temperature-dependent viscosity) of the glass former; Bose peaks are generally prominent in strong glass formers but are much less prominent in fragile glass formers [1]. A strong connection evidently exists between the high frequency dynamics reflected by the presence of the Bose peak and the long-time macroscopic dynamics of glass formers [2].

Despite the large number of studies devoted to the Bose peak and its evident importance, the microscopic details of its origin remain unclear. There is a general consensus that the redistribution of vibrational mode density arises from strong scattering of acoustic modes from the disordered lattice. According to the Ioffe-Regel criterion, a propagating wave spontaneously localizes when its wavelength becomes comparable to its mean free path L_F. For acoustic modes, this leads to the condition that the Bose peak should appear at a frequency $f_{Bose} = V_S/L_F$, where V_S is the sound velocity. However, the nature of the scattering mechanism which gives rise to this localization is unknown. Numerous mechanisms have been proposed, involving collective interactions on length scales ranging from a few molecular units to several nanometers [3-6]. Currently there is no consensus among these, in large part because there have been few experiments which have been able to distinguish among the various proposals.

Here, we demonstrate that nanometer-scale confinement can induce a strong Bose peak in a most fragile liquid, room temperature water. We further show that, by systematically controlling the size of the confinement, we can gain new understanding about the length scale associated with the anomalous vibrational modes. This permits us to experimentally distinguish between several of the proposed microscopic explanations for the phenomenon.

We have used terahertz time-domain spectroscopy to measure the far-infrared dielectric function of water pools at the core of inverse micelles. These systems have been extensively characterized using a number of different techniques, and it is by now well established that this water pool is essentially identical to the bulk liquid, at least with respect to its local (i.e., nearest neighbor) structure [7,8]. Unlike most other techniques, far infrared spectroscopy provides an indication of the structure and dynamics on mesoscopic, rather than microscopic length scales. It is therefore a probe of collective effects in the liquid.

These measurements consist of transmission spectroscopy, using a variable path length liquid cell. Measured spectra are deconvolved using the known spectrum of the background solvent (heptane), and of the surfactant shell which segregates the water pool. This procedure relies on a core-shell dielectric model commonly used for these and similar systems [9]. This permits a determination of the absorption coefficient of the water pool, over a broad spectral range.

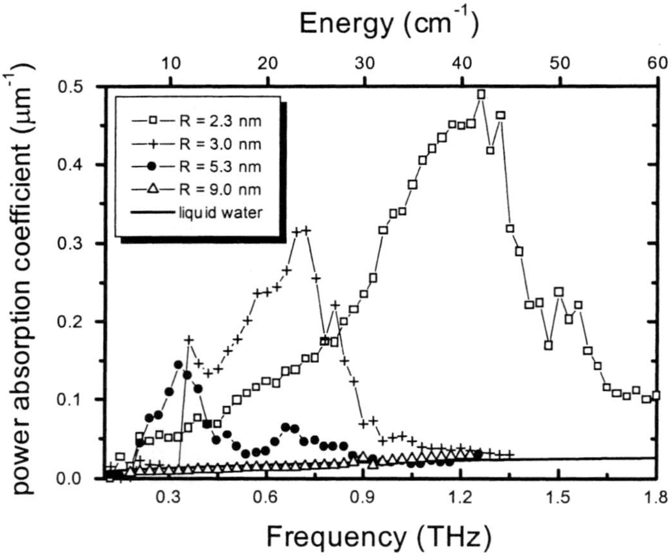

Fig. 1. Measured absorption spectra of representative micelle samples, with different water pool radii shown. The solid line is the double-Debye model representing the absorption of the bulk liquid. Note that the absorption in the largest micelles converges to the bulk limit.

Figure 1 shows absorption spectra of several different micelle samples. An important feature to note is that, for the larger micelles, the absorption spectrum converges to that of ordinary bulk water, illustrated by the slowly varying solid line. This validates the core-shell model used for the data analysis. In addition, it confirms that the water pool at the core of sufficiently large micelles behaves identically to the bulk liquid. As the micelle size is reduced, the absorption

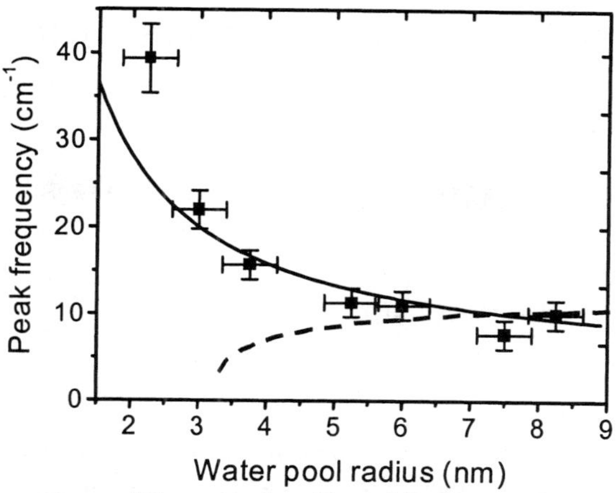

Fig. 2. The dependence of the spectral position of the Bose peak on water pool radius. The solid and dashed lines shows the prediction of two models; the solid line assumes a frequency-independent mean free path, while the dashed line assumes that Rayleigh scattering (for which $L_F \sim \lambda^4$) plays a dominant role.

spectrum develops a pronounced peak that grows and shifts to higher frequency with decreasing water pool radius. Figure 2 shows the evolution of the spectral location of this peak as a function of the water pool radius.

These results permit us to narrow the range of possible explanations for the nature of the interaction which gives rise to the Bose peak. First, the fact that the peak shifts even for radii as large as a few nanometers indicates that the modes involved cannot be highly localized as proposed in [3], but must be extended over mesoscopic length scales. Further, the blue shift with decreasing size is consistent with a frequency-independent mean free path as in [4,5] (solid line), and not with a mean free path varying as λ^4, as in Rayleigh scattering or the soft potential model [6] (dashed line). These data represent, for the first time, a systematic variation of the properties of the Bose peak as a function of an experimentally controllable parameter, the size of the water pool. They are also the first report of observations of the Bose peak in liquid water, and may have a bearing on the recently proposed fragile-to-strong phase transition in this system [10]

[1] A. P. Sokolov, *et al.*, *Phys. Rev. Lett.* **69**, 1540 (1992).
[2] M. D. Ediger, C. A. Angell, S. R. Nagel, *J. Phys. Chem.* **100**, 13200 (1996).
[3] T. Nakayama, *Physica B* **263-264**, 243 (1999).
[4] V. N. Novikov, *Phys. Rev. B* **58**, 8367 (1998).
[5] S. R. Elliott, *Europhys. Lett.* **19**, 201 (1992).
[6] D. A. Parshin, *Phys. Rev. B* **49**, 9400 (1994).
[7] A. Maitra, *J. Phys. Chem.* **88**, 5122 (1984).
[8] G. Onori, A. Santucci, *J. Phys. Chem.* **97**, 5430 (1993).
[9] T. Hanai, T. Imakita, N. Koizumi, *Coll. Polym. Sci.* **260**, 1029 (1982).
[10] K. Ito, C. T. Moynihan, C. A. Angell, *Nature* **398**, 492 (1999).

Interaction of terawatt laser pulses with neat water

S. Pommeret[1], F. Gobert[1], M. Mostafavi[2], I. Lampre[2], P. Pernot[2], R. Haïdar[1], S. Buguet[1], G. Vigneron[1] and J.-C. Mialocq[1]

[1] CEA/Saclay, DSM/DRECAM/SCM/URA 331 CNRS, 91191 Gif-sur-Yvette Cedex, France, E-mail: Stanislas.Pommeret@cea.fr

[2] Laboratoire de Chimie-Physique UMR 8610, CNRS/Université Paris-Sud, Bâtiment 350, 91405 Orsay Cedex, France, E-mail: Mehran.Mostafavi@lcp.u-psud.fr

Abstract. Femtosecond time-resolved experiments in neat water under terawatt ultraviolet laser pump power density are reported. For the first time, the formation of a decimolar concentration of hydrated electron is demonstrated. The contribution of non-geminate recombination is discussed.

1. Introduction

Nowadays the use of femtosecond lasers in surgical applications is increasingly popular [1]. The study of the interaction of ultra intense laser pulses with neat water is of interest since the human body is essentially composed of water. However, water has not been studied as intensively as other dielectric media such as fused silica [2,3].

It is known that the interaction of intense ultraviolet laser pulses with neat water produces via a multiphoton absorption mechanism, hydrated electrons, hydroxyl radicals and hydronium cations [4-6]. Then subsequent reactions occur on the picosecond, nanosecond and microsecond time scale, mainly reactions of hydrated electrons with hydronium cations and hydroxyl radicals [5-7]. Those pioneering studies have been performed at "low" power density (0.01-0.1 TW/cm^2). The present experiments using very intense laser pulses (0.1-1.0 TW/cm^2) allow the formation of those primary species at high concentration of particular interest for the radiation chemistry community. Indeed, when an ionizing particle interacts with a liquid, it creates inhomogeneously distributed spurs containing highly concentrated primary species.

2. Experimental Methods

The femtosecond absorption spectroscopy experiments have been performed using a commercial kilohertz Ti:Sa laser system delivering at 800 nm laser pulses of 50 fs duration with an energy of 700 μJ. The pump beam at 266 nm with an energy up to 100 μJwas focussed in a 70 μm thick water jet. The 180 μm diameter of the pump beam was measured using a CCD camera. The probe pulse was delayed using a motorized optical delay line and then spatially filtered using a 50 μm pinhole inside a Keplerian beam expander of a magnification of 1 and overall length of 0.4 m. The measured diameter of the probe beam was within the resolution of our measurement system (50 μm). The probe and the reference beam were detected using silicon photodiodes.

For nanosecond and microsecond time scale experiments, the 266 nm beam and a He:Ne laser beam (632.8 nm) were used for excitation and probing, respectively.

Water had a resistivity greater than 18 MΩ cm and was acidified -when necessary- using perchloric acid to scavenge hydrated electrons. The perchlorate anion is considered inert with respect to electron photodetachment.

3. Results and Discussion

In pure water, at 1.5 TW/cm^2 laser power density, the initial 632.8 nm absorbance at 1 ns is approximately 0.2. Its decay follows a second order kinetics as expected from the hydrated electron in the nanosecond time range. Moreover, the absorbance decays more and more rapidly with increasing the acid concentration. In strongly acidic solutions, the hydrated electron disappears in the sub-nanosecond time range. The 0.2 initial absorbance in pure water corresponds to a concentration of about 2 10^{-3} M assuming an homogenous distribution.

To get a better understanding of the primary events under laser excitation with high power densities, experiments were performed in the femto- and picosecond time range. The results shown in figure 1 have been obtained at 1.5 TW/cm^2 and the absorbances at 2 ps delay after excitation are roughly equal to 1.0. Confirming the nanosecond observations, in strongly acidic solutions (1.5 M) the 800 nm absorbance disappears within 300 ps. Assuming that at 2 ps time delay the absorption is mainly due to the fully hydrated electron, the 1.0 initial absorbance corresponds to a concentration of about 10^{-2} M. To our knowledge, such a high concentration of hydrated electrons produced by a single laser pulse has never been reported. Moreover, the calculation of the energy deposit assuming a two-photon absorption law enables us to demonstrate that the ratio between the peak concentration and the measured concentration is 15. Therefore, the hydrated electron peak concentration is 0.15 M. Such a high concentration never reported is comparable to the expected concentrations inside radiolysis spurs. Within our experimental conditions we have been able to demonstrate that shapes of the kinetics are dependent on the excitation power density, demonstrating for the first time the role of non-geminate recombination in the picosecond time scale when creating the hydrated electron with intense UV laser pulses.

Finally we observed that after a few microseconds the continuous analyzing light (He:Ne laser) is no longer transmitted through the jet due to light scattering. We also note the presence of water vapor around the jet and a noise is heard due to the laser induced volume expansion and/or vaporization. As there is no evidence of plasma formation, this phenomenon might be a consequence of the H atoms peak concentration (0.15 M) which is well above the solubility limits. Such a high concentration will imply a violent volume expansion in the microsecond time range such as that we observed and that already reported in other liquids [8].

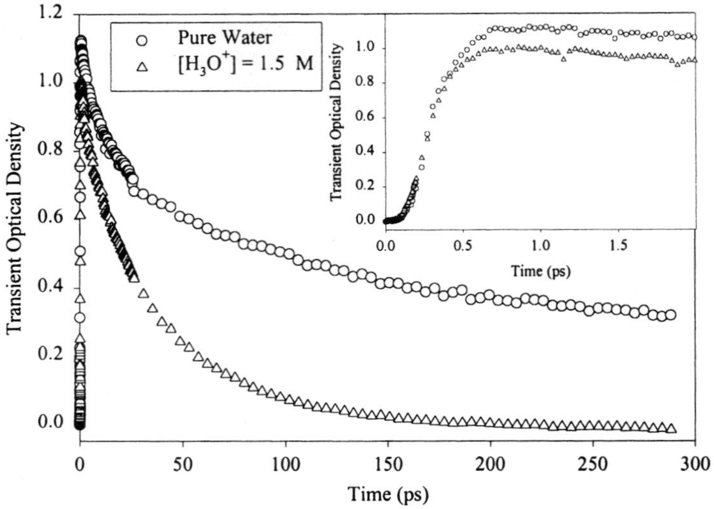

Fig. 1. Decay kinetics of the hydrated electron observed at 800 nm after terawatt excitation of water at 266 nm, in the absence and in the presence of perchloric acid (1.5 M).

To conclude, the formation of decimolar concentrations of hydrated electron under high laser pump power density at 266 nm has been demonstrated in neat water. Non-geminate recombination plays an important role in the picosecond time range.

References

1 T. Juhasz, F.H. Loesel, R.M. Kurtz, C. Horvath, J. F. Bille and G. Mourou, IEEE J. Sel. Top. Quantum Electron. **5**, 902, 1999.

2 A.-C. Tien, S. Backus, H. Kapteyn, M. Murnane and G. Mourou, Phys. Rev. Lett. **82**, 3883, 1999 and references therein.

3 F. Quéré, S. Guizard, P. Martin, G. Petite, O. Gobert, P. Meynadier and M. Perdrix, Appl. Phys. B **68**; 459, 1999.

4 A. Migus, Y. Gauduel, J.L. Martin and A. Antonetti, Phys. Rev. Lett **58**, 1559, 1987.

5 D.N. Nikogosyan, A.A. Oraevsky and V.I. Rupasov, Chem. Phys. **77**, 131, 1983.

6 R. A. Crowell and D.M. Bartels, J. Phys Chem. **100**, 17940, 1996.

7 Y. Gauduel, S. Pommeret, A. Migus and A. Antonetti, J. Phys. Chem. **93**, 3880, 1993.

8 Y. Tsuboi, K. Hatanaka, H. Fukumura, H. Masuhara, J. Phys. Chem. A **102**, 1661, 1998

Picosecond Infrared Spectroscopy of Carboxylic Acid Dimers in Liquid Solution and Water-in-Oil Microemulsions

G. Seifert, T. Patzlaff, M. Janich and H. Graener

Martin-Luther-University Halle, Physics Institute, Hoher Weg 8, D-06099 Halle, Germany
E-mail: g.seifert@physik.uni-halle.de

Abstract. The picosecond dynamics of acetic and stearic acid dimers in liquid solution was studied using infrared double resonance spectroscopy after vibrational excitation of stretching vibrations in the dimer ring and the C-H stretching modes, respectively. The same technique was used in combination with dynamic light scattering to investigate the ultrafast dynamics of ternary liquid mixtures of water, Aerosol OT and octane (microemulsions) containing nanoscaled water droplets.

1. Introduction

Carboxylic acids with longer hydrocarbon chains such as stearic acid (so-called lipid molecules) are important constituents of human skin. In this work, the vibrational and structural dynamics of hydrogen-bonded dimers of stearic acid in liquid solution were studied using picosecond infrared double resonance spectroscopy. As the 3000 cm^{-1} region of both infrared and Raman spectrum of these dimers is dominated by the hydrocarbon chain, additionally acetic acid dimers were investigated to get information about the dynamics of the doubly hydrogen-bonded C–O–H ring.

Ternary liquid mixtures of hydrophobic liquid, surfactant and water are called microemulsions; these mixtures can be used as drug-carrier systems or nanoreactors. A typical water-in-oil microemulsion is a mixture of water, AOT (Sodium di-(2-ethyl-hexyl) sulfosuccinate) and n-octane. The picosecond dynamics after vibrational excitation of the water molecules in this system has also been studied using the infrared double resonance technique. Additional information was obtained by help of dynamic light scattering.

2. Results and Discussion

The carboxylic acids were dissolved in CCl_4, where they form rather stable dimers [1]. The transient transmission changes $\ln(T/T_0)$ produced by excitation of the CH_2 stretching modes of 0.1 mol/L stearic acid in carbon tetrachloride are given in Fig. 1 at three different delay times. The dashed spectrum refers to excitation of the antisymmetric CH_2 stretching vibration, the solid curves gives spectra obtained after excitation of the symmetric mode. As in both cases symmetric and antisymmetric mode are bleached already at zero delay, a fast energy redistribution between these modes can be deduced directly from the data of Fig. 1. Tracing the amplitude of the high-frequency shoulders, e.g. the one at 2960 cm^{-1} referring to the antisymmetric CH_3 vibration, energy migration along the hydrocarbon chain (16 CH_2 groups) within a few picoseconds can be concluded. Analyzing time-resolved data a unique effective lifetime of approximately 17 ps for the C–H stretching modes was found. So the CH_3 endgroups seem to play an important role for the relaxation of vibrational energy into the solvent. Similar results were obtained in the past for polyalkenes in solution [2], indicating that in the case of stearic acid the dimer ring is not important for the relaxation of C-H vibrational energy into the solvent.

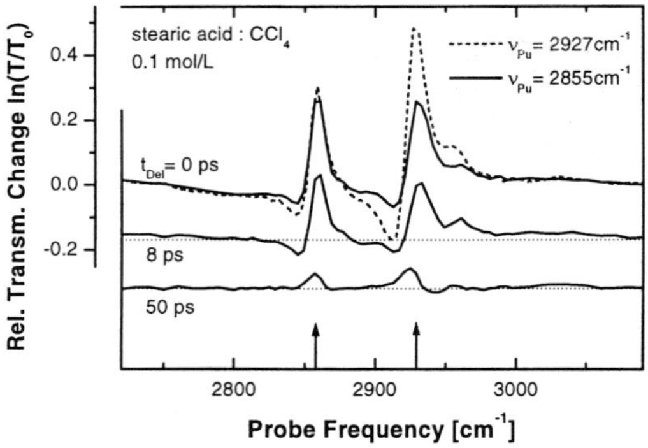

Fig. 1. Transient transmission changes of a liquid solution of 0.1 mol/L stearic acid in CCl_4 after excitation at 2927 cm^{-1} (dashed curve) and 2855 cm^{-1} (solid curves, three different delay times)

From the experiments on 0.5 mol/L acetic acid in CCl_4 it could be concluded that the vibrational modes of the dimer ring are strongly coupled leading to a fast redistribution (time scale 1 ps) of the vibrational energy among the manifold of ring modes [3,4]. An effective lifetime of ≈ 4 ps is found for this population evaluating the (red-shifted) excited state absorptions. But also a blue-shifted induced absorption indicating a modification of the hydrogen bonds was observed; analyzing the time evolution in this region, a time constant of 15-20 ps was obtained for the rearrangement of the H-bond system.

The experiments on microemulsions were performed on ternary liquid mixtures of octane, the surfactant AOT, and 0.28 mol/L H_2O at room temperature. In these systems water droplets with a typical radius of several nanometers are formed; the size of these droplets depends on the content of AOT, usually characterized by the concentration ratio $\mu = [H_2O]/[AOT]$. Dynamic light scattering exhibits an increase of the radius with increasing μ (a linear fit of the data was used to scale the upper abscissa in Fig. 2). Picosecond IR spectroscopy after excitation within the broad O–H stretching band of water yields at very early times several narrow induced absorptions on the high-frequency side indicating the presence of quasi-free O–H groups [6]. At later times, only a broad bleaching and blue-shifted induced absorption are observed, which can be explained by the thermal line shift of the water O-H stretching band. This features show a quite slow, clearly non-exponential decay, which is getting slower with increasing μ. As the vibrational relaxation and thermalization of bulk water is known to be finished within a few picoseconds [6-8], this behavior can be assigned to thermalization of the excess energy within the microemulsion. Fitting a series of data using stretched exponential functions $y \propto \exp(-t/\tau)^{\beta}$ typically yields increasing values of the parameters τ and β with increasing μ, as can easily be seen for the example given in Fig. 2. A more detailed analysis reveals that most probably the variation of τ and β is determined by the average radius of the reverse micelles and the polydispersity of their sizes.

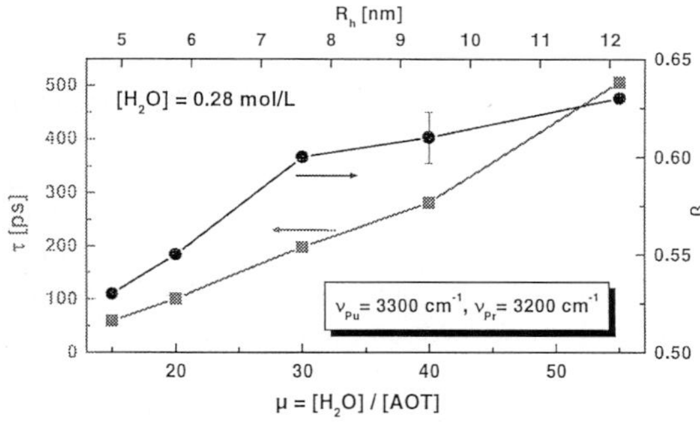

Fig. 2: Parameters of stretched exponential fits to a series of time resolved data with constant water concentration, but varying AOT content μ.

References

1 Y. Marechal and A. Witkowski, J. Chem Phys. **48**, 3697 (1968).
2 H. Graener, T.-Q. Ye, and A. Laubereau, J. Phys. Chem. **93**, 7044 (1989).
3 G. Seifert, H Graener, Bull. Pol. Acad. Sci. Chem. **47**, 397 (1999).
4 G. Seifert, T. Patzlaff, H. Graener, *to be published.*
5 T. Patzlaff, M. Janich, G. Seifert and H. Graener, *submitted.*
6 R. Laenen, C. Rauscher and A. Laubereau, Phys. Rev. Lett. **80**, 2622 (1998).
7 G. M. Gale, G. Gallot, F. Hache, N. Lascoux, S. Bratos and J.-Cl. Leicknam, Phys. Rev. Lett. **82**, 1068 (1999).
8 H.-K. Nienhuys, S. Woutersen, R. A. van Santen and H. J. Bakker, J. Chem Phys. **111**, 11494 (1999).

Ultrafast coherent response of hydrogen bonds

Jens Stenger[1], Dorte Madsen[2], Jens Dreyer[1], Erik T. J. Nibbering[1], Peter Hamm[1], and Thomas Elsaesser[1]

[1] Max-Born-Institut für Nichtlineare Optik und Kurzzeitspektroskopie, Max-Born-Strasse 2A, D-12489 Berlin, Germany.
E-mail: dreyer@mbi-berlin.de
[2] Department of Chemistry, Aarhus University, Langelandsgade 140, 8000 Aarhus C, Denmark.

Abstract. Femtosecond pump-probe spectroscopy in the mid-infrared demonstrates the coherent oscillatory modulation of the hydrogen bond distance and its strength in an intramolecular hydrogen bond, giving evidence of anharmonic coupling between high- and low-frequency modes.

Hydrogen bonding is responsible for structural and functional properties of numerous molecular systems. The dynamics of proton motion in a hydrogen bond, however, is far from being understood. Vibrational spectra give detailed information about hydrogen bonding. The high-frequency stretching band of a X–H hydrogen donor group in a X–H...Y hydrogen bond is strongly affected by hydrogen bonding with the most evident effects being the red shift of the ν(X–H) frequency compared to a free X–H group, its intensity increase and broadening, often accompanied by a peculiar bandshape with substructure. Anharmonic coupling to low-frequency modes, which modulate the hydrogen bond distance, is generally considered to be of major importance for ultrafast hydrogen motion and for explaining the vibrational lineshape [1]. So far, however, direct insight into those phenomena has not been attained.

As a representative example of a medium strong intramolecular hydrogen bond we studied phthalic acid monomethylester (PMME, Fig. 1A) in the nonpolar solvent C_2Cl_4. To avoid the interference of C–H stretching modes within the high-frequency stretching band, our study is focused on the O–D stretching band of the carboxy-deuterated derivative (PMME-D). Mid-infrared pulses of 130 fs duration, 150 cm^{-1} bandwidth, and more than 1 μJ energy/pulse were generated by parametric difference frequency mixing [2]. Pump and probe measurements were performed with excitation and probe pulses centered at 2100 and 2300 cm^{-1} (Fig. 1B). The probe and reference pulses were spectrally dispersed after interaction with the sample and broad-band detected using a dual 16-channel HgCdTe detector array.

The broad stationary O–D stretching band of PMME-D in C_2Cl_4 is composed of two main subbands covering a hardly visible fine structure (Fig. 1B). Time resolved transients are shown in Figure 2 for excitation at 2100 cm^{-1} (A) and 2300 cm^{-1} (B). For negative delay times and around zero delay, the transients are strongly influenced by perturbed free induction decay and cross phase modulation. However, the transients observed for the pure solvent (Fig. 2B - dashed line) show an amplitude small compared to the overall transmission change for delay times longer than 300 fs.

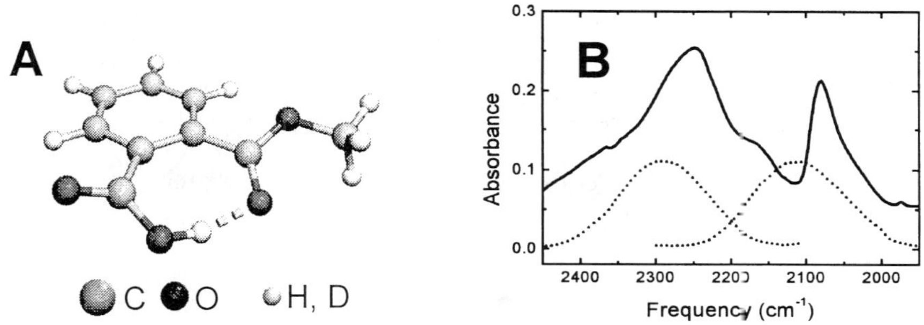

Fig. 1. A. Molecular structure of PMME-D. **B.** Linear O–D stretching absorption band (solid line) and power spectrum of the pump pulses centered at 2100 and 2300 cm^{-1} (dashed line).

In Figure 2A, an absorption increase is found at 2028 cm^{-1}, whereas a decrease of absorption, i.e. bleaching, occurs at higher probe frequencies. The enhanced absorption at 2028 cm^{-1} is due to the $v_{OD} = 1$ to $v_{OD} = 2$ transition and decays with a time constant of $T_1 \sim 430$ fs, the lifetime of the $v_{OD} = 1$ state. The bleaching at higher probe frequencies shows a slow nonexponential decay (~ 20 ps). The general behavior for excitation at 2300 cm^{-1} is similar (Fig. 2B). After 500 fs, induced absorption appears in the blue wing (2315 cm^{-1}) decaying within about 20 ps indicating an overall blue shift of the band. Those transients reflect intramolecular vibrational energy redistribution and energy dissipation to the solvent [3].

The pronounced oscillatory contributions display a period of 330 fs in all measurements and persist up to a delay time of about 1 ps accompanied by distinct phase shifts. The Fourier transform of the oscillatory signal at 2028 cm^{-1} shows a maximum at 100 cm^{-1} with a spectral width of 25 cm^{-1} (Fig. 2C). The oscillations observed at 2208 cm^{-1} are damped with a time constant of ~ 450 fs, very close to the $v_{OD} = 1$ decay time T_1.

In our experiments, the pump pulse creates a low-frequency vibrational wavepacket in the $v_{OD} = 1$ state and – via an impulsive Raman-like process – in the $v_{OD} = 0$ ground state. The transient absorption at a probe frequency of 2028 cm^{-1} is due to the $v_{OD} = 1$ to $v_{OD} = 2$ transition and the oscillatory signal is mainly due to the motion of the wavepacket in the $v_{OD} = 1$ state. This motion is damped with a time constant of 430 fs, the decay time of the $v_{OD} = 1$ population. In contrast, the negative signals are dominated by bleaching and stimulated emission on the $v_{OD} = 0$ to $v_{OD} = 1$ transition and here the wavepackets in both the $v_{OD} = 0$ and $v_{OD} = 1$ states contribute. The variation of the oscillation amplitudes and phases with probe frequency suggests that the spectral position of the ν(O–D) band is blue-shifted and modulated by the coherent motion along the low-frequency coordinate.

The observed oscillation frequency of 100 cm^{-1} is attributed to a low-frequency mode contributing to an absorption band in the far-infrared around ~ 115 cm^{-1} (Fig. 2C). Normal mode calculations on PMME-D indicate that this mode is

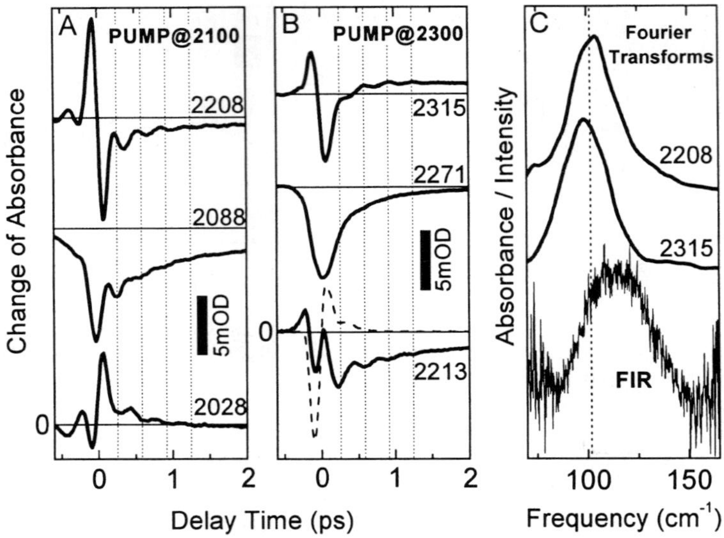

Fig.2. Change of absorption after excitation at 2100 (A) and 2300 cm^{-1} (B). Pure solvent signal for 2213 cm^{-1} (dashed line). C. Fourier transfrom of the oscillatory signal at 2208 and 2315 cm^{-1}, together with the far-infrared vibrational absorption of PMME-D.

connected with out-of-plane motions of PMME-D, strongly modulating the length of the hydrogen bond. Anharmonic coupling of this mode to the fast O–D stretching motion allows the impulsive excitation of low-frequency wavepackets by the pump pulse, resulting in the observed oscillatory absorption changes.

In conclusion, we report the first observation of coherent wavepacket motion along a hydrogen bond giving direct evidence of anharmonic coupling between the high-frequency stretching and low-frequency motions. The spectral position of the high-frequency band is modulated by a coherent motion corresponding to a periodic variation of the hydrogen bond distance and its strength.

References

1 O. Henri-Rousseau, P. Blaise, *Adv. Chem. Phys.* **103**, 1, 1998.

2 P. Hamm, M. Lim, R. M. Hochstrasser, *J. Chem. Phys.* **107**, 10523, 1998.

3 H.-K. Nienhuys, S. Woutersen, A. A. van Santen, H. J. Bakker, *J. Chem. Phys.* **111**, 1494, 1999.

Phase-sensitive detection for polarization-selective femtosecond Raman spectroscopy

M. Khalil, O. Golonzka, N. Demirdöven, and A. Tokmakoff

Department of Chemistry, Massachusetts Institute of Technology, Cambridge, MA 02139
E-mail: tokmakof@mit.edu

Abstract: Phase-sensitive detection of polarization-selective femtosecond nonresonant Raman spectroscopy is demonstrated using a two-diffractive-optic interferometer. Additionally, this method allows for direct phase calibration and removal of unwanted signal contributions.

Introduction. Phase control of input fields and in heterodyne detection is of increaisng interest for the development of optical analogs to NMR. We present a new method of phase sensitive heterodyne detection for nonresonant femtosecond (fs) Raman experiments which employs two diffractive optics (DO). The use of DOs has been previously demonstrated for heterodyne-detection [1-4]. The two DO set-up allows us to vary the phase, ϕ, between our signal ($\mathbf{E}_s$) and local oscillator ($\mathbf{E}_{LO}$) fields to selectively measure the in-quadrature birefringent ($\mathbf{E}_{bir}$) and the in-phase dichroic ($\mathbf{E}_{dic}$) contributions to the third-order nonlinear polarization. Additionally, we can suppress two-pulse signal contributions generated in the same phase-matched direction. In the limit that $\mathbf{E}_{LO} \gg \mathbf{E}_s(\tau) = \mathbf{E}_{bir}+\mathbf{E}_{dic}$, only the desired signal is detected:

$$I(\tau,\phi) \approx \frac{nc}{2\pi}\left[E_{LO}(\tau)E_{bir}(\tau)\sin(\phi) + E_{LO}(\tau)E_{dic}(\tau)\cos(\phi)\right] \quad (1)$$

This method also allows us to independently vary the polarization of the input beams in order to measure the tensor components of the third-order nonlinear response functions. Previous fs nonresonant experiments have predominantly measured the anisotropic response ($R_{ANISO} = R_{ZZZZ}-R_{ZZYY}$), while the isotropic response ($R_{ISO} = R_{ZZZZ}+2R_{ZZYY}$), which suppresses reorientational motion, has largely been ignored. We measure the R_{ISO} response function of CS_2 using the phase-sensitive interferometer and demonstrate the selectivity of the R_{ANISO} and R_{ISO} response functions of CCl_4 using a nonresonant pump-probe experiment.

Experimental Methods. Experiments were performed with 50 fs pulses at 650 nm with the interferometer illustrated in Figure 1(a). The pump and probe beams were focussed on DO1 and the +1 and –1 diffraction orders from each pulse were imaged onto the sample using the familiar "box car" geometry shown in Figure 1(b). The $\mathbf{E}_S$ and the $\mathbf{E}_{LO}$ ($\mathbf{k}_3$) were recombined on DO2 and the spot corresponding to the overlap of the +1 and –1 diffraction orders of the two beams was sent to a silicon photodiode.

The relative phase (ϕ) between the signal and LO was calibrated by taking an interference pattern between the two probe beams with the pump beam blocked. Polarization-selective pump-probe measurements were made by rotating the angle

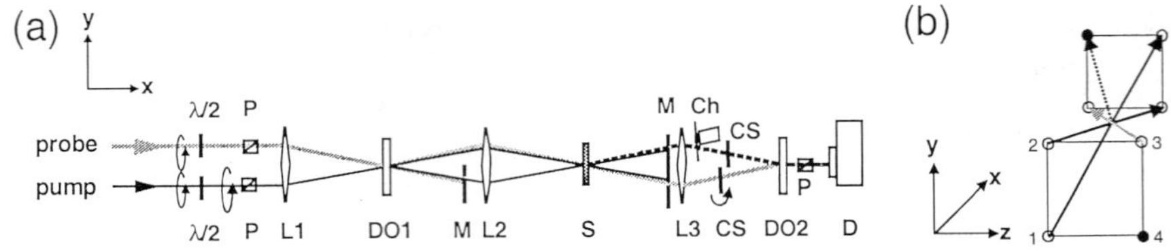

Fig. 1. (a) Dual-diffractive-optic interferometer for phase-sensitive detection. λ/2: half-wave plate, P: polarizer, L: lens, DO: diffractive optic, S: sample, M: mask, Ch: chopper, CS: cover slips and D: detector. In the actual experiment L2 and L3 represent a concave imaging mirror and a flat folding mirror. (b) The 'box-car' phase matching geometry showing the third-order signal being generated along $\mathbf{k}_4$.

of the pump polarizer (θ_{pump}), while the angles of the probe polarizer (θ_{probe}) and the analyzer (θ_{anal}) were kept constant at 0°.

Results and Discussion. Figure 2 demonstrates the phase-sensitive detection of the nonresonant response of liquid CS_2 and CH_3CN where the birefringent response ($\phi=\pi/2$, $3\pi/2$...) is appreciably greater than the dichroic response ($\phi=\pi$, 2π...). Along with $\mathbf{E}_S$ and $\mathbf{E}_{LO}$, there are pump-probe (PP) signals radiated along $\mathbf{k}_3$: $\mathbf{PP}_3'(\tau, \mathbf{k}_3 = \mathbf{k}_1\text{-}\mathbf{k}_1\text{+}\mathbf{k}_3)$ and $\mathbf{PP}_3''(\tau, \mathbf{k}_3 = \mathbf{k}_2\text{-}\mathbf{k}_2\text{+}\mathbf{k}_3)$ which are removed by chopping the signal before recombining it with the LO on the second DO. A phase-sensitive measurement technique such as this allows phase-cycling methods for data acquisition to be designed. Analogous to phase-cycling in NMR, we can remove unwanted signal contributions, such as pump-probe and homodyne contributions, and baseline imperfections by subtracting two data scans where the LO is π-phase-shifted between scans. This principle becomes particularly important in the development of two-dimensional nonresonant techniques.

A comparison of the three traces in Figure 3 (left) reveal that the R_{ZZZZ} and R_{ZZYY} response functions of CS_2 show the characteristic 1.6 ps long time relaxation. However, the isotropic dynamics damp away rapidly suggesting that the relaxation at early time results from a combination of isotropic and anisotropic dynamics and is dominated by reorientational motion after 2 ps.

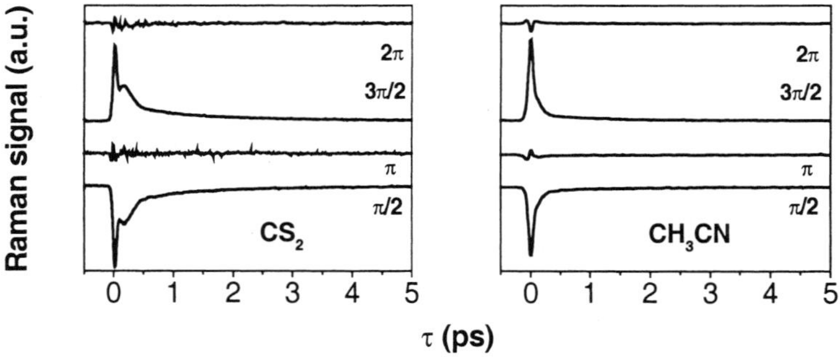

Fig. 2. Phase-sensitive detection of the third-order polarization in CS_2 (left) and CH_3CN (right) for the R_{ZZZZ} tensor element for different φ.

To demonstrate polarization selectivity as an important tool for separating reorientational and isotropic motions in high frequency intramolecular vibrational dephasing, we studied CCl_4 using a two-pulse fs Raman pump-probe experiment with control over the field polarizations [5]. The $Im[R_{ZZZZ}(\omega)]$ and $Im[R_{ZZYY}(\omega)]$ spectra in Figure 3 (right) shows that the amplitude and vibrational phase of the depolarized asymmetric vibrational modes at 217 (ν_2) and 314 cm^{-1} (ν_4) are dependent on the probe polarization while the strongly polarized ν_1 mode at 461 cm^{-1} is invariant to it. The modes at 217 and 314 cm^{-1} are absent from the isotropic spectrum, but they are the only modes present in the $Im[R_{ANISO}(\omega)]$ spectrum.

While previous experiments have demonstrated polarization selectivity and/or heterodyning capabilities, the two DO method allows for both of these capabilities with the additional discrimination against contaminating pump-probe signals. This technique holds great promise for resonant non-linear experiments by separating contributions from excited and ground states with the added advantages of passively phase-locked heterodyne detection and the ability to design phase cycles for data acquisition.

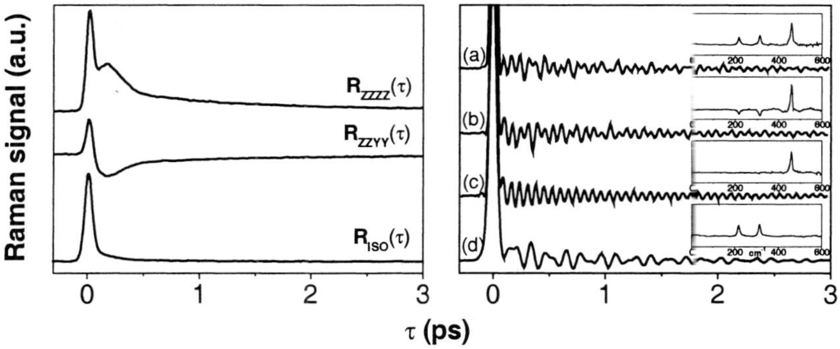

Fig.3. (Left) Variable polarization measurements of the intermolecular polarizability relaxation of CS_2 using phase-sensitive detection ($\phi = \pi/2$). (Right) Time-domain response functions measured by the nonresonant fs pump-probe for four polarization geometries with their respective deconvolved imaginary Fourier spectra: (a) R_{ZZZZ} (θ_{pump}=0°, $\theta_{anal}=\theta_{probe}$=0°) (b) R_{ZZYY} (θ_{pump}=90°, $\theta_{anal}=\theta_{probe}$=0°) (c) R_{ISO} (θ_{pump} = 54.7°, $\theta_{anal}=\theta_{probe}$=0°) (d). R_{ANISO} (θ_{pump}= 45°, θ_{anal}=90°; θ_{probe}=0°)

References

1 G. D. Goodno, G. Dadusc, and R. J. D. Miller, *J. Opt. Soc. Am. B*. 15, 1791 (1998).
2 G. D. Goodno, V. Astinov, and R. J. D. Miller, in *Ultrafast Phenomena IX*, Edited by T. Elsaesser (Springer-Verlag, 1998).
3 A. A. Maznev, T. F. Crimmins, and K. A. Nelson, *Opt. Lett.*, 23, 1378 (1998).
4 A. A. Maznev, K. A. Nelson and J. A. Rogers, *Opt. Lett.*, 23, 1319 (1998).
5 M. Khalil, O. Golonzka, N. Demirdöven, C. Fecko, and A. Tokmakoff, *Chem. Phys. Lett.*, 321, 231 (2000).

Studies of the Third and Fifth Order Non-Linear Susceptibilities of Optical Materials using Femtosecond Transient Absorption Spectroscopy

Katrin Ekvall[1], Cecilia Lundevall[2], and Peter van der Meulen[2]

[1] Royal Institute of Technology, Department of Physics, S-100 44 Stockholm, Sweden
E-mail: ekvall@atom.kth.se
[2] University of Stockholm, Department of Physics, S-113 85 Stockholm, Sweden
E-mail: cia@physto.se, meulen@physto.se

Abstract. The real and imaginary parts of the third order non-linear susceptibility of various glasses and liquids are determined using femtosecond transient absorption pump-probe spectroscopy. Some evidence for the interference with the fifth order contribution is obtained.

1. Introduction

Femtosecond transient absorption pump-supercontinuum probe spectroscopy is a powerful tool to study chemical reaction dynamics in the condensed phase [1]. In a typical experiment, the reaction is initiated by the absorption of a photon from a strong pump pulse and its progress is monitored by recording the absorption ($\Delta OD = -\log(I_{probe}/I_{reference})$) of a weak supercontinuum probe that can be delayed with respect to the pump. In order to maintain frequency resolution, the supercontinuum probe is dispersed by a spectrograph located behind the sample and is then detected by a CCD camera. However, it has recently been demonstrated that such a transient absorption set-up cannot only be employed to elucidate reaction dynamics, but can also be used to study the (third order) non-linear response of liquids or glasses [1-3]. In this contribution we have used the femtosecond transient absorption pump-supercontinuum probe technique to investigate the third and possibly the fifth order non-linear response of three optical glasses and one liquid: UV grade fused silica, BK7 optical glass, BS7 optical glass and ethylene glycol.

2. Experimental Methods

As a pump we used the second harmonic of a Ti:sapphire regenerative amplifier (Clark CPA 2001) at 387 nm. The energy of the pump pulses was varied between 5 μJ/pulse and 150 μJ/pulse. The duration of the second harmonic pulse was about 170 fs FWHM. The peak pump intensity at the sample position was estimated to lie between approximately 5 GW/cm^2 and 200 GW/cm^2, depending on the pump pulse energy. The supercontinuum probe was generated by focussing a small fraction of the Ti:sapphire fundamental in a 2.5 mm thick fused silica plate. The chirp rate of the supercontinuum was measured to be about $3x10^{-3}$ fs^{-2}. All samples had a thickness of only 200 μm to reduce the importance of propagation effects in the sample [2].

3. Results and Discussion

In Fig. 1 the pump-induced changes in the optical density for three different wavelengths of the supercontinuum probe are shown for all four nominally transparent samples. These results were obtained at pump energies between 5 and 60 μJ/pulse, i.e. at a peak pump intensity of less than 100 GW/cm^2. In Fig. 1a the signal from the UV grade fused silica plate is shown. This signal is due to cross-phase modulation (xpm) only [1-4], since UV grade fused silica absorbs at wavelengths below 170 nm and two-photon absorption (tpa) is thus not energetically allowed. In Fig. 1b, the signal from a free flowing jet of ethylene glycol is shown. The signals at 425 nm and 496 nm also originate from xpm, while the strong negative signal at 438 nm corresponds to (stimulated) Raman scattering from the C-H and/or O-H intra-molecular stretching vibrations. In Fig. 1c the experimental signal for BK7 optical glass is shown. In this case, the signal is a combination of xpm and tpa. Two-photon absorption is important here because BK7 glass absorbs at wavelengths below 350 nm. For BS7 optical glass (Fig. 1d) mostly tpa is observed. Note that the one-photon absorption edge for this material is at about 370 nm so tpa is indeed expected to be dominant.

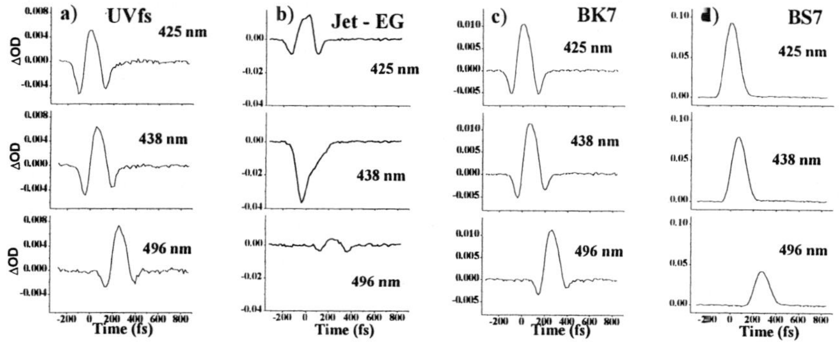

Fig. 1. Experimental ΔOD for three different wavelengths of the supercontinuum probe: a) UV grade fused silica b) free-flowing jet of ethylene glycol c) BK7 optical glass d) BS7 optical glass

Provided the experimental conditions (pulse duration, chirp rate, etc.) are known, a comparison of these experimental data with theoretical calculations (not shown) allow for a determination of both the real (xpm) and the imaginary (tpa) of the third order non-linear susceptibility ($\chi^{(3)}$) over a wide range of probe wavelengths. Alternatively, these measurements can be used to characterize both the pump and the continuum pulses.

At higher pump intensities the signal of the UV grade fused silica sample changes qualitatively, as shown in Fig. 2a where all curves have been normalized to the same maximum value.

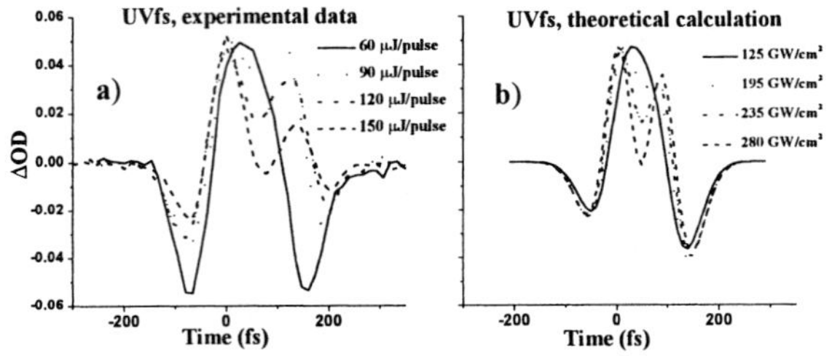

Fig. 2. Normalized experimental (a) and calculated (b) ΔOD at 425 nm for a 0.2 mm thick sample of UV grade fused silica for different pump intensities

The change in appearance observed at higher intensities suggests that a higher order non-linearity becomes important, and we tentatively propose that the variation in the shape of the signal is due to an interference between the $\chi^{(3)}$ and the $\chi^{(5)}$ (or perhaps cascaded $\chi^{(3)}$) contributions to the response [5]. The results of preliminary simulations including this fifth order response are shown in Fig. 2b. It can be seen that the change in shape of the signal with increasing intensity is well reproduced, although the agreement is certainly not perfect. Improvements of the calculations are currently in progress.

4. Conclusions

Femtosecond transient absorption spectroscopy with supercontinuum probing appears to be a promising technique to study the third and the fifth order non-linear response of optical materials.

References

1 S.A. Kovalenko, A.L. Dobryakov, J. Ruthmann, and N.P. Ernsting, Phys. Rev. A **59**, 2369, 1999.
2 K. Ekvall, P. van der Meulen, C. Dhollande, L.-E. Berg, S. Pommeret, R. Naskrecki, and J.-C. Mialocq, J. Appl. Phys. **87**, 2340, 2000.
3 J.K. Wang, T.-L. Chiu, C.-H. Chi, and C.-K. Sun, J. Opt. Soc. Am. B **16**, 651, 1999.
4 R.R. Alfano, *The Supercontinuum Laser Source*, Springer, New York, 1989.
5 S. Wu, X.-C. Zhang, and R.L. Fork, Appl. Phys. Lett. **61**, 919, 1992.

Nuclear Dynamics in the Ground and Excited Electronic States for a Molecule in Liquid

June-Sik Park and Taiha Joo

Division of Molecular and Life Science, Department of Chemistry, Pohang University of Science and Technology (POSTECH), Pohang, 790-784, KOREA
E-mail: thjoo@postech.ac.kr

Abstract. Ground state spectral hole and excited state particle dynamics are observed for a molecule in liquid by spectrally resolved four wave mixing. Detailed information on the intramolecular vibrations and solvation processes in each electronic state is obtained.

1. Introduction

In a typical resonant third order nonlinear spectroscopy such as pump/probe transient absorption, both the ground state bleach (GB) and the excited state stimulated emission (SE) components contribute to the signal. A full set of frequency-time resolved data may allow the ground state hole and the excited state particle dynamics to be observed separately, since they usually appear at different frequencies due to the Stokes shift. Many key questions regarding condensed phase nuclear dynamics, such as vibrational frequencies, dephasing times, and the spectral densities responsible for the solvation dynamics for each electronic state, can be answered, if GB and SE contributions to a transient spectra are resolved. Detailed molecular dynamics of an elementary chemical reaction can also be studied by observing the nuclear wave packet motion along the reaction coordinate in the excited state.

We have measured time resolved GB and SE spectra of a dye indocyanine green (IR125) in methanol by spectrally resolved four wave mixing (SRFWM) [1]. High quality time resolved transient spectra can be obtained by this technique. In addition, the three-pulse scheme provides the possibility of controlling wave packet formation by varying the time delay between the two pump pulses [2].

2. Experimental Methods

The light source was a cavity-dumped Kerr lens mode-locked Ti:sapphire laser that generates 18 fs pulses centered at 795 nm. Optical layout was similar to a standard three pulse transient grating (TG) setup in a 'box' geometry [1]. Two pump pulses ($\mathbf{k}_1$ and $\mathbf{k}_2$) with a time delay τ and a probe pulse ($\mathbf{k}_3$), delayed by T from $\mathbf{k}_2$, were focused into a sample solution. The third order nonlinear signal into $-\mathbf{k}_1+\mathbf{k}_2+\mathbf{k}_3$ phase matching direction was spatially filtered and directed to a spectrograph equipped with a CCD detector. Spectra of the probe pulses were measured simultaneously by using a two-leg fiber. The third order signal was normalized by the probe pulse spectrum to give an SRFWM signal.

3. Results and Discussion

A representative SRFWM at T=80 ps (τ = 0 fs) and two log-normal functions fit are shown in Fig. 1a. For time delays of T > 80 fs, the SRFWM can be well separated into GB and SE components. Inter- and intra-molecular nuclear dynamics manifest in the change of center frequency and line width of the GB and SE spectra [1]. The center frequencies of the GB and SE components obtained from the two log-normal function fits are shown in Fig. 1b and 1c. It is demonstrated that the center frequencies can be measured with an exceptional precision (±5 cm^{-1}).

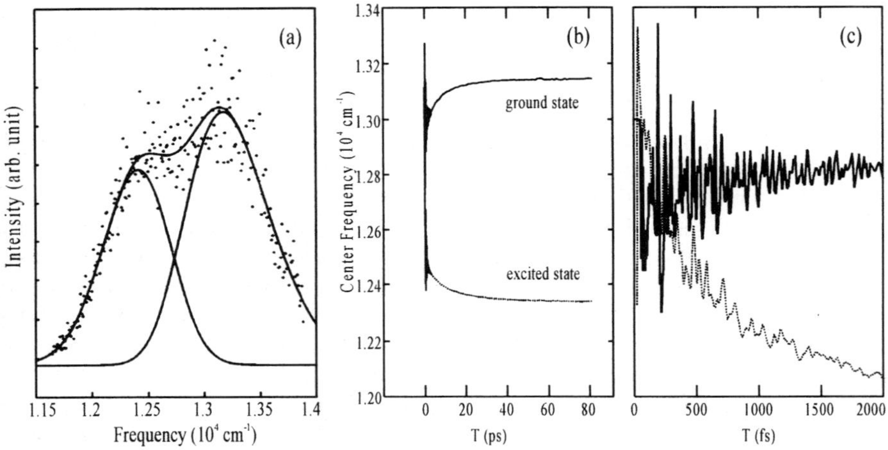

Fig. 1. (a) Spectrally resolved four wave mixing of indocyanine green in methanol and two log-normal functions fit, T = 80 ps and τ = 0 fs. The center frequencies vs. T for the SE and GB components are shown in 80 ps (b) and 2 ps (c) ranges.

As T increases, the SE spectrum shifts to the red due to solvation dynamics, while the GB spectrum shifts to the blue. The blue shift of the GB spectrum is due to the ground state spectral hole filling process by the energy fluctuations of the molecules in the ground state. The separate observation of the dynamic Stokes shift by the dissipation process while the molecules reside in the excited state and the hole-filling process by the energy fluctuation of the molecules in the ground state provides unique opportunity to test the validity of the usually assumed linear response approximation. The red-shift time constants of the SE and the blue-shift time constants of the GB are similar, although the GB shows a slightly faster initial time constant. The change of the dipole moment upon electronic transition is rather small in IR125, which justifies the linear response of the molecule in methanol [3].

Intramolecular vibrational wave packet motions in the excited and ground electronic states can also be resolved. The center frequencies of the GB and SE oscillate due to the vibrational wave packet dynamics in each electronic state.

Fourier transforms of the center frequencies are shown in Fig. 2. For comparison, Fourier transform of the transient grating signal is also shown. The frequencies in GB are generally higher by ~10 cm^{-1} with narrower line width than those in SE, which are consistent with the usual expectation that vibrations in the excited state have lower frequencies and dephase faster. Dephasing times of the vibrations in the excited states are 500-700 fs, slower than usually assumed ~100 fs, while they are around 1 ps in the ground state. This sets the lower bound for the vibrational relaxation time in the excited state to be a few hundred fs.

When the time delay between the pump pulses is varied, the wave packet formation can be controlled. SRFWM also provides an opportunity to observe the wave packet formation in the excited and in the ground state independently. Fig. 2b shows the Fourier transform of the center frequencies of the GB and SE spectra when τ is set to 20 fs. Vibrations having higher frequencies are enhanced, while they are suppressed when τ is 10 fs (not shown).

In conclusion, SRFWM is shown to be a powerful tool for the detailed study of the inter- and intra- molecular dynamics in condensed phases.

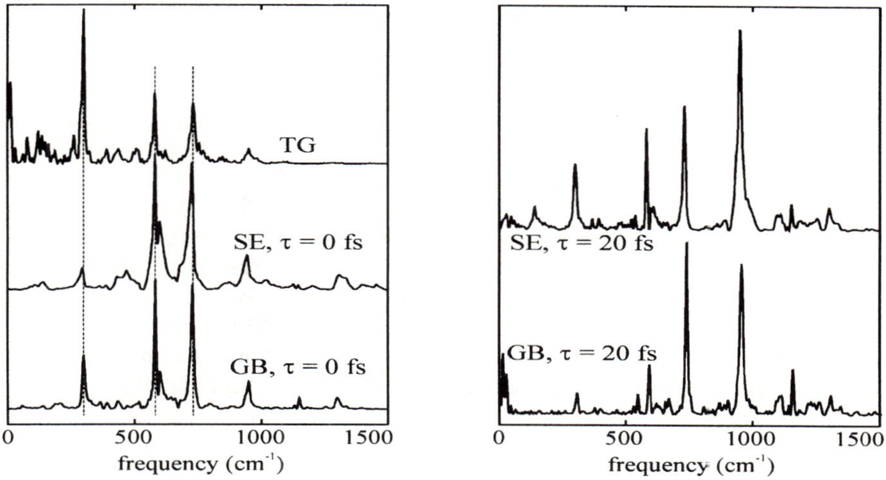

Fig. 2. Fourier transforms of the GB and SE center frequencies in SRFWM for $\tau = 0$ fs (a) and $\tau = 20$ fs (b). Fourier transform of a TG signal is shown for comparison.

References

1 S.-H. Lee, J.-S. Park, and Taiha Joo, J. Phys. Chem. *in press*.

2 E. J. Brown, I. Pastirk, B. I. Grimberg, V. V. Lozovoy, and M. Dantus, J. Chem. Phys. **111**, 3779, 1999.

3 P. V. Kumar and M. Maroncelli, J. Chem. Phys. **103**, 3038, 1995.

Ultrafast Solute Vibrational Spectral Evolution During the Solvation Process

John B. Asbury, Yongqiang Wang, and Tianquan Lian

Department of Chemistry, Emory University, Atlanta, GA 30322
Phone: (404)727-6649, Email: tlian@emory.edu

Abstract. Using femtosecond infrared spectroscopy, the rate and degree of frequency shift in the CO stretch of Re(dmbpy)$(CO)_3$Cl in its MLCT excited state is found to vary with the polarity and solvation time in a series of solvents.

1. Introduction

The dynamics of solvent response to the change in solute charge and charge distribution plays an important role in determining chemical reaction dynamics in solution. Some previous studies of solvation dynamics used Dynamic Fluorescence Stokes Shifts[1,2] to monitor change in the electronic energy level of the solute. Through these and other studies, a comprehensive understanding of solvation dynamics has been obtained. However, the change of solute vibrational spectrum during solvation has not been carefully studied. Since vibrational spectral features can often be assigned to specific groups or bonds, their changes during solvation may provide detailed insights of specific interactions between solvent and solute molecules. In this paper, we report a study of the spectral evolution of the excited state CO stretching mode of Re(dmbpy)$(CO)_3$Cl in a series of solvents. The sample preparation and femtosecond broadband detection mid-IR spectrometer used in the study have been described elsewhere[3].

2. Results

Transient IR spectral change was measured for Re(dmbpy)$(CO)_3$Cl [dmbpy=2,2'-bipyridine-4,4'dimethylacetate] in dimethylformamide(DMF), methanol(MeOH), ethanol(EtOH), iso-propanol(iPrOH), and iso-butanol(iBuOH). Re(dcbpy)$(CO)_3$Cl [dcbpy=2,2'-bipyridine-4,4'carboxylate] was studied on a nanocrystalline ZrO_2 thin film surface. The complexes have three ground state CO stretching bands. Only the highest frequency band at ~2024 cm^{-1} is analyzed because it is free of spectral interference observed in the lower frequency modes. Fig. 1 shows transient IR spectra of Re(dmbpy)$(CO)_3$Cl in MeOH and iPrOH. The transient spectra show bleach of the ground state and a partially overlapping, time-evolving excited state peak. The spectrum at each delay time is fitted with two Gaussian bands. The band representing the ground state bleach is fixed in height, FWHM and position, while these parameters are allowed to vary for the excited state band.

Figure 2 shows the time evolution of the excited state peak shift relative to its position at 100 fs. The shifts from the ground state to the final excited state

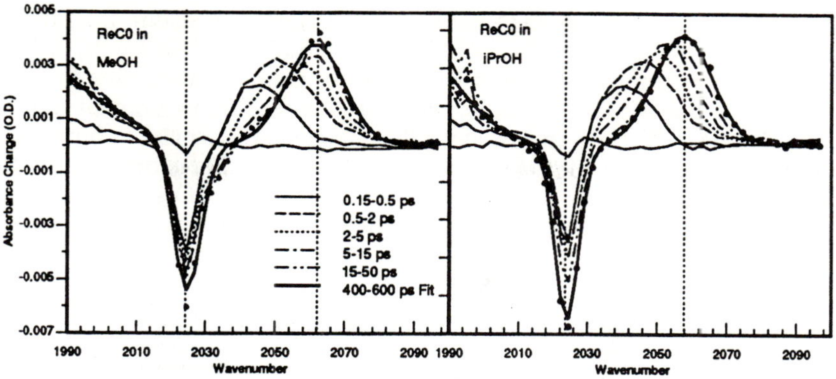

Fig. 1: Transient IR spectra of $Re(dmbpy)(CO)_3Cl$ in a) MeOH and b) in iPrOH

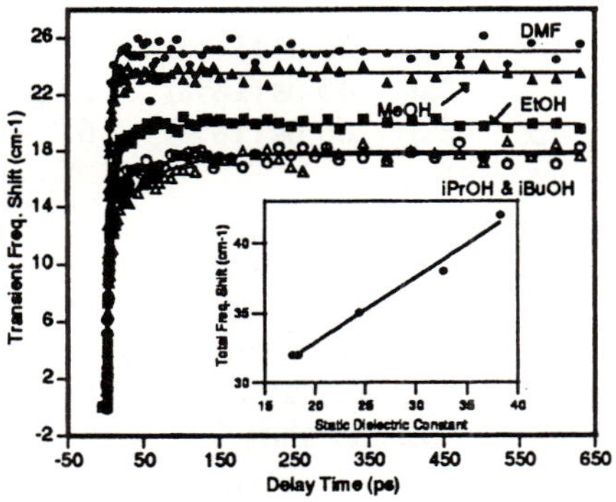

Fig. 2: The excited state CO peak position shift over time. Inset shows correlation of shift and solvent dielectric constant.

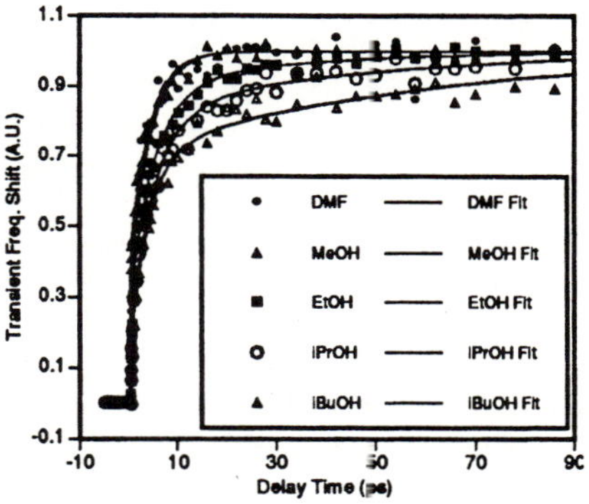

Fig. 3: The rate of evolution can be compared by normalizing the final shift.

position are listed in table 1 as well as the solvent static dielectric constant[4]. The inset of Fig. 2 shows a linear relation between these two values.

The rate of excited state spectral evolution decreases in the series of solvents going from DMF to iBuOH. Figure 3 shows this by normalizing the total shift to the same final value. The solid lines in Fig. 2 and 3 are fits to the data with multi-exponential rise functions. The fit parameters are shown in Table 1 along with solvation times measured by Dynamic Fluorescence Stokes Shift[1].

Similar spectra were also measured for $Re(dcbpy)(CO)_3Cl$ on ZrO_2 thin films surface (results not shown). The ground state high frequency CO stretching band is located at $2048cm^{-1}$. The excited state band shifts to 2072 cm^{-1} at > 10 ps. The shift of the peak is well fitted by a single-exponential rise with 4 ps time constant.

3. Discussion

Excitation at 400 nm produces a ^{1}MLCT state ~10000 cm^{-1} above the long-lived ^{3}MLCT state minimum. The transition promotes an electron from the Re center to bipyridine, oxidizing the Re. Due to back electron donation from the Re to the CO π^* orbital, the CO stretch frequency should increase in the MLCT state. Although not previously measured in the Re complex, relaxation from the ^{1}MLCT state to the ^{3}MLCT state occurred in ≤100 fs for $Ru(bpy)_3$ complex[5]. In the ^{3}MLCT state, a dipole is created along the direction of Re center to bipyridine. The solvent will rearrange in response to the created dipole. Therefore, the spectral evolution observed at >100fs can be viewed as change of the CO stretch frequency in the ^{3}MLCT state from solvation and vibrational relaxation.

Table 1. Comparison of spectral shift and dynamics in different solvents

Solvent	Shift (cm^{-1})	Static dielectric const[4]	Fitting parameter			Avg. Solv. Time[1]
			t_1(100fs)	t_2	t_3	
DMF	42	38.25	32%	4ps (68%)	-	~1 ps
MeOH	38	32.66	40%	4ps (60%)	-	~5 ps
EtOH	35	24.3	30%	5ps (59%)	30 ps (11%)	
IPrOH	32	18.3	20%	5ps (62%)	45 ps (18%)	33 ps
IBuOH	32	17.7	17%	4ps (54%)	60 ps (18%)	61 ps
Surface	24	N/A	0%	4ps (100%)	-	N/A
Matrix	20	N/A				N/A

For the Re complex on the surface of dry ZrO_2, the observed spectral shift is attributed to vibrational relaxation because there is no solvent. Table 1 shows a 4–5 ps component in all the samples studied. This suggests a similar vibrational relaxation rate in different media, dominated by intra-molecular pathways. In low temperature matrix, where solvent molecules are fixed, the shift from ground to excited state is only 20 cm^{-1}. A similar result was observed on dry ZrO_2 surface.

The amount of shift increases with solvent static dielectric constant. In more polar solvents, there is a stronger interaction of the CO dipole with the solvent molecules resulting in a bigger shift when the solvent reorients. As solvent molecules orient themselves to minimize the interaction energy with the dipole of the ^{3}MLCT state, they might adopt a configuration that is not optimal for interaction with the CO groups. Excluding the contribution from vibrational relaxation, the timescale of the spectral shift should reflect the rate of solvent reorientation. Fig. 3 and Table 1 show that the spectral shift is slower in solvents with longer solvation times. It is notable that there is a fast component with a time constant of ~100 fs in all the solvents studied. This is attributed to the inertial component of the solvent response[2]. The microscopic solvent motion responsible for the ensuing spectral shift is not yet clear.

References

1 P. F. Barbara and W. Jarzeba, Adv. Photochem. **15**, 1 (1990).

2 S. J. Rosenthal, X. L. Xie, M. Du *et al*, J. Chem. Phys. **95**, 4715-4718 (1991).

3 Y.Q. Wang, J. B. Asbury, and T. Lian, J. Phys. Chem. A **104**, 4291 (2000).

4 D. R. Lide, "CRC Handbook of Chemistry and Physics," (CRC press, Boca Raton, 1996).

5 N. H. Damrauer, G. Cerullo, A. Yeh *et al.*, Science **275**, 54 (1997).

Separating Inertial and Diffusive Rotation and Solvation for a Nonpolar Solute

Yunhan Zhang, Jianwen Jiang and Mark A. Berg

University of South Carolina, Department of Chemistry and Biochemistry, Columbia, SC 29208
E-mail: berg@mail.chem.sc.edu

Abstract: Femtosecond heterodyned dichroism measurements of anthracene solutions cleanly separate times for diffusive and inertial rotation and diffusive solvation. Comparisons are made to theory and to other solvation experiments.

1. Introduction

A persistent problem in measuring ultrafast dynamics in liquids is separating the contributions from the various processes that can contribute to a given spectroscopic observation. We present femtosecond dichroism measurements on anthracene in benzyl alcohol and toluene in which a good identification of all the decay components is made. Because the vibronic structure of anthracene is sparse, vibrational dynamics and oscillations are largely eliminated. Nonetheless, the results show as many as four distinct decay components. We apply a definition of diffusive and inertial dynamics that depends on the behavior with changing viscosity – diffusive rates scale directly with the viscosity of the liquid; inertial rates are viscosity independent [1]. Using this definition, inertial rotation of a single molecule is identified. An intermediate decay time (~0.5 ps) in toluene [2] is identified as a mixture of diffusive solvation and inertial rotation.

2. Dichroism Measurements

Dichroism signals $S(\tau)$ contain contributions from both the time dependent anisotropy $r(\tau)$ and the time dependent absorption $\Delta A(\tau)$, i.e., $S(\tau) = \Delta A(\tau) r(\tau)$. The anisotropy $r(\tau)$ is directly related to single-molecule rotation of the solute. The absorption decay $\Delta A(\tau)$ is due to hole burning and Stokes shifting of the absorption-bleach and stimulated-emission spectra. Within the cumulant approximation, all four-wave mixing measurements are determined by a single, one-dimensional frequency correlation function, $M(t) = \langle \delta\omega(t)\delta\omega(0)\rangle$. Thus dichroism contains complete information on both solvation and rotation.

We use a standard polarization spectroscopy experiment, with a frequency doubled Ti:sapphire (382 nm) source. Heterodyne detection selects the dichroism signal and rejects birefringent effects. The autocorrelation trace and the frequency spectrum of our pulses are both fit well with a 40 fs FWHM, chirp-free Gaussian.

3. Results

On the longest time scale (~100 ps in benzyl alcohol), the dichroism shows an exponential decay, which is assigned to rotational diffusion (Fig. 1a). The hydrodynamic volume implied by our measurements is consistent with other measurements of anthracene's rotation, with computer simulations [3] and with the van der Waals volume.

Figure 1a also shows that in benzyl alcohol, there is another 5-10 ps decay component. This decay time also scales with the solvent viscosity, and therefore represents a diffusive relaxation. It will be assigned to diffusive solvation below. A biexponential representing these two components fits the data very well from 1-200 ps (Fig. 1a). Below 1 ps, the fits deviate from the data, revealing a third, subpicosecond decay (inset to Fig. 1b). The fits were subtracted from the data to isolate this decay (Fig. 1b). Its decay time does not change as the viscosity is changed, so it represents an inertial relaxation.

This decay is assigned to inertial rotation. Computer simulations of anthracene rotation by Jas, et al. [3] were analyzed in the same way as our data, and the inertial component was compared to our data (Fig. 1b). The subpicosecond decay component has both the amplitude and decay time expected for inertial rotation.

The 5-10 ps diffusive decay in benzyl alcohol (see Fig. 1a) is assigned to diffusive solvation. It cannot represent rotational motion. The total anisotropy amplitude of 0.4 is fully accounted for by the ~100 ps and ~0.5 ps decays. Furthermore, the decay time is a factor of ten smaller than hydrodynamic predictions of the rotational time. However, it is within the range of times estimated from the viscoelastic model of nonpolar solvation dynamics [4].

Measurements were also made in toluene, which is approximately ten times less

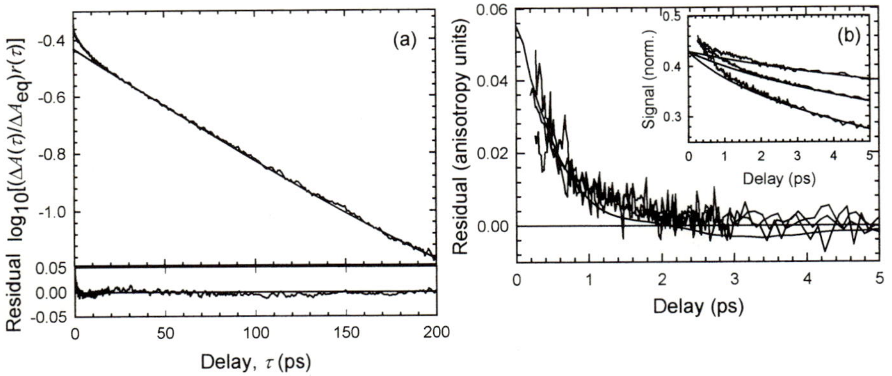

Fig. 1. Dichroism measurements on anthracene in benzyl alcohol. (a). Beyond 1 ps, the decays have two components: diffusive rotation (line), and diffusive solvation. Residuals from a biexponential fit of both processes (solid curves in inset) are shown below (a). (b) Inset: Below 1 ps, there is an additional decay component above the biexponential fit (solid). Top to bottom: 11.4 cP (4.0 °C), 5.0 cP (28.0 °C), 2.5 cP (56 °C). Main: Residuals from the biexponential fit are independent of viscosity, i.e., they represent an inertial process. The decays (noisy curves) match the inertial rotation decay extracted from the computer simulation of Jas, et al. [3] (smooth curve).

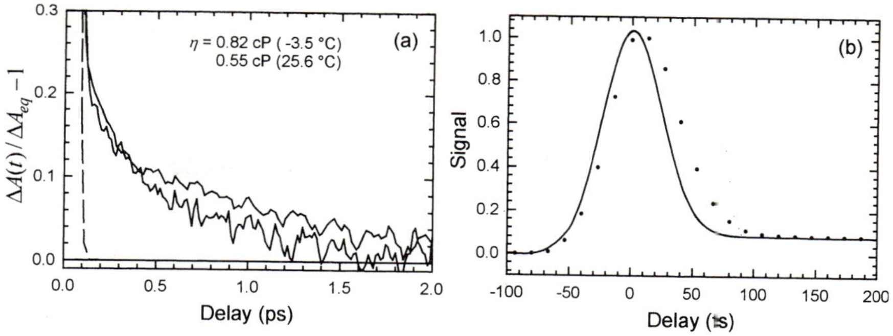

Fig. 2. Dichroism measurements of anthracene in toluene. a. The diffusive rotational contribution is removed (solid). The remaining decay depends on the viscosity (0.82 cP, upper curve; 0.55 cP, lower curve) and is assigned to a combination of diffusive solvation and inertial rotation. b. At short times, a large peak is seen. A combination of two-level dynamics and a non-resonant interaction (solid curve) cannot account for its width. Excited state absorption is believed to be responsible.

viscous than benzyl alcohol. The viscoelastic theory predicts that the diffusive solvation time will be subpicosecond at these viscosities, overlapping with the inertial rotation. By fitting the diffusive rotation decay, and by assuming that the amplitude and decay time of the inertial rotation are the same as in benzyl alcohol, the pure solvation component of the toluene data was extracted (Fig. 2a). The $1/e$ time of 500 fs at room temperature is close to the 650 fs decay component reported by Larsen et al. from photon echo experiments in toluene [2]. We do not see evidence for the 3 ps or 120 ps decays they reported. Our experiments identify the ~500 fs decay as diffusive solvation and support the general features of the viscoelastic theory of solvation in nonpolar solvents.

Acknowledgement. This work was supported by the National Science Foundation under Grant No. CHE-9809719.

References

1. M. Berg, J. Phys. Chem. A **102,** 17 (1998).
2. D. S. Larsen, K. Ohta and G. R. Fleming, J. Chem. Phys. **111,** 8970 (1999).
3. G. S. Jas, Y. Wang, S. W. Pauls, C. K. Johnson and K. Kuczera, J. Chem. Phys. **107,** 8800 (1997).
4. J. Ma, J. T. Fourkas, D. A. Vanden Bout and M. Berg, *Multiple Time Scales in the Nonpolar Solvation Dynamics of Supercooled Liquids,* J. T. Fourkas, D. Kivelson, U. Mohanty and K. A. Nelson (American Chemical Society, Washington, DC).

Solute Dependence of three pulse photon echo peak shift measurement

Yutaka Nagasawa[1], Ayako Watanabe[1], and Tadashi Okada[1,2]

[1] Department of Chemistry, Graduate School of Engineering Science
[2] Research Center for Materials Science at Extreme Conditions,
Osaka University, Toyonaka, Osaka 560-8531, Japan
E-mail: okada@chem.es.osaka-u.ac.jp

Abstract. We have measured three pulse photon echo peak shifts (3PEPS) of some dyes in methanol at room temperature. The peak shifts were smaller and 3PEPS decays were faster for dyes with longer peak absorption wavelengths.

1. Introduction

Recently, three pulse photon echo peak shift (3PEPS) measurement has become a powerful tool to study solvation dynamics in condensed phase [1, 2]. We have measured 3PEPS of some different dyes in methanol. The aim of this study is to understand how the 3PEPS signal depends on solute molecules. Our expectation was to observe a correlation between 3PEPS signal and the spectral bandwidth of the electronic transition because the bandwidth is closely related to the mechanism of optical dephasing. Despite our expectation, the 3PEPS signal exhibited a correlation with the peak absorption wavelength and not with the bandwidth. The peak shifts were smaller and 3PEPS decays were faster for dyes with longer maximum absorption wavelengths.

2. Experimental Methods

A home-made cavity-dumped Kerr lens mode-locked Cr: forsterite laser was used for the experiment. The cavity-dumped IR beam was focused into a LBO crystal to generate the visible second harmonic at 635 nm. The measured FWHM of the pulse was about 33 fs. The experiment is a three pulse stimulated echo in which both echo signals at phase-matching conditions, $\boldsymbol{k}_1 - \boldsymbol{k}_2 + \boldsymbol{k}_3$ and $-\boldsymbol{k}_1 + \boldsymbol{k}_2 + \boldsymbol{k}_3$ are simultaneously recorded. Since there are three pulses, there are three time periods to consider, the first time period during which the system is in an electronic superposition state is called τ. The second time period in which the system is in a diagonal population state is called T. The third pulse creates the final superposition which leads to rephasing and echo formation. This last period is integrated to record the echo intensity as a function of τ. The observable of interest is the location of the echo maximum with respect to zero delay for different fixed values of T. The shift from zero delay, $\tau^*(T)$, we refer to as the peak shift, and by measuring both phase matched echoes simultaneously, this can be recorded with extremely high precision. A plot of $\tau^*(T)$ vs. T constitutes a 3PEPS data set.

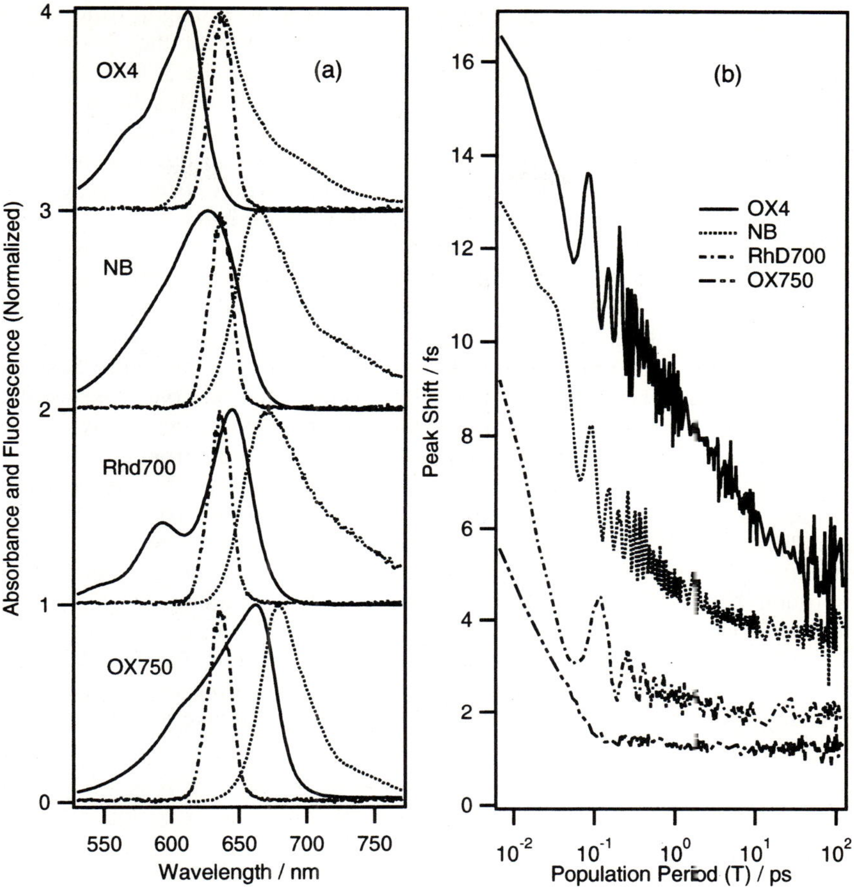

Figure 1. (a) Normalized absorption and fluorescence spectra of oxazine 4, Nile blue, rhodamine 700, and oxazine 750 in methanol solution. Spectrum of the excitation pulse is also shown for comparison. (b) The 3PEPS signals of oxazine 4, Nile blue, rhodamine 700, and oxazine 750 in methanol solution.

3. Results and Discussion

Figure 1 (a) shows the absorption and fluorescence spectra of dyes, oxazine 4 (OX4), Nile blue (NB), rhodamine 700 (RhD700), and oxazine 750 (OX750), in methanol. All the dyes have a similar rigid xanthene structure. Spectrum of the excitation laser pulse is also shown for comparison. It can be seen that the absorption and fluorescence spectra shift to longer wavelength from top to bottom. The excitation pulse has the largest overlap with the fluorescence spectrum of OX4 while the smallest with that of OX750. The 3PEPS signals of the dyes in methanol are shown in Figure 1 (b). The total peak shifts and decay time constants decrease in the order of OX4 > NB > RhD700 > OX750. When the spectra are compared with the 3PEPS signal, it can be easily noticed that the molecules with longer peak absorption wavelengths exhibit faster decay. The 3PEPS signal depends on which part of the spectrum is excited.

When the blue side of the absorption band is excited, it can be easily understood that the electronic transition will occur to the higher vibrational levels of the excited state. IVR in the excited state is considered to be extremely fast and can affect the earlier part of the 3PEPS signal. Multiple excitation of the high frequency modes will also lead to ultrafast initial decay of the 3PEPS signal. However, this mechanism may not fully explain the observed phenomena. Previous 3PEPS experiments have shown that the picosecond decay can be assigned to the diffusive solvation process because it depends strongly on solvent. The picosecond decay times are 8.4 ps and 3.8 ps for OX4 and NB, respectively, and there is no picosecond decay for RhD700 and OX750. To explain such a large difference in picosecond decay, excitation dependence of the solvation process may also have to be taken into account. When the blue side of the absorption band is excited, the population will be generated away from the minimum on the excited state free energy surface. In this case, both population shift and broadening have to take place for thermal relaxation. On the other hand, when the red side of the absorption is excited, population will be generated somewhere near the minimum and only the broadening will take place.

4. Conclusions

We have carried out solute dependence of three pulse photon echo peak shift (3PEPS) measurements in methanol. Solute molecules with longer absorption wavelength displayed faster 3PEPS decay. We think that this effect can be attributed not only to IVR in the excited state but also to excitation wavelength dependence of the solvation process. Our conclusion is based on two major assumptions; (1) contribution of the spectral shift and broadening to the 3PEPS decay change with excitation wavelength, and (2) presence of nonlinear solvation process which causes the shift to occur faster than the broadening. As a future plan to confirm this assumption, we have to carry out an experiment on a single solute-solvent system with changing excitation wavelength.

References

1 T. Joo, J. Jia, J.-Y. Yu, M. J. Lang, and G. R. Fleming, *J. Chem. Phys.* **104,** 6089 1996.

2 W. P. de Boeij, M. S. Pshenichnikov, and D. A. Wiersma, *J. Phys. Chem.* **100,** 11806 1996.

Ultrafast Dynamics in DNA

Eric B. Brauns[1], Mihaela M. Madaras[2], Robert S. Coleman[2], Catherine J. Murphy[1], and Mark A. Berg[1]

[1]University of South Carolina, Department of Chemistry and Biochemistry, Columbia, SC 29208
E-mail: berg@mail.chem.sc.edu, murphy@mail.chem.sc.edu
[2]Ohio State University, Department of Chemistry, Columbus, Ohio 43210
E-mail: coleman@chemistry.ohio-state.edu

Abstract. A DNA base pair has been replaced by a coumarin, and the DNA's "solvation" dynamics monitored by time-resolved Stokes shift measurements. The shift is logarithmic in time, indicating an extremely broad distribution of relaxation times.

1. Introduction

There are few experimental measurements of fast dynamics of DNA. Computer simulations of DNA show significant fluctuations about its equilibrium structure with dynamics extending over a wide range of time scales from <1 ps to 100's of ps or longer [1]. Unfortunately, the poor photophysics of native DNA bases has made ultrafast spectroscopy of DNA difficult. Anisotropy measurements of intercalated dyes are sensitive to global motions of an oligonucleotide [2], but have poor sensitivity to localized motions of individual bases, sugars, phosphates, etc. We have developed a method for adapting standard ultrafast "solvation" experiments on liquids to use in DNA [3-5]. A base-pair in an oligonucleotide is synthetically replaced by a Coumarin-102 group (Fig. 1), and the dynamics of the DNA's reorganization around the excited state are measured through the time-resolved fluorescence Stokes shift.

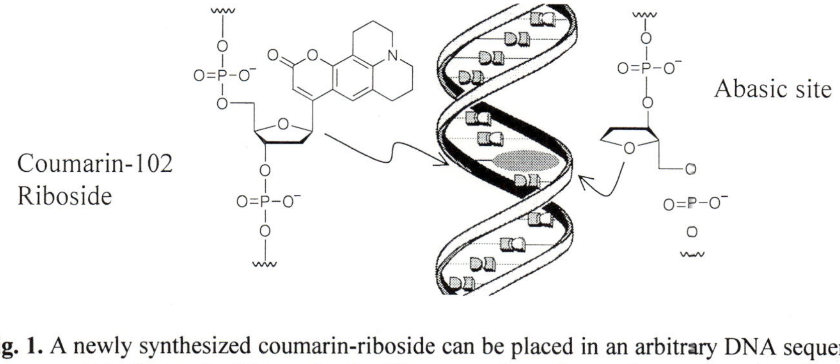

Fig. 1. A newly synthesized coumarin-riboside can be placed in an arbitrary DNA sequence by automated oligonucleotide synthesis. The coumarin replaces a base-pair and probes the local "solvation" dynamics in the interior of DNA.

2. Experimental Methods

Figure 1 shows how a probe suitable for ultrafast spectroscopy was incorporated into DNA. Coumarin-102 and its variants are well established as excellent probes of solvation dynamics in simple fluids. We synthesized a Coumarin-102 riboside which can be incorporated into an arbitrary DNA sequence using an automated DNA synthesizer. The coumarin is covalently attached in the same position as a normal DNA base and has a size and shape similar to an entire base-pair. The complementary strand is synthesized with a vacant (abasic) site opposite the coumarin. In the annealed double helix, the coumarin sits in the same place as a native DNA base-pair and samples the fluctuating electric field in the interior of DNA. Molecular modeling, melting studies, and CD spectroscopy all indicate that the modified DNA adopts a near normal structure. Once an appropriate probe chromophore was placed in DNA, standard methods developed for measuring ultrafast solvation dynamics were applied to the DNA [5].

3. Results and Discussion

Figure 2 shows a plot of the time-resolved Stokes shifts for three different 17-base-pair DNA sequences. The most dramatic feature is that none of them show a special relaxation time. Rather, the Stokes shifts are all nearly linear on a logarithmic time axis. This result is not consistent with a simple kinetic model involving 2-3 well defined substates. Instead, it implies that there are a wide variety of substructures with interconversion rates spanning a wide time range.

A model that explains the logarithmic time dependence of the Stokes shift $S(t)$ is based on a distribution of relaxing elements. The number of elements $A(\tau)$ with relaxation time τ decreases as the inverse of τ

$$S(t) = S_0 + \int_0^{\infty} A(\tau) e^{-t/\tau} \, \mathrm{d}\tau \qquad (1)$$

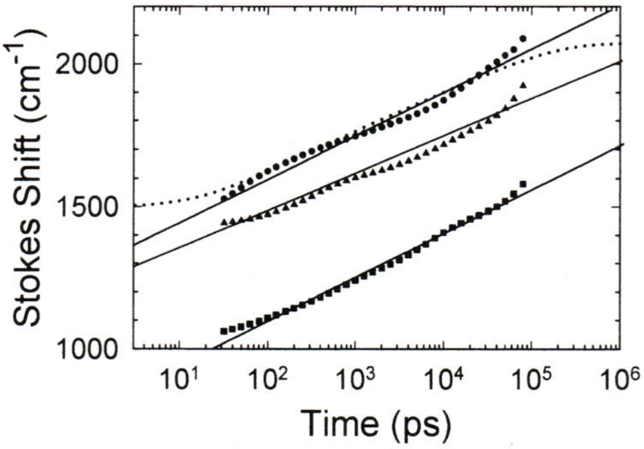

Fig. 2. Stokes shift relative to cyclohexane of three different DNA sequences. Solid lines - logarithmic time dependence. Dotted curve - Eqs. 1 and 2. ● = 5'-GCATGCGC*CGCGTACG-3'; ▲ = 5'-GCAAAAAA*AAAAAACG-3'; ■ = 5'-GCTATATA*ATATATCG-3'; * = coumarin

$$A(\tau)=\begin{cases}1/\tau\ ; & \tau_{\min}<\tau<\tau_{\max}\\ 0\ ; & \text{otherwise}\end{cases} \qquad (2)$$

A fit to $S(t)$ with $\tau_{\min}$ = 20 ps and $\tau_{\max}$ = 100 ns is shown as the dotted curve in Fig. 2. The values of $\tau_{\min}$ and $\tau_{\max}$ are only upper and lower bounds, respectively.

In addition to the basic logarithmic time dependence, each sequence shows small deviations ("wiggles") about the logarithmic dependence. A purely empirical model that accurately reproduces these wiggles is a double stretched exponential, $0.37\exp[-(t/47\ \text{ps})^{0.5}]+0.63\exp[-(t/63\ \text{ns})^{0.5}]$ (fit not shown). Regardless of the model used, the dynamics must have a broad distribution of rates.

There is intense interest in the possibility that different DNA sequences have different physical properties that aid in their recognition by proteins. Our data show that there are subtle but measurable differences in the dynamics of the three sequences studied. The slopes of the best logarithmic fits are 145 (●), 130 (▲), and 131 (■) cm^{-1}/decade. In addition, the "wiggles" around the logarithmic dependence differ from one sequence to the next. Future modeling and simulation will reveal if these differences are large enough to alter the biochemical interactions of different sequences.

Although the current data are for times >30 ps, there are strong indications that even more important processes lie at shorter times. The shifts in Fig. 2 are measured relative to the spectra in cyclohexane, which has a nearly zero Stokes shift. The large value of the Stokes shifts seen at early times shows that as much as 70% of the relaxation has already occurred by 30 ps. Moreover, extending the logarithmic time dependence to cover the 0.1-20 ps range does not account for the early amplitude. Thus, there must be a qualitatively different relaxation dynamic at shorter times. Techniques for extending our current measurements to this time regime are readily available.

Acknowledgements. This work was supported by the National Science Foundation under Grant No. CHE-9809719 to MAB and by the National Institutes of Health under Grant Nos. GM-47991 to RSC and GM-55566 to CJM.

References

1. M. A. Young, G. Ravishanker and D. L. Beveridge, Biophys. J. **73,** 2313 (1997).
2. J. Schurr, B. S. Fujimoto, P. Wu and L. Song, *Topics in Fluorescence Spectroscopy Volume 3: Biochemical Applications,* J. R. Lakowicz (Plenum Press, New York and London).
3. R. S. Coleman and M. L. Madaras, J. Org. Chem. **63,** 5700 (1998).
4. E. B. Brauns, C. J. Murphy and M. A. Berg, J. Am. Chem. Soc. **120,** 2449 (1998).
5. E. B. Brauns, M. L. Madaras, R. S. Coleman, C. J. Murphy and M. A. Berg, J. Am. Chem. Soc. **121,** 11644 (1999).

Femtosecond Photophysics of DNA and RNA Nucleosides in Aqueous Solution

Jean-Marc Pecourt, Jorge Peon, and Bern Kohler

The Ohio State University, Department of Chemistry, 100 W. 18th Ave., Columbus OH 43210, USA. Tel.: (614) 688-3944, Fax. (614) 292-1685
E-mail: kohler@chemistry.ohio-state.edu

Abstract. Excited state absorption of the DNA nucleosides in aqueous solution is reported for the first time. Femtosecond transient absorption measurements show that the S_1 states of the bases undergo internal conversion to the ground electronic state in 300-700 fs.

1. Introduction

Recent research activity directed at understanding photooxidative damage in nucleic acids has refocused attention on the unusual photophysics of the DNA bases. While it has been known for several decades that the nucleobases exhibit extremely low fluorescence quantum yields in room temperature solution [1], efforts to directly observe the singlet excited states of the bases have been largely unsuccessful. Accurate values of the S_1 lifetimes and even crude information about the deactivation pathways of the singlet excited states are still lacking. We report here the first excited state absorption study of the DNA bases. Using femtosecond pump-probe techniques we have characterized the $S_1 \rightarrow S_N$ absorption spectra and observed the subpicosecond decay of the S_1 states.

2. Experimental Setup

Aqueous solutions of the nucleosides were studied by pump-probe methods using 150 fs, 265-nm pump pulses derived from an amplified, kHz titanium:sapphire laser system. Probe pulses were obtained by continuum generation in water using the laser fundamental. The polarizations of pump and probe pulses were set to the magic angle. Millimolar solutions were recirculated through a 300-μm thick jet. The pump intensity was controlled by a waveplate/polarizer combination in order to study the signal intensity dependence.

3. Results and Discussion

The purine nucleosides, adenosine and guanosine, show enhanced fluorescence at low pH [2]. Under these conditions, the S_1 lifetime is considerably longer than at neutral conditions, and excited state absorption was readily detected. Protonated Guo ($GuoH^+$) absorbs in a broad band centered near 600 nm at pH 2. Following

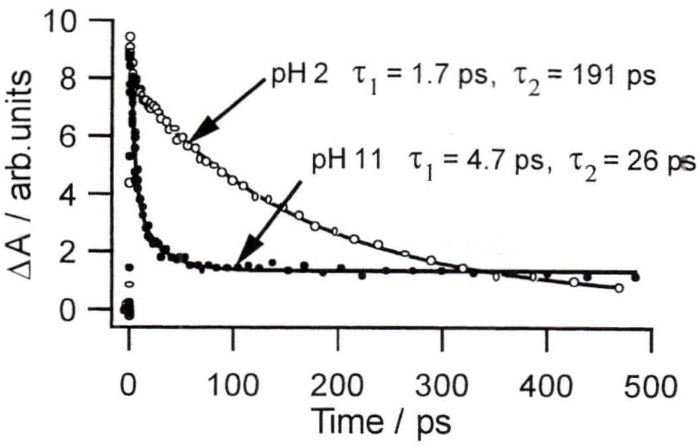

Fig. 1. Transient absorption at 600 nm of protonated guanosine in acidic (pH 2) and basic (pH 11) aqueous solution

an instrument-limited rise, a biexponential decay is observed (Figure 1). Signals recorded at five separate probe wavelengths have identical kinetics, and were globally fit to the sum of two exponentials. The exponential time constants, $\tau_1 = 1.7 \pm 0.3$ ps and $\tau_2 = 191 \pm 4$ ps, yield an excellent fit at all probe wavelengths. The slower time constant agrees very well with the emission lifetime of 196 ± 7 ps reported in a streak camera study of $GuoH^+$ [3], confirming that the signal seen here is due to excited state absorption by the S_1 state. The faster time constant is due to the presence of a small concentration of unprotonated Guo molecules.

At neutral pH the transient absorption shows only a subpicosecond decay (τ = 460 fs) that nearly recovers to the baseline (Figure 2). This result demonstrates the acute sensitivity of the guanosine chromophore to protonation. Figure 2 also shows the transient for the pyrimidine base, thymidine. In this case, an enhanced long-time offset is observed. Transients for adenosine and cyticine show similar offsets. The ratio of the offset to the maximum signal at very early times increases strongly as the pump pulse intensity is varied from 2-50 GW/cm^2. This intensity-dependent reshaping occurs because the long-time signal arises from a two-photon process in contrast to the one-photon nature of the short-time decay.

The long-time signal is most intense near 720 nm, and is assigned to the hydrated electron produced by two-photon ionization. To identify the source of the photoionization, transients were recorded at constant pump intensity for different solute concentrations of 5' guanosine monophosphate. As the solute concentration was decreased, the signal amplitude at early time ($t < 500$ fs) decreased linearly with concentration, proving that the short-time signal arises from one-photon excitation of the nucleobase. In contrast, the long-time ($t > 4$ ps) signal increased slightly as the solute concentration was decreased, indicating that solvated electron formation occurs through two-photon ionization of the abundant water molecules. The strong two-photon absorption of water in the UV and the weak absorption of the base excited states make it very difficult to completely eliminate the photoionization signal from bulk water. As Figure 2 indicates, it was slightly easier to reduce the contribution of the offset for the purine bases than for the pyrimidine bases, due to the weaker excited state absorption of the latter.

Decay times for all nucleosides studied are summarized in Table 1. These values are considerably shorter than previous estimates [3-6].

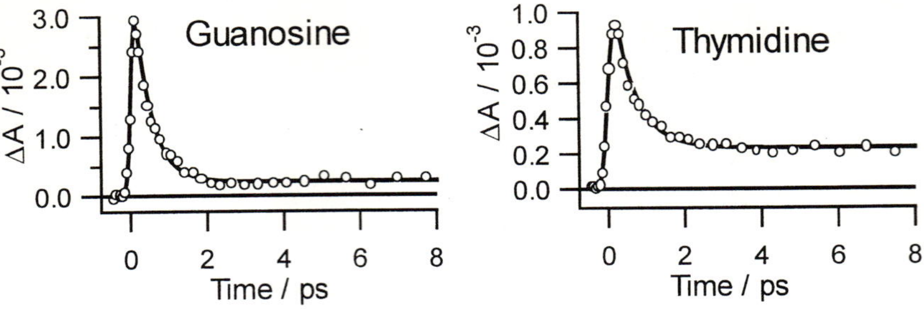

Fig. 2. Transient absorption at 610 nm of guanosine and thymidine in pH 7 aqueous solutions

Table 1. Nucleoside S_1 lifetimes in neutral water solution (± 40 fs)

	Adenosine	Guanosine	Cytidine	Thymidine
τ / fs	290	460	720	540

4. Conclusions

We have detected weak, excited state absorption from DNA and RNA nucleosides in aqueous solution. The absorption cross-sections for the first excited singlet state are sufficiently small that two-photon ionization of bulk water is unavoidable. This two-photon process dominates at high excitation intensities and obscures the electronic relaxation of the solute. It was nonetheless possible to accurately determine the S_1 lifetimes of the nucleosides by the disappearance of the excited state absorption at low pump intensity. These results provide direct evidence that internal conversion occurs in less than 1 ps.

References

1 P. R. Callis, Ann. Rev. Phys. Chem. **34**, 329, 1983.
2 H. C. Börresen, Acta Chem. Scand. **17**, 921, 1963.
3 T. Fujiwara, Y. Kamoshida, R. Morita, and M. Yamashita, J. Photochem. Photobiol. **41**, 114, 1997.
4 D. N. Nikogosyan, D. Angelov, B. Soep, and L. Lindqvist, Chem. Phys. Lett. **252**, 322, 1996.
5 T. Häupl, C. Windolph, T. Jochum, O. Brede, and R. Hermann, Chem. Phys. Lett. **280**, 520, 1997.
6 A. Reuther, D. N. Nikogosyan and A. Laubereau, J. Chem. Phys. **100**, 5570, 1996.

Investigation of Coherent Dynamics in Charge-Transfer Reaction of Plastocyanin

Satoru Nakashima[1], Yutaka Nagasawa[1], Tadashi Okada[1], Maki Sato[2], and Takamitsu Kohzuma [2]

[1] Department of Chemistry, Graduate School of Engineering Science and Research Center for Materials Science at Extreme Conditions, Osaka University, Toyonaka, 560-8531 Japan
E-mail: okada@chem.es.osaka-u.ac.jp
[2] Faculty of Science, Ibaraki University, Bunkyo, Mito, 310-8512 Japan

Abstract. Protein dynamics associated with optical excitation and radiationless deactivation of its charge-transfer state were studied by ultrafast pump-probe spectroscopy. A coherent low-frequency (33 cm^{-1}) vibrational mode that was not observed in resonance Raman spectrum, was discovered. The observed coherency of this low frequency mode appears via the excited state deactivation and this mode most probably can be identified to the delocalized mode including the protein matrix motion.

1. Introduction

Electron transfer reactions in proteins use the motion of the protein matrix surrounding their active sites in order to proceed the reactions efficiently. Observation of the photo induced reaction by an ultrafast laser pulse reveals such non-dissipative coherent nuclear motions [1-3]. Elucidation of the mechanism by which protein motion plays a key role in such efficient reaction may be possible.

Plastocyanin transports an electron between the cytochrome b_6f complex to P700 of photosystem I in oxygenic photosynthesis. The copper atom is ligated with two histidines, one cystein, and one methionine in a distorted tetrahedral geometry. Plastocyanin in the oxidized form shows intense absorption band that is assigned to the S(cys) to Cu(II) ligand-to-metal charge-transfer (LMCT), at visible region around 600 nm [4]. The Cu-S bond has been suggested to play a key role in physiological electron transfer reaction. Resonance Raman spectroscopic studies by the excitation at the LMCT band have provided the vibrational structure around the active site and the dynamics on the excited-state [5,6].

In this study, we used the homemade cavity-dumped Cr:Forsterite laser having a center frequency of 635 nm and a pulse duration of 30 fs. This laser system can excite the LMCT state directly and the coherence of all the localized modes around the active site, which have been measured in resonance Raman spectra can be observed. A coherent low-frequency (33 cm^{-1}) vibrational mode was observed. The assignment of the observed low-frequency mode and the origin of the coherency will be discussed.

2. Experimental Methods

Synechococcus sp. PCC 7942 plastocyanin was expressed in *E. coli*, and the expressed plastocyanin was purified and used in this experiment. The details of the laser system are described elsewhere. The output of the second harmonics of the cavity-dumped Cr:Forsterite laser, centered at 635 nm with 30 fs duration, was divided into and used as pump and probe beams. The sample was placed into the custom-made rotational cuvette. The probe beam was detected directly by a Si photodiode (wavelength integral) or after spectrometer (wavelength resolved). Low-frequency resonance Raman spectra were taken by a double monochromater attached with PMT.

For data analysis, non-linear least squares convolution fitting (using the Levenberg-Marquardt algorithm) was performed using various theoretical functions. For high-frequency mode, resonance Raman data was used for initial parameters and cosine functions including dephasing were applied. In order to analyze the high-frequency vibrational mode, FFT was performed on the signal that has been deconvoluted using the instrument response function.

3. Results and Discussion

The Fig. 1 (a) shows the pump-probe signal of plastocyanin excited at LMCT band in ambient temperature. The signal contains the decay componen ts with the superimposed coherent oscillations. At a glance, the logarithmic plot of the signal intensity shows a rapid fall indicating a difficulty in fitting the decay by a simple sum of exponential components. For decay components, we have tried various types of fitting functions. Only a critically dumped oscillation with a frequency of 33 cm^{-1}, a dephasing time of 360 fs gave the satisfactory result. (Fig. 1 (b)) Our analysis demonstrates that a low-frequency coherent motion is involved in the recovery of the ground-state population

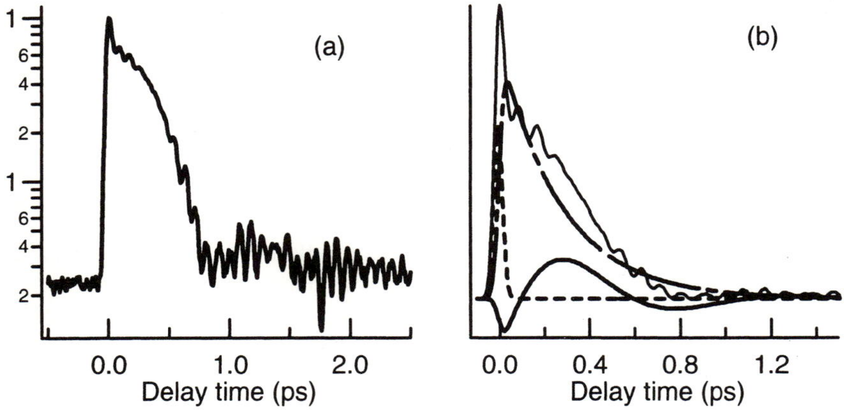

Fig. 1. (a) Logarithmic plot of the pump-probe signal of plastocyanin (b) Fitting results.

The identification of this low frequency mode becomes the greatest question. Since all high-frequency modes that localize around the active site were observed in resonance Raman spectrum, and our pump-probe data were consistent with these modes, it is suggested that this low-frequency mode can be regarded to delocalized mode that includes a protein matrix motion.

The feature that must be emphasized here is the phase of the low-frequency mode. The fit of the low-frequency mode to a cosine function gave phase shifts of approximately -120 degrees suggesting that the oscillation is not due to impulsive stimulated Raman scattering. Actually resonance Raman spectrum (> 10 cm^{-1}) showed that high-frequency (> 200 cm^{-1}) modes are consistent with pump-probe data, but the low-frequency mode (33 cm^{-1}) did not appear. In the case of wavelength resolved measurements, the dephasing time of this low-frequency mode became faster when probe wavelength became longer. From these findings, the most likely mechanism is that the observed coherency of this low frequency mode appears via the excited state deactivation. Such correlation between coherence and reaction in condensed phase has been studied intensively, and the present findings might provide insight into the fundamental mechanism of the chemical reactions.

4. Conclusions

In conclusion, we have measured LMCT band (635 nm) excitation pump-probe signal of a blue copper protein, plastocyanin, by 30 fs pulse and followed the protein dynamics after photo-excitation. A coherent low-frequency (33 cm^{-1}) vibrational mode of which the phase was shifted from those of the other localized Raman modes was observed. This mode most probably can be identified to the delocalized mode including the protein skeletal motion. The observed coherency of this low frequency mode appears via the excited state deactivation, rather than arising merely from impulsive stimulated Raman.

Acknowledgements. I would like to thank Prof. T. Kitagawa (IMS), Prof. S. Kinoshita and Prof. J. Watanabe (Osaka Univ.) for helping the Raman measurements.

References

1 U. Liebl, G. Lipowski, M. Negrerie, J.-C. Lambry, J.-L. Marin, M. H. Vos, Nature **401** 181, 1999.

2 Q. Wang, R. W. Schoenlein, L. A. Peteanu, R. A. Mathies, C. V. Shank, Science **266** 422, 1994.

3 L. D. Book, D. C. Arnett, H. Hu, N. F. Scherer, J. Phys. Chem. **102** 4350, 1998.

4 E. I. Solomon, M. J. Baldwin, M. D. Lowery, Chem. Rev. **92** 521, 1992.

5 D. Qiu, s. Dasgupta, P. M. Kozolwski, W. A. Goddard III , T G. Spiro, J. Am. Chem. Soc. **120** 12791, 1998.

6 G. R. Loppnow, E. Fraga, J. Am. Chem. Soc. **119** 896 ,1997.

Part XIII

Polymers, Aggregates, and Macromolecular Systems

Vibrational Dynamics in Molecules and Polymers Revealed by Sub-5-fs Real-time Spectroscopy

Takayoshi Kobayashi

Department of Physics, University of Tokyo, 7-3-1 Hongo, Bunkyo, Tokyo 113-0033, Japan
E-mail: kobayashi@phys.s.u-tokyo.ac.jp

Abstract. Pump-probe experiment was performed for spin-coated films of conjugated polymers (substituted polydiacetylene), J-aggregates of a porphyrin derivative, and dye molecules (cresyl violet) by using a sub-5fs visible laser. Real-time spectra were obtained for the systems and new dynamical processes have been observed, i.e. dynamic mode coupling, dynamic intensity borrowing, and dynamic Dushinsky rotation.

1. Introduction

Femtosecond pulses shorter than 10 fs have become available and enabled the study of real-time vibrational dynamics of various molecular systems. Especially, femtosecond wavepacket dynamics has been recently studied to realize the possibility of controlling molecular dynamics using tailored ultrashort pulse. Information about the dynamic coupling among vibrational modes in molecules in condensed phase is needed for the realization of such selective chemical reaction. Dynamic mode coupling is also important for the clarification of the mechanism of structural relaxation in the excited state of molecules. Recently sub-5-fs visible pulses have been generated in our group on the base of a novel optical parametric amplifier with a non-collinear configuration i.e. non-collinear optical parametric amplifier (NOPA) [1,2]. The laser system generating such short pulses has enabled real-time observation of the transmittance change associated with molecular vibrations. Real-time spectroscopy has been applied to conjugated polymers, J -aggregates of porphyrin, and dye molecules. Energy migration among the modes of C-C, C=C, and C≡C stretching and other low frequency out-of-plane bending in polymers through mode coupling have been observed by using a sub-5fs visible laser [3]. Dynamic intensity borrowing has also been observed for the first time using a Frenkel exciton system of J-aggregates of porphyrin. Coupling with an out-of-plane bending mode in cresyl violet doped in polyvinyl alcohol and in polymethylmethacrylate is found to induce the phase and amplitude modulation of the breathing mode, resulting in the Dynamic Dushinsky rotation.

2. Experimental

The pulse-front-matched NOPA generates transform-limited (TL) 4.7-fs, 5-μJ 1-kHz pulses centered at 650nm with a 250-nm bandwidth [1,2]. It has two sets of appropriately designed chirped mirrors for the compensation of group-velocity dispersion [1,2]. One of the two sets is for pre-compensation to obtain temporal overlap of seed pulse and pump pulse in the BBO crystal for efficient parametric amplification. The other set of chirped mirrors is for optimization of the amplified pulses. Pulse-front-matching (PFM) geometry essential for transform-limited (TL) pulse generation was utilized to obtain visible sub-5-fs pulses and tunable pulses in a sub-10-fs regime are realized by the same NOPA system with and without the chirped mirrors, respectively. The NIR sub-10-fs pulses by angular dispersion compensation were also obtained. All of the samples of polymers of diacetylene, J- aggregates of porphyrin and cresyl violet were prepared in the form of spin-coated film.

3. Polydiacetylenes

Ultrafast dynamics of conjugated polymers such as polydiacetylenes (PDA) has been attracting many scientists because they are model systems of quasi-one dimensional semiconductors with a strong electron-phonon coupling. From our extensive studies [3-6], the initial kinetics after photo-excitation can be explained by a relaxation of a free exciton (FE) to a self-trapped exciton (STE) within 150fs. The recent achievement of sub-5-fs visible pulse generation based on a novel noncollinear optical parametric amplification (NOPA)[1,2] has enabled the real-time observation of the formation of a STE in a PDA for the first time. The stretching-mode vibration(C-C, 27 fs, C=C, 23 fs and C≡C, 16 fs) coupled with the excitonic transition drives the coherent relaxation of the backbone structure from an acetylene (A)-like $(=CR-C\equiv C-CR'=)_n$ to a butatriene (B)-like $(-CR=C=C=CR'-)_n$ configuration.
The real-time spectrum i.e. the transient differential transmittance clearly (Fig. 1) shows a coherent molecular motion composed of several modes over the whole probe spectral range. From the complex features, it is also evident that more than two modes are contributing to the time-dependent transmittance change. In spite of the complexity, there are systematic features in the phases of the oscillations probed at various probe photon energies. For example, there are eight peaks at 23, 44, 90, 154, 180, 200, 260, and 308 fs among the eleven prominent peaks in the delay time range of real time spectrum representing the trace of transmittance change at 2.12 eV. The signal intensity is minimum at these delay times in the trace of 1.75 eV. This result of the coincidence indicates that there are constant phase relation among the observed modes and that the coherence times of the relevant modes are substantially longer than 300fs.

Previous papers [7,8] reported that only the C=C stretching among the three stretching modes and ~700-cm^{-1} modes were observed and that they remained only for the initial 100 fs. This quick damping was explained by

dephasing due to anharmonicity. The behavior observed in the present study probed around the exciton band has similarity to the previous results; but there is a striking difference, namely weak oscillations continue beyond 400-fs delay even after the population decay. The underlying signal due to the population change also decays quickly along with the oscillating component due to wave-packet motion. This fast decay is observed in the transients for probe photon energy $E_{probe} < \sim 2.0$ eV. In the region $E_{probe} > \sim 2.0$ eV, on the other hand, the amplitude of the oscillations is maintained or even increases slightly in the initial positive signal range. The coherence is maintained as long as ~1 ps in the delay-time range of negative transmittance change due to photo-induced absorption.
The integrated Fourier analysis of the real-time spectrum reveals two intense modes of C=C(~1455cm^{-1}) and C≡C(~2080cm^{-1}) stretching, in addition to weaker modes around 220, 700, and 1220cm^{-1}, of which intensities are dependent on the probe wavelength. These modes are also observed in the static Raman spectrum. The A->B isomerization and following thermalization of STE is visualized by a real-time frequency analysis using a spectrogram [9].

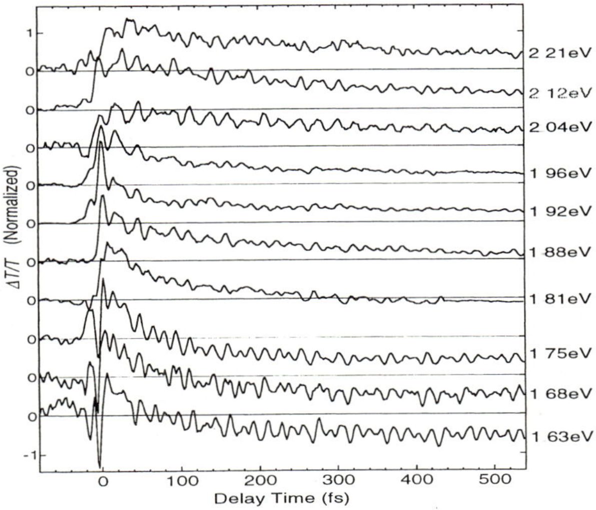

Fig.1. Coherent wavepacket motion in the transient differential transmittance. The probe photon energy is marked on the right with a 5-nm resolution.

From the calculations the Fourier-sine and –cosine transform of the transmittance change, the wavelength dependence of the phases of molecular vibration was found to be same for three modes of C-C C=C, and C≡C stretching. This indicates that these three modes are excited impulsively in a common phase. Energy migration among the three stretching modes and other low frequency mode presumably due to out-of-plane bending through mode coupling has been observed. This is the first observation of dynamic mode coupling.
The spectrogram was calculated using various window functions and widths. Care must be taken for the mixing of the signal of Fourier power spectrum of

neighboring modes between which frequency difference is smaller than the inverse of the width of window function used. Using broad enough width, modulation of frequencies and amplitudes of three high-frequency modes of carbon-carbon stretching could are observed. Some of the modulations are correlated with the same low modulation frequency.

4. J-aggregates

Fourier spectra of the time dependence of absorbance change observed in porphyrin J-aggregates clearly show a strong peak at 244 cm^{-1}, which is also observed in the stationary Raman spectrum. This mode is drastically enhanced by aggregation by a factor of more than 30 [10]. A π-phase jump is clearly observed around 1.77eV. This result can be explained in terms of dynamical intensity borrowing. The intensity of absorption associated with the transition of the ground state to the Q-band is borrowed from that of B-band by the configuration-interaction mechanism. The degree of the intensity borrowing is modified by vibrational motion of the ruffling mode by the exciton-phonon coupling.

5. Dye molecule: cresyl violet

Sub-5-fs real-time spectroscopy has been performed for a dye molecular system of cresyl violet (CV) doped in poly (methylmetaacrylate) (PMMA) The sample was prepared by casting solution of CV and PMMA in chloroform on a glass substrate. The breathing mode of frequency of 590 cm^{-1}, which corresponds to 60 fs, is found to be modulated in both the amplitude and frequency. The sizes of modulations of amplitudes and frequencies are compared with each other. The maximum fractional changes in the amplitude are as large as 10% as probed at 600 nm. On the other hand, those in the frequency $\delta(\Delta\omega/\omega)$ are only 4 %. If the force constant change that induces frequency change as mentioned above is the only one origin of the amplitude modulation, then it is expected that this size of the frequency modulation be calculated to be 2%. In the molecular systems time dependent configuration change induced by possibly an out-of-bending motion leads to the change in the potential-minimum coordinate and the zero-zero transition energy in the stretching configuration space. The effect in cresyl violet is much larger than that due to force constant change. The mode coupling can be modeled by including the coupling term in the force constant coefficient phenomenological estimated from the amount of the modulation. The result is expressed as a Dushinsky rotation.

6.Metal-halogen mixed valence compound

The time dependence of the reflectivity change ($\Delta R / R$) of a single crystal of $[Pt(en)_2][Pt(en)_2I_2](ClO_4)$ at 620 nm. The negative signal is observed and it can be separated into two contributions. The non-oscillatory contribution decreases exponentially with a time constant of 1200±100 fs at 620 nm. This signal reflects the relaxation process of STEs. The other oscillatory contribution is described by the sum of the two dumped sinusoidal functions,

$$\left(\frac{\Delta R(t)}{R}\right)_{osc} = A_1 \cdot \exp\left(-\frac{t}{\tau_1}\right) \cdot \cos(\omega_1 t + \theta_1) + A_2 \cdot \exp\left(-\frac{t}{\tau_2}\right) \cdot \cos(\omega_2 t + \theta_2) \quad (1).$$

The parameters are determined as τ_1=1100±100 fs, $2\pi/\omega_1$=270±10 fs (124 cm^{-1}), and θ_1=-85±6° for slowly dumped oscillatory component and τ_2=290 ±10 fs, $2\pi/\omega_2$=310±10 fs (107 cm^{-1}), and θ_2=240±6° for rapidly dumped oscillatory component. The slowly and rapidly dumped components are attributed to the wave packet motions in the ground state and STE state, respectively.

7.Conclusion

The present results will offer a valuable information about the geometrical relaxation and encourage the theoretical investigations of the election-phonon and electron-electron interactions in non-degenerate conjugated polymers. The experimental technique of real–time spectroscopy with sub-5 fs pulses can reveals the dynamical processes in the excited states and intermediate states through the oscillating features appearing in the transmittance change

Acknowledgement

The author is grateful to Drs. Shirakawa and Fuji, and Messrs. Kanou, Matsuzawa, and Saito, and Prof. Nakanishi for collaboration. He also thanks Profs. Toyozawa, Abe, and Okamoto for valuable discussion. This work was partly supported by Research for the Future of Japan Society for the Promotion of Science (JSPS-RFTF-97P-00101).

References

[1] A. Shirakawa, I. Sakane, and T. Kobayashi, Opt. Lett. **23**, 1292(1998).

[2]A. Shirakawa, I. Sakane, M. Takasaka, and T. Kobayashi, Appl. Phys. Lett. **74**, 2268 (1999).

[3] T. Kobayashi, A. Shirakawa, H. Matsuzawa, H. Nakanishi, Chem. Phys. Lett. 321, 385 (2000).

[4] E.g., T. Kobayashi, M. Yoshizawa, U. Stamm, M. Taiji, and M. Hasegawa, J. Opt. Soc. Am. B. **7**, 1558(1990).

[5] M. Yoshizawa, Y. Hattori, and T. Kobayashi, Phys. Rev. B **123**, 121(1994).

[6] T. Kobayashi, M. Yasuda, S. Okada, H. Matsuda, H. Nakanishi, Chem. Phys. Lett. **267**, 472(1997).

[7] T. A. Pham, A. Daunois, J. C. Merle, J. Le Moigne and J. Y. Bigot, Phys. Rev. Letters 74 (1995) 904.

[8] J. Y. Bigot, T. A. Pham and T. Barisien, Chem. Phys. Letters 259 (1996) 469.

[9] M. J. J. Vrakking, D. M. Villeneuve, and A. Stolow, Phys. Rev. A **54**, 37(1996).

[10] D. L. Akins, H. -R. Ahu, and C. J. Guo, J. Phys. Chem., 98, 17159 (1995).

Early events in the photoexcitation of π-conjugated chains studied with sub-10-fs time resolution

G. Cerullo, M. Zavelani-Rossi, G. Lanzani, S. De Silvestri

Istituto Nazionale per la Fisica della Materia, CEQSE-C.N.R., Dipartimento di Fisica, Politecnico, Milano (Italy), e-mail: giulio.cerullo@polimi.it

Abstract. Using sub-10-fs pulses, pump-probe experiments were carried out on prototypical π-conjugated chains belonging to the class of carotenoids and semiconducting polymers. The internal-conversion rate and the subsequent excited state thermalization are directly measured.

1. Introduction

From the extensive theoretical and spectroscopic investigation on polyenes [1], it is known that the lowest singlet excited state (S_1) is a covalent excitation of "gerade" symmetry, i.e. a dark state with negligible optical coupling. Optical excitation in the visible range populates the higher lying singlet state (S_2), with ionic character and "ungerade" parity. The underlying physics associated to the initial ultrafast internal conversion (IC) from the bright state to the dark state is of great interest because it is this process which ultimately controls the fate of the energy deposited in the molecule by the optical field. Energy transfer reactions, non linear responses and charge photo-generation are examples of phenomena depending on that process.

In this work we investigate IC ($S_2 \rightarrow S_1$) following photoexcitation in prototype conjugated molecules by using sub-10-fs pulses in the visible range, which allow carrying out pump-probe experiments with unprecedented time resolution. We present results on polydiacetylene and β-carotene. The former, a non-luminescent semiconducting polymer, is a model system for studying one-dimensional exciton dynamics. The latter is a wide spread natural pigment, with vital functions in living systems [2]. Our results can help enlightening the long-standing discussion on photoexcitation relaxation in this class of materials.

2. Experimental Methods

The excitation source was a recently developed optical parametric amplifier (OPA) based on non-collinear phase matching in *β*-barium borate followed by pulse compression using chirped mirrors of novel design [3]. The pulses have a broad band extending from 500 nm to 720 nm, energy up to 1 μJ and are compressed to a near transform-limited duration of 5-6 fs. In the experiments we used typically few tens of nJ in a non-collinear pump-probe geometry.

We studied solutions of poly[1,6-bis(3,6-dihexadecyl-N-carbazolyl)-2,4-hexadiyne] (polyDCHD-HS) dissolved in benzene and β-carotene dissolved in cyclohexane. PolyDCHD-HS has very narrow absorption line-width (about 500 cm^{-1}) and exists in benzene solution as isolated chain. β-carotene was purchased by Aldrich and used without further purification. In both samples the excitation reaches the vibrational manifold of the first optically allowed singlet state (S_2).

3. Results and Discussion

In Fig.1 we plot transmission difference in polyDCHD-HS as a function of pump-probe delay at 580 nm. The positive signal is assigned to stimulated emission (SE) of $S_2 \rightarrow S_0$ transitions. It shows an ultrafast build-up, reaching a maximum at about 40 fs after photoexcitation, followed by a decay into photoinduced absorption (negative ΔT). We assign the initial process to intramolecular vibrational relaxation, directly measured for the first time, and the subsequent decay to IC ($S_2 \rightarrow S_1$), with a time constant of about 70 fs.

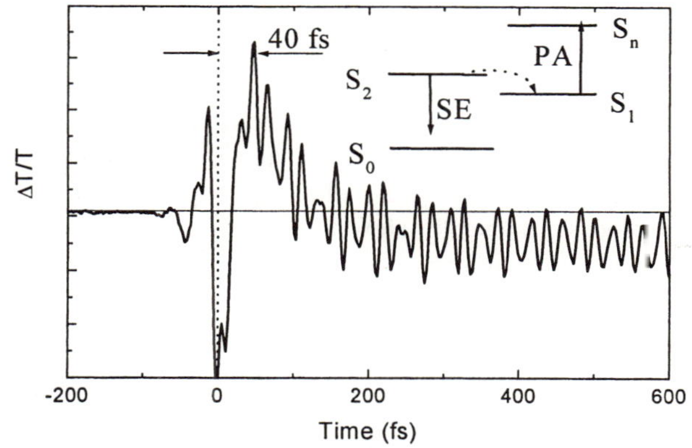

Fig. 1. Transmission difference as a function of pump-probe delay in poly-DCHD-HS at 580 nm. The level scheme shows the internal conversion in S_2 following photoexcitation.

At all probing wavelengths an oscillatory component is superimposed to the transient transmission, with larger amplitudes in the SE region. Its Fourier power spectrum contains two main frequencies at 1510 cm^{-1} and 2115 cm^{-1}, assigned to collective vibrational coherence initiated by the short pump pulse.
The oscillations are slowly damped in time due to dephasing among the ensemble and persist longer than 1 ps. Since the lifetime of S_2 is estimated to be about 70 fs, this rules out the hypothesis of excited state nuclear motion. In addition the observed frequencies correspond to those observed in cw resonant Raman spectra and assigned to the double and triple carbon-carbon stretching vibration. Both

observations lead us to conclude that they are caused by ground state wavepacket motion. The loss of excited state coherence during IC is most likely due to dephasing as the system evolves on an anharmonic potential energy surface.

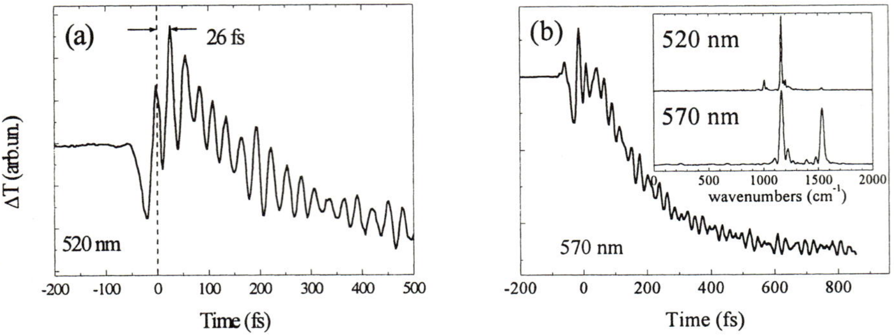

Fig. 2. (a) Transmission difference kinetics in β-carotene at 520 nm. SE evolves into PA. (b) Buildup of PA at 570 nm. The inset shows the Fourier power spectrum of both data.

β-carotene behaves in a similar manner as polyDCHD-HS, but shows a slower kinetics of IC. In Fig. 2 we show ΔT/T signals at 520 and 570 nm: in both traces the kinetics of IC is clearly seen, giving a time constant of S_2 depletion of about 170 fs, in agreement with previous measurements using photoluminescence up-conversion [4]. In addition a full spectral analysis of the transient (not reported here) reveals the thermalization of S_1, which is associated to a blue-shift of PA with time. The process, unusually slow for such a large molecule, lasts for few picoseconds after IC. This information is relevant to the modeling of carotenoids for chlorophyll energy transfer. Our data demonstrate that there is indeed a bottleneck effect that slows down vibrational relaxation in S_1, thus pointing to the role of each vibrational sub-level in the energy transfer.

Again impulsive collective vibrational coherence is evident almost everywhere in the transient spectrum as periodic modulation of the signal. An example of Fourier power spectra is reported in the inset of Fig.2 (b). Vibrational coherence is assigned to ground state motion, based on observations that retrace closely the discussion for polyDCHD-HS.

References

1 B. S. Hudson, B. E. Kohler, K. Schulten, in *Excited States*, Edited by E. C. Lim, Academic, New York, 1982.
2 A. Damjanovic, T. Ritz, K. Schulten, Phys. Rev. B **59**, 3293 (1999).
3 G. Cerullo, M. Nisoli, S. Stagira, S. De Silvestri, G. Tempea, F. Krausz, K. Ferencz, Opt. Lett. **24**, 1529 (1999)
4 H. Kandori, H. Sasabe, M. Mimuro , J. Am. Chem. Soc. **116**, 2671 (1994).

Direct Probe of Spectrally Line-Narrowed Emission from Cooperative Optical State in π-conjugated Polymeric Thin Films

Sae Chae Jeoung and Dongho Kim

National Creative Research Initiatives Center for Ultrafast Optical Characteristics Control and Spectroscopy Laboratory, Korea Research Institute of Standards and Science, Taejon 305-600. Korea
e-mail:dongho@kriss.re.kr

Abstract: New results of time-resolved emission spectroscopy of π-conjugated polymeric thin film revealed an apparent induction time for the appearance of spectrally narrowed emission (SNE) in contrast with an instantaneous build-up of normal and stimulated emission.

1. Introduction

The realization of a new solid-state plastic laser has gained prospects after reports on lasing action of polymers in various forms. Furthermore, SNE was reported for various polymeric films under high photoexcitation density[1,2]. But the underlying mechanism for this peculiar phenomenon has been the subject of some speculation. Cooperative processes such as superradiance (SR) or superfluorescence (SF) were initially proposed to explain SNE from conjugated polymers. However, most spectroscopic results to elucidate the mechanism for SNE have been interpreted in terms of amplified spontaneous emission (ASE) and/or stimulated emission (SE) of incoherent excitations including exciton-exciton annihilation. More specifically, transient absorption experiments exhibit a gradual decrease in SE lifetimes with increasing excitation density, which seems to be fairly consistent with a model utilizing ASE. However, this behavior is no longer a direct evidence for SE or ASE. A direct way to clarify the disputes regarding the mechanism for SNE is the time-resolved measurement of SNE under various excitation densities because its temporal profiles are expected to exhibit markedly different behaviors depending on its origins[2].

2. Results and Discussion

We measured the emission temporal profiles of a PPV derivative, poly[2,5-bis(cyclohexylmethoxy)-1,4-phenylenevinylene] (PDCHPV), film with an excitation density of 1 GW/cm^2 by using fluorescence up-conversion technique with femtosecond pump and gate pulses at 400 and 800 nm, respectively. And the constructed time-resolved emission spectra were displayed in 3-dimensional and contour plots in Fig. 1. Two emission bands at 600 and 650 nm are responsible for

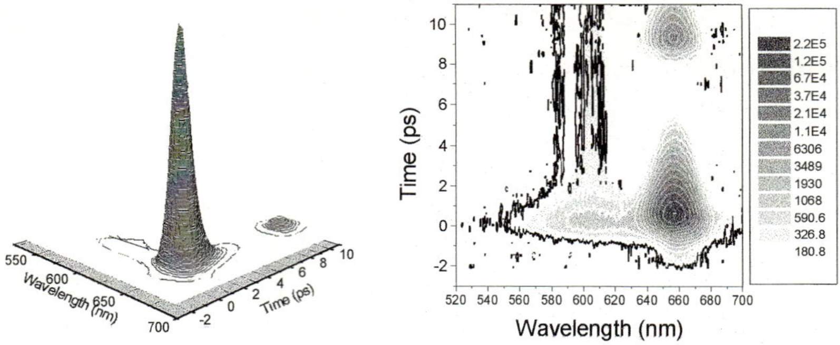

Fig. 1. Three-dimensional (left) and its contour plot (right) of time-resolved emission spectra of PDCHPV film upon photoexcitation at 400 nm by fluorescence up-conversion method.

normal emission and SNE, respectively. The distinct observation is a train of events for SNE, which is separated by ~ 8 ps judging from peak-to-peak positions and persists longer than 30 ps in delay time. Another interesting finding is a time lag (~0.5 ps) in the appearance of SNE in contrast with an instantaneous build-up of normal emission. This time lag is strongly dependent on the excitation power density.

Fig. 2 presents the temporal profiles of normal emission and SNE under five different excitation densities marked as I-V. At low excitation density (Fig. 2(I)), the temporal profile for normal emission fits well to a double-exponential decay with time constants of about 15 ps and >100 ps. The temporal profiles at 600 and 650 nm exhibit almost the same features, indicating that identical emissive species are involved. On the other hand, a gradual increase of the excitation power density changes the SNE dynamics dramatically (Fig. 2(II)-(V)). More importantly, so-called "induction time", a time lag in the appearance of SNE, was clearly observed. The induction time becomes shorter with an increase in the excitation power density. We measured the SNE dynamics with varying the illuminated spot-size from 0.7 to 2 mm in diameter keeping the excitation density constant. If the optical excitations (exciton) are localized and have a coherence length of only a few monomer units, it is difficult to observe remarkable changes in decay dynamics by varying the illuminated spot-size in millimeter range. However, the dynamics, for instance an induction time, illustrates an evident spot-size dependence.

We also investigated the excitation power dependence of the SE dynamics at 650 nm by the transient absorption spectroscopy (Fig. 3) under the five different excitation densities as marked in Fig. 2. At low power density, the SE temporal profile is identical to the normal emission dynamics shown in Fig. 2(I). More interestingly, the dynamics under higher excitation power densities exhibit faster non-exponential decay with an apparent induction time, during which SE signal remains almost constant. After the induction time, the SE signal starts to decrease abruptly. An increase in the excitation power density leads to a gradual shortening of the induction time (Fig. 3).

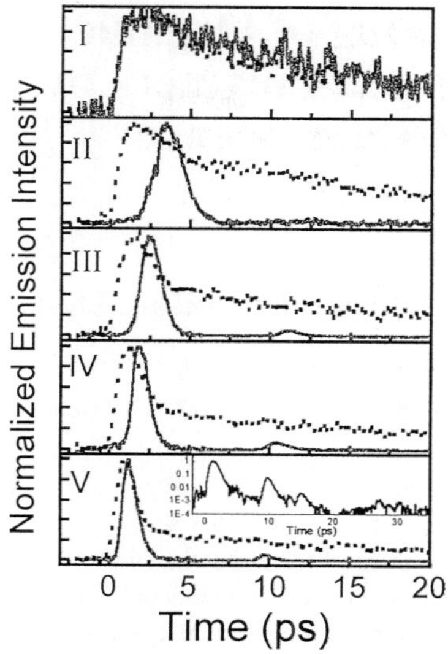

Fig. 2 Normalized temporal profiles of Normal emission at 600 nm and SNE at 650 nm (solid lines) under the excitation densities marked as I, II, III, IV and V.

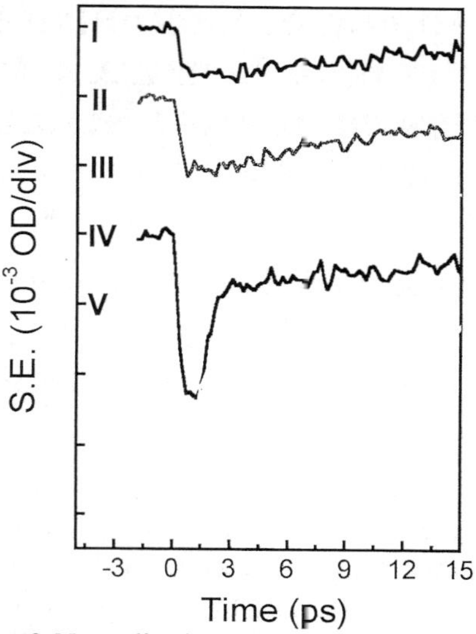

Fig. 3 Normalized temporal profiles for SE from pump-probe experiment under the same photoexcitation densities marked in Fig.2.

3. Conclusions

The direct probe of SNE temporal profiles of π-conjugated polymer films clearly reveals the induction time in the appearance of the SNE relative to the normal emission and their oscillation behaviors which persist up to ~30 ps. The induction time is also well correlated with the duration time where the SE signal remains constant in the SE temporal profiles probed by the transient absorption experiment. Our experimental observations indicate that that the SNE might originate from cooperative optical states, where the overall dynamics is governed by the dephasing dynamics of these newly formed states, instead of collective processes such as ASE.

Acknowledgements. This work has been financially supported by the National Creative Research Initiatives of the Ministry of Science & Technology of Korea (DK)

References

1. N. Tessler, G. J. Denton and R. H. Friend, Nature **382**, 695 (1996).
2. F. Hide *et al.* Science **273**, 1833 (1997).

Mapping of Molecular Docking of Organic Cations on Negatively Charged Polymers by Picosecond Time Resolved Fluorescence Spectroscopy.

Dina Pines, Jacob Cohen, Y. Zhao, Vitaliy Khutorsky, Zvi Priel and Ehud Pines.

Department of Chemistry, Ben-Gurion University of the Negev, P.O.B. 653, Beer-Sheva 84105, Israel.
E-mail: epines@begumail.ac.il

Abstract. Picosecond time resolved spectroscopy is used to probe the binding between positively charged diaminostilbene derivatives and negatively charged polyions. Several binding modes of roughly equal stability but of different polarity and environment were found; implying a compensation mechanism between the hydrophobic and electrostatic interactions.

1. Introduction

Host-guest complexes in biological systems exhibit a basic structural fit and are stabilized by both electrostatic and hydrophobic interactions. This results in remarkable high degree of chemical selectivity.

In our studies we have modeled such host-guest complexes by the binding of charged amino-stilbene derivatives to negatively charged polymers of moderately high molecular weights, 10,000 < (M.W.) < 200,000. We have found out that binding affects both the trans-cis photoisomerization rate of the stilbene derivatives and the relaxation of their fluorescence polarization anisotropy. In some cases the binding process was found to be selective in respect to either the cis or the trans isomer of the stilbene derivatives [1].

A remarkable backbone fit was found by molecular mechanics simulations between the positively charged 4-amino-4-trimethylamino stilbene ($TAAS^+$) and poly(styrene sulfonate) (PSS). In this case the structural backbone fit allows both electrostatic and hydrophobic interactions to contribute to the total binding interactions.

In the present study we report on the mapping of the molecular docking of $TAAS^+$ on PSS^+ by combination of picosecond-time-resolved lifetime measurements with picosecond-time-resolved fluorescence polarization anisotropy measurements.

2. Experimental Methods

The time–resolved fluorescence measurements were done by using an ultrafast single photon counting setup which yields lifetime values comparable to up-conversion techniques down to about 2 ps with at least two order of magnitude greater sensitivity [2].

3. Results and Discussion

We were able to identify three equilibrated modes of binding of $TAAS^+$ to PSS, which differ by both the trans-cis photoisomerization rate and the relaxation of their fluorescence polarization anisotropy. (figures 1 and 2). The correlation between the time behavior of the two processes is shown in Fig.2 and demonstrates that each of the binding mode belongs to a different micro-environment: A purely aqueous binding environment via the head-on coulombic interaction between the charged end groups of the polymer and TASS, a territorial binding mode which is moderately non-polar and a dominantly non-polar binding mode which penetrates the inner aromatic skeleton of PSS.

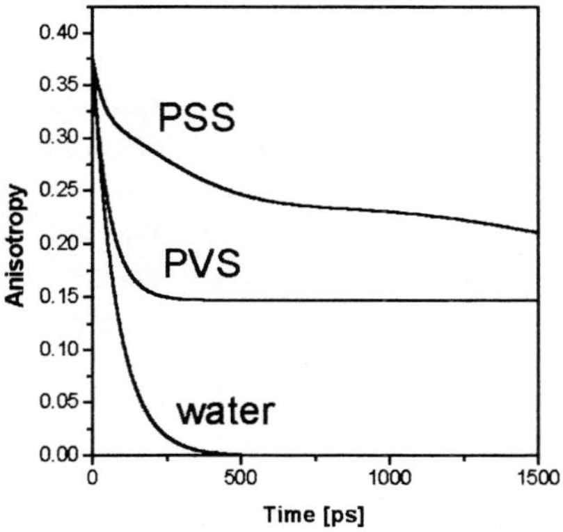

Fig. 1. Best fit curves of experimental polarization data snowing the anisotropy decay of TAAS+ bound to PSS, PVS and as a free rotor in water solution. The magic angle photoisomerization lifetimes are: PSS (τ_1 = 74ps (34%), τ_2 = 230ps (57%), τ_3 = 690 ps (9%)), PVS (74ps), water (74ps). The rotation relaxation times and the order parameters are: PSS (τ_1 = 58ps, S=0.63), τ_2 = 470ps, S=0.57), τ_3 = 2830ps, S=0), PVS (58ps, S=0.57), water (82ps).

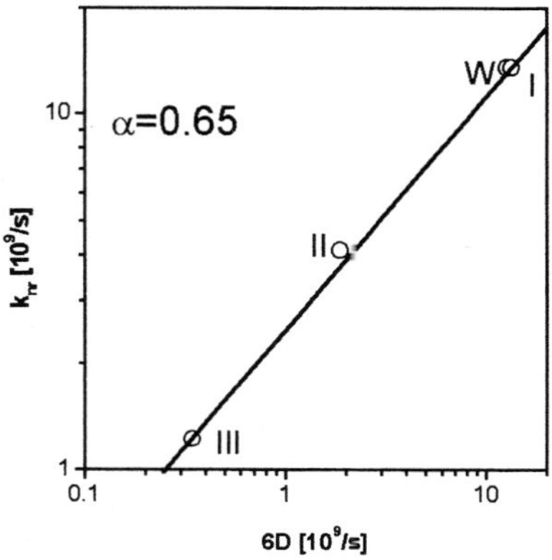

Fig. 2. The estimated isomerization rates are plotted against the 6D values extracted from the rotation relaxation data. (W)= water, and I, II, II are the 3 $TAAS^+$ binding sites. The straight line is $k_{nr} = (6D)^{\alpha}$ with α=0.65. The value of α is in agreement with the analysis of Fleming et al [3] of the viscosity dependence of the cis-trans photoisomerization of stilbene.

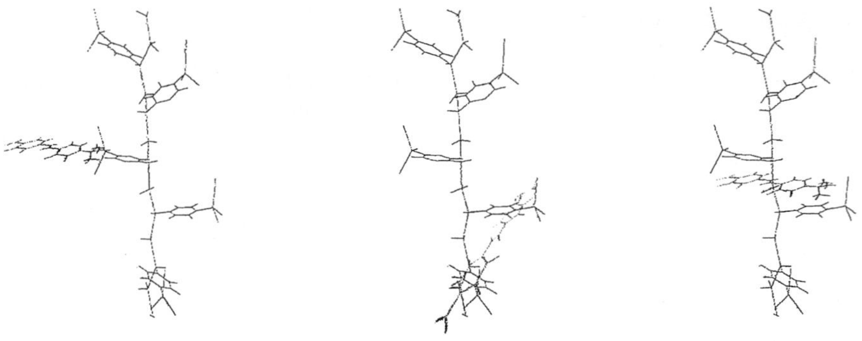

I II III

Fig. 3. The three equilibrated binding modes of $TAAS^+$-PSS complex.

The three binding modes (Fig.3) were found to be in equilibrium with each other. When complexed with poly(vinylsulfonate) (PVS) which lacks π systems, TAAS+ retained only the purely electrostatic aqueous head-on binding mode.

4. Conclusions

The overall free-energy of binding of $TAAS^+$ to PSS was found to be about 10kcal/mole, while the free-energy differences between the three equilibrated binding modes of $TAAS^+$ to PSS were less than one kcal/mole. It follows that even in a simple high-affinity host-guest complex, such as in our case, which combines a structural backbone fit with electrostatic and hydrophobic interactions, there are several energetically close-laying binding modes which largely differ by their polarity and local environment. This points out the formidable task of simulating the exact structure of biological host-guest complexes from energy minimization procedures. In fact, this work may imply th at in many cases such efforts may be inherently fruitless.

Acknowledgements. Ehud Pines acknowledges James Franck German-Israel Binational Program

References

1. Y. Zhao, J. Cohen, V. Khutorsky, N. Strashnikova, E. Pines, Z.Priel, J. Photochem. Photophys. **A** 2000.
2. E. Pines, D. Pines, T. Barak, B-Z Magnes, L.M. Tolbert, & J.E. Haubrich, Ber. Bunsen. fur Phys. Chem. **102**, 511, 1988
3. S.H. Courtney and G.R. Fleming, J. Chem.Phys. **83**, 215 (1985)

Direct observation of the ultrafast electron transfer process in a polymer/fullerene blend

Ch. J. Brabec[1], G. Zerza[1], N.S. Sariciftci[1], G. Cerullo[2], G. Lanzani[2], S. De Silvestri[2], J.C. Hummelen[3]

[1] Physical Chemistry, Johannes Kepler University of Linz, Altenbergerstr. 69, A-4040 Linz
E-mail: cristoph.brabec@jk.uni-linz.ac.at
[2] Istituto Nazionale per la Fisica della Materia, CEQSE-C.N.R., Dipartimento di Fisica, Politecnico, Milano (Italy)
[3] Stratingh Institute and Materials Science Center, University of Groningen, Nijenborgh 4, 9747 Groningen (The Netherlands)

Abstract. Optical studies on conjugated polymer-fullerene blends are performed with sub-10-fs temporal resolution. The photoinduced electron transfer process is directly monitored in the time domain, obtaining a forward electron transfer time constant of 45 fs.

1. Introduction

Photoinduced electron transfer in organic molecules is an intensively investigated topic in physics, chemistry and biology. A very efficient electron transfer from conjugated polymers to Buckminsterfullerene (C_{60}) has been reported [1], with promising applications to photovoltaics, non-linear optics and artificial photosynthesis. While the forward electron transfer time is remarkably fast, the backtransfer is heavily hindered, resulting in metastable charge separated states with lifetimes in the μs timescale. Up to now only an upper limit of the electron forward transfer time of $\approx$ 100 fs has been reported [2], but a detailed time resolution of the process is still missing and is highly needed to shed light on the photophysics of the charge transfer process.

In this work we perform ultrafast experiments on conjugated polymer/fullerene blends resonantly excited with sub-10-fs light pulses. Monitoring the formation of the absorption band of the charged polymer, which provides a spectral signature of the charge transfer, we are able to time resolve for the first time the process, obtaining a forward electron transfer time constant of $\approx$ 45 fs.

2. Experimental Methods

The excitation used in the experiments is a visible optical parametric amplifier [3], based on noncollinear phase-matching in β-barium borate, followed by a pulse compressor using chirped mirrors. This source provides ultrabroadband pulses, with spectrum extending from 500 to 720 nm, compressed to an almost transform limited duration of 5-6 fs. Pump-probe experiments are performed in a standard noncollinear configuration. We study films of poly [2-methoxy, 5-(3',7' – dimethyl-ocyloxy)]-p-phenylene vinylene (MDMO-PPV) blended with [6,6]-Phenyl C_{61} – butyric acid methyl esther (PCBM), cast onto thin quartz substrates, with a 1:1 ratio in molecular concentration. At this high PCBM concentration diffusion does not influence the dynamics and we measure the intrinsic charge transfer process. We also study, for comparison, pure MDMO-PPV films.

3. Results and Discussion

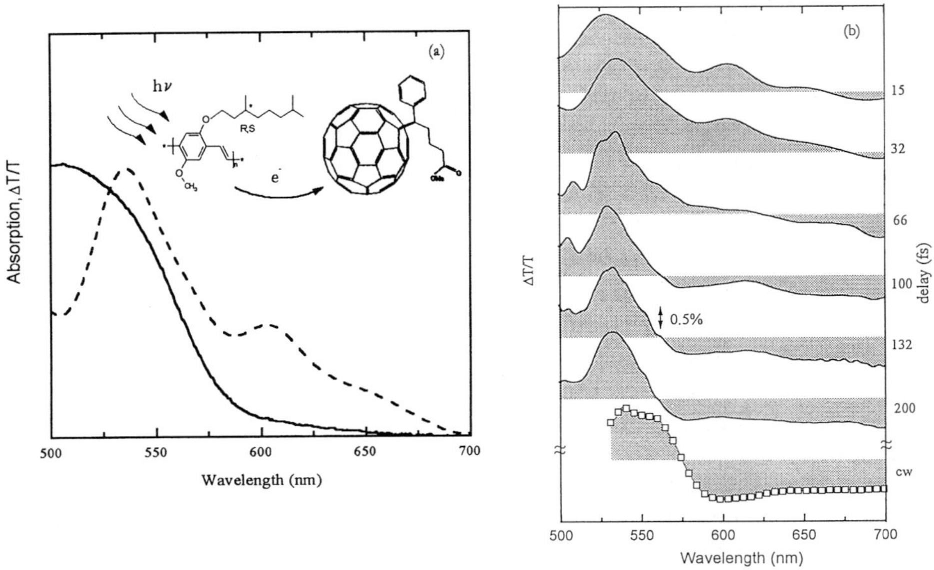

Fig. 1. (a) Absorption spectrum of a pure MDMO-PPV film (solid line) and ΔT/T spectrum at 200 fs pump-probe delay (dashed line). The inset shows a scheme of the MDMO-PPV/PCBM electron transfer process (b) ΔT/T spectra of a MDMO-PPV/PCBM blend at various delays after excitation by a sub-10-fs pulse. The cw spectrum (□) was measured at 100 K.

Fig. 1a shows as a solid line the absorption spectrum of MDMO-PPV and as a dashed line the ΔT/T spectrum at 200 fs pump-probe delay. In agreement with previous results [4], the ΔT/T signal can be attributed to a superposition of photobleaching of the ground state absorption and stimulated emission (SE) from the photoexcited state. The shape of the ΔT/T spectrum does not show any major evolution in the first few ps from photoexcitation, apart from an overall decay. By adding PCBM to the polymer matrix, the excited state evolution scenario changes dramatically, as shown in Fig. 1(b). At early time delays (see the 10 fs and the 30 fs data) the spectrum closely resembles that of pure MDMO-PPV, confirming the predominant excitation of this molecule. The SE band from MDMO-PPV rapidly gives way to a photoinduced absorption (PA) feature, the formation of which is completed within 200 fs. After this initial quick evolution the ΔT/T spectrum remains stationary on the timescale of the experiment (40 ps). This PA feature is assigned to the PPV cation radical (positive polaron) by comparison to the near steady state ΔT/T spectrum of MDMO-PPV/PCBM composite shown as hollow squares. This band provides a direct signature for the charge transfer process.

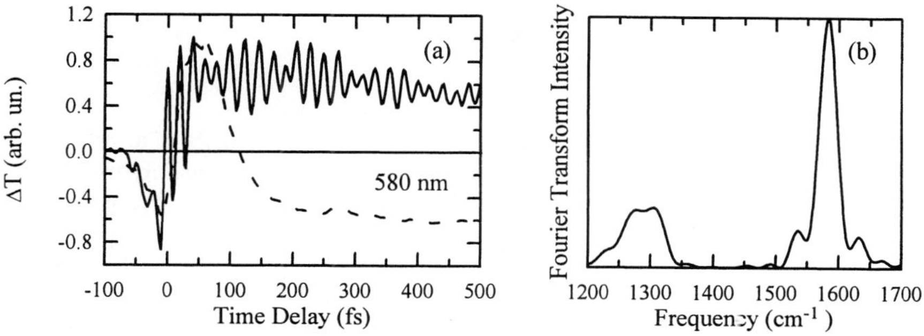

Fig. 2. (a) Differential transmission dynamics of pure MDMO-PPV (solid line) and PPV/PCBM (dashed line) at 580 nm probe wavelength; (b) Fourier transform of the oscillatory component of the MDMO-PPV signal.

Fig. 2(a) compares the dynamics of the blend at 580 nm probe wavelength (dashed line) to that of the pure polymer (solid line). A fast rise to positive values of ΔT/T, due to SE from the polymer, is followed immediately by a fast decay until the signal stabilises on negative ΔT/T, indicating the PA of the charge transferred state. By an exponential fitting of the PA risetime, a time constant for the electron transfer process of 45±3 fs can be calculated. Note that the coherently excited oscillations observed in the pure polymer signal (see Fourier spectrum in Fig. 2(b)) are almost completely quenched in the blend, indicating evolution of the system on an anharmonic charge transfer potential energy surface, formed immediately after photoexcitation.

References

1 N. S. Sariciftci, L. Smilowitz, A.J. Heeger, and F. Wudl, Science **258**, 1474 (1992).

2 B. Kraabel, D. McBranch, N.S. Sariciftci, D. Moses, and A.J. Heeger, Phys. Rev. B **50**, 18543-18552 (1994).

3 G. Cerullo, M. Nisoli, S. Stagira, S. De Silvestri, G. Tempea, F. Krausz, K. Ferencz, Opt. Lett. **24**, 1529 (1999).

4 S.V. Frolov, W. Gellermann, M. Ozaki, K. Yoshino, Z.V. Vardeny, Phys. Rev. Lett. **78**, 729 (1997).

The Controlled Coherent Excitation of Vibrational Modes in the Electronic Ground State of Polydiacetylenes

T. Chen, A. Vierheilig, W. Kiefer, and A. Materny

Institut für Physikalische Chemie der Universität Würzburg,
Am Hubland, D-97074 Würzburg, Germany
E-mail: amaterny@phys-chemie.uni-wuerzburg.de

Abstract. Vibrational modes in the electronic ground state of polymers of diacetylene are selectively excited and then probed by means of a femtosecond time-resolved CARS (coherent anti-Stokes Raman scattering) scheme. Control is achieved by varying the timing and the phase shape (chirp) of the exciting laser pulses.

The control of chemical reactions is one of the main goals of chemists. One way to achieve this goal is photochemistry where the energy to break bonds is added in form of light [1]. The absorption of light results in internal vibration as well as translational or rotational motion. When molecules are heated indiscriminately the energy will be distributed statistically among all degrees of freedom; no selectivity can be achieved, the weakest bonds will break first and therefore, only a limited set of reactions is possible.

In our recent work we have investigated the possibility of selectively exciting specific vibrational modes in molecules which are characterized by multidimensional potential energy surfaces (PES) [2]. In the present contribution we demonstrate that by the application of fs time-resolved coherent anti-Stokes Raman scattering (CARS) we are able to selectively excite ground state vibrations and also detect the time response of these modes as well as neighboring modes.

The experiments were performed on diacetylene single crystals containing less than 1% polymer molecules. For the polydiacetylenes (PDAs) there exist two different mesomeric structures of their backbones which can be described as "acetylene-type" $(=RC-C\equiv C-CR'=)_n$ and "butatriene-type" $(-RC=C=C=CR'-)_n$. While the ground state of the considered PDAs (TS6 and FBS, see e.g. Ref. 2) is described by the acetylenic structure, the excitonic state is characterized by a geometric relaxation model. Pham *et al.* [3] assumed that the resonant optical excitation of the PDAs takes them to their free excitonic state (FE) from where a formation of self-trapped excitons (STE) takes place. FE and STE backbones have butatriene and acetylene type structure, respectively. Fig. 1a shows a scheme suggested by Kobayashi and coworkers [4] where the different electronic (excitonic) states of the PDAs are depicted. Fig. 1b gives the energy scheme of the femtosecond time-resolved CARS process. The dynamics of the different states can be accessed by varying the delays between Stokes and probe (τ_2) or pump and Stokes pulses (τ_1). Positive τ_1 means that the pump laser pulse interacts with the sample earlier than the Stokes laser pulse. In the experiments discussed here, we scan τ_2 for different fixed τ_1. Additionally, we can distinguish signals arising from different phonon modes detecting the frequency-resolved anti-Stokes signal.

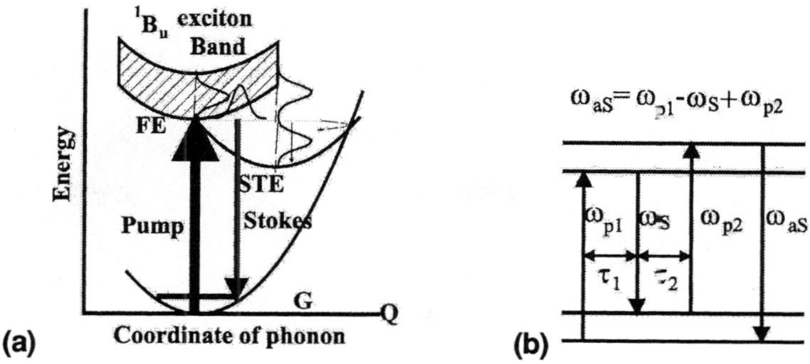

Fig. 1. (a) Scheme of the energy relaxation process in PDAs (compare Ref. 4); (b) the energy diagram of the femtosecond time-resolved CARS experiment.

The general idea of the experiments discussed in the following was to optimize the excitation of different vibrational modes in the electronic ground state of the PDA molecules. This goal can be achieved by varying the time dependent phase of the exciting pump and /or Stokes lasers as well as making use of the dynamics on the excitonic PES.

In our first experiment, we have minimized the chirp of the laser pulses in order to remove any influence of the phase shape of the lasers. The excitation process was influenced by a variation of the delay time τ_1 between pump and Stokes pulses. In this case mainly the dynamics in the excited excitonic states are determining the outcome of the pump-dump process. As a function of τ_1, Franck-Condon windows are opened at different positions of the PESs. The geometric relaxation process is responsible for a dramatic change in the transition conditions for the Stokes laser. As suggested by Pham *et al.* [3] a transition from the butatrienic mesomeric chain structure to the acetylenic one takes place on a femtosecond time-scale, where the STE is self-trapped in the C=C double bonds. The transition to the electronic ground state (with acetylenic backbone structure) therefore is expected to favor the $\nu_2/\nu_{2'}$ modes (C=C stretching and Fermi coupled sidegroup vibration, respectively).

Fig. 2a shows the results of an experiment, where the delay time τ_2 was fixed at $\tau_2 = 0$ fs and the delay time τ_1 between pump and Stokes pulses was varied. The relative populations corresponding to $\nu_2/\nu_{2'}$ and ν_3 (single bond stretching) are given for comparison. At positive τ_1 delay times the vibrational motion belonging to the $\nu_2/\nu_{2'}$ mode is strongly driven; at the same time, the vibrational motion of ν_3 is suppressed. This result is in accordance with the model discussed above.

The second experiment performed was of more qualitative character. In this experiment, τ_1 was fixed at $\tau_1 = 0$ fs, the pump and Stokes pulses were overlapping in time. The transients shown in panel b of Fig. 2 were taken as a function of signal wavenumber and delay time τ_2. Now, the excitation process was influenced by a variation of the chirp of the exciting laser pulses (pump and Stokes). The first transient (upper chart) was obtained for minimized chirp and serves as a reference. At $\tau_2 = 0$ fs all vibrational modes contribute which are excited by the spectrally broad pump and Stokes lasers (Raman resonance for the central wavenumbers with the ν_3 mode was chosen). In a recent publication, we have analyzed this transient in more detail [5].

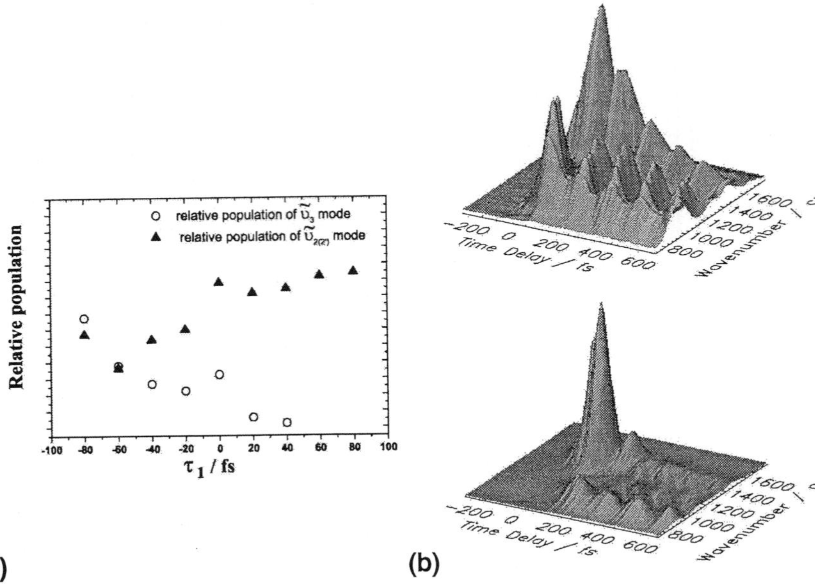

Fig. 2. (a) The relative population of $\tilde{\nu}_2/\tilde{\nu}_{2'}$ and $\tilde{\nu}_3$ modes as a function of τ_1. (b) The influence of the exciting lasers' chirp on the mode excitation; upper transient: minimal chirp; lower transient: laser chirp was introduced in order to selectively enhance the $\tilde{\nu}_2/\tilde{\nu}_{2'}$ contributions.

By adding chirp to the exciting laser pulses, the behaviour can be dramatically changed. Without optimizing the phases of pump or Stokes lasers we were able to achieve a nearly exclusive excitation of $\nu_2/\nu_{2'}$ at $\tau_2 \approx 0$ fs. This is clearly demonstrated with the lower transient in Fig.2b.

In summary, we have discussed two ways of influencing the coherent excitation of vibrational modes in the electronic ground state of polymers of diacetylene. By making use of the dynamics in the excited excitonic states of the PDAs as well as by varying the linear chirp of the exciting lasers of the CARS process, a selective excitation of modes could be achieved.

References

1 A. H. Zewail and R. Bernstein, *Chem. Eng. News* , 66, 24 , 1988.

2 T. Chen, A. Vierheilig, P. Waltner, M. Heid, W. Kiefer, and A. Materny, *Chem. Phys. Lett.*, in press.

3 T. A. Pham, A. Daunois, J.-C. Merle, J. Le Moigne, and J. Y. Bigot, *Phys. Rev. Lett.*, 74, 904, 1995.

4 T. Kobayashi, M. Yoshizama, U. Stamm, M. Taiji, and M. Hasegawa, *J. Opt. Soc. Am.*, 7, 1558, 1990.

5 A. Vierheilig, T. Chen, P. Waltner, W. Kiefer, A. Materny, and A. H. Zewail, *Chem. Phys. Lett.*, 312, 349, 1999.

Molecular Dynamics Simulations of Collective Electronic and Nuclear Modes in Conjugated Systems

Matteo Tommasini, Vladimir Chernyak and Shaul Mukamel

Department of Chemistry, University of Rochester, Rochester NY, 14627
Tel.: (716) 275-3080, Fax: (716) 473-6889, e-mail: mukamel@chem.rochester.edu

Abstract. A new time dependent algorithm based on the Collective Electronic Oscillators (CEO) method for computing excited state potential surfaces is used for simulating femtosecond nuclear dynamics in polydiacetylene oligomers.

1. Introduction

Impulsive pump-probe experiments performed by Kobayashi and coworkers on polydiacetylene (PDA) using 5 fs pulses showed, among other features, the details of the geometry relaxation on the excited state [1]. Similar measurements has been performed on different conjugated systems by De Silvestri and coworkers [2]. To simulate the nuclear dynamics probed by these experiments it is necessary to compute excitation energies and their gradients with respect to nuclear coordinates. This is the main bottleneck in molecular dynamics simulations of photoinduced processes [3]. Computing the excited state wavefunction is prohibitively expensive, in particular when it needs to be repeated for many nuclear configurations along the trajectory. In this work we carried out simulations using a molecular dynamics Collective Electronic Oscillators (CEO) algorithm which avoids the direct computation of excited state many-electron wavefunctions [4]. It is based on a reduced description of the electronic structure which uses a minimal amount of relevant excited state information by focusing on the time-dependent single-electron density matrix of the molecule. The CEO associated with the molecular excited state $|\nu\rangle$ with energy E_ν is obtained as the eigenvector $|\xi_\nu(\mathbf{x}))$ of a Liouville space eigenvalue problem ($\mathbf{x}$ represents the set of nuclear coordinates):

$$L(\mathbf{x})\ |\xi_\nu(\mathbf{x})) = \Omega_\nu(\mathbf{x})\ |\xi_\nu(\mathbf{x})). \tag{1}$$

The electronic normal modes $|\xi_\nu(\mathbf{x}))$ represent the transition density matrices between the ground $|0\rangle$ and excited $|\nu\rangle$ states, and the eigenvalues $\Omega_\nu(\mathbf{x})$ give the excitation frequencies. Solution of Eq. (1) at various nuclear configurations $\mathbf{x}$ yields the geometry dependent excitation energies Ω_ν. The excited-state adiabatic surface $E_\nu(\mathbf{x})$ is then obtained by adding $\Omega_\nu(\mathbf{x})$ to the ground-state adiabatic surface $E_0(\mathbf{x})$, which is an input to this calculation. Vibronic coupling enters Eq. (1) through the dependence of the electronic hamiltonian, and thus of L, on nuclear geometry. The excitation energy gradient can be written as $\partial\Omega_\nu/\partial x = (\xi_\nu|\partial L/\partial x|\xi_\nu)$ (here x stands for some generic nuclear coordinate). The gradient of the excited state hypersurface $E_\nu = E_0 + \Omega_\nu$ is thus given by $\partial E_0/\partial x + \partial\Omega_\nu/\partial x$. The non adiabatic couplings which are essential for simulating processes occurring near conical intersections, such as photoisomerization reactions, can be computed as well [5]. Our algorithm can be used to perform MD simulations of ultrafast nuclear relaxation on the excited state. We do not need to select a region on the excited state surface prior to the simulation: the surface is computed "on the fly" during the simulation.

2. Computations and discussion

We have applied the molecular dynamics CEO code to explore the dynamics of the lowest excited state of a PDA oligomer containing five carbon-carbon triple bonds (see Fig. 1). The left panel shows the ground state density matrix: notice that the five triple and the four double CC bonds can be identified along the diagonal (the carbons are numbered along the chain). The right panel shows the transition density matrix for the first excited state; the two arrows point to the major change in the density upon excitation corresponding to the central triple bond and the two adjacent double bonds. This reflects the transition from the acetylenic structure of the ground state to a butatrienic structure for the excited state, as discussed in [1] and [6]. The transition density matrix also shows that the main electronic change upon ecitation is localized in the center of the molecule. We have simulated an ideal vertical excitation from

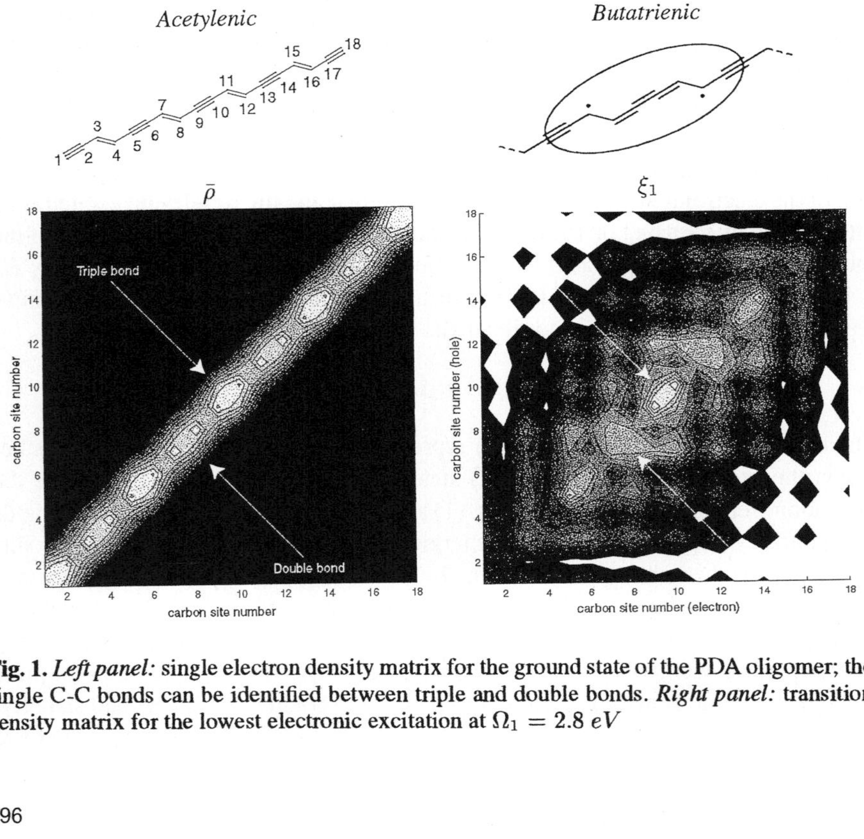

Fig. 1. *Left panel:* single electron density matrix for the ground state of the PDA oligomer; the single C-C bonds can be identified between triple and double bonds. *Right panel:* transition density matrix for the lowest electronic excitation at $\Omega_1 = 2.8\ eV$

the equilibrium ground state geometry to the lowest CEO mode. During the first few fs the triple bonds are lengthening, while the single bonds are shortening (see Fig. 2, left panel). This is a signature of the transition to the butatrienic form of the excited state: the single and the triple bonds become double bonds in the excited state so that the bond lengths change as shown in Fig. 2. The high frequency CH stretching vibrations are decoupled from other vibrational degrees of freedom. In contrast the time evolution of carbon-carbon bond lenghts shows the effect of couplings among CC stretchings and with bendings along the chain. A Fourier analysis of the time domain data shows the coupling between stretching and bending vibrations [7]. As can be seen in the right panel of Fig. 2, the main peak of the C-C=C bending vibration shows up in the spectrum of the CC stretchings (in particular 2 and 3, as marked by the two solid arrows). The linear bending defined on the single and triple bond does not interact with the CC stretchings but interacts with the C-C=C bending (blank arrow). The same type of coupling has been observed in the experiments [1].

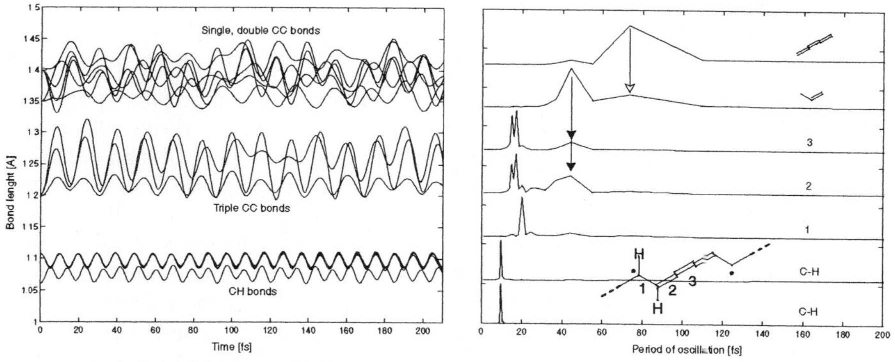

Fig. 2. *Left panel:* time evolution of the various bond lengths on the excited state surface. *Right panel:* Fourier transform of the time evolution of the various relevant bond lengths and bond angles at the center of the oligomer

Acknowledgements. The support of the National Science Foundation is gratefully acknowledged.

References

1. T. Kobayashi, A. Shirakawa, H. Matsuzawa, and H. Nakanishi, Chem. Phys. Lett., **321**, 385-393 (2000).
2. G. Cerullo, G. Lanzani, L. Pallaro, and S. De Silvestri, J. Mol. Struct., **521**, 261-270 (2000).
3. W. Domcke, and G. Stock, Advances in Chemical Physics, **100**, 1-169 (1997).
4. E. V. Tsiper, V. Chernyak, S. Tretiak, and S. Mukamel, J. Chem. Phys. **110**, 8328-8337 (1999).
5. V. Chernyak and S. Mukamel, J. Chem. Phys. **112**, 3572-3579 (2000).
6. M. Turki, T. Barisien, J.Y. Bigot, and C. Daniel, J. Chem. Phys. **112**, 10526-10537 (2000).
7. M. Tommasini, V. Chernyak, and S. Mukamel, in preparation

Theoretical investigation of the excited states dynamics of the polydiacetylene backbone.

T. Barisien[1], M. Turki[2], C. Daniel[2], and J.-Y. Bigot[1]

[1] Institut de Physique et Chimie des Matériaux de Strasbourg, Unité Mixte 7504 CNRS-ULP-ECPM, 23 rue du Loess BP20, 67037 Strasbourg Cedex, France
E-mail: Jean-Yves.Bigot@ipcms.u-strasbg.fr
[2] Laboratoire de Chimie Quantique UMR 7551 du CNRS et de l'Université Louis Pasteur, Institut Le Bel, 4 rue Blaise Pascal, 67000 Strasbourg, France
E-mail: daniel@quantix.u-strasbg.fr

Abstract. The geometrical relaxation from the acetylenic to the butatrienic conformation in polydiacetylene is studied by simulating the dynamics of wave packets on the excited state potential energy curves of a cyclic model of the chain.

The theoretical description of the electronic structure of conjugated polymers remains of major interest since the efficient optimization of the electronic properties (namely the linear susceptibilities or the photo- and electro-luminescence yields) is based on the ability to understand first, how the mechanisms intrinsic to the conjugated backbone such as vibronic coupling, affect the dynamics of the primary excitations. Among the great diversity of conjugated materials, the polydiacetylene chain thanks to its quasi-unidimensional structure represents a model system particularly appropriate to address the influence of the molecular motion which can be initiated experimentally in a coherent manner on the potential energy surfaces associated to the first excited states of the molecule. In the present paper the geometrical relaxation of isolated chains of polydiacetylene in response to an optical ultra-short pulse is investigated by determining the potential energy curves (PEC) which connect the mesomeric structures of the chain. The propagation of vibrational wave packets on the coupled PECs allows us to draw the main frame of the molecular motion observed in classical spectrally resolved pump-probe experiments in which the pump and probe pulses have a duration of 10 femtoseconds.

1. Model and calculations

From a general point of view the important delocalization of electrons occurring in conjugated polymers makes the exploration of the excited states organization particularly tricky. The latter indeed requires an extensive treatment of the correlation effects including both the interactions between electrons of the π conjugated system and interactions between π electrons and those from the core cloud (σ electrons). In this work the wave functions and excitation energies of the two lowest singlet excited states are determined systematically in a family of linear oligomers (fig. 1a) as well as in the folded model of the chain introduced in

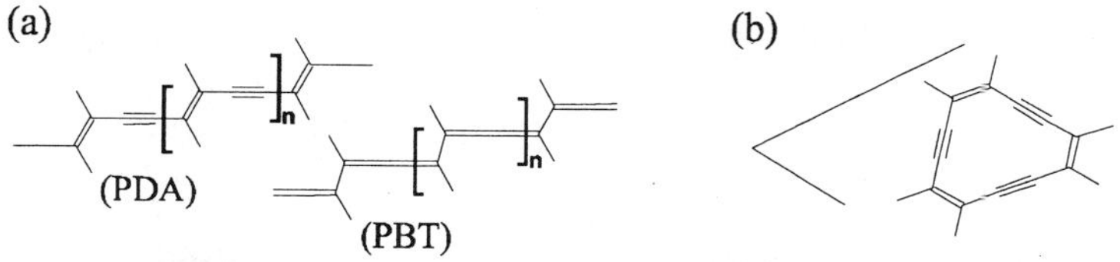

Fig. 1. (a) Ground state structure of the chain in the polydiacetylenic (PDA) and polybutatrienic (PBT) configurations. The substituents have been replaced by hydrogen atoms to enable ab initio calculations. The energy and character of the first excited states have been investigated for n=1, 2, 3.5. (b) Cyclic model of the polydiacetylene backbone.

the early work of Karpfen [1] and designed to describe the PECs connecting the ground and excited states of the polydiacetylenic and polybutatrienic conformations of the chain (fig 1b).

The CASSCF/CASPT2 calculations indicate in particular that the $2A_g$ dark state of the PDA structure which correlates with the $1A_g$ ground state of the PBT chain still lies above the first excited optically active 1^1B_u state in the trimer compound. The picture of the potential energy curves associated to the cyclic molecule (figure 2) points out the presence of an intersystem crossing between the 2^1A_g and 1^1B_u states next to the Franck Condon region compatible with the description of the relaxation process via a self trapping of the elementary excitations in the $2A_g$ potential well [2]. It also shows that, when the effects of correlation are taken into account, the PBT structure no longer corresponds to a stationary configuration of the chain in the ground state and raises the question of the identification of the coupling mode between the 1^1B_u and 2^1A_g states, orthogonal in the adiabatic representation of figure 2. At the CASPT2 level of calculation the energetic separation between the 2^1A_g and 1^1A_g states in the region of the avoided crossing is estimated to be about 0.35 eV.

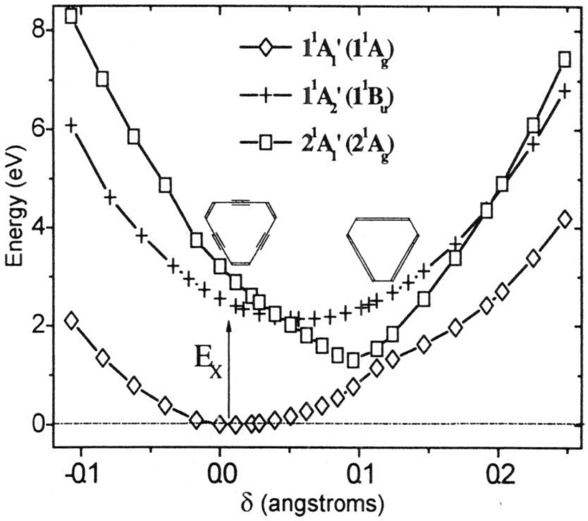

Fig. 2. PECs connecting the PDA and PBT structures of the folded chain. The x axis corresponds to the stretching of the double C=C bond when passing from the PDA to the PBT structure. Vertical excitation energy E_x is 2.4 eV and must be compared to the value of the excitonic resonance in the blue phase polymer (~2 eV). Note that the calculations have been performed in the C_{2v} group. The symmetry indicated in parenthesis corresponds to the symmetry of the state which has the same structure in the C_{2h} group, characteristic of the linear molecule in the crystal.

2. Excited states dynamics

In figure 3a, we have represented the time dependent overlap (autocorrelation function) between a bra wave packet whose position is frozen in the Franck Condon region and a ket wave packet propagating on the potential energy curve of the 1^1B_u excited state. The oscillatory behaviour of the function can unambiguously be assigned to the coherent vibrational motion associated to the C=C stretching modes previously detected in the spectro-temporal variations of the pump-probe signal [3] as a quantum beat with a period of 23 fs (figure 3c). The simulated dynamics on the coupled 1^1B_u – 2^1A_g states indicates that internal conversion occurs with the characteristic time of about 130 fs (figure 3b) and is accompanied by the geometrical relaxation of the chain from the PDA to the PBT conformation. Let us note that the theoretical value of the rate constant is typically the one reported in recent time resolved fluorescence measurements [4]. This study also provides meaningful information concerning the subsequent stages of the relaxation mechanism. In particular it stresses the important effect of thermalization on the rate of the internal conversion process to the 1^1A_g ground state [5].

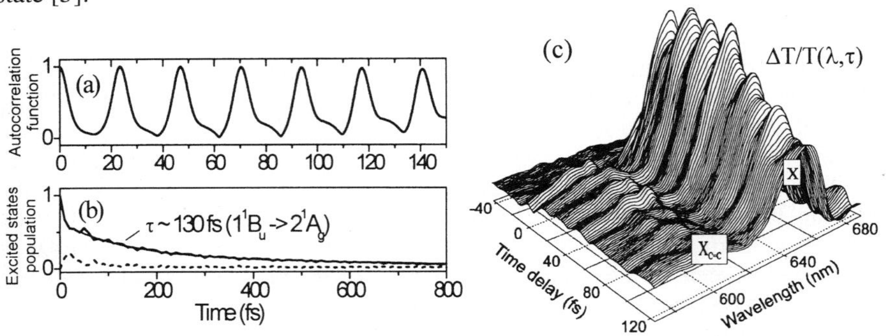

Fig. 3. (a) Time dependent overlap between the vibrational wave packet at t=0 on the 1^1B_u state and a wave packet which propagates on the potential energy curve of this same state. (b) Simulated decay associated to the internal conversion process between the 1^1B_u and 2^1A_g states. In order to reproduce realistic simulations, the wave packet of the 2^1A_g state is absorbed by an attractive potential before it repopulates the 1^1B_u state. The dashed line of figure (b) represents the population of the 2^1A_g state during the simulation. (c) Spectro-temporal variations of the pump-probe signal measured in pDCH polydiacetylene (X is for the main excitonic transition and $X_{C=C}$ corresponds to the replica of the C=C double bond).

References

1 A. Karpfen, J. Phys. C: Solid St. Phys. **13**, 5673, 1980.
2 T. Kobayashi, Pure & Appl. Chem. **67**, 387, 1995.
3 J.-Y. Bigot, T.-A. Pham, T. Barisien, Chem. Phys. Lett. **259**, 469, 1996.
4 S. Haacke, J. Berrehar, C. Lapersonne Meyer, M. Schott, Chem. Phys. Lett. **308**, 363, 1999.
5 M. Turki, T. Barisien, J.-Y. Bigot, C. Daniel, J. Chem. Phys. **112**, 10526, 2000.

Ultrafast Energy Relaxation Processes of Long Molecular Wires Based on Zinc(II)porphyrins

Yong Hee Kim[1], Dae Hong Jeong[1], Sae Chae Jeoung[1], Dongho Kim[1], Naoki Aratani[2] and Atsuhiro Osuka[2]

[1]National Creative Research Initiatives Center for Ultrafast Optical Characteristics Control, KRISS, Taejon 305-600, Korea,
E-mail: dongho@kriss.re.kr
[2]Department of Chemistry, Kyoto University, Kyoto 606-8502, Japan
E-mail: osuka@kuchem.kyoto-u.ac.jp

Abstract. Ultrafast excition migration processes in long molecular wires up to 128 porphyrin units joined directly together were investigated by using various time-resolved spectroscopic techniques

1. Introduction

The motivation for the molecular design of directly *meso-meso* linked porphyrin arrays as a new model of light-harvesting antenna as well as a molecular photonic wire is stemming from facts that the porphyrins must be brought closer to permit rapid energy transfer yet without formation of any energy sink which disrupts the energy flow along the arrays. The linear rod-like shape and orthogonal connection of the neighboring porphyrins are ideal for this purpose. Zinc(II) 5,15-bis(3,5-dioctyloxyphenyl)porphyrin (Z1), its directly *meso-meso* linked porphyrin arrays up to Z128 (Zn, n represents the number of porphyrins) were synthesized and their photophysical properties have been investigated in order to gain further insight into the electronic coupling energy between the neighboring porphyrins in the ground-state, S_1 state (Q band), and S_2 state (B band), the exciton migration, the energy flow mechanism upon the increase in the number of porphyrins[1,2].

2. Results and Discussion

The absorption spectra of these porphyrin arrays change in a systematic manner with an increase in the number of porphyrins; the high-energy absorption bands of the split Soret bands remain at nearly the same wavelength (413-414 nm), while the low-energy absorption bands are gradually red-shifted, resulting in a progressive increase in the exciton splitting energy which is nicely correlated with the values of $\cos[\pi/(N + 1)]$ according to the Kasha's exciton coupling theory. The increasing red-shifts were also observed for the Q bands but these spectral changes were rather modest. The steady-state fluorescence excitation anisotropy spectra of the porphyrin arrays show that the photoexcitation of the high-energy Soret band induces a large angle difference between absorption and emission dipoles in contrast with the photoexcitation of the low-energy exciton split Soret and Q bands. This result indicates that the high-energy exciton Soret band exhibits the localized monomeric character with its dipole moment deviated from the array chain

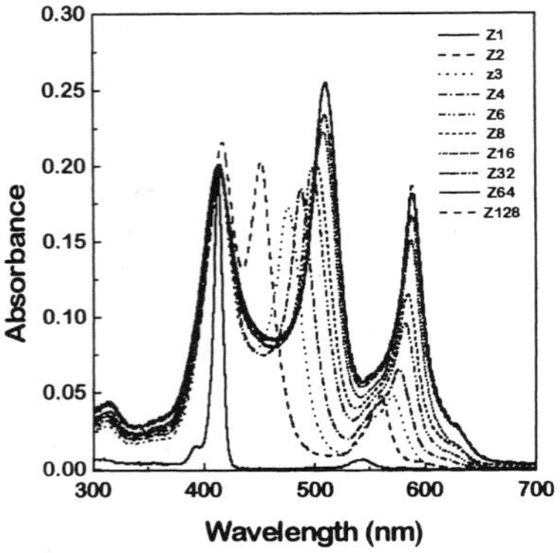

Fig. 1. Absorption Spectra of Porphyrin ararrays

direction, while the low-energy exciton split Soret band is delocalized through the array chain.

Ultrafast fluorescence decay measurements show that the $S_2 \rightarrow S_1$ internal conversion process occurs in less than 1 ps in the porphyrin arrays due to the existence of exciton split band as a ladder-type deactivation channel, while this process is relatively slow in Z1 (~1.6 ps). The rate of this process seems to follow the energy gap law, which is mainly determined by the energy gap between the exciton split Soret bands of the porphyrin arrays. In addition, the decay components with approximately 20-30 ps time constant were observed in the fluorescence decay profiles. These components are believed to arise from the conformational change in the excited states mainly induced by the dihedral angle distribution in these arrays[3].

From the singlet state lifetimes, photoexcitation density dependence on the fluorescence decay profiles and systematic frequency shift of some Raman modes, the exciton delocalization length (coherent length) was estimated to be 6-8 porphyrin units in these porphyrin arrays. According to the formalism developed by Kakitani et al.,[4] the coherent length (N_c) can be estimated by eq. (1).

$$Nc = \sqrt{\frac{3\pi^2|\Delta E_0|}{\gamma}} - 1 \qquad (1)$$

where γ represents the homogeneous broadening in the exciton level and can be estimated for chromophores in very similar environments by eq. (2),

$$\gamma = \frac{1}{\pi(FC)_o} = \frac{\sqrt{4\pi\lambda k_B T}}{\pi} \qquad (2)$$

On the basis of $\Delta E_o = 4300cm^{-1}$ for the Soret-band (S_2-state), the coupling strength for Q-band (S_1-state) has been estimated to be 270 cm^{-1}. Adopting the full width at half height of the maximum absorption (FWHM), 236 cm^{-1} for the Q-band as the value of γ, we obtained $N_C = 5$ for the S_1-state. When γ is estimated as a half of the Stokes shift in eq. (2), we obtained $N_C = 7$ for the S_1-state. These estimations are nicely fit with the experimental results. Applying this numerical calculation to the Soret-band, we obtained $N_C = 15$ and 21 on the basis of FWHM and the Stokes

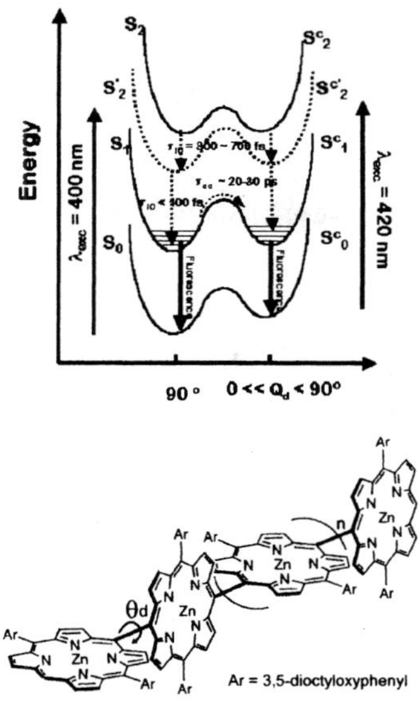

Fig. 2. Schematic potential energy diagram of porphyrin arrays

shift, respectively. Thus, we can expect larger coherent length for the upper excited S_2-state.

It is noteworthy that the energy transfer rate of $(\sim 0.55\ \mathrm{ps})^{-1}$ in the directly linked Zn(II) and free-base porphyrin dimer is quite similar to that from B800 to B850 in LH2 complexes ($(\sim 0.65\ \mathrm{ps})^{-1}$). Considering this fact, the directly linked porphyrin arrays would be the closest artificial light harvesting array to the natural photosynthetic organisms, providing that the directly linked porphyrin arrays are the most suitable synthetic molecular modules for the realization of molecular photonic devices based on the porphyrin arrays

Acknowledgements. This work has been financially supported by the National Creative Research Initiatives of the Ministry of Science & Technology of Korea (DK).

References

1. N. Aratani , A. Osuka, D. Kim, Y. H. Kim, D. H. Jeong, Angew. Chem. **39,** 1458, 2000.
2. H. S. Cho, N.W. Song, Y. H. Kim, S. C. Jeoung, D. Kim, S. K. Kim, N. Yoshida, Osuka, J. Phys. Chem. B **104,** 3287, 2000.
3. Y. H. Kim, D. H. Jeong, H. S. Cho, S. J. Hahn, N. W. Song, S. C. Jeoung, S. K. Kim, N. Aratani, A. Osuka, J. Am. Chem. Soc. submitted.
4. T. Kakitani, A. Kimura, H. Sumi, J. Phys. Chem. B **103**, 3720, 1999.

Dynamic intensity borrowing in porphyrin J-aggregates revealed by sub-5-fs spectroscopy

Hideaki Kano, Takashi Saito, and Takayoshi Kobayashi

Department of Physics, University of Tokyo, Hongo 7-3-1, Bunkyo, Tokyo 113-0033, JAPAN
E-mail: kanou@femto.phys.s.u-tokyo.ac.jp

Abstract. Sub-5-fs spectroscopy of porphyrin J-aggregates reveals for the first time coherent molecular vibration coupled to the Frenkel exciton. The oscillations with the frequency of 244cm^{-1} are described by a plus-cosine function for bleaching and a minus-cosine function for induced absorption. The coherent oscillation is explained by a modulated transition dipole moment, which is due to the transfer of an oscillator strength from the intense B-band to the weak Q-band through the ruffling mode with 244cm^{-1}-frequency.

1. Introduction

Molecular J-aggregates have attracted much attention because of the large nonlinearity and ultrafast response. The excitonic transition of the J-aggregates is characterized by a sharp absorption band, called the J-band, of which transition energy is lower than that of monomers by the value of twice the dipole-dipole interaction. Among J-aggregates composed of various aromatic or macrocyclic compounds, tetraphenylporphine tetrasulfonic acid (TPPS) J-aggregates are of special interest since they are model substances for aggregates of the light-harvesting antenna chlorophyll. The absorption spectrum of TPPS aggregates, which is depicted in Fig. 1, shows a relatively weak band in the visible region and a strong peak in the near ultraviolet, denoted as Q- and B-bands, respectively, composing a quasi-three-band Frenkel exciton system. J-aggregates are also known as the system in which excitons interact only weakly with molecular vibrations. It is indicated by the small Stokes shift in stationary fluorescence spectrum.

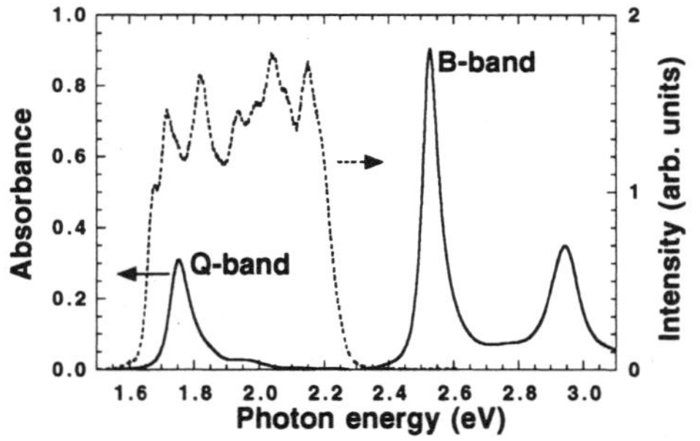

Fig. 1. Linear absorption spectrum of porphyrin J-aggregates (solid line) and laser spectrum (dashed line).

Although several studies have been made of incoherent coupling of excitons and phonons [1], no investigation of the coherent exciton-vibration coupling in J-aggregates have ever been performed in real-time domain because the excitons interact only weakly with vibrations and very high time-resolution is required for direct measurement of intramolecular vibrations. In the present study, we apply for the first time the sub-5-fs real-time spectroscopy to study the coherent exciton-vibration coupling in the Frenkel exciton system. We observed for the first time the coherent molecular vibration coupled to the Frenkel exciton.

2. Results and Discussion

A unidirectionally oriented film of porphyrin J-aggregates was prepared using a vertical spin-coating technique developed by our group [2]. The pump pulse energy and photon density are about 14nJ and 5.9×10^{14} photons/cm^2, respectively. Both the polarization of the pump and probe are parallel to the aggregate axis. The time dependence of the transmittance change in porphyrin J-aggregates at 3 different probe photon energies is shown in Fig.2 (a). The predominant feature common to the traces is high-frequency oscillations in addition to the underlying slow-dynamics transient components. To the best of our knowledge, this is the first observation of the coherent molecular vibration in J-aggregates. The positive and negative signals in $\Delta T/T(t)$ are associated with the bleaching and induced absorption (IA) due to the transition from the one-exciton state to two-exciton states, respectively. Figure 2 (b) shows the Fourier-power spectra calculated from the oscillating components. The Fourier spectra clearly show intense peaks at 244cm^{-1} in the whole spectral region of the Q-band. We also performed resonant Raman experiment under resonant excitation of the B-band and observed a peak at the same frequency. This peak is assigned to a ruffling mode and the Raman signal of this mode is drastically enhanced by a factor of more than 30 [3]. The phase of the oscillation can be evaluated by the complex Fourier transformation. Figure 3

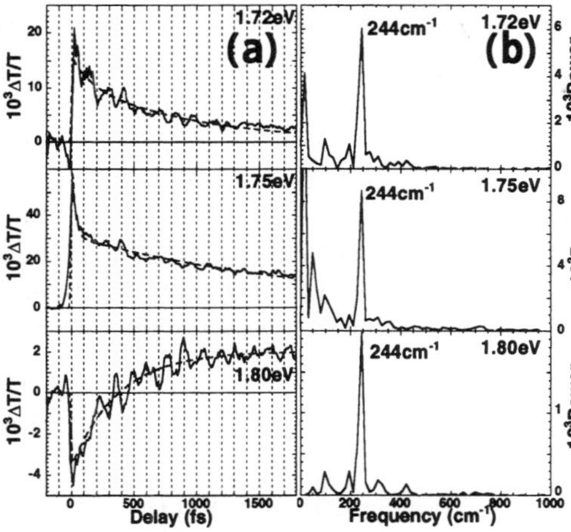

Fig. 2. (a) Time-dependence of the transimttance change at 3 photon energies marked on the right. (b) Fourier-power spectra of real-time spectra.

shows the probe-photon energy dependence of the phase and amplitude of the $244cm^{-1}$-ruffling mode. The phase of the oscillations are constant in the whole spectral region of the Q-band and the sign of the amplitude is reversed around 1.77eV. Around 1.77eV, the transient signal changes from the bleaching (positive signal) to the IA (negative signal) and the oscillation amplitude is extremely decreased. The constant phase is different from expected continuously varying phase change along the probe photon energy. The plus-cosine oscillation in the bleaching and the minus-cosine oscillation in the IA indicate that the bleaching and IA increase (or decrease) synchronously. This unexpected result can be explained in terms of dynamic intensity borrowing (DIB). The intensity of absorption associated with the transition of the ground state to the Q-band is borrowed from that of B-band by the configuration-interaction mechanism. The degree of the intensity borrowing is modified by vibrational motion of the ruffling mode by the exciton-phonon coupling. From the experimental transmittance change, the intensity of the vibronic coupling is estimated to $H_{vib} = 2.8meV$ from $\delta(\mu^2)/\mu^2 = 4.4 \pm 1.0\%$.

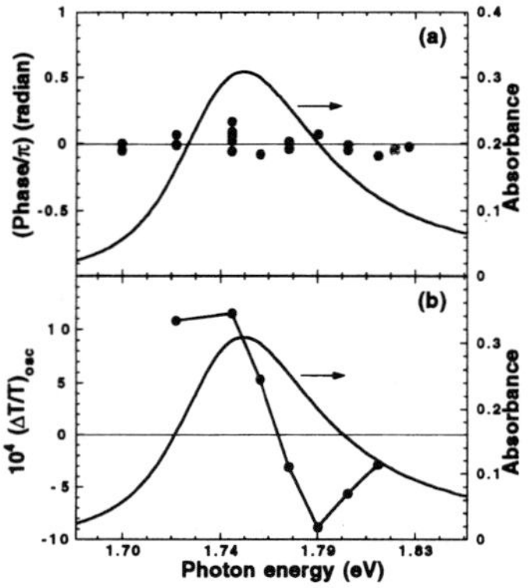

Fig. 3. Probe-photon energy dependence of the phase (a) and amplitude (b) of the oscillation in the transient signals.

3. Conclusions

The coherent molecular vibration coupled to the Frenkel exciton is observed for the first time. The experimental result is interpreted by the transition dipole moment due to DIB. DIB is a new mechanism of ultrafast nonlinearity.

References

1 S. De Boer and D. A. Wiersma, Chem. Phys. **131** , 135 (1989)
2 K. Misawa, H. Ono, K. Minoshima, and T. Kobayashi, Appl. Phys. Lett. **63**, 577 (1993).
3 D. L. Akins, H. -R. Ahu, and C. J. Guo, J. Phys. Chem., 100, 14390 (1995).

Wavepacket motion of self-trapped exciton under excitation by femtosecond chirped pulses in one dimensional metal dimethyl-glyoxime complex

S. Iwai[1], T. Kamata[1], S. Murata[1], and K. Yamamoto[2]
[1] National Institute of Materials and Chemical Research, 1-1 Higashi, Tsukuba , JAPAN
E-mail: iwai@nimc.go.jp
[2] Institute for Molecular Science, Myodaiji Okazaki, JAPAN

Abstract. Wavepacket motion of self-trapped excitons (STE) was observed under excitation by positively and negatively chirped pulses in molecular aggregate of one-dimensional metal complex. Negatively chirped excitation enhances the ocsillative signal due to the wavepacket motion. This shows that the time evolution of the STE wavepacket is affected by the chirp of the excitation pulses.

1. Introduction

To understand the relaxation or photo-chemical reaction processes in solids or molecular aggregates, it is essential to know which vibrational modes couple with the electronic excited state strongly among a lot of inter- and intramolecular vibrations. Recent progerss in the technique of femtosecond spectroscopy enables us to observe the excited state wavepacket motion directly in many systems. This gives useful information about the configurational coordinate and structures of APES of the excited state. Furthermore, wavepacket evolution on the excited state APES which strongly depends on the structure of APES can be modulated by chirped excitation pulses. So, this also presents more precise information about excited state APES.

One dimentional molecular aggregates of d^8 metal dimethylglyoxime complexes ($M(dmg)_2$) have large third order optical non-linearlity and it decays extremely fast [1,3]. Ultrafast decay of the nonlinearity is due to the relaxation of 1-D exciton to self-trapped exciton (STE) followed by thermalization of STE [3,4]. However, it has not been revealed yet which vibrational mode is related to the formation of STE. In this study, to obtain the information about the structure of APES including free and self-trapped exciton, the wavepacekt motion of STE was observed under excitation by chirped pulses.

2. Experimental

The sample of the thin film of Pt(dmg)$_2$ (bis(dimethylglyoxime) platinum(II)) used in this study is the same as that used in a previous work [3,4]. In the pump-probe measurements, a regenerative amplifier system of Ti:Al_2O_3 laser (Spectra Physics Tsunami-35 fs and sub-50 fs Super Spitfire system) with an OPA (Spectra Physics OPA800) operating at 1KHz was used as a light source. The center wavelengths of pump and probe pulses are 800 nm (pump) and 920nm, 1033 nm (probe), respectively. The response function of our measurement system and the excitation density are about 60 fs and 0.1 mJ/cm^2, respectively. The pump and probe pulses are precompensated separetely by two distinct prism pairs and lead to ordinarly two-color pump-probe optics. The pump-probe measurements were made by changing the chirp of the excitation pulse, while the probe pulse was fixd in transform limited pulse.

3. Results and Discussions

Fig.1 shows the time evolution of the transient absorption of Pt(dmg)$_2$ observed at (a) 1.20 eV and (b) 1.35 eV under excitation by non-chirped (transform limited) pulses. The temperature of the sample was 80K. Positive absorption change observed in this energy region has been assigned to the induced absorption due to the STE [3]. Underdamped oscillatory structure whose period and damping time are 120 fs and 170 fs can be seen in the time profile of the induced absorption. These temporal behaviors of the induced absorption can be explained by the wavepacket motion of the STE during thermalization on the APES surface. In addition to the oscillatory structure, the decay (at 1.20 eV) and rise (at 1.35 eV) components with a time constant of about 0.2 ps were observed.

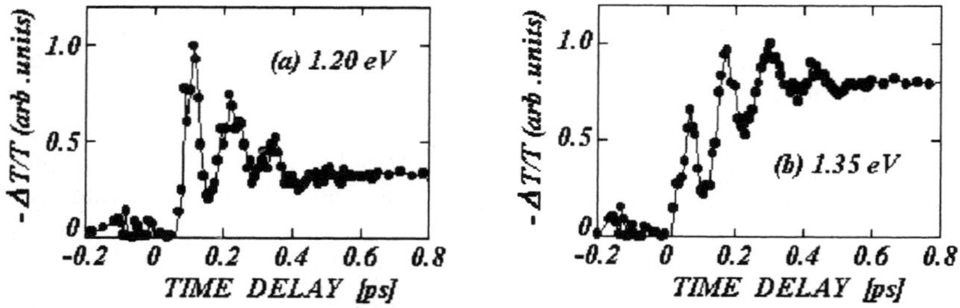

Fig.1 : Time evolution of the induced absorption of the STE observed at (a) 1.20 and (b) 1.35 eV under excitation by non-chirped (transform limitted) 1.55 eV pulses. Excitation density is 0.1mJ/cm^2. The response function of the measurement system is 60 fs.

The difference between the time profiles observed at 1.20 and 1.35 eV can be ascribed to the defference in the probing position on the APES surface. The position corresponding to 1.35 eV is closer to the minimum of the potential surface than that corresponding to 1.20 eV. As the thermalization (= vibrational relaxation and dephasing) of STE proceeds, the population observed at 1.20 eV decreases, while that observed at 1.35 eV increases.

The period of oscillation can be used to consider which vibrational mode is coupled strongly with the optical transition of STE. The period of 120 fs corresponds to 270 cm^{-1}. On the other hand, the Raman peaks due to intramolecular bending modes with A_g symmetry which are expected to couple with the 1-D electronic state are observed between 200 and 500 cm^{-1} of the spectrum of $Pt(dmg)_2$. The frequencies of other A_g modes such as intramolecular breathing mode (>1000 cm^{-1}) or intermolecular stretching mode (<100cm^{-1}) are far from 290 cm^{-1}. Thus, the STE in $Pt(dmg)_2$ is related to the intramolecular bending mode whose frequency is 270 cm^{-1}. This intramolecular distortion is considered to break the translational symmetry along the 1-D chain and induce the localization of the exciton.

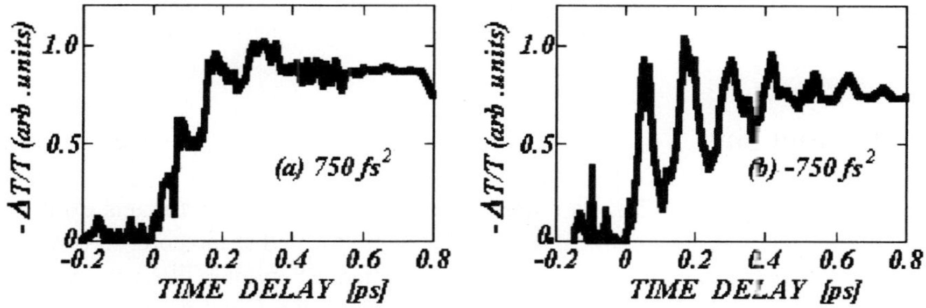

Fig.2: Time evolutions of induced absorption observed at 1.20 eV under excitation by chirped pulses. The chirp parameters of the excitation pulses are approximately described as the second order group velocity dispersion (DG [fs^2]). ((a): DG=750 fs^2 , (b): DG=-750 fs2)

Fig.2 shows the time evolutions of induced absorption observed at 1.20 eV under excitation by chirped pulses. Excitation density is 0.1mJ/cm^2. The chirp parameters of the excitation pulse are approximately described as the second order group velocity dispersion (D_G [fs^2]). ((a): D_G=750 fs^2 , (b): D_G=-750 fs^2). In positively chirped excitation (a), the oscillation is not observed very clearly. However, in negatively chirped excitation (b), an oscillatory structure is apparently larger than in the case of the excitation by transform limited or positively chirped pulses. This shows that the time evolution of the STE wavepacket is affected by the chirp of the excitation pulses.

4. References

[1] T.Kamata, T. Fukaya, M. Mizuno, H. Matuda, and F. Mizukami, Chem. Phys. Lett. **221**,194-198 (1994).

[2] K. Yamaoto, T. Kamata, S. Iwai, S. Kazaoui, N. Minami, F. Mizukami, K. Misawa, T. Ohta, and T. Kobayashi, Chem. Phys. Lett. **302**,609-614 (1999).

[3] S. Iwai, T. Kamata, S. Murata, T. Fukaya, T. Kodaira, F. Mizukami, M. Tachiya, K. Yamamoto, T. Ohta, J. Chem. Phys. **110**,8687-8695 (1999).

[4] S. Iwai, T. Kamata, S. Murata, K. Yamamoto, T. Ohta, J. Chem. Phys. **111**, 5402-5407 (1999).

Energy funneling in the dendrimeric nanostar probed by time-resolved nonlinear spectroscopies

Jason C. Kirkwood, Christoph Scheurer, Tatsuya Minami, Vladimir Chernyak and Shaul Mukamel

Department of Chemistry, University of Rochester, Rochester NY, 14627
Tel.: (716) 275-3080, Fax: (716) 473-6889, e-mail: mukamel@chem.rochester.edu

Abstract. The complex excitonic dynamics of a four-generation phenylacetylene dendrimer with a perylene acceptor and its signatures in time- and frequency-gated fluorescence and pump-probe spectroscopies are simulated using density matrix Liouville space wavepackets and the Doorway/Window representation of sequential optical signals.

1. Introduction

The nanostar (Fig. 1) exhibits an efficient energy gradient, which funnels excitons from the periphery to the chemical sensor (perylene) located at the core. Recent experimental [1] and theoretical [2] studies of poly-phenylacetylene (PPA) dendrimers have shown that optical excitations are completely localized within the linear segments. The meta branching at the benzene rings between the individual PPA units disrupts the charge transfer between linear segments. Exciton migration proceeds via Coulombic interaction and can be described by the Frenkel exciton Hamiltonian.

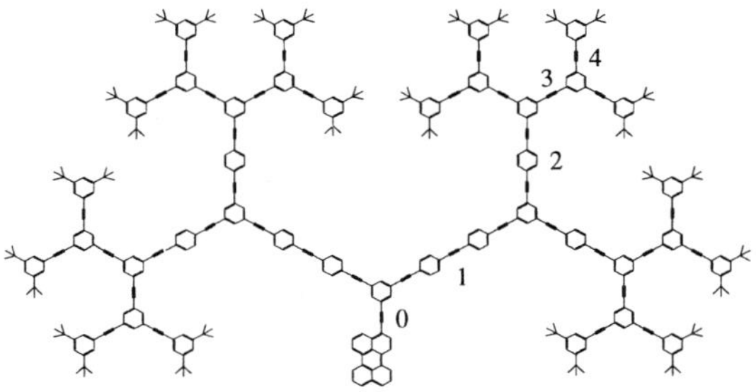

Fig. 1. The nanostar dendrimer ($C_{460}H_{424}$) comprised of 4 generations 1,2,3,4 (as indicated) and a perylene acceptor at the core, 0.

Owing to its self-similarity and high symmetry, the nanostar has a relatively simple excitonic band structure, originating from the three different sizes of linear segments that comprise the molecule. The lengths of the linear segments increase and the excitation energy decreases with generation, as one moves from the periphery to the core. Upon excitation of the most energetic band (corresponding to the shortest linear segments which lie on the periphery), excitons migrate from the periphery, through the intermediate regions, ultimately residing on the locus of the dendrimer, where the perylene moiety acts as an energy trap.

2. Probing exciton migration by time-resolved spectroscopy

We have simulated the exciton transport and its probing by time- and frequency-resolved fluorescence and pump-probe spectroscopy. Exciton transport is described by the Redfield equations for the exciton density matrix. The signals were recast using the Doorway-Window representation [3,4]. In this picture, the system interacts twice with the pump, creating exciton populations and coherences. The resulting exciton density matrix (the *Doorway wavepacket*) then propagates during the time-delay between pulses, where the exciton-phonon coupling induces exciton relaxation. The signal is finally formed by the Liouville space overlap of a second, *Window wavepacket*, created by the probe with the propagated Doorway.

The dispersed fluorescence signals at various delay times are displayed in Fig. 2, where the pump pulse is tuned to 29044 cm^{-1} to coincide with the highest exciton. At 0 s, the peak intensity from the highest exciton is largest, with a minor contribution from intermediate excitons. At 30 ps, exciton migration into the intermediate regions becomes appreciable. Finally at 90 ps, the system has nearly reached thermal equilibrium, whereby the propagated wavepacket ultimately resides on the perylene trap.

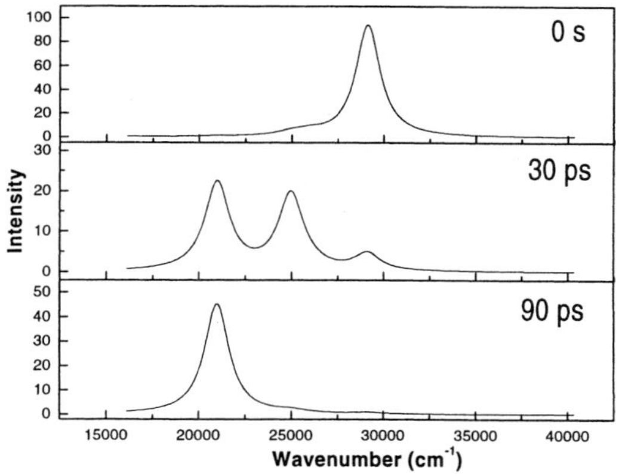

Fig. 2. Computed dispersed fluorescence signals from the nanostar at three time delays.

To explore the effects of energy funneling and thermal relaxation in greater detail, we have examined the propagated Doorway wavepacket. The absolute magnitude of the Doorway in the site representation is displayed in Fig. 3. Initially, coherences exist between various linear segments, which dephase and vanish by ~10 ps. Subsequent evolution of the population towards the perylene moiety is apparent in the right panel of Fig. 3. Further illustration of the energy migration effects of the nanostar is shown in Fig. 4. The diameter of the circles corresponds to the value of the diagonal elements of the propagated Doorway wavepacket in the site representation. At zero delay, the peripheral segments are preferentially populated, while at 10 ps, thermal relaxation towards the inner region has occurred. By 100 ps, the excitons have migrated fully to the perylene trap.

Analysis of the propagated density matrix wavepacket in both the exciton and site representations, and calculation of the participation ratio associated with both its

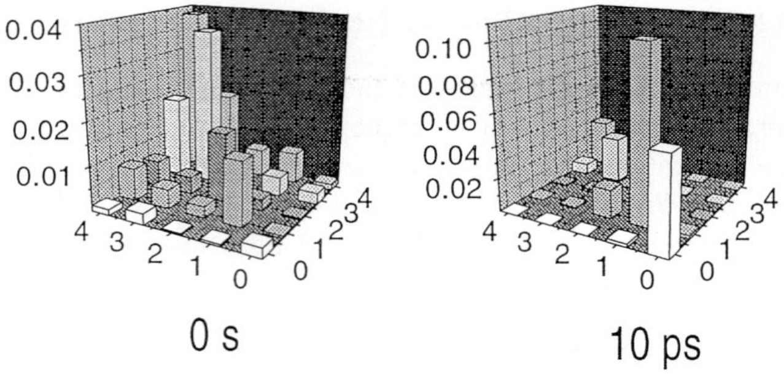

Fig. 3. The time-dependent Doorway wavepacket in the site representation. The numbering of the axes in the x-y plane refers to the individual generations (linear segments) as given in Fig. 1.

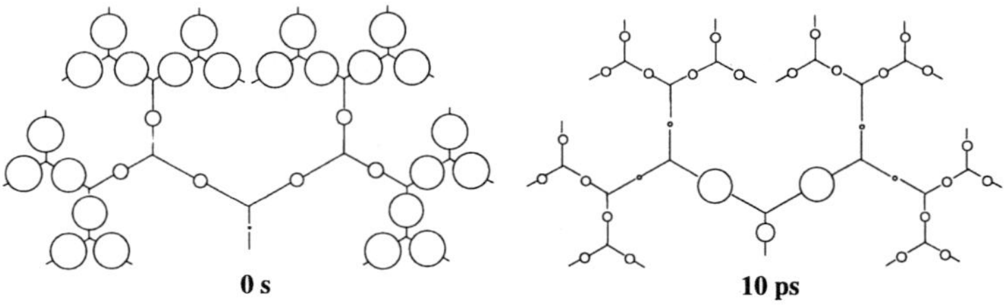

Fig. 4. The diagonal elements of the propagated Doorway wavepacket in the site representation.

diagonal and off-diagonal elements reveal the complex nature of exciton migration and energy funneling [5,6].

Acknowledgement. The support of the National Science Foundation is gratefully acknowledged.

References

1. M. Shortreed, S. Swallen, Z-Y Shi, W. Tan, Z. Yu, C. Devadoss, J. Moore and R. Kopelman, J. Phys. Chem. B **101**, 6318-6322 (1997); R. Kopelman, M. Shortreed, Z-Y Shi, W. Tan, Z. Xu, J.Moore, A. Bar-Haim and J. Klafter, Phys. Rev. Lett. **78**, 1239-1242 (1997).
2. V. Chernyak, E. Poliakov, S. Tretiak and S. Mukamel, J. Chem. Phys. **111**, 4158-4168 (1999).
3. M. Dahlbom, T. Minami, V. Chernyak, T, Pullertis, V. Sundstrom and S. Mukamel, J. Phys. Chem. B **104**, 3976-3983 (2000); V. Chernyak, T. Minami and S. Mukamel, J. Chem. Phys. **112**, 7953-7963 (2000).
4. S. Mukamel, *Principles of Nonlinear Optical Spectroscopy*, Oxford University Press, New York, 1995.
5. J.C. Kirkwood, C. Scheurer, V. Chernyak and S. Mukamel, (in preparation).
6. S. Raychaudhuri, Y. Shapir, V. Chernyak and S. Mukamel, Phys. Rev. Lett. **85**, 282-285 (2000).

Ultrafast Investigations of Intra-molecular Interactions and Energy Transfer in Dendrimers and Polymeric Materials

Theodore Goodson III[1]

[1] Department of Chemistry, 59 Chemistry Building, Wayne State University, Detroit, MI, 48202, USA E-mail: tgoodson@chem.wayne.edu

Abstract The dendritic architecture allows for the control of energy transfer. Ultra-fast excited state dynamics of organic dendrimers have been studied with both polarized and depolarized femtosecond fluorescence spectroscopy. The dynamics of novel dendrimer-metal nanocomposite systems is also discussed.

1. Introduction

Recently there has been an increasing amount of attention devoted to the study of dendrimers, which are highly branched macromolecular systems [1-3]Along with the synthesis and chemical properties, the photo-physics of these systems is attracting rising interest in connection with their applications as artificial antenna systems.[4] However, the nature of electronic excitations and specific mechanisms of intramolecular energy transport in dendrimers are not well understood.[3-5] In the A-DSB system the nitrogen branching centers[5] can cause a disruption of π-electron conjugation of linear building blocks. In this case the optical excitation creates the electron-hole pairs localized on small segments[4]. However, these localized electron-hole pairs can contribute either to *collective* optical excitations (excitons)[3] or to the incoherent hopping[4] due to inter-segment interactions. The goal of our research is to identify the mechanism of energy transfer and intermolecular interactions in these systems.

2. Experimental Methods

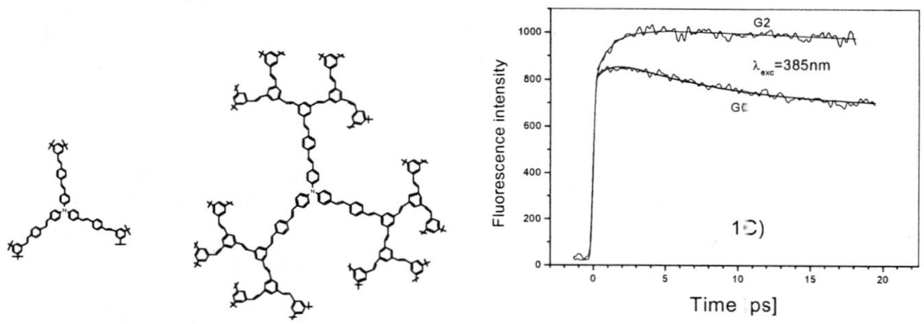

1a) GO 1b) G2

Fig. 1: a) and b) The chemical structures of the A-DSB dendrimers generations zero and two, and c) the time-resolved emission of A-DSB.

The structure of the nitrogen-cored distyrylbenzene-stilbene (A-DSB) and the second generation dendrimer is shown in Fig.1. The linear absorption spectrum shows two prominent features with maxima at 410 nm and 320 nm. Synthesis and spectroscopic features of all the structures have been described elsewhere.[5,6] Solutions in chloroform were used in this study. Femtosecond upconversion spectroscopy was employed to resolve temporally the polarized fluorescence. Our optical arrangement for the upconversion experiments has been described previously.[7-9] The laser had an average pulse width of 100 fs tuned at 790-860 nm and a repetition rate of 82 MHz.

3. Results and Discussion

The fluorescence isotropic decay of A-DSB on the pico-second time scale was found to be non-exponential and dependent on the emission wavelength (Fig. 1C). The effect of fluorescence rise-time is clearly detected in the "red" regions of the fluorescence spectrum. This fluorescence time-behavior can be assigned either to intra-molecular interactions in the dendrimer (excimer formation, energy transfer) or to local relaxation processes in the chromophore (conformational changes, solvent rearrangement).

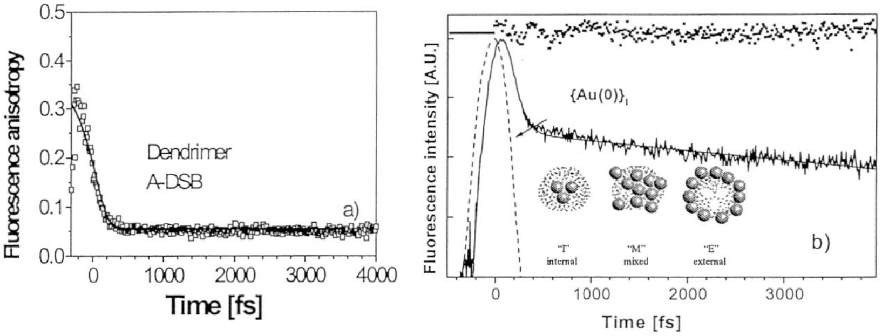

Fig. 2: a) Anisotropy decay of A-DSB dendrimer. b) Fluorescence decay of Au dendrimer nanocomposite (Au-PAMAM internal type).

The decay curves in Fig. 1c have been normalized by the peak absorption of the dendrimer core at 420nm. It is clearly seen from Fig. 1c that the fluorescence dynamics of G2 differ from those of G0 when fluorescence is excited at 385nm. Additional rise-time component in G2-dynamics can be attributed to the dendrons-to-core energy transfer process.

To gain a further insight into energy transfer (energy migration) processes in A-DSB dendrimer molecules we investigated the fluorescence anisotropy dynamics in G0 and G2. The result for G2 is shown in Fig. 2a. The Fluorescence anisotropy r(t) was calculated from the decay curves for the intensities of fluorescence polarized parallel $I_{par}(t)$ and perpendicularly $I_{per}(t)$ to the polarization of the excitation light. It is clearly seen in Fig.2a that the fluorescence anisotropy (FA) of A-DSB decays to a value close to zero in less than 200fs after excitation. The initial spike at about zero delay is comparable in duration with the time resolution of our system. Best fit of a difference signal with convolution with IRF gave the anisotropy decay time of .57±7fs. The dendrimer core (G0) is a star-like molecule consisting of three DSB –segments bonded to a central nitrogen atom. The interaction between branches could lead either to the formation of coherent excitonic states[10] or hopping-type[11] relaxation of excitations localized on one DSB-branch.

The dynamics of novel dendrimer nanocomposites systems have also been investigated. The PAMAM dendrimer with silver and gold transition metals have been investigated using time-resolve luminescence and the result for gold internal nanocomposite structure is shown in figure 2b. The isotropic fluorescence decay was found to be much slower than the depolarization rate. The proposed mechanism of the emission decay is due to electron-electron and electron-surface scattering processes. We found that the emission decay was shorter than the relaxation associated with a phonon coupling process (~1 ps)[12].

4. Conclusions

In conclusion, fast fluorescence depolarization within 200fs was found for A-DSB dendrimer. These results demonstrate the ultra-fast transition dipole reorientation due to inter-segment interactions in dendrimer. The ultra-fast measurements with a novel dendrimer metal nanocomposite showed fast fluorescence decay related to the interaction of the PAMAM dendrimer and the metal centers.

Acknowledgements. This research was supported in part by the American Chemical Society Petroleum Research Fund.

References

1. D.A. Tomalia, A.N. Naylor, W.A. Goddard III, Angew.Chem.Int.Ed.Engl. **29,** 138 (1990).
2. J.M.J. Frechet, Science **263,** 1710 (1994).
3. S. Tretiak, V. Chernyak, S. Mukamel, J. Chem Phys. **110,** 8161 (1999).
4. R. Kopelman, M. Shortreed, Zhong-You Shi, W. Tang, Zhifu Xu, J. Moore, A. Bar-Haim, J.Klafter, Phys.Rev.Lett. *78*, 1239 (1997).
5. J.N.G. Pillow, M. Halim, J.M. Lupton, P. Burn, I.D.W. Samuel, Macromolecules **32**, 5985 (1999).
6. J.M. Lupton, I.D.W. Samuel, R. Beavington, P.L. Burn, H. Bassler (submitted).
7. O. Varnavski, L. Liu , J. Takacs, T. Goodson III, J.Phys.Chem.. **104**, 179 (2000).
8. O. Varnavski, A. Leanov, L. Liu, J. Takacs, T. Goodson III, Phys. Rev. B, **61,** 12732 (2000).
9. O. Varnavski and T. Goodson III* Chem.Phys.Lett. **320**, 688 (2000).
10. R. Kumble, R.M. Hochstrasser J. Chem Phys. **109,** 855 (1998).
11. L. Larettini, G. DeBelder, , G. Schweitzer, M. Van der Auweraer, F.C. DeSchryver Chem. Phys. Lett. **295,** 11 (1998).
12. J.-Y. Bigot, V. Halté, J.-C. Merle, A. Daunois, Chem. Phys. **251**, 181 (2000).

Systematic Fluorescence Upconversion Investigation of Intramolecular Processes in Peryleneimide Dendrimers containing One to Four Chromophores

Gerd Schweitzer, G. De Belder, S. Jordens, S. Mitra, M. Lor and F.C. De Schryver

Katholieke Universiteit Leuven, Molecular Dynamics and Spectroscopy, Celestijnenlaan 200F, 3001 Heverlee, Belgium. E-mail: Gerd.Schweitzer@chem.kuleuven.ac.be

Abstract. First generation dendrimers bearing 1,2,3 or 4 peryleneimide chromophores at the rim have been investigated using fluorescence upconversion detection with 250 fs time resolution. For all samples, intramolecular vibrational reorganisation and vibrational relaxation processes could be attributed. Moreover, in multichromophoric samples, an additional 10 ps time constant was found and attributed to singlet singlet annihilation.

1. Introduction and Experimental Methods

Dendrimers are highly branched macromolecular systems whose structure can be defined on a molecular level [1]. In this study, a first generation tetraphenyl carbon core dendrimer (Fig. 1, insert) has been decorated systematically with 1, 2, 3 or 4 peryleneimide chromophores at the rim to study their influence selectively. The fluorescence of all samples was measured at 15 different wavelengths between 530

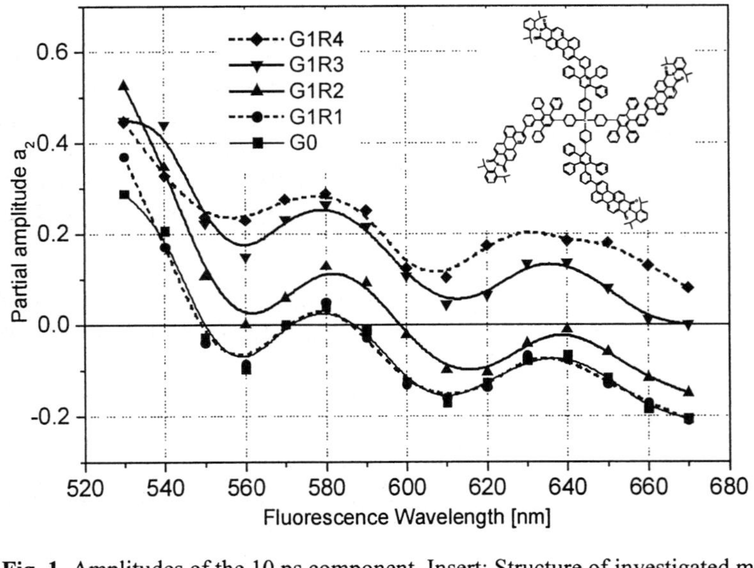

Fig. 1. Amplitudes of the 10 ps component. Insert: Structure of investigated molecules

nm and 670 nm under magic angle conditions using an upconversion detection scheme [2] with a time resolution of 250 fs. The excitation was done at 495 nm with 350 nJ pulses obtained from a regeneratively amplified fs laser system [3]. All measurements were done at room temparture in 1 mm cuvettes at an optical density of 0.5 / mm solutions in toluene.

2. Results and Discussion

Employing global analysis of all sets, a sum of four exponentials has been found necessary to fit the data. The longest component of 4.2 ns time constant is attributed to the fluorescence emission, the second one ranging from 100 ps to 300 ps was found at a small concentration dependent partial amplitude and thus partly attributed to an intermolecular process. The time constant of fastest component was found to be wavelength dependent giving values rising from 0.5 ps (detected at 530 nm) to 2 ps (670 nm). The related amplitude, however, was negative at all wavelengths corresponding to a rise term in the decays. This kinetic component is attributed to a fast Intramolecular Vibrational Reorganisation (IVR) process within

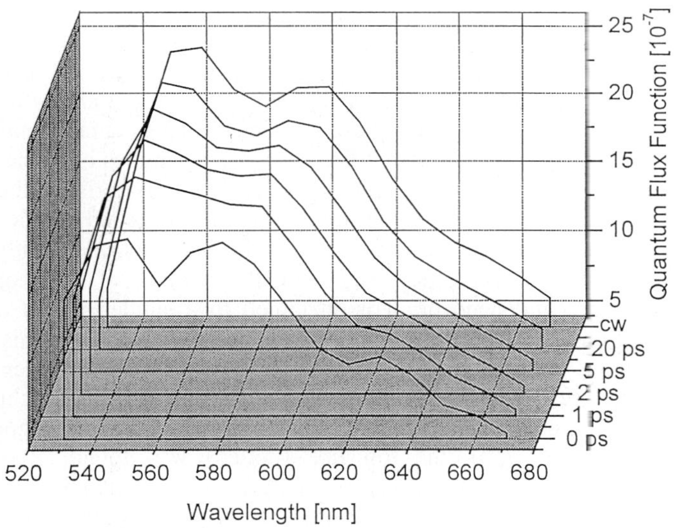

Fig. 2. Caclulated fluorescence spectra for various delays

the peryleneimide. This behaviour can be seen in Fig. 2 displaying fluorescence emission spectra calculated from the data sets for various delay times after excitation, which clearly shows a ca. 10 nm red shift in the spectra appearing in the course of time. Moreover, it is apparent that the spectral shift is finished to a large extent within the first 10 ps after excitation, thus all the intramolecular redistribution yielding the change in spectrum is completed within that time scale. The resulting amplitude spectra for the 10 ps component for all samples is shown in Fig. 1. The two bottom lines resulting from the two monochromophoric samples G0 (model compound) and G1R1 are identical and going from positive to negative

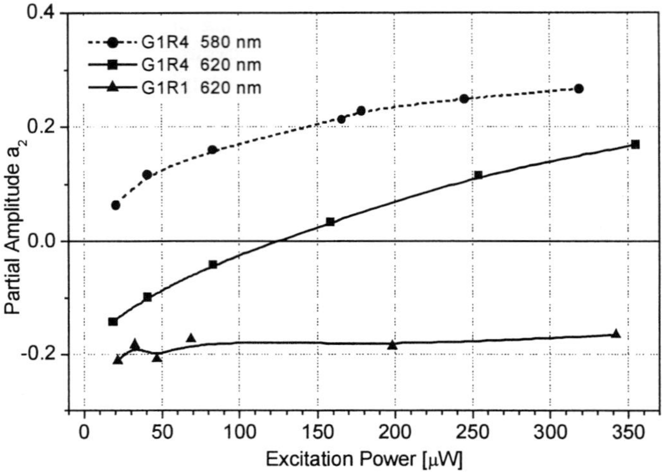

Fig. 3. Dependence of the amplitude of the 10 ps component from the excitation power

amplitudes upon increasing fluorescence detection wavelength. This kinetic component is attributed to a vibrational relaxation going on in the electronically excited state of the peryleneimide chromophore. The curves for the multichromophoric sample exhibit a different behaviour (top 3 lines, Fig. 1): although similar in shape and spectral distribution, they are all shifted to higher partial amplitude values. Thus, for these samples the 10 ps component is assumed to actually be a mixture of two parts: first, also these samples have the vibrational relaxation going on in each single chromophore, which gives rise to the shape and spectral amplitude distribution. Superimposed on that, a second kinetic component also having a time constant of ca. 10 ps is found. This latter one is only present in multichromophoric samples, increasing with the number of chromophores and almost wavelength independent. It is attributed to a singlet singlet annihilation process going on in the multichromophoric dendrimer molecules. This attribution could further be substantiated by measuring the power dependence of this component in three specific situations (Fig. 3): first, for a multichromophoric sample detected at a wavelength where the vibrational relaxation component has a zero amplitude (580 nm), the top line (dotted) was found. Repeating this measurment at 620 nm detection, a mixture of vibrational relaxation and annihilation (middle line) is seen. Finally, for a monochromophoric sample as expected no variation with the sample excitation power is determined.

Acknowledgements. Support through FWO (GOA 1/96), the EC (TMR Sisitomas) and DWTC (IUAP-IV-11) is gratefully acknowledged.

References

1 J. M. Frechet, *Science* **1994,** *263*, 1710.
2 Y. Karni, S.Jordens, G.De Belder, G. Schweitzer, J. Hofkens, T. Gensch, M. Maus, F. C. De Schryver, A. Herrmann, K. Mullen, *Chem. Phys. Lett.* **1999,** *310*, 73.
3 G. Schweitzer, L. Xu, B. Craig, F. C. De Schryver, *Opt. Commun.* **1997,** *142*, 283.

Part XIV

Ultrafast Chemistry

Experimental and theoretical study of chirped pulse excitation

Boris D. Fainberg[1,2], Dan Huppert[2], and Jonathan Segal[2]

[1] Holon Academic Technological Institute, 52 Golomb St., Holon 58102, Israel
E-mail: fainberg@barley.cteh.ac.il
[2] Raymond and Beverly Sackler Faculty of Exact Science, School of Chemistry, Tel-Aviv University, Tel-Aviv 69978, Israel
E-mail: huppert@chemosf1.tau.ac.il

Abstract. We used positive and negative ultrashort intense linear chirped pulses to excite cryptocyanine dye in liquid solutions. We found that the chirp of the pulse has a profound effect on the absorption spectrum.

1. Introduction

The possibility of controlling molecular dynamics using properly tailored pulses has been the subject of intensive studies in the last few years [1-3]. Several control schemes have been proposed that make use of the coherent nature of the laser radiation. Ultrashort pulses intrinsically consist of a broad range of frequency components. The relative phase of these frequency components can be systematically changed by introducing positive or negative linear chirp. Chirped pulses can selectively excite coherent wave packet motion either on the ground electronic potential energy surface of a molecule or on the excited electronic potential energy surface. In particular, a negatively chirped (NC) pulse creates a nonstationary ground state component, while a positively chirped (PC) pulse does not [2].

We present a systematic study of the effect of ultrashort chirped pulses on the absorption spectrum of large dye molecules in liquid solution.

2. Theory

We have used the theory, and maintained the same conditions, which are described in detail in ref. [1] to calculate the transmission spectrum of the chirped pulse. The absorption spectrum at frequency $\omega' + \omega$ can be written in the form:

$$\alpha(\omega') \sim \omega \operatorname{Im} P(\omega') / E(\omega') = \omega \operatorname{Im} \frac{\int_{-\infty}^{\infty} dt P^{+}(t) \exp(i\omega' t)}{\int_{-\infty}^{\infty} E(t) \exp(i\omega' t) dt} \tag{1}$$

where the field of a phase modulated pulse can be presented as $(1/2)E(t)\exp(-i\omega t)+c.c.$, $E(t)=\mathcal{E}(t)\exp(i\varphi(t))$, $\mathcal{E}(t)$ and $\varphi(t)$ are real functions of time. $P(\omega')$ and $E(\omega')$ are the Fourier transforms of the positive frequency component of the polarization $P^{+}(t)$ and the field amplitude $E(t)$, respectively; $\omega(t)=\omega - d\varphi(t)/dt$ is the instantaneous pulse frequency.

For relatively strong chirped pulses when the pulse duration t_p is much longer than the corresponding transform-limited one, the integrals in Eq.(1) can be evaluated using the method of stationary phase. As a result, we obtain

$$\alpha(\omega') = \alpha[\omega(t_{sp}) - \omega] \sim \omega \operatorname{Im} \chi[\omega(t_{sp}), t_{sp}] \tag{2}$$

where $\chi[\omega(t),t]$ is the susceptibility determined by Eq.(22) of Ref. [1], and t_{sp} is the phase point determined by the following equation: $\omega' = \omega(t_{sp}) - \omega$. Thus, the dependence of the imaginary part of the susceptibility on the instantaneous frequency $\omega(t) - \omega$ describes the absorption spectrum of comparatively strong chirped pulses.

3. Experimental

Laser pulses of 40 fs duration, centered near 800 nm, with a pulse energy of ~600 μJ and repetition rate of 1 kHz, were generated by a Ti:Sapphire-based oscillator and amplified by a multipass Ti:Sapphire amplifier. These pulses were used to pump an OPA which generates tunable short intense and coherent pulses of 50fs with ~25 nm bandwidth in the spectral range of 600-770 nm (SHG).

Negative group-velocity dispersion (GVD) is achieved by a pair of prisms. The chirp rate μ is determined by the phase term [1,2]:

$$\mu = -4\Phi''(\omega)[\tau_{p0}^4 + 4\Phi''^2(\omega)]^{-1} \tag{3}$$

where τ_{p0} determines the duration of transform limited pulse, and $\Phi''(\omega) = \Phi''(\nu)/(4\pi^2)$ is the phase term.

Positive GVD is achieved by passing the beam through a known thickness of quartz material. The maximum negative chirp is controlled by the distance between the prisms and the angle of incidence of the laser beam on the first prism.

The sample was irradiated with pulses of ~1 μJ focused to about 200 μm spot. The pulse intensity was attenuated by neutral density filters. The fluorescence and transmission of samples of 1mm optical path length were measured at several laser intensities and chirp rates.

4. Results and discussion

In this study, we varied the intensity and the chirp of ultrashort pulses used to excite cryptocyanine, a large organic molecule, in room temperature liquids. We measured the absorption and the integrated fluorescence, which is directly proportional to the amount of the excited state population. We found in this study that the chirp of the pulse has a profound effect on the excited state population and transient spectroscopy of the molecular system.

Figure 1 shows the experimental and calculated absorption spectra of positively chirped pulses, centered at 710 nm, after passing through the sample cell containing cryptocyanine dye in methanol solution. The low power absorbance at the peak (707 nm) is 1.3. The negative chirp rate corresponds to the phase term $\Phi''(\nu)$=-12000 fs^2.

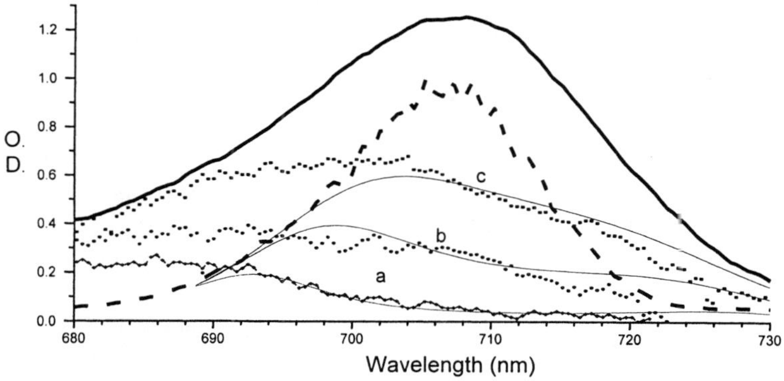

Fig .1. Absorption spectra for negative chirped pulse for cryptocyanine in methanol .

The calculated (solid lines) and the experimental (dots) absorption spectra are shown for three laser energies of 1 (a), 0.25 (b) and 0.06 (c) μJ. The calculation is based on the theory of Ref [1]. In the figure we also plotted the low power absorption spectrum of the cryptocyanine sample (bold solid line) as well as the laser spectrum in arbitrary units (dashed line).

As it can be seen, we obtain a rather good fit of the experimental spectra by the calculated signal. The input parameters for the calculation are: the cryptocyanine absorption cross-section $\sigma(\lambda)$, the laser intensity and spectrum and the chirp rate. For the highest intensity used, the calculated excited state population is 0.45 for the negative chirp. Thus, the experimental data support the theory that the chirp of the pulse has a profound effect on the absorption spectrum.

References

1 B.D. Fainberg, "Nonpertubative analytic approach to interaction of intense chirped pulses with molecules in solution: Picture of "moving" potentials," J. Chem. Phys. **109**, 4523-4532 (1998).

2 G. Cerullo, C.J. Bardeen, Q. Wang and C.V. Shank, "High-power femtosecond chirped pulse excitation of molecules in solution," Chem. Phys. Lett. **262**, 362-368 (1996).

3 J.L. Krause, R.M. Whitnel, K.R. Wilson, Y.J. Yan and S. Mukamel, "Optical control of molecular dynamics: molecular cannons, reflectrons and wave-packet focusers," J. Chem. Phys. **99**, 6562-6578 (1993).

Probing CTTS Dynamics of Na^- in THF: Novel Relaxation and Polarization Effects

Zhaohui Wang, Ophir Shoshana, and Sanford Ruhman.

Dept. of Physical Chemistry, and the Farkas Center for Light Induced Processes, the Hebrew University, Jerusalem 91904, Israel. E-mail: sandy@fh.huji.ac.il

Abstract. Ultrafast photoselective photolysis of sodide demonstrates for the first time that its CTTS transition dipole is oriented in the solvent, and the related absorption band is inhomogeneously broadened. Continuous evolution of absorption and its anisotropy reflecting solvent dynamics, are observed.

Electron photodetachment dynamics following excitation into CTTS absorption bands of solvated anions has been the focus of recent interest [1]. In a pioneering study, Schwartz and coworkers investigated CTTS dynamics of Na^- in THF via tunable pump-probe spectroscopy with ~200 fsec time resolution [2]. Appearance of solvated electrons was directly detected, and shown to be delayed with respect to the excitation by nearly a picosecond. Their visible transmission data was interpreted with a three state kinetic model whereby excitation generates a CTTS excited state with a constant absorption spectrum peaked at ~600 nm. The metastable excited state then decays quantitatively within ~700 fsec to produce a solvated electron, and sodium cation–solvated electron contact pair. Geminate recombination was shown to take place within ~3 psec, with a strongly wavelength dependent efficiency. No inherently dynamic continuous spectral shifts or direct evidence for involvement of solvent relaxation was recovered.

Here we report preliminary results of a renewed investigation of sodide in THF using a laser source capable of following the solvent response on all its timescales, in order to capture details of the underlying solvent controlled dynamics. Na^- in THF was prepared by vacuum distillation of all ingredients into a pyrex vessel equiped with a 1 mm quartz cell attached to a side arm. Data were typically obtained with OD 2 (730 nm) solutions. Excitation pulses were derived directly from a 30 fsec multipass amplified Ti;sapphire laser system working at 1 kHz (pulse spectrum in insert to figure 1). Probe pulses were interference filtered from a white supercontinuum. Time zero was precisely determined by conducting OKE scans in a thin quartz flat.

Figure 1 presents transient transmission data at 790, both for VV and VH relative pulse polarizations. An instantaneous increase of transmission is observed, which is nearly 3 times more intense in the VV than in VH. The rise is symmetrically situated with respect to the zero of delay, suggesting it is not a coherence spike resulting from the degeneracy of pump and probe frequencies. A very rapid decrease in the bleach follows, which is complete within 200 fsec, and is accompanied with a reduction of the VV/VH signal ratio from 3 to ~1.2. A much slower continued regeneration of absorption follows, and is accompanied with a gradual coalescence of the VV and VH data at ~ 2.5 psec.

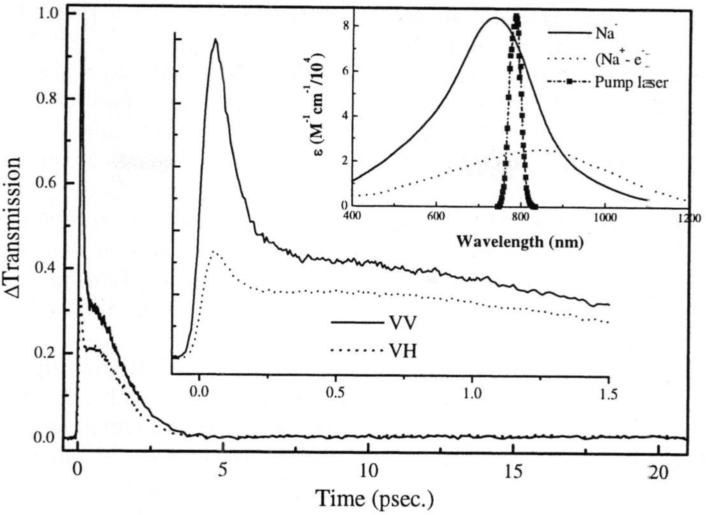

Fig 1. Transient transmission at 790 nm, with first 1.5 psec presented on expanded scale. The insert presents spectra of sodide, of Na^+-e^- contact pairs, and of the laser.

A partial sample of data obtained at other probe frequencies is presented in figure 2. Sufficiently removed from the peak of the transient absorption band an initial bleach appears, and decays presumably both due to delayed photoejection, and geminate recombination. In the vicinity of the excitation wavelength the bleach appears instantaneously, and exhibits a rapid initial decay as observed in figure1.

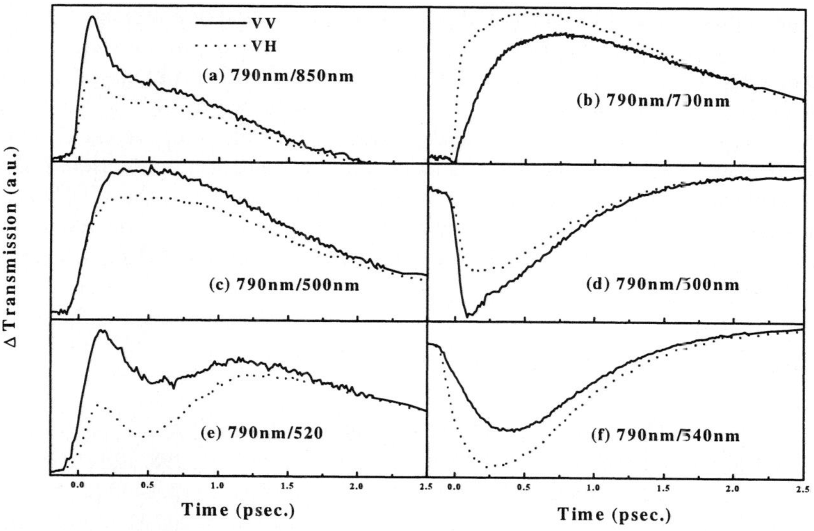

Fig 2. VV/VH transient transmission scans at various probing wavelengths following excitation at 790 nm.

In contrast, at the other side of the band at 500 nm, the bleach rises slowly, attaining its peak intensity only near the end of the initial decay in the 790 nm data. At 600 nm, the peak of the transient absorption, an instantaneous absorption appears, and indications of the fast initial relaxation process are observed as well. The 850 nm scans closely resemble the 790 nm results, demonstrating again that the fast process of bleach recovery is not associated with a coherence spike. The 700nm, 520nm, and 540 nm data are particularly revealing. Aside from the unusual reversals in sign of the anisotropy, they exhibit nonmonotonic transmission changes in the course of relaxation. A gradual rise in the 700 nm bleach is matched by a gradual absorption rise at 520 and 540. Both trends then turn around beyond 0.5 psec. These results permit us to interpret the underlying dynamics as following:

1) The similar kinetics and bleach anisotropies observed at 790 and 500 nm at t>200 fsec should indicate that these signals are not appreciably affected by residual absorptions from the CTTS excited state. We therefore tentatively assign the initial bleach decay at the former and delayed rise at the later to spectral inhomogeneity of the CTTS band, which undergoes spectral diffusion and equilibration within 200 fsec.

2) The striking anisotropy in these signals suggests that the sodide ions are preferentially oriented, either due to interactions with the solvent, or with other species in the solution. The unusual polarization behavior at wavelengths dominated by transient absorption further indicate that the CTTS excited state is also absorbing anisotropically.

3) Finally, the concerted 0.5 psec loss of OD at the red edge of the transient absorption (700 nm), and increase at the blue (540, 520 nm) must represent a continuous spectral shifting of this band following the impulsive excitation. It is most likely due to solvation of the CTTS excited state, and is not amenable to representation with kinetic models [2].

This interpretation raises new riddles. It is tempting to associate the fast and slow relaxation stages of absorption anisotropy to the inertial and collective stages of the solvent response [3]. How then is it that spectral diffusion seems to be sensitive only to the former? The polarization changes with probing wavelength are inconsistent with superposition of two transitions whose VV/VH ratio is constant across the band. Is it possible that under the envelope of the CTTS band numerous electronic levels exist, which differ in their transition orientations? This is consistent with findings that despite the substantial lifetime of the CTTS state the spatial distribution of ejected electrons depends crucially upon excitation wavelengths. Finally – the mechanism for orientation of the CTTS transition dipole must be identified.

References

1. W. Sheu and P.J. Rossky, *J. Pyhs. Chem.* **100**, 1295 (1996). J.A. Kloepfer, V.H. Vilchiz, V.A. Lenchenkov, S.E. Bradforth, *Chem. Phys. Lett.*, **298**, 120 (1998).
2. E. R. Bartel, I. G. Martini, B. J. Schwartz, *J. Chem. Phys.*, **112**, 9433 (2000).
3. L. Reynolds, J. A. GardeckiS. J. V. Frankland, M. L. Horng, and M. Maroncelli, *J. Phys. Chem.*, **100**, 10337 (1996).

The Ultrafast Dissociation Dynamics of Iodine Embedded in Microporous Crystalline Porosils of Different Structure

G. Flachenecker and A. Materny

Institut für Physikalische Chemie der Universität Würzburg,
Am Hubland, D-97074 Würzburg, Germany
E-mail: amaterny@phys-chemie.uni-wuerzburg.de

Abstract. The dissociation and recombination dynamics of iodine embedded in porosils is investigated by femtosecond time-resolved pump-probe spectroscopy. The caging process is studied for porosils of different dimensionality.

The photoinduced reaction dynamics of iodine inside a matrix is strongly influenced by the interaction with the surroundings. This can nicely be demonstrated for the dissociation reaction of iodine going from the isolated molecule to a liquid-like environment simulated by high-pressure rare gas [1]. In the present contribution, we discuss results obtained by femtosecond pump-probe spectroscopy on iodine, embedded in a microporous matrix formed by polycrystalline porosils [2,3]. Porosils are zeolites having a neutral corner-sharing $SiO_{4/2}$ tetrahedra framework of different structure (Fig.1). The host-guest interactions are assumed to be dominated by collisions; the weak van-der-Waals interactions usually are neglected. The porosils form cavities, in which the iodine molecules are inserted via the gas phase [2,3]. In the case of the DDR porosil (Deca-dodecasil 3R) the guest components reside isolated from each other inside cages having a volume of approximately 350 Å^3. In contrast to the quasi zero-dimensional structure of DDR, the TON porosil (silica-ZSM-22) forms channel-like cavities in which the iodine molecules are not isolated from each other. In addition to the host-guest interaction an interaction between the guest molecules along the channel axis takes place. The microchannels of TON are assumed to be isolated from each other, so that the interaction between iodine molecules takes only place in one dimension. Our experiments were performed on iodine in TON and are compared to former results obtained from the iodine-DDR system [4].

After excitation of iodine into the B state with the pump pulse, a predissociation induced by collisions occurs along repulsive potential branches which cross the B state. After this initial predissociation of the iodine molecules, a recombination of the atomic fragments is forced by collisions with the surrounding cavity and in the case of TON additionally by neighboring iodine molecules ("caging"). The recombination processes take place via the lower energetic weakly bound A/A' states and the X state. The different dynamics are probed by a time delayed second laser pulse, which further excites the iodine molecules into their ion pair states. The response is a transient absorption signal which varies with the wavelength of the probe laser and reflects predissociation, recombination as well as vibrational relaxation processes at different internuclear distances and on different electronic potential energy surfaces. As mentioned

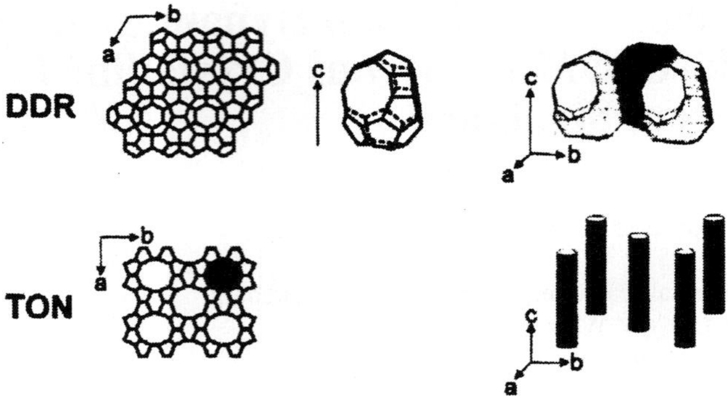

Fig. 1. The structures of the porosils DDR and TON [3]. Left and center: view along crystallographic axes (each line corresponds to an Si-O-Si bridge); right: schematic representation of the pore structures.

above, in TON additionally to the interaction with the cage, an interaction between the iodine molecules themselves have to be taken into account. This results in a change of the potential energy curves of the iodine molecules. Compared to DDR, where the iodine potential energies (except the ion pair states) are not much influenced by the crystalline matrix, considerable spectral shifts were observed for iodine embedded in TON [2,3]. From Raman spectroscopic investigations, we know that also the electronic ground state is influenced, which is reflected by a Raman shift to lower wavenumbers, pointing to a decrease of the molecular binding energy [2,3]. The change of the absorption spectrum mainly points to a shift of the excited B state, but also the weakly bound A/A' and the repulsive states crossing the B state might be affected which should result in a change of the dissociation and recombination dynamics compared to the system I_2-DDR.

As expected, a significantly different dynamical behaviour can be observed in the femtosecond time-resolved pump-probe experiments on iodine in DDR and TON. As an example, in panel (a) of Fig. 2 we show transients taken of iodine in the different environments for same probe wavelengths and pump energy close to resonant B-state excitation. The DDR transient in panel (a) of Fig. 2 starts with a rapid rise, which reflects the onset of the recombination dynamics probed close to the bottom of the A/A' states. The slow decay reflects the vibrational relaxation out of the probe window. The constant signal after about 40 ps is due to the recombined iodine molecules which have relaxed in the weakly bound A/A' states. The rise and decay of the B state signal from iodine in DDR is a rapid process, which is shown in panel (b) of Fig. 2. The slow decay of the I_2-DDR transient in panel (a) is therefore exclusively caused by the dynamics on the A/A' states after the dissociation and recombination of iodine [4]. The I_2-TON transient shown in panel (a) of Fig. 2 is clearly different. The dynamics seen in the beginning is faster than that observed for I_2-DDR and might be due to B state dynamics contributing to the transient.

After this, the transient signal is further decreasing and about 150 ps later the signal is vanishing. No residuum signal like for I_2-DDR can be observed. It

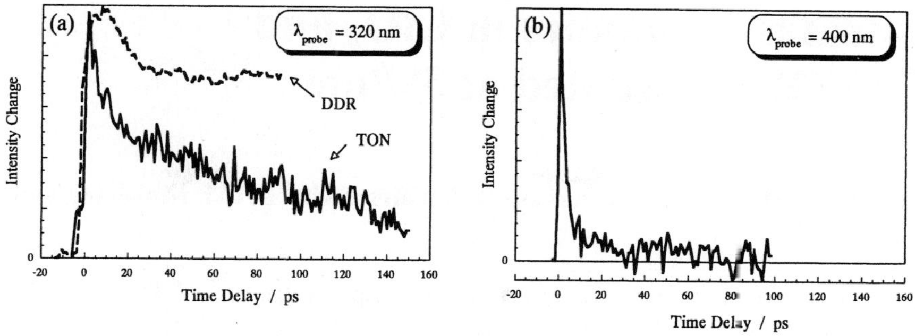

Fig. 3. Transients of I_2 in DDR and TON observed for a pump laser wavelength of 535 and 548 nm, respectively. (a) λ_{probe} = 320 nm: the I_2-DDR transient results from the recombination dynamics along the A/A' states. The remaining signal intensity is due to vibrationally relaxed molecules stabilized in these states. The I_2-TON transient is considerably different which is caused by the iodine-iodine interactions along the microchannels in the TON crystal structure. (b) λ_{probe} = 400 nm: the signal decay due to predissociation of the B state dominates the I_2-DDR transient.

becomes obvious that the iodine-iodine interaction strongly influences the reaction dynamics of the dissociation and recombination. Due to the solvent shift of the ion pair states, a probe laser with a wavelength of 320 nm should probe the vibrationally relaxed A/A' states also for I_2-TON. One therefore would expect a constant signal from the population "trapped" in the weakly bound A/A' states if caging takes place. Several possibilities have to be considered for an explanation of this phenomenon. It might be that the A/A' states are strongly changed by the I_2 – I_2 interaction. Nongeminate recombination processes with considerable excess energies forming instable "intermediate molecules" are very likely to take place and have to be taken into consideration.

In summary, we present for the first time femtosecond time-resolved investigations of the dissociation and recombination dynamics of iodine in the quasi one-dimensional channels of the porosil TON. We compare these results to former results obtained for iodine molecules embedded in quasi zero-dimensional cages of the porosil DDR. Considerably different dynamics can be observed which is due to the strong iodine-iodine interaction in the TON porosils and the restriction of the rotational motion of the iodine molecules.

References

1 A. Materny, C. Lienau, and A. H. Zewail, *J. Phys. Chem.* Vol. 100, 18650, 1996.
2 G. Wirnsberger, H.P. Fritzer, A. Popitsch, G. van de Goor, and P. Behrens, *Angew. Chem,* 108, 2951,1996.
3 G. Wirnsberger, H.P. Fritzer, G. van de Goor, B. Pillep, P. Behrens, and A. Popitsch, *J. Mol. Struct.*, 123, 410-411, 1997.
4 G. Flachenecker, P. Behrens, G. Knopp, M. Schmitt, T. Siebert, A. Vierheilig, G. Wirnsberger, and A. Materny, *J. Phys. Chem. A* , 103, 3854, 1999.

Ultrafast dynamics in ClO_2 and $ClO_2(H_2O)_n$ excited at 398nm

V. Stert, H.-H. Ritze, P. Farmanara, F. Tschirschwitz, E.T. J. Nibbering and W. Radloff

Max-Born-Institut für Nichtlineare Optik und Kurzzeitspektroskopie
Max-Born-Strasse 2a, D-12489 Berlin, Germany
E-mail: radloff@mbi-berlin.de

Abstract: The different relaxation and fragmentation channels of ClO_2 and $ClO_2(H_2O)_n$ excited at 398nm to the ($^2A_2$, v_1=6) state have been studied in pump-probe experiments with 50fs laser pulses by photoion and photoelectron detection.

1. Introduction

Earlier time-resolved studies of the ClO_2 molecule excited at 386nm to the vibrationally excited $^2A_2$ state have demonstrated a two-step fragmentation process in which a relatively slow (4.6ps time constant) internal conversion to the $^2A_1$ and $^2B_2$ states precedes the fast dissociation to ClO+O as well as to Cl+O_2 [1]. For excitation to higher vibrational energies of the $^2A_2$ state at 329nm besides this two-step channel (with smaller time constants) the ultrafast direct dissociation (50fs) of ClO_2 has been observed [2].

We have conducted experiments with much shorter laser pulses of 50fs at a lower excitation energy of 3.1eV in order to study the different reaction channels with improved time resolution. Furthermore, the change of the reaction dynamics for $ClO_2(H_2O)_n$ clusters with respect to ClO_2 has been measured and compared with that of ClO_2 in the liquid solution.

2. Experimental Results and Discussion

In our pump-probe experiments we used the second (398nm) and the third harmonic (265nm) of a Ti:sapphire laser as pump and probe beam, respectively. The pulse duration τ_L=50fs and the delay time zero were determined by nonresonant multiphoton ionization of Xe atoms. The ClO_2 molecules and $ClO_2(H_2O)_n$ clusters were cooled in a molecular beam by adiabatic expansion through a pulsed conical nozzle. Both the ion and the electron signals were analyzed in time-of-flight spectrometers. In Fig. 1 the time-dependent ion signals are shown for the parent molecule and the main fragmentation product ClO. The thin solid lines are fit curves obtained by our theoretical model which bases on optical Bloch equations for the coherent resonant excitation process coupled with rate equations for the secondary relaxation steps. Different contributions $\rho_n^{(m)}$ arising from distinct excited states are added to the resulting signal (heavy solid line). The lifetime τ_1=7ps of the contribution $\rho_2^{(1)}$ corresponds to the transition from the originally excited ($^2A_2$, v_1=6) state to the $^2A_1$, $^2B_2$ states which is followed by fast dissociation to ClO+O and Cl+O_2, respectively, proceeding within τ_2=250fs. The long-time transition is due to spin-orbit and vibrational coupling between the corresponding states. Due to this long lifetime a coherent modulation of the ClO_2^+ signal with a period of 250fs is observable which reflects the simultaneous excitation of the two vibrational combination bands (6, 0, 0) and (5, 0, 2) with a distance of about 140cm^{-1} (Fig. 1). The very fast contribution $\rho_2^{(2)}$ near τ=0 resembling the cross correlation curve of the laser pulses results from nonresonant probe plus pump photon excitation as has been confirmed by our photoelectron spectroscopy. The contribution $\rho_2^{(3)}$ with a decay time of 250fs is tentatively assigned to the dissociation of the $^2A_1$ state which probably can be directly reached from the electronic ground state by the pump photon with low probability. This channel, not observed in previous studies [1, 2], is suggested by resonance Raman measurements of ClO_2 [3]. Preliminary results of our photoelectron spectroscopy seem to confirm this suggestion.

The ClO^+ signal contains two strong contributions $\rho_3^{(1)}$ and $\rho_2^{(3)}$ resulting from dissociative ionization of ClO_2. The third additional term $\rho_5^{(1)}$ represents the ClO fragment formation out of the $^2A_1$, $^2B_2$ states within about 7ps followed by three probe photon ionization. The analysis of the Cl^+ signal (not shown here) reveals significant differences with respect to the time dependence of the ClO^+ signal which indicate distinct pathways of the two dissociation channels.

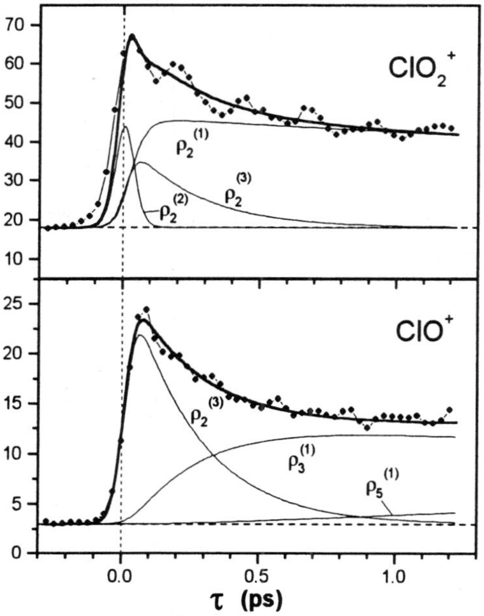

Fig. 1. Parent and fragment ion signals vs. delay time τ ($\rho_n^{(m)}$ denote different contributions, see text)

In Fig. 2 the time-dependent ion signals $ClO_2(H_2O)^+$ and $ClO_2(H_2O)_9^+$ are compared to ClO_2^+. The fit procedure reveals quite similar contributions as described above for ClO_2^+, except for the shorter transition time τ_1=2ps for ClO_2 bound to water clusters in comparison to τ_1=7ps for the free ClO_2 molecule. This lifetime of 2ps has been also found for ClO_2 in aqueous solution at identical excitation conditions. However, no $ClO(H_2O)_n^+$ ion signals were observed in the gas phase. Possibly, due to the high excess energy these cluster fragments dissociate completely leading to ClO^+ or $(H_2O)_mH^+$ ions.

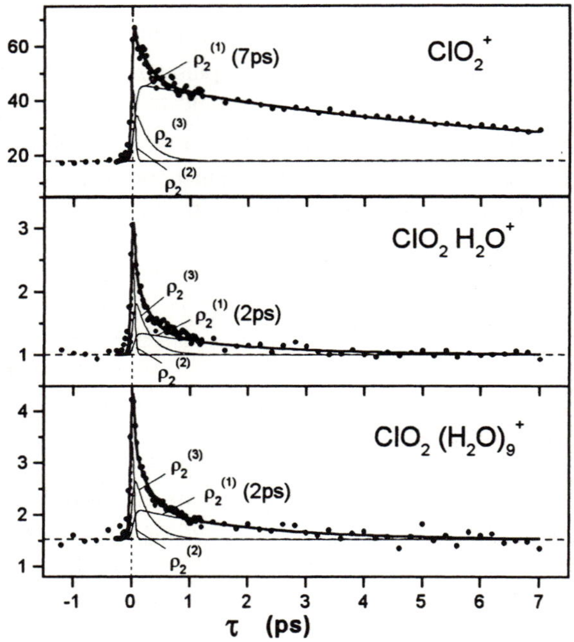

Fig. 2. Time-dependent ion signals of ClO_2 and $ClO_2(H_2O)_{1,9}$ for 398nm pump and 265nm probe wavelengths

References

[1] P. Ludowise, M. Blackwell, Y. Chen, Chem. Phys. Lett. 273, 211, 1997
[2] T. Baumert, J.L. Herek, and A.H. Zewail, J. Chem. Phys. 99, 4430, 1993
[3] P. J. Reid, A.P. Esposito, C.E. Foster, and R.A. Beckman, J. Chem. Phys. 107, 8262, 1997

Vibrational Mode-Specific Ultrafast Photodissociation Dynamics of Chlorine Dioxide in Solution

H. Fidder[1], F. Tschirschwitz[2], O. Dühr[2], and E. T. J. Nibbering[2]

[1]Department of Physical Chemistry, Uppsala Universitet,
Box 532, S-751 21 Uppsala, Sweden. henk.fidder@fki.uu.se
[2]Max Born Institut für Nichtlineare Optik und Kurzzeitspektroskopie, Max Born Strasse 2A, D-12489 Berlin, Germany. nibberin@mbi-berlin.de

Abstract. We show with femtosecond pump-probe measurements that the photodissociation of OClO in solution shows vibrational mode-specific features.

Much effort has been put in unraveling the photochemistry of OClO in gas and condensed phases, since the outcome could have implications for stratospheric chemistry [1]. From gas-phase work it is known that excitation to the A^2A_2 state of OClO can lead to dissociation in two ways, either yielding ClO + O or Cl + O_2, with the pathway ClO + O as the main dissociation channel. In the condensed phase additional features like cage effects, fragment recombination, and relaxation to the ground state potential energy surface, can occur [2-4].

Recently Ludowise *et al.* [5] combined femtosecond spectroscopy with mass and photoelectron spectroscopy to determine lifetimes of optically prepared states and dark intermediate states in the dissociation process. They found lifetimes for optically accessible vibronic states near 386 nm to be 2-4 ps. Experimental data on OClO in solution gave dissociation times of optically excited states of a few hundred femtoseconds [2-4]. Here we present femtosecond pump-probe results from one-color experiments at 403 nm with 55 fs temporal resolution, and two-color experiments with pumping at 400 nm and probing at 800 nm or 267 nm with >100 fs time resolution. We conclude that the photodissociation dynamics of OClO in solution is dependent on which vibrational level is excited within the A^2A_2 state.

The absorption spectrum of OClO in solution consists of a number of vibrational progressions in the symmetric stretch mode. The vibronic series connected to the ground state to excited state transition are $(\nu_1,0,0)$, $(\nu_1,0,2)$, $(\nu_1,1,0)$, and $(\nu_1,1,2)$, where (ν_1,ν_2,ν_3) denote the number of vibrational quanta in the symmetric stretch, bending, and asymmetric stretch mode, respectively. The positions of the vibronic transitions are red-shifted compared to the gas phase by 275 cm^{-1}, 150 cm^{-1}, and 0 cm^{-1} for cyclohexane, water and acetonitrile, respectively. In our experiments we excite with femtosecond laser pulses near 400 nm the (6,0,0), (5,0,0), (6,1,0), (5,1,0), (5,0,2), (4,0,2) and (4,1,2) states.

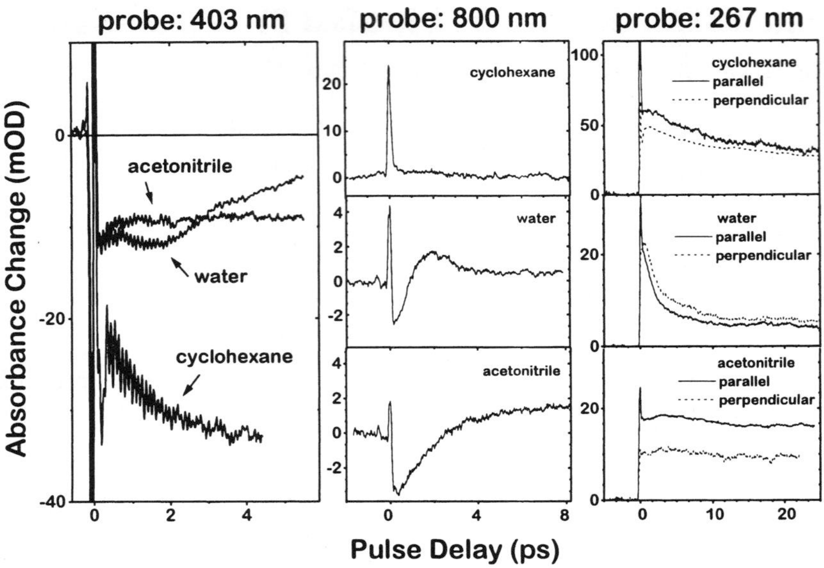

Fig. 1. Time-resolved pump-probe signals of OClO in different solvents for various probe wavelengths.

Figure 1 shows the results on OClO in cyclohexane, water or acetonitrile at the probe wavelengths 403, 800 and 267 nm. A coherent coupling signal dominates during pulse overlap. Probe measurements at 403 nm show initially a decrease in bleach in all solvents for about 0.4 ps. At longer times the dynamics depend on the solvent. In cyclohexane the bleach increases again with a time constant of 1.3 ps, and then remains more or less constant. For OClO in water a small decrease in absorption occurs up to 2 ps, after which the absorbance change drops to zero due to vibrational relaxation to the electronic ground state [3]. In acetonitrile no changes are observed after the first picosecond up to long time delays, in accordance with previous reports [4]. In all solvents the one–color 403 nm pump-probe signals are strongly modulated with a period of 100 fs, which corresponds to a frequency of 330 cm^{-1} (Fig.2). The dephasing time of this oscillating contribution in cyclohexane is identical to the 1.3 ps time constant of the bleach increase. Probing at 800 nm we see an instantaneous increase in transmission for water and acetonitrile, followed by a rising transient absorption at longer times. In cyclohexane only a strong coherent artifact was observed at this wavelength. Probe measurements at 267 nm show for all solvents a rapid increase of absorption, followed by a solvent dependent decay at longer times.

Signal contributions can be assigned to: <u>400 nm</u>, bleach of the OClO ground state X^2B_1 and stimulated emission from the A^2A_2 state; <u>800 nm</u>, transient

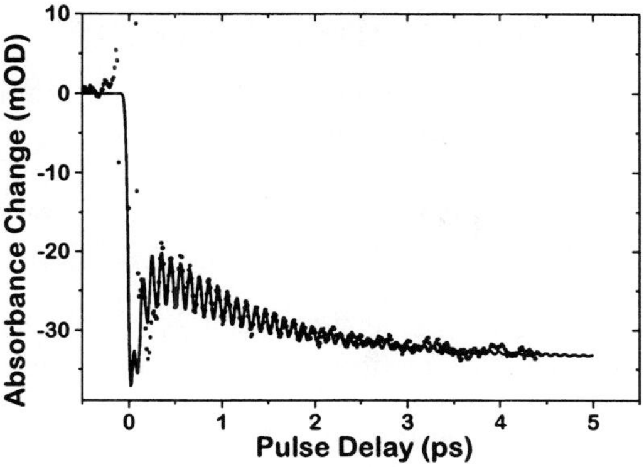

Fig. 2. Expanded view of the time-resolved pump-probe signals of OClO in cyclohexane recorded at 403 nm (dots), showing an oscillatory behaviour with a 100 fs period. The solid trace has been calculated using a kinetic model.

absorption from the vibrationally hot ground state X^2B_1 [3]; 267 nm, transient absorption related to the radical reaction products ClO and Cl, as well as the isomer ClOO [2,3]. Other groups [2,3] have concluded that in water, after excitation near 400 nm, all OClO dissociates and recombines within a picosecond. With this in mind no explanation can be given for the initial transmission increase at 800 nm, and the observed features at 403 nm in the first 3 ps. We can describe the observed features at all wavelengths if vibrational levels with different degrees of excitation of symmetric stretch, bend and asymmetric stretch in the A^2A_2 state exhibit state-dependent dynamics. We obtain a closed description of all observed features if we assume that the bend mode releases a vibrational quantum to the solvent and relaxes in 1.3-1.5 ps mainly through the pathway (6,1,0) → (6,0,0). Excitation of the asymmetric stretch leads to a direct fast dissociation to ClO + O. On the other hand, states where only the symmetric stretch is excited exhibit longer solvent dependent lifetimes: in water ~3 ps, whereas in acetonitrile and cyclohexane these lifetimes exceed 10 picoseconds. The features at 267 nm we ascribe to formation of ClO and recombination.

References:

1. V. Vaida and J. D. Simon, Science **268**, 1443 (1995).
2. Y. J. Chang, and J. D. Simon, J. Chem. Phys. **100**, 6406 (1996).
3. J. Aa. Poulsen *et al.*, J. Chem. Phys. **108**, 8461 (1998).
4. M. J. Philpott, S. Charalambous and P. J. Reid, Chem. Phys. Lett. **281**, 1 (1997).
5. P. Ludowise, M. Blackwell and Y. Chen, Chem.Phys.Lett. **273**, 211 (1997).

Ultrafast Infrared Studies of Ligand Rearrangement at Coordinatively Unsaturated Transition Metal Centers

K.T. Kotz[1], H. Yang[2], P.T. Snee[1], C.K. Payne[1], and C.B. Harris[1]

[1] Department of Chemistry, University of California, Berkeley, CA 94720, USA and Chemical Sciences Division, E.O. Lawrence Berkeley National Laboratory, Berkeley, CA 94720, USA
E-mail: harris@socrates.berkeley.edu
[2] Department of Chemistry and Chemical Biology, Harvard University, 12 Oxford St., Cambridge, MA 02138, USA

Abstract. Femtosecond infrared spectroscopy has been applied to the study of ligand rearrangement dynamics of coordinatively unsaturated transition metal complexes. A new site-diffusion model based on the Langevin equation is proposed to describe the observed rearrangement dynamics.

1. Introduction

The absorption of light by group 6 transition metal carbonyls results in the loss of a CO ligand, forming a highly reactive coordinatively unsaturated transition metal center. Alcohol solvents can form a strongly-bound complex to the vacant metal site through the hydroxyl OH group or a weakly-bound complex through the C-H bonds of the alkyl chain of the alcohol. The rearrangement of the weakly-bound kinetic product to the strongly-bound thermodynamic product in short-chain alcohols occurs on the ultrafast timescale and is dependent upon the alcohol chain length. This rearrangement process has been previously studied for the transition metal compound, $Cr(CO)_6$, using picosecond UV-Vis spectroscopy[1]. An intra-molecular chain-walk mechanism was used to describe the observed kinetics, in which an initially formed alkyl complex rearranges to the thermodynamic hydroxyl product through a series of jumps between carbon atoms along the alkyl backbone of the bound alcohol. In order to better understand these reactive intermediates which have important implications in synthetic organometallic chemistry, we have re-examined the rearrangement dynamics following the photogeneration of $Cr(CO)_5$ in alcohol solutions using femtosecond infrared techniques[2].

2. Results and Discussion

The ultrafast transient difference spectra following photodissociation of CO from $Cr(CO)_6$ dissolved in 1-propanol and 2-propanol are shown in Figure 1. The rearrangement in 1-propanol and in 2-propanol occurs with similar timescales of 160 ps and 100 ps respectively. Furthermore, preliminary data for the

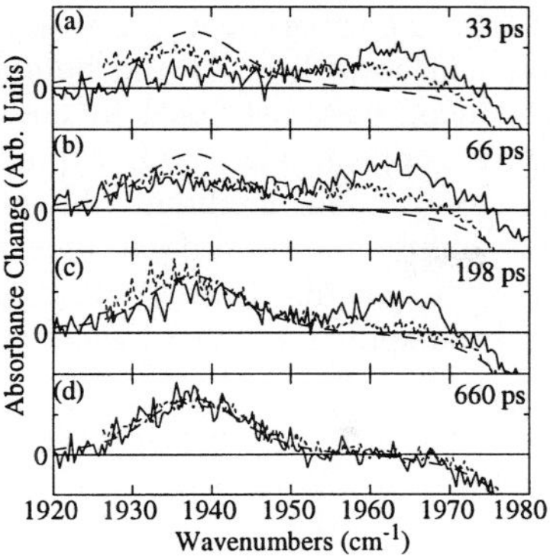

Figure 1. Ultrafast time-resolved infrared difference absorption spectra of Cr(CO)5 in 1-propanol (solid line), 2-propanol (short-dashed line). A static FTIR for the hydroxyl bound product (long-dashed line) is overlaid as a guide to the reader. The numbers in the upper right hand corner of each spectrum is the time delay between the pump and probe laser pulses. The CO stretching bands which appear at 1937 cm^{-1} are due to the hydroxyl bound pentacarbonyl fragment while the corresponding alkyl bound moieties appear at 1964 cm^{-1}.

rearrangement in ethanol show a rearrangement time of < 60 ps and the absence of the expected alkyl bound peak. The difference in re-arrangement times seen in ethanol and 2-propanol as well as the similarity between propanol kinetics are difficult to explain using an intramolecular mechanism.

In order to better understand our results, a new theoretical model has been developed and applied to the data. We considered the rearrangement process as a site-diffusion process by which the $Cr(CO)_5$ fragment "hops" between different C-H binding sites before binding to the final hydroxyl site. The trajectory of each one of these fragments was described as a particle moving along a one-dimensional potential of mean force (PMF). The potential was modeled as a series of cosine wells representing either intermediate CH sites along the alkyl backbone or the terminating OH site. The model used can include effects due to different binding energies at different binding sites. The depths at each unique binding site were based on the binding energies calculated by ab initio and DFT methods. The dynamics of this simplified one-dimensional system are assumed to conform to the Langevin equation subjected to stochastic fluctuation $F(t)$ at room temperature T_R. $F(t)$ is assumed to have a Gaussian distribution with mean $\langle F(t)\rangle=0$ and is subjected to white noise. The initial velocities of simulated particles were taken from a Boltzmann distribution at T_R and the initial placement of particles in potential wells simulated random solvation. The equation of motion was then integrated using standard techniques[3].

Figure 2 displays the normalized population decays of particles residing at intermediate carbon sites for the alcohols 1-propanol and 2-propanol. The simulated kinetic traces can be described by the stretched exponential, or Williams-Watts function. With a heterogeneous rearrangement potential, where each alkyl site is assigned a unique well depth, both traces exhibit apparent non-single-exponential behavior ($\beta ww < 1$), consistent with the non-single-exponential kinetics reported by Simon and coworkers. In the homogeneous potential case, where all of the alkyl well depths are set equal to one another, the simulated kinetic traces shown in Figure 2 appear to be single-exponential ($\beta_{WW} \sim 1$), as one would have expected for a homogeneous medium. In both cases, the simulated $(CO)_5Cr$ / 1-propanol system exhibits a mean decay time $<\tau_{WW}>$ twice as long as that of the $(CO)_5Cr$ / 2-propanol system. Current efforts are underway in our lab to further develop this model and apply it to other ligand rearrangement reactions.

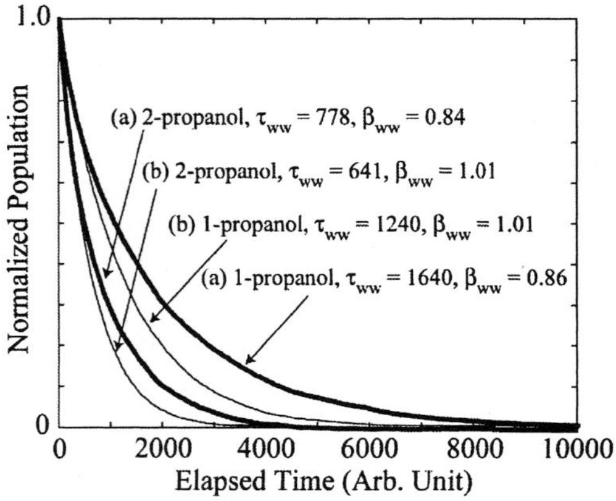

Figure 2. Normalized population decay curves for alkyl bound particles in 1-propanol and 2-propanol determined with (a) an heterogeneous PMF (bold curves) and with (b) an homogeneous PMF (thin curves). The values, τ_{ww} and β_{ww}, are derived from fits of the decay curves to the Williams-Watt stretched exponential function.

Acknowledgements. This work was supported by a grant from the National Science Foundation. All computations were carried out on workstations supported by the Office of Basic Energy Science, Chemical Science Division, U.S. Department of Energy contract # DE-AC03-76FS00098. We also thank Professor David Chandler for fruitful discussions in the initial phase of this work.

References

1 X. Xie, J.D. Simon, "Picosecond time-resolved absorption studies of the solvation of Cr(CO)5 in alcohols: A unimolecular kinetic model for the formation of Cr(CO)5 (OHR) from photogenerated Cr(CO)5(ROH)," *J. Am. Chem. Soc.* **112**, 1130-1136, 1990.

2 S.E. Bromberg, H. Yang, M.C. Asplund, T. Lian, B.K. McNamara, K.T. Kotz, J.S. Yeston, M. Wilkens, H. Frei, R.G. Bergman, C.B. Harris, "The mechanism of a C-H bond activation reaction in room-temperature alkane solution," *Science* **278**, 260-263, 1997.

3 M.P. Allen and D.J. Tildesley, *Computer Simulation of Liquids* (Oxford University Press, New York, NY, 1994).

Ultrafast fragmentation dynamics of the dicarbonyl(cyclopentadienyl)iron halides

M. Bergt, N. Damrauer, C. Dietl, B. Kiefer, and G. Gerber

Physikalisches Institut, Universität Würzburg, D-97074 Würzburg, Germany
Phone: +49 (0)931 8885715, Fax: +49 (0)931 8884906
Email: gerber@physik.uni-wuerzburg.de

Abstract. We present the first systematic experimental study of the influence of metal-ligand bonding properties on the ultrafast photo-fragmentation dynamics of organometallic molecules in the gas phase.

1. Introduction

Organometallic molecules are widely used as photocatalysts in synthetical chemistry [1]. The catalytic activity of such systems is often based on the activation of a stable, saturated complex by primary dissociation events. Subsequent dynamics of the metal-containing species influence its utility in affecting desired photo-chemical transformations. For understanding the mechanisms of primary relaxation and fragmentation processes, the influence of the electronic properties of the metal-ligand bonds on the dissociation dynamics have to be taken into account.

We report the influence of different halogen ligands X (X = Cl, Br, I) on the photodissociation mechanisms of the transition metal complexes of the form $CpFe(CO)_2X$. Femtosecond pump-probe spectroscopy in combination with a molecular beam and reflectron time-of-flight ion mass spectrometry is the ideal tool to study those primary photochemical processes [2-4].

3.. Results and Discussion

In an initial experiment we performed single pulse multi-photon ionization (MPI) experiments in order to determine the fragmentation patterns of the parent ions $[CpFe(CO)_2X]^+$. An example is shown in Fig. 1 for the $CpFe(CO)_2Cl$ molecule, where the mass spectrum is obtained with an 80 fs laser pulse at a center wavelength of 800 nm. The spectrum is dominated by the peak of the parent ion $[CpFe(CO)_2Cl]^+$ and a series of smaller ionic fragments down to Fe^+. The double peaks result from the isotopes of the chlorine atom. The spectrum shows only one dissociation channel of subsequent CO bond cleavage leading to $[CpFeCl]^+$. The absence of the $[CpFe(CO)_2]^+$ and $[CpFe(CO)]^+$ fragments suggests that the breaking of the iron-halogen bond is not involved in the primary steps of fragmentation. This scheme changes upon photodissociation of the $[CpFe(CO)_2I]^+$ species. In this case, the peaks at lower masses dominate the spectrum.

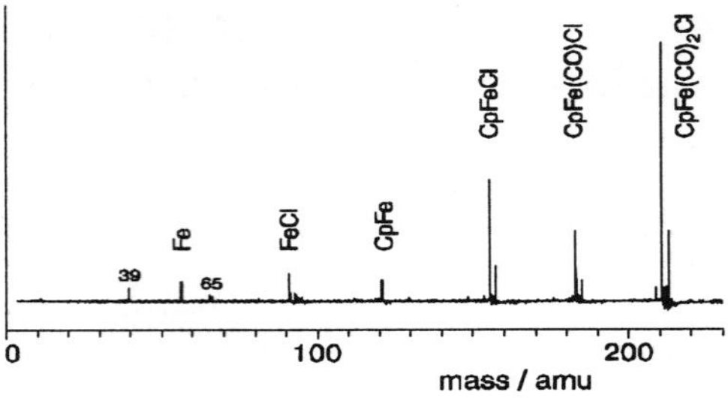

Fig. 1. MPI mass spectrum of $CpFe(CO)_2Cl$

Additionally, the $[CpFe(CO)_2]^+$ and $[CpFe(CO)]^+$ peaks are observed, giving evidence that iron-halogen bond breaking is important in the primary mechanism of photofragmentation.

In order to study the photofragmentation dynamics of the electronically excited neutral species, we performed femtosecond pump-probe experiments with a 400 nm pump laser pulse. The excited-state dynamics is initiated, while a time-delayed 800 nm probe laser pulse ionizes the neutral photofragments. The ions are then observed in the mass spectrometer. The pump laser pulse is attenuated so that no ions are formed. Figure 2 shows the transient ionization spectra of $[CpFe(CO)_2Cl]^+$

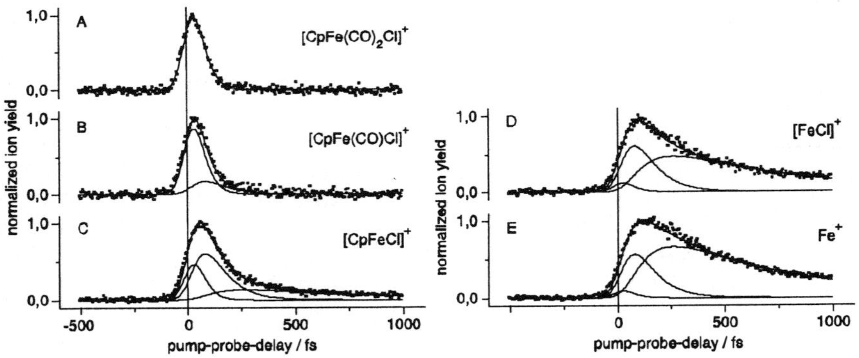

Fig. 2. Transient ionization spectra of $CpFe(CO)_2Cl$ (A) and its fragments (B-E)

(A) and its fragments (B-E). Each transient shows the temporal variation of the detected ion yield of a particular mass.

Our interests are in determining the time-scales of the formation of the neutral intermediates involved in the sequential fragmentation of the photo-excited parent. In order to extract this information, a kinetic model has been applied to the data employing rate equations. This model also accounts for additional fragmentation of ionic species which are introduced by the probe pulse but contain no additional kinetic information.

The dissociation model extracted from the analysis of the $CpFe(CO)_2Cl$ data provides lifetimes of the intermediates and is shown in Fig. 2. The initial excited state $[CpFe(CO)_2Cl]^*$ undergoes ultrafast fragmentation by loss of the first CO ligand in $\tau \approx$ 40 fs. Subsequent loss of the second CO occurs in $\tau \approx$ 90 fs resulting in $[CpFeCl]^*$. On a longer time scale of $\tau \approx$ 800 fs $[CpFeCl]^*$

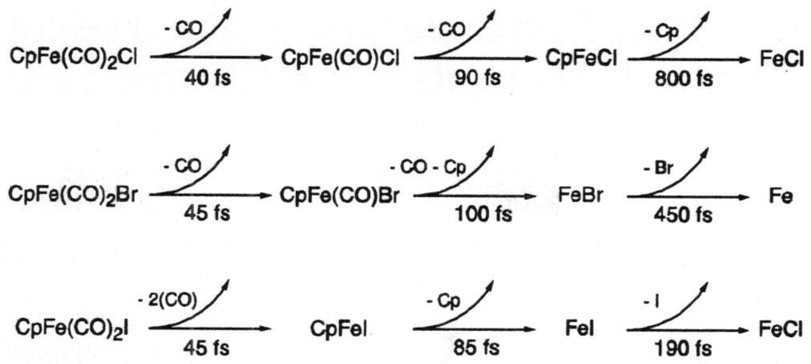

Fig. 3. Assumed dissociation model of the $CpFe(CO)_2X$, X = Cl, Br, I. Only the fragmentation channel characterized by an initial CO loss is shown

dissociates into FeCl and Cp. Using a similar methodology, the transient data collected for $CpFe(CO)_2Br$ and $CpFe(CO)_2I$ can be analyzed. The $CpFe(CO)_2Br$ photofragmentation also follows a mechanism involving a sequential loss of CO fragments. The main difference between $CpFe(CO)_2Cl$ and $CpFe(CO)_2Br$ is that in the second case the Cp ring dissociates simultaneously with the second CO group. For the fragmentation of $CpFe(CO)_2I$ a sequential cleavage of the Fe-CO bonds is not seen. In this case the two CO ligands dissociate in a concerted manner. A second mechanism is involved only in the $CpFe(CO)_2I$ species which involves the primary breaking of the metal-halogen bond.

3. Conclusion

The tools of ultrafast spectroscopy and mass spectrometry can be combined to provide detailed mechanistic information about the photo-fragmentation of complex molecules in the gas phase. We have performed such an experiment on the $CpFe(CO)_2X$, (X = Cl, Br, I) species. Our data indicate that the rate of complete fragmentation of the molecule is slowest for the X = Cl species ($\tau \approx 1$ ps) and increases throughout the series of compounds (for the X = Br species $\tau \approx 600$ fs and for the X = I species ($\tau \approx 130$ fs). Furthermore, while sequential dissociation steps are observed for the X = Cl species, concerted and more complex mechanisms are seen for species containing the larger halogens. These observations suggest that in gas-phase photo-reactions, the time-scales and mechanisms of fragmentation can be modified and possibly even tuned by changing the electronic properties of the metal-halogen bond.

References

1 M. S. Wrighton et. al, Pure Appl. Chem. **41**, 671 1975
2 L. Banares et. al., J. Chem. Phys **108**, 5799 1998
3 M. Bergt et. al., J. Chem. Phys. A **103**, 10381 1999
4 M. Bergt et. al., J. Mol. Struct. **480**, 207 1999

Time-Resolved Photoelectron Spectroscopy of Excited State Intramolecular Proton Transfer Dynamics

Stefan Lochbrunner[1], Michael Schmitt[2], James P. Shaffer[2], Thomas Schultz[2], and Albert Stolow[2]

[1] Sektion Physik, Ludwig-Maximilians-Universität, Oettingenstr. 67, D-80538 München, Germany
E-mail: Stefan.Lochbrunner@physik.uni-muenchen.de
[2] Steacie Institute for Molecular Sciences, National Research Council of Canada, Ottawa, Ontario, Canada K1A 0R6
E-mail: Albert.Stolow@nrc.ca

Abstract: Time-resolved photoelectron spectroscopy of excited state intramolecular proton transfer is presented for the first time. In o-hydroxybenzaldehyde, it occurs within 100 fs on an almost barrierless potential. The excess energy dependence of the subsequent internal conversion is presented.

1. Introduction

Excited state intramolecular proton transfer (ESIPT) is ideal for the study of bond breaking and bond formation. The reaction center is a chelate ring in which the bond between an H atom and a hetero atom switches to another hetero atom in the excited S_1-state [1]. o-Hydroxybenzaldehyde (OHBA) plays an important role as a model compound in both experiment [2-4] and theory [5] because it is the smallest aromatic system which exhibits ESIPT. We investigated ESIPT in OHBA using time-resolved photoelectron spectroscopy. This technique reveals information on both nuclear and electronic changes [6] and allows us to directly observe the dynamics at the keto-minimum of the S_1-state which is populated by the ESIPT.

2. Experimental Results and Discussion

Fig. 1 shows the electronic states involved in the pump-probe photoelectron experiments. We excited jet-cooled OHBA from its ground state, S_0, to the first excited singlet state S_1 with tunable fs-pump pulses. The excited molecules are ionized after a time delay, Δt, using probe pulses centered at 208 nm.

Fig. 2 displays time-dependent photoelectron spectra of OHBA for various time delays Δt between the pump and probe pulses. For $\Delta t = 0$ the photoelectron spectrum has a sharp peak at 0.9 eV. This peak is due to ionization of the S_1 enol-state into the ground state, D_0, of the enol-cation. The S_1 excited molecule changes via ESIPT from its enol- to its keto-form ($S_1 \rightarrow S_1'$ conversion) leading to the broad feature centered at 0.4 eV.

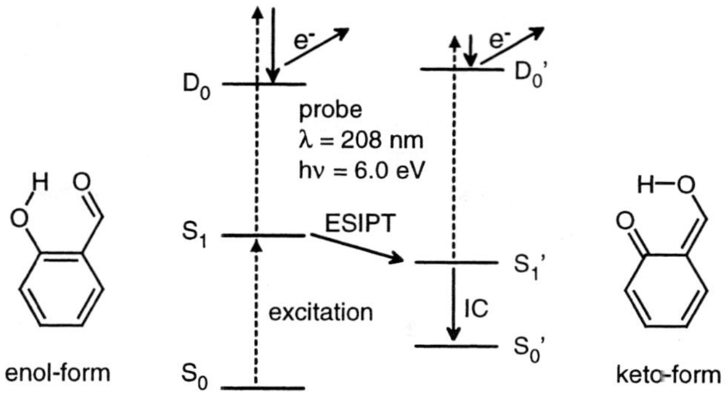

Fig. 1. Enol- and keto-form of o-hydroxybenzaldehyde and detection scheme of the time resolved fs-photoelectron spectroscopy. For details see text.

For non-overlapping lasers the photoelectron spectrum shows one feature at 0.3 eV. This feature is due to ionization of S_1-keto form (S_1') to the keto form of the ion (D_0'). The sharp enol peak decays within the crosscorrelation of the pump and probe pulses. Thus the proton transfer time has to be considerably faster than 100 fs.

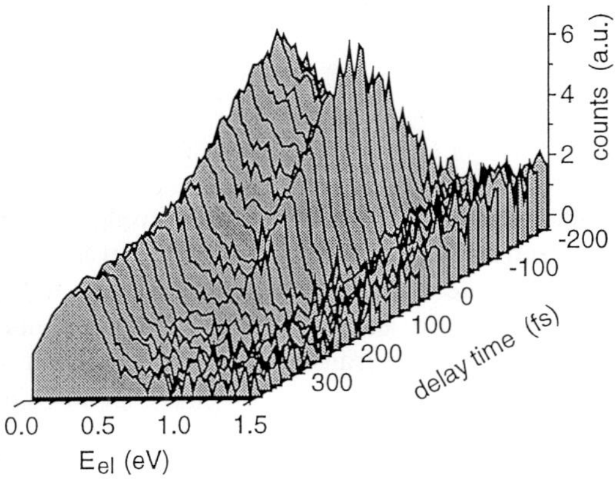

Fig. 2. Photoelectron spectra of OHBA excited at 336 nm in dependence of the delay time.

The photoelectron peak at 0.3 eV monitors the population in the S_1-keto minimum. It decays within several picoseconds (see Fig. 3a) indicating an efficient internal conversion pathway from the S_1-keto minimum to the electronic ground state. This observation is consistent with fluorescence yield measurements [4]. Photoelectron spectra were recorded for various pump wavelengths between 286 nm and 346 nm all showing qualitatively the same features as the spectrum discussed above. However, the internal conversion rate increases with decreasing pump wavelengths. In Fig. 3b the rate is shown as a function of excitation energy.

The same experiments were performed where the transferred proton was replaced with deuterium. The time-resolved photoelectron spectra of the deuterated molecule show a deuterium transfer time of less than 100 fs. Since no deuterium effect was observed for the proton transfer, any barrier on the proton transfer coordinate must be small. The internal conversion rate was found to be similar to those found in the case of OHBA.

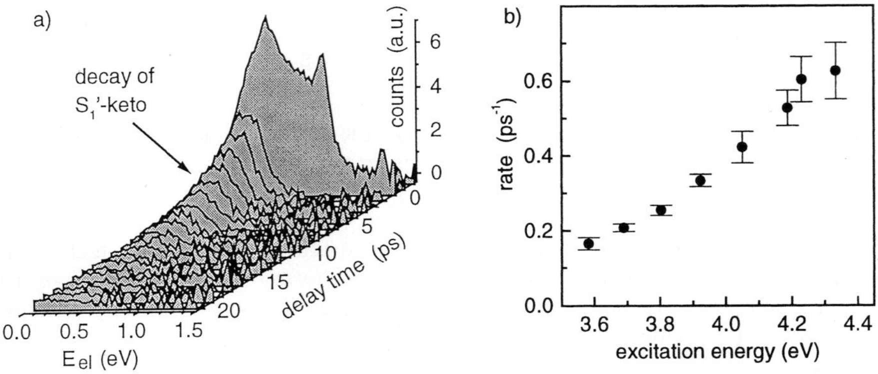

Fig. 3. a) Disappearance of the photoelectron spectrum on a ps-time scale at a pump wavelength of 336 nm due to internal conversion. b) Rate of the internal conversion in dependence of the excitation energy.

3. Conclusions

The ESIPT of OHBA and the subsequent internal conversion were investigated by time resolved photoelectron spectroscopy. The ESIPT was found to occur within less than 100 fs while the rate of the internal conversion is in the order of 0.4 ps^{-1} and increases with increasing excess energy. Experiments with deuterium substitution indicate that the proton transfer barrier is small.

4. References

1 A. Weller, Z. Elecktrochem. **60**, 1144 (1956).
2 J. Catalán, F. Toribio, and A. U. Acuña, J. Phys. Chem. **86**, 303 (1982).
3 S. Nagaoka, N. Hirota, M. Sumitani, and K. Yoshihara, J. Am. Chem. Soc. **105**, 4220 (1983).
4 S. Nagaoka, Y. Shinde, K. Mukai, and U. Nagashima, J. Phys. Chem. A **101**, 3061 (1997).
5 A. L. Sobolewski and W. Domcke, in *The Reaction Path in Chemistry: Current Approaches and Perspectives*, edited by D. Heidrich, Kluwer Academic Publishers, Netherlands, 1995, pp. 257
6 V. Blanchet, M. Z. Zgierski, T. Seideman and A. Stolow, Nature **401**, 52 (1999).

Does the proton move during ultrafast excited state intramolecular proton transfer?

E. Riedle[1], S. Lochbrunner[1], A. J. Wurzer[1], V. de Waele[2], and R. de Vivie-Riedle[2]

[1] Sektion Physik, Ludwig-Maximilians-Universität, Oettingenstr. 67, D-80538 München
E-mail: Eberhard.Riedle@physik.uni-muenchen.de
[2] Max-Planck-Institut für Quantenoptik, Hans-Kopfermannstr. 1, D-85748 Garching

Abstract. The coherent wavepacket dynamics observed in 35 fs crosscorrelation pump-probe spectra of 2-(2'-hydroxyphenyl)benzothiazole render a novel and detailed picture of the excited state intramolecular proton transfer, the contributing skeletal deformations and the electronic configuration changes.

Intermolecular and intramolecular proton transfer are fundamental reactions in chemistry and molecular biology, that occur at largely varying rates. The excited state intramolecular proton transfer (ESIPT) has been shown to proceed in less than 100 fs in prototype molecules [1-3]. The simplest notion of ESIPT is that the proton is dislocated within the molecule by a large distance after absorption of a photon. What is experimentally observed quite readily is a very large Stokes shift [4]. It is, however, not clear how much the change in electronic configuration, the deformation of the whole molecule and the actual motion of the proton contribute to the overall process.

We have performed measurements of ESIPT with a cross correlation of 35 fs that allow us for the first time the direct observation of the relevant nuclear motions. We observe strongly oscillating pump-probe signals due to vibronic wavepacket motion during and after the ESIPT process. Their analysis together with high level quantum chemistry calculations leads to a novel picture of the proton transfer.

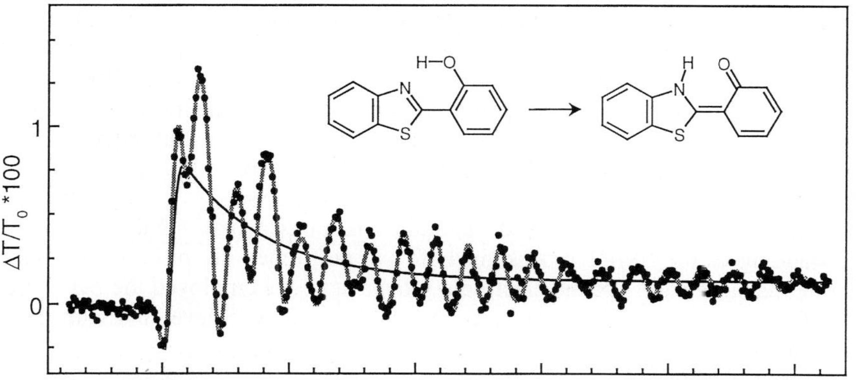

Fig. 1. Transmission change at 502 nm after excitation of HBT at 347 nm. The exponential contribution to the signal is shown separately from the complete fit.

UV pulses of 347 nm center wavelength and 25 fs duration were used to excite 2-(2'-hydroxyphenyl)benzothiazole (HBT) dissolved in cyclohexane. The subsequent evolution of the sample was probed with 20 fs pulses tuned between 450 and 687 nm, the full cw fluorescence band. The ultrashort pulses were generated in noncollinearly phasematched optical parametric amplifiers [5]. A typical pump-probe trace at a detection wavelength of 502 nm is shown in Fig. 1 (dots). In addition to the typical exponential signal contributions (solid black line) the trace is highly modulated.

Use of a sliding window Fourier transformation allows the identification of the 4 contributions at 118, 254, 289 and 529 cm^{-1}. The time evolution of the Fourier transform and selected cuts are depicted in Fig. 2). They show that the oscillations are damped with time constants of about 1 ps.

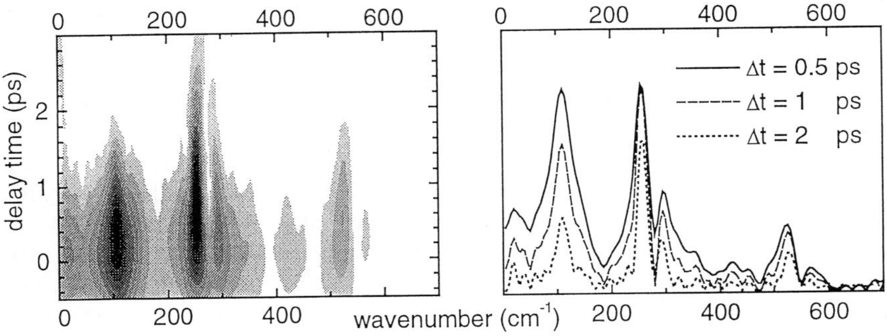

Fig. 2. Fourier transformation of the transmission change at 598 nm (1 ps sliding time window). The non-oscillatory contributions have been subtracted prior to transformation.

With these 4 oscillations all 13 experimental traces can be well fitted. The high degree of agreement between fit (gray line) and experimental data (dots) is shown in Fig. 1. Due to the large Stokes shift in HBT of about 7500 cm^{-1} we exclusively observe wavepacket motion in the excited electronic state. The stimulated emission signal sets in with a detection wavelength dependent delay of around 60 fs, a time that can be taken as a rough measure of the proton transfer time. The signal initially decreases due to rapid relaxation in many vibrational degrees of freedom, resulting in a red shift of the emission spectrum within 300 fs. Subsequently the signal decays with the known S_1 life time of more than 100 ps [6,7].

In order to identify the observed vibrational motions and to obtain a detailed understanding of the ESIPT process in HBT we performed ab initio calculations along the reaction path. For the electronic ground state we employed the MP2 and the DFT method. The excited state was calculated on the CIS level and with time dependent DFT.

The major findings of the calculations are: 1) The 4 vibrations observed are in-plane motions with a large distortion of the central chelate ring in HBT. The chelate ring is shown in detail in Fig. 3 and contains the two heteroatoms (O and N) that form the alternate binding sites of the proton. 2) The energy difference between the excited enol and keto form only amounts to about one third of the Stokes shift. The major part of the Stokes shift is due to the deformation of the molecule in the ground state. 3) There is at most a very small barrier for the proton transfer in the S_1 state. 4) Along the S_1 reaction path the electronic configuration changes abruptly in the chelate ring, corresponding to the breaking of the OH bond and the formation of the NH one.

From the experimental observations with unprecedented temporal resolution and the quantum chemistry calculations we find the novel picture of the ESIPT process depicted schematically in Fig. 3. After the $\pi\pi^*$ photoexcitation (I) the O and the N atom start moving closer together, leading to a significant tightening of the chelate ring (II). About halfway along the reaction path the electrons start to move along the chelate ring in a cyclic fashion toward the nitrogen atom (III). Finally the ring opens again partially, however, the whole molecule is now vibrating in those 4 modes that correspond to the deformation of the chelate ring (IV). The coherent vibration of the product keto-HBT is detected in the pump-probe signal as strong modulations. Their detailed analysis gives the experimental proof of the outlined model.

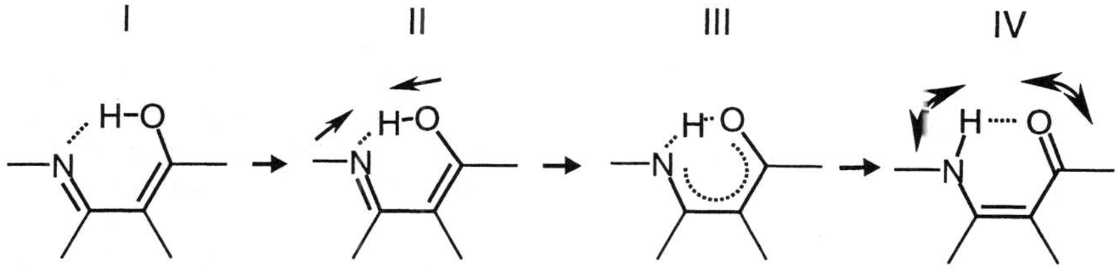

Fig. 3. Sequence of the ESIPT process according to the new model. For details see text.

Nearly all of the available vibrational energy (about 2500 cm^{-1}) is concentrated in the 4 skeletal modes. The proton follows the whole process adiabatically and the resulting NH mode is not excited. This means that the change in electronic configuration and the deformation of the overall molecule are the important aspects in the proton transfer. The motion of the proton itself does not significantly contribute to the energetics and the experimentally observable signatures.

References

1 J. L. Herek, S. Pedersen, L. Bañares, A. H. Zewail, J.Chem.Phys **97**, 9046 (1992)
2 T. Arthen-Engeland, T. Bultmann, N. P. Ensting, M. A. Rodriguez, W. Thiel, Chem.Phys. **163**, 43 (1992)
3 C. Chudoba, E. Riedle, M. Pfeiffer, T. Elsaesser, Chem.Phys.Lett. **263**, 622 (1996)
4 A. Weller, Z. Elecktrochem. **60**, 1144 (1956)
5 T. Wilhelm, J. Piel, and E. Riedle, Opt. Lett. **22**, 1494 (1997)
6 P. F. Barbara, L. E. Brus, P. M. Rentzepis, J. Am. Chem. Soc. **102**, 5631 (1980)
7 F. Laermer, T. Elsaesser, and W. Kaiser, Chem. Phys. Lett. **148**, 119 (1988)

Direct Observation of Ultrafast Excited State Dynamics in Condensed Phase Photochemistry and Photobiology.

Roseanne J. Sension, Allwyn G. Cole, Neil A. Anderson, and Joseph J. Shiang

Department of Chemistry, University of Michigan, Ann Arbor, MI, USA 48109-1055
E-mail: rsension@umich.edu

Abstract. Tunable, ultrashort, sub 40 fs, pulses have been used to investigate the excited state photochemistry and photophysics of vitamin B_{12} and the B_{12} coenzymes 5'-deoxyadenosylcobalamin and methylcobalamin.

1. Introduction

In the mammalian enzyme methylmalonyl-CoA mutase, and related bacterial enzymes such as glutamate mutase, Coenzyme B_{12} (5'-deoxyadenosylcob(III)-alamin) undergoes thermal homolysis of the carbon-cobalt bond, forming cob(II)alamin and catalyzing radical rearrangement reactions [1]. A second B_{12} coenzyme, methylcobalamin, catalyzes a methyl transfer reaction in the mammalian enzyme methionine synthase and related bacterial enzymes. Thermal bond homolysis is not expected in these methyl transfer reactions [2].

Photoexcitation of alkylcobalamins in the visible and near ultraviolet region of the spectrum results in bond homolysis with high yield (10% to 50%) [3], suggesting that photons may be used to trigger enzyme relevant chemistry under the right circumstances. Femtosecond transient absorption experiments performed in our laboratory have demonstrated wavelength dependent photochemistry (400 nm to 520 nm) for several cobalamins including the biologically active coenzyme methylcobalamin [4,5]. By contrast the photochemistry of adenosylcobalamin is independent of excitation wavelength [4,6]. In the work discussed here, tunable ultrashort visible pulses, 23 - 40 fs at 470-600 nm, are being used to investigate the ground and excited state dynamics of B_{12} cofactors. These studies are aimed at exploring ultrafast internal conversion processes and the partitioning between reaction pathways, as well as the time scales for the initial bond cleavage process.

2. Experimental Methods

Transient absorption measurements were performed using a femtosecond laser system and experimental method as described previously [5,6], with a few important modifications. A home-built Titanium Sapphire oscillator/regenerative amplifier, was used to provide 400 μJ, 70 fs pulses at a repetition rate of 1 kHz. Tunable pump pulses were generated by sending the 800 nm laser pulse into a home-built optical parametric amplifier constructed according to the design of Wilhelm *et al.*[7] For the present set of measurements a prism pair was used to compress the output pulses, resulting in 20-40 fs pulses with pulse energies of 0.5 to 1.5 μJ/pulse. Pump and probe pulse timing and instrument response functions were determined by optical Kerr effect measurements in carbon tetrachloride.

3. Results and Discussion

Typical transient absorption signals are shown in Fig. 1. The signal obtained following excitation of methylcobalamin exhibits a rapid internal conversion to a long-lived electronically excited state. This state decays on a 1 ns time scale, resulting in bond homolysis with a yield of 14% [4,5]. Excitation of cyanocobalamin (vitamin B_{12}) results in a signal which exhibits a growth on a time scale of 80 fs followed by a rapid biexponential (190 fs and 6.9 ps) decay reflecting internal conversion to the ground electronic state. In contrast to the relatively simple signals exhibited by these cofactors, adenosylcobalamin exhibits several fast components and a clear non-exponential, oscillatory signal (See Fig. 1B). The initial deep bleach observed at 520 nm exhibits a rapid recovery (92 ± 8 fs) followed by the appearance of a cob(II)alamin-like difference spectrum on a time scale of a few hundred femtoseconds. The oscillation observed in Fig. 1 may provide additional insight into the ground and excited state dynamics. Although these data are not of sufficient quality to permit a complete characterization, a preliminary analysis suggests contributions from at least two vibrational modes, 295 ± 10 cm^{-1} and 88 ± 5 cm^{-1}, with an exponential damping constant of *ca.* 250 fs.

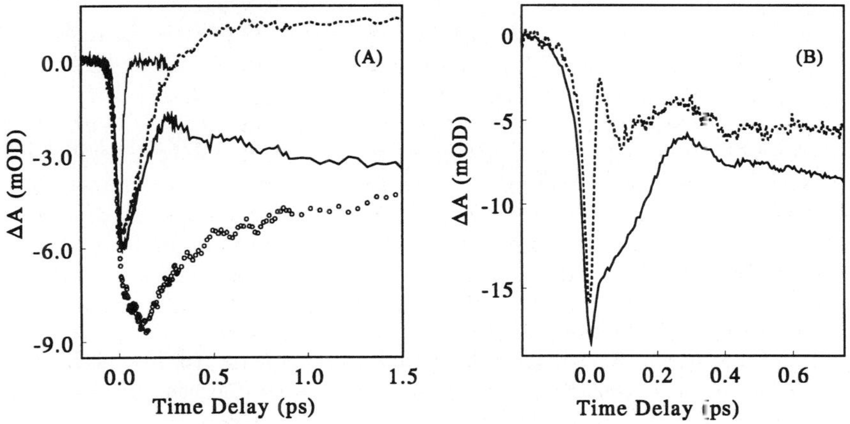

Fig. 1. (A) Transient absorption traces obtained following the excitation of cyanocobalamin (open circles), adenosylcobalamin (solid line) and methylcobalamin (dashed line) with a 28 fs pulse at 520 nm (Autocorrelation is shown in the figure.). The probe wavelength was also 520 nm. (B) Transient absorption traces of adenosylcobalamin obtained following excitation with a ca. 33 fs pulse at 535 nm. The probe pulses were centered at 535 nm (solid line) and 555 nm (dashed line). The oscillations observed in these traces are dominated by a 380 fs component.

For the cobalamin species studied here, overall reorientational diffusion is expected to be slow (i.e. >>100 ps). Therefore fast anisotropy decays reflect the conversion between species with different polarization properties. This information can be used to gain further insight into the time scale for photoinduced bond homolysis in adenosylcobalamin. The anisotropy decay observed following the excitation of adenosylcobalamin at 535 nm is displayed in Fig. 2. The anisotropy starts near 0.4 and exhibits a rapid (<100fs) rise to a value of 0.52, followed by a decay, on a time scale of *ca.* 0.4 ps, to a constant value of 0.26. The plateau persists for at least 100 ps [6].

The rapid changes in anisotropy observed for time delays less than ca. 600 fs reflect changes in the average transition dipole for $S_n \rightarrow S_m$ absorption brought about by internal conversion in the excited state manifold followed by bond dissociation. At longer times the anisotropy is constant, although the relative magnitudes of the ground state bleach and the photoproduct absorption continue to evolve. This plateau in the anisotropy reflects the similar polarization behavior expected for the $\pi\pi^*$ transitions of cob(II)- and cob(III)alamins [8], while the continuing evolution in the magic angle signal reflects cob(II)alamin relaxation. These observations support the conclusion that bond dissociation is complete within 600 fs.

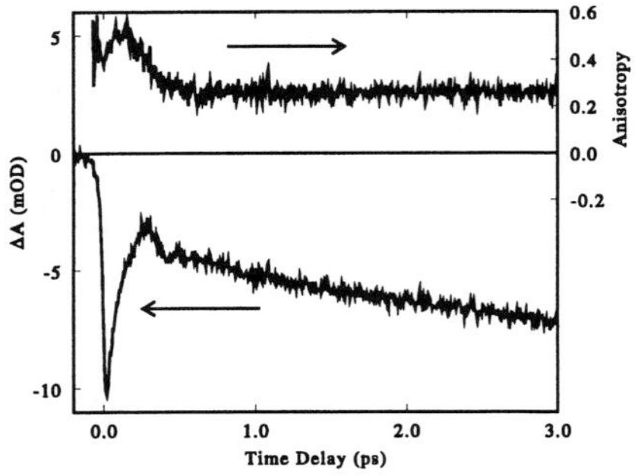

Fig. 2. Anisotropy (top) and magic angle (bottom) decays obtained following excitation of 5'-deoxyadenosylcobalamin at 535 nm. The anisotropy was calculated from data obtained with parallel and perpendicular pump and probe polarization geometries as $(I_{PARA}-I_{PERP})/(I_{PARA}+2I_{PERP})$. The magic angle decay was calculated as $(I_{PARA}+2I_{PERP})/3$.

Acknowledgements. This work was supported by the NIH DK53842 and the Center for Ultrafast Optical Science NSF STC PHY 8920108. .

References

1 E. N. G. Marsh, BioEssays, **17**, 431, 1995.
2 M. L Ludwig and R. G. Matthews, Annual Review Biochem., **66**, 269, 1997.
3 R. T. Taylor, L. Smucker, M. L. Hanna, J. Gill, Arch. Biochem. Biophys., **156**, 521, 1973.
4 J.J. Shiang, L.A. Walker II, N.A. Anderson, A.G. Cole, and R.J. Sension, J. Phys. Chem. B, **103** 10532, 1999.
5 L.A. Walker II, J.T. Jarrett, N.A. Anderson, S.H. Pullen, R.G. Matthews, and R.J. Sension, J. Am. Chem. Soc., **120**, 3597, 1998.
6 L.A. Walker II, J.J. Shiang, N.A. Anderson, S.H. Pullen, and R.J. Sension, J. Am. Chem. Soc., **120**, 7286, 1998.
7 T. Wilhelm, J. Piel, E. Riedle, Opt. Lett., **22**, 1494, 1997.
8 R.W. Fugate, C.-A. Chin, and P.-S. Song, Biochim. Biophys. Acta, **421**, 1, 1976.

Part XV

Biology

Disorder in Photosynthetic Light Harvesting Complexes: A Three-Pulse Photon Echo Peak Shift Study of the B800 band of the LH2 complex

Ritesh Agarwal, Mino Yang, and Graham R. Fleming

Department of Chemistry, University of California at Berkeley, and Physical Bioscience Division, Lawrence Berkeley National Laboratory, Berkeley, CA 94720, USA
E-mail: grfleming@lbl.gov

Abstract. Three-Pulse Echo Peak Shift (3PEPS) data on the B800 band of LH2 complex reveals energy transfer on a time-scale of 500 fs within the band and is modeled properly by combing the appropriate optical response functions with the population kinetics obtained via the solution of the master equation. Our model suggests the presence of two-levels of disorder in the B800 ring.

Light Harvesting Complexes (LHC's) absorb light and transfer this excitation energy to the reaction center with a quantum efficiency close to 100%. [1] The mechanism and time-scales of energy transfer strongly depend on the electronic coupling between the pigments, the coupling of the pigments to the protein fluctuations (electron-phonon coupling) and the distribution of electronic excitation energies of the chromophores due to differences in the protein environment (static-disorder). [1,2] Three-Pulse Photon Echo Peak Shift (3PEPS) spectroscopy has been shown to elucidate details of the electron-phonon coupling and the extent of disorder in chromophores dissolved in condensed phase [3] and for electronically coupled systems (e.g. LHC's). [4,5] 3PEPS experiments [4,5] and theory [6] have shown that the peak shift is also very sensitive to energy transfer between pigments that absorb within the laser spectral window (intra-band energy transfer). Hence, interpretation of 3PEPS data on LHC's requires a detailed modeling of the system with proper inclusion of electron-phonon coupling, disorder and its effect on energy transfer.

Here we report the results of a 3PEPS experiment on the B800 ring of the B800-850 complex of purple bacterium *Rps. acidophila* (strain 10050). We combine the experimental results with an extension of a theory developed by Yang and Fleming [5] that incorporates the effect of a distribution of energy transfer rates due to static disorder. The third-order optical response functions for weakly couped systems take into account the interaction of the chromophores with the laser pulses, whereas the energy transfer dynamics are incorporated by solving the master equation for the whole ring. The effect of disorder is taken into account by a Monte-Carlo sampling procedure that selects out transition energies from a Gaussian distribution of site energies. The spectral overlaps are then calculated using the homogeneous line-shapes centered at the site energy which are randomly chosen by the Monte Carlo procedure. The peak shift is calculated by combining the ensemble averaged population kinetics with the appropriate response functions.

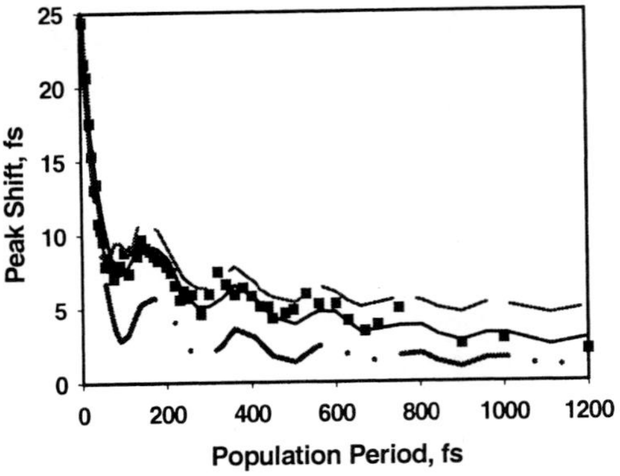

Fig. 1. 3PEPS data of the B800 band of the LH2 complex of *Rps. acidophila* (squares). The dash-dot-dot line is the simulated peak shift with only one level of disorder (σ= 100 cm^{-1}; Σ = 0 cm^{-1}); The dashed line is a simulation with $\sigma = \Sigma =$ 70 cm^{-1}; the solid line is a simulation with σ = 90 cm^{-1} and Σ = 50 cm^{-1}

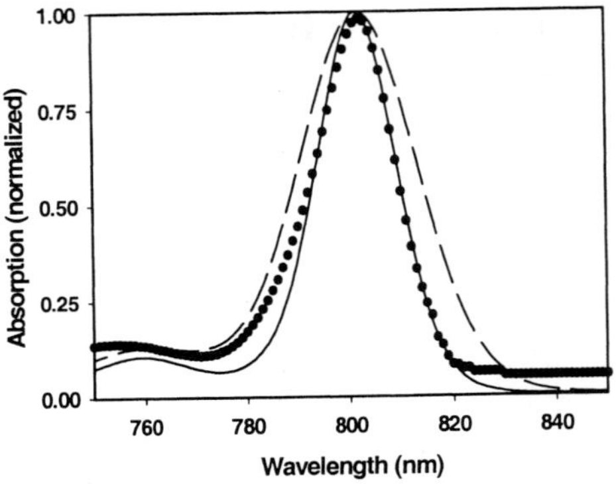

Fig. 2 The absorption spectrum of the B800 band after removing the contribution from the B850 band (solid circles). The simulated absorption spectrum with disorder width = 150 cm^{-1} is shown in dashed line and with disorder width = 100 cm^{-1} in solid line.

Figure 1 shows the experimental peak shift obtained for the B800 band. To model the energy transfer kinetics, the nearest neighbor electronic coupling value used in our calculations is 27 cm^{-1} and the next to nearest neighbor coupling is 4 cm^{-1}. The disorder value used is 160 cm^{-1} (standard deviation). The energy transfer kinetics are solved for the whole ring of B800 pigments with each pigment transferring energy to the B850 band on a time-scale of 800 fs. The initial excitation is placed on one chromophore (randomly selected) and the other B800 chromophores are treated as acceptors. We then sum over the acceptor populations. This is because in our model, energy transfer to an acceptor that absorbs within the laser pulse leads to a free induction decay signal and therefore to a decay in peak shift. Energy transfer out of the band (i.e. to the B850 band) has very little effect on peak shift. [6] The calculation is then averaged over 3000 rings and coupled with the optical response functions to calculate the peak shift. This model simulates the peak shift data well but fails to simulate the absorption spectrum. In Fig. 2, we show the B800 absorption spectrum (after removing the contribution from the B850 band) and the B800 simulated spectrum with our model (dashed line). A static disorder of ~ 100 cm^{-1} simulates the absorption spectrum reasonably well but it cannot simulate the peak shift (fig. 1, dashed-dot-dot curve). It underestimates the value of long time peak shift. Clearly, something is missing in our model.

The above discussion takes into account the distribution of static energies within a ring of chromophores only, and assumes that the distribution of static energies is the same for all the rings in the ensemble. However, it may be possible that the mean of the distribution of static energy within a ring can be different from other rings, i.e., there can be a distribution of the mean static energy between different rings.

Therefore, to model these two-levels of disorder, we assume an ensemble of disordered rings. When such an ensemble is coherently excited, part of the phase memory due to energy transfer/exciton-relaxation within the individual rings is lost, whereas phase memory due to distribution of transition energies between the rings is retained for a very long time. To model the effect of two-levels of disorder in the peak shift, we have two levels of averaging in our theory. The first average is taken over the distribution of site energies within a ring given by σ, and the second over the distribution of energies of different rings given by Σ. In order to simulate both the peak shift data and the absorption spectrum, we fix the total static disorder width at 100 cm^{-1} and partition it into the two distributions. For a given value of σ, we calculate the ensemble averaged population kinetics (as described before) and combine it with the modified response functions that properly accounts for two-levels of disorder. [6] The population dynamics do not change drastically by changing the value of σ. Fig. 1 shows peak shift simulations for different values of σ and Σ and the best fit is obtained for $\sigma = 90\ cm^{-1}$ and $\Sigma = 50\ cm^{-1}$ (solid line). Therefore, long time behaviour of the peak shift can be simulated with a second level of disorder even though the static disorder value is smaller than that required to get a satisfactory fit with only one level of disorder.

We have presented a resasonably complete microscopic model of the B800 band that simultaneously simulates our peak shift data and the absorpion spectrum. It is remarkable that the experimental peak shift decay can be simulated by a kinetic model with no explicit rates of energy transfer and with almost no adjustable parameters for the energy transfer kinetics. This study also shows that energy transfer takes place within the B800 band in competition with energy transfer to the B850 band on a time-scale of 500 fs. Our model suggests the presence of two-levels of disorder in the B800 band. In addition, two-levels of disorder in light harvesting complexes have also been suggested recently in the analysis single molecule spectroscopy at low temperatures. [7]

Acknowledgements

This work was supported by the Director, Office of Science, office of Basic Energy Sciences, Chemical Sciences Divison, of the U.S. Department of Energy under Contract No. DE-AC03-76SF00098

References

1Sundstrom, V., Pullerits, T., van Grondelle, R. J. Phys. Chem. B, **103**, 2327(1999).
2Scholes, G. D., Fleming, G. R. J. Phys. Chem. B, **104**, 1854 (2000).
3Fleming, G. R., Cho, M. Annual Rev. Phys. Chem. **47**, 109 (1996).
4Jimenez, R., van Mourik, F., Yu, J-Y, Fleming, G. R. J. Phys. Chem. B, **101**, 7350 (1997).
Yu, J-Y., Nagasawa, Y., van Grondelle, R. Fleming, G. R. Chem. Phys. Lett., **280**, 404 (1997).
5Agarwal, R., Krueger, B. P., Scholes, G. D., Yang, M., Yom, J., Mets, L., Fleming G. R. J. Phys. Chem. B, **104**, 2908 (2000).
6Yang, M., Fleming, G.R. J. Chem. Phys. **111**, 27 (1999).
Yang M., Fleming G.R. J. Chem. Phys. *in press*.
7 Van Oijen, A. M., Ketelaars, M., Kohler, J., Aartsma, T. J., Schmidt, J. Science, **285**, 400 (1999).
Van Oijen, A. M., Ketelaars, M., Kohler, J., Aartsma, T. J., Schmidt, J. Biophys. J., **78**, 1570 (2000).

Highly unusual growing-in of coherence and strong oscillations in the bacterial FMO antenna complexes, observed in photon echo kinetics

Valentin I.Prokhorenko[1], Alfred R.Holzwarth[1], Frank R.Nowak[2], and Thijs J.Aartsma[2]

[1] Max-Planck Institut für Strahlenchemie, Stiftstr. 34-36, D-45470, Mülheim a.d. Ruhr, Germany
E-mail: prokh@mpi-muelheim.mpg.de
[2] Department of Biophysics, Huygens Laboratory, Leiden University, P.O. Box 9504, 2300 RA Leiden, The Netherlands
E-mail: aartsma@biophys.leidenuniv.nl

Abstract. A very unusual free induction decay in the FMO antennae was monitored by femtosecond photon echo spectroscopy, characterized by strong non-damped oscillations near the red absorption edge and a growing-in of coherence at early times after excitation.

1. Introduction

The bacteriochlorophyll *a*-protein Fenna-Matthew-Olson complexes (FMO) of the green sulfur bacteria *Chlorobium tepidum* and *Prosthecochloris aestuarii* have a similar structural organization. Both contain tree identical subunits, arranged into C_3 trimer, and each subunit contains seven relatively weakly coupled BChl *a* molecules, in which pigment #3 can be considered, in accordance with excitonic calculations [1,2], as an essentially isolated pigment with degree of excitation energy localization of 80~90%. This chromophore determines mainly the well resolved long-wave absorption band at 825 nm, and can be used as a model of a single isolated pigment in a protein environment for the study of pigment-protein interaction.

2. Experimental Methods

The free-induction decay (FID) of excitation has been investigated at 1.27 K in the two FMO complexes by one-color two-pulse photon echo (2PE) techniques in the range of 795-835 nm with ~90 fs resolution and low excitation density of 3-$6 \cdot 10^{12}$ photons/cm^2/pulse. The repetition rate of applied laser pulses was 80 kHz and the FWHM of the light laser pulse was ~11 nm. For noise suppression a double lock-in technique has been used.

3. Results and Discussion

At short excitation wavelengths ≤ 800 nm the 2PE signal displays a relatively fast FID with ~150-250 fs average lifetime, mainly due to fast energy transfer, without any oscillations. The fast oscillations in the 2PE kinetics appear only in the long wavelength absorption band, and they exhibit very large amplitudes (up to 40-60%) near the red edge of the absorption (Figure.1A). Monitoring of the

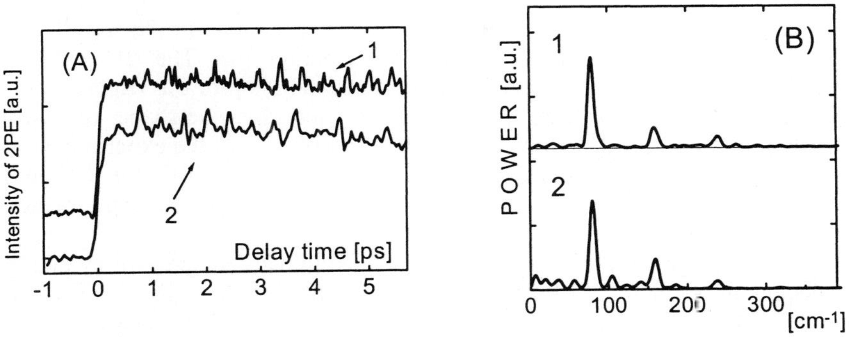

Fig. 1. (A) 2PE traces at 827 nm for FMO *Cl. tepidum* (1) and *Pr. aestuarii* (2) and Fourier spectra of oscillations (B).

2PE signal in a long delay window shows that these oscillations are still not substantially damped up to 300 ps. The Fourier transform of the oscillations clearly displays the presence at least 3 approximately equally spaced components with frequencies of ~80, ~160, and ~240 cm^{-1} for both FMO complexes (Figure 1B). The highest modes could not be resolved due to the relatively narrow spectrum of the laser pulses. These frequencies were not observed in the site selective fluorescence spectra [3] which allows us to propose that they originate from the BChl vibrational states in the excited (S_1) state.

Besides these oscillations the 2PE signal shows at initial delays a rising component in the time evolution of FID. Figure 2A shows a plot of the measured

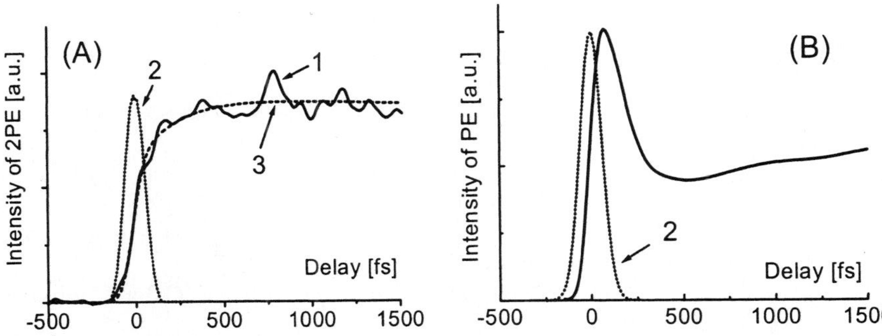

Fig. 2. (A) Measured 2PE trace at 827 nm for FMO *aestuarii* (1), measured cross-correlation function of laser pulses (2), and (3) fit with rise component of 230 fs (-32%) and decay components of 28 ps (+107%) and 920 ps (+25%). (B) Modeled 2PE trace.

2PE for FMO *aestuarii* at early delay times containing a rise time of 230 fs. For the FMO *tepidum* the rising of the FID occurs with a time of ~150 fs. For both FMOs the observed 2PE signal in the red edge could not be fitted without a rising component. The maximum of the amplitude distribution corresponds to the spectral location of BChl *a* #3. The presence of this rising component in the FID is also supported by accumulated photon echo investigations, performed in a completely different setup.

A rise of the 2PE signal indicates that the coherence is growing after impulsive excitation. That means that a molecule develops a coherent state due to interaction with its environment. This unusual effect can not be explained in the framework of generally accepted theory of coherence formation [4]. The 2PE signal, calculated on the basis of the theory of the PE formation [4] for our experimental conditions, is shown in Figure 2B. We have used the line-shape function, constructed on the spectral density of bath states derived from the experimentally measured phonon profile [3]. This modeling shows that the 2PE signal should contain a positive peak around zero delay which disagrees with the experimental observations.

However, a growing-in of coherence can be qualitatively understood by taking into account a correlation between the bath variables in the ground (S_0) and excited (S_1) states. This gives a rising component in the induced polarization at essentially low system-bath interaction, characterized by a small Huang-Rhys factor (~0.3 for FMOs).

4. Conclusions

We have experimentally discovered an effect of growing-in of coherence of BChl *a* molecules, immersed in a protein environment. The rising of coherence occurs with a characteristic time of 150-250 fs after impulsive excitation due to interaction with the protein bath.

Acknowledgements. This work was supported by Deutsche Forschungsgemeinschaft SFB # 189, and by the EU TMR Project on "Green bacterial photosynthesis".

References

1 R.J.W. Louwe, J. Vrieze, A.J Hoff, and T.J. Aartsma, J. Phys. Chem. B **101**, 11280, 1997.

2 S.I.E. Vulto, M.A. de Baat, R.J. Louwe, H.P. Permentier, T. Neef, M. Miller, H. van Amerongen, and T.J. Aartsma, J. Phys. Chem. B **102**, 9577, 1998.

3 S.I.E. Vulto, M.A. de Baat, S. Neerken, F.R. Nowak, H. van Amerongen, J. Amesz, and T.J. Aartsma, J. Phys. Chem. B **103**, 8153, 1999.

4 S. Mukamel, *Principles of nonlinear optical spectroscopy*, Oxford University Press, New York, 1995.

Time-Dependent Exciton Delocalization Size in the B850 Band of Purple Bacterial LH2

L. D. Book[1], A. E. Ostafin[1], N. Ponomarenko[1], J. R. Norris[1], S. Mukamel[2] and N. F. Scherer[1]

[1] Department of Chemistry, the James Franck Institute and the Institute for Biophysical Dynamics, University of Chicago, Chicago IL 60637, USA
E-mail: nfschere@uchicago.edu
[2] Department of Chemistry, University of Rochester, Rochester NY 14627, USA

Abstract. Several ultrafast coherence spectroscopies are employed to study initially delocalized excitons and their subsequent sub-100 fs localization dynamics in LH2. Rigorous simulations of the exciton relaxation process allow for determination of the time-dependent exciton size.

Characterization of exciton dynamics in the light harvesting complexes of photosynthetic purple bacteria is an issue of great current interest [1]. One aspect of the photophysics of these systems that has engendered considerable debate is the extent to which the B850 exciton of light harvesting complex II (LH2) is delocalized over the aggregates's 18 bacteriochlorophyll (BChl) chromophores. For example, two-photon absorption studies [2] have suggested that the exciton is delocalized over the majority the B850 ring, while measurements of the rate of superradiant emission indicated a coherence length of 3-4 BChl's [3]. In any measurement of the delocalization size, the timescale of the experiment must be taken into account for proper interpretation of the result, as shown here and more comprehensively in Ref. [4]. Delocalization is of importance in the optical spectroscopy these antenna complexes because the transition dipole moment of a molecular aggregate is enhanced over that of the constituent monomer by a factor equal to the square root of the exciton delocalization size [5]. The delocalization size can be rigorously defined as the inverse participation ratio, $P(t)$, of the one-exciton density matrix [6]:

$$P(t) = \frac{\left[\sum_{mn} |N_{mn}(t)|\right]^2}{L \sum_{mn} |N_{mn}(t)|^2} \qquad (1)$$

where the N_{mn}'s are elements of the one exciton density matrix in the site (i.e., real-space) representation and L is the total number of chromophores in the aggregate. Since the density matrix elements are time-dependent, *the delocalization size is time-dependent.*

Fig. 1 shows the magic angle pump-probe (MAPP) and pump-probe (PPA) anisotropy signals from B850. Note the steady decay in the value of the MAPP signal. This decay cannot be due to population relaxation to the ground state or a fast Stokes shift, as the fluorescence lifetime of B850 is almost a nanosecond and the Stokes shift is less than 150 cm^{-1} [3], well within the laser spectral width. The value of the PPA begins at a little below 0.4, but it drops to a

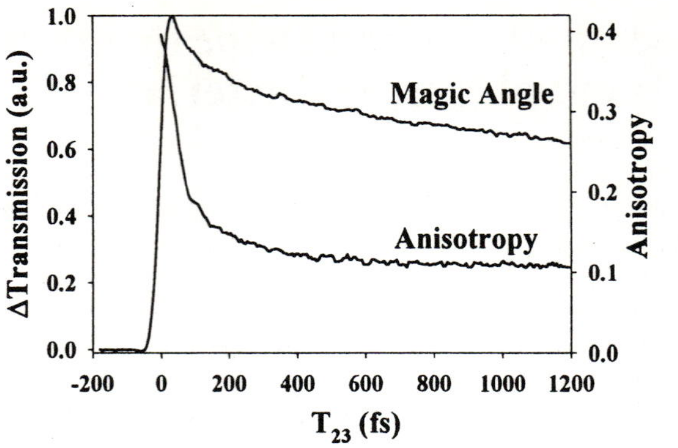

Fig. 1. Magic angle pump-probe and pump-probe anisotropy of the B850 band of LH2.

bit above 0.1 with a time constant of about 50 fs, reflecting ultrafast transition dipole moment reorientation.

The decay of the MAPP and PPA signals can be best explained by dephasing-induced contraction and reorientation of the B850 transition dipole moment. At the instant when B850 absorbs a photon, the magnitudes of the off-diagonal elements of the one-exciton density matrix will be at their peak values, meaning that at this time the delocalization size and the transition dipole strength of the exciton will achieve their highest values. As the delay time in an experiment is increased beyond zero, homogeneous and inhomogeneous dephasing processes manifest themselves and the density matrix becomes less and less coherent. Thus, the delocalization size and transition dipole moment of the exciton decrease. Because a pump-probe signal is proportional to the transition dipole of the system at the time of interaction with the probe pulse, we interpret the decay of the MAPP signal as a direct observation of exciton localization. When an exciton in B850 is highly delocalized, the transition moment is degenerate in the *x-y* plane of the ring. As the exciton localizes, it becomes confined to an increasingly smaller area of the ring and the "two-dimensional" character of the transition moment decreases. These dynamics are reflected in the decay of the pump-probe anisotropy, and thus this measurement has the same interpretation as the MAPP result.

These experimental data provide essential constraints for the theoretical simulation of exciton dynamics in B850. Proper modeling of ultrafast four-wave mixing in molecular aggregates requires the incorporation of the one- and two-exciton band structure, exciton-phonon coupling and inhomogeneous broadening. Mukamel and co-workers have developed a theory that rigorously includes all of these elements [7]. The exciton-phonon interaction is modeled with arbitrary spectral densities between all pairs of electronic levels in the aggregate. As a starting point for the simulation of our B850 data, it is assumed that the nuclear fluctuations acting on different chromophores are identical and uncorrelated. The form for these spectral densities is taken to be that of an overdamped Brownian oscillator. Assigning the coupling strength and relaxation time of the Brownian oscillator to be 240 cm^{-1} and 130 fs, respectively, and assuming an inhomogeneous width of 530 cm^{-1}, a good fit is made to the linear absorption spectrum. Fig. 2 shows experimental and calculated frequency-resolved pump-probe signals for B850 at delay times 0, 50, 100, 500, 1000 and 1500 fs. The calculated signals at the three longest delay times are in good agreement with experiment. The disparity between experiment and theory for the shorter delay times are due to the neglect of finite pulse widths, "improperly time-ordered" Feynman pathways, and monomer excited-state absorption. Note that exciton localization is the only mechanism in the calculations that can cause a decline in absolute signal level.

Further refinement of these calculations should enable the determination of the delocalization size as a function of time elapsed since creation of the exciton.

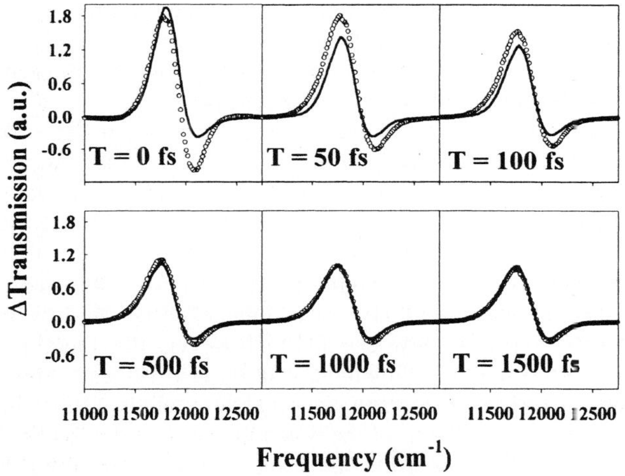

Fig. 2. Experimental (points) and calculated (lines) frequency-resolved pump-probe signals from the B850 band of LH2. Experimental and simulated results are normalized to the T = 1000 fs data.

References

1 V. Sundström *et al*, in *J. Phys. Chem B*, Vol. 103, 2327, 1999.
2 D. Leupold *et al*, in *Phys. Rev. Lett.*, Vol. 77, 4675, 1996.
3 R. Monshouwer *et al*, in *J. Phys. Chem. B*, Vol. 101, 7241, 1997.
4 L. D. Book *et al*, in *J. Phys. Chem. B*, Vol. 104, xxxx, 2000.
5 F. C. Spano and S. Mukamel, in *Phys. Rev. A*, Vol. 40, 5783, 1989.
6 T. Meier *et al*, in *J. Phys. Chem. B*, Vol. 101, 7332, 1997.
7 T. Meier *et al*, in *J. Chem. Phys.*, Vol. 107, 8759, 1997.

8 Acknowledgment: NIH (R01-8M57768), DOE (DEFG0296ER14675) and NSF.

The Kinetics of Primary Electron Transfer steps in PS I Reaction Centers

Sergei Savikhin[1], Peter Martinson[2], Parag Chitnis[3], Wade Johnson[3], Wu Xu[3] and Walter Struve[1,2]

[1] Ames Laboratory, Ames IA 50011
E-mail: sergei@ameslab.gov, wstruve@amealb.gov
[2] Department of Chemistry, 0207 Gilman, Iowa State University, Ames IA 50011
[3] Biochemistry & Biophysics, Molecular Biology, Iowa State University, Ames, IA 50011
E-mail: chitnis@iastate.edu, wxu@iastate.edu

Abstract. The primary electron transfer kinetics were determined for the first time in core antenna - reaction center complexes from photosystem I of wild-type *Synechocystis* sp. PCC 6803 containing about 100 chlorophylls, by comparing femtosecond pump-probe profiles for samples with open and closed reaction centers.

1. Introduction

The primary events triggered by light absorption in the Chl *a* core antenna of photosystem I (PS I) complexes have been intensively studied by many groups since the mid-1980's (for review see [1]). However, the kinetic sequence of electron transfer events in PS I following oxidation of P700 is not well established. The considerable overlap of antenna and reaction center (RC) Chl absorption spectra makes selective excitation of the reaction center cofactors like P700 or A_0 impossible, and signals arising from the electron transfers among the six Chl's within the RC are masked by signals reflecting energy transfer among ~100 antenna pigments. In a recent review of indirect and cumulative kinetic measurements, Brettel [2] suggested that electron transfer from P700* to the primary acceptor A_0 occurs with 1-3 ps kinetics, followed by 20-50 ps electron transfer from A_0 to the phylloquinone secondary acceptor A_1.

2. Experimental Methods

The PS I trimers were isolated from wild-type *Synechocystis* sp. PCC 6803. The samples contained 20 mM sodium ascorbate. The femtosecond optical pump-probe system as well as the steady-state absorption difference spectrometer have been described in detail elsewhere [3].
Hastings et al. [4] proposed a strategy for isolating the reaction center kinetics by obtaining time-resolved absorption difference profiles for PS I complexes with open and closed RC's. Subtraction between these profiles would cancel the antenna kinetics, if the latter are unaffected by the P700 oxidation state. Savikhin et al. [3]

justified this by showing that the kinetics of antenna excitation trapping at P700 and $P700^+$ are essentially identical in *Synechocystis* sp.
In this work we used an external light source to control the RC oxidation state. As shown earlier [3], samples in experiments conducted in total darkness contained predominantly open reaction centers; continuous illumination of the sample cell by a 3V flashlight bulb yielded samples in which the reaction centers were almost exclusively closed.

3. Results and Discussion

Figure 1 (top) shows the transient absorption difference *(ΔA)* profiles observed at 690 nm after excitation with femtosecond pulses at 660 nm. The solid line represents the case for open RC's, while the dashed line corresponds to closed RC's. The bottom panel of figure 1 shows the difference in signal *(ΔΔA)* between these two cases. Since the antenna kinetics for open and closed RC's, they cancel in the *ΔΔA* profiles; the signal diffrence carries information about processes occuring within RC. The 690 nm probe wavelength is of particular interest, because the absorption at this wavelength increases when $P700^+$ is formed. In order to interpret these data, we have measured the steady-state absorption difference spectra of PS I between open and closed RC's (Figure 2, noisy solid line) and fitted it with the sum of four gaussians (smooth solid line). The fit reveals two major bands – a broad photobleaching band centered at ~700 nm (long-dash), and a much narrower additional absorbance at 690 nm (short-dash).

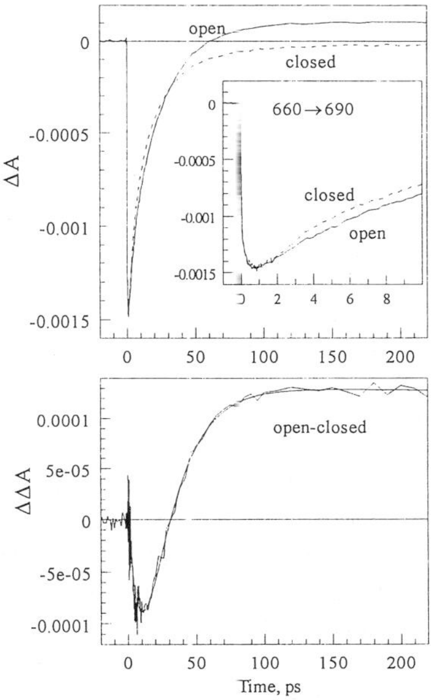

Fig. 1. Top: PS I transient *ΔA* spectum for open and closed RC's. Bottom: difference between open and closed profiles, fitted according to the model described in text.

According to our scenario, excitation is transferred from antenna to the P700 in 23 ps, forming electronically excited P700*. As a result the broad P700 exciton absorption band (Fig. 2, long dashed line) bleaches, causing a drop in the *ΔΔA* signal. P700* then transfers an electron to A_0 with 10 ps kinetics, and the photobleaching becomes dominated by the emerging absorption band of the non-ionized Chl within the oxidized special pair $P700^+$. We were able to fit the *ΔΔA* profile within the described scenario (smooth line on top of the $\Delta\Delta A$ profile in Fig. 1) with only one lifetime as a free

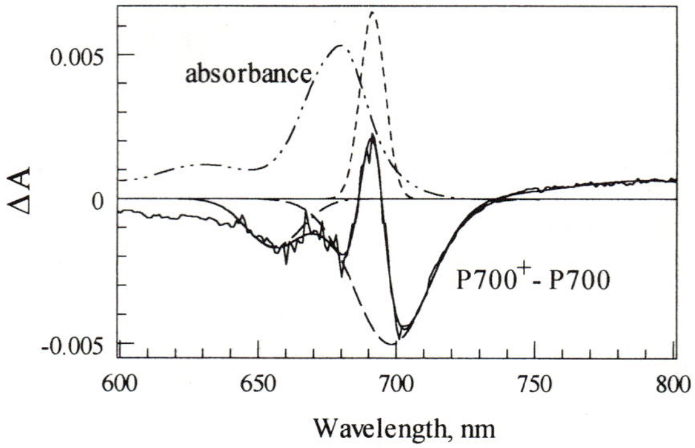

Fig 2. Steady state absorption difference spectrum of PS I between cases where the reaction centers are open and closed (noisy solid line), superimposed on a multiple Gaussian fit. Two major gaussian components are shown by dashed curves. The steady state PS I absorbance spectrum (dot-dot-dashed curve, scaled down by a factor of 100) is shown for comparison.

parameter; the amplitudes were provided independently by the steady state absorption difference spectrum (Fig. 2). This scenario was also supported by other data observed in our experiments at different pump-probe wavelength combinations (not shown). Our measurements strongly suggest that the timescale for the primary charge separation is 7-10 ps, significantly longer than previously estimated [2, 4].

Acknowledgements. Research at the Ames Laboratory was supported by the Division of Chemical Sciences, Office of Basic Energy Sciences, U. S. Department of Energy. The preparation of PS I mutants was supported by NSF grant MCB970023 and MCB0078264 (to PRC).

References:

1. Van Grondelle, R., J. P. Dekker, T. Gillbro, and V. Sundström, Biochim. Biophys. Acta **1187**, 1-65 (1994).

2. Brettel, K Biochim. Biophys. Acta **1318**, 322-373 (1997).

3. Savikhin, S., W. Xu, P. Martinsson, P. R. Chitnis, and W. S. Struve, Biophys. J. in press (2000).

4. Hastings, G., F. A.M. Kleinherenbrink, S. Lin, and R. E. Blankenship, Biochemistry **33**, 3193-3200 (1994).

Ultrafast charge separation in the plant photosystem I reaction center

Shigeichi Kumazaki[1], Isamu Ikegami[2], Shuichiro Yasuda[1] and Keitaro Yoshihara[1]

[1] School of Materials Science, Japan Advanced Institute of Science and Technology, Tatsunokuchi, Ishikawa 923-1292, Japan.
E-mail: kumazaki@jaist.ac.jp
[2] Faculty of Pharmaceutical Sciences, Teikyo University, Sagamiko, Kanagawa, 199-0195, Japan

Abstract. The primary charge separation in photosystem I reaction center of plant at room temperature is found to occur intrinsically on a subpicosecond time scale, which is faster than that of purple bacterial reaction center.

1. Introduction

Plants and cyanobacteria contains a pigment-protein complex called photosystem I (PSI) reaction center (RC), which generates the strongest reductant among all the biological systems and leads to the reduction of carbon dioxide. The arrangement of the electron carrier chlorophylls of PSI RC has been recently clarified [1]. However, the details of the electronic excited state and charge separation associated with the chlorophylls are not yet understood. This is mainly due to the large number of antenna chlorophylls (about 90) in the core RC complex.

We have prepared PSI RC particles with a significantly reduced number of antenna chlorophylls [2]. These are designated P700-enriched RC, where P700 is the primary electron donor chlorophyll. The P700-enriched RCs retain the original electron-transfer capability from the excited state of P700 (P700*) to the electron acceptor A_0 [3]. The electron acceptor phylloquinone has been removed from the P700-enriched RC particles, by which one can accumulate the radical pair of chlorophylls ($P700^+A_0^-$) [3, 4]. Since the so-called long-wavelength-absorbing chlorophylls" with their absorption peaks redder than that of P700 are also removed [5], P700 can be excited preferentially. The focus of our study is on the absorbance changes in the 710 – 757 nm region, where chlorophyll radicals give an absorption band [4], but the absorption of excited chlorophylls is small [3]. The high relative population of P700* on the excitation and the direct detection of the chlorophyll radical pair with a femtosecond time-resolution have enabled us to observe the primary charge separation of PS I RC with minimal interference by excitation transfers.

2. Experimental Methods

The P700-enriched RC particles with a chlorophyll/P700 ratio of 12 - 13 were prepared from spinach as previously reported [2]. The transient absorbance changes were recorded under the magic angle polarization conditions by using femtosecond pump and probe pulses at 1 kHz repetition rate. The instrument-response functions (IRFs) had a FWHM of 0.12 – 0.15 ps in the single-wavelength-probe method and 0.30-0.35 ps in the transient-spectrum method. The sample was put in a cylindrical cuvette with a path length of 1.5 mm. This cuvette was cooled ($\approx$ 280 K) and rotated at $\approx$ 5 Hz during the measurement. The Qy band peak wavelength of the sample was checked before and after each measurement, and the change was at most 1.0 nm. We estimated that at most each pump pulse excited 20 % of the reaction center particles.

3. Results and Discussion

The transient signals under the P700-neutral conditions (Fig. 1(a,b)) are fitted by the sum of the three exponential components: 0.1 ps (1.04×10^{-3} absorbance decay), 0.7 ps (1.02×10^{-3} absorbance rise) and 9 ps (1.30×10^{-3} absorbance rise). The rising rate of the signal under the P700-preoxidized conditions was much slower and the amplitudes are smaller than under the P700-neutral conditions. The transient spectra in the 710 – 757 nm region (Fig. 1(c, d)) are analyzed by a global curve fitting method, which supported that there are two rising components as in Fig. 1(a, b). The three spectra in Fig. 1(d) are similarly flat in the 725 – 750 nm region, but the amplitude around 755 nm of the 0.2 ps spectrum is relatively small.

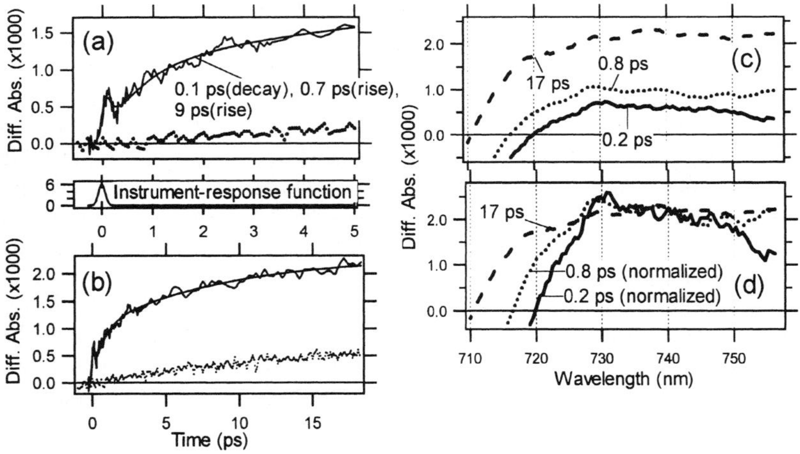

Fig. 1. (a) and (b): Kinetic trace of transient absorption at 738 nm. Lines connecting the data points: under the P700-neutral conditions, excited at 699 nm. Smooth lines: fitted three-exponential curve. Dots: under the P700-preoxidized conditions, excited at 692 nm. (c) and (d): Transient absorption spectra of P700-enriched RC under the P700-neutral conditions.

The instrument-response-limited rise and the 0.1 ps decay (Fig. 1(a)) are ascribed to the dynamics of P700*. The relatively strong transient absorption of P700* in the 725 – 750 nm region is due to the larger excited state absorption and/or weaker vibrational progression in the fluorescence spectrum compared with monomeric chlorophyll. Since the 0.2 ps spectrum in Fig. 1(c, d) is substantially determined by the absorption of P700*, the similarity between the 0.2 ps and 17 ps spectra is intriguing. This may indicate that P700* can be described as a superposition of excitonic state(s) and internal charge-transfer state(s) of the pair of chlorophylls, as proposed for the special pair of purple bacterial RCs [6].

The two rising components in Fig. 1(a, b) indicate that the charge separation is biphasic: 0.7 (50 %) and 9 ps (50%), which approximately correspond to the charge separation during and after the excitation equilibration, respectively. Computer simulations to reproduce the overall charge separation kinetics yielded 0.5 – 0.8 ps for the intrinsic time constant of the primary charge separation. Given the empirical relation between the edge-to-edge distance of the pigments and the time constant of the electron transfer [7], the 0.7 ps rise in Fig. 1(a) seems to be associated with the charge separation between P700 and the nearest acceptor chlorophyll (the chlorophyll between P700 and A_0)[1]. The intrinsic time constant of the charge separation of PSI RC is thus shorter than those of purple bacterial and photosystem II RCs (~3 ps) [6,8].

5. Conclusions

This study has given evidence that the primary charge separation in PSI RC is one of the fastest electron transfer ever observed in photosynthetic systems. The quantum yield of the charge separation in PSI RC is maintained very high by the intrinsically subpicosecond charge separation reaction.

References

1 Klukas, O., Schubert, W.-D., Jordan, P., Krauß, N., Fromme, P., Tobias, H. & Saenger, W., *J. Biol. Chem.* **274**, 7361 – 7367 (1999).

2 Ikegami, I. & Katoh, S. *Biochim. Biophys. Acta* **376**, 588 – 592 (1975).

3 Kumazaki S., Kandori H., Petek H., Yoshihara K. & Ikegami I. *J. Phys. Chem.* **98,** 10335 - 10342 (1994).

4 Mathis, P., Ikegami, I. & Sétif, P. *Photosynth. Res.* **16**, 203 – 210 (1988)..

5 Iwaki, M., Mimuro, M. & Itoh, S. *Biochim. Biophys. Acta*, **1100**, 278 – 284 (1992).

6 Parson W. W. In *Protein Electron Transfer* Bendall D.S., ed. 125 - 160 (BIOS Scientific Publishers 1996), Chap. 6.

7 Moser, C. C., Keske, J. M., Warncke, K., Farid, R. S. & Dutton, P. L. *Nature* **355**, 796 – 802 (1992).

8 Klug D. R., Durrant J. R. & Barber J. *Phil. Trans. R. Soc. Lond. A*. **356**, 449 - 464 (1998).

Direct observation of the S_1 level of the carotenoid spheroidene using near-infrared femtosecond spectroscopy

Tomáš Polívka[1], Donatas Zigmantas[1], Jennifer L. Herek[1], James A. Bautista[2], Harry A. Frank[2] and Villy Sundström[1]

[1]Dept. of Chemical Physics, Lund University, Box 124, S-221 00 Lund, Sweden
[2]Dept. of Chemistry, University of Connecticut, U-60, 55 North Eagleville Road, Storrs, CT 06269-3060 USA

Abstract. In this work, we have determined the energy of the S_1 state of the carotenoid spheroidene. The energy of this state is 13,400 ± 90 cm^{-1} at both 293 K and 186 K, showing that there is no temperature-induced shift of the S_1 level. A discrepancy of about 800 cm^{-1} between the S_1 energy determined here and that obtained from previous fluorescence and resonance Raman measurements is explained in terms of the different conformational species co-existing in the S_1 excited state.

Introduction

The S_1 energy level of carotenoids is suggested to play an important role in various functions of carotenoids, yet the corresponding mechanisms are poorly understood due to the long-standing difficulty in assigning the location of this energy level. In carotenoids, the S_1 state is spectroscopically "dark", as the transition from the ground state S_0 to the first excited S_1 state is symmetry forbidden [1]. Recently, we presented a general method for measuring the S_1 state of carotenoids directly using methods of femtosecond spetroscopy [2]. Briefly, the experiment can be described as follows: a femtosecond pulse excites the carotenoid to the well-characterized S_2 state of carotenoid, which then relaxes to the forbidden S_1 state within approximately 200 fs. Given the time resolution of our laser system, we are able to detect excited state absorption from the S_1 state to the S_2 state by probing in the spectral region 5,000 – 12,500 cm^{-1}. Then, with knowledge of the S_1-S_2 and S_0-S_2 transitions in hand, we can easily locate the position of the S_1 level of carotenoids by simple subtraction $E(S_0\text{-}S_2) - E(S_1\text{-}S_2)$. Here we present an extension of these experiments to another biologically important carotenoid spheroidene, which is found in light-harvesting complexes and reaction centers of the purple bacterium *Rhodobacter sphaeroides*.

Experimental

Spheroidene, extracted and purified according to the method described in [3], was dissolved in n-hexane and contained in 2-mm optical path cuvette. The femtosecond spectrometer consists of an amplified Ti:Sapphire laser system that provides pulses with ~100 fs duration at a repetition rate of 5 kHz. The output was divided to pump two independent parametric amplifiers for generation of the excitation and probe pulses. The parametric amplifier producing the probe beam was computer driven enabling scanning of the probe beam in the spectral region 5,000 – 12,500 cm^{-1}. The S_2 state of spheroidene was excited to the low-energy edge of the absorption spectrum (20,400 cm^{-1} at room temperature and 20,100 cm^{-1} at 186 K) to avoid any contribution from vibrational relaxation within the S_2 state.

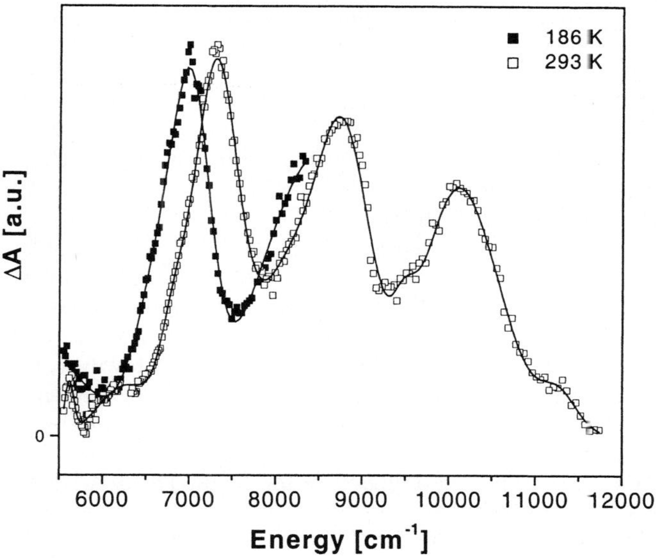

Fig. 1. Transient absorption spectra of carotenoid spheroidene recorded 3 ps after excitetion at 293 K and 186 K

Results and discussion

The transient absorption spectra of spheroidene recorded 3 ps after the excitation pulse at both room and low temperature are shown in Fig. 1. Both spectra clearly exhibit the three-peak structure characteristic of the S_2 state of carotenoids. The low temperature spectrum (only the profile of the first peak was measured) is shifted approximately 400 cm^{-1} to the lower energy, similar to the shift of the steady-state absorption spectrum, suggesting that the position of the S_1 level of spheroidene is not affected by temperature. The maxima of the red-most peak, which is believed to be the 0-0 transition from the S_1 state to the S_2 state, have been determined to be 7,310 cm^{-1} (293 K) and 6,980 cm^{-1} (186 K). Decay kinetics measured at the maximum of the this band exhibit single-exponential decays with decay times 8.1 ps (293 K) and 9.9 ps (186 K). The 8.1 ps decay measured at 293 K matches well with that determined via probing the S_1-S_N transition [3], supporting the conclusion that the observed transient absorption spectrum indeed reflects the S_1-S_2 transition. The S_1 lifetime of spheroidene has not been previously measured at low temperature, however, a slightly longer lifetime measured at 186 K is in agreement with measurements of the temperature dependence of the S_1 lifetime in β-carotene [4]. Simple subtraction (E_{S0-S2} - E_{S1-S2}), where E_{S0-S2} is given by steady-state absorption spectrum, provides the energy of the S_1 level of spheroidene: 13,400±100 cm^{-1} at both 293 K and 186 K.

The location of the S_1 energy level observed here by direct measurements is substantially lower (~800 cm^{-1}) than that predicted by the energy gap law and measured by fluorescence and resonance Raman spectroscopies [5,6]. The experimental results described above can be fully explained by a model in which the S_1 potential surface of spheroidene has a more complex structure corresponding to the presence of different conformers. These conformers appear in the S_1 state as a result of excitation of torsional modes by the S_2-S_1 internal conversion process. Then, after vibrational relaxation in the S_1 state, a statistical distribution of these conformers will co-exist in the S_1 state. Some of the conformations, however, deviate from the idealized C_{2h} symmetry of all-trans spheroidene, making the forbidden S_1-S_0 transition more allowed than that of the all-trans species. In such a case, the S_1-S_0 transition belonging to the conformers that deviate from the idealized C_{2h} symmetry will dominate in spectroscopic methods relying on the forbidden transition (such as fluorescence), as it is more allowed than the S_1-S_0 transition of the all-trans species. However, this is not the case for the near-infrared femtosecond approach presented here, because we explore the allowed S_1-S_2 transition. It is therefore natural to ascribe the S_1 energy of 13,400 cm^{-1} to the all-trans spheroidene, while the value of 14,200 cm^{-1} extracted from fluorescence experiments is the S_1 energy of twisted species that deviate from C_{2h} symmetry.

References

1. Frank, H. A. & Cogdell R. (1996) Photochem. Photobiol. 63, 257–264.
2. Polívka, T., Herek, J. L., Zigmantas, D., Åkerlund, H.-E. & Sundström, V. (1999) Proc. Natl. Acad. Sci. U.S.A. 96, 4914-4917.
3. Frank, H. A., Desamero, R. Z. B., Chynwat, V., Gebhard, R., van der Hoef, I., Jansen, J. F., Lugtenburg, J., Gosztola, D. & Wasielewski, M. R. (1997) J. Phys. Chem. A 101, 149-157.
4. Wasielewski, M. R., Johnson, D. G., Bradford, E. G. & Kispert, L. D. (1991) J. Chem. Phys. 91, 6691-6697.
5. Sashima, T., Shiba,, M., Hashimoto, H., Nagae, H. & Koyama, Y. (1998) Chem. Phys. Lett. 290, 36-42.
6. Fujii, R., Onaka, K., Kuki, M., Koyama, Y. & Watanabe, Y. (1998) Chem. Phys. Lett. 288, 847-853

Fs-transient Absorption and Fluorescence Upconversion after Two-Photon Excitation of Carotenoids in Solution and in LHC II

P.J. Walla, P.A. Linden, and G.R. Fleming

Department of Chemistry, University of California Berkeley, and Physical Biosciences Division, Lawrence Berkeley National Laboratory, Berkeley CA 94720, USA
E-mail: grfleming@lbl.gov

Abstract: We present the results of time-resolved fs-transient absorption and fluorescence upconversion measurements after two-photon excitation of the forbidden S_1 states of β-Carotene, Lutein and the energy transferring carotenoids in LHC II. The direct excitation of the carotenoid S_1 state enabled us to measure the subsequent dynamics on a fs time scale without any interference from higher excited states and to determine the two-photon excitation spectrum of the S_1 state.

1. Introduction

Carotenoids (Cars) such as Lutein and β-Carotene are important photoprotective and light harvesting pigments in the light-harvesting proteins of higher plants and green algae [1]. However, many aspects of the Cars are still poorly known, because their first excited singlet state, S_1, is optically dipole-forbidden and has a significant double-triplet character. To understand this, the electronic states of carotenoids are often described by analogy to polyenes (Fig. 1): Their ground state (S_0) and first excited state (S_1) both posses A_g symmetry, their second excited state (S_2) possesses B_u symmetry in the idealized C_{2h} point group. Therefore only the $S_0 \leftrightarrow S_2$ transition is optically dipole-allowed and contributes significantly to the absorption spectrum of Cars. However, after excitation into S_2 Cars undergo very rapid (~100-200 fs) internal conversion (IC) to S_1 [2].

An elegant way to directly populate S_1 is by two-photon excitation (TPE) because transitions between states of A_g symmetry are two-photon allowed. In contrast to one-photon techniques [3,4], direct excitation of the Car S_1 state enables the measurement of subsequent dynamics on a fs time scale without interference from higher excited states and the estimation of the S_1 state energy.

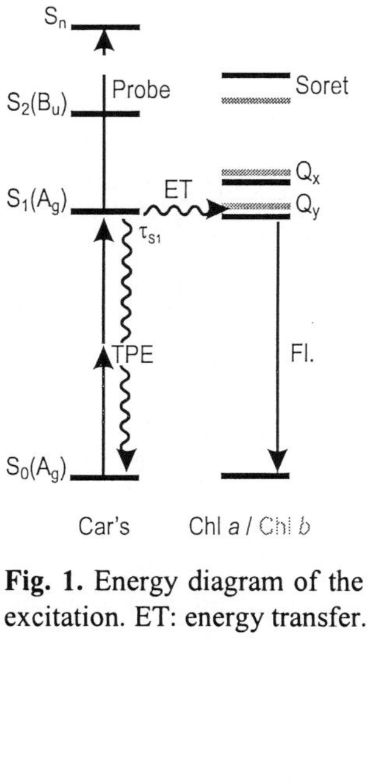

Fig. 1. Energy diagram of the carotenoids and chlorophylls in LHC II. TPE: Two-photon excitation. ET: energy transfer. Fl.: Fluorescence.

If there is energy transfer (ET) from S_1 in light-harvesting proteins, the chlorophyll (Chl) fluorescence can be used to monitor the TPE of the Cars. The best way to detect the S_1 state of non-energy transferring carotenoids is via its very strong excited-state absorption, because the S_1 state does not fluoresce. In this work we present TPE pump-probe and TPE fluorescence upconversion experiments to explore the S_1 state of Cars in solution and proteins.

2. Results and Discussion

In Fig. 2a the time dependence of the TPE pump probe signal is shown for β-Carotene and Lutein. A monoexponential fit to the two-photon data of β-Carotene gave S_1 lifetimes known from the literature [5]. To assure that we really observe signals, S_{TPE}, due to TPE we measured the power-and the polarization dependencies. We observed a power-dependence of $S_{TPE} \propto P^{2.2}$. In TPE spectroscopy, the signal is different for linear, circular or any elliptical polarization of the pump-beam. The polarization ratio, $\Omega = S_{TPE}^{circ.} / S_{TPE}^{lin.}$, is dependent on the symmetries of the excited and ground state wave functions. We found a value of $\Omega=0.84\pm0.07$, which is identical to the value of $\Omega\sim0.8$ given in the literature for all-trans diphenyloctatetraene and diphenylhexatriene [6].

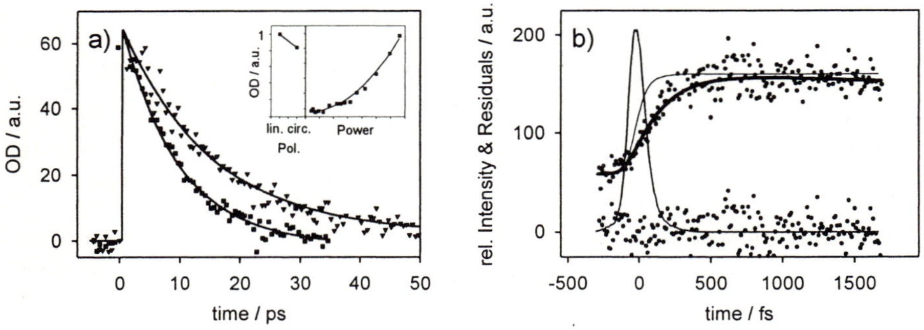

Fig. 2. a) Time dependence of excited state absorption (λ_{det}=550 nm) of β-Carotene in octane (squares) and Lutein in 1-octanol (triangles) after TPE (λ_{exc}=1310 nm). Solid lines are monoexponential fits (β-Carotene: 9±0.2 ps; Lutein: 15±0.5 ps. The insets show the polarization dependence (Ω=0.84±0.07) and the power dependence of excited state absorption. b) Upconversion data (λ_{det}=685 nm) after TPE (λ_{exc}=1200 nm) of LHC II (circles), along with exponential fit (thick line); residuals (circles), instrument response function (medium line) and calculated curve corresponding to instantaneous rise (thin line).

Fig. 2b shows the upconversion data and fits for the Chl Q_y emission after TPE of LHC II at 1220 nm. We deconvoluted the data with a triexponential function and found a 250±50 fs rise and two decay times of 3 ps (~15 %) and >100 ps (~85%). Although the first 1-2 ps can be affected by spectral evolution, IC of Chls or ET between Chls, the ET time constant, τ_{ET}, cannot be much slower than the observed rise, because the ET is the limiting factor for the observation of any Chl-fluorescence. The very fast Car $S_1 \rightarrow$Chl ET is quite remarkable because usually time constants in the ps range are observed. However, in the particular case of LHC II the fast ET can be understood by the fact, that its carotenoids are in van der Waals contact with a large number (~4) of acceptors (Chls).

In Fig. 3a and 3b the TPE pump probe signal of β-Carotene and Lutein is shown as a function of the TPE wavelength. The result of this measurement is a direct TPE spectrum of the Car S_1 state which enables us to estimate the 0-0-transition of β-Carotene (~14300 ± 300 cm^{-1}) and Lutein (~14350 ± 350 cm^{-1}). In Fig. 3c the TPE spectrum of LHC II, observed by monitoring the chlorophyll fluorescence, is plotted. The onset of this spectrum is ~ 15100 ± 300 cm^{-1}.

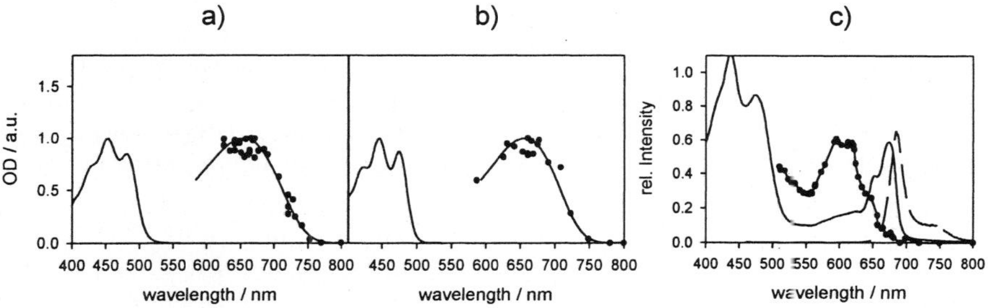

Fig. 3. Absorption spectra (solid lines) and TPE spectra (dots with solid lines) of a) β-Carotene, b) Lutein and c) LHC II of *Chlamydomonas Reinhardtii* (dashed line = Fl.)

The last result has very important implications for the role of these carotenoids in light-harvesting proteins of plants. First of all, the data of Fig 3c provides clear evidence that there is a Car $S_1 \rightarrow$Chl ET in LHC II. However, it can be seen from the comparison with Fig. 3a and 3b that the energy transferring Cars states in LHC II have significantly higher energy than the most abundant (~ 60 % in LHC II) carotenoid, Lutein (10 double bonds). If the 0–0-transition of Lutein in LHC II is not shifted to very high energies, either the major part of Car $S_1 \rightarrow$Chl ET occurs predominatly from higher vibronic S_1 states or the ET is due to Neoxanthin or Violaxanthin (~30-40% in LHC II, 9 double bonds).

Acknowledgements. This work was supported by the DOE (DE-AC03-76SF 00098). P.J.W. is grateful for support from the Deutsche Forschungsgemeinschaft.

References

1 R. van Grondelle, J. P. Dekker, T. Gillbro, and V. Sundström, *Biochim. Biophys. Acta* 1187, 1, 1994.
2 A. N. Macpherson, and T. Gillbro, *J. Phys. Chem. A* 102, 5049, 1998.
3 T. Polívka, J. L. Herek, D. Zigmantas, H. E. Åkerlund, and V. Sundström, *Proc. Natl. Acad. Sci.* 96, 4914, 1999.
4 K. Onaka, R. Fujii, H. Nagae, M. Kuki, Y. Koyama, and Y. Watanabe, *Chem. Phys. Lett.* 315, 75, 1999.
5 V. Chynwat, and H. A. Frank, *Chem. Phys.* 194, 237, 1995.
6 G. R. Holtom, and W. M. McClain, *Chem. Phys. Lett.* 44, 436, 1976.

Quantitative modeling of ultrafast energy transfer dynamics in the photosynthetic apparatus

Xanthipe J. Jordanides, Gregory D. Scholes, Graham R. Fleming

Department of Chemistry, University of California Berkeley and Physical Biosciences Division, Lawrence Berkeley National Laboratory, Berkeley, California 94720-1460
E-mail: GRFleming@lbl.gov

Abstract. The dynamics of energy transfer processes in photosynthetic reaction centers and the light harvesting antenna LH2 are elucidated quantitatively using a new prescription for applying Förster theory to complex molecular aggregates. We conclude that the design of highly efficient light harvesting antenna systems generally cannot be predicted by pairwise Förster theory.

1. Introduction

The remarkable efficiencies of solar energy conversion attained by photosynthetic organisms derives in part from the robustness of the light-harvesting (e.g. LH2) and trapping apparatus (the reaction center). Despite many wide-ranging studies, truly quantitative elucidation and prediction of function for such multichromophoric assemblies has only just been attained. This is because, although Förster theory [1] for electronic energy transfer (EET) has proved successful for predicting rates of energy transfer based on the overlap of donor emission and acceptor absorption spectra, we have found that when several donor and/or acceptor chromophores are arranged in a confined geometry, the pairwise energy hopping model breaks down [2]. High resolution structural models of reaction centers and light-harvesting complexes, combined with advances in nonlinear spectroscopy and electronic structure calculations have now enabled us to integrate all the relevant data to examine quantitatively the *mechanism* of energy transfer in any complex multichromophoric system.

The concepts underpinning our approach are (i) explicit separation of electronic and nuclear factors, (ii) a proper calculation of the electronic couplings, and (iii) a detailed calculation of the nuclear spectral overlap factor. Our result is that the rate of energy transfer from donor (D) states δ to acceptor (A) states α is given by eq 1,

$$k = \frac{2\pi}{\hbar}\left\langle \int_0^{\infty} \mathrm{d}\varepsilon \sum_{\delta,\alpha} P_\delta |V_{\delta\alpha}|^2 J_{\delta\alpha}(\varepsilon) \right\rangle \qquad (1)$$

where $V_{\delta\alpha}$ is the electronic coupling between D and A and P_δ is a normalized population factor. The angle brackets denote that an ensemble average is taken to account for static disorder in the monomer site energies. The spectral overlap $J_{\delta\alpha}$ is defined in terms of donor and acceptor densities of states (DOS) [2].

2. Results and Discussion

Energy transfer in reaction centers. The photosynthetic reaction center (RC) of purple bacteria is a pigment-protein complex comprised of a bacteriochlorophyll dimer called "the special pair" (P), two monomeric "accessory" bacteriochlorophylls (B), and two bacteriopheophytins (H). Upon excitation of H or B energy is transferred to P on an ultrafast time scale. An experiment on *Rb. sphaeroides* resolved EET times from B to P of 120 fs at 200K and 160 fs at 85 K, indicating a weak temperature dependence of the energy transfer.[3] It has proved very difficult to reproduce these rates theoretically.[4]

Although construction of a complete energy transfer model in the reaction center poses a great theoretical challenge, we believe our treatment incorporates all the essential details for calculating EET rates between the RC pigments: pigment arrangement, electronic couplings between pigments, electron phonon coupling, energetic disorder, and vibrational modes. Through our reformulated Förster theory of Eq. 1, we find that the principal acceptor state in energy transfer from B to P is the upper exciton state of the special pair dimer, P_+. The calculated energy transfer rate from B to P at 298 K for *Rb. sphaeroides* is 154 fs, which compares favorably to the experimental time of 116 fs [3]. At 77 K and 15 K, the EET times are 147 fs and 144 fs, respectively. Hence, we have calculated rates that predict the weak temperature dependence evident in experiments, quantitatively match experimental values, and clarify the mechanism of the EET.

Energy funneling in antennae. Light absorbed by the B800 ring of the peripheral antenna LH2 of *Rb. sphaeroides* is transferred to the B850 ring within 650 fs at 298 K and 1.2 ps at 77 K. Förster theory provides an unsatisfactory estimate of this time scale, and the insensitivity to temperature. Our approach to model B800 to B850 EET, described in reference [2], correctly predicts the time scales and temperature dependence of the energy transfer. We find that disorder plays an important role in accelerating energy transfer and producing an almost temperature independent rate.

3. Conclusions

We believe that the calculations reported in this work capture the essential details and principles of the light harvesting mechanisms in the RC and LH2 of purple bacteria. The calculations of the energy transfer times also reinforce the quantitative success of this modeling. The pictures of RC and LH2 that emerge are of complex molecular systems whose function cannot be readily predicted by examining the components individually. Instead, the true behavior emerges only after the whole system is included in the model. Table I on the following page compares and contrasts the two photosynthetic apparatuses.

Table 1. Summary of the energy transfer concepts in the photosynthetic apparatus.

Purple bacterial reaction center (RC)	
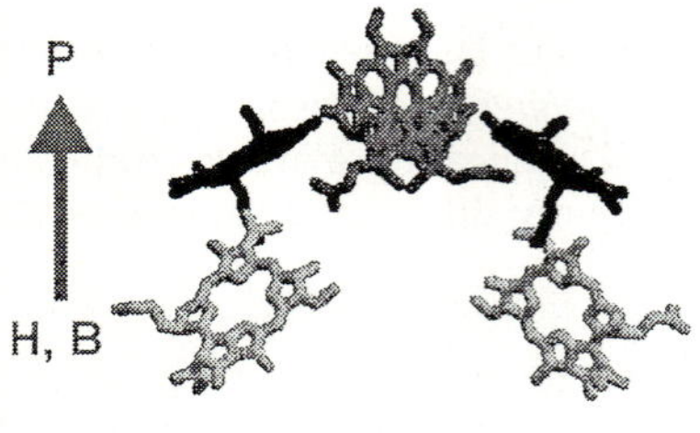	Multiple coupled donors and acceptors. Electronic couplings reflect pigment arrangement rather than electronic spectra. Thus Forster theory fails to capture the essential dynamics. Spectral overlap between DOS permits weakly allowed transitions(e.g. P_+) to have significant overlap. Small role played by disorder.
Purple bacterial light-harvesting complex (LH2)	
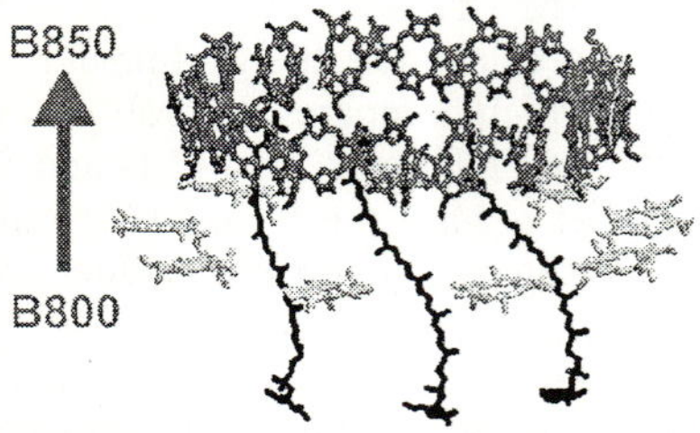	Strongly interacting acceptors, weakly interacting donors. Complex acceptor DOS available as an energy sink. Important role played by disorder.

Acknowledgments. This work was supported by the Director of the U. S. Department of Energy (Office of Science, Office of Basic Energy Sciences, Chemical Sciences Division) under Contract No. DE-AC03-76SF00098.

References

1 Th. Förster, "Zwischenmolekulare Energiewanderung und Fluoreszenz," Ann. Phys. **6**, 55-75 (1948).

2 G. D. Scholes and G. R. Fleming, "On the mechanism of light-harvesting in photosynthetic purple bacteria: B800 to B850 energy transfer," J. Phys. Chem. **104**, 1854-1868 (2000).

3 R. J. Stanley, B. King, and S. G. Boxer, "Excited state energy transfer pathways in photosynthetic reaction centers .2. Heterodimer special pair," J. Phys. Chem. **100,** 12052-12059 (1996).

4 J. M. Jean, C. -K. Chan, and G. R. Fleming, "Electronic Energy Transfer in Photosynthetic Bacterial Reaction Centers," Isr. J.Chem. **28,** 169-175 (1988).

Primary Reactions of Sensory Rhodopsins: Two Proteins with Vastly Different Dynamics

I. Lutz[1], A. Sieg[1], I. Boche[2], M. Otsuka[2], D. Oesterhelt[2], J. Wachtveitl[1], W. Zinth[1]

[1] Sektion Physik, Ludwig-Maximilians-Universität München, Oettingenstr. 67, 80538 München, Germany
E-mail: Wolfgang.Zinth@physik.uni-muenchen.de
[2] Max-Planck-Institut für Biochemie, am Klopferspitz 18a, 82152 Martinsried, Germany

Abstract. The first steps in the photocycles of the photoreceptors sensory rhodopsin I and II from archaea have been investigated. The formation of red-shifted photoproducts with widely different kinetics was observed.

1. Introduction

Halobacterium salinarum contains a series of retinal based photoactive chromoproteins. The ion pumps bacteriorhodopsin (BR) and halorhodopsin (HR) allow photosynthetic growth and defined membrane potentials whereas the two sensory rhodopsins sSRI and sSRII control phototaxis and give the bacteria a kind of "color-vision" [1, 2]. In the past, the ion pumps have attracted most attention. It is only recently that sufficient sample quantities of the sensory rhodopsins are available for spectroscopic investigations. In this paper we present the first study of the primary reaction dynamics of these two different photosensors.

2. Experimental Methods

The two sensory rhodopsins were isolated and purified similar to the procedure given in [3]. A thin fused silica cuvette allowed exchange of the excited sample volume between two laser shots. The laser system was operated at 1 kHz, excitation pulses were generated by a noncollinear optical parametric amplifier (NOPA) at λ=480 nm (sSRII) and λ=580 nm (sSRI and BR) as described in [4]. White-light probing pulses were generated in a sapphire plate. Multichannel (cross-correlation width $\tau_{cl} \approx 90$ fs) and single channel detection ($\tau_{cl} \approx 30$ fs) have been used.

3. Results and Discussion

The two sensory rhodopsins exhibit different absorption spectra (Fig. 1). While sSRI has a BR-like spectrum peaking at 585 nm, sSRII is strongly blue shifted with the absorption peak at 490 nm. Fig. 2 shows the results from transient absorption measurements performed in the blue part of the spectrum: in bacteriorhodopsin (BR) strong excited state absorption decaying with 500 fs and formation of the

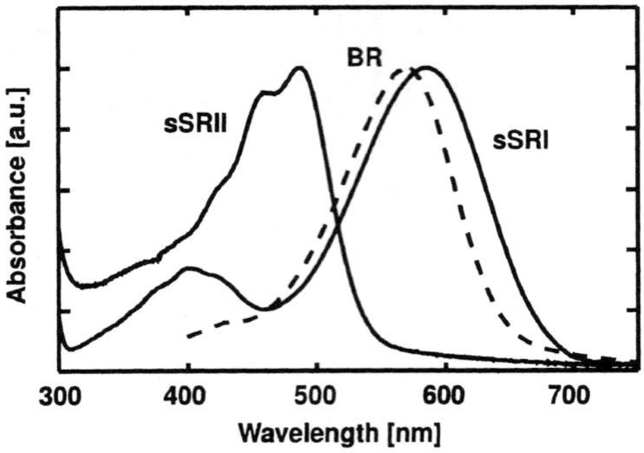

Fig. 1. Absorption spectra of bacteriorhodopsin (dashed line) and sensory rhodopsin I and II from *Halobacterium salinarum* (solid lines).

product state J [5, 6], are seen at a probing wavelength of 496 nm. In sSRI we again observe strong excited state absorption which decays (apart from a weak ~150 fs contribution) much slower with 4 ps and 33 ps. The same time constants dominate the whole observed spectral region. The data on sSRI can be summarized as follows: After a first S_1 reaction in the 150 fs time domain the excited electronic state decays with the two time constants of 4 ps and 33 ps into a red-shifted photoproduct. The remaining absorbance change at late delay times is very weak. This can be explained if the product state has an absorption spectrum similar to the ground state spectrum of sSRI or if the quantum efficiency for product formation is small.

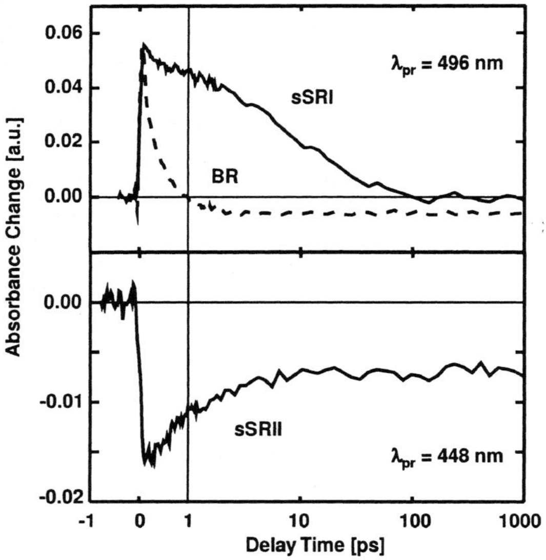

Fig. 2. Transient absorbance changes of BR (dashed line) and sSRI (solid line) at 496 nm (top). Transient absorbance changes of sSRII at 448 nm (bottom).

The reaction dynamics of sSRII is completely different: at 448 nm in the blue wing of the ground state absorption we still find a reduced absorbance which decays with $\tau_1 = 400$ fs and $\tau_2 = 4$ ps to a long lasting absorption bleach. The frequency dependence of the amplitudes (decay associated spectra see Fig. 3) in combination with single channel gain measurements at 890 nm suggests the following reaction model: after optical excitation and initial motions on the S_1 potential surface the initial red-shifted product state "sSRII$_J$" is formed within 400 fs. The secondary transition (within $\tau_1 = 4$ ps) corresponds to a state "sSRII$_K$" with a weaker red-shift. This state persists for the rest of the observation time until $\tau_D = 1$ ns.

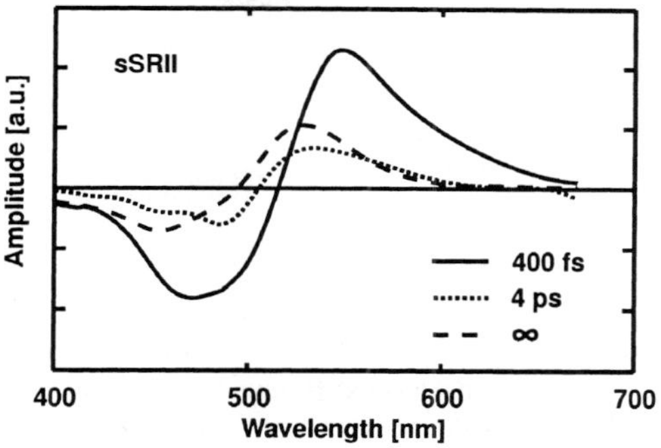

Fig. 3. Decay associated spectra of sSRII resulting from a multiexponential global fitting procedure.

4. Conclusions

The experiments present for the first time the primary reactions of the halobacterial photosensors SRI and SRII. In both systems there is a fast formation of a red-shifted photoproduct state. While in sSRI this reaction takes place within 4 and 33 ps, a faster initial step is observed in sSRII which closely resembles the primary reaction in BR [7]. It is interesting to note, that the reaction of sSRI has the same structure –namely slow and biexponential decay of the excited electronic state- as HR and mutants of BR, where the polar amino acid D85 has been replaced by uncharged residues. The common feature of these retinal proteins, the lack of a negative charge at position 85 is repeated in sSRI, which contains a protonated (i.e. uncharged) residue D85 at pH6.0 used in our experiments. The results indicate very similar underlying reaction mechanisms in the different retinal containing proteins yielding important information for a unifying theoretical description.

References

1 D. Oesterhelt, Curr. Opin. Struct. Biol. **8**, 489, 1998.
2 J. L. Spudich, D. N. Zacks, R. A. Bogomolni, Isr. J. Chem. **55**, 495, 1995.
3 M. P. Krebs, E. N. Spudich, J. L. Spudich, Protein Expres. Purif. **6**, 780, 1995.
4 T. Wilhelm, J. Piel, E. Riedle, Opt. Lett. **22**, 1474, 1997.
5 H.-J. Polland, M. A. Franz, W. Zinth, W. Kaiser, D. Oesterhelt, Biophys. J. **49**, 651, 1986.
6 K. C. Hasson, Feng Gai, P. A. Anfinrud, Proc. Natl. Acad. Sci. USA **93**, 15124, 1996
7 J. Dobler, W. Zinth, W. Kaiser, Chem. Phys. Lett. **144**, 215, 1988.

Femtosecond spectroscopy and model calculations for an understanding of the primary reaction in bacteriorhodopsin

W. Zinth[1], A. Sieg[1], P. Huppmann[1], T. Blankenhorn[1], D. Oesterhelt[2] and M. Nonella[1]

[1] Sektion Physik, Ludwig-Maximilians-Universität München, Oettingenstr. 67, 80538 München, Germany
E-mail: zinth@physik.uni-muenchen.de
[2] Max-Planck-Institut für Biochemie, 82152 Martinsried, Germany

Abstract. Detailed investigations on the initial reaction dynamics of native and modified bacteriorhodopsin combined with extensive model calculations give new insights into primary reaction and optimization of this important proton pump.

Bacteriorhodopsin (BR) was the first molecule, where reactive motions on the S1-surface could be followed in realtime [1, 2]. These early experiments have demonstrated that the optically prepared Franck-Condon state is rapidly left [2] and that a fast (~200 fs) initial decrease of short wavelength stimulated emission occurs prior to the 500 fs transition to the groundstate [1]. In the meantime additional studies on native and modified BR samples gave important additional information showing that: (i) there is no finite build-up time for the long wavelength stimulated emission and that (ii) BR containing locked retinal molecules - unable to undergo photoisomerization - show similar initial kinetics [3, 4]. In combination with new theoretical concepts for the primary isomerization, this calls for a more detailed experimental and theoretical study of the primary reaction [5, 6]

In this paper we studied BR from wild type (WT-BR) of *Halobacterium salinarum* and the mutated system BR-D85T [7], where the aspartate D85 is replaced by a threonine changing strongly the electrostatics in the surrounding of the Schiffs base and modifying the later reaction steps [8]. The experiments were performed with independently tunable light pulses generated by two non collinear parametric amplifier (NOPA) [9] yielding a cross correlation of below 30fs.

Results from both samples measured at a probing wavelength of 890 nm and an excitation wavelength of 570 nm are shown in Fig. 1. We find a fast instantaneous rise of stimulated emission for 70% of the total amplitude. The remaining part of this long wavelength stimulated emission is build up within 50 fs. During the delayed rise one observes wavepacket motion with a high frequency of $\nu = 23$ THz, ($\nu/c = 770$ cm^{-1}). In WT-BR the signal decays later on with a $\approx$ 500 fs time constant. For the mutant D85T the same wavepacket motions are observed in the signal rise. However, the decay of the stimulated emission is much slower. The signal relaxes biexponentially with the two time constants of $\approx$ 3 ps (10%) and $\approx$18 ps (90%) (see Fig. 2).

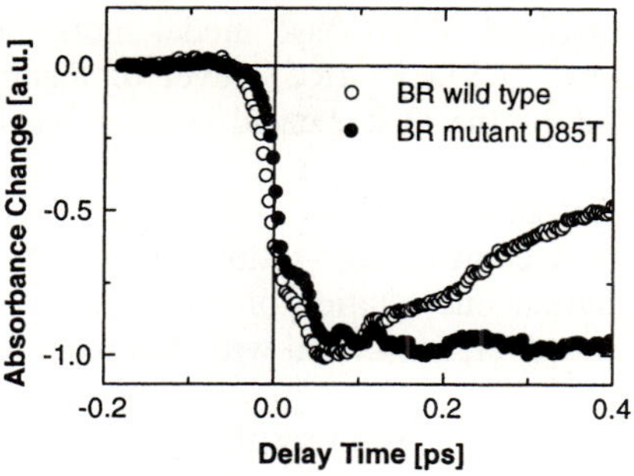

Fig. 1. The initial reaction dynamics of wild type BR (open circles) and of the mutant D85T (points).

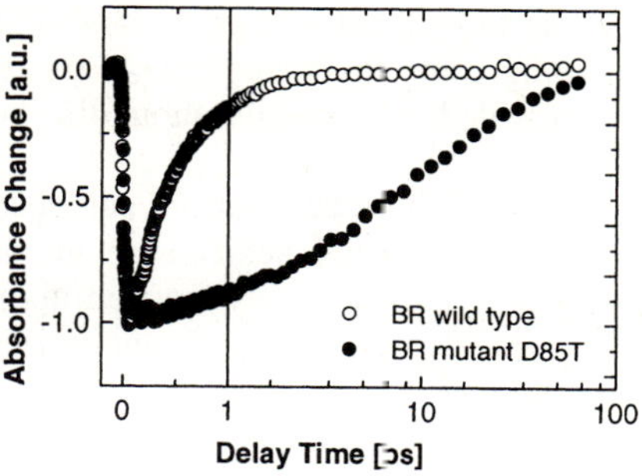

Fig. 2. Decay of the stimulated emission of wild type BR (open circles) and of the mutant D85T (points).

Probing in the spectral range λ = 660-730 nm (data not shown) we observe in both samples short lived (decay time ≈ 150 fs) stimulated emission and a pronounced wavepacket motion with frequencies up to ≈800 cm^{-1}. Even if the two BR samples show very different temporal dependence at later times their initial behavior during the first 200 fs is very similar. A qualitative interpretation of the primary reaction is given in Fig. 3. It will be supported by the calculation presented below: Optical excitation promotes the system to the Franck-Condon region (FC) on the excited state potential surface. From here the S_1 reaction proceeds along different coordinates (for simplicity we show only two). The experimentally observed 150 fs reaction together with highly damped vibrational motions suggests a fast downhill motion along high frequency coordinates. (The model calculation supports this interpretation since the isomerizational motion is only possible after a relaxation of C-C distances.) All these motions promote the system towards a valley from where the further motion along low frequency (torsional) coordinates proceeds. During this motion the transition to the photoproduct with the 500 fs (WT-BR) occurs presumably via a conical intersection. The slow reaction in BR-D85T can be understood within this picture by a modified potential in this range of the coordinate space, changing the accessibility or the existence of the conical intersection. This interpretation is strongly supported by high level model calculations [10].

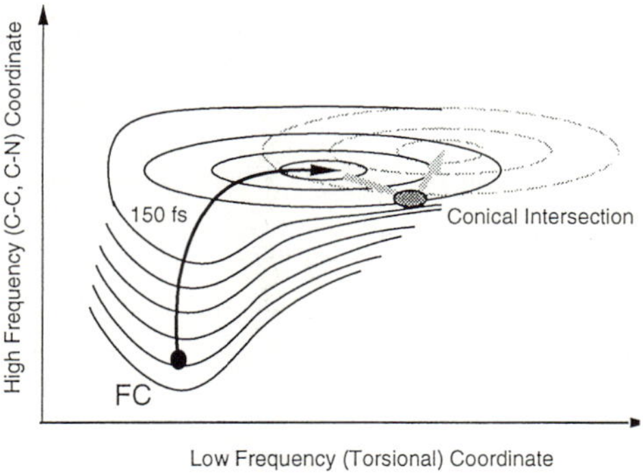

Fig. 3. Scheme of the potential energy surface for the initial motions after optical excitation to the Franck-Condon (FC) region. For simplicity only one and one low frequency coordinate are shown.

These new ab initio calculations of a protonated Schiff base model molecule containing four conjugated double bonds at the CASSCF/MRCI level of theory have provided new details about the behavior of the first excited state of such molecules.

The results of these calculations can be summarized as follows:

- For a protonated Schiff base in vacuum a conical intersection between the states S_1 and S_0 is only reached upon simultaneous rotation of the bonds C_{13}-C_{14} and C_{14}-C_{15}. A significant barrier is, however, associated with this reaction coordinate.
- Adding a counter ion in the vicinity of the -NH^+- group results in a conical intersection also along the reaction coordinate of a single isomerization. The barrier for this reaction coordinate is significantly smaller than the one for a double isomerization.
- The small barrier connected with such a single isomerization is lost after relaxation of the bond lengths of the C-C and C-N bonds in the excited state. The reaction coordinate for a double isomerization is still associated with a large barrier.

These new results demonstrate, that electrostatic interactions not only affect the **shape** of potential energy surfaces but also the **character** of the wavefunctions. The distribution of charged groups as found in the wild type of Bacteriorhodopsin is most likely optimized for allowing a fast all-trans $\rightarrow$ 13-cis isomerization. A perturbation, for example introduced through a point mutation as in the case of the D85T mutant, can modify the region of the conical intersection and thereby reduces the probability for the system to reach this region. Nature itself has used electrostatics to tune both, the absorption properties and the reaction dynamics. The different retinal proteins of *Halobacterium salinarum* are prominent examples.

References

1 J. Dobler, W. Zinth, W. Kaiser, and D. Oesterhelt, Chem. Phys. Lett. 144, 215, 1988.

2 R. Mathies, B. Cruz, W. Pollard, and C. Shank, Science **240**, 777, 1988.

3 T. Ye, N. Friedman, G. H. Atkinson, M. Sheves, M. Ottolenghi, and S. Ruhman, J. Phys. Chem. B **103**, 5122, 1999.

4 K. C. Hasson, F. Gai, and P. A. Anfinrud, Proc. Natl. Acad. **93**, 15124, 1996.

5 K. Schulten, W. Humphrey, I. Logunov, M. Sheves, and D. Xu, Israel J. Chem. **35**, 447, 1995.

6 M. Garavelli, F. Negri, and M. Olivucci, J. Am. Chem. Soc. **121**, 1023, 1999.

7 J. Tittor, U. Schweiger, D. Oesterhelt, and E. Bamberg, Biophys. J. **67**, 1682, 1994.

8 S. Logunov, M. El-Sayed, L. Song and J. Lanyi, J. Phys. Chem. **100**, 2391, 1996.

9 T. Wilhelm, J. Piel, and E. Riedle, Opt. Lett. **22**}, 1494 1997.

10 M. Nonella, J. Phys. Chem. to be published 2000.

Resolving the primary dynamics of bacteriorhodopsin, and Locked analogs in the reactive excited state.

Tong Ye[1], Bixue Hou[1], Erez Gershgoren[1], Michael Ottolenghi[1], Noga Friedman[2], Mudi Sheves[2], Sanford Ruhman[1].

[1] Department of Physical Chemistry, and the Farkas Center for Light Induced Processes, the Hebrew University, Jerusalem 91904, Israel. E-mail: sandy@fh.huji.ac.il.
[2] Department of Organic Chemistry, The Weizmann Institute of Science, Rehovot 76100, Israel. E-Mail: cosheves@wicc.weizmann.ac.il

Abstract. Stimulated emission from excited bacteriorhodopsin is compared with that from isomerize. Ultrafast Stokes shifting of emission, and coherent vibrations in the excited state are observed in both for the first time. Biological relevance of coherent motions is tested by comparing to similar motions in free protonated Sciff base (PSB) of retinal.

Deciphering the primary light induced events in Bacteriorhodopsin (BR), and pinpointing the stage of isomerization known to take place in the retinal prosthetic group, have been the subject of active investigation. Light absorption by this protein, which pumps protons across the membrane of photosynthetic bacteria, has been demonstrated to unleash dramatic subpicosecond spectral changes, consisting of the appearance of a near IR (NIR) stimulated emission band, and a blue absorption band, peaking at 850 and 460 nm respectively [1]. These features decay concertedly within ~1
presumably on the ground electronic surface.

Recent studies in this lab provided compelling evidence that *all-trans* $\rightarrow$ *13-cis* retinal isomerization is not the primary consequence of light absorption in BR. Comparing the ultrafast spectroscopy of BR with that of an artificial pigment BR5.12 which is prohibited synthetically from isomerizing around the active double bond, demonstrated identical spectral changes in both (up to ~200 fsec). The absorption and emission bands associated with the reactive excited state rose instantaneously with photoexcitation in both proteins [2]. Despite this, emission in both proteins is strongly
BR is absent in BR5.12, suggesting that this spectral change mirrors the stage of isomerization. No spectral evolution was observed in the transient bands of emission and absorption from the reactive excited state throughout the duration of its existence. The objective of this investigation is: a) To fully resolve the rise of

dynamics of rapid motion out of the impulsively excited Franck Condon state, with superior time resolution. b) To seek evidence for coherent nuclear dynamics on the reactive excited potential surface. And c) to compare results from both

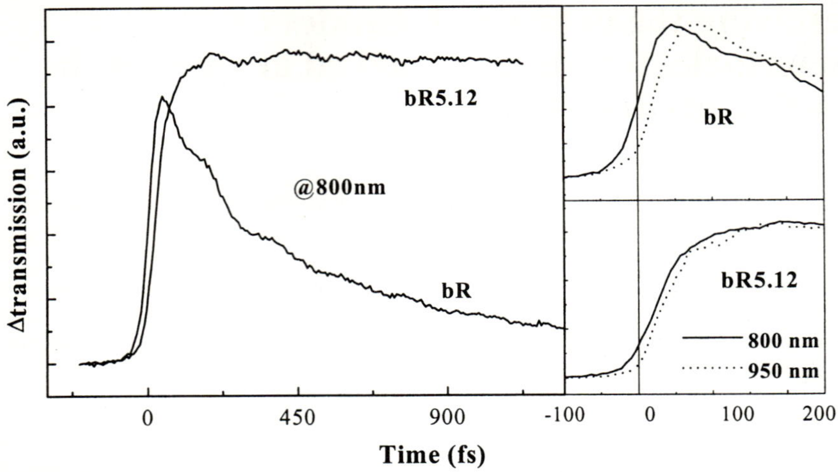

Figure 1. Stimulated emission scans of native and locked bR. See text for details

proteins to reveal involvement of torsion along the active double bond in the observations reported in a) and b).

A home made multipass amplified Ti;sapphire laser system coupled to a commercial OPA (TOPAS) producing ~20 fsec pulses throughout the visible was 550 nm, and probed at 800, and 950 nm. The zero of time was determined in the sample cell before and after each run by an OKE scan in water. PSB in the *all-trans* configuration was dissolved in ethanol, and flowed through the same 100 μm sample cell used for the BR experiments. PSB is pumped at 400nm, and results are presented for 600 nm probing pulses.

Magic angle transient transmission scans at 800 and 950 nm are presented in figure 1, both for BR and BR5.12. All the scans are normalized to make the rapid stage of signal rise roughly equal. In the left panel stimulated emission scans are shown at 800 nm for both proteins. While in BR the emission decays rapidly to zero, in BR5.12, following a rapid rise, a secondary slower rise component is clearly discernable over a period of a few hundred fsec, after which further changes are arrested. In both proteins, a periodical oscillation component in the transmission is clearly observable. Oscillations were detected in emission scans at 950 nm as well, but were much weaker. The differential rise in probe transmission over the first few hundred fsec is depicted in the right panel of figure 1. The rise of stimulated emission at 950 nm, is clearly delayed by 10-20 fsec with respect to the related rise at 800 nm in both. For BR, the rise of stimulated emission at 800nm is almost perfectly symmetrical with respect to t=0, indicating that it arises nearly instantaneously with light absorption. In BR5.12 the delay is less impressive, and even at 800 nm the rise is not symmetric about t=0. None the less the observed relative delay at the two frequencies was checked a number of times and found to be significant and reproducible.

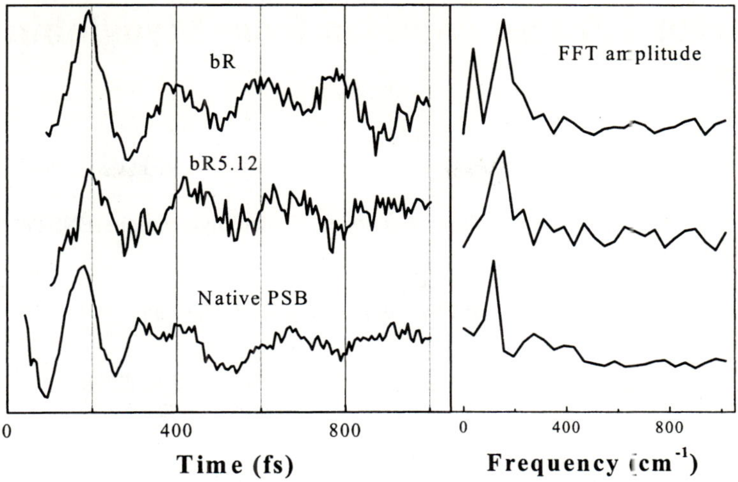

Figure 2. Spectral modulations obtained from 800 nm emission scans in BR and BR5.12, along with those obtained from PSB at 600 nm. Fourier analysis is presented to right.

The first conclusion we draw from the results presented [3] is that appearance of emission at the red edge of the band is delayed with respect to time zero, and its rise at 800 nm, for BR and BR5.12 as shown in figure 1. This result demonstrates that time consuming structural reorganization leads to
excited state emission. It also shows that the structural relaxation is extremely rapid, and is not dependent on torsion around the $_{13}=C_{14}$ double bond.
The second conclusion is that impulsive excitation of the native and locked

evidenced by the spectral modulations. Preliminary study of the isolated PSB was aimed to assess whether these coherences involve the protein and the prosthetic group, and whether they are related directly to the biological activity of the protein. The results clearly implicate vibrations of the pigment primarily. The residuals and their Fourier analysis does however show that the average period of modulations following the intense first feature in BR is markedly shorter than that in BR5.12, or native PSB (notice the π phase shift in PSB modulations). Also, this
vibrational motion. Future work will be dedicated to appreciating whether these distinctions are related to the possible concurrent isomerization being underway in the BR, and not in the other samples.

References

1 G. G. Kochendoerfer and R. A. Mathies, Isr. J. Chem. **35**, 211(1995).

2 T. Ye, N. Friedman, Y. Gat, G. H. Atkinson, M. Sheves, M. Ottolenghi, S. Ruhman, J. Phys. Chem. B, **103**, 5122 (1999).

3 T. Ye et.al., Chem. Phys. Lett., **314,** 429-434 (1999).

Coherent infrared emission from Myoglobin crystals

Marie-Louise Groot[1,*], Marten H. Vos[1], Ilme Schlichting[2] and Jean-Louis Martin[1]

[1] INSERM U451, Laboratoire d'Optique Appliquee, Ecole Polytechnique-ENSTA, 91761 Palaiseau Cedex, France
E-mail: vos, martin@enstay.ensta.fr
[2] Max Planck Institute for Molecular Physiology, Dortmund, Germany

Abstract. We have measured coherent infrared emission from myoglobin crystals liganded with CO.

1. Introduction

Coherent infrared emission interferometry is a new spectroscopic technique which directly measures the electrical field emitted by oscillating dipoles in the infrared, by letting it interfere with a short reference-electrical field. In this way one can determine frequency, phase, dephasing time and polarization of the oscillating dipoles

The reference electrical field is generated by optical rectification [1], which results in our case in quasi single-cycle pulses of ~30 fs with a corresponding spectrum extending to 2000 cm^{-1} (or 60 Thz, 5 μm). We have explored the possibilities of this technique in the study of protein dynamics. In a crystal where the proteins are aligned, coherently excited vibrations of the haem will lead to net oscillating dipoles which emit light in the infrared. Excitation of the haem with a short 12-15 fs visible pulse may also lead to an ultrafast response of the protein, and the impulsive (<50 fs) photodissociation process in myoglobin must lead to a fast reaction of the nearby protein residues, in particular of the proximal and distal histidines. These may therefore also result in coherent excitation of protein vibrations and consequently emit a protein electrical field. Since these experiments are performed on crystals in order to have a net dipole, a direct relation of the observed dynamics with static X-ray structures can be made.

2. Experimental Methods

In order to have the dipoles of the protein aligned a crystal is needed. We used crystals of spacegroup $P2_1$, typical dimensions 200x50μm, thickness ~8μm, the crystals were kept in $(NH_4)_2SO_4$. A 1 kHz Ti:saph. amplifier (Spectra Physics), seeded by a home-built oscillator, was used to pump a non-collinear OPA [2].

* Present adress: Dept. of Biophysics, Vrije Universiteit, Amsterdam, The Netherlands, E-mail: marloes@nat.vu.nl

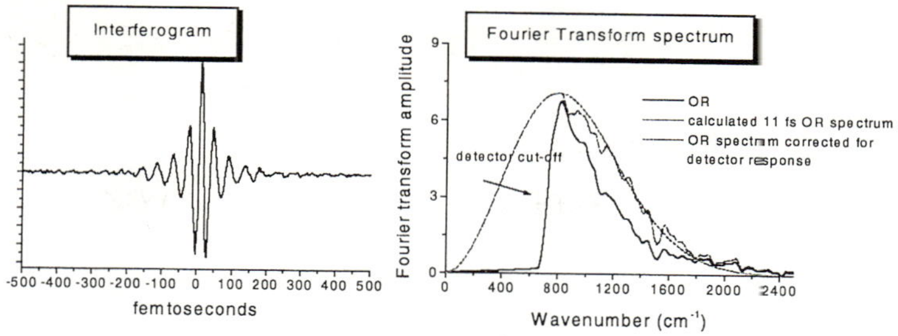

Fig. 1. Optical rectification of the 10-12 fs NOPA output centered at ~580 nm results in a short, almost single cycle, infrared pulse. The interferogram of two infrared pulses is shown, and in the right panel the corresponding spectrum which extends to 2000 cm^{-1}. The strong detector cut-off is responsible for the 'ringing' in the interferogram.

From the NOPA ~50 nJ was splitted off to excite the myoglobin crystal at ~580 nm 100 nJ was used to generate broadband infrared in GaAs with an efficiency of ~10^{-6}, see fig. 1. The reference GaAs-generated infrared pulse travels through the sample to facilitate overlap on the HgCdT detector with the induced infrared emission of the crystal. An interferogram is measured by scanning the delay line in one of the two arms of the interferometer.

3. Results and Discussion

Infrared emission is induced by excitation of the cofactor in myoglobin, a haem, (liganded with CO) at its lowest electronic transition at 580 nm with a short ~12-15 fs pulse. This results in excitation of vibrations coupled to the excited state, or via the Impulsive Stimulated Raman Mechanism, in excitation of vibrations on the ground state. The electrical field emitted by the coherently oscillating vibrational dipoles (see fig. 2) is detected by letting it interfere with the reference infrared electrical field induced in GaAs. Fourier-Transform analysis reveals the presence of 5-6 different modes in the protein E-field (not shown).

4. Conclusions

We have for the first time directly measured the electrical field emitted by an orientated protein complex. Measurements as a function of crystal rotation angle (not shown) and a careful analysis taking into account the dispersion of the reference E-field, will enable us to make a relation with static X-ray structures. With the phase information from our data we will have a real-time view of vibrational motion of the haem and possibly nearby protein residues.

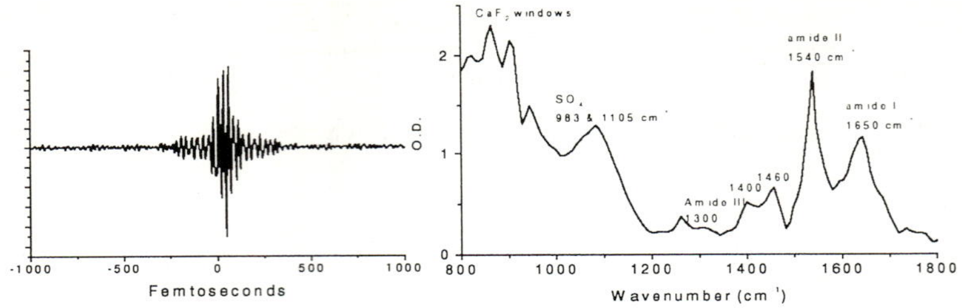

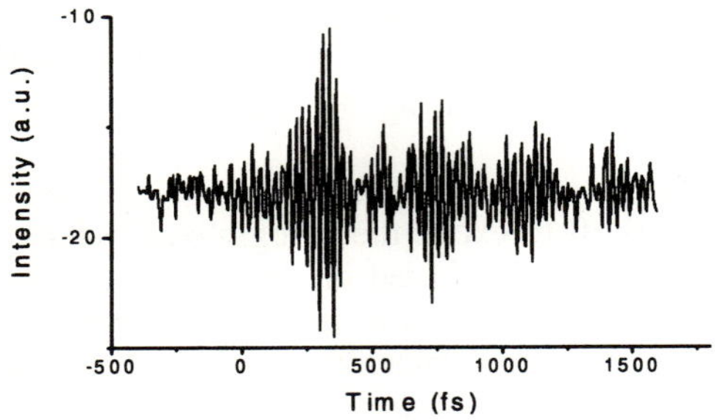

Fig. 2. (top left) Interferogram of Mb-CO crystal transmission, (top right) corresponding O.D. spectrum and (left) the Mb-CO infrared emission interferogram. The dephasing time of the protein emission is 0.6-0.8 ps, the intensity is ~200 photons / shot.

Acknowledgements. We thank dr. Andrea Mozarelli for providing us with myoglobin crystals which were used in an early stage of the experiments, Jacques Breton for use of equipment, Lisa Miller and Mark Chance for sending us the absorption spectrum of 4-MeIm, Jean-Christoph Lambry for writing the measuring program and Manuel Joffre and Antigoni Alexandrou for stimulating discussions.

References

1 Bonvalet, A., Nagle, J., Berger, V., Migus, A., Martin, J.-L. and Joffre, M. (1996) *Phys. Rev. Lett.* 76, 4392-4396

M. Joffre, A. Bonvalet, A. Migus, J.-L. Martin (1996) *Opt. Lett.* **21**, 964

Bonvalet, A., Joffre, M., Martin, J.-L. and Migus, A. (1995) *Appl Phys Lett.* **67**, 2907-2909

2 G. Cerullo, M. Nisoli, S. Stagira, S. De Silvestri *Opt. Lett.* **23**, 1283 (1998)

A. Shirakawa, I. Sakane, T. Kobayashi *Opt. Lett.* **23**, 1292 (1998)

T. Wilhelm, J. Piedl, E. Riedle *Opt. Lett.* **22**, 1494 (1997)

G. Cerullo, M. Nisoli, S. Stagira, S. De Silvestri *Appl. Phys. Lett.* **71**, 3616 (1997)

A. Shirakawa, T. Kobayashi *Appl. Phys. Lett*, **72**, 147(1998)

Femtosecond fluorescence dynamics of porphyrins; internal conversion, energy relaxation, and quantum beats

Y. Shibata[1,2], H. Chosrowjan[2], N. Mataga[2], N. Yoshida[3], and A. Osuka[3]

[1] Precursory Research for Embryonic Science and Technology (PRESTO), Japan Science and Technology Corporation (JST), E-mail: yshibata@ile.osaka-u.ac.jp
[2] Institute for Laser Technology, Hyogo 661-0974, Japan
[3] Department of Chemistry, Graduate School of Science, Kyoto University, Kyoto 606-8502, Japan

Abstract. Fluorescence dynamics of some porphyrins have been studied by up-conversion technique to elucidate the mechanism of internal conversion, subsequent intramolecular energy relaxation, coherent motion of vibrational excited states, and the competing electron-transfer reaction.

1. Introduction

Recent progress in fs laser technology has enabled us to study ultrafast intramolecular energy-relaxation processes from the highly excited state. Various types of porphyrins now available are quite effective samples for such investigation because their S_2 state lies around 400 nm which can be excited by a conventional fs-laser systems. Systematic studies on porphyrin family will shed light on the old, but still unresolved problems; 1) What kind of vibrational modes accept the excess energy in the course of IC process? For example, why is S_2 state of free-base (Fb) porphyrin nonfluorescent, while Zn-derivative often emits rather strong S_2 fluorescence? 2) Does any coherence remain in highly excited vibrational states populated just after the IC? 3) How do the ultrafast energy-relaxation processes affect the competing ultrafast photoinduced reactions such as electron transfer (ET)? To clarify these problems, in the present study, we have carried out detailed investigation of fluorescence up-conversion spectroscopy on various porphyrin derivatives.

2. Experimental Methods

The experimental setup of the fluorescence up-conversion apparatus has been given in the previous paper in detail [1]. All experiments were made at room temperature. The exciting wavelength is 405 nm which lies in the blue edge of the Soret band for all samples examined and the typical excitation power is about

15 mW. The temporal resolution of the apparatus is determined from the cross-correlation time profile between the pump and fundamental beam, and is estimated to be ~ 220 fs.

3. Results and Discussion

In Fig. 1 the fluorescence dynamics of Zn tetraphenylporphyrin (TPP) in ethanol (a) and FbTPP in tetrahydrofuran (THF) (b) at several monitoring wavelengths are compared. Here, the intensity of the signal is normalized to unity at the peak. As shown in the previous paper [2, 3], the wavelength-dependent fluorescence dynamics of ZnTPP can be interpreted as follows. The same time constant between the decay of S_2 fluorescence and the rise of S_1 fluorescence indicates that main fraction of molecules relaxes to the vibrational ground state within S_1 immediately after IC because of a very fast energy redistribution. The weak fluorescence in the intermediate spectral region of 500-580 nm, on the other hand, implies that small fraction of molecules relaxes through relatively slow relaxation path, and is reasonably interpreted as emitted from vibrational excited states of S_1. Thus, the whole fluorescence dynamics of ZnTPP can be explained by considering multi-path relaxation.

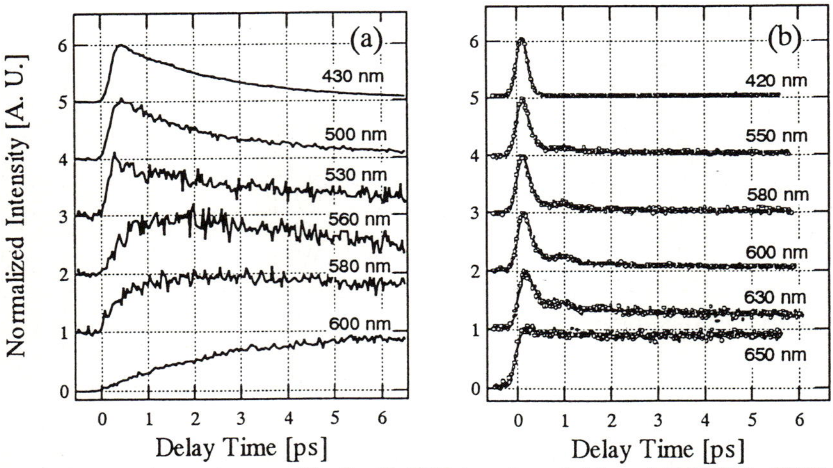

Fig. 1. Fluorescence time profile for ZnTPP in ethanol (a) and FbTPP in THF (b).

In the case of FbTPP in THF, the fluorescence from S_2 monitored at 420 nm shows very rapid decay resulting in slightly broader time profile than the instrumental response. This clearly reflects its very high IC rate. As in the case of ZnTPP, the rise time of S_1 fluorescence at 650 nm seems almost identical to the decay time of S_2 fluorescence, suggesting an immediate energy redistribution after IC process. In the case of FbTPP, again fluorescence signal has been detected in the intermediate spectral region. The fluorescence dynamics monitored in the spectral region of 550-630 nm conspicuously shows low-frequency quantum beat. These fluorescence may be emitted from the vibrational excited states within S_1 state, and their quantum beat implies coherent wave-packet motion of intramolecular low-frequency vibrational modes, which is probably responsible for accepting the excess energy accumulated in the course of IC. We have observed no apparent isotope effect on this oscillatory fluorescence component. Namely, temporal behavior was almost the same for FbTPP in ethanol and in hevy ethanol. We have also confirmed that the fluorescence time profile is same for V/V and V/H configuration.

To analyze the fluorescence dynamics of FbTPP in THF including the oscillatory component in more detail, we fitted the observed fluorescence time profile to the form, the sum of two exponentially decaying components and one damping-oscillation component, $\exp(-kt)\cos(\omega t-\phi)$. The solid lines in Fig. 1 (b) are the fitting curves convoluted with the instrumental response, which are in good agreement with the observed data. According to this analysis, both the decay time of S_2 fluorescence and the rise time of S_1 fluorescence were determined to be 120 fs. The period of the oscillation of the fluorescence quantum beat is estimated to be about 1 ps, corresponding to frequency of about 30~40 cm^{-1}. A peak is also seen around 30-40 cm^{-1} in the spectrum obtained from Fourier transform of the fluorescence time profile. Thus, the origin of the quantum beat is suggested to be very low-frequency motion. Our analysis has also shown that there is no apparent wavelength dependence of the phase ϕ of the oscillation. This suggests that the fluorescence quantum beat observed here arise not from the temporal oscillation of the peak position of the fluorescence spectrum due to the wave-packet motion, but from the periodic modulation of the fluorescence intensity.

In addition to the above results, we have made investigation of the ultrafast photoinduced ET dynamics [2]. In this case, we have used the samples of porphyrin derivatives where electron acceptors are directly combined to the meso-position of porphyrins. Our preliminary experiments have clarified ultrafast fluorescence quenching from S_2 state of Zn-porphyrin derivative, because of ET reaction to the attached acceptor. The energy-gap dependence of the estimated ET rate showed clear-cut inverted effect, which is observed for the first time for the charge separation process.

Acknowledgements. H. C. thanks Monbu-sho for financial support.

References

1 H. Chosrowjan, N. Mataga, N. Nakashima, Y. Imamoto, and F. Tokunaga, Chem. Phys. Lett. **270**, 267, 1997.

2 Y. Shibata, H. Chosrowjan, N. Mataga, N. Yoshida, and A. Osuka, J. Lumin. **87-89**, 757, 2000.

3 N. Mataga, Y. Shibata, H. Chosrowjan, N. Yoshida, and A. Osuka, J. Phys. Chem. B, **104**, 4001, 2000.

Femtosecond Pump-Probe Studies of Radiationless Decay Dynamics in Arthropod and Mollusc Hemocyanins

Maurice D. Edington[1], James S. Floyd[2], Nicole M. Haralampus-Grynaviski[2], Tong Ye[2], Bo Zheng[2], and John D. Simon[2]

[1]Department of Chemistry, Florida A&M University, Tallahassee, FL 32307
[2]Department of Chemistry, Duke University, Durham, NC 27708

Abstract. Femtosecond pump-probe spectroscopy has been used to measure the dynamics following excitation of the ligand-to-metal charge-transfer states in arthropod and molluscan hemocyanins isolated from *Limulus polyphemus* and *Busycon canaliculatum*, respectively. These results are discussed in terms of the thermal elimination and geminate recombination of oxygen following rapid nonradiative relaxation in the protein. The two hemocyanins exhibit the same dynamics, consistent with the structural similarity of their active sites.

1. Introduction

Hemocyanins are large multi-subunit copper-containing proteins that serve as the dioxygen carriers for a variety of molluscs and arthropods. Hemocyanins from molluscs and arthropods have similar active-site electronic structures but they differ dramatically in quaternary protein structures [1,2]. Molluscan hemocyanins are composed of 350-400 kDa subunits that are each comprised of 7-8 50-kDa domains. Each domain contains one binuclear copper site and consists of ~400 amino acid residues; the subunits assemble as decamers or di-decamers in the blood. Arthropod hemocyanins can exist as hexamers or multiple hexamers of 75 kDa subunits. Each subunit contains ~650 amino acid residues and a single binuclear copper site.

A detailed picture of the active-site geometry of hemocyanin has been obtained from X-ray crystallographic studies (Figure 1) [3-5]. In deoxy hemocyanin from *Panulirus interruptus,* two Cu(I) atoms are each coordinated to two histidine ligands with Cu-N distances of 2.0 Å and N-Cu-N bond angles of ~108°. A third histidine ligand is coordinated to each Cu atom at a distance of ~2.7 Å. The Cu-Cu distance is 3.6 Å. In deoxy hemocyanin from *Limulus polyphemus*, the Cu-Cu distance is 4.5 Å and each Cu(I) atom is coordinated to three histidine ligands with Cu-N distances of 2.0 Å.

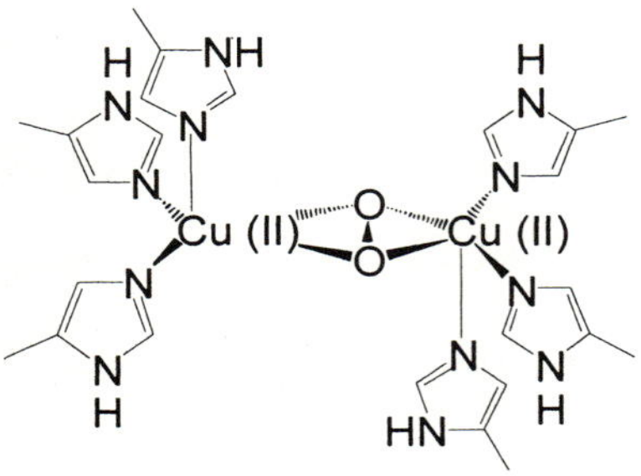

Fig. 1. Schematic of the active site of hemocyanin. The oxygen bridges the two copper centers.

The results of numerous theoretical and experimental investigations on model compounds and oxyhemocyanins from various organisms have led to a detailed understanding of the spectroscopic features observed [6-9]. Molecular oxygen binds symmetrically as peroxide to both copper atoms in a side-on (μ-η^2:η^2) bridging configuration. Significant overlap between the peroxo π–orbitals and copper d-orbitals results in the appearance of two absorption bands in the oxy spectrum that are absent in the deoxy spectrum. These bands correspond to ligand-to-metal charge-transfer (LMCT) transitions and are centered at 345 nm and 570 nm.

2. Experimental

Hemocyanin was isolated as described elsewhere [10]. Femtosecond pulses were generated by a commercial regeneratively-amplified Ti:Sapphire laser system (Spectra Physics). The amplified pulses (80 fs, 1 kHz, 1 mJ/pulse) seed an optical parametric oscillator (OPA), which is tuned from 300-400 nm. Pulses from the OPA (200 fs, 100 nJ/pulse) are split into pump and probe beams, and attenuated with neutral density filters. Excitation energies on the order of 50 nJ/pulse and probe pulse energies of 1 nJ/pulse are used.

3. Results and Discussion

We recorded degenerate pump-probe transients at 340 nm on hemocyanin from *Limulus polyphemus* and *Busycon canaliculatum.* Figure 2 shows the data obtained at 340 nm for hemocyanin isolated from *Busycon canaliculatum.*

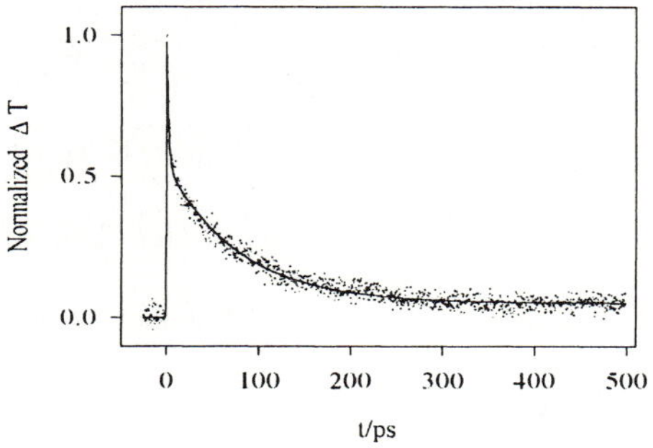

Fig. 2. Transient absorption signal for the 340 nm degenerate pump-probe experiment on hemocyanin isolated from *Busycon canaliculatum.*

Following the formation of an instantaneous bleach, the transient exhibits a decay that is well described by a two-exponential model with time constants of 1.7 ± 0.2 ps and 80 ± 10 ps [10]. The transient exhibits a residual bleach of ~4% of the initial bleach when the probe delay is extended to 500 ps. The value of the offset was observed to increase when large pulse energies were used. Identical results were observed for hemocyanin from *Limulus polyphemus.*

Busycon and *Limulus* hemocyanins exhibited no detectable emission when excited at 340 nm [10]. In addition, temperature-dependent photoacoustic studies reveal that all of the absorbed energy is released as heat on the subnanosecond time scale. These data indicate that there must be a rapid nonradiative relaxation back to the ground state following photoexcitation. Based on these data, the following model is proposed to account for the time-resolved absorption data shown in Figure 2. The fast recovery of the transient bleach (1.7 ± 0.2 ps) is assigned to a nonradiative relaxation from the electronically excited protein to its ground state. This rapid relaxation produces a "hot" active site and thermal dissociation of oxygen competes with vibrational cooling. The longer component of the bleach recovery (80 ± 10 ps) is assigned to the dynamics associated with the geminate recombination of molecular oxygen to the dicopper binding site. Finally the observation that 4% of the initial bleach persists onto long time scales (> 500 ps) suggests that a small fraction of the dissociated oxygen escapes the active site and diffuses into the protein.

Busycon and *Limulus* hemocyanins have different quarternary structure. However, time-resolved data reveal that the dynamics in the vicinity of the active site are identical for the two systems. This is consistent with the similarity in structure of the active sites among the hemocyanins.

Acknowledgements. This work was supported by the National Institutes of Health MBRS program, GM08111(MDE), Duke University, and the North Carolina Biotechnology Center. We thank Dr. Celia Bonaventura of the Duke University Marine Laboratories for providing a sample of hemocyanin from *Limulus polyphemus.*

References

1. J. Lamy, S. Compin, J. N. Lamy, *Arch. Biochem. Biophys.* **1983**, 223, 584-603.
2. K. E. Van Holde, K. I. Miller, *Quart. Rev. Biophys.* **1982**, 15, 1-129.
3. K. A. Magnus, B. Hazes, H. Ton-That, C. Bonaventura, J. Bonaventura, W. G. J. Hol, *Proteins* **1994**, 19, 302-309
4. B. Hazes, K. A. Magnus, C. Bonaventura, J. Bonaventura, Z. Dauter, K. H. Kalk, G. J. Hol, *Prot. Sci.* **1993**, 2, 597-619.
5. M. E. Cuff, K. I. Miller, K. E. van Holde, W. A. Hendrickson, *J. Mol. Biol.* **1998**, 278, 855-870.
6. M. Baldwin, D. Root, J. Pate, K. Fujisawa, N. Kitajima, E. Solomon, *J. Am. Chem. Soc.* **1992**, 114, 10421-10431.
7. P. Ross, E. Solomon, *J. Am. Chem. Soc.* **1991**, 113, 3246-3259.
8. E. Solomon, F. Tuczek, D. Root, C. Brown, *Chem. Rev.* **1994**, 94, 827-856.
9. E. Solomon, M. Lowery, *Science* **1992**, 259, 1575-1581.
10. M. D. Edington, J. S. Floyd, N. M. Haralampus-Grynaviski, T. Ye, J. D. Simon, *J. Phys. Chem. B.*, in preparation (2000).

Initial Conformational Dynamics in Cyclic Azobenzene Peptides

J. Wachtveitl[1], S. Spörlein[1], W. Zinth[1], C. Renner[2], R. Behrendt[2], L. Moroder[2]

[1] Sektion Physik, Ludwig-Maximilians-Universität München, Oettingenstr. 67, 80538 München, Germany
E-mail: Josef.Wachtveitl@physik.uni-muenchen.de
[2] Max-Planck-Institut für Biochemie, am Klopferspitz 18a, 82152 Martinsried, Germany

Abstract. Conformational transitions in small cyclic chromopeptides are photoinduced and verified by NMR-analysis. Femtosecond UV/vis-spectroscopic analysis of a cyclic azo-octapeptide reveals the sub-ps photoisomerization of the light-switch and the slower photoresponse of the peptide backbone.

1. Introduction

Femtosecond time resolved spectroscopy of model-peptide-compounds is the most direct approach to study the mechanisms underlying the fast reactions in protein folding. The ultrafast photoisomerization of azobenzene [1-3] can be utilized to switch between structural motifs in cyclic peptides. In a monocyclic molecule containing the photochromic (4-amino)phenylazobenzoic acid (APB) and the active site fragment of thioredoxin reductase (H-Ala-Cys-Ala-Thr-Cys-Asp-Gly-Phe-OH), the light induced reversible *trans/cis*-isomerization leads to a conformational transition from a very rigid, highly constrained *trans*-APB-peptide to a more flexible *cis*-APB-peptide (Fig. 1, [4]).

Azobenzene rather than the pseudo-amino acid APB should be compared to the APB-peptides, since the same n-π^* transition should be excited in all samples (λ_{exc} = 480 nm; Fig. 2, top). The electronic properties of APB, however, lead to an overlap of π-π^* and n-π^* states, preventing a selective n-π^* photoexcitation [3].

The global analysis over a broad spectral range reveals the main steps of the fast, reversible photoswitch and the response of the peptide indicating fast conformational dynamics of the entire molecule.

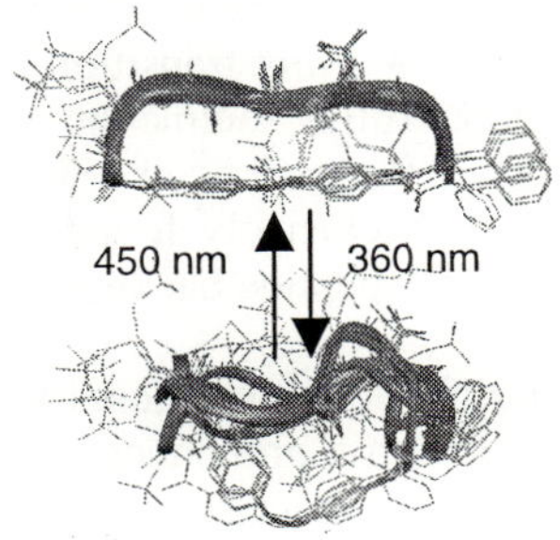

Fig. 1. Structural changes upon photoisomerization of a cyclic azopeptide. The ten structures of lowest energy are shown for the *trans*- (top) and *cis*- (bottom) configuration of cyclo(Ala-Cys(StBu)-Ala-Thr-Cys(StBu)-Asp-Gly-Phe-APB). The ribbons are a smoothed representation of the peptide backbone structure.

2. Experimental Methods

The pump-probe setup including the multi-channel detection system is described in detail elsewhere [5]. A 1 kHz regenerative chirped pulse amplifier together with a noncollinear optical parametric amplifier (NOPA) [6] provided excitation pulses with a pulse length of 20-30 fs (FWHM) at a central wavelength of 480 nm. In combination with a sapphire white light continuum used for probing a cross correlation width of about 80 fs was achieved.

The three samples (azobenzene, linear and cyclic APB-peptides) dissolved in DMSO were measured in a 0.5 mm fused silica flow cuvette to assure exchange of the sample volume between the laser pulses. A stable isomer ratio is maintained by cw-illumination with 360 nm or 450 nm light, respectively.

3. Results and Discussion

In Fig. 2 the transient absorption dynamics at two different probing wavelengths are depicted for both isomers of the three samples. For azobenzene (Fig. 2, first row) the experiments reveal a fast *cis* → *trans* photoisomerization. For the *cis*-isomer the excited state decays within 150 fs, the subsequent slower time constant (here $\tau = 4.1$ ps) is assigned mainly to cooling processes in the ground state [1]. For *trans*-azobenzene time constants of 320 fs and 3.8 ps are observed. Due to the weak ground state absorption at 504 nm the photoproduct signal is very small, but can be clearly identified at shorter wavelengths ($\lambda_{pr} = 368$ nm). The initial photoreaction is similar to reported values obtained for an excitation at a shorter wavelength of $\lambda_{pr} = 435$ nm [1], the weak deceleration of the reaction dynamics agrees well with the dependence of this reaction on excitation wavelength [2].

The linear APB-peptide is the ideal system for a comparison with the cyclic APB-peptide, since the electronic properties (Fig. 2, top) and the size of both molecules are very similar. For the linear APB-peptides the ps-components dominate (1.4 ps and 9 ps for the *cis* → *trans* and 2.3 ps and 11 ps for the *trans* → *cis* reaction), the fast component is weak for both isomerization directions. The clearly increased lifetime of the excited state may reflect a slower photoisomerization reaction due to the larger molecular mass of the molecule.

The *trans* → *cis* photoisomerization of the cyclic peptide leads to a transition from a rigid molecule into a much more relaxed conformation with many accessible structures (Fig. 1). Interestingly, this reaction occurs faster than in the linear peptide (with dominating time constants 0.6 ps and 7.8 ps). In contrast, the restriction of the conformational space, occuring during the *cis* → *trans* transition, leads to a drastically slower reaction dynamics. Within a global fitting routine the curves can be approximated by four time constants (0.18 ps, 2.9 ps, 45 ps and 800 ps). The presence of very slow components ($\tau > 100$ ps) is exclusively found in the cyclic system and may be indicative for the response of the peptide backbone upon azobenzene isomerization.

In conclusion, we observe for the first time an ultrafast conformational transition of cyclic and linear chromophore-peptide-systems with a pronounced isomer dependence. The data clearly indicate a slowing down of the reaction upon restriction of the accessible conformational space.

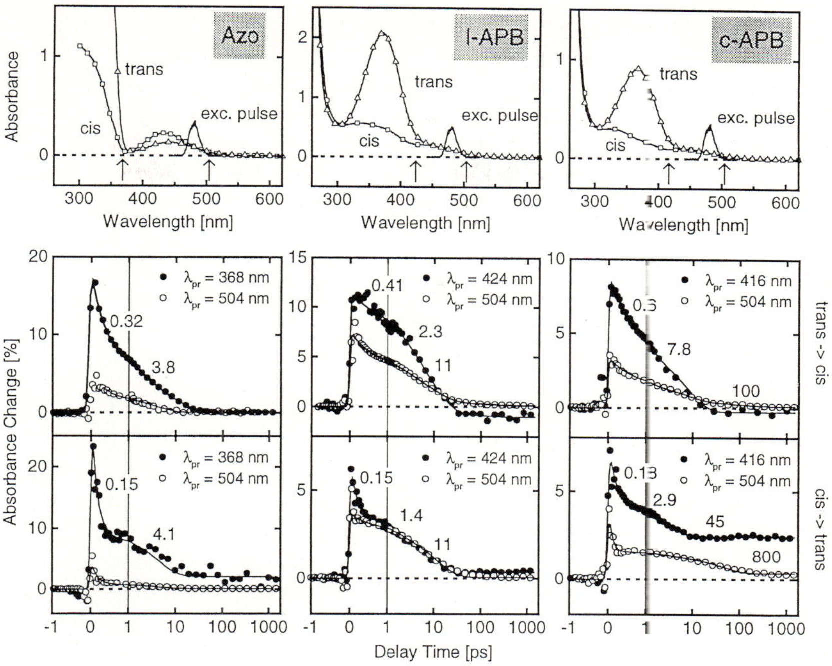

Fig. 2. top: Spectra for the *trans*- and *cis*- configuration of azobenzene (left), linear- (middle) and cyclic APB-peptides (right). All three systems were bistable, i.e. the established photostationary *cis/trans* balance persisited for several hours. Transient absorbance changes of *trans*- (middle row) and *cis*-isomers (lower row) for two representative probing wavelengths. Especially the restriction of conformational space upon the *cis* → *trans* transition in the cyclopeptide (lower right plot) leads to the appearance of slow components.

Our approach can be extended to construct various photoresponsive cyclopeptides based on conformational library informations. The conformational dynamics of various structural motifs can be investigated from fs to s and compared with the structural data. This should improve the understanding of important biological processes like the dynamics of ligand binding or protein folding.

References

1 T. Nägele, R. Hoche, W. Zinth and J. Wachtveitl. Chem. Phys. Lett. **272**, 439, 1997.

2 I.K. Lednev, T.Q. Ye, M. Matousek, M. Towrie, P. Foggi, F.V.R. Neuwahl, S. Umapathy, R.E. Hester and J.N. Moore, Chem. Phys. Lett. **290**, 68, 1998.

3 J. Wachtveitl, T. Nägele, B. Puell, W. Zinth, M. Krüger, S. Rudolph-Böhner, D. Oesterhelt and L. Moroder, Photochem. Photobiol. **105**, 283, 1997.

4 R. Behrendt, C. Renner, M. Schenk, F. Wang, J. Wachtveitl, D. Oesterhelt and L. Moroder, Angew. Chem. Int. Ed. Engl., **38**, no.18, 2771, 1999.

5 M. Seel, E. Wildermuth and W. Zinth, Meas. Sci. Technol. **8**, 449, 1997.

6 T. Wilhelm, J. Piel and E. Riedle, Opt. Lett. **22**, 1474, 1997.

Index of Contributors

Springer Series in Chemical Physics

Editors: Vitalii I. Goldanskii Fritz P. Schäfer J. Peter Toennies

1 **Atomic Spectra and Radiative Transitions** By I. I. Sobelman

2 **Surface Crystallography by LEED** Theory, Computation and Structural Results. By M. A. Van Hove, S. Y. Tong

3 **Advances in Laser Chemistry** Editor: A. H. Zewail

4 **Picosecond Phenomena** Editors: C. V. Shank, E. P. Ippen, S. L. Shapiro

5 **Laser Spectroscopy** Basic Concepts and Instrumentation By W. Demtröder 3rd Printing

6 **Laser-Induced Processes in Molecules** Physics and Chemistry Editors: K. L. Kompa, S. D. Smith

7 **Excitation of Atoms and Broadening of Spectral Lines** By I. I. Sobelman, L. A. Vainshtein, E. A. Yukov

8 **Spin Exchange** Principles and Applications in Chemistry and Biology By Yu. N. Molin, K. M. Salikhov, K. I. Zamaraev

9 **Secondary Ion Mass Spectrometry SIMS II** Editors: A. Benninghoven, C. A. Evans, Jr., R. A. Powell, R. Shimizu, H. A. Storms

10 **Lasers and Chemical Change** By A. Ben-Shaul, Y. Haas, K. L. Kompa, R. D. Levine

11 **Liquid Crystals of One- and Two-Dimensional Order** Editors: W. Helfrich, G. Heppke

12 **Gasdynamic Laser** By S. A. Losev

13 **Atomic Many-Body Theory** By I. Lindgren, J. Morrison

14 **Picosecond Phenomena II** Editors: R. M. Hochstrasser, W. Kaiser, C. V. Shank

15 **Vibrational Spectroscopy of Adsorbates** Editor: R. F. Willis

16 **Spectroscopy of Molecular Excitions** By V. L. Broude, E. I. Rashba, E. F. Sheka

17 **Inelastic Particle-Surface Collisions** Editors: E. Taglauer, W. Heiland

18 **Modelling of Chemical Reaction Systems** Editors: K. H. Ebert, P. Deuflhard, W. Jäger

19 **Secondary Ion Mass Spectrometry SIMS III** Editors: A. Benninghoven, J. Giber, J. László, M. Riedel, H. W. Werner

20 **Chemistry and Physics of Solid Surfaces IV** Editors: R. Vanselow, R. Howe

21 **Dynamics of Gas-Surface Interaction** Editors: G. Benedek, U. Valbusa

22 **Nonlinear Laser Chemistry** Multiple-Photon Excitation By V. S. Letokhov

23 **Picosecond Phenomena III** Editors: K. B. Eisenthal, R. M. Hochstrasser, W. Kaiser, A. Laubereau

24 **Desorption Induced by Electronic Transitions DIET I** Editors: N. H. Tolk, M. M. Traum, J. C. Tully, T. E. Madey

25 **Ion Formation from Organic Solids** Editor: A. Benninghoven

26 **Semiclassical Theories of Molecular Scattering** By B. C. Eu

27 **EXAFS and Near Edge Structures** Editors: A. Bianconi, L. Incoccia, S. Stipcich

28 **Atoms in Strong Light Fields** By N. B. Delone, V. P. Krainov

29 **Gas Flow in Nozzles** By U. G. Pirumov, G. S. Roslyakov

30 **Theory of Slow Atomic Collisions** By E. E. Nikitin, S. Ya. Umanskii

31 **Reference Data on Atoms, Molecules, and Ions** By A. A. Radzig, B. M. Smirnov

32 **Adsorption Processes on Semiconductor and Dielectric Surfaces I** By V. F. Kiselev, O. V. Krylov

33 **Surface Studies with Lasers** Editors: F.R. Aussenegg, A. Leitner, M.E. Lippitsch

34 **Inert Gases** Potentials, Dynamics, and Energy Transfer in Doped Crystals. Editor: M. L. Klein

35 **Chemistry and Physics of Solid Surfaces V** Editors: R. Vanselow, R. Howe

36 **Secondary Ion Mass Spectrometry, SIMS IV** Editors: A. Benninghoven, J. Okano, R. Shimizu, H. W. Werner

37 **X-Ray Spectra and Chemical Binding** By A. Meisel, G. Leonhardt, R. Szargan

38 **Ultrafast Phenomena IV** By D. H. Auston, K. B. Eisenthal

39 **Laser Processing and Diagnostics** Editor: D. Bäuerle

Springer Series in Chemical Physics

Editors: Vitalii I. Goldanskii Fritz P. Schäfer J. Peter Toennies

Managing Editor: H. K.V. Lotsch

40 **High-Resolution Spectroscopy of Transient Molecules**
By E. Hirota

41 **High Resolution Spectral Atlas of Nitrogen Dioxide 559–597 nm**
By K. Uehara and H. Sasada

42 **Antennas and Reaction Centers of Photosynthetic Bacteria**
Structure, Interactions, and Dynamics
Editor: M. E. Michel-Beyerle

43 **The Atom-Atom Potential Method**
Applications to Organic Molecular Solids
By A. J. Pertsin and A. I. Kitaigorodsky

44 **Secondary Ion Mass Spectrometry SIMS V**
Editors: A. Benninghoven, R. J. Colton, D. S. Simons, and H. W. Werner

45 **Thermotropic Liquid Crystals, Fundamentals**
By G. Vertogen and W. H. de Jeu

46 **Ultrafast Phenomena V**
Editors: G. R. Fleming and A. E. Siegman

47 **Complex Chemical Reaction Systems**
Mathematical Modelling and Simulation
Editors: J. Warnatz and W. Jäger

48 **Ultrafast Phenomena VI**
Editors: T. Yajima, K. Yoshihara, C. B. Harris, and S. Shionoya

49 **Vibronic Interactions in Molecules and Crystals**
By I. B. Bersuker and V. Z. Polinger

50 **Molecular and Laser Spectroscopy**
By Zu-Geng Wang and Hui-Rong Xia

51 **Space-Time Organization in Macro molecular Fluids**
Editors: F. Tanaka, M. Doi, and T. Ohta

52 **Multiple-Photon Laser Chemistry**
By. R. V. Ambartzumian, C. D. Cantrell, and A. Puretzky

53 **Ultrafast Phenomena VII**
Editors: C. B. Harris, E. P. Ippen, G. A. Mourou, and A. H. Zewail

54 **Physics of Ion Impact Phenomena**
Editor: D. Mathur .

55 **Ultrafast Phenomena VIII**
Editors: J.-L. Martin, A. Migus, G. A. Mourou, and A. H. Zewail

56 **Clusters of Atoms and Molecules**
Solvation and Chemistry of Free Clusters, and Embedded, Supported and Compressed Clusters
Editor: H. Haberland

57 **Radiationless Transitions in Polyatomic Molecules**
By E. S. Medvedev and V. I. Osherov

58 **Positron Annihilation in Chemistry**
By O. E. Mogensen

59 **Soot Formation in Combustion**
Mechanisms and Models
Editor: H. Bockhorn

60 **Ultrafast Phenomena IX**
Editors: P. Barbara, W. H. Knox, G. A. Mourou, and A. H. Zewail

61 **Gas Phase Chemical Reaction Systems**
Experiments and Models 100 Years After Max Bodenstein
Editors: J. Wolfrum, H.-R. Volpp, R. Rannacher, and J. Warnatz

62 **Ultrafast Phenomena X**
Editors: P. F. Barbara, J. G. Fujimoto, W. H. Knox, and W. Zinth

Druck: Strauss Offsetdruck, Mörlenbach
Verarbeitung: Schäffer, Grünstadt